T0257124

CLYMER® MANUALS

YAMAHA

V-STAR 1100 • 1999-2009

WHAT'S IN YOUR TOOLBOX?

YouTube

Disclaimer

There are risks associated with automotive repairs. The ability to make repairs depends on the individual's skill, experience and proper tools. Individuals should act with due care and acknowledge and assume the risk of performing automotive repairs.

The purpose of this manual is to provide comprehensive, useful and accessible automotive repair information, to help you get the best value from your vehicle. However, this manual is not a substitute for a professional certified technician or mechanic.

This repair manual is produced by a third party and is not associated with an individual vehicle manufacturer. If there is any doubt or discrepancy between this manual and the owner's manual or the factory service manual, please refer to the factory service manual or seek assistance from a professional certified technician or mechanic.

Even though we have prepared this manual with extreme care and every attempt is made to ensure that the information in this manual is correct, neither the publisher nor the author can accept responsibility for loss, damage or injury caused by any errors in, or omissions from, the information given.

More information available at haynes.com
Phone: 805-498-6703

J H Haynes & Co. Ltd.
Haynes North America, Inc.

ISBN-10: 1-59969-298-8
ISBN-13: 978-1-59969-298-2
Library of Congress: 2009932593

Author: *Clymer Staff*
Technical Photography: *Clymer Staff*
Technical Illustrations: *Errol McCarthy*
Tools and Equipment: *K & L Supply Co. (www.klsupply.com)*
Cover: *Mark Clifford Photography, Los Angeles, California (www.markclifford.com)*

M281-4, 8V1, 17-416

Chapter One
General Information 1

Chapter Two
Troubleshooting 2

Chapter Three
Lubrication, Maintenance and Tune-up 3

Chapter Four
Engine Top End 4

Chapter Five
Engine Lower End 5

Chapter Six
Clutch and Primary Drive Gear 6

Chapter Seven
Transmission, Shift Mechanism and Middle Gear 7

Chapter Eight
Air/Fuel, Exhaust and Emission Control Systems 8

Chapter Nine
Electrical System 9

Chapter Ten
Wheels and Tires 10

Chapter Eleven
Front Suspension and Steering 11

Chapter Twelve
Rear Suspension and Final Drive 12

Chapter Thirteen
Brakes 13

Chapter Fourteen
Frame 14

Index 15

Wiring Diagrams 16

Common spark plug conditions

NORMAL

Symptoms: Brown to grayish-tan color and slight electrode wear. Correct heat range for engine and operating conditions.
Recommendation: When new spark plugs are installed, replace with plugs of the same heat range.

WORN

Symptoms: Rounded electrodes with a small amount of deposits on the firing end. Normal color. Causes hard starting in damp or cold weather and poor fuel economy.
Recommendation: Plugs have been left in the engine too long. Replace with new plugs of the same heat range. Follow the recommended maintenance schedule.

CARBON DEPOSITS

Symptoms: Dry sooty deposits indicate a rich mixture or weak ignition. Causes misfiring, hard starting and hesitation.
Recommendation: Make sure the plug has the correct heat range. Check for a clogged air filter or problem in the fuel system or engine management system. Also check for ignition system problems.

ASH DEPOSITS

Symptoms: Light brown deposits encrusted on the side or center electrodes or both. Derived from oil and/or fuel additives. Excessive amounts may mask the spark, causing misfiring and hesitation during acceleration.
Recommendation: If excessive deposits accumulate over a short time or low mileage, install new valve guide seals to prevent seepage of oil into the combustion chambers. Also try changing gasoline brands.

OIL DEPOSITS

Symptoms: Oily coating caused by poor oil control. Oil is leaking past worn valve guides or piston rings into the combustion chamber. Causes hard starting, misfiring and hesitation.
Recommendation: Correct the mechanical condition with necessary repairs and install new plugs.

GAP BRIDGING

Symptoms: Combustion deposits lodge between the electrodes. Heavy deposits accumulate and bridge the electrode gap. The plug ceases to fire, resulting in a dead cylinder.
Recommendation: Locate the faulty plug and remove the deposits from between the electrodes.

TOO HOT

Symptoms: Blistered, white insulator, eroded electrode and absence of deposits. Results in shortened plug life.
Recommendation: Check for the correct plug heat range, over-advanced ignition timing, lean fuel mixture, intake manifold vacuum leaks, sticking valves and insufficient engine cooling.

PREIGNITION

Symptoms: Melted electrodes. Insulators are white, but may be dirty due to misfiring or flying debris in the combustion chamber. Can lead to engine damage.
Recommendation: Check for the correct plug heat range, over-advanced ignition timing, lean fuel mixture, insufficient engine cooling and lack of lubrication.

HIGH SPEED GLAZING

Symptoms: Insulator has yellowish, glazed appearance. Indicates that combustion chamber temperatures have risen suddenly during hard acceleration. Normal deposits melt to form a conductive coating. Causes misfiring at high speeds.
Recommendation: Install new plugs. Consider using a colder plug if driving habits warrant.

DETONATION

Symptoms: Insulators may be cracked or chipped. Improper gap setting techniques can also result in a fractured insulator tip. Can lead to piston damage.
Recommendation: Make sure the fuel anti-knock values meet engine requirements. Use care when setting the gaps on new plugs. Avoid lugging the engine.

MECHANICAL DAMAGE

Symptoms: May be caused by a foreign object in the combustion chamber or the piston striking an incorrect reach (too long) plug. Causes a dead cylinder and could result in piston damage.
Recommendation: Repair the mechanical damage. Remove the foreign object from the engine and/or install the correct reach plug.

CONTENTS

QUICK REFERENCE DATA IX

CHAPTER ONE
GENERAL INFORMATION 1

Manual organization
Model names and numbers
Warnings, cautions and notes
Safety
Serial numbers
Fasteners
Shop supplies
Tools
Precision measuring tools
Electrical system fundamentals
Basic service methods
Storage
Serial numbers
Model names and numbers
Decimal and metric equivalents
Specifications
Conversion formulas
Technical abbreviations
Metric tap and drill size

CHAPTER TWO
TROUBLESHOOTING 34

Operating requirements
Starting the engine
Starting difficulties
Engine performance
Engine noises
Engine lubrication
Engine leakdown test
Clutch
Gearshift linkage
Transmission
Final drive
Carburetor
Carburetor heater
Fuel pump
Electrical testing
Electrical test equipment
Basic electric test procedures
Electrical system
Front suspension and steering
Brakes

CHAPTER THREE
LUBRICATION, MAINTENANCE AND TUNE-UP 56

- Fuel type
- Maintenance intervals
- Cylinder numbering
- Engine rotation
- Tune-up
- Spark plugs
- Engine oil
- Fuel and exhaust systems
- Control cables
- Transmission
- Sidestand
- General lubrication
- Battery
- Tires
- Front suspension
- Steering head
- Rear suspension
- Final drive
- Brakes
- Fasteners
- Maintenance schedule
- Recommended lubricants and fluids
- Specifications

CHAPTER FOUR
ENGINE TOP END 87

- Servicing engine in the frame
- Cylinder head covers
- Cylinder head
- Camshaft
- Rocker arms
- Cam chain and cam chain drive assembly
- Valves and valve components
- Cylinder
- Pistons and piston rings
- Specifications

CHAPTER FIVE
ENGINE LOWER END 133

- Servicing the engine in the frame
- Engine
- Alternator cover
- Stator and pickup coils
- Flywheel and starter clutch
- Oil pump
- Crankcase
- Crankshaft
- Connecting rods
- Break-in
- Connecting rod bearing insert selection
- Specifications

CHAPTER SIX
CLUTCH AND PRIMARY DRIVE GEAR 168

- Clutch cover
- Clutch
- Clutch release mechanism
- Clutch cable replacement
- Primary drive gear
- Specifications

CHAPTER SEVEN
TRANSMISSION, SHIFT MECHANISM AND MIDDLE GEAR 185

- Shift pedal/footrest assembly
- External shift mechanism
- Internal shift mechanism
- Transmission
- Middle drive shaft assembly
- Middle driven shaft assembly
- Checking middle gear backlash
- Specifications

CHAPTER EIGHT
AIR/FUEL, EXHAUST AND EMISSION CONTROL SYSTEMS 203

Fuel tank
Fuel valve
Air filter housing
Surge tank
Carburetor
Pilot screw
Fuel level
Idle speed adjustment
Throttle cable adjustment
Throttle position sensor
Fuel cut solenoid valve
Carburetor heater
Throttle cable
Choke cable
Fuel filter
Fuel pump
Air induction system
Evaporative emission control (California models)
Exhaust system
Specifications

CHAPTER NINE
ELECTRICAL SYSTEM 241

Charging system
Stator
Voltage regulator/rectifier
Ignition system
Spark plug cap
Ignition coil
Pickup coil
Diode
Ignitor unit
Starting system
Starter
Starting circuit cutoff relay
Starter relay
Lighting system
Headlight
Meter assembly
Taillight/brake light
Signal system
Horn
Turn signals
Neutral indicator light
Oil level indicator light
Speed sensor
Switches
Self-diagnostic system
Fuses
Wiring connectors
Replacement bulbs
Specifications

CHAPTER TEN
WHEEL AND TIRES 283

Motorcycle stand
Brake rotor protection
Wheel inspection
Bearing removal
Bearing installation
Front wheel
Front hub
Rear wheel
Rear hub
Laced wheel service
Wheel balance
Tire changing
Specifications

CHAPTER ELEVEN
FRONT SUSPENSION AND STEERING. 308

Front fork
Handlebar
Steering head
Steering head bearing races
Lower bearing replacement
Specifications

CHAPTER TWELVE
REAR SUSPENSION AND FINAL DRIVE 330

Shock absorber
Suspension linkage
Swing arm
Final drive assembly
Specifications

CHAPTER THIRTEEN
BRAKES 348

Brake service
Preventing brake fluid damage
Brake bleeding
Brake fluid draining
Brake pads
Front caliper
Front master cylinder
Rear brake caliper
Rear brake master cylinder
Brake pedal/footrest assembly
Brake disc
Brake hose replacement
Specifications

CHAPTER FOURTEEN
FRAME 381

Seats
Ignitor panel
Left side cover
Toolbox cover
Toolbox panel
Battery cover
Right side cover
Battery box
Frame neck cover
Rider floorboards
Passenger footpegs
Sidestand
Front fender
Rear fender
Windshield
Specifications

INDEX. 395

WIRING DIAGRAMS 400

QUICK REFERENCE DATA

MODEL:______________________________ YEAR:______________

VIN NUMBER:______________________________

ENGINE SERIAL NUMBER:______________________________

CARBURETOR SERIAL NUMBER OR I.D. MARK:______________________________

TIRE INFLATION PRESSURE[1]

XVS1100 models	
0-90 kg (0-198 lb.) load[2]	
Front	200 kPa (28.5 psi)
Rear	225 kPa (32 psi)
90-200 kg (198-441 lb.) load[2]	
Front	225 kPa (32 psi)
Rear	250 kPa (36 psi)
XVS1100A models	
0-90 kg (0-198 lb.) load[2]	
Front	225 kPa (32 psi)
Rear	225 kPa (32 psi)
90-200 kg (198-441 lb.) load[2]	
Front	225 kPa (32 psi)
Rear	250 kPa (36 psi)
Maximum load[2]	200 kg (441 lb.)

1. Tire inflation pressures apply to original equipment tires only. Aftermarket tires may require different pressures. Refer to the tire manufacturer's specifications.
2. Load equals the total weight of rider, passenger, accessories and all cargo.

RECOMMENDED LUBRICANTS AND FLUIDS

Fuel	Regular unleaded
Octane	86 [(R + M)/2 method] or research octane of 91 or higher
Capacity	17 liter (4.49 U.S. gal.)
Reserve	4.5 liter (1.19 U.S. gal.)
Engine oil	
API classification	SE, SF or SG
Viscosity	
5° C (40° F) or above	SAE 20W40
15° C (60° F) or below	SAE 10W30
Capacity	
Oil change only	3.0 L (3.2 U.S. qt.)
Oil and filter change	3.1 L (3.3 U.S. qt.)
When engine completely dry	3.6 L (3.8 U.S. qt.)

(continued)

RECOMMENDED LUBRICANTS AND FLUIDS (continued)

Final gear oil	
Viscosity	
Single grade	SAE 80 hypoid gear oil
Multigrade	SAE 80W-90 hypoid gear oil
Grade	API GL-4, GL-5 or GL-6
Capacity	200 cc (6.8 U.S. oz.)
Brake fluid	DOT 4
Battery	Maintenance free
Fork oil	
Viscosity	SAE 10W fork oil
Capacity per leg	
1999-2003 all models	464 cc (15.7 U.S. oz.)
2004-on XVS1100 models	464-481 cc (15.8-16.3 U.S. oz.)
2004-on XVS1100A models	488 cc (16.5 U.S. oz.)
Oil level (measured from top of the fully compressed fork tube with the fork spring removed)	
1999-2003 models	108 mm (4.25 in.)
2004-on XVS1100 models	99-105 mm (3.89-4.13 in.)
2004 XVS1100A models	99 mm (3.89 in.)

MAINTENANCE AND TUNE-UP SPECIFICATIONS

Item	Specification
Recommended spark plug	NGK BPR7ES, Denso W22EPR-U
Spark plug gap	0.7-0.8 mm (0.028-0.031 in.)
Idle speed	950-1050 rpm
Pilot screw (Europe and Australia models)	3 turns out
Valve clearance	
Intake	0.07-0.12 mm (0.0028-0.0047 in.)
Exhaust	0.12-0.17 mm (0.0047-0.0067 in.)
Compression pressure (at sea level)	
Standard	1000 kPa (142 psi)
Minimum	900 kPa (128 psi)
Maximum	1100 kPa (156 psi)
Ignition timing	10° BTDC @ 1000 rpm
Vacuum pressure (at idle)	34.7-37.3 kPa (260-289 mm Hg [10.2-11.4 in. Hg])
Front brake pad wear limit	0.8 mm (0.03 in.)
Rear brake pad wear limit	0.5 mm (0.020 in.)
Brake pedal height	
XVS1100 models (above footpeg)	81.8 mm (3.22 in.)
XVS1100A models (above floorboard)	98.5 mm (3.88 in.)
Throttle cable free play (at the flange)	4-6 mm (0.16-0.24 in.)
Brake free play (at lever end)	5-8 mm (0.20-0.31 in.)
Clutch cable free play (at lever end)	5-10 mm (0.20-0.39 in.)
Shift rod length	
XVS1100 models	114.7 mm (4.52 in.)
XVS1100A models	114.9 mm (4.53 in.)
Rim runout service limit	
Radial	1.0 mm (0.04 in.)
Axial	0.5 mm (0.02 in.)

MAINTENANCE AND TUNE-UP TORQUE SPECIFICATIONS

Item	N•m	in.-lb.	ft.-lb.
Air filter cover bolts	2	18	–
Air filter housing bolts	10	89	–
Brake pedal height adjuster locknut	16	–	12
Cam sprocket cover bolts	10	89	–
Clutch adjuster locknut	12	106	–
Final gearcase drain bolt	23	–	17
Final gearcase oil filler bolt	23	–	17
Fork bottom Allen bolt*	30		22
Fork cap bolt	23	–	17
Front axle	59	–	43
Front axle clamp bolt	20	–	15
Front brake caliper mounting bolt	40	–	30
Front caliper retaining bolt			
1999-on XVS1100 models	23	–	17
All models except 1999-on XVS1100 models	27	–	20
Lower fork bridge clamp bolt	30	–	22
Oil drain bolt	43	–	32
Oil filter outer cover bolt	10	89	–
Rear axle clamp nut	23	–	17
Rear axle nut	107	–	79
Rear brake adjuster locknut	16	–	12
Rear caliper bracket bolt	40	–	30
Rear caliper mounting bolt	40	–	30
Rocker arm bolt	38	–	28
Sidestand nut	56	–	41
Spark plugs	20	–	15
Steering head nut	110	–	81
Steering stem adjuster nut			
First stage	52	–	38
Second stage	18	–	13
Upper fork bridge clamp bolt	20	–	15
Valve adjuster locknut	27	–	20
Valve cover bolts	10	89	–

*Apply Loctite 242 (blue) or an equivalent medium strength threadlocking compound.

CHAPTER ONE

GENERAL INFORMATION

This detailed and comprehensive manual covers Yamaha XVS1100 and XVS1100A models from 1999-2009. The text provides complete information on maintenance, tune-up, repair and overhaul. Hundreds of original photographs and illustrations created during the complete disassembly of the motorcycle guide the reader through every job. All procedures are in step-by-step form and designed for the reader who may be working on the motorcycle for the first time.

MANUAL ORGANIZATION

A shop manual is a tool and as in all Clymer manuals, the chapters are thumb tabbed for easy reference. Main headings are listed in the table of contents and the index. Frequently used specifications and capacities from the tables at the end of each individual chapter are listed in the *Quick Reference Data* section at the front of the manual. Specifications and capacities are provided in U.S. standard and metric units of measure.

During some of the procedures there will be references to headings in other chapters or sections of the manual. When a specific heading is called out in a step it will be *italicized* as it appears in the manual. If a sub-heading is indicated as being "in this section" it is located within the same main heading. For example, the sub-heading ***Handling Gasoline Safely*** is located within the main heading ***SAFETY***.

This chapter provides general information on shop safety, tools and their usage, service fundamentals and shop supplies. Refer to **Tables 1-4** at the end of this chapter for model numbers, general motorcycle information and general shop technical information.

Subsequent chapters describe troubleshooting, maintenance, engine, transmission, clutch, fuel and exhaust systems, suspension and brake procedures.

MODEL NAMES AND NUMBERS

Yamaha has used several names to identify the motorcycles covered in this manual. For example, XVS1100 models have been named V-Star 1100, V-Star 1100 Custom, V-Star 1100 Custom Midnight and V-Star 1100 Custom Flame. The XVS1100A models have been named V-Star 100 Classic and V-Star 1100 Silverado.

In the interest of clarity, this manual refers to the models by their number, either XVS1100 or XVS1100A.

WARNINGS, CAUTIONS AND NOTES

The terms WARNING, CAUTION and NOTE have specific meanings in this manual.

A WARNING emphasizes areas where injury or even death could result from negligence. Mechanical damage may also occur. WARNINGS *should be taken seriously.*

A CAUTION emphasizes areas where equipment damage could result. Disregarding a CAUTION could cause permanent mechanical damage, though injury is unlikely.

A NOTE provides additional information to make a step or procedure easier or clearer. Disregarding a NOTE could cause inconvenience but would not cause equipment damage or personal injury.

SAFETY

Professional mechanics can work for years and never sustain a serious injury or mishap. Follow these guidelines and practice common sense to safely service the motorcycle.

1. Do not operate the vehicle in an enclosed area. The exhaust gasses contain carbon monoxide, a poisonous gas that is odorless, colorless, and tasteless. Carbon monoxide levels build quickly in a small enclosed area, and it can cause unconsciousness and death in a short time. Make sure the work area is properly ventilated or operate the vehicle outside.
2. *Never* use gasoline or any extremely flammable liquid to clean parts. Refer to *Cleaning Parts* and *Handling Gasoline Safely* in this section.
3. *Never* smoke or use a torch in the vicinity of flammable liquids, such as gasoline or cleaning solvent.
4. When welding or brazing on the motorcycle, remove the fuel tank, carburetor and shocks to a safe distance at least 50 ft. (15 m) away.
5. Use the correct type and size tool to avoid damaging fasteners.
6. Keep tools clean and in good condition. Replace or repair worn or damaged equipment.
7. When loosening a tight fastener, be guided by what would happen if the tool slips.
8. When replacing fasteners, make sure the new fasteners are of the same size and strength as the original ones.
9. Keep the work area clean and organized.
10. Wear eye protection *any time* the safety of your eyes is in question. This includes procedures involving drilling, grinding, hammering, compressed air and chemicals.
11. Wear the correct clothing for the job. Tie up or cover long hair so it cannot be caught in moving equipment.
12. Do not carry sharp tools in clothing pockets.
13. Always have an approved fire extinguisher available. Make sure it is rated for gasoline (Class B) and electrical (Class C) fires.
14. Do not use compressed air to clean clothes, the motorcycle or the work area. Debris may be blown into your eyes or skin. *Never* direct compressed air at yourself or someone else. Do not allow children to use or play with any compressed air equipment.
15. When using compressed air to dry rotating parts, hold the part so it can not rotate. Do not allow the force of the air to spin the part. The air jet is capable of rotating parts at extreme speed. The part may be damaged or disintegrate, causing serious injury.
16. Do not inhale the dust created by brake pad and clutch wear. In most cases, these particles contain asbestos. In addition, some types of insulating materials and gaskets may contain asbestos. Inhaling asbestos particles is hazardous to health.
17. Never work on the vehicle while someone is working under it.
18. When placing the vehicle on a stand, make sure it is secure before walking away.

Handling Gasoline Safely

Gasoline is a volatile, flammable liquid and is one of the most dangerous items in the shop.

Because gasoline is used so often, many people forget that it is hazardous. Only use gasoline as fuel for gasoline internal combustion engines. When working on a vehicle, remember that gasoline is always present in the fuel tank, fuel lines and carburetors. To avoid a disastrous accident when working around the fuel system, observe the following precautions:

1. *Never* use gasoline to clean parts. See *Cleaning Parts* in this section.
2. When working on the fuel system, work outside or in a well-ventilated area.

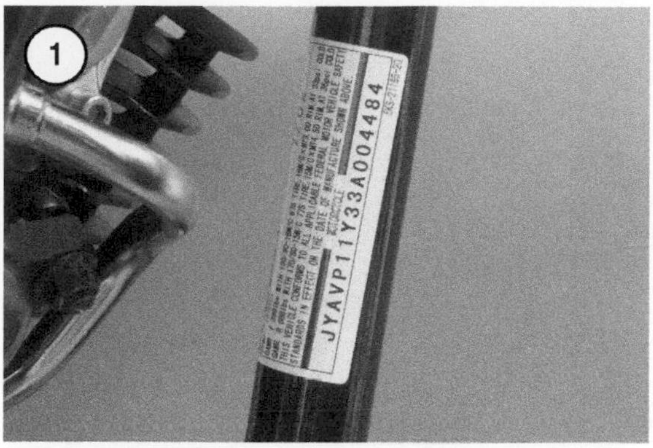

3. Do not add fuel to the fuel tank or service the fuel system while the vehicle is near open flames, around sparks or near someone who is smoking. Gasoline vapor is heavier than air; it collects in low areas and is more easily ignited than liquid gasoline.

4. Allow the engine to cool completely before working on any fuel system component.

5. When draining the carburetor, catch the fuel in a plastic container and then pour it into an approved gasoline storage container.

6. Do not store gasoline in glass containers. If the glass breaks, a serious explosion or fire may occur.

7. Immediately wipe up spilled gasoline with rags. Store the rags in a metal container with a lid until they can be properly disposed of, or place them outside in a safe place for the fuel to evaporate.

8. Do not pour water onto a gasoline fire. Water spreads the fire and makes it more difficult to extinguish. Use a class B, BC or ABC fire extinguisher to put out a gasoline fire.

9. Always turn off the engine before refueling. Do not spill fuel onto the engine or exhaust system. Do not overfill the fuel tank. Leave an air space at the top of the tank to allow room for the fuel to expand due to temperature fluctuations.

Cleaning Parts

Cleaning parts is one of the more tedious and difficult service jobs performed in the home garage. There are many types of chemical cleaners and solvents available for shop use. Most are poisonous and extremely flammable. To prevent chemical exposure, vapor buildup, fire and serious injury, observe each product warning label and note the following:

1. Read and observe the entire product label before using any chemical. Always know what type of chemical is being used and whether it is poisonous and/or flammable.

2. Do not use more than one type of cleaning solvent at a time. If mixing chemicals is called for, measure the proper amounts according to the manufacturer's instructions.

3. Work in a well-ventilated area.

4. Wear chemical-resistant gloves.

5. Wear safety glasses.

6. Wear a vapor respirator if the instructions call for it.

7. Wash hands and arms thoroughly after cleaning parts.

8. Keep chemical products away from children and pets.

9. Thoroughly clean all oil, grease and cleaner residue from any part that must be heated.

10. Use a nylon brush when cleaning parts. Metal brushes may cause a spark.

11. When using a parts washer, only use the solvent recommended by the manufacturer. Make sure the parts washer is equipped with a metal lid that will lower in case of fire.

Warning Labels

Most manufacturers attach information and warning labels to the vehicle. These labels contain instructions that are important to personal safety when operating, servicing, transporting and storing the vehicle. Refer to the owner's manual for the description and location of labels. Order replacement labels from the manufacturer if they are missing or damaged.

SERIAL NUMBERS

Serial numbers and model codes are stamped into the motorcycle and/or printed on labels.

The VIN or vehicle identification number (**Figure 1**) appears on a label on the inside of the left fork leg. This number is also stamped on the right side of the steering head (**Figure 2**). Yamaha's primary ID number is a variation of the VIN number.

The model code (**Figure 3**) appears on a label on the side of the right main frame tube.

The engine number (**Figure 4**) is stamped on the right side of the upper crankcase.

Record these numbers in the *Quick Reference Data* section at the front of the manual. Have these numbers available when ordering parts.

FASTENERS

Proper fastener selection and installation is important to ensure that the vehicle operates as designed and can be serviced efficiently. Original equipment fasteners are designed for their specific applications. Make sure replacement fasteners meet all the same requirements as the originals.

Threaded Fasteners

Threaded fasteners secure most of the components on the vehicle. Most are tightened by turning them clockwise (right-hand threads). If the normal rotation of the component would loosen the fastener, it may have left-hand threads. If a left-hand threaded fastener is used, it is noted in the text.

Two dimensions are required to match the thread size of the fastener: the number of threads in a given distance and the outside diameter of the threads.

Two systems are currently used to specify threaded fastener dimensions: the U.S. standard system and the metric system (**Figure 5**). Pay particular attention when working with unidentified fasteners. Mismatching thread types can damage threads.

CAUTION

To ensure the fastener threads are not mismatched or cross-threaded, start all fasteners by hand. If a fastener is hard to start or turn, determine the cause before tightening it with a wrench.

The length (L, **Figure 6**), diameter (D) and distance between thread crests (pitch) (T) classify metric screws and bolts. A typical bolt may be identified by the numbers, 8—1.25 × 130. This indicates the bolt has diameter of 8 mm, the distance between thread crests is 1.25 mm and the length is 130 mm. Always measure bolt length as shown in **Figure 6** to avoid purchasing replacements of the wrong length.

The numbers located on the top of the fastener (**Figure 6**) indicate the strength of metric screws and bolts. The higher the number, the stronger the fastener. Unnumbered fasteners are the weakest.

Many bolts and studs are combined with nuts to secure particular components. To indicate the size of a nut, manufacturers specify the internal diameter and the thread pitch.

The measurement across two parallel flats on a nut or bolt head indicates the wrench size that fits the fastener.

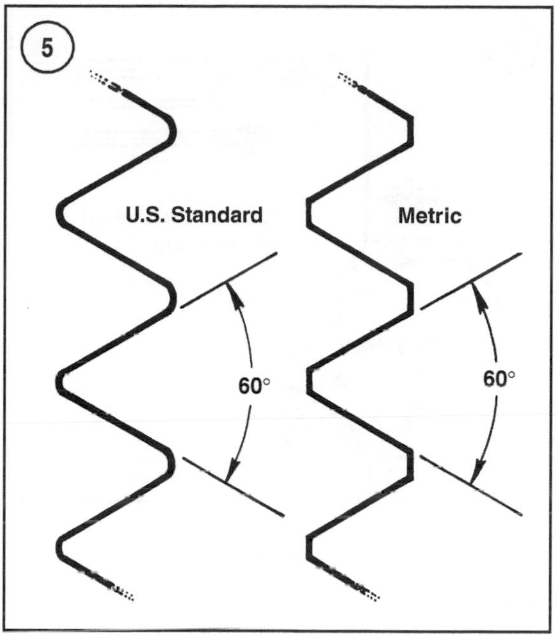

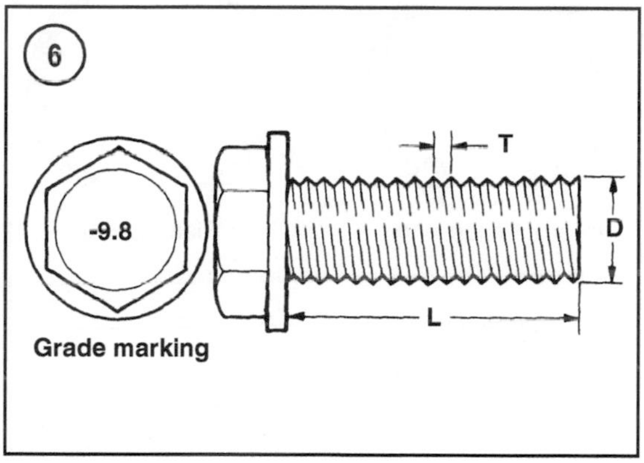

WARNING
Do not install fasteners with a strength classification lower than what was originally installed by the manufacturer. Doing so may cause equipment failure and/or damage.

Torque Specifications

The components of the motorcycle may be subjected to uneven stresses if the fasteners of the various subassemblies are not installed and tightened correctly. Fasteners that are improperly installed or that work loose can cause extensive damage. Use an accurate torque wrench when tightening fasteners, and tighten each fastener to its specified torque.

Torque specifications for specific components appear in the procedures and at the end of the appropriate chapters. Specifications for torque are provided in Newton-meters (N•m), foot-pounds (ft.-lb.) and inch-pounds (in.-lb.). Refer to **Table 3** for torque conversion factors and to **Table 2** for torque recommendations. To use the general torque table, first determine the size of the fastener as described in *Threaded Fasteners* in this section. Locate that size fastener in **Table 2**, and tighten the fastener to the indicated torque. Torque wrenches are described under *Basic Tools* in this chapter.

Self-Locking Fasteners

Several types of bolts, screws and nuts use various means to create interference between the threads of two fasteners to prevent the fasteners from loosening. The most common types are the nylon-insert nut or a dry adhesive coating on the threads of a bolt.

Self-locking fasteners improve resistance to vibration and therefore provide greater holding strength than standard fasteners. Most self-locking fasteners cannot be reused. The materials used to form the lock become distorted after the initial installation and removal. Always discard and replace self-locking fasteners after their removal. Do not replace self-locking fasteners with standard fasteners.

Washers

There are two basic types of washers: flat washers and lockwashers. Flat washers are simple discs with a hole for a screw or bolt. Flat washers help distribute fastener load, they protect components from fastener damage, and they can be used as spacers and seals. Lockwashers are used to prevent a fastener from working loose.

When replacing washers, make sure the replacements are of the same design and quality as the originals.

Cotter Pins

A cotter pin is a split metal pin inserted into a hole or slot to prevent a fastener from working loose. In certain applications, such as the rear axle on an ATV or motorcycle, a cotter pin and castellated (slotted) nut is used.

To use a cotter pin, first make sure the pin's diameter is correct for the hole in the fastener. After correctly tightening the fastener and aligning the holes, insert the cotter pin through the hole and bend the ends over the fastener (**Figure 7**). Unless instructed to do so, never loosen a torqued fastener to align the holes. If the holes do not align, tighten the fastener just enough to achieve alignment.

Cotter pins are available in various diameters and lengths. Measure length from the bottom of the head to the tip of the shortest pin.

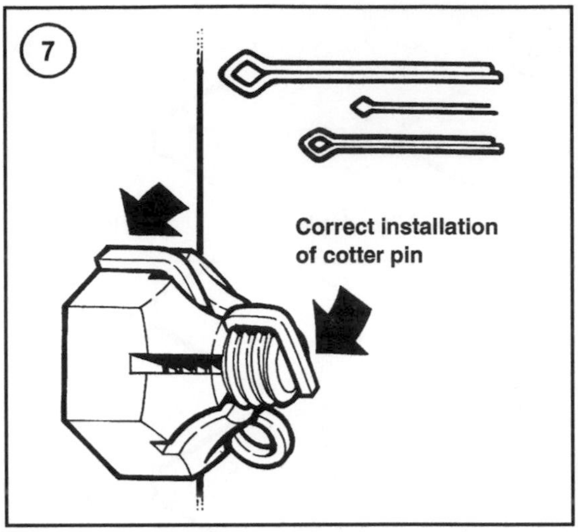

Snap Rings and E-clips

Snap rings (**Figure 8**) are circular-shaped metal retaining clips. They secure parts and gears onto shafts, pins or rods. External type snap rings are used to retain items on shafts. Internal type snap rings secure parts within housing bores. In some applications, in addition to securing the component(s), snap rings of varying thickness also determine endplay. These are usually called selective snap rings.

Two basic types of snap rings are used: machined and stamped snap rings. Machined snap rings (**Figure 9**) can be installed in either direction since both faces have sharp edges. Stamped snap rings (**Figure 10**) are manufactured with a sharp edge and a round edge. When installing a stamped snap ring in a thrust application, install the sharp edge facing away from the part producing the thrust.

E-clips are used when it is not practical to use a snap ring. Remove E-clips with a flat blade screwdriver by prying between the shaft and E-clip. To install an E-clip, center it over the shaft groove, and push or tap it into place.

Observe the following when installing snap rings:

1. Remove and install snap rings with snap ring pliers. See *Snap Ring Pliers* in this chapter.
2. In some applications, it may be necessary to replace snap rings after removing them.
3. Compress or expand snap rings only enough to install them. If overly expanded, they lose their retaining ability.
4. After installing a snap ring, make sure it seats completely.
5. Wear eye protection when removing and installing snap rings.

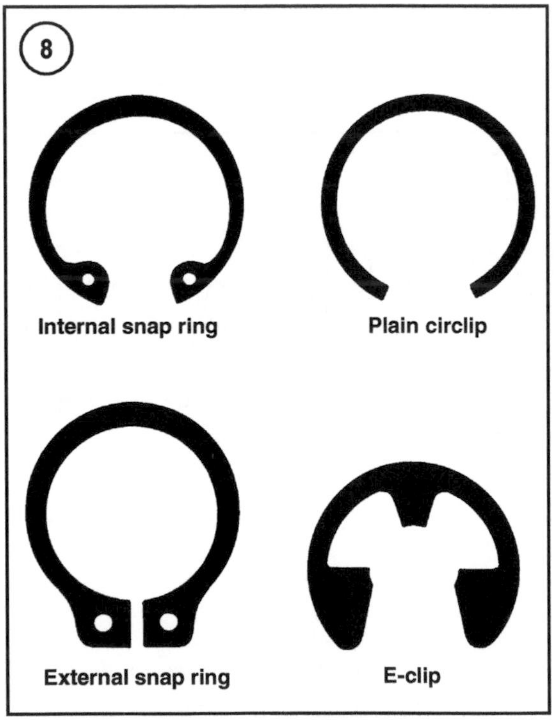

SHOP SUPPLIES

Lubricants and Fluids

Periodic lubrication helps ensure a long service life for any type of equipment. Using the correct type of lubricant is as important as performing the lubrication service, although in an emergency the wrong type of lubricant is better than none. The following section describes the types of lubricants

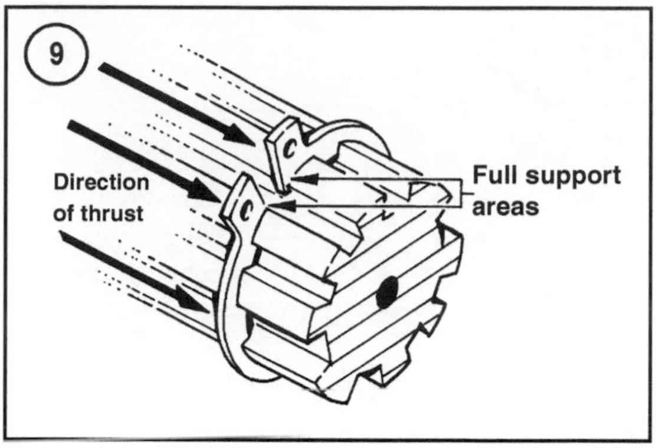

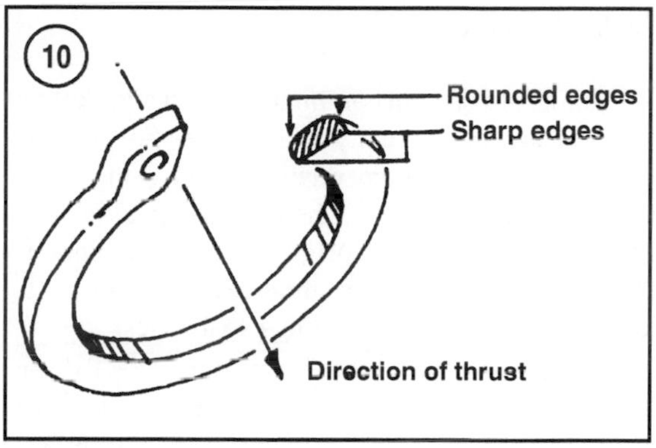

most often required. Make sure to follow the manufacturer's recommendations for lubricant types.

Engine oils

Generally, all liquid lubricants are called oil. They may be mineral-based (including petroleum bases), natural-based (vegetable and animal bases), synthetic-based, or emulsions (mixtures).

Engine oil is classified by two standards: the American Petroleum Institute (API) service classification and the Society of Automotive Engineers (SAE) viscosity rating. This information is on the oil container label. Two letters indicate the API service classification (SF, SG, etc.). The number or sequence of numbers and letter (10W-40 for example) is the oil's viscosity rating. The API service classification and the SAE viscosity index are not indications of oil quality.

The service classification indicates that the oil meets specific lubrication standards. The first letter in the classification (*S*) indicates that the oil is for gasoline engines. The second letter indicates the standard the oil satisfies.

Always use an oil with a classification recommended by the manufacturer. Using an oil with a different classification can cause engine damage.

Viscosity is an indication of the oil's thickness. Thin oils have a lower number while thick oils have a higher number. A *W* after the number indicates that the viscosity testing was done at low temperature to simulate cold-weather operation. Engine oils fall into the 5- to 50-weight range for single-grade oils.

Most manufacturers recommend multigrade oil. Multigrade oils (10W-40, for example) are less viscous (thinner) at low temperatures and more viscous (thicker) at high temperatures. This allows the oil to perform efficiently across a wide range of engine operating conditions. The lower the number, the better the engine will start in cold climates. Higher numbers are usually recommended when operating an engine in hot weather.

Greases

Grease is an oil to which a thickening base has been added so the end product is semi-solid. Grease is often classified by the type of thickener added, such as lithium soap. The National Lubricating Grease Institute (NLGI) grades grease. Grades range from No. 000 to No. 6, with No. 6 being the thickest. Typical multipurpose grease is NLGI No. 2. For specific applications, manufacturers may recommend a water-resistant type grease or one with an additive such as molybdenum disulfide (MoS_2).

Brake fluid

Brake fluid is the hydraulic fluid used to transmit hydraulic pressure (force) to the wheel brakes. Brake fluid is classified by the Department of Transportation (DOT). Current designations for brake fluid are DOT 3, DOT 4 and DOT 5. This classification appears on the fluid container.

Each type of brake fluid has its own definite characteristics. Do not intermix different types of brake fluid. DOT 5 fluid is silicone-based. DOT 5 is not compatible with other fluids or in systems for which it was not designed. Mixing DOT 5 fluid with other fluids may cause brake system failure. When adding brake fluid, *only* use the fluid recommended by the vehicle manufacturer.

Brake fluid will damage plastic, painted or plated surfaces. Use extreme care when working with brake fluid. Immediately clean up any spills with soap and water. Rinse the area with plenty of clean water.

Hydraulic brake systems require clean and moisture-free brake fluid. Never reuse brake fluid.

Brake fluid absorbs moisture, which greatly reduces its ability to perform correctly. Keep brake fluid containers and reservoirs properly sealed. Purchase brake fluid in small containers, and discard any small leftover quantities properly. Do not store a container of brake fluid with less than 1/4 of the fluid remaining. This small amount absorbs moisture very rapidly.

WARNING

Never put a mineral-based (petroleum) oil into the brake system. Mineral oil will cause rubber parts in the system to swell and break apart, resulting in complete brake failure.

Cleaners, Degreasers and Solvents

Many chemicals are available to remove oil, grease and other residue from the vehicle. Before using cleaning solvents, consider how they will be used and disposed of, particularly if they are not water-soluble. Local ordinances may require special procedures for the disposal of various cleaning chemicals. Refer to *Safety* and *Cleaning Parts* in this chapter for more information on their use.

Generally, degreasers are strong cleaners used to remove heavy accumulations of grease from engine and frame components.

Use brake parts cleaner to clean brake system components when contact with petroleum-based products will damage seals. Brake parts cleaner leaves no residue.

Use electrical contact cleaner to clean electrical connections and components without leaving any residue.

Carburetor cleaner is a powerful solvent used to remove fuel deposits and varnish from fuel system components. Use this cleaner carefully, as it may damage finishes.

Most solvents are designed to be used in a parts washing cabinet for individual component cleaning. For safety, use only nonflammable or high flash point solvents.

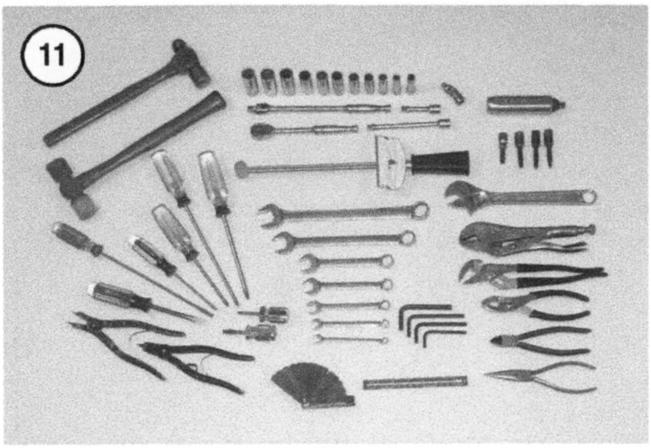

Gasket Sealant

Sealants are used in combination with a gasket or seal and are occasionally used alone. Follow the manufacturer's recommendation when using sealants. Use extreme care when choosing a sealant other than the type originally recommended. Choose sealants based on their resistance to heat, various fluids and their sealing capabilities.

One of the most common sealants is RTV, or room temperature vulcanizing sealant. This sealant cures at room temperature over a specific time period. It allows the repositioning of components without damaging gaskets.

Moisture in the air causes the RTV sealant to cure. Always install the tube cap as soon as possible after applying RTV sealant. RTV sealant has a limited shelf life and will not cure properly if the shelf life has expired. Keep partial tubes sealed, and discard them if they have surpassed the expiration date.

Applying RTV sealant

Clean all old gasket residue from the mating surfaces. Remove all gasket material from blind threaded holes; it can cause inaccurate bolt torque. Spray the mating surfaces with aerosol parts cleaner, and then wipe them with a lint-free cloth. The area must be clean for the sealant to adhere.

Apply RTV sealant in a continuous bead, 2-3 mm (0.08-0.12 in.) thick. Circle all the fastener holes unless otherwise specified. Do not allow any sealant to enter these holes. Assemble and tighten the fasteners to the specified torque within the time frame recommended by the RTV sealant manufacturer.

12

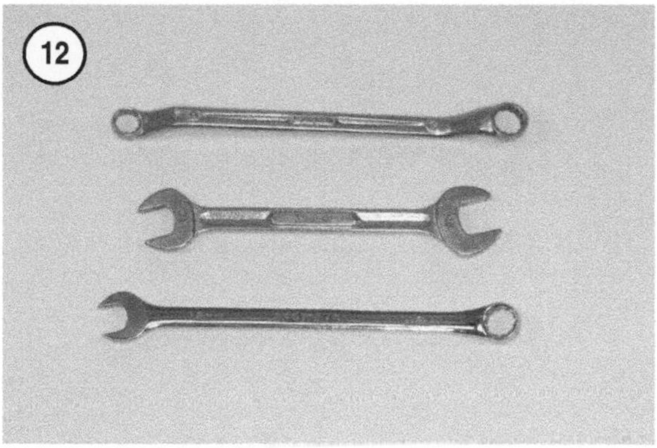

Gasket Remover

Aerosol gasket remover can help remove stubborn gaskets. This product can speed up the removal process and prevent damage to the mating surface that may be caused by a scraping tool. Most of these products are very caustic. Follow the gasket remover manufacturer's instructions for use.

Threadlocking Compound

A threadlocking compound is a fluid applied to the threads of fasteners. After tightening the fastener, the fluid dries and becomes a solid filler between the threads. This makes it difficult for the fastener to work loose from vibration or heat expansion and contraction. Some threadlocking compounds also provide a seal against fluid leaks.

Before applying threadlocking compound, remove any old compound from both thread areas and clean them with aerosol parts cleaner. Use the compound sparingly. Excess fluid can run into adjoining parts.

Threadlocking compounds are available in various strength, temperature and repair applications. Follow the manufacturer's recommendations regarding compound selection.

TOOLS

Most of the procedures in this manual can be carried out with simple hand tools and test equipment familiar to the home mechanic. Always use the correct tools for the job at hand. Keep tools organized and clean. Store them in a tool chest with related tools organized together.

Quality tools are essential. The best are constructed of high-strength alloy steel. These tools are light, easy to use and resistant to wear. Their working surface is smooth, and the tool is carefully polished. They have an easy-to-clean finish and are comfortable to use. Quality tools are a good investment.

When building a new tool kit, consider purchasing a basic set (**Figure 11**) from a large tool supplier. These sets contain a variety of commonly used tools, and they provide substantial savings when compared to individually purchased tools. As one becomes more experienced and tasks become more complicated, specialized tools can be added.

Some of the procedures call for special tools. In most cases, the tool is illustrated in use. It may be possible to substitute similar tools or fabricate a suitable replacement. However, in some cases, the specialized equipment may make it impractical for the home mechanic to perform the procedure. When necessary, such operations are identified in the text. It may be less expensive to have a professional perform these tasks, especially if the cost of the equipment is high.

In the case of Yamaha special tools, two part numbers are listed. The part number for special tools sold in the United States begin with a two letter prefix, for example: YM-90069. Part numbers for tools sold in Europe and Australia markets begin with a number, for example: 90890-01290.

Screwdrivers

Screwdrivers of various lengths and types are mandatory for the simplest tool kit. The two basic types are the slotted tip (flat blade) and the Phillips tip. These are available in sets that often include an assortment of tip sizes and shaft lengths.

As with all tools, use a screwdriver designed for the job. Make sure the size of the tip conforms to the size and shape of the fastener. Use them only for driving screws. Never use a screwdriver for prying or chiseling metal. Repair or replace worn or damaged screwdrivers. A worn tip may damage the fastener, making it difficult to remove.

Wrenches

Box-end, open-end and combination wrenches (**Figure 12**) are available in a variety of types and sizes.

The number stamped on the wrench refers to the distance between the work areas. This must match the distance across two parallel flats on the bolt head or nut.

The box-end wrench is an excellent tool because it grips the fastener on all sides. This reduces the chance of the tool slipping. The box-end wrench is designed with either a 6- or 12-point opening. For stubborn or damaged fasteners, the 6-point provides superior holding ability by contacting the fastener across a wider area at all six edges. For general use, the 12-point works well. It allows the wrench to be removed and reinstalled without moving the handle over such a wide arc.

An open-end wrench is fast and works best in areas with limited overhead access. Because it contacts the fastener at only two points, an open-end wrench is subject to slipping under heavy force or if the tool or fastener is worn. A box-end wrench is preferred in most instances, especially when applying considerable force to a fastener.

The combination wrench has a box-end on one end and an open-end on the other. This combination makes it a very convenient tool.

Adjustable Wrenches

An adjustable wrench or Crescent wrench (**Figure 13**) fits nearly any nut or bolt head that has clear access around its entire perimeter. An adjustable wrench is best used as a backup wrench to hold a large nut or bolt while the other end is being loosened or tightened with a box-end or socket wrench.

Adjustable wrenches contact the fastener at only two points, which makes them more subject to slipping off the fastener. The fact that one jaw is adjustable and may loosen only aggravates this shortcoming. Make certain the solid jaw is the one transmitting the force.

Socket Wrenches, Ratchets and Handles

Sockets that attach to a ratchet handle (**Figure 14**) are available with 6-point (A, **Figure 15**) or 12-point (B) openings and different drive sizes. The drive size indicates the size of the square hole that accepts the ratchet handle. The number stamped on the socket is the size of the work area and must match the fastener head.

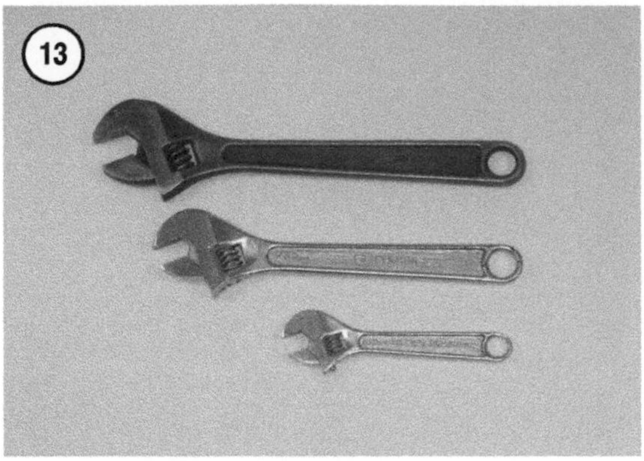

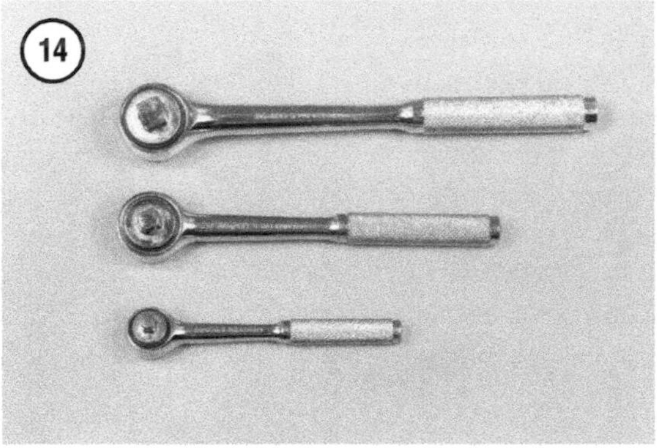

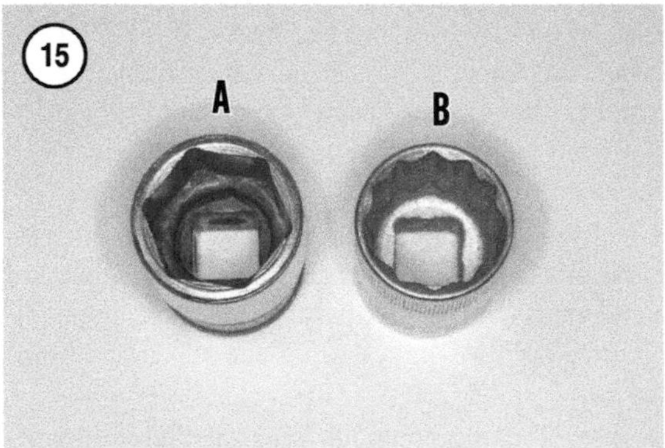

As with wrenches, a 6-point socket provides superior-holding ability, while a 12-point socket needs to be moved only half as far to reposition it on the fastener.

Sockets are designated for use with either hand. Impact sockets are made of thicker material for more durability. Compare the size and wall thickness of a 19-mm hand socket (A, **Figure 16**) and the 19-mm impact socket (B). Use impact sockets when using an impact driver or air tools. Use hand sockets with hand-driven attachments.

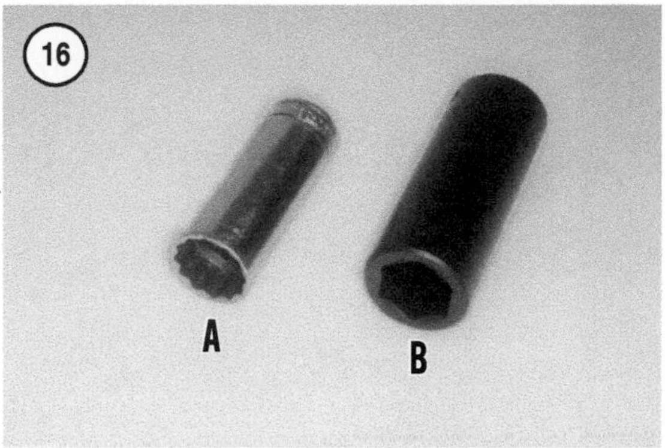

16

17

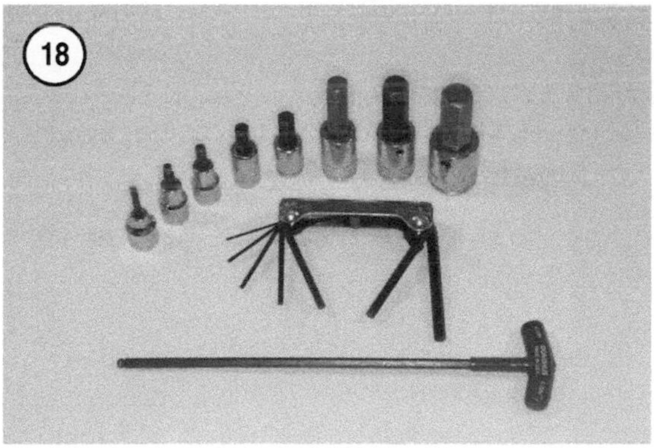

18

WARNING
Do not use hand sockets with air or impact tools. They may shatter and cause injury. Always wear eye protection when using impact or air tools.

Various handles are available for sockets. The speed handle is used for fast operation. Flexible ratchet heads in varying lengths allow the socket to be turned with varying force and at odd angles. Extension bars allow the socket setup to reach difficult areas. The ratchet is the most versatile wrench. It allows the user to install or remove the nut without removing the socket.

Sockets combined with any number of drivers make them undoubtedly the fastest, safest and most convenient tool for fastener removal and installation.

Impact Driver

An impact driver provides extra force for removing fasteners by converting the impact of a hammer into a turning motion. This makes it possible to remove stubborn fasteners without damaging them. Impact drivers and interchangeable bits (**Figure 17**) are available from most tool suppliers. When using a socket with an impact driver, make sure the socket is designed for impact use. Refer to *Socket Wrenches, Ratchets and Handles* in this section.

WARNING
Do not use hand sockets with air or impact tools. They may shatter and cause injury. Always wear eye protection when using impact or air tools.

Allen Wrenches

Allen or setscrew wrenches (**Figure 18**) are used on fasteners with hexagonal recesses in the fastener head. These wrenches are available in L-shaped bars, sockets and T-handles. A metric set is required when working on most vehicles made by Japanese and European manufacturers. Allen bolts are sometimes called socket bolts.

Torque Wrenches

A torque wrench is used with a socket, torque adapter or similar extension to tighten a fastener to a measured torque. Torque wrenches come in several drive sizes (1/4, 3/8, 1/2 and 3/4) and use various methods of reading the torque value. The drive size indicates the size of the square drive that accepts the socket, adapter or extension. Common methods of reading the torque value are the deflecting beam (A, **Figure 19**), the dial indicator (B) and the audible click (C).

When choosing a torque wrench, consider the torque range, drive size and accuracy. The torque

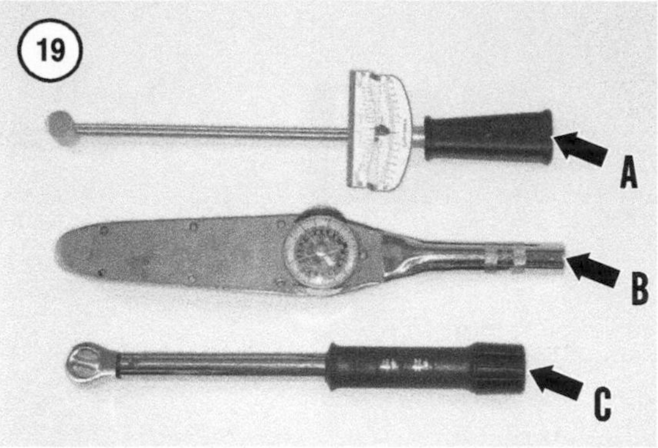

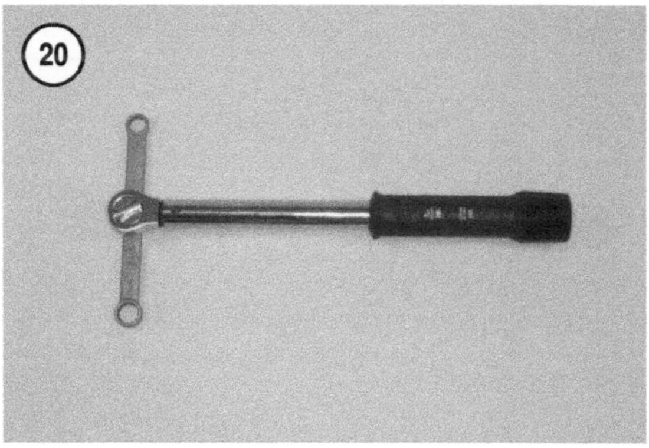

specifications in this manual provide an indication of the range required.

A torque wrench is a precision tool that must be properly cared for to remain accurate. Store torque wrenches in cases or separate padded drawers within a toolbox. Follow the manufacturer's instructions for their care and calibration.

Torque Adapters

Torque adapters extend or reduce the reach of a torque wrench. The torque adapter shown in **Figure 20** is used to tighten a fastener that cannot be reached due to the size of the torque wrench head, drive, and socket. Since a torque adapter changes the effective lever length (**Figure 21**) of a torque wrench, the torque reading on the wrench does not equal the actual torque applied to the fastener. It is necessary to calculate the adjusted torque reading on the wrench to compensate for the change of lever length. When a torque adapter is used at a right angle to the drive head, calibration is not required, since the effective length has not changed.

To calculate the adjusted torque reading when using a torque adapter, use the following formula.

$$TW = \frac{TA \times L}{L + A}$$

TW is the torque setting or dial reading on the wrench.

TA is the torque specification and the actual amount of torque that will be applied to the fastener.

A is the amount that the adapter increases (or in some cases reduces) the effective lever length as measured along the centerline of the torque wrench from the center of the drive to the center of adapter box end (**Figure 21**).

L is the lever length of the wrench as measured from the center of the drive to the center of the grip.

The effective length of the torque wrench is measured along the centerline of the torque wrench as is the sum of *L* and *A*. For example:

To apply 20 ft.-lb. to a fastener using an adapter as shown in the top example in **Figure 21**.

TA = 20 ft.-lb.

A = 3 in.

L = 14 in.

$$TW = \frac{20 \times 14}{14 + 3} = \frac{280}{17} = 16.5 \text{ ft. lb.}$$

In this example, a click-type torque wrench would be set to the calculated torque value (TW = 16.5 ft.-lb.) . When using a dial or beam-type torque wrench, tighten the fastener until the pointer aligns with 16.5 ft.-lb. In either case, although the torque wrench reads 16.5 ft.-lb., the actual torque applied to the fastener is 20 ft.-lb.

Pliers

Pliers come in a wide range of types and sizes. Pliers are useful for holding, cutting, bending, and crimping. Do not use them to turn fasteners. **Figure 22** shows several types of useful pliers. Each design has a specialized function. Slip-joint pliers are general-purpose pliers used for gripping and bending. Diagonal cutting pliers cut wire and can be used to remove cotter pins. Adjustable pliers can be adjusted to hold different size objects. The jaws remain parallel so they grip around objects such as pipe or tubing. Needlenose pliers are used to hold or bend small objects. Locking pliers (**Figure 23**), sometimes called vise-grips, are used to hold objects very tightly. They have many uses ranging from holding two parts together to gripping the end of a broken stud. Use caution when using locking

21

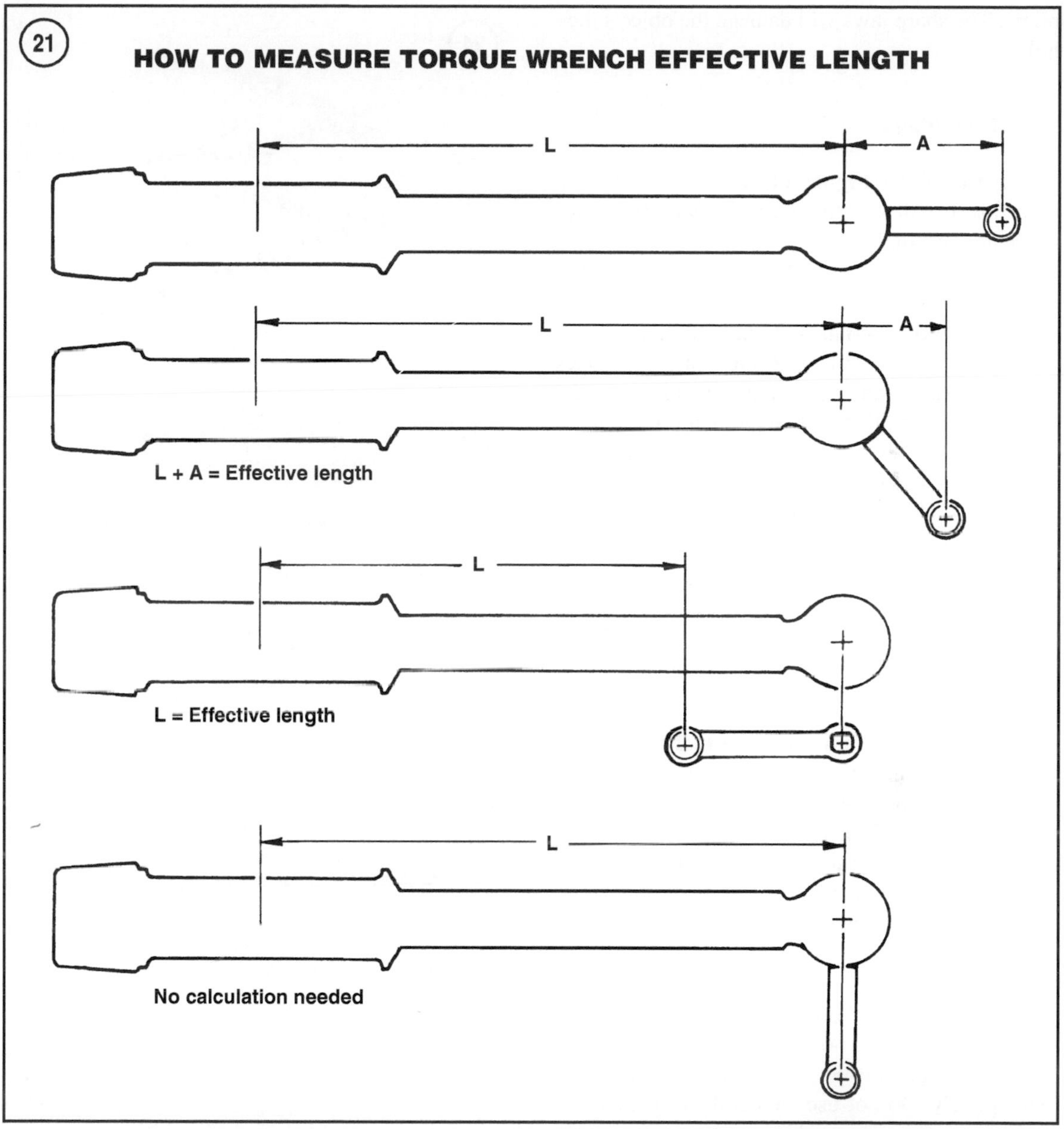
HOW TO MEASURE TORQUE WRENCH EFFECTIVE LENGTH
L
A
L
A
L + A = Effective length
L
L = Effective length
L
No calculation needed

22

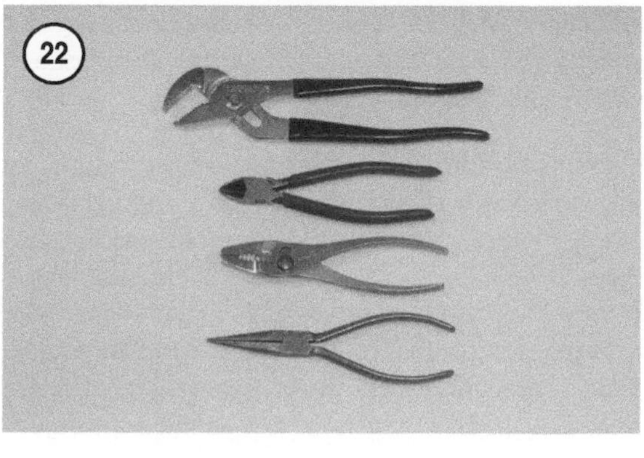

23

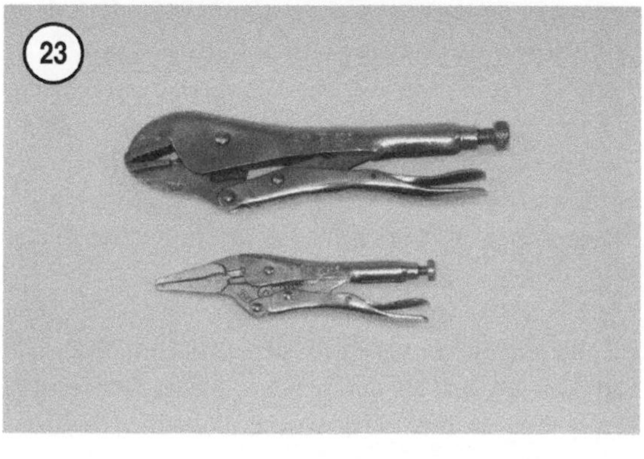

pliers. The sharp jaws will damage the objects they hold.

Snap Ring Pliers

Snap ring pliers (**Figure 24**) are specialized pliers with tips that fit into the ends of snap rings to remove and install them.

Snap ring pliers are available with a fixed action (either internal or external) or convertible (one tool works on both internal and external snap rings). They may have fixed tips or interchangeable ones of various sizes and angles. For general use, select convertible type pliers with interchangeable tips.

WARNING
Snap rings can slip and fly off during removal and installation. In addition, the tips may break. Always wear eye protection when using snap ring pliers.

Hammers

Various types of hammers (**Figure 25**) are available to fit a number of applications. A ball-peen hammer is used to strike another tool, such as a punch or chisel. Soft-faced hammers are required when a metal object must be struck without damaging it. *Never* use a metal-faced hammer on engine or suspension components. Damage will occur in most cases.

Always wear eye protection when using hammers. Make sure the hammer face is in good condition and the handle is not cracked. Select the correct hammer for the job and make sure to strike the object squarely. Do not use the handle or the side of the hammer to strike an object.

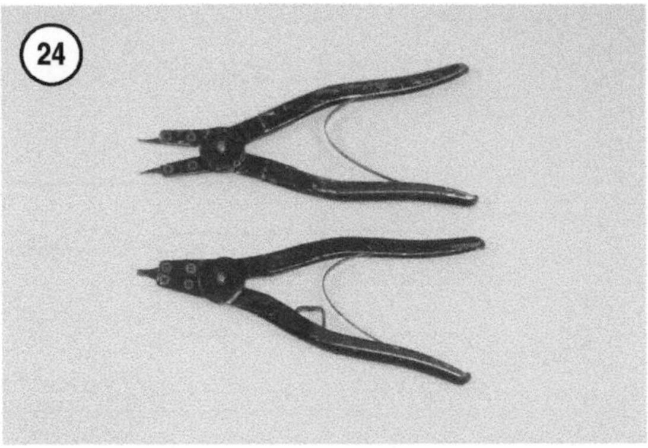

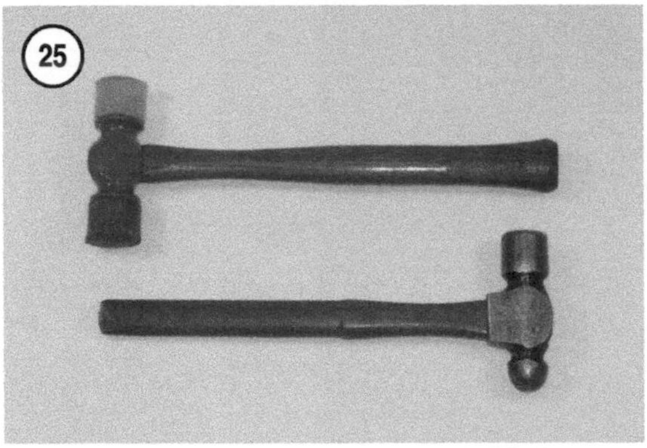

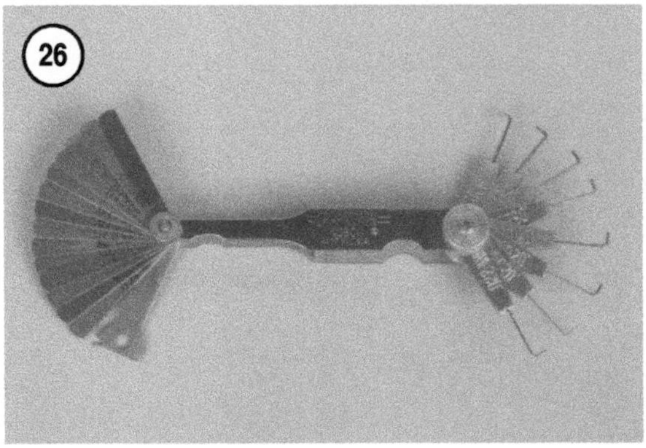

PRECISION MEASURING TOOLS

Each type of measuring instrument is designed to measure a dimension with a particular degree of accuracy and within a certain range. When selecting a measuring tool, make sure it is applicable to the task.

As with all tools, measuring tools provide the best results if they are cared for properly. Improper use can damage the tool and result in inaccurate results. If any measurement is questionable, verify the measurement using another tool. A standard gauge is usually provided with measuring tools to check accuracy and calibrate the tool.

Precision measurements can vary according to the experience of the person taking the measurement. Accurate results are only possible if the mechanic possesses a feel for using the tool. Heavy-handed use of measuring tools produces less accurate results than if the tool is handled gently. Grasp precision measuring tools with your fingertips so the point at which the tool contacts the object

27

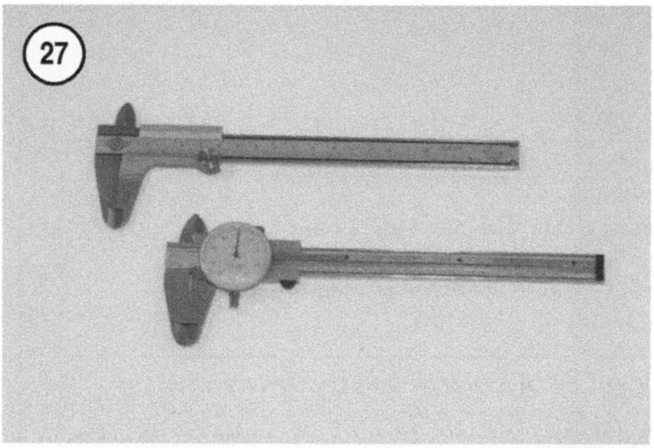

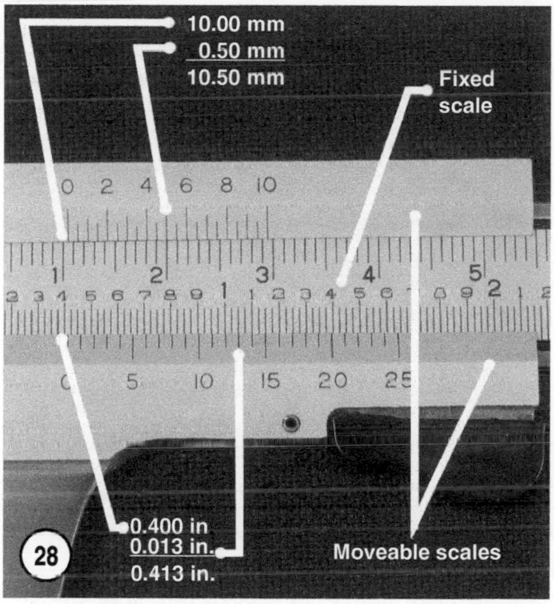

28

is easily felt. This feel for the equipment produces consistently accurate measurements and reduces the risk of damaging the tool or component. Refer to the following sections for a description of various measuring tools.

Feeler Gauge

The feeler or thickness gauge (**Figure 26**) is used for measuring the distance between two surfaces. A common use for a feeler gauge is to measure valve clearance. Wire (round) type gauges are used to measure spark plug gap.

A feeler gauge set consists of an assortment of steel strips of graduated thicknesses. Each blade is marked with its thickness. Blades can be of various lengths and angles for different procedures.

Calipers

Calipers (**Figure 27**) are excellent tools for obtaining inside, outside and depth measurements. Although not as precise as a micrometer, they allow reasonable precision, typically to within 0.05 mm (0.001 in.). Most calipers have a range up to 150 mm (6 in.).

Calipers are available in dial, vernier or digital versions. Dial calipers have a dial, which is easy to read. Vernier calipers have marked scales that must be compared to determine the measurement. The digital caliper uses an LCD display to show the measurement.

Properly maintain the measuring surfaces of the caliper. There must not be any dirt or burrs between the tool and the object being measured. Never force the caliper closed around an object. Close the caliper around the highest point so it can be removed with a slight drag. Some calipers require calibration. Always refer to the manufacturer's instructions when using a new or unfamiliar caliper.

Figure 28 shows a measurement taken with a metric vernier caliper. To read the measurement, note that the fixed scale is graduated in centimeters, which is indicated by the whole numbers 1, 2, 3 and so on. Each centimeter is then divided into millimeters, which are indicated by the small line between the whole numbers. (1 centimeter equals 10 millimeters). The movable scale is marked in increments of 0.05 (hundredths) mm. The value of a measurement equals the reading on the fixed scale plus the reading on the movable scale.

To determine the reading on the fixed scale, look for the line on the fixed scale immediately to the left of the 0-line on the movable scale. In **Figure 28**, the fixed scale reading is 1 centimeter (or 10 millimeters).

To determine the reading on the movable scale, note the one line on the movable scale that precisely aligns with a line on the fixed scale. Look closely. A number of lines will seem close, but only one aligns precisely with a line on the fixed scale. In **Figure 28**, the movable scale reading is 0.50 mm.

To calculate the measurement, add the fixed scale reading (10 mm) to the movable scale reading (0.50 mm) for a value of 10.50 mm.

(29)

DECIMAL PLACE VALUES*

0.1	Indicates 1/10 (one tenth of an inch or millimeter)
0.010	Indicates 1/100 (one one-hundreth of an inch or millimeter)
0.001	Indicates 1/1,000 (one one-thousandth of an inch or millimeter)

*This chart represents the values of figures placed to the right of the decimal point. Use it when reading decimals from one-tenth to one one-thousandth of an inch or millimeter. It is not a conversion chart (for example: 0.001 in. is not equal to 0.001 mm).

Micrometers

A micrometer is an instrument designed for linear measurement using the decimal divisions of the inch or meter (**Figure 29**). While there are many types and styles of micrometers, most of the procedures in this manual call for an outside micrometer. The outside micrometer is used to measure the outside diameter of cylindrical forms and the thickness of materials.

A micrometer's size indicates the minimum and maximum size it can measure. The usual sizes (**Figure 30**) are 0-1 in. (0-25 mm), 1-2 in. (25-50 mm), 2-3 in. (50-75 mm) and 3-4 in. (75-100 mm).

Micrometers that cover a wider range of measurement are available. These use a large frame with interchangeable anvils of various lengths. This type of micrometer offers a cost savings; however, its overall size may make it less convenient.

Reading

When reading a micrometer, read numbers from different scales and add them together.

For accurate results, properly maintain the measuring surfaces of the micrometer. There must not be any dirt or burrs between the tool and the measured object. Never force the micrometer closed around an object. Close the micrometer around the highest point so it can be removed with a slight drag.

The standard metric micrometer (**Figure 31**) is accurate to one one-hundredth of a millimeter (0.01-mm). The sleeve line is graduated in millimeter and half millimeter increments. The marks on the upper half of the sleeve line equal 1.00 mm. Every fifth mark above the sleeve line is identified with a number. The number sequence depends on the size of the micrometer. A 0-25 mm micrometer, for example, has sleeve marks numbered 0 through 25, in 5 mm increments. This numbering sequence continues with larger micrometers. On all metric micrometers, each mark on the lower half of the sleeve equals 0.50 mm.

(30)

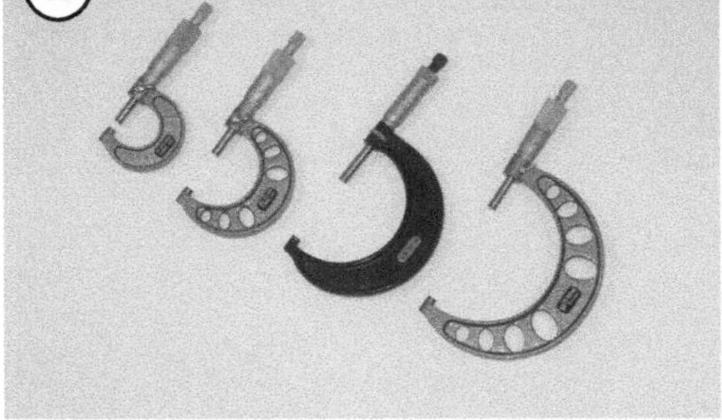

The tapered end of the thimble has fifty lines marked around it. Each mark equals 0.01 mm.

One complete turn of the thimble aligns its 0 mark with the first line on the lower half of the sleeve line or 0.50 mm.

When reading a metric micrometer, add the number of millimeters and half-millimeters on the sleeve line to the hundredths of a millimeter shown on the thimble. Perform the following steps and refer to **Figure 32**.

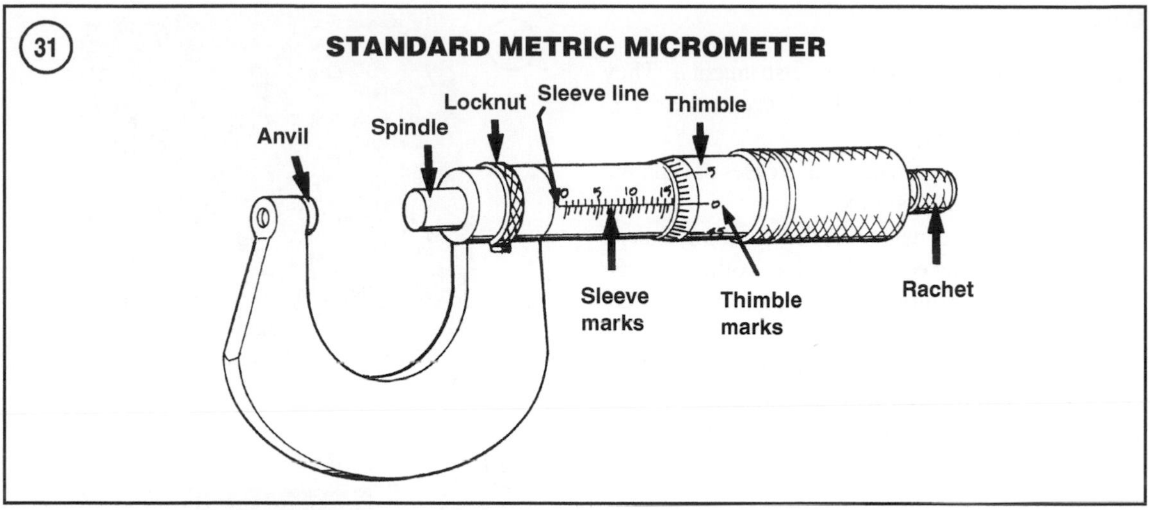

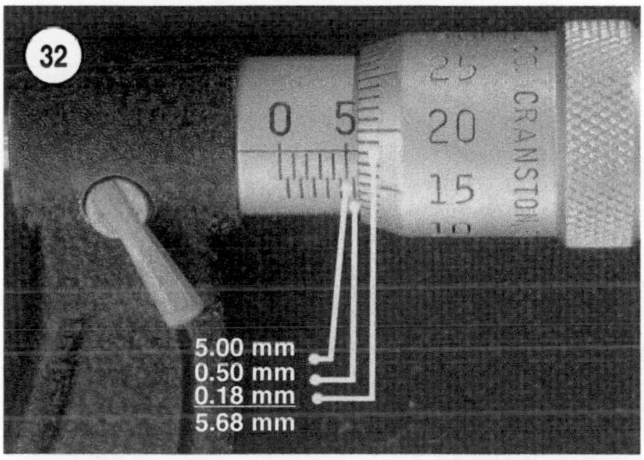

1. Read the upper half of the sleeve line and count the number of lines visible. Each upper line equals 1 mm.

2. See if the half-millimeter line is visible on the lower sleeve line. If so, add 0.50 to the reading in Step 1.

3. Read the thimble mark that aligns with the sleeve line. Each thimble mark equals 0.01 mm.

NOTE

If a thimble mark does not align exactly with the sleeve line, estimate the amount between the lines. For accurate readings to two-thousandths of a millimeter (0.002 mm), use a metric vernier micrometer.

4. Add the readings from Steps 1-3.

Adjustment

Before using a micrometer, check its adjustment as follows:

1. Clean the anvil and spindle faces.

2A. To check a 0-1 in. or 0-25 mm micrometer:

a. Turn the thimble until the spindle contacts the anvil. If the micrometer has a ratchet stop, use it to ensure the proper amount of pressure is applied.

b. The adjustment is correct if the 0 mark on the thimble aligns exactly with the 0 mark on the sleeve line. If the marks do not align, the micrometer is out of adjustment.

c. Follow the manufacturer's instructions to adjust the micrometer.

2B. To check a micrometer larger than 1 in. or 25 mm, use the standard gauge supplied by the manufacturer. A standard gauge is a steel block, disc or rod that is machined to an exact size.

a. Place the standard gauge between the spindle and anvil, and measure its outside diameter or length. If the micrometer has a ratchet stop, use it to ensure the proper amount of pressure is applied.

b. The adjustment is correct if the 0 mark on the thimble aligns exactly with the 0 mark on the sleeve line. If the marks do not align, the micrometer is out of adjustment.

c. Follow the manufacturer's instructions to adjust the micrometer.

Care

Micrometers are precision instruments. They must be used and maintained with great care.

Note the following:

1. Store micrometers in protective cases or separate padded drawers in a toolbox.
2. When in storage, make sure the spindle and anvil faces do not contact each other or any other objects. If they do, temperature changes and corrosion may damage the contact faces.
3. Do not clean a micrometer with compressed air. Dirt forced into the tool will cause wear.
4. Lubricate micrometers with WD-40 to prevent corrosion.

Telescoping and Small Bore Gauges

Use telescoping gauges (**Figure 33**) and small bore gauges (**Figure 34**) to measure bores. Neither gauge has a scale for direct readings. An outside micrometer must be used to determine the reading.

To use a telescoping gauge, select the correct size gauge for the bore. Compress the movable post and carefully insert the gauge into the bore. Carefully move the gauge in the bore to make sure it is centered. Tighten the knurled end of the gauge to hold the movable post in position. Remove the gauge, and measure the length of the posts with a micrometer. Telescoping gauges are typically used to measure cylinder bores.

To use a small-bore gauge, select the correct size gauge for the bore. Carefully insert the gauge into the bore. Tighten the knurled end of the gauge to carefully expand the gauge fingers to the limit within the bore. Do not overtighten the gauge, as there is no built-in release. Excessive tightening can damage the bore surface and damage the tool. Remove the gauge and measure the outside dimension (**Figure 35**). Small bore gauges are typically used to measure valve guides.

Dial Indicator

A dial indicator (A, **Figure 36**) is a gauge with a dial face and needle used to measure variations in dimensions and movements. Measuring brake rotor runout is a typical use for a dial indicator.

Dial indicators are available in various ranges and graduations. They use three basic types of

33

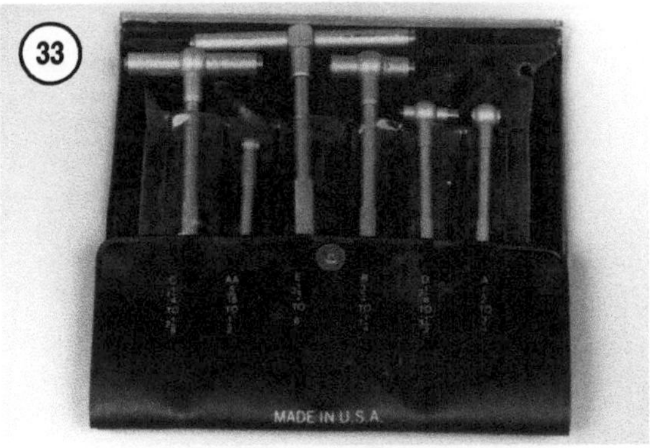

34

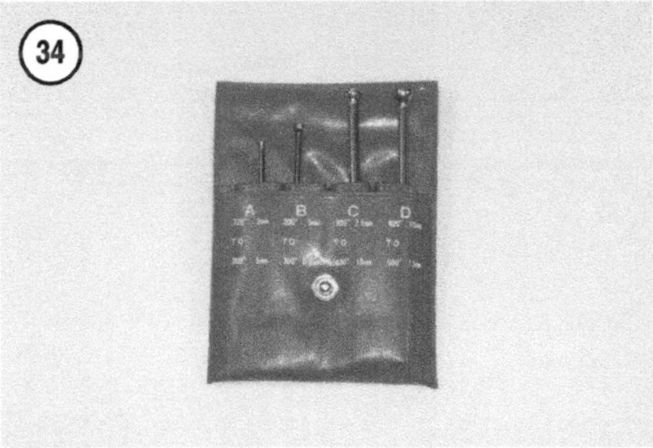

35

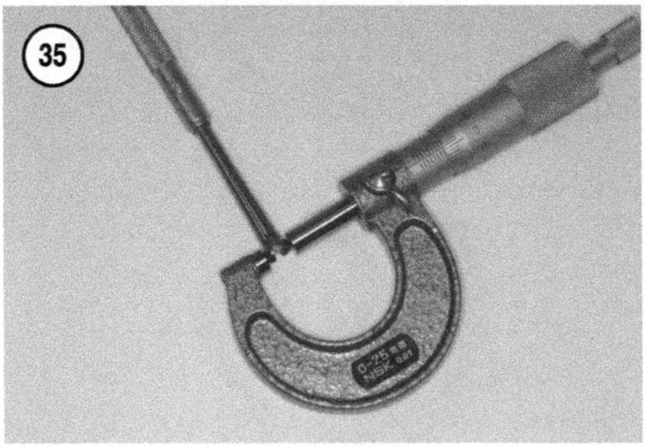

36

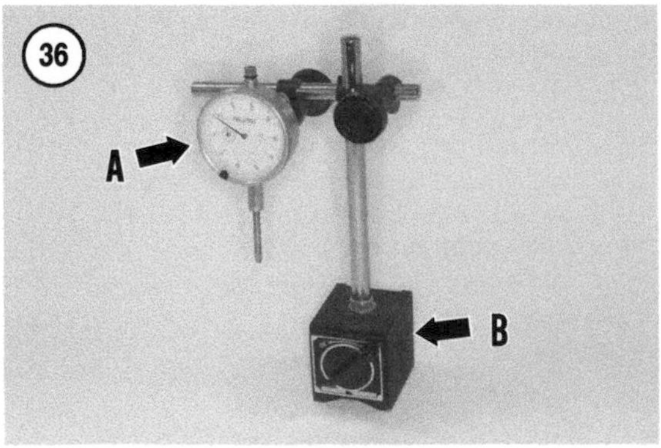

37

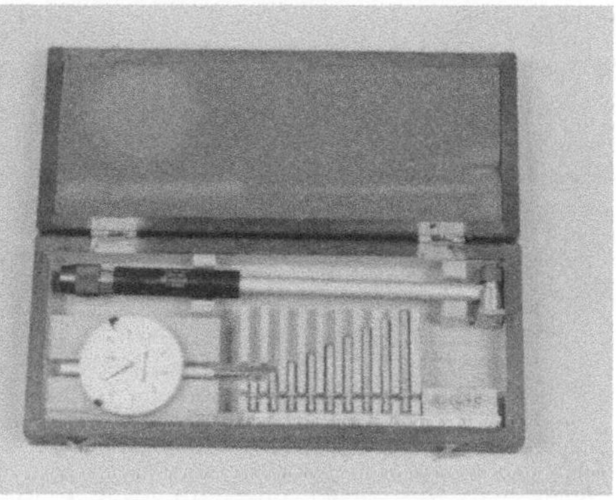

38

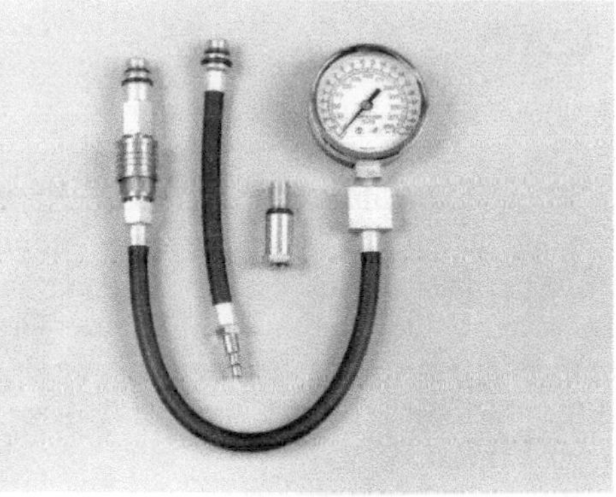

39

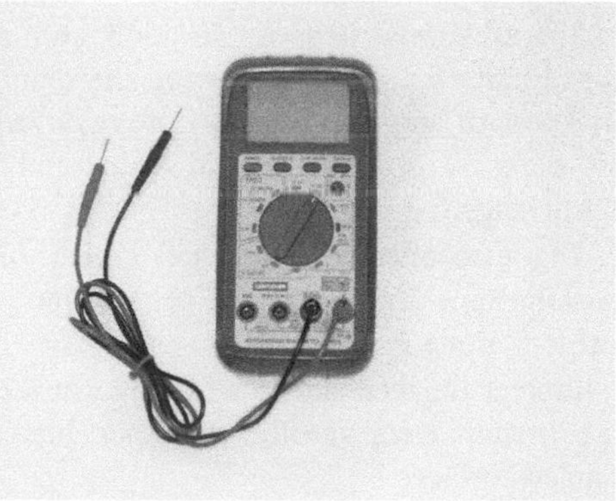

mounting bases: magnetic, clamp, or screw-in stud. When purchasing a dial indicator, select the magnetic stand type (B, **Figure 36**) with a continuous dial.

Cylinder Bore Gauge

A cylinder bore gauge is similar to a dial indicator. The gauge set shown in **Figure 37** consists of a dial indicator, handle, and different length adapters (anvils) to fit the gauge to various bore sizes. The bore gauge is used to measure bore size, taper and out-of-round. When using a bore gauge, follow the manufacturer's instructions.

Compression Gauge

A compression gauge (**Figure 38**) measures combustion chamber (cylinder) pressure, usually in psi or kg/cm^2. The gauge adapter is either inserted or screwed into the spark plug hole to obtain the reading. Disable the engine so it will not start and hold the throttle in the wide-open position when performing a compression test. An engine that does not have adequate compression cannot be properly tuned. See Chapter Three.

Multimeter

A multimeter (**Figure 39**) is an essential tool for electrical system diagnosis. The voltage function indicates the voltage applied or available to various electrical components. The ohmmeter function tests circuits for continuity and measures the resistance of a circuit.

Some test specifications for electrical components are based on results using a specific test meter. Results may vary if a meter not recommend by the manufacturer is used. Such requirements are noted when applicable.

Ohmmeter (analog) calibration

Each time an analog ohmmeter is used or if the scale is changed, the ohmmeter must be calibrated. Digital ohmmeters do not require calibration.

1. Make sure the meter battery is in good condition.
2. Make sure the meter probes are in good condition.
3. Touch the two probes together and watch the needle. It should align with the 0 mark on the scale.
4. If necessary, rotate the set-adjust knob until the needle points directly to the 0 mark.

ELECTRICAL SYSTEM FUNDAMENTALS

A thorough study of the many types of electrical systems used in today's vehicles is beyond the scope of this manual. However, an understanding of electrical basics is necessary to perform simple diagnostic tests.

Voltage

Voltage is the electrical potential or pressure in an electrical circuit and is expressed in volts. The more pressure (voltage) in a circuit, the more work can be performed.

Direct current (DC) voltage means the electricity flows in one direction. All circuits powered by a battery are DC circuits.

Alternating current (AC) means that the electricity flows in one direction momentarily then switches to the opposite direction. Alternator output is an example of AC voltage. This voltage must be changed or rectified to direct current to operate in a battery powered system.

Resistance

Resistance is the opposition to the flow of electricity within a circuit or component and is measured in ohms. Resistance causes a reduction in available current and voltage.

Resistance is measured in an inactive circuit with an ohmmeter. The ohmmeter sends a small amount of current into the circuit and measures how difficult it is to push the current through the circuit.

An ohmmeter, although useful, is not always a good indicator of a circuit's actual ability under operating conditions. This is due to the low voltage (6-9 volts) that the meter uses to test the circuit. The voltage in an ignition coil secondary winding can be several thousand volts. Such high voltage can cause the coil to malfunction, yet the fault may not be detected during a resistance test.

Resistance generally increases with temperature. Perform all tests with the component or circuit at room temperature. Resistance tests performed at high temperatures may indicate high resistance readings and result in the unnecessary replacement of a component.

Amperage

Amperage is the unit of measurement for current within a circuit. Current is the actual flow of electricity. The higher the current, the more work can be performed. However, if the current flow exceeds the circuit or component capacity, the system will be damaged.

Electrical Tests

Refer to Chapter Two for a description of various electrical tests.

BASIC SERVICE METHODS

Most of the procedures in this manual are straightforward and can be performed by anyone reasonably competent with tools. However, consider personal capabilities carefully before attempting any operation involving major disassembly of the engine.

1. Front, in this manual, refers to the front of the vehicle. The front of any component is the end closest to the front of the vehicle. The left and right sides refer to the position of the parts as viewed by the rider sitting on the seat facing forward. For example, the throttle control is on the right side of the handlebar.
2. Whenever servicing an engine or suspension component, secure the vehicle in a safe manner.
3. Tag all similar parts for location, and mark all mated parts for position. Record the number and thickness of any shims as they are removed. Identify parts by placing them in sealed and labeled plastic bags.
4. Tag disconnected wires and connectors with masking tape and a marking pen. Do not rely on memory alone.
5. Protect finished surfaces from physical damage or corrosion. Keep gasoline and other chemicals off painted surfaces.
6. Use penetrating oil on frozen or tight bolts. Avoid using heat where possible. Heat can warp, melt or affect the temper of parts. Heat also damages the finish of paint and plastics.
7. When a part is a press fit or requires a special tool for removal, the information or type of tool is identified in the text. Otherwise, if a part is difficult to remove or install, determine the cause before proceeding.
8. To prevent objects or debris from falling into the engine, cover all openings.
9. Read each procedure thoroughly and compare the illustrations to the actual components before

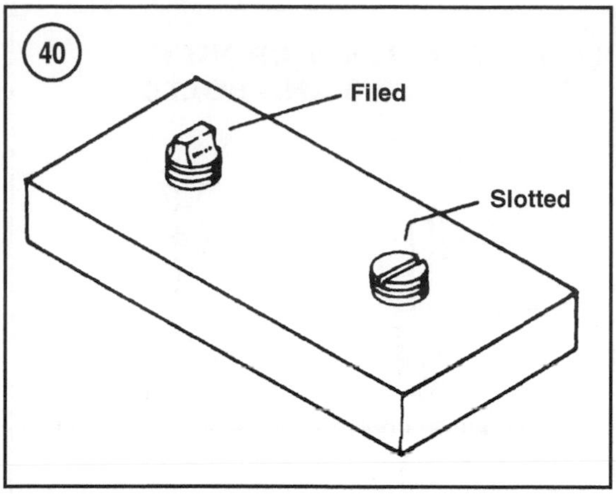

starting the procedure. Perform each procedure in sequence.

10. Recommendations are occasionally made to refer service to a dealership or specialist. In these cases, the work can be performed more economically by the specialist than by the home mechanic.

11. The term *replace* means to discard a defective part and install a new part in its place. *Overhaul* means to remove, disassemble, inspect, measure, repair and/or replace parts as required to recondition an assembly.

12. Some operations require the use of a hydraulic press. If a press is not available, have these operations performed by a shop equipped with the necessary equipment. Do not use makeshift equipment that may damage the vehicle.

13. Repairs are much faster and easier if the vehicle is clean before starting work. Degrease the vehicle with a commercial degreaser; follow the directions on the container for the best results. Clean all parts with cleaning solvent as they are removed.

CAUTION

Do not direct high-pressure water at steering bearings, carburetor hoses, wheel bearings, suspension and electrical components. The water will force the grease out of the bearings and possibly damage the seals.

14. If special tools are required, have them available before starting a procedure. When special tools are required, they will be described at the beginning of the procedure.

15. Make diagrams of similar-appearing parts. For instance, crankcase bolts are often not the same lengths. Do not rely on memory alone. It is possible that carefully laid out parts will become disturbed, making it difficult to reassemble the components correctly without a diagram.

16. Make sure all shims and washers are reinstalled in the same location and position.

17. Whenever a rotating part contacts a stationary part, look for a shim or washer.

18. Use new gaskets if there is any doubt about the condition of old ones.

19. If self-locking fasteners are used, replace them with new ones. Do not reuse a self-locking fastener. Also, do not install standard fasteners in place of self-locking ones.

20. Use grease to hold small parts in place if they tend to fall out during assembly. However, do not apply grease to electrical or brake components.

Removing Frozen Fasteners

If a fastener cannot be removed, several methods may be used to loosen it. First, apply penetrating oil such as Liquid Wrench or WD-40. Apply it liberally, and let it penetrate for 10-15 minutes. Rap the fastener several times with a small hammer. Do not hit it hard enough to cause damage. Reapply the penetrating oil if necessary.

For frozen screws, apply penetrating oil as described. Insert a screwdriver in the slot, and rap the top of the screwdriver with a hammer. This loosens the rust so the screw can be removed in the normal way. If the screw head is too damaged to use this method, grip the head with locking pliers and twist the screw out.

Avoid applying heat unless specifically instructed, as it may melt, warp or remove the temper from parts.

Removing Broken Fasteners

If the head breaks off a screw or bolt, several methods are available for removing the remaining portion. If a large portion of the remainder projects out, try gripping it with locking pliers. If the projecting portion is too small, file it to fit a wrench or cut a slot in it to fit a screwdriver (**Figure 40**).

If the head breaks off flush, use a screw extractor. To do this, center punch the remaining portion of the screw or bolt. Drill a small hole in the screw and tap the extractor into the hole. Back the screw out with a wrench on the extractor (**Figure 41**).

Repairing Damaged Threads

Occasionally, threads are stripped through carelessness or impact damage. Often the threads can be repaired by running a tap (for internal threads on nuts) or die (for external threads on bolts) through the threads (**Figure 42**). To clean or repair spark plug threads, use a spark plug tap.

If an internal thread is damaged, it may be necessary to install a Helicoil or some other type of thread insert. Follow the manufacturer's instructions when installing their insert.

If it is necessary to drill and tap a hole, refer to **Table 9** for metric tap and drill sizes.

Stud Removal/Installation

A stud removal tool is available from most tool suppliers. This tool makes the removal and installation of studs easier. If one is not available, thread two nuts onto the stud and tighten them against each other. Remove the stud by turning the lower nut (**Figure 43**).

1. Measure the height of the stud above the surface.
2. Thread the stud removal tool onto the stud and tighten it, or thread two nuts onto the stud.
3. Remove the stud by turning the stud remover or the lower nut.
4. Remove any threadlocking compound from the threaded hole. Clean the threads with an aerosol parts cleaner.
5. Install the stud removal tool onto the new stud or thread two nuts onto the stud.
6. Apply threadlocking compound to the threads of the stud.
7. Install the stud and tighten it with the stud removal tool or the top nut.
8. Install the stud to the height noted in Step 1 or torque it to specification.
9. Remove the stud removal tool or the two nuts.

41 REMOVING BROKEN SCREWS AND BOLTS

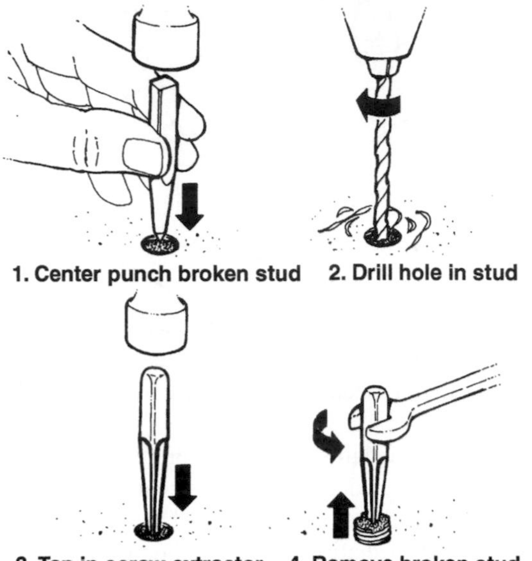
1. Center punch broken stud
2. Drill hole in stud
3. Tap in screw extractor
4. Remove broken stud

42

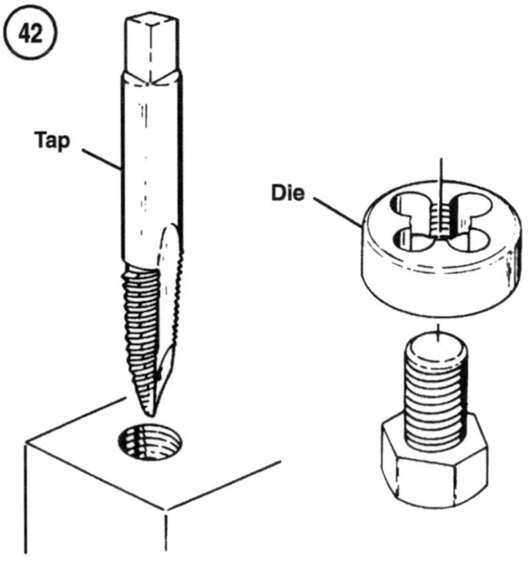
Tap
Die

Removing Hoses

When removing stubborn hoses, do not exert excessive force on the hose or fitting. Remove the hose clamp and carefully insert a small screwdriver or pick tool between the fitting and hose. Apply a spray lubricant under the hose and carefully twist the hose off the fitting. Use a wire brush to clean any corrosion or rubber hose material from the fitting. Clean the inside of the hose thoroughly. Do not use any lubricant when installing the hose (new or old). The lubricant may allow the hose to come off the fitting even with the clamp secure.

Bearings

Bearings are used in the engine and transmission assembly to reduce power loss, heat and noise re-

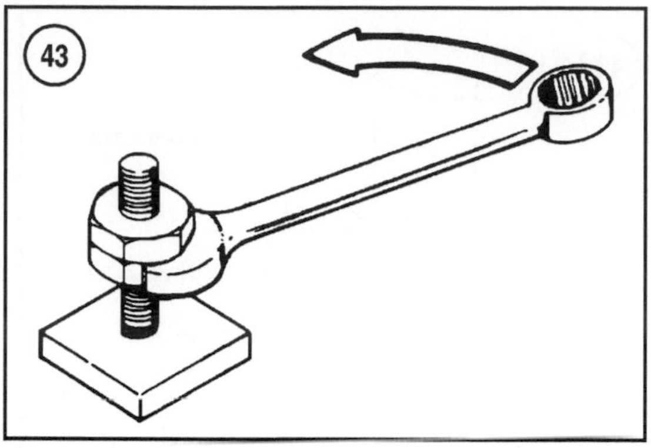

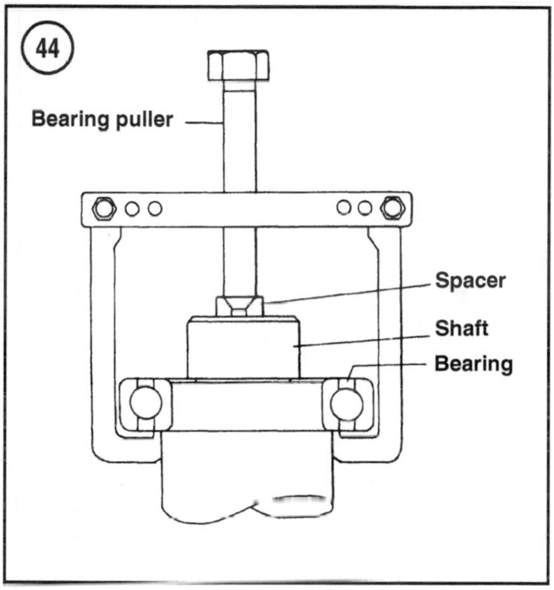

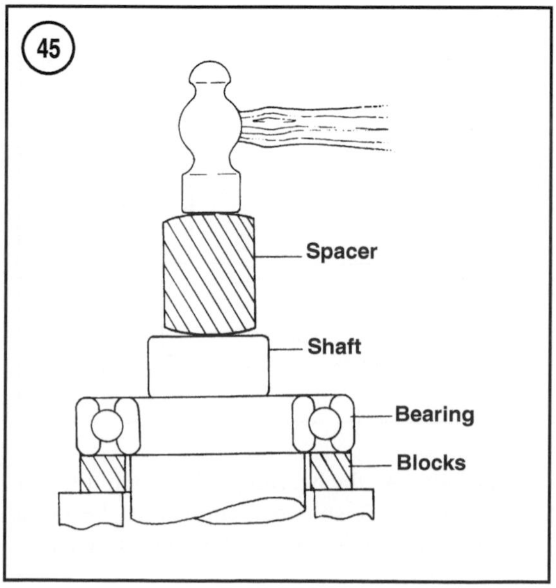

sulting from friction. Because bearings are precision parts, they must be properly lubricated and maintained. If a bearing is damaged, replace it immediately. When installing a new bearing, take care to prevent damaging it. Bearing replacement procedures are included in the individual chapters where applicable; however, use the following sections as a guideline.

NOTE
Unless otherwise specified, install bearings with the manufacturer's mark or number facing out.

Removal

While bearings are normally removed only when damaged, there may be times when a good bearing must be removed. Improper bearing removal will damage the bearing and maybe the shaft or case half. Note the following when removing bearings:

1. When using a puller to remove a bearing from a shaft, take care that shaft is not damaged. Always place a piece of metal between the end of the shaft and the puller screw. In addition, place the puller arms next to the inner bearing race. See **Figure 44**.
2. When using a hammer to remove a bearing from a shaft, do not strike the hammer directly against the shaft. Instead, use a brass or aluminum spacer between the hammer and shaft (**Figure 45**) and make sure to support both bearing races with wooden blocks as shown.
3. A hydraulic press is the ideal tool for bearing removal. Note the following when using a press:
   a. Always support the inner and outer bearing races with a suitable size wooden or aluminum ring (**Figure 46**). If only the outer race is supported, pressure applied against the balls and/or the inner race will damage them.
   b. Always make sure the press arm (**Figure 46**) aligns with the center of the shaft. If the arm is not centered, it may damage the bearing and/or shaft.
   c. The moment the shaft is free of the bearing, it will drop to the floor. Secure or hold the shaft to prevent it from falling.

Installation

1. When installing a bearing into a housing, apply pressure to the *outer* bearing race (**Figure 47**).

When installing a bearing onto a shaft, apply pressure to the *inner* bearing race (**Figure 48**).

2. When installing a bearing as described in Step 1, some type of driver is required. Never strike the bearing directly with a hammer or the bearing will be damaged. When installing a bearing, use a piece of pipe or a driver with a diameter that matches the bearing race. **Figure 49** shows the correct way to use a driver and hammer to install a bearing onto a shaft.

3. Step 1 describes how to install a bearing in a case half or over a shaft. However, when installing a bearing over a shaft and into a housing at the same time, a tight fit is required for both outer and inner bearing races. In this situation, install a spacer underneath the driver tool so pressure is applied evenly across both races. See **Figure 50**. If the outer race is not supported as shown in **Figure 50**, the balls will push against the outer bearing race and damage it.

Interference fit

1. Follow this procedure when installing a bearing over a shaft. When a tight fit is required, the bearing inside diameter will be smaller than the shaft. In this case, driving the bearing onto the shaft using normal methods may cause bearing damage. Instead, heat the bearing before installation. Note the following:
   a. Secure the shaft so it is ready for bearing installation.
   b. Clean all residue from the bearing surface of the shaft. Remove burrs with a file or sandpaper.
   c. Fill a suitable pot or beaker with clean mineral oil. Place a thermometer rated above 120° C (248° F) in the oil. Support the thermometer so it does not rest on the bottom or side of the pot.
   d. Remove the bearing from its wrapper and secure it with a piece of heavy wire bent to hold it in the pot. Hang the bearing in the pot so it does not touch the bottom or sides of the pot.
   e. Turn the heat on and monitor the thermometer. When the oil temperature rises to approximately 120° C (248° F), remove the bearing from the pot and quickly install it. If necessary, place a socket on the inner bearing race and tap the bearing into place (**Figure 49**). As the bearing chills, it will tighten on the shaft so installation must be done quickly. Make sure the bearing is installed completely.

46

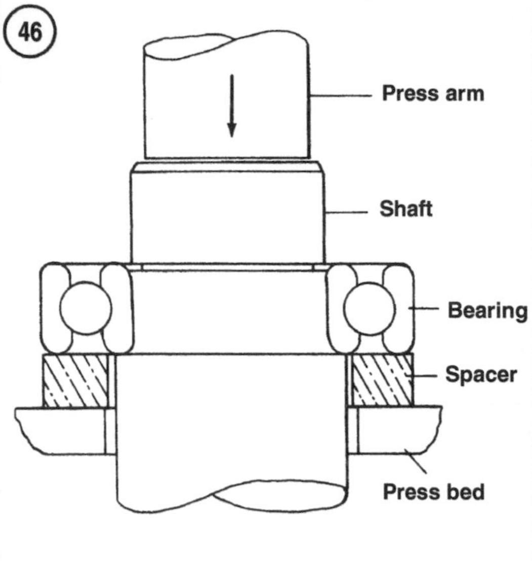

47

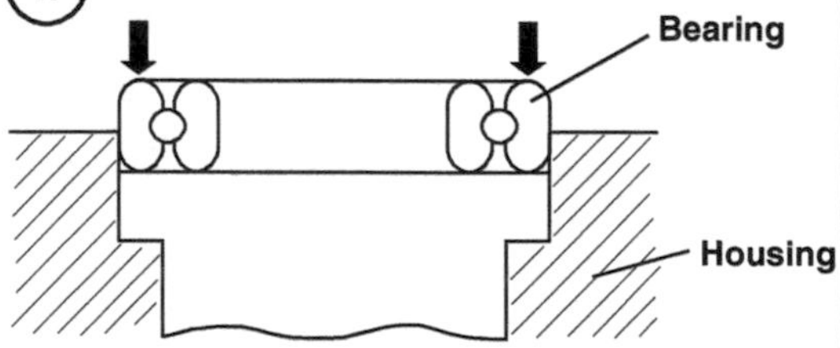

48

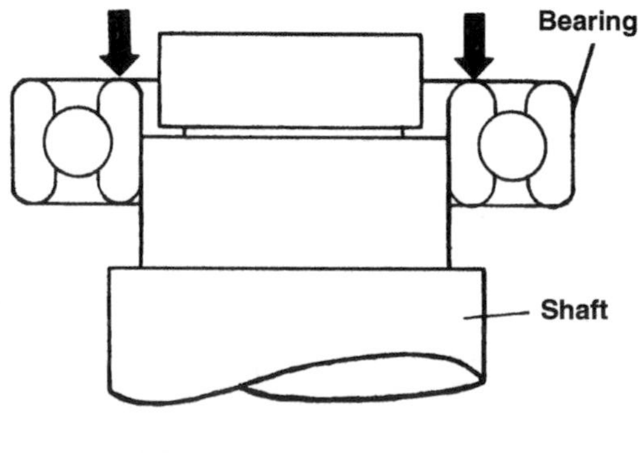

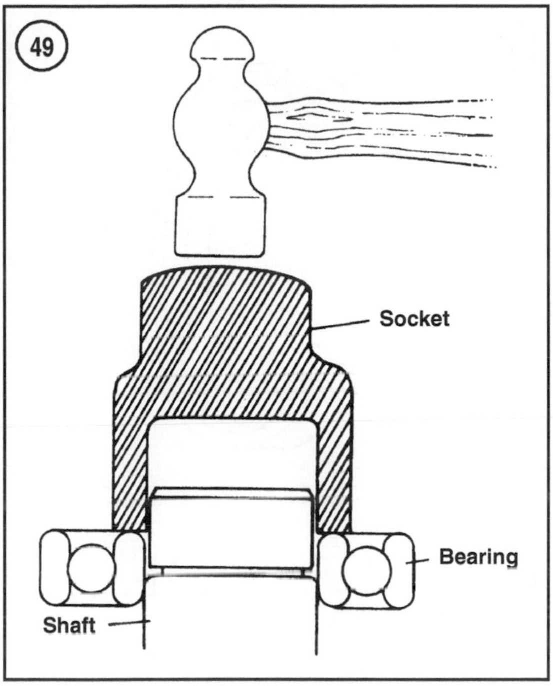

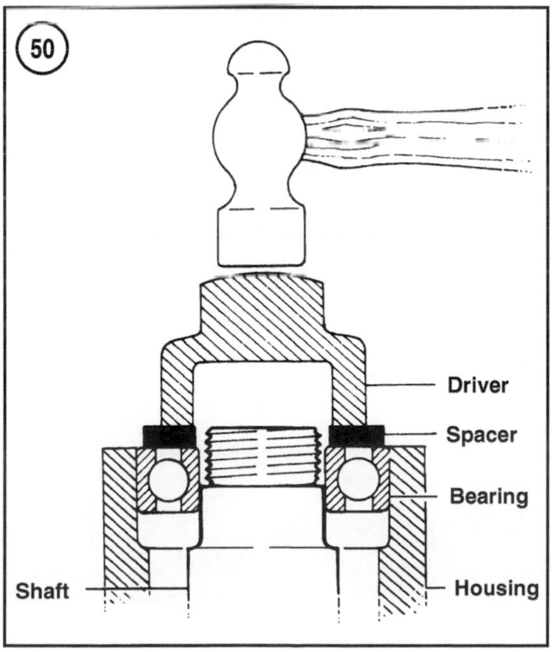

2. Follow this step when installing a bearing in a housing. Bearings are generally installed in a housing with a slight interference fit. Driving the bearing into the housing using normal methods may damage the housing or cause bearing damage. Instead, heat the housing before the bearing is installed. Note the following:

CAUTION

Before heating the housing, wash the housing thoroughly with detergent and water. Rinse and rewash the cases as required to remove all traces of oil and other chemical deposits.

a. Heat the housing to approximately 100° C (212° F) in an oven or on a hot plate. To check the housing temperature, fling tiny drops of water onto the housing. If they sizzle and evaporate immediately, the temperature is correct. Heat only one housing at a time.

CAUTION

Do not heat the housing with a propane or acetylene torch. Never bring a flame into contact with the bearing or housing. The direct heat will destroy the case hardening of the bearing and will likely warp the housing.

b. Remove the housing from the oven or hot plate, and hold onto the housing with a kitchen potholder, heavy gloves or heavy shop cloth. It is hot!

NOTE

Remove and install the bearings with a suitable size socket and extension.

c. Hold the housing with the bearing side down and tap the bearing out. Repeat for all bearings in the housing.
d. Before heating the bearing housing, place the new bearing in a freezer. Chilling a bearing slightly reduces its outside diameter while the heated bearing housing assembly is slightly larger due to heat expansion. This will make bearing installation easier.

NOTE

Always install bearings with the manufacturer's mark or number facing outward.

e. While the housing is still hot, install the new bearing(s) into the housing. Install the bearings by hand, if possible. If necessary, lightly tap the bearing(s) into the housing with a socket placed on the outer bearing race (**Figure 47**). Do not install new bearings by driving on the inner bearing race. Install the bearing(s) until it seats completely.

Seal Replacement

Seals (**Figure 51**) are used to contain oil, water, grease or combustion gasses in a housing or shaft. Improper removal of a seal can damage the housing or shaft. Improper installation of the seal can damage the seal. Note the following:

1. Prying is generally the easiest and most effective method for removing a seal from a housing. However, always place a rag underneath the pry tool (**Figure 52**) to prevent damage to the housing.
2. Pack waterproof grease in the seal lips before the seal is installed.
3. Install seals with the manufacturer's numbers or marks facing out.
4. Install seals with a socket placed on the outer circumference of the seal as shown in **Figure 53**. Drive the seal squarely into the housing. Never install a seal by striking the top of the seal with a hammer.

51

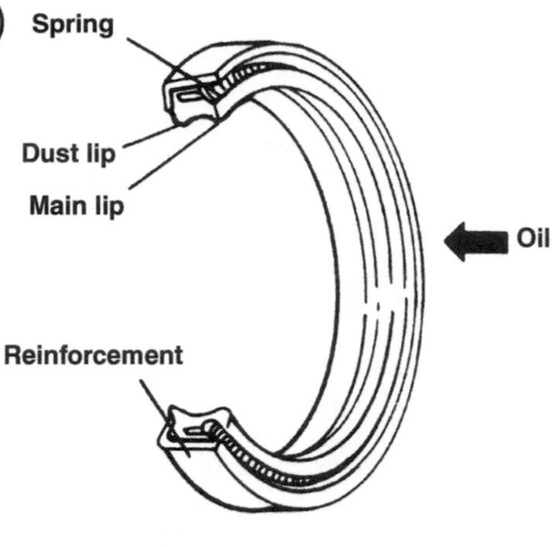

52

53

STORAGE

Several months of non-use can cause a general deterioration of the motorcycle. This is especially true in areas of extreme temperature variations. This deterioration can be minimized with careful preparation for storage. A properly stored motorcycle will be much easier to return to service.

Storage Area Selection

When selecting a storage area, consider the following:

1. The storage area must be dry. A heated area is best, but not necessary. It should be insulated to minimize extreme temperature variations.
2. If the building has large window areas, mask them to keep sunlight off the motorcycle.
3. Avoid buildings in industrial areas where corrosive emissions may be present. Avoid areas close to saltwater.
4. Consider the area's risk of fire, theft or vandalism. Check with an insurer regarding motorcycle coverage while in storage.

Preparing the Motorcycle for Storage

The amount of preparation a motorcycle should undergo before storage depends on the expected length of non-use, storage area conditions and personal preference. Consider the following list the minimum requirement:

1. Wash the motorcycle thoroughly. Make sure all dirt, mud and road debris are removed.

2. Start the engine and allow it to reach operating temperature. Drain the engine oil regardless of the

riding time since the last service. Fill the engine with the recommended type of oil.

3. Drain all fuel from the fuel tank, run the engine until all the fuel is consumed from the lines and carburetor.

4. Remove the spark plugs and ground the plug wire. Pour a teaspoon of engine oil into each cylinder. Cover the openings and crank the engine to distribute the oil in the cylinders.

5. Remove the battery. Store the battery in a cool and dry location.

6. Cover the exhaust and intake openings.

7. Reduce the normal tire pressure by 20%.

8. Apply a protective substance to the plastic and rubber components. Make sure to follow the manufacturer's instructions for each type of product being used.

9. Place the motorcycle on the center stand and support the front of the engine with a jack or wooden blocks so the wheels are off the ground. This is especially important on a motorcycle as large as the XVS1100 or XVS1100A. The tires can easily develop flat spots if the motorcycle is stored while resting on them.

10. Cover the motorcycle with old bed sheets or something similar. Do not cover it with plastic material that will trap moisture.

Returning the Motorcycle to Service

The amount of service required when returning a motorcycle to service after storage depends on the length of non-use and storage conditions. In addition to performing the reverse of the above procedure, make sure the brakes, clutch, throttle and engine stop switch work properly before operating the motorcycle. Refer to Chapter Three and evaluate the service intervals to determine which areas require service.

Table 1 METRIC, INCH AND FRACTIONAL EQUIVALENTS

mm	in.	Nearest fraction	mm	in.	Nearest fraction
1	0.0394	1/32	26	1.0236	1 1/32
2	0.0787	3/32	27	1.0630	1 1/16
3	0.1181	1/8	28	1.1024	1 3/32
4	0.1575	5/32	29	1.1417	1 5/32
5	0.1969	3/16	30	1.1811	1 3/16
6	0.2362	1/4	31	1.2205	1 7/32
7	0.2756	9/32	32	1.2598	1 1/4
8	0.3150	5/16	33	1.2992	1 5/16
9	0.3543	11/32	34	1.3386	1 11/32
10	0.3937	13/32	35	1.3780	1 3/8
11	0.4331	7/16	36	1.4173	1 13/32
12	0.4724	15/32	37	1.4567	1 15/32
13	0.5118	1/2	38	1.4961	1 1/2
14	0.5512	9/16	39	1.5354	1 17/32
15	0.5906	19/32	40	1.5748	1 9/16
16	0.6299	5/8	41	1.6142	1 5/8
17	0.6693	21/32	42	1.6535	1 21/32
18	0.7087	23/32	43	1.6929	1 11/16
19	0.7480	3/4	44	1.7323	1 23/32
20	0.7874	25/32	45	1.7717	1 25/32
21	0.8268	13/16	46	1.8110	1 13/16
22	0.8661	7/8	47	1.8504	1 27/32
23	0.9055	29/32	48	1.8898	1 7/8
24	0.9449	15/16	49	1.9291	1 15/16
25	0.9843	31/32	50	1.9685	1 31/32

Table 2 TORQUE RECOMMENDATIONS

Fastener size or type	N•m	in.-lb.	ft.-lb.
5 mm screw	4	35	–
5 mm bolt and nut	5	44	–
6 mm screw	9	80	–
6 mm bolt and nut	10	88	–
6 mm flange bolt			
(8 mm head, small flange)	9	80	–
6 mm flange bolt			
(10 mm head) and nut	12	106	–
8 mm bolt and nut	22	–	16
8 mm flange bolt and nut	27	–	20
10 mm bolt and nut	35	–	26
10 mm flange bolt and nut	40	–	29
12 mm bolt and nut	55	–	41

Table 3 CONVERSION FORMULAS

Multiply:	By:	To get the equivalent of:
Length		
Inches	25.4	Millimeter
Inches	2.54	Centimeter
Miles	1.609	Kilometer
Feet	0.3048	Meter
Millimeter	0.03937	Inches
Centimeter	0.3937	Inches
Kilometer	0.6214	Mile
Meter	3.281	Feet
Fluid volume		
U.S. quarts	0.9463	Liters
U.S. gallons	3.785	Liters
U.S. ounces	29.573529	Milliliters
Imperial gallons	4.54609	Liters
Imperial quarts	1.1365	Liters
Liters	0.2641721	U.S. gallons
Liters	1.0566882	U.S. quarts
Liters	33.814023	U.S. ounces
Liters	0.22	Imperial gallons
Liters	0.8799	Imperial quarts
Milliliters	0.033814	U.S. ounces
Milliliters	1.0	Cubic centimeters
Milliliters	0.001	Liters
Torque		
Foot-pounds	1.3558	Newton-meters
Foot-pounds	0.138255	Meters-kilograms
Inch-pounds	0.11299	Newton-meters
Newton-meters	0.7375622	Foot-pounds
Newton-meters	8.8507	Inch-pounds
Meters-kilograms	7.2330139	Foot-pounds
Volume		
Cubic inches	16.387064	Cubic centimeters
Cubic centimeters	0.0610237	Cubic inches
Temperature		
Fahrenheit	(°F – 32) × 0.556	Centigrade
Centigrade	(°C × 1.8) + 32	Fahrenheit

(continued)

Table 3 CONVERSION FORMULAS (continued)

Multiply:	By:	To get the equivalent of:
Weight		
Ounces	28.3495	Grams
Pounds	0.4535924	Kilograms
Grams	0.035274	Ounces
Kilograms	2.2046224	Pounds
Pressure		
Pounds per square inch	0.070307	Kilograms per square centimeter
Kilograms per square centimeter	14.223343	Pounds per square inch
Kilopascals	0.1450	Pounds per square inch
Pounds per square inch	6.895	Kilopascals
Speed		
Miles per hour	1.609344	Kilometers per hour
Kilometers per hour	0.6213712	Miles per hour

Table 4 TECHNICAL ABBREVIATIONS

A	Austria model
ABDC	After bottom dead center
AIS	Air induction system
ATDC	After top dead center
API	American Petroleum Institute
BBDC	Before bottom dead center
BDC	Bottom dead center
BTDC	Before top dead center
C	Celsius (centigrade)
Ca	California model
cc	Cubic centimeters
CDI	Capacitor discharge ignition
Cdn	Canada model
cid	Cubic inch displacement
cu. in.	Cubic inches
D	Germany
F	Fahrenheit
Fin	Finland model
ft.	Feet
ft.-lb.	Foot-pounds
gal.	Gallons
H/A	High altitude
hp	Horsepower
in.	Inches
in.-lb.	Inch-pounds
I.D.	Inside diameter
kg	Kilograms
kgm	Kilogram meters
km	Kilometer
kPa	Kilopascals
L	Liter
m	Meter
MAG	Magneto
ml	Milliliter
mm	Millimeter
N•m	Newton-meters

(continued)

Table 4 TECHNICAL ABBREVIATIONS (continued)

O.D.	Outside diameter
oz.	Ounces
psi	Pounds per square inch
PTO	Power take off
pt.	Pint
qt.	Quart
rpm	Revolutions per minute
SCCR	Starting circuit cutoff relay
TCI	Transistor controlled ignition
TPS	Throttle position sensor
USA	49-state model

NOTES

NOTES

NOTES

CHAPTER TWO

TROUBLESHOOTING

Begin any troubleshooting procedure by defining the symptoms as precisely as possible. Gather as much information as possible to aid diagnosis. Never assume anything and do not overlook the obvious. Make sure there is fuel in the tank, and the fuel valve is in the on position. Make sure the engine stop switch is in the run position and the spark plug wires are attached to the spark plugs.

If a quick check does not reveal the problem, turn to the troubleshooting procedures described in this chapter. Identify the procedure that most closely describes the symptoms, and perform the indicated tests.

If the meter on the instrument panel is flashing, refer to the *Self-Diagnostic System* section in Chapter Nine to begin troubleshooting.

In most cases, expensive and complicated test equipment is not needed to determine whether repairs can be performed at home. A few simple checks could prevent an unnecessary repair charge. On the other hand, be realistic and do not attempt repairs beyond your capabilities. Many service departments will not take work that involves the reassembly of damaged or abused equipment. If they do, expect the cost to be high.

OPERATING REQUIREMENTS

An engine needs three basic requirements to run properly: correct air/fuel mixture, compression and a spark at the proper time. If any element is missing, the engine will not run. Four-stroke engine operating principles are described in **Figure 1**.

If the machine has been sitting for any length of time and refuses to start, check and clean the spark plugs and then look to the fuel delivery system. This includes the fuel tank, fuel valve, fuel pump and fuel lines to the carburetor. Gasoline deposits may have gummed up the carburetor jets and air passages.

Gasoline tends to lose its potency after standing for long periods. Condensation may contaminate the fuel with water. Drain the old fuel (fuel tank, fuel lines and carburetor) and try starting with a tank of fresh gasoline.

STARTING THE ENGINE

Starting Preliminaries

1. A sidestand ignition cutoff system is used. The position of the sidestand affects engine starting. Note the following:
   a. The engine cannot start when the sidestand is down and the transmission is in gear.
   b. The engine can start when the sidestand is down and the transmission is in neutral. The engine will stop, however, if the transmission is put in gear while the sidestand is down.

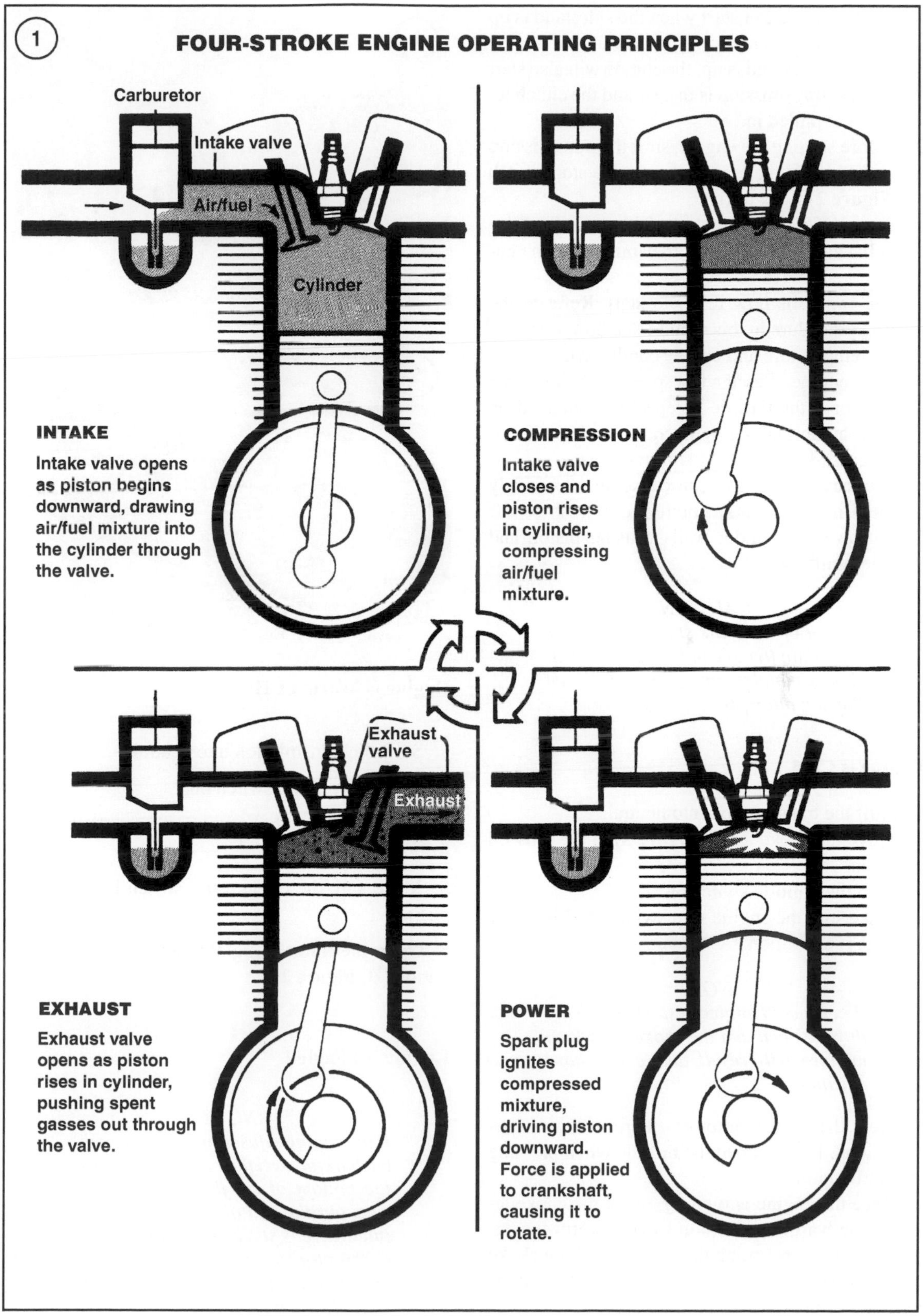
1
FOUR-STROKE ENGINE OPERATING PRINCIPLES
Carburetor
Intake valve
Air/fuel
Cylinder
INTAKE
Intake valve opens as piston begins downward, drawing air/fuel mixture into the cylinder through the valve.
COMPRESSION
Intake valve closes and piston rises in cylinder, compressing air/fuel mixture.
Exhaust valve
Exhaust
EXHAUST
Exhaust valve opens as piston rises in cylinder, pushing spent gasses out through the valve.
POWER
Spark plug ignites compressed mixture, driving piston downward. Force is applied to crankshaft, causing it to rotate.

c. The engine can start when the sidestand is up and the transmission is in neutral.

d. If the sidestand is up, the engine will also start if the transmission is in gear and the clutch lever is pulled in.

2. Before starting the engine, shift the transmission into neutral and make sure the engine stop switch (A, **Figure 2**) is set to run.

3. Turn the ignition switch on. The neutral indicator light should be on (when transmission is in neutral).

4. The engine is now ready to start. Refer to the starting procedure in this section that best describes the air temperature and engine conditions.

5. If the engine idles at a fast speed for more than 5 minutes or if the throttle is repeatedly snapped on and off at normal air temperatures, the exhaust pipes may discolor.

6. Excessive choke use can cause an excessively rich fuel mixture. This condition can wash oil off the piston and cylinder walls causing piston and cylinder scuffing.

CAUTION

Do not operate the starter motor for more than five seconds at a time. Wait approximately ten seconds between starting attempts.

Engine is Cold

1. Shift the transmission into neutral.

2. Move the choke lever (**Figure 3**) to the fully on position.

3. Turn the ignition switch on.

4. Make sure the engine stop switch (A, **Figure 2**) is in the run position.

NOTE

When a cold engine is started with the throttle open and the choke on, a lean mixture will result and cause hard starting.

5. Press the starter button (B, **Figure 2**) and start the engine. Do not open the throttle when pressing the starter button.

6. Once the engine is running, move the choke lever to the warm up position to help warm the engine. Continue warming the engine until the choke can be turned to the fully off position and the engine responds to the throttle cleanly.

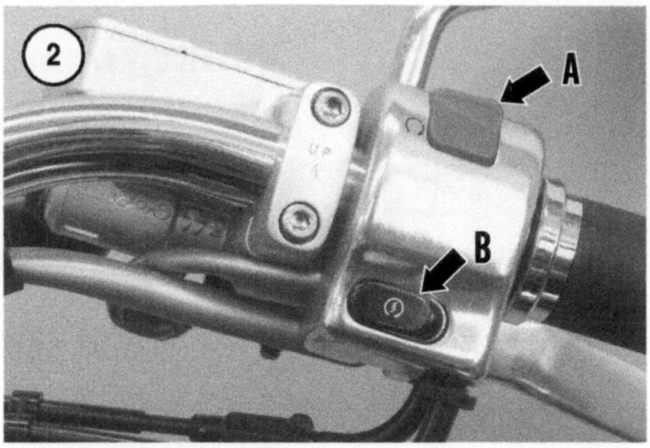

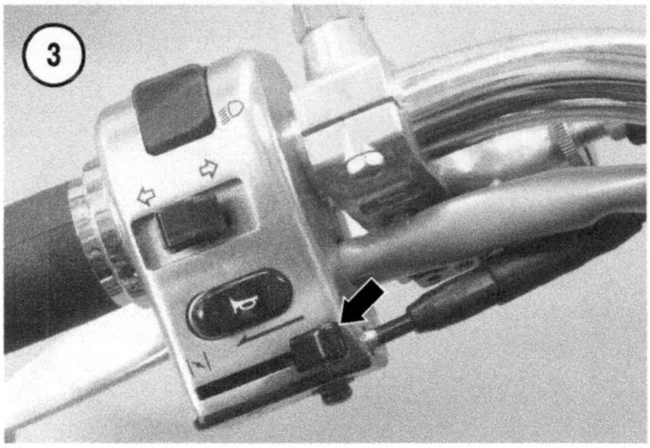

Engine is Warm or Hot

1. Shift the transmission into neutral.

2. Turn the ignition switch on.

4. Make sure the choke lever (**Figure 3**) is in the fully off position.

5. Make sure the engine stop switch is in the run position.

6. Open the throttle slightly and press the starter button (B, **Figure 2**).

Engine is Flooded

NOTE

If the engine refuses to start, check the carburetor overflow hose attached to the bottom of the float bowl. If fuel runs out the end of the hose, the fuel inlet valve is stuck open, allowing the carburetor to overfill. Remove the carburetor and correct the problem. Refer to Chapter Eight.

4

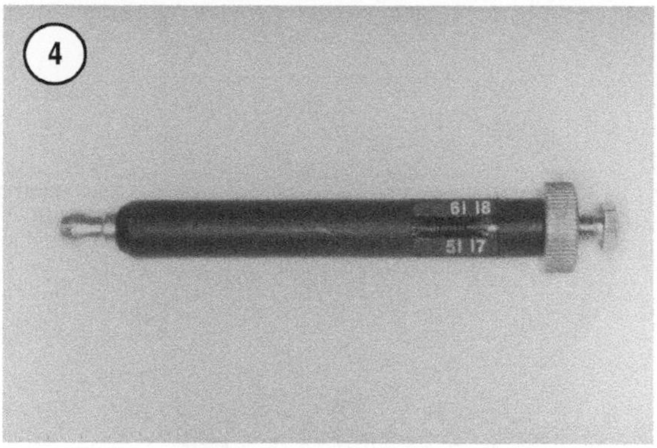

If you smell gasoline after attempting to start the engine, the engine is probably flooded. To start a flooded engine:

1. Move the choke lever (**Figure 3**) to the fully off position.
2. Turn the engine stop switch to the run position (A, **Figure 2**).
3. Open the throttle completely and press the starter button (B, **Figure 2**). If the engine starts, close the throttle quickly. If necessary, operate the throttle to keep the engine running until it smoothes out. If the engine does not start, wait 10 seconds and then try to restart the engine by following normal starting procedures. If the engine will not start, refer to *Starting Difficulties* in this chapter.

STARTING DIFFICULTIES

Perform each step while remembering the engine operating requirements described in this chapter. If the engine still will not start, refer to the appropriate troubleshooting procedures in this chapter.

1. Make sure the choke lever (**Figure 3**) is in the correct position.

WARNING
Do not use an open flame to check fuel in the tank. A serious explosion is certain to result.

2. Make sure the engine stop switch (A, **Figure 2**) moves freely and works properly. Also confirm that the switch wire is not broken or shorting out. If necessary, test the switch as described in Chapter Nine.
3. Make sure the sidestand is up and the sidestand switch operates properly. If necessary, test the switch as described in Chapter Nine.
4. Make sure both spark plug wires are on tight. Push each spark plug cap and slightly rotate it to clean the electrical connection between the plug and the connector.
5. Test the integrity of the ignition system by performing the *Spark Test* described in this section.
6. If the test produces a good spark, proceed with Step 7. If the spark is weak or if there is no spark, troubleshoot the ignition system as described in this chapter.
7. Check engine compression as described in Chapter Three. If the compression is good, perform Step 8. If the compression is low, check for one or more of the following:
   a. Leaking cylinder head gasket(s).
   b. Cracked or warped cylinder head(s).
   c. Worn piston rings, pistons and cylinders.
   d. Valve stuck open.
   e. Worn or damaged valve seat(s).
   f. Incorrect valve timing.
8. Perform the fuel pump operational test described in Chapter Eight. If the fuel flow is good, check for one or more of the following:
   a. Clogged fuel filter or fuel line.
   b. Stuck or clogged carburetor float valve.

Spark Test

The ignition spark test checks the integrity of the ignition system. The greater the air gap that a spark will jump, the stronger the ignition system. Use the Yamaha Ignition Checker (part No. YM-34487 or 90890-06754), the Motion Pro Ignition System Tester (part No. 08-122), or a similar tool (**Figure 4**) to perform this test. Connect the tester and perform the test as described by the tool manufacturer's instructions. The ignition system is working properly if a spark jumps a gap equal to or greater than the ignition minimum spark gap of 6 mm (0.24 in.).

If the above equipment is not available, perform the test with a new spark plug.

1. Remove one of the spark plugs as described in Chapter Three.
2. Make sure the spark plug gap is correct and insert the spark plug into the spark plug cap. Touch the plug base to a good engine ground (**Figure 5**). Position the plug so you can see the electrode.

WARNING
Make sure the spark plug is away from the spark plug hole in the cylinder so the spark cannot ignite the mixture in that cylinder. If the engine is flooded, do not perform this test. The firing of the spark plug can ignite fuel ejected from the opened spark plug hole.

WARNING
During the next step, do not hold the spark plug, plug wire or connector with your fingers. Serious electrical shock may result.

3. Turn the main switch on, and crank the engine over with the starter. A fat blue spark should be evident across the plug terminals. Repeat this test for the other cylinder.
4. Repeat Steps 1-3 for the other spark plug.
5. If the spark is good at each spark plug, the ignition system is functioning properly. Check for one or more of the following possible malfunctions:
   a. Obstructed fuel line or fuel filter.
   b. Damaged fuel pump system.
   c. Low compression or engine damage.
   d. Flooded engine.
6. If the spark was weak or if there was no spark at one or more plugs, note the following:
   a. If there is no spark at both plugs, there may be a problem in the input side of the ignition system, ignitor unit, sidestand switch or neutral switch. Troubleshoot the ignition system as described this chapter.
   b. If there is no spark at only one spark plug , the spark plug is probably faulty or there is a problem with that spark plug's wire, plug cap or ignition coil. Retest it with a spark tester or use a new spark plug. If there is still no spark at that plug, make sure the spark plug cap is installed correctly. Check the spark plug wiring and cap, and test the related ignition coil as described in Chapter Nine.

Engine is Difficult to Start

1. Check for fuel flow to the carburetors. If fuel is reaching the carburetors, go to Step 2. If not, check for one or more of the following possible malfunctions:

5

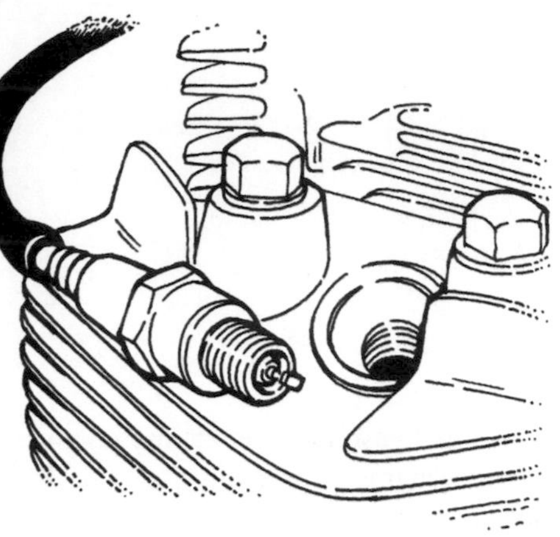

   a. Clogged fuel hose and/or fuel filter.
   b. Clogged fuel tank breather hose.
   c. Damaged fuel pump.
   d. Damaged starting circuit cutoff relay.
   e. Loose or disconnected starting circuit cutoff relay connector.
2. Perform the *Spark Test* described in this section. Note the following:
   a. If the spark plugs are wet, go to Step 3.
   b. If the spark is weak or if there is no spark, go to Step 4.
   c. If the spark is good, go to Step 5.
3. If the plugs are wet, the engine may be flooded. Check the following:
   a. Flooded carburetors.
   b. Dirty air filter.
   c. Throttle valve(s) binding or stuck open.
   d. Incorrect choke operation.
   e. Needle valve in carburetor stuck open.
4. If the spark is weak or if there is no spark, check the following:
   a. Fouled spark plug(s).
   b. Damaged spark plug(s).
   c. Loose or damaged spark plug wire(s).
   d. Loose or damaged spark plug cap(s).
   e. Damaged ignitor unit.
   f. Damaged ignition coil.
   g. Damaged engine stop switch.
   h. Damaged ignition switch.
   i. Damaged sidestand switch.
   j. Dirty or loose terminals.

5. If the spark is good, check the following:
   a. Try starting the engine by following normal starting procedures. If the engine does not start, go to Step 6.
   b. If the engine starts but then stops, check for an inoperative choke, incorrect carburetor adjustment, leaking intake manifolds, improper ignition timing or contaminated fuel.
6. If the engine turns over but does not start, the engine compression is probably low. Check for the following possible malfunctions:
   a. Leaking cylinder head gasket.
   b. Valve clearance too tight.
   c. Bent or stuck valve.
   d. Incorrect valve timing.
   e. Improper valve-to-seat contact.
   f. Worn cylinders and/or piston rings.

Engine Does Not Crank

If the engine will not turn over, check for one or more of the following:
1. Blown fuse.
2. Discharged battery.
3. Defective starter, starter relay, starting circuit cutoff switch, or starter switch.
4. Faulty starter clutch.
5. Seized pistons(s).
6. Seized crankshaft bearings.
7. Broken connecting rod.
8. Locked-up transmission or clutch assembly.
9. Defective starter clutch.

ENGINE PERFORMANCE

In the following checklists, it is assumed that the engine runs, but is not operating at peak performance. This section serves as a starting point from which to isolate a performance problem. Where ignition timing is mentioned as a problem, remember that the ignition timing cannot be adjusted. If the ignition timing is incorrect, a part within the ignition system is faulty. Check the individual ignition system components, and replace any faulty part.

Engine Will Not Idle

1. Carburetors not synchronized.
2. Incorrect idle speed adjustment.
3. Fouled or improperly gapped spark plug(s).
4. Leaking head gasket(s) or vacuum leak.
5. Incorrect valve clearance or valve timing.
6. Obstructed fuel line or fuel valve.
7. Low engine compression.
8. Starter valve (choke) stuck in the open position.
9. Incorrect pilot screw adjustment.
10. Clogged pilot jet or pilot air jet.
11. Clogged air filter element.
12. Faulty fuel pump.
13. Incorrect throttle cable free play.
14. Faulty ignition system component.

Low or Poor Engine Power

1. Securely support the motorcycle with the rear wheel off the ground, and spin the rear wheel by hand. If the wheel spins freely, perform Step 2. If the wheel does not spin freely, check for the following conditions:
   a. Dragging rear brake.
   b. Excessive rear axle torque.
   c. Worn or damaged rear wheel bearings.
   d. Worn or damaged final gear bearings.
2. Check the tire pressure. If pressure is normal, perform Step 3. If pressure is low, the tire valve is faulty.
3. Ride the motorcycle, and accelerate rapidly from first to second gear. If the engine speed reduces when the clutch is released, perform Step 4. If the engine speed does not change when the clutch is released, check for the following:
   a. Slipping clutch.
   b. Worn or warped clutch plates or friction discs.
   c. Weak clutch spring.
   d. Incorrect clutch cable free play.
   e. Check the engine oil for additives.
4. Ride the motorcycle and accelerate lightly. If the engine speed increases relative to throttle operation, perform Step 5. If engine speed does not increase, check for the following:
   a. Choke valve opened.
   b. Clogged air cleaner.
   c. Fuel flow restricted.
   d. Clogged muffler.
   e. Faulty fuel pump.
   f. Faulty starting circuit cutoff relay.
5. Check for one of the following:

a. Incorrect ignition timing due to a malfunctioning ignition component.
b. Improperly adjusted valves or worn valve seats.
c. Low engine compression.
d. Clogged carburetor jet(s).
e. Fouled spark plugs.
f. Incorrect spark plug heat range.
g. Oil level too low or too high.
h. Contaminated oil.
i. Worn or damaged valve train assembly.
j. Engine overheating. See *Engine Overheating* in this section.

6. If the engine knocks when accelerating or when running at high speed, check for the following:
a. Incorrect type of fuel.
b. Lean carburetor jetting.
c. Advanced ignition timing caused by malfunctioning ignition component.
d. Excessive carbon buildup in the combustion chamber.
e. Worn pistons and/or cylinder bores.

Poor Idle Speed or Low Speed Performance

1. Check the valve clearance. Adjust the valves as necessary.
2. Check for damaged intake manifolds, or loose carburetor, air filter housing, or surge tank clamps.
3. Perform the spark test described in this chapter. Note the following:
a. If the spark is good, perform to Step 4.
b. If the spark is weak, test the ignition system as described in this chapter.
4. Check the ignition timing as described in Chapter Three. If the ignition timing is incorrect, troubleshoot the ignition system as described in this chapter. If the ignition timing is correct, check the carburetor and fuel system.

Poor High Speed Performance

1. Check the ignition timing as described in Chapter Three. If the ignition timing is correct, perform Step 2. If the timing is incorrect, test the following ignition system components as described in Chapter Nine:
a. Pickup coil.
b. Ignition coils.
c. Ignitor unit.
3. Perform the fuel pump operational test described in Chapter Eight. If the fuel flow is acceptable, check for one or more of the following:
a. Clogged fuel line.
b. Stuck or clogged carburetor float valve.
4. Remove the carburetor assembly as described in Chapter Eight. Then remove the float bowl(s) and check for contamination and plugged jets. If there is any contamination, disassemble and clean the carburetor(s). Also pour out and discard the remaining fuel in the fuel tank and flush the tank thoroughly. If there was no contamination and the jets were not plugged, perform Step 4.
5. Incorrect valve timing and worn or damaged valve springs can cause poor high-speed performance. If the valve timing was set just before the onset of this type of problem, the valve timing may be incorrect. If the valve timing was not recently set or changed, remove the cylinder head and inspect the valve train assembly.
6. Check the carburetor and fuel system. Pay particular attention to the following:
a. A faulty diaphragm in the carburetors.
b. Improperly set fuel level.
c. Clogged or loose main jet.
d. Faulty fuel pump.
e. Clogged air filter.

Engine Overheating

1. Incorrect spark plug gap.
2. Improper spark plug heat range.
3. Faulty ignitor unit.
4. Incorrect carburetor adjustment or jet selection.
5. Incorrect fuel level.
6. Clogged air filter.
7. Heavy engine carbon deposits in combustion chamber.
8. Low oil level.
9. Incorrect oil viscosity.
10. Oil not circulating properly.
11. Valves leaking.
12. Dragging brake(s).
13. Clutch slipping.

Engine Runs Roughly

1. Clogged air filter element.

2. Carburetor adjustment incorrect; mixture too rich.
3. Choke not operating correctly.
4. Contaminants in the fuel.
5. Clogged fuel line.
6. Spark plugs fouled.
7. Ignition coil defective.
8. Ignitor unit or pickup coil defective
9. Loose or defective ignition circuit wire.
10. Short circuit from damaged wire insulation.
11. Loose battery cable connection(s).
12. Valve timing incorrect.

Engine Lacks Acceleration

1. Carburetor mixture too lean.
2. Clogged fuel line.
3. Improper ignition timing.
4. Dragging brake(s).
5. Slipping clutch.

Engine Backfires

1. Improper ignition timing.
2. Carburetor(s) improperly adjusted.
3. Lean fuel mixture.

Engine Misfires During Acceleration

1. Improper ignition timing.
2. Lean fuel mixture.
3. Excessively worn or defective spark plug(s).
4. Ignition system malfunction.
5. Incorrect carburetor adjustment.

ENGINE NOISES

Often the first evidence of an internal engine problem is a strange noise. A new knocking, clicking or tapping sound may be an early sign of trouble. While engine noises can indicate problems, they are difficult to interpret correctly. They can seriously mislead inexperienced mechanics.

Professional mechanics often use a special stethoscope to isolate engine noises. A home mechanic can do nearly as well with a length of dowel or a section of small hose. Place one end in contact with the area in question and the other end to the front of your ear (not directly on your ear) to hear the sounds emanating from that area. At first, this will be a cacophony of strange noise. Distinguishing a normal noise from an abnormal one can be difficult. If possible, have an experienced mechanic help you sort out the noises.

Consider the following when troubleshooting engine noises:

1. A knocking or pinging during acceleration is usually caused by the use of a low octane fuel. It may also be caused by poor fuel, a spark plug of the wrong heat range or carbon buildup in the combustion chamber. Refer to *Spark Plugs* and *Compression Test* in Chapter Three.
2. Slapping or rattling noises at low speed or during acceleration may be caused by excessive piston-to-cylinder wall clearance (piston slap).

NOTE
Piston slap is easier to detect when the engine is cold and before the pistons have expanded. Once the engine has warmed up, piston expansion reduces piston-to-cylinder clearance.

3. A knocking or rapping while decelerating is usually caused by excessive connecting rod bearing clearance.
4. A persistent knocking and vibration that occurs every crankshaft rotation is usually caused by worn connecting rod or main bearing(s). Can also be caused by broken piston rings or damaged piston pins.
5. A rapid on-off squeal may indicate a compression leak around cylinder head gasket or spark plug(s).
6. If valve train noise is evident, check for the following:
   a. Excessive valve clearance.
   b. Excessively worn or damaged camshaft.
   c. Damaged cam chain tensioner.
   e. Worn or damaged valve lifters and/or shims.
   f. Damaged valve bore(s) in the cylinder head.
   g. Valve sticking in guide.
   h. Broken valve spring.
   i. Low oil pressure.
   j. Clogged cylinder oil hole or oil passage.
   k. Excessively worn or damaged timing chain.
   l. Damaged timing chain sprockets.

ENGINE LUBRICATION

An improperly operating engine lubrication system will quickly lead to engine seizure. Check the engine oil level before each ride, and top off the oil as described in Chapter Three. Oil pump service is described in Chapter Five.

Oil Consumption High or Engine Smokes Excessively

1. Worn valve guides.
2. Worn or damaged piston rings.

Excessive Engine Oil Leaks

1. Clogged air filter breather hose.
2. Loose engine parts.
3. Damaged gasket sealing surfaces.

Black Smoke

1. Clogged air filter.
2. Incorrect carburetor fuel level (too high).
3. Choke stuck open.
4. Incorrect main jet (too large).

White Smoke

1. Worn valve guide.
2. Worn valve oil seal.
3. Worn piston ring oil ring.
4. Excessive cylinder and/or piston wear.

Low Oil Pressure

1. Low oil level.
2. Damaged oil pump.
3. Clogged oil strainer screen.
4. Clogged oil filter.
5. Internal oil leak.
6. Incorrect type of engine oil being used.
7. Oil pressure relief valve stuck open.

High Oil Pressure

1. Incorrect type of engine oil being used.
2. Plugged oil filter, oil gallery or metering orifices.

6 LEAKDOWN TESTER

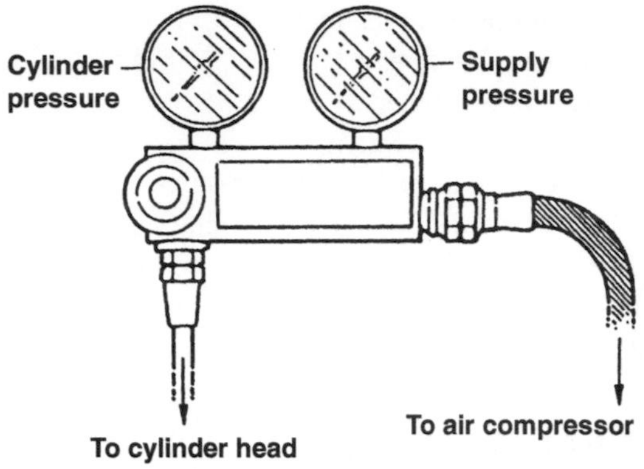

3. Oil pressure relief valve stuck closed.

No Oil Pressure

1. Damaged oil pump.
2. Excessively low oil level.
3. Damaged oil pump drive shaft.
4. Damaged oil pump drive sprocket.
5. Incorrect oil pump installation.

Low Oil Level

1. Oil level not maintained at correct level.
2. Worn piston rings.
3. Worn cylinder.
4. Worn valve guides.
5. Worn valve stem seals.
6. Piston rings incorrectly installed during engine overhaul.
7. External oil leak.
8. Oil leaking into the cooling system.

Oil Contamination

1. Blown head gasket allowing coolant to leak into the engine.
2. Water contamination.
3. Oil and filter not changed at specified intervals or when operating conditions demand more frequent changes.

ENGINE LEAKDOWN TEST

Perform an engine leakdown test to pinpoint engine problems caused by compression leaks. While a compression test (Chapter Three) can identify a weak cylinder, a leakdown test can determine where the leak occurs. A cylinder leakdown test is made by applying compressed air through the cylinder head (with the valves closed), then measuring the leak rate as a percentage. Under pressure, air leaks past worn or damaged parts. You will need a cylinder leakdown tester (**Figure 6**) and an air compressor to perform this test.

1. Start and run the engine until it is warm. Then turn the engine off.
2. Remove the air filter housing and surge tank as described in Chapter Eight. Open and secure the throttle at its wide-open position.
3. Set the No. 1 cylinder (rear) to top dead center on its compression stroke as described in *Valve Clearance* in Chapter Three.
4. Remove the spark plug from the No. 1 cylinder.
5. Thread the tester's 10 mm adapter into the No. 1 cylinder spark plug hole following the manufacturer's instructions. Then connect the leakdown tester onto the adapter. Connect an air compressor hose onto the tester's fitting. **Figure 6** shows a typical leakdown tester and its hose connections.

WARNING
To prevent the engine from turning over as compressed air is applied to the cylinder, shift the transmission into fifth gear and have an assistant apply the rear brake. Remove any tools attached to the end of the crankshaft.

6. Apply compressed air to the leak tester and perform a cylinder leakdown test following the manufacturer's instructions. Read the percent of leak on the gauge. Note the following:
   a. For a new or rebuilt engine, a leak rate of 0 to 5 percent per cylinder is desirable. A leak rate of 6 to 14 percent is acceptable and means the engine is in good condition.
   b. For a used engine, the critical rate is not the leak percent for each cylinder but the difference between the cylinders. On a used engine, a leak rate of 10 percent or less between cylinders is satisfactory.
   c. A leak rate exceeding 10 percent between cylinders points to an engine that is in very poor condition and requires further inspection and possible engine repair.
7. After measuring the percent of leak, and with air pressure still applied to the combustion chamber, listen for air escaping from the following areas:

NOTE
Use a mechanic's stethoscope to help listen for air leaks in the following areas.

   a. Air leaking through the exhaust pipe indicates a leaking exhaust valve.
   b. Air leaking through the carburetor indicates a leaking intake valve.
   c. Air leaking through the crankcase breather suggests worn piston rings or a worn cylinder bore.
8. Remove the leakdown tester, and repeat these steps for each cylinder.

CLUTCH

Basic clutch troubles and their causes are listed in this section. Clutch service procedures are in Chapter Six.

Clutch Lever Hard to Pull In

If the clutch lever has become hard to pull in, check the following:

1. Clutch cable requires lubrication.
2. Clutch cable improperly routed or bent.
3. Damaged clutch lifter bearing.

Clutch Lever Soft or Spongy

1. Air in the hydraulic system.
2. Low fluid level.
3. Leaking hydraulic system.

Rough Clutch Operation

1. Excessively worn, grooved or damaged clutch hub and clutch housing slots.
2. Worn friction disc tangs.

Clutch Slips

If the engine speed increases without an increase in motorcycle speed, the clutch is probably slipping. Some main causes of clutch slipping are:

1. Incorrect clutch cable adjustment.
2. Weak clutch springs.
3. Worn clutch or friction plates.
4. Damaged pressure plate.
5. Clutch release mechanism.
6. Incorrectly assembled clutch.
7. Loose clutch nut.
8. Improper oil level.
9. Improper oil viscosity.
10. Engine oil additive being used (clutch plates contaminated).

Clutch Drag

If the clutch will not disengage or if the bike creeps with the transmission in gear and the clutch disengaged, the clutch is dragging. Some main causes of clutch drag are:

1. Incorrectly assembled clutch.
2. Uneven clutch spring tension.
3. Warped clutch plates or pressure plate.
4. Bent push rod.
5. Damaged clutch boss.
6. Damaged primary driven gear bushing.
7. Swollen friction discs.
8. Clutch marks not properly aligned.
9. Engine oil level too high.
10. Incorrect oil viscosity.
11. Engine oil additive being used.

GEARSHIFT LINKAGE

The gearshift linkage assembly connects the shift pedal to the shift drum (internal shift mechanism). The external shift mechanism can be examined after the clutch has been removed. The internal shift mechanism can only be examined once the engine has been removed and the crankcase split.

Common gearshift linkage troubles and their checks are listed below.

Transmission Jumps Out of Gear

1. Incorrect shift pedal position.
2. Bent or worn shift fork.
3. Bent shift fork shaft.
4. Gear groove worn.
5. Damaged stopper bolt.
6. Weak or damaged stopper arm spring.
7. Loose or damaged shift cam.
8. Worn gear dogs or slots.
9. Damaged shift drum grooves.
10. Weak or damaged gearshift linkage springs.

Difficult Shifting

1. Improperly assembled clutch or clutch master cylinder.
2. Incorrect oil viscosity.
3. Bent shift fork shaft(s).
4. Bent or damaged shift fork(s).
5. Worn gear dogs or slots.
6. Damaged shift drum grooves.
7. Weak or damaged gearshift linkage springs.

Shift Pedal Does Not Return

1. Bent shift shaft.
2. Weak or damaged shift shaft spindle return spring.
3. Shift shaft incorrectly installed.
4. Improper shift pedal linkage adjustment.
5. Bent shift fork shaft.
6. Damaged shift fork.
7. Seized transmission gear.
8. Improperly assembled transmission.

TRANSMISSION

Transmission symptoms are sometimes hard to distinguish from clutch symptoms. Common transmission troubles and their checks are listed below. Refer to Chapter Seven for transmission service procedures. Before working on the transmission, make sure the clutch and gearshift linkage assemblies are working properly.

Difficult Shifting

1. Incorrect clutch adjustment.
2. Incorrect clutch operation.
3. Bent shift fork shaft.
4. Damaged shift fork guide pin(s).

5. Bent or damaged shift fork.
6. Worn gear dogs or slots.
7. Damaged shift drum grooves.

Jumps Out of Gear

1. Loose or damaged shift drum stopper arm.
2. Bent or damaged shift fork(s).
3. Bent shift fork shaft(s).
4. Damaged shift drum grooves.
5. Worn gear dogs or slots.
6. Broken shift linkage return spring.
7. Improperly adjusted shift lever position.

Incorrect Shift Lever Operation

1. Bent shift lever.
2. Stripped shift lever splines.
3. Damaged shift lever linkage.
4. Improperly adjusted shift pedal rod.

Excessive Gear Noise

1. Worn bearings.
2. Worn or damaged gears.
3. Excessive gear backlash.

FINAL DRIVE

Excessive Final Drive Noise

1. Low oil level.
2. Worn or damaged pinion and ring gears.
3. Excessive pinion-to-ring gear backlash.
4. Worn or damaged pinion gear and splines.
5. Scored driven flange and wheel hub.
6. Scored or worn ring gear and driven flange.

Excessive Rear Wheel Backlash

1. Worn driveshaft or pinion splines.
2. Loose gearcase bearing(s).
3. Worn driven flange and ring gear splines.
4. Excessive ring gear-to-pinion gear backlash.

Final Gearcase Oil Leak

1. Loose or missing cover bolts.
2. Damaged final gearcase seals.
3. Clogged breather.
4. Oil level too high.

CARBURETOR

Test the indicated items when experiencing a particular symptom.

Engine Will Not Start

Check the following items if the engine will not start and the electrical and mechanical systems are working correctly.

1. If there is no fuel going to the carburetors, check for the following:
   a. Clogged fuel tank breather.
   b. Clogged fuel tank-to-carburetor line.
   c. Clogged fuel filter.
   d. Faulty fuel pump.
   e. Faulty starting circuit cutoff relay.
   f. Incorrect float adjustment.
   g. Stuck or clogged float valve in carburetor(s).
2. If the engine is flooded (too much fuel), check the following:
   a. Flooded carburetor(s).
   b. Float valve in carburetor(s) stuck open.
   c. Clogged air filter element.
3. A faulty emission control system (if equipped) can cause fuel problems. Check for a loose, disconnected or plugged emission control system hoses.
4. If you have not located the problem in Steps 1-3, check for the following:
   a. Contaminated or deteriorated fuel.
   b. Intake manifold air leak.
   c. Clogged pilot or choke circuit.

Engine Starts but Idles and Runs Poorly or Stalls Frequently

An engine that idles roughly or stalls may have one or more of the following problems:

1. Clogged air cleaner.
2. Contaminated fuel.
3. Incorrect pilot screw adjustment.
4. Incorrect idle speed.
5. Loose, disconnected or damaged fuel and emission control vacuum hoses.
6. Intake air leak.

7. Incorrect air/fuel mixture.
8. Plugged carburetor jets.
9. Partially plugged fuel tank breather hose.
10. Faulty fuel pump.
11. Faulty starting circuit cutoff relay.

Incorrect Fast Idle Speed

A fast idle speed can be due to one of the following problems:
1. Idle adjust screw incorrectly set.
2. Incorrect carburetor synchronization.
3. Stuck choke valve.

Poor Fuel Mileage and Engine Performance

Poor fuel mileage and engine performance can be caused by infrequent engine tune-ups. Check your records against the recommended tune-up intervals in Chapter Three. If your last tune-up was within the specified service intervals, check for one or more of the following problems:
1. Clogged air filter.
2. Clogged fuel system.
3. Loose, disconnected or damaged fuel and emission control vacuum hoses.
4. Ignition system malfunction.

Rich Fuel Mixture

A rich carburetor fuel mixture can be caused by one or more of the following conditions:
1. Clogged or dirty air filter.
2. Worn or damaged fuel valve and seat.
3. Clogged air jets.
4. Incorrect float level (too high).
5. Flooded carburetor(s).
6. Damaged vacuum piston.

Lean Fuel Mixture

A lean carburetor fuel mixture can be caused by one or more of the following conditions:
1. Clogged carburetor jet(s).
2. Clogged fuel filter.
3. Restricted fuel line.
4. Intake air leak.
5. Incorrect float level (too low).

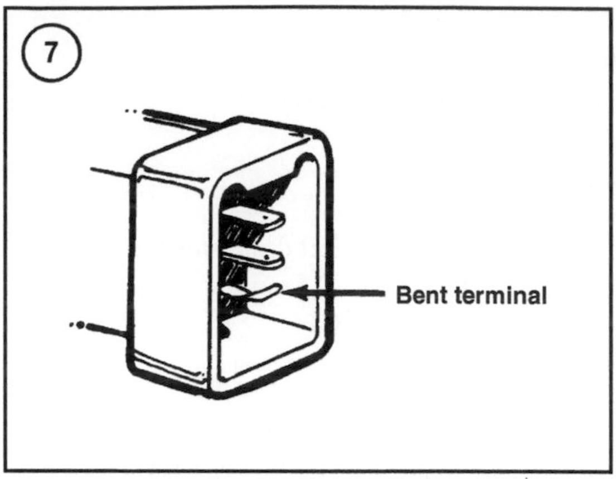

6. Worn or damaged float valve.
7. Faulty throttle valve.
8. Faulty vacuum piston.

Engine Backfires

1. Lean fuel mixture.
2. Incorrect carburetor adjustment.

Engine Misfires During Acceleration

When there is a pause before the engine responds to the throttle, the engine is misfiring. An engine misfire can occur when starting from a dead stop or at any speed. An engine misfire may be due to one of the following:
1. Lean fuel mixture.
2. Faulty ignition coil secondary wires. Check for cracking, hardening or bad connections.
3. Faulty vacuum hoses. Check for kinks, splits or bad connections.
4. Vacuum leaks at the carburetor(s) and/or intake manifold(s).
5. Fouled spark plug(s).
6. Low engine compression, especially at one cylinder only. Check engine compression as described in Chapter Three. Low compression can be caused by worn engine components.
7. Faulty fuel pump.

CARBURETOR HEATER

Whenever there is a problem with the carburetor heater system, follow the carburetor heater trouble-

8

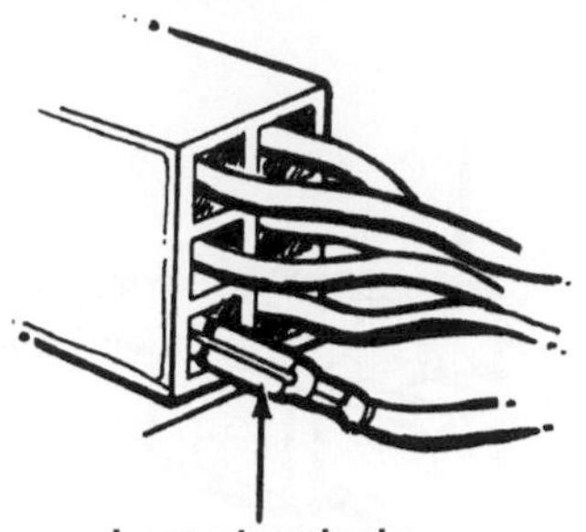

shooting procedures listed below. Refer to the wiring diagram at the end of this book.

Perform the troubleshooting test procedures in the listed sequence. Each test presumes the components tested in the earlier steps are working properly. The tests can yield invalid results if they are performed out of sequence. If a test indicates that a component is working properly, reconnect the electrical connections and proceed to the next step.

1. Inspect the main fuse and the carburetor fuse as described in Chapter Nine.

2. Check the battery as described in Chapter Three.

3. Check the continuity of the main switch (Chapter Nine).

4. Check the continuity of the neutral switch (Chapter Nine).

5. On 1999-2003 models, test the carburetor heater relay (Chapter Eight).

6. Test the thermoswitch (Chapter Eight).

7. Test the carburetor heater (Chapter Eight).

8. Check the carburetor heating system wiring and connectors. Repair as necessary.

FUEL PUMP

When troubleshooting a fuel pump problem, first perform the fuel pump operational test described in Chapter Eight. If the fuel pump is operational, the problem is in the fuel pump system circuit. Troubleshoot the circuit by following the fuel pump circuit troubleshooting procedures listed below. Refer to the wiring diagram at the end of this book.

Fuel Pump Circuit Troubleshooting

Perform these test procedures in the listed sequence. Each test presumes the components tested in the earlier steps are working properly. The tests can yield invalid results if they are performed out of sequence. If a test indicates that a component is working properly, reconnect the electrical connections and proceed to the next step.

1. Inspect the main fuse and the ignition fuse as described Chapter Nine.
2. Check the battery (Chapter Three).
3. Check the continuity of the main switch (Chapter Nine).
4. Check the continuity of the engine stop switch (Chapter Nine).
5. Check the fuel pump relay in the starting circuit cutoff relay (Chapter Eight).
6. Check the resistance of the fuel pump (Chapter Eight).
7. Check the fuel system wiring and all connectors. Repair as necessary.

ELECTRICAL TESTING

This section describes the basics of electrical testing and the use of test equipment.

Preliminary Checks and Precautions

Prior to starting any electrical troubleshooting perform the following:

1. Check the main fuse (Chapter Nine). If the fuse is blown, replace it.
2. Check the individual fuses mounted in the fuse box (Chapter Nine). Inspect the suspected fuse, and replace it if it is blown.
3. Inspect the battery (Chapter Three). Make sure it is fully charged, and the battery leads are clean and securely attached to the battery terminals.
4. Disconnect each electrical connector in the suspect circuit and make sure there are no bent metal terminals inside the electrical connector (**Figure 7**). A bent terminal will not connect to its mate in the other connector, causing an open circuit.
5. Make sure the terminals on the end of each wire (**Figure 8**) are pushed all the way into the connector. If not, carefully push them in with a narrow blade screwdriver.

2

6. Check all electrical wires where they join with the individual metal terminals in both the male and female connectors.
7. Make sure all electrical terminals within the connectors are clean and free of corrosion. Clean them, if necessary, and pack the connectors with dielectric grease.
8. Push the connector halves together. Make sure they are fully engaged and locked together (**Figure 9**).
9. Never pull the electrical wires when disconnecting an electrical connector. Only pull the connector plastic housing.

NOTE
Scan the QR code or search for "Clymer Manuals YouTube Tech Tips" to see an overview on electrical troubleshooting with a wiring diagram.

ELECTRICAL TEST EQUIPMENT

Test Light or Voltmeter

A test light can be constructed from a 12-volt light bulb with a pair of test leads carefully soldered to the bulb. To check for battery voltage in a circuit, attach one lead to ground and the other lead to various points along the circuit. The bulb lights when battery voltage is present.

A voltmeter is used in the same manner as the test light to find out if battery voltage is present in any given circuit. The voltmeter, unlike the test light, also indicates how much voltage is present at each test point. When using a voltmeter, attach the positive test lead to the component or wire to be checked and the negative test lead to a good ground (**Figure 10**).

Ammeter

An ammeter measures the flow of current (amps) in a circuit (**Figure 11**). When connected in series in a circuit, the ammeter determines if current is flowing through the circuit and if that current flow is excessive because of a short in the circuit. Current flow is often referred to as current draw. Comparing actual current draw in the circuit or component to the manufacturer's specified current draw provides useful diagnostic information.

9

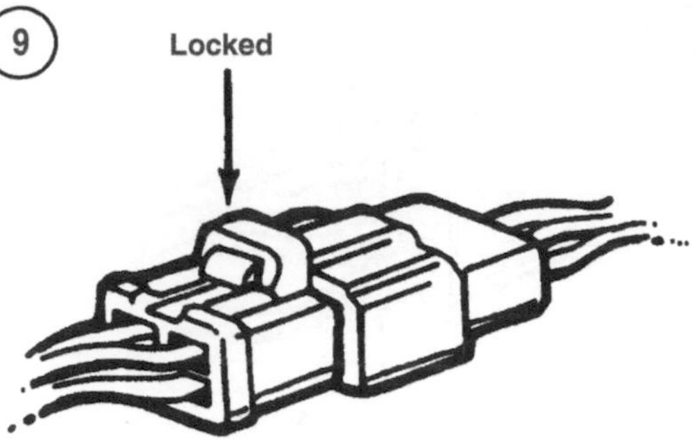

10

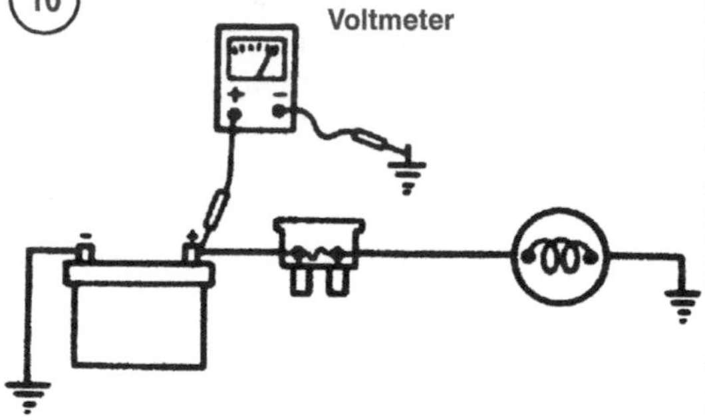

Self-powered Test Light

A self-powered test light can be constructed from a 12-volt light bulb, a pair of test leads and a 12-volt battery. When the test leads are touched together the light bulb should go on.

Use a self-powered test light as follows:

1. Touch the test leads together to make sure the light bulb turns on. If it does not, correct the problem before using the test light in a test procedure.
2. Disconnect the motorcycle's battery or remove the fuse(s) that protects the circuit to be tested.

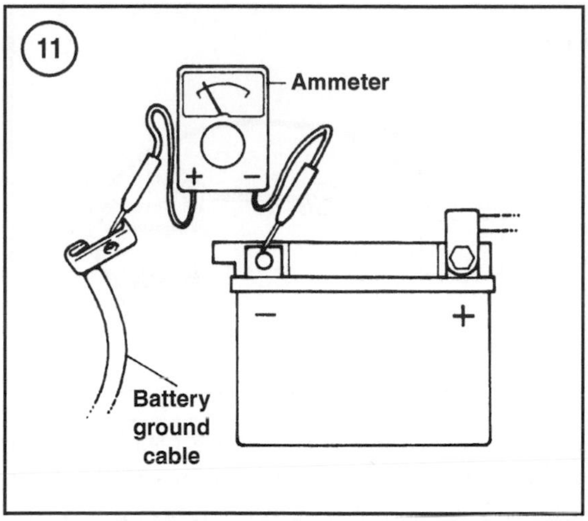

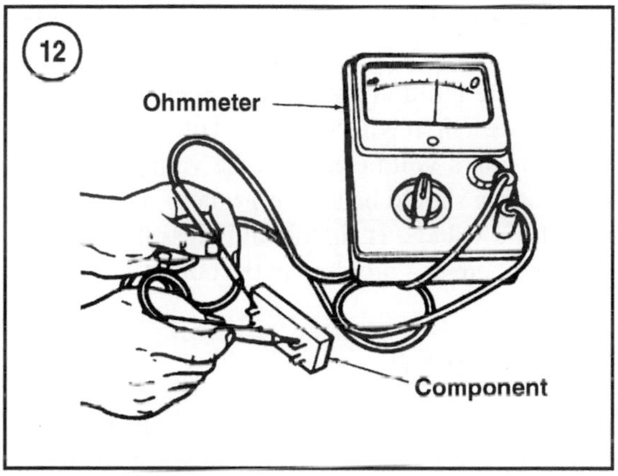

3. Select two points within the circuit where there should be continuity.

4. Attach one lead of the self-powered test light to each point.

5. If there is continuity, the self-powered test light bulb will turn on.

6. If there is no continuity, the self-powered test light bulb will not come on indicating an open circuit.

7. Never use a self-powered test light on circuits that contain solid-state devices. The solid-state device may be damaged.

Ohmmeter

An ohmmeter measures the resistance in ohms to current flow in a circuit or component. Like the self-powered test light, an ohmmeter contains its own power source and should not be connected to a live circuit.

Ohmmeters may be analog (needle scale) or digital (LCD or LED readout). Both types of ohmmeters have different ranges of resistance for accurate readings. The analog ohmmeter also has a set-adjust control which is used to zero or calibrate the meter; digital ohmmeters do not require calibration.

Connect an ohmmeter test leads to the terminals or leads of the circuit or component to be tested (**Figure 12**). When using an analog meter, calibrate it by touching the test leads together and turning the set-adjust knob until the meter needle reads zero. When the leads are uncrossed, the needle should move to the other end of the scale indicating infinite resistance.

During a continuity test, a reading of infinity indicates that there is an open in the circuit or component. A reading of zero indicates continuity, which means there is no measurable resistance in the circuit or component being tested. If the meter needle falls between the two ends of the scale, this indicates the actual resistance to current flow that is present. To determine the resistance, multiply the meter reading by the ohmmeter scale. For example, a meter reading of 5 multiplied by the R × 1000 scale is 5000 ohms of resistance.

CAUTION
Never connect an ohmmeter to a circuit which has power applied to it. Always disconnect the battery negative lead before using an ohmmeter.

Jumper Wire

A jumper wire is a simple way to bypass a potential problem and isolate it to a particular point in a circuit. If a faulty circuit works properly with a jumper wire installed, an open exists between the two jumper points in the circuit.

To troubleshoot with a jumper wire, first use the wire to determine if the problem is on the ground side or the load side of a device. In the example shown in **Figure 13**, test the ground (A) by connecting a jumper between the lamp and a good ground. If the lamp comes on, the problem is the connection between the lamp and ground. If the lamp does not come on with the jumper installed, the lamp's connection to ground is good so the problem is between the lamp and the power source.

To isolate the problem, connect the jumper between the battery and the lamp (B, **Figure 13**). If it comes on, the problem is between these two points. Next, connect the jumper between the battery and the fuse side of the switch. If the lamp comes on, the switch is good. By moving the jumper from one point to another, the problem can be isolated to a particular place in the circuit.

Pay attention to the following when using a jumper wire:

1. Make sure the jumper wire gauge (thickness) is the same as that used in the circuit being tested. Smaller gauge wire will rapidly overhead and could melt.
2. Install insulated boots over alligator clips. This prevents accidental grounding, sparks or possible shock when working in cramped quarters.
3. Jumper wires are temporary test measures only. Do not leave a jumper wire installed as a permanent solution. This creates a severe fire hazard that could easily lead to complete loss of the motorcycle.
4. When using a jumper wire, always install an inline fuse/fuse holder (available at most automotive supply stores or electronic supply stores) to the jumper wire.
5. Never use a jumper wire across any load (a component that is connected and turned on). This would result in a direct short and will blow the fuse(s).

BASIC ELECTRIC TEST PROCEDURES

Voltage Testing

Make all voltage tests with the electrical connectors still connected unless otherwise specified. Insert the test leads into the backside of the connector and make sure the test lead touches the electrical wire or metal terminal within the connector housing. Touching the wire insulation will yield a false reading.

Always check both sides of the connector as one side may be loose or corroded thus preventing electrical flow through the connector. This type of test can be performed with a test light or a voltmeter. A voltmeter gives the best results.

NOTE
When using a test light, either lead can be attached to ground.

13

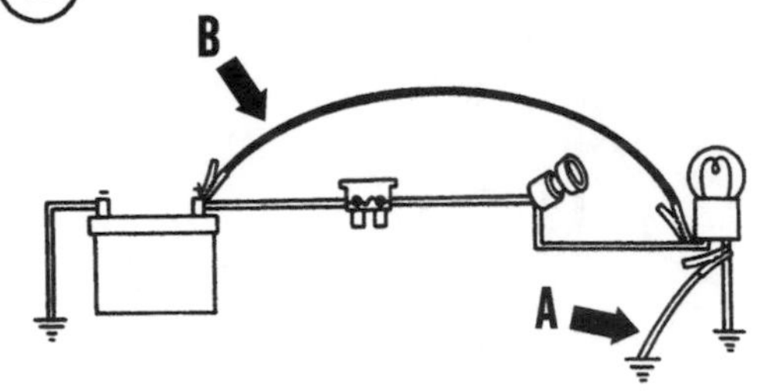

1. Attach the voltmeter negative test lead to a good ground. Make sure the part used for ground is not insulated with a rubber gasket or rubber grommet.
2. Attach the voltmeter positive test lead to the point (electrical connector, etc.) to be checked (**Figure 10**).
3. Turn the ignition switch on. When using a test light, the test light will come on if voltage is present. When using a voltmeter, note the voltage reading. The reading should be within 1 volt of battery voltage. If the voltage is significantly less than battery voltage, there is a problem in the circuit.

Voltage Drop Test

Since resistance causes voltage to drop, a voltmeter can be used to determine resistance in an active circuit. This is called a voltage drop test. A voltage drop test measures the difference between the voltage at the beginning of the circuit and the available voltage at the end of the circuit while the circuit is operating. If the circuit has no resistance, there is no voltage drop so the voltmeter indicates 0 volts. The greater the resistance in a circuit, the greater the voltage drop reading. A voltage drop of 1 or more volts indicates that a circuit has excessive resistance.

Remember a 0 reading on a voltage drop test is good. Battery voltage, on the other hand, indicates an open circuit. A voltage drop test is an excellent way to check the condition of solenoids, relays, battery cables and other high-current electrical components.

1. Connect the voltmeter positive test lead to the end of the wire or device closest to the battery.
2. Connect the voltmeter negative test lead to the ground side of the wire or device (**Figure 14**).
3. Turn the components on in the circuit.

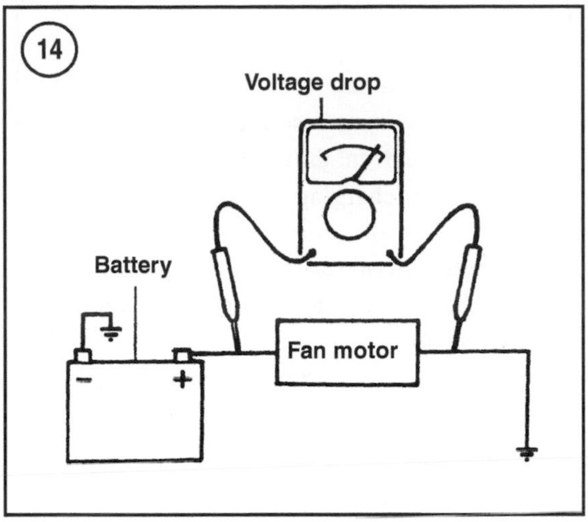

4. The voltmeter should indicate 0 volts. If there is a drop of 1 volt or more, there is a problem within the circuit. A voltage drop reading of 12 volts indicates an open in the circuit.

Continuity Test

A continuity test is used to determine the integrity of a circuit, wire or component. A circuit has continuity if it forms a complete circuit; if there are no opens in either the electrical wires or components within the circuit. A circuit with an open has no continuity.

This type of test can be performed with a self-powered test light or an ohmmeter. An ohmmeter gives the best results. When using an analog ohmmeter, calibrate the meter by touching the leads together and turning the calibration knob until the meter reads zero.

1. Disconnect the battery negative lead.

2. Attach one test lead (test light or ohmmeter) to one end of the part of the circuit to be tested.

3. Attach the other test lead to the other end of the part or the circuit to be tested.

4. The self-powered test light comes on if there is continuity. An ohmmeter reads 0 or very low resistance if there is continuity. A reading of infinite resistance indicates no continuity; the circuit has an open.

Testing for a Short with a Self-powered Test Light or Ohmmeter

1. Disconnect the battery negative lead.
2. Remove the blown fuse from the fuse panel.
3. Connect one test lead of the test light or ohmmeter to the load side (battery side) of the fuse terminal in the fuse panel.
4. Connect the other test lead to a good ground. Make sure the part used for a ground is not insulated with a rubber gasket or rubber grommet.
5. With the self-powered test light or ohmmeter attached to the fuse terminal and ground, wiggle the wiring harness of the suspect circuit at 15.2 cm (6 in.) increments. Start next to the fuse panel and work away from the fuse panel.
6. Watch the self-powered test light or ohmmeter as you progress along the harness. If the test light blinks or the needle on the ohmmeter moves when the harness is wiggled, there is a short-to-ground at that point in the harness.

Testing For a Short with a Test Light or Voltmeter

1. Remove the blown fuse from the fuse panel.
2. Connect the test light or voltmeter across the fuse terminals in the fuse panel. Turn the ignition switch on and check for battery voltage.
3. With the test light or voltmeter attached to the fuse terminals, wiggle the wiring harness of the suspect circuit at 15.2 cm (6 in.) intervals. Start next to the fuse panel and work away from the panel.
4. Watch the test light or voltmeter as you progress along the harness. If the test light blinks or if the needle on the voltmeter moves when the harness is wiggled, there is a short-to-ground at that point in the harness.

ELECTRICAL SYSTEM

Electrical troubleshooting can be very time-consuming and frustrating without proper knowledge and a suitable plan. Refer to the wiring diagrams at the end of the book for component and connector identification. Use the wiring diagrams to trace the current paths from the power source through the circuit components to ground. Also check any circuits that share the same fuse, ground or switch. If the other circuits work properly and the shared wiring is good, the cause must be in the wiring used only by the suspect circuit. If all related circuits are faulty at the same time, the probable cause is a poor ground connection or a blown fuse(s).

As with all troubleshooting, analyze typical symptoms in a systematic manner. Never assume anything, and do not overlook the obvious, like a blown fuse or an electrical connector that has separated. Test the simplest and most obvious items first and try to make tests at easily accessible points on the motorcycle.

The troubleshooting procedures for various electrical systems are listed below. Whenever there is a problem with the electrical system, perform the troubleshooting procedures for the affected system. Start with the first inspection in the list and perform the indicated check(s). If a test indicates that a component is working properly, reconnect the electrical connections and proceed to the next step. Systematically work through the troubleshooting checklist until the problem is found. Repair or replace faulty parts as described in the appropriate section of the manual.

Perform these procedures in the listed sequence. Each test presumes that the components tested in the earlier steps are working properly. The tests can yield invalid results if they are performed out of sequence.

Electrical Component Replacement

Most motorcycle dealerships and parts suppliers will not accept the return of any electrical part. If you cannot determine the *exact* cause of any electrical system malfunction, have a Yamaha dealership retest that specific system to verify your test results. If you purchase a new electrical component(s), install it, and then find that the system still does not work properly, you will probably be unable to return the unit for a refund.

Consider any test results carefully before replacing a component that tests only *slightly* out of specification, especially resistance. A number of variables can affect test results dramatically. These include: the testing meter's internal circuitry, ambient temperature and conditions under which the machine has been operated. All instructions and specifications have been checked for accuracy; however, successful test results depend to a great degree upon individual accuracy.

Charging System

1. Check the main fuse.
2. Check the battery as described in Chapter Three.
3. Perform the current draw test (Chapter Nine).
4. Perform the charging voltage test (Chapter Nine).
5. Test the stator coil resistance (Chapter Nine).
6. Check the wiring and connections in the entire charging system.
7. Replace the voltage regulator/rectifier.

Ignition System

1. Inspect the main fuse and the ignition fuse as described in Chapter Nine.
2. Check the battery (Chapter Three).
3. Check the condition of each spark plug (Chapter Three).
4. Perform the ignition spark test in this chapter.
5. Check the resistance of each spark plug cap (Chapter Nine).
6. Check the ignition coil resistance (Chapter Nine).
7. Check the pickup coil resistance (Chapter Nine).
8. Check the continuity of the main switch (Chapter Nine).
9. Check the continuity of the engine stop switch (Chapter Nine).
10. Check the continuity of the neutral switch (Chapter Nine).
11. Check the continuity of the sidestand switch (Chapter Nine).
12. Check the ignition system diode by performing the *SCCR Diode Test* described in Chapter Nine.
13. Check the connections in the entire ignition system.
14. Have a dealership or qualified shop check the ignitor unit.

Starting System

1. Check the main fuse and the ignition fuse.
2. Check the battery as described in Chapter Three.
3. Perform the starter operational test (Chapter Nine).

4. Test the starting circuit cutoff relay (SCCR) by performing the continuity test described in Chapter Nine.
5. Test the starting system diode by performing the SCCR diode test described in Chapter Nine.
6. Test the continuity of the starter relay (Chapter Nine).
7. Check the continuity of the main switch (Chapter Nine).
8. Check the continuity of the engine stop switch (Chapter Nine).
9. Check the continuity of the neutral switch (Chapter Nine).
10. Check the continuity of the sidestand switch (Chapter Nine).
11. Check the continuity of the clutch switch (Chapter Nine).
12. Check the continuity of the start switch (Chapter Nine).
13. Test the continuity of the diode as described in *Ignition System* of Chapter Nine.
14. Check the wiring and each connector in the starting circuit.

Lighting System

The lighting system consists of the headlight, taillight, high beam indicator light, meter illumination lights and front turn signal/position light.

1. Check the affected bulb.
2. Check the main fuse and the headlight fuse.
3. Check the battery as described in Chapter Three.
4. Check the continuity of the main switch (Chapter Nine).
5. Check the continuity of the dimmer switch (Chapter Nine).
6. Check the continuity of the pass switch (Chapter Nine).
7. On models with a pass light, check the wiring and each connector in the lighting circuit (Chapter Nine).
8. Locate the symptom among the following descriptions and perform the indicated tests.

The headlight and high beam indicator do not turn on

1. Check the continuity of the headlight bulb and socket.
2. Perform the headlight voltage test in Chapter Nine.

A meter light does not turn on

1. Check the continuity of the affected meter bulb and socket.
2. Perform the meter indicator light test in Chapter Nine.

The taillight does not turn on

NOTE

*Refer to **Signal System** if the brake light does not turn on.*

1. Check the continuity of the taillight bulb and socket.
2. Perform the taillight test in Chapter Nine.

Signal System

The signal system includes the horn, turn signal lights, brake light and indicator lights (except the high beam indicator, which is part of the lighting system).

1. Check the main fuse and the signal system fuse.
2. Check the battery as described in Chapter Three.
3. Check the continuity of the main switch (Chapter Nine).
4. Check the wiring and each connector in the signal system circuit.
5. Locate the symptom among the following descriptions and perform the indicated tests.

The horn does not sound

Perform the horn circuit test in Chapter Nine.

The brake light does not turn on

1. Check the continuity of the brake light bulb and socket.
2. Check the continuity of the affected brake light switch (Chapter Nine).
3. Perform the brake light test in *Taillight/Brake Light* in Chapter Nine.

A turn signal light and/or turn signal indicator fails to flash

1. Check the continuity of the affected turn signal bulb and socket.
2. Check the continuity of the bulb and socket for the turn signal indicator light.
3. Check the continuity of the turn signal switch (Chapter Nine).
4. Perform the turn signal flash test (Chapter Nine).

The neutral indicator light does not turn on

1. Check the continuity of the bulb and socket for the neutral indicator light.
2. Check the continuity of the neutral switch (Chapter Nine).
3. Perform the neutral indicator circuit test in Chapter Nine.

The oil level light does not turn on

1. Check the continuity of the bulb and socket for the oil level indicator light.
2. Check the continuity of the oil level switch (Chapter Nine).
3. Perform the oil level indicator test (Chapter Nine).

FRONT SUSPENSION AND STEERING

Poor handling may be caused by improper tire pressure, a damaged/bent frame or front steering components, a worn front fork assembly, worn wheel bearings or dragging brakes.

Steering is Sluggish

1. Incorrect steering stem adjustment (too tight).
2. Improperly installed upper or lower fork bridge.
3. Damaged steering head bearings.
4. Tire pressure too low.
5. Worn or damaged tire.

Steering to One Side

1. Bent front or rear axle.
2. Bent frame or fork.
3. Worn or damaged wheel bearings.
4. Worn or damaged swing arm pivot bearings.
5. Damaged steering head bearings.
6. Bent swing arm.
7. Incorrectly installed wheels.
8. Front and rear wheels are not aligned.
9. Uneven front fork adjustment.
10. Fork legs positioned unevenly in the fork bridges.

Front Suspension Noise

1. Loose mounting fasteners.
2. Damaged fork or rear shock absorber.
3. Low fork oil capacity.

Wheel Wobble/Vibration

1. Loose front or rear axle.
2. Loose or damaged wheel bearing(s).
3. Damaged wheel rim(s).
4. Damaged tire(s).
5. Loose swing arm pivot bolt.
6. Unbalanced tire and wheel.
7. Loose spokes.

Hard Suspension (Front Fork)

1. Insufficient tire pressure.
2. Damaged steering head bearings.
3. Incorrect steering head bearing adjustment.
4. Bent fork tubes.
5. Binding slider.
6. Incorrect weight fork oil.
7. Plugged fork oil passage.
8. Worn or damaged fork tube bushing or slider bushing.
9. Damaged damper rod.

Hard Suspension (Rear Shock Absorber)

1. Excessive rear tire pressure.
2. Bent or damaged shock absorber.
3. Incorrect shock adjustment.
4. Damaged shock absorber bushing(s).
5. Damaged swing arm pivot bearings.
6. Poorly lubricated suspension components.

Soft Suspension (Front Fork)

1. Insufficient tire pressure.
2. Insufficient fork oil level or fluid capacity.
3. Incorrect oil viscosity.
4. Weak or damaged fork springs.

Soft Suspension (Rear Shock Absorbers)

1. Insufficient rear tire pressure.
2. Weak or damaged shock absorber spring.
3. Damaged shock absorber.
4. Incorrect shock absorber adjustment.
5. Leaking damper unit.

BRAKES

The front and rear brake units are critical to riding performance and safety. Inspect the front and rear brakes frequently, and repair any problem immediately. When adding or changing the brake fluid, use only DOT 4 brake fluid from a closed container. See Chapter Thirteen for additional information on brake fluid selection and disc brake service.

When checking brake pad wear, make sure the brake pads in each caliper contact the disc squarely. If one of the brake pads is wearing unevenly, suspect a warped or bent brake disc or damaged caliper.

Brake Drag

1. Clogged brake hydraulic system.
2. Sticking caliper pistons.
3. Sticking master cylinder piston.
4. Incorrectly installed brake caliper.
5. Warped brake disc.
6. Incorrect wheel alignment.
7. Contaminated brake pad and disc.
8. Excessively worn brake disc or pad.
9. Caliper not sliding correctly.

Brakes Grab

1. Contaminated brake pads and disc.
2. Incorrect wheel alignment.
3. Warped brake disc.
4. Caliper not sliding correctly.

Brake Squeal or Chatter

1. Contaminated brake pads and disc.
2. Incorrectly installed brake caliper.
3. Warped brake disc.
4. Incorrect wheel alignment.

Soft or Spongy Brake Lever or Pedal

1. Low brake fluid level.
2. Air in brake hydraulic system.
3. Leaking brake hydraulic system.
4. Clogged brake hydraulic system.
5. Worn brake caliper seals.
6. Worn master cylinder seals.
7. Sticking caliper piston.
8. Sticking master cylinder piston.
9. Damaged front brake lever.
10. Damaged rear brake pedal.
11. Contaminated brake pads and disc.
12. Excessively worn brake disc or pad.
13. Warped brake disc.

Hard Brake Lever or Pedal Operation

1. Clogged brake hydraulic system.
2. Sticking caliper piston.
3. Sticking master cylinder piston.
4. Worn caliper piston seal.
5. Glazed or worn brake pads.
6. Damaged front brake lever.
7. Damaged rear brake pedal.
8. Caliper not sliding correctly.

CHAPTER THREE

LUBRICATION, MAINTENANCE AND TUNE-UP

This chapter describes lubrication, maintenance and tune-up procedures.

Minor problems found during periodic inspections are generally simple and inexpensive to correct. However, they could lead to major problems if not corrected promptly.

Before servicing the motorcycle, become acquainted with the tools and parts available at motorcycle dealerships and parts supply houses. Also pay attention to the various service and maintenance products such as engine oil, brake fluid, threadlocking compounds and greases. Refer to Chapter One.

If this is your first experience with motorcycle maintenance, start by doing simple tune-up, lubrication and maintenance procedures. Attempt more complicated jobs as you gain experience.

When inspecting the components mentioned in this chapter, compare any measurements to the maintenance and tune-up specifications at the end of this chapter. Replace any part that is damaged, worn or out of specification. During assembly, torque fasteners to the specified torque.

FUEL TYPE

The engine is designed to use gasoline with a pump octane rating of 86 or higher (or a research octane of 91 or higher). The pump octane rating ([R + M]/2) is normally displayed at the service station fuel pump. Using gasoline with a lower octane rating can cause pinging or spark knock. Either condition can lead to engine damage.

When adding fuel, note the following:

1. Do not overfill the fuel tank. There should be no fuel in the filler neck, which is located between the fuel cap and the tank.
2. When using an oxygenated fuel, make sure it meets the minimum octane rating.
3. Oxygenated fuels can damage plastic and painted parts. Be careful not to spill these fuels on the motorcycle when refueling.
4. An ethanol (ethyl or grain alcohol) gasoline that contains more than 10 percent ethanol (by volume) may cause engine starting and performance problems.

MAINTENANCE INTERVALS

Refer to **Table 1** for maintenance intervals. Strict adherence to these recommendations helps ensure a long service life from the motorcycle. If the motorcycle is operated in an area of high humidity, the lubrication services must be performed more frequently to prevent rust and corrosion.

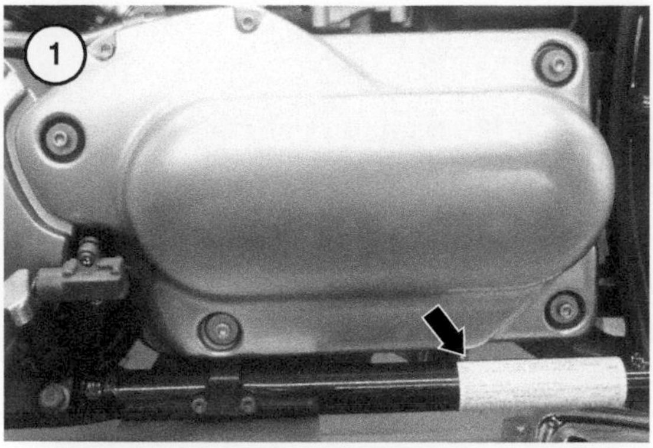

For convenience, most of the procedures listed in **Table 1** are described in this chapter. Procedures that require more than minor disassembly or adjustment are covered in the appropriate chapter in the manual. Refer to the *Table of Contents* or *Index* to locate a particular procedure.

CYLINDER NUMBERING

The rear cylinder is the No. 1 cylinder; the front cylinder No. 2.

ENGINE ROTATION

Engine rotation is clockwise when viewed from the left side. Use the flywheel nut to rotate the crankshaft manually, and always turn the crankshaft clockwise.

TUNE-UP

A complete tune-up restores performance that is lost due to normal wear and deterioration. Because engine wear occurs over a combined period of time and mileage, perform the engine tune-up procedures at the intervals specified in **Table 1**. More frequent tune-ups may be required if the motorcycle is primarily operated in stop-and-go traffic.

The Tune-Up Specification label provides tune-up specifications. This label is on the lower frame member just beneath the left side cover (**Figure 1**). Always refer to the specifications on this label when servicing the motorcycle. If the specifications on the label differ from those in this manual, use the specifications from the label.

When performing a tune-up, service the following items as described in this chapter:

1. Air filter.
2. Spark plugs.
3. Engine compression.
4. Engine oil and filter.
5. Ignition timing.
6. Valve clearance.
7. Carburetor adjustment.
8. Brake system.
9. Suspension components.
10. Tires and wheels.
11. Fasteners.

Air Filter Replacement

The air filter removes dust and abrasive particles from the incoming air before it enters the carburetors and the engine. Without an air filter, very fine particles could enter the engine and rapidly wear the piston rings, cylinders and bearings. These particles could also clog small passages in the carburetors. Never run the motorcycle without the air filter element installed.

The air filter is a dry-element type; no oiling is required. Proper air filter servicing can do more to ensure long service from your engine than almost any other single item. Remove and clean the air filter element at the interval listed in **Table 1**. Replace the air filter if it is soiled, severely clogged or broken in any area.

1. Make sure the ignition switch is off.
2. Securely support the motorcycle on level ground.
3. Remove the cover screws (A, **Figure 2**), and remove the air filter cover (B).
4. Remove the air filter element from the posts (**Figure 3**) on the housing.

5. Wipe the interior of the air filter housing with a shop rag dampened with cleaning solvent. Remove any debris that may have passed through a broken element.

6. Inspect the air filter element for tears or other damage that would allow unfiltered air to pass into the engine. Also check the element gasket (**Figure 4**) for tears. Replace the element if necessary.

7. Gently tap the air filter element to loosen the dust.

CAUTION

In the next step, do not direct compressed air directly toward the inside surface of the element. This will force the dirt and dust into the pores of the element, thus restricting air flow.

8. Apply compressed air toward the *outside surface* of the element (**Figure 5**) to remove all loosened dirt and dust from the element.

9. Installation is the reverse of removal. Note the following:

   a. Install the air filter so the notches on the filter engage the posts (**Figure 3**) on the air filter housing.
   b. Torque the air filter cover bolts (A, **Figure 2**) to 2 N•m (18 in.-lb.).

Compression Test

An engine compression test is one of the quickest ways to check the condition of the rings, head gasket, piston and cylinder. Record the compression reading during each tune-up in the maintenance log at the end of the manual. Compare the current reading with those taken during earlier tune-ups. This will help to identify any developing problems.

Use a screw-in type compression gauge with a flexible adapter (**Figure 6**) when performing this test. Check the rubber gasket on the end of the adapter before each use. This gasket seals the cylinder to ensure accurate compression readings.

Before starting this test, confirm that the cylinder head bolts are tightened to the specified torque (Chapter Four), the valves are properly adjusted as described in this chapter, and the battery is fully charged to ensure proper cranking speed.

1. Warm the engine to normal operating temperature, and turn the engine off.

3

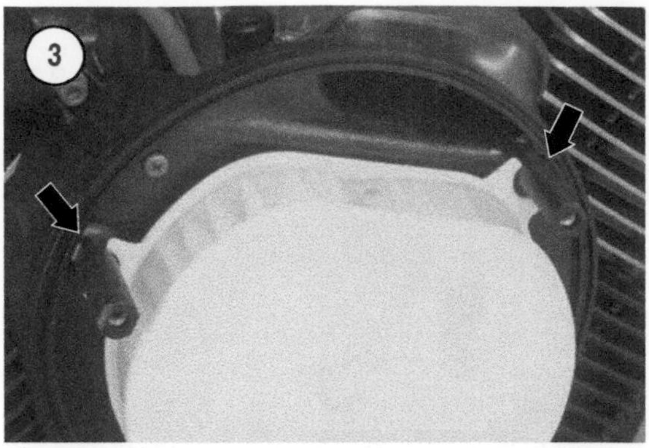

4

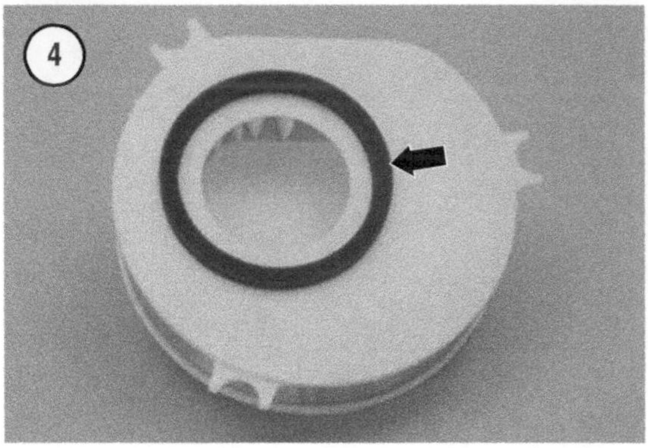

5

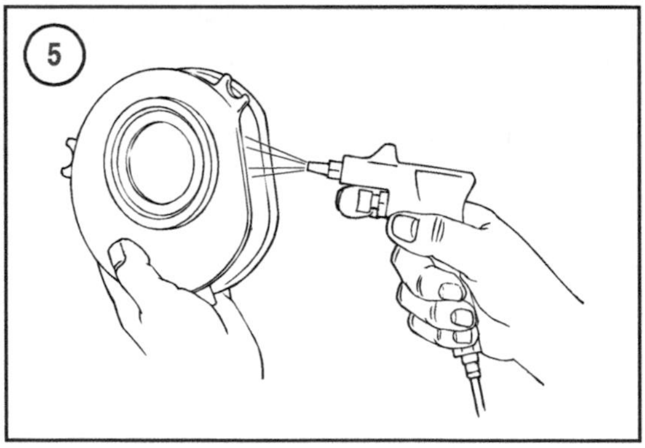

6

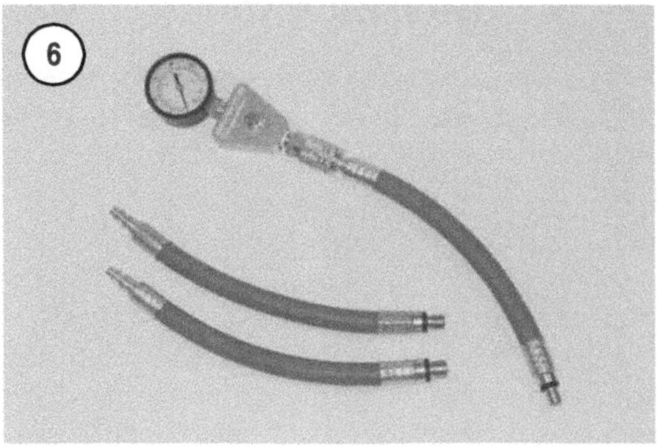

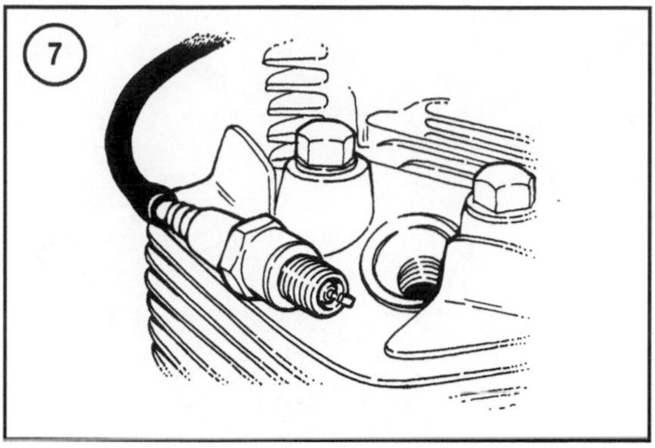

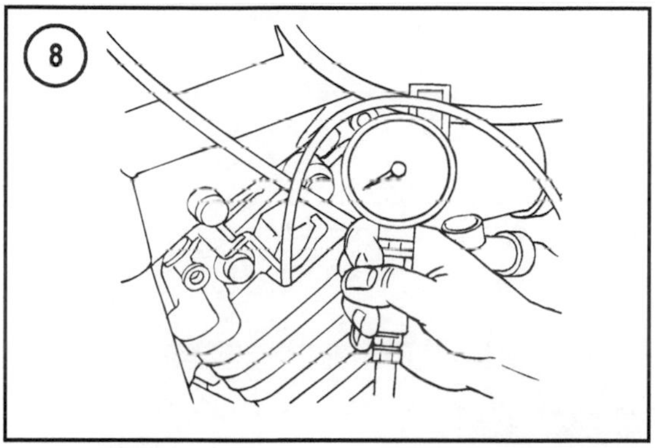

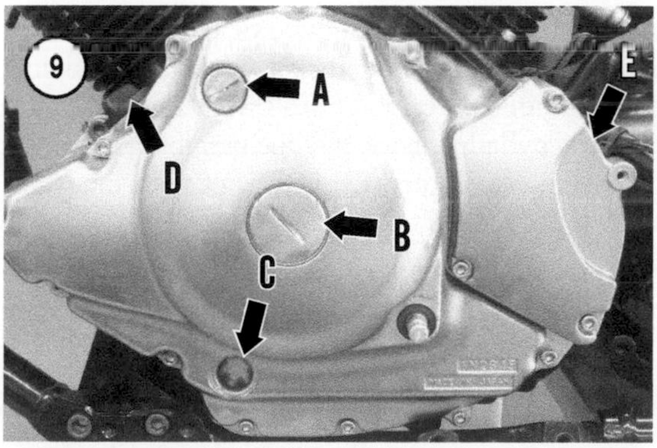

2. Remove the spark plugs as described in this chapter.
3. Insert each spark plug into its spark plug cap, and ground the spark plug against the cylinder head (**Figure 7**).
4. Turn the compression gauge into one cylinder following the manufacturer's instructions (**Figure 8**). Make sure the gauge is properly seated in the cylinder head.
5. Completely open the throttle, and crank the engine until there is no further rise in pressure. Remove the gauge and record the reading.
6. Repeat Steps 3-5 for the other cylinder.
7. Standard compression pressure is specified in **Table 5**. When interpreting the results, note the difference between the cylinders. Large differences indicate worn or broken rings, leaky or sticky valves, blown head gasket or a combination of all.
   a. If the compression reading between cylinders does not differ by more than 10%, the rings and valves are in good condition.
   b. If the reading is low (10% or more) on one of the cylinders, it indicates valve or ring trouble.
   c. To determine which, pour about a teaspoon of engine oil through the spark plug hole onto the top of the piston. Turn the engine over once to distribute the oil, then take another compression test and record the reading. If the compression returns to normal, the rings are worn or defective. If compression does not increase, the valves are leaking.

NOTE
If the compression is low, the engine cannot be tuned to maximum performance. The worn parts must be replaced.

Ignition Timing

Ignition timing is not adjustable. However, the timing can be checked to make sure all ignition components are operating correctly.
1. Start the engine and let it reach normal operating temperature. Shut the engine off.
2. Remove the timing inspection cover (A, **Figure 9**) from the alternator cover.
3. Connect a portable tachometer following the manufacturer's instructions.
4. Connect a timing light to the No. 1 spark plug wire (rear cylinder) following the manufacturer's instructions.
5. Start the engine and let it idle at the idle speed listed in **Table 5**.
6. Aim the timing light at the timing window. The timing is correct if the timing mark (A, **Figure 10**) on the flywheel aligns with the cutout in the timing window (B).
7. If the timing is incorrect, there is a problem in the ignition system. Follow the ignition system troubleshooting procedures listed in Chapter Two. The ignition timing cannot be adjusted.

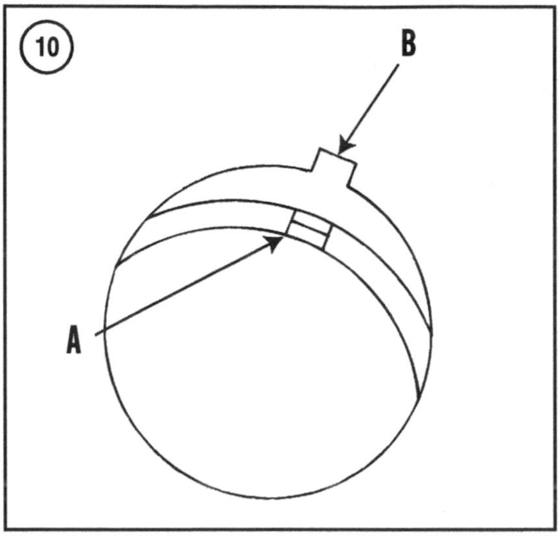

8. Shut off the engine, and disconnect the timing light and portable tachometer. Install the timing inspection cover (A, **Figure 9**).

Valve Clearance

Valve clearance measurement and adjustment must be performed with the engine at room temperature (below 35° C [95° F]).

The exhaust valve is located on the rear side of the rear cylinder and on the front side of the front cylinder.

NOTE
This procedure is shown with the engine removed from the frame for photographic clarity. The valves can be adjusted with the engine in the frame.

1. Remove the rider and passenger seats as described in Chapter Fourteen.
2. Remove the fuel tank, air filter housing and the surge tank (Chapter Eight).
3. Remove the cylinder head covers (Chapter Four).
4. Remove both the intake (**Figure 11**) and exhaust valve covers from each cylinder head.
5. Remove the timing inspection cover (A, **Figure 9**) and flywheel nut cover (B) from the alternator cover.

NOTE
A breather plate is not used in the rear cylinder head.

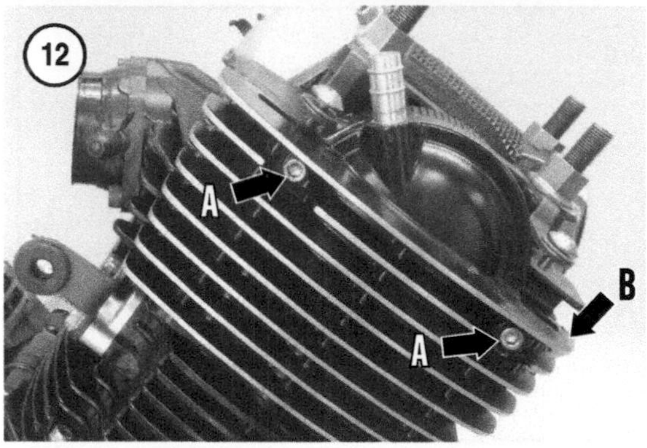

6. Remove the mounting screws (A, **Figure 12**) and remove the cam sprocket cover (B) and its O-ring from each cylinder head. When working on the front cylinder, make sure the breather plate and its O-ring (A, **Figure 13**) come out with the cam sprocket cover (B).
7. Remove both spark plugs. This makes it easier to rotate the engine.

NOTE
When a cylinder is set to TDC on the compression stroke, the timing mark on its cam sprocket may not precisely align with pointer on the cylinder head. On some models, the mark could be off by as much as 1/2 tooth.

8. Use the flywheel nut to turn the crankshaft clockwise until the rear cylinder is at top dead center on the compression stroke. The *rear cylinder* is at TDC on the compression stroke when the T-mark (A, **Figure 14**) on the flywheel aligns with the cutout in the alternator cover (B), and when the timing mark on the rear cam sprocket (A, **Figure 15**) aligns with the pointer on the rear cylinder head (B).

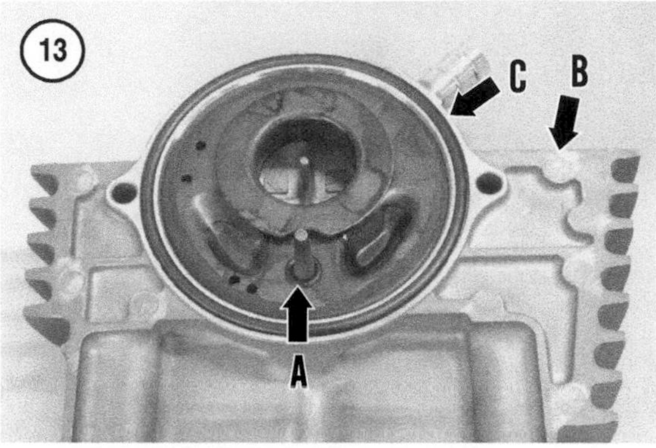

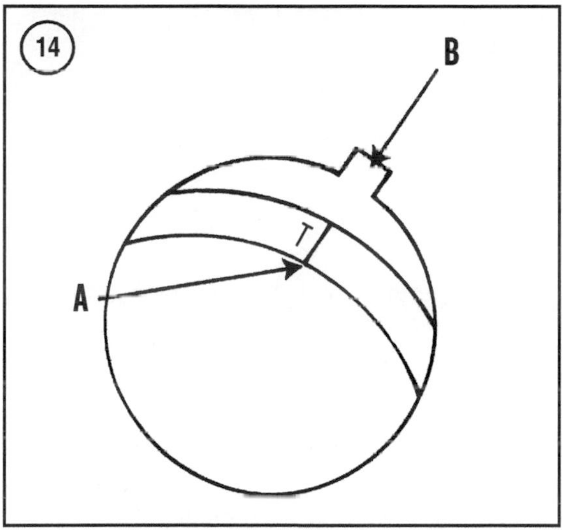

NOTE
A cylinder at TDC on the compression stroke has free play in both rocker arms, which indicates that both valves are closed.

9. Make sure the cylinder is at TDC by pressing each rocker arm. The intake and exhaust rocker arms should have free play. If both rocker arms do not have free play, rotate the engine an additional 360° until they do.

10. Check the clearance of both the intake and exhaust valves by performing the following. Refer to **Figure 16**.

   a. Insert a feeler gauge between the valve stem and the adjuster end.
   b. The clearance is correct if there is a slight drag on the feeler gauge when it is inserted and withdrawn. The correct valve clearance for the intake and exhaust valves is listed in **Table 5**.

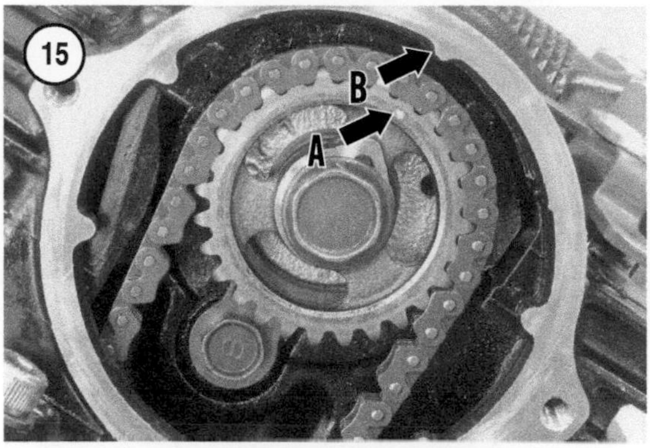

11. If necessary, adjust the valve clearance by performing the following:

   a. Loosen the locknut on the valve adjuster.
   b. With the feeler gauge between the valve stem and the adjuster end, turn the valve adjuster to obtain the specified clearance.
   c. Hold the adjuster to prevent it from turning, and torque the valve adjuster locknut to 27 N•m (20 ft.-lb.).

12. Use the flywheel nut to turn the crankshaft clockwise until the front cylinder is at top dead center on the compression stroke.

   a. Set the rear cylinder to TDC by rotating the engine clockwise until the T-mark (A, **Figure 14**) on the flywheel aligns with the alternator cover cutout (B).
   b. Rotate the crankshaft another 290° clockwise until the I-mark on the flywheel (A, **Figure 17**) aligns with the cutout in the alternator cover (B).
   c. Check the timing mark on the front cam sprocket plate (A, **Figure 18**). It should align with the pointer on the front cylinder head (B).

17

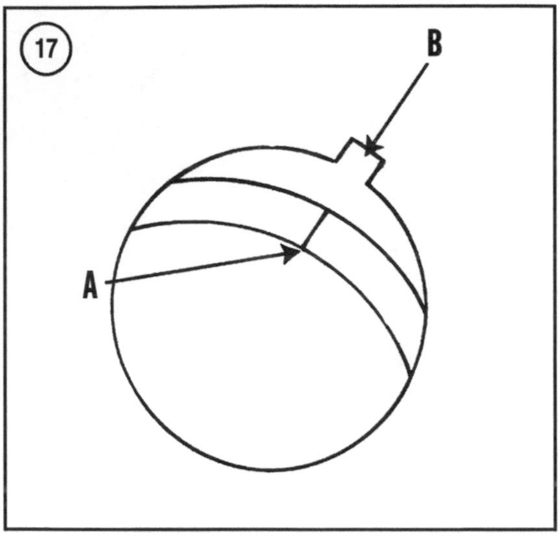

13. Perform Steps 10 and 11 to check and adjust each valve in the front cylinder.
14. When the clearance of each valve is within specification, reinstall the removed parts by reversing the removal procedure. Note the following:
   a. Before installing the front-cylinder cam sprocket cover, install a new O-ring (A, **Figure 13**) onto the breather plate and set the breather plate in place in the cam sprocket cover (B).

NOTE
A breather plate is not used on the rear cam sprocket cover.

   b. Use a new O-ring (C, **Figure 13**) when installing each cam sprocket cover onto the cylinder head. Torque the cam sprocket cover bolts (A, **Figure 12**) to the 10 N•m (89 in.-lb.).
   c. Torque the valve cover bolts to 10 N•m (89 in.-lb.) and the spark plugs to the 20 N•m (15 ft.-lb.).
   d. Install the cylinder head covers as described in Chapter Four.

SPARK PLUGS

Removal

A spark plug can be used to help determine the operating condition of its cylinder when it is properly read. As each spark plug is removed, label it with its cylinder number.

18

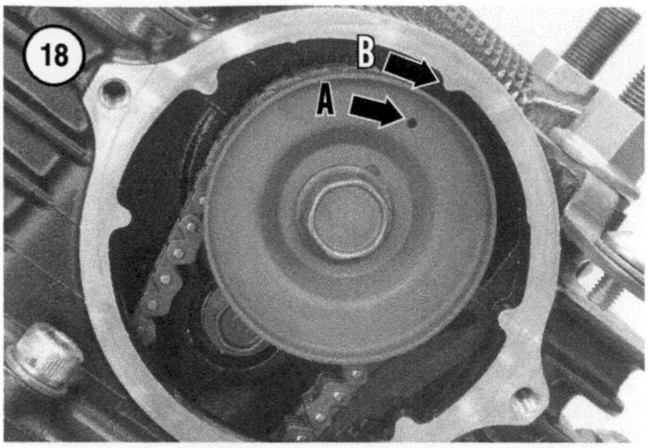

19

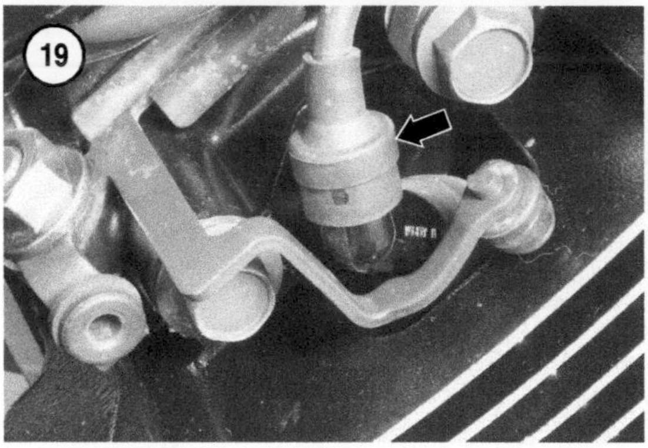

1. Remove the cylinder head covers as described in Chapter Four.
2. Grasp the spark plug lead (**Figure 19**) as near to the plug as possible and pull the lead off the plug.

CAUTION
Whenever a spark plug is removed, dirt around it can fall into the plug hole. This can cause serious engine damage.

3. Blow away any dirt that has accumulated in the spark plug well.

NOTE
If a plug is difficult to remove, apply penetrating oil, like WD-40 or Liquid Wrench, around base of plug and let it soak in for about 10-20 minutes.

4. Remove the spark plug (**Figure 20**) with a 5/8-in. spark plug wrench with a rubber insert to hold the plug. Label the spark plug by cylinder number.
5. Repeat for the remaining spark plug.

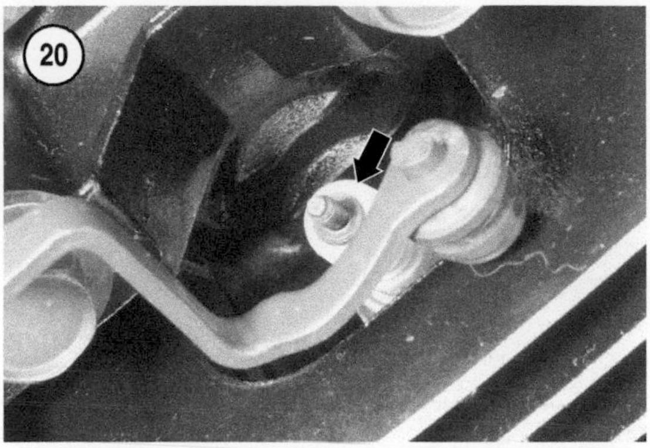

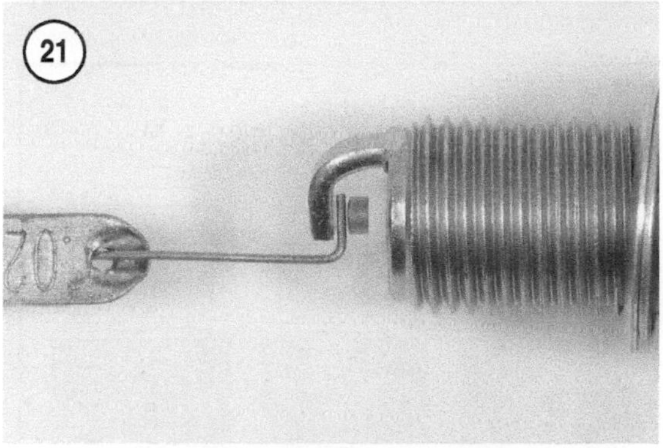

CAUTION
Do not clean the spark plugs with a sand-blasting device. This type of cleaning may leave abrasive material on the plug, which can enter the cylinder and cause damage.

6. Inspect the spark plug carefully. Look for a plug with broken center porcelain, excessively eroded electrodes, and excessive carbon or oil fouling. Replace a damaged plug. If deposits are light, the plug may be cleaned in solvent with a wire brush. Regap the plug as described in this section.
7. Inspect the spark plug cap and wires for cracks, hardness or other damage. If necessary, test the spark plug cap as described in Chapter Nine.

Gap and Installation

1. Remove the new spark plugs from the boxes. If the terminal nut is installed, unscrew it from the end of the plug. This adapter is not used.
2. Insert a wire feeler gauge between the center and side electrodes of the plug (**Figure 21**). The specified gap is listed in **Table 5**. The gap is correct if there is a slight drag as the wire is pulled through. If there is no drag or if the gauge will not pass through, bend the side electrode with a gaping tool (**Figure 22**) and set the gap to specification.
3. Apply a light coat of antiseize compound onto the threads of the spark plug before installing it. Do not use engine oil on the plug threads.

CAUTION
The cylinder head is aluminum. Cross-threading the spark plug can easily damage the spark plug hole threads.

4. Screw the spark plug in by hand until it seats. Very little effort is required. If force is necessary, the plug is cross-threaded. Unscrew it and try again.

NOTE
Do not overtighten the spark plug. This will only distort the gasket and destroy its sealing ability.

5. Torque the spark plugs to 20 N•m (15 ft.-lb.). If a torque wrench is not available, turn the plug until the gasket makes contact with the head, then tighten the plug an additional quarter to a half turn. When reinstalling old, regapped plugs with the old gasket, tighten the plug an additional quarter turn.

CAUTION
Do not use a plastic hammer or any type of tool to tap the plug cap assembly onto the spark plug. The assembly will be damaged.

NOTE
Be sure to push the plug cap all the way down to make full contact with

the spark plug post. If the cap does not completely contact the plug, the engine may falter and cut out at high engine speeds.

6. Install each plug cap onto the correct spark plug. Press the cap onto the spark plug and rotate the assembly slightly in both directions. Make sure it is attached to the spark plug and to the sealing surface of the cylinder head cover.
7. Install the cylinder head cover as described in Chapter Four.

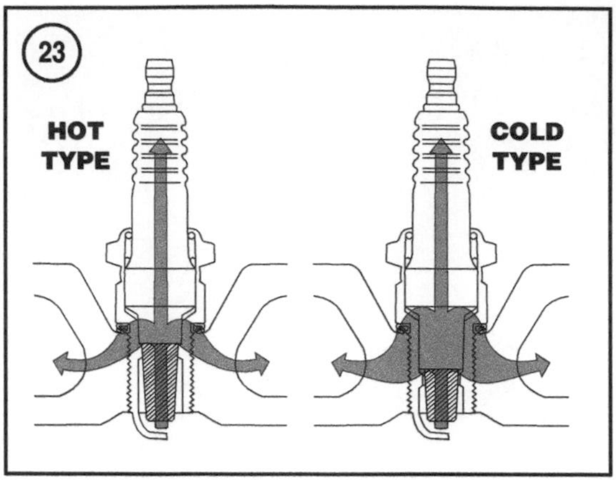

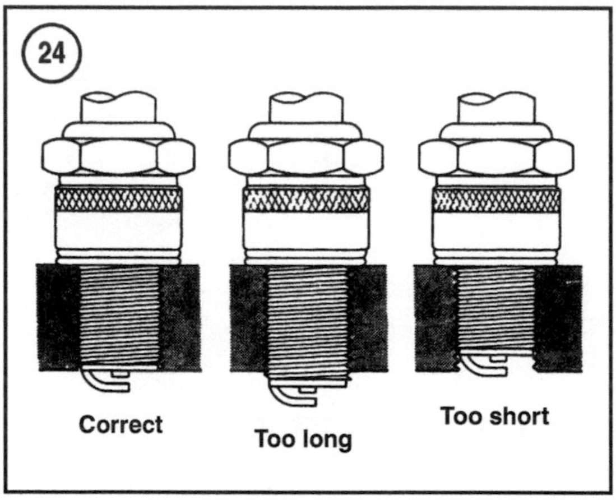

Heat Range

Spark plugs are available in various heat ranges that are hotter or colder than the plugs originally installed by the manufacturer.

Select a plug with a heat range designed for the loads and conditions under which the motorcycle will be operated. A plug with an incorrect heat range can foul, overheat and cause piston damage.

In general, use a hot plug for low speeds and low temperatures. Use a cold plug for high speeds, high engine loads and high temperatures. See **Figure 23**. The plug should operate hot enough to burn off unwanted deposits, but not so hot that it is damaged or causes preignition. To determine if plug heat range is correct, remove each spark plug and examine the insulator.

Do not change the spark plug heat range to compensate for adverse engine or carburetion conditions.

When replacing plugs, make sure the reach (**Figure 24**) is correct. A longer than standard plug could interfere with the piston and cause engine damage.

Refer to **Table 5** for recommended spark plugs.

Reading

Reading the spark plugs can provide information about spark plug operation, air/fuel mixture composition and engine conditions (such as oil consumption or pistons). Before checking the spark plugs, operate the motorcycle under a medium load for approximately 6 miles (10 km). Avoid prolonged idling before shutting off the engine. Remove the spark plugs as described in this section. Examine each plug and compare it to those in **Figure 25**. Refer to the following sections to determine the operating conditions.

If the plugs are being inspected to determine if carburetor jetting is correct, start with new plugs and operate the motorcycle at the load that corresponds to the jetting information desired. For example, if the main jet is in question, operate the motorcycle at full throttle, shut the engine off and coast to a stop.

Normal condition

A light tan- or gray-colored deposit on the firing tip and no abnormal gap wear or erosion indicate good engine, ignition and air/fuel mixture conditions. A plug with the proper heat range is being used. It may be serviced and returned to use.

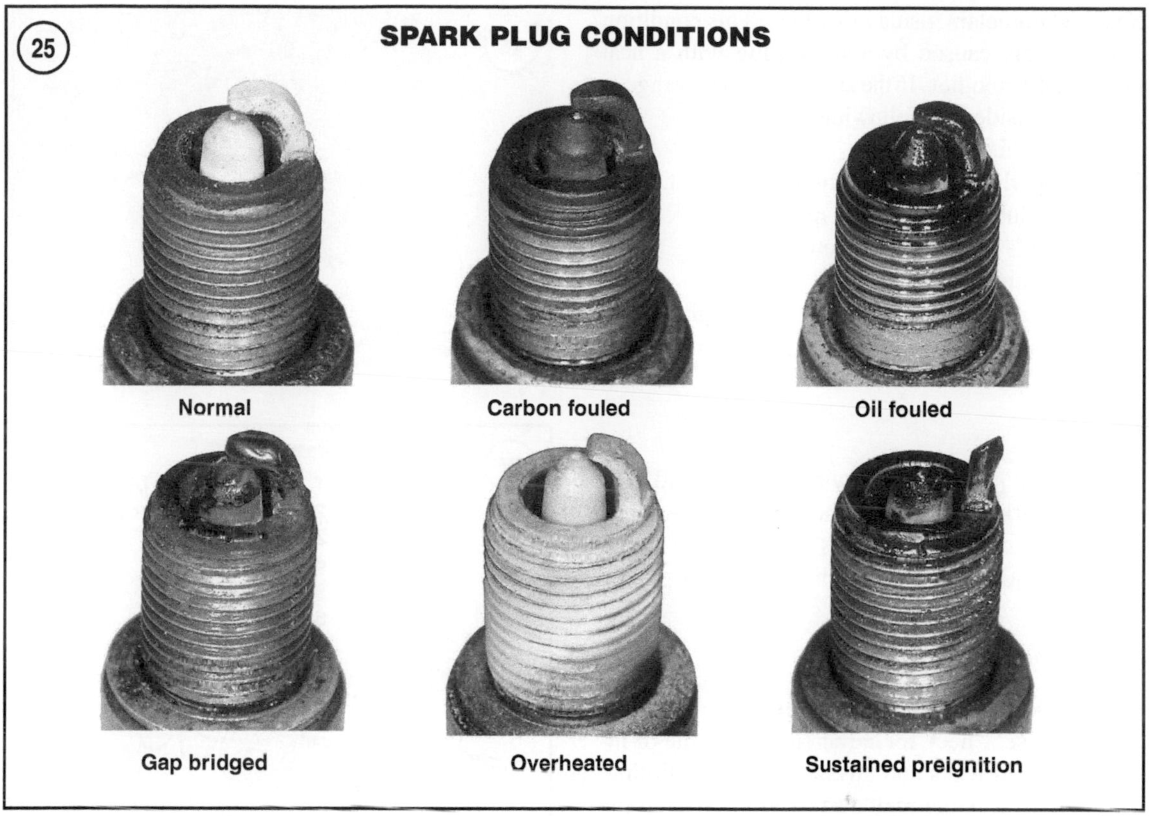

Carbon fouled

Soft, dry, sooty deposits covering the entire firing end of the plug are evidence of incomplete combustion. Even though the firing end of the plug is dry, the deposits decrease the plug's insulation. The carbon forms an electrical path that bypasses the electrodes resulting in a misfire. One or more of the following conditions can cause carbon fouling:

1. Air/fuel mixture too rich.
2. Spark plug heat range too cold.
3. Clogged air filter.
4. Improperly operating ignition component.
5. Ignition component failure.
6. Low engine compression.
7. Prolonged idling.

Oil fouled

An oil fouled plug has a black insulator tip, a damp oily film over the firing end and a carbon layer over the entire nose. The electrodes are not worn. Common causes for this condition are:

1. Incorrect air/fuel mixture.
2. Low idle speed or prolonged idling.
3. Ignition component failure.
4. Spark plug heat range too cold.
5. Engine still being broken in.
6. Valve guides worn.
7. Piston rings worn or broken.

Oil fouled spark plugs may be cleaned in an emergency, but it is better to replace them. It is important to correct the cause of fouling before the engine is returned to service.

Gap bridging

Plugs with this condition have deposits building up between the electrodes. The deposits reduce the gap and eventually close it entirely. If this condition is encountered, check for excessive carbon or oil in the combustion chamber. Be sure to locate and correct the cause of this condition.

Overheating

Badly worn electrodes and premature gap wear are signs of overheating, along with a gray or white blistered porcelain insulator surface. This condition is commonly caused by a spark plug with a heat range that is too hot. If the spark plug heat range is correct, consider the following causes:

1. Lean air/fuel mixture.
2. Improperly operating ignition component.
3. Engine lubrication system malfunction.
4. Cooling system malfunction.
5. Engine air leak.
6. Improper spark plug installation (over-tightening).
7. No spark plug gasket.

Worn out

Corrosive gases formed by combustion and high voltage sparks have eroded the electrodes. A spark plug in this condition requires more voltage to fire under hard acceleration. Install a new spark plug.

Preignition

If the electrodes are melted, preignition is probably the cause. Check for carburetor mounting or intake manifold leaks and advanced ignition timing. The plug heat range may also be too hot. Find the cause of the preignition before returning the engine into service. For additional information on preignition, refer to Chapter Two.

ENGINE OIL

Engine Oil Level Check

Check the engine oil level through the sight glass (C, **Figure 9**) in the alternator cover.

1. Start the engine and let it reach normal operating temperature.
2. Stop the engine and let the oil settle.

CAUTION
If the bike is not parked correctly, the oil level reading will be incorrect.

3. Have an assistant hold the motorcycle so it stands straight up and level.
4. The oil level should be between the maximum and minimum window marks. If necessary, remove the oil filler cap (D, **Figure 9**) from the top of the alternator cover and add enough of the recommended oil (see **Table 4**) to raise the oil to the proper level. Do not overfill the crankcase.

26

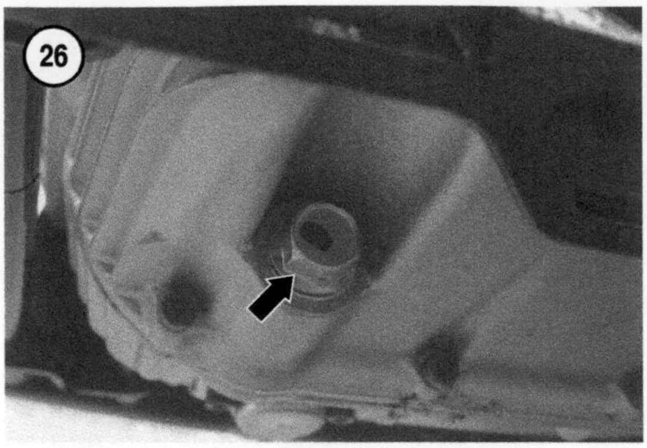

27

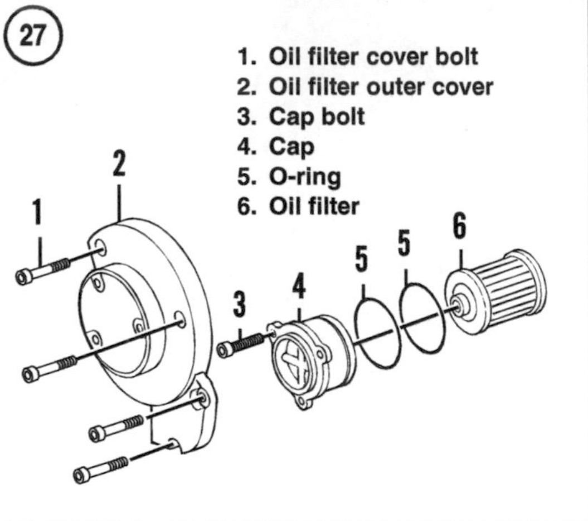

5. Reinstall the oil filler cap and tighten it securely.

Engine Oil and Filter Change

The recommended oil and filter change interval is specified in **Table 1**. This assumes that the motorcycle is operated in moderate climates. If a motorcycle is operated under dusty conditions, the oil will get dirty more quickly and should be changed more frequently than recommended.

Select an engine oil with a API service classification of SE, SF or SG. The classification is on the container label. Try to use the same brand of oil at each oil change and avoid the use of oil additives. They may cause clutch slippage. Refer to **Table 4** and select the correct oil viscosity for the anticipated ambient air temperatures (not the engine oil temperature).

CAUTION
Do not use engine oil classified as ***Energy Conserving.*** *These types of*

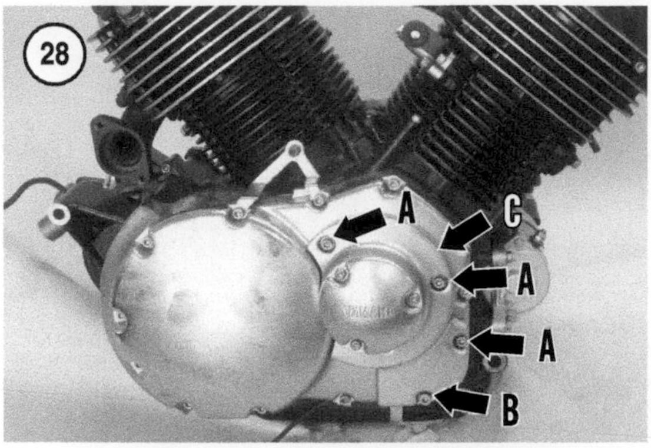

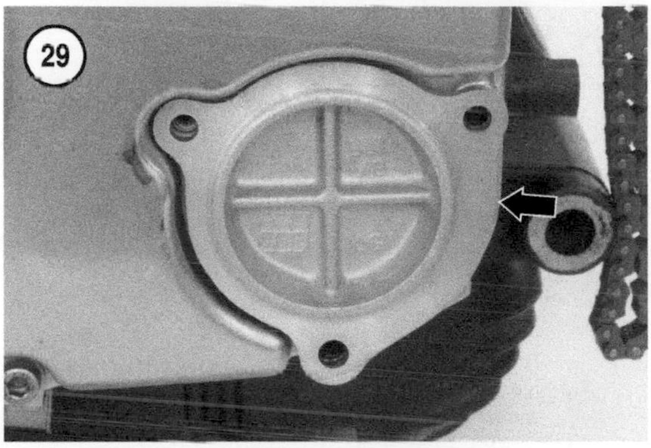

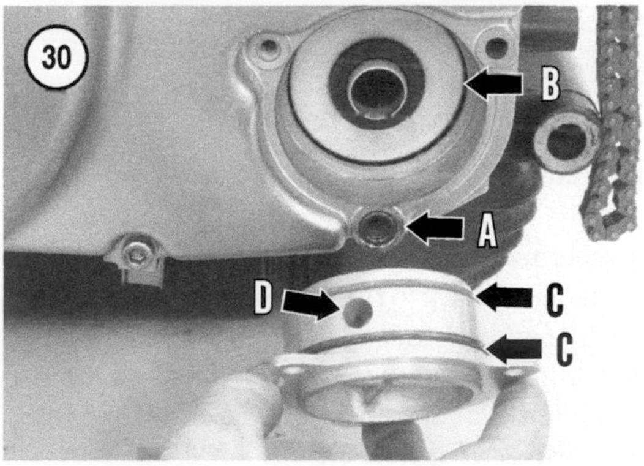

oils are designed specifically for automotive applications. The additives added to these oils may cause engine and/or clutch damage in motorcycle applications.

NOTE

Never dispose of engine oil in the trash, on the ground or down a storm drain. Many service stations and oil retailers accept used oil for recycling. Do not combine other fluids with engine oil for recycling. To locate a recycler, contact the American Petroleum Institute (API) at ***www.recycleoil.org****.*

1. Start the engine and run it until it reaches normal operating temperature, then turn the engine off.
2. Securely support the bike on a level surface.
3. Place a drain pan under the crankcase, and remove the oil drain bolt (**Figure 26**) from the left side of the crankcase.
4. Let the oil drain for at least 15-20 minutes.
5. Inspect the sealing washer on the crankcase drain plug. Replace the washer if its condition is in doubt.
6. Install the oil drain bolt and washer. Torque the bolt to 43 N•m (32 ft.-lb.).

WARNING

The exhaust pipes will be ***hot****! Wear thick work gloves when removing the exhaust pipes.*

7. Replace the oil filter (6, **Figure 27**) as follows:
   a. Remove the mufflers and front exhaust pipe as described in Chapter Eight.

   NOTE

   The emblem does not have to be removed when the oil filter is changed.

   a. Remove the outer cover bolts (A and B, **Figure 28**) and remove the outer cover (C) from the clutch cover. Note that the shouldered bolt is installed in the lower cover hole (B, **Figure 28**). It must be reinstalled here during assembly.
   b. Remove the oil filter cap (**Figure 29**). Watch for the O-ring (A, **Figure 30**) in the lower mount.
   c. Remove the oil filter (B, **Figure 30**) from the oil filter cavity.
   d. Apply fresh oil to the grommet on a new oil filter and install the filter so the end with the grommet faces out as shown in **Figure 31**.
   e. Check the O-rings for tears or brittleness. Replace them as necessary.
   f. Apply fresh engine oil to the two O-rings on the oil filter cap (C, **Figure 30**) and the small O-ring in the lower mount (A).
   g. Install the oil filter cap so its oil delivery hole (D, **Figure 30**) faces up.
   h. Install the oil filter outer cover (C, **Figure 28**). Install the shouldered bolt (B, **Figure 28**)

in the hole with the small O-ring. See A, **Figure 30** and B, **Figure 28**. Torque the oil filter outer cover bolts to 10 N•m (89 in.-lb.).

9. Remove the oil filler cap (D, **Figure 9**) and insert a funnel into the oil filler hole. Fill the crankcase with the correct weight and quantity of oil (**Table 4**). Screw on the oil filler cap securely.

10. Check the oil pressure as follows:
   a. Slightly loosen the intake rocker arm shaft bolt on the rear cylinder.
   b. Start the engine and let it idle. Oil should seep from the loosened rocker arm bolt. If it does not do so within one minute, turn the engine off and inspect the oil lines, oil filter and oil pump for damage.
   c. Turn the engine off and torque the rocker arm bolt to 38 N•m (28 ft.-lb.).

11. Check for oil leaks. Once the oil has settled, check the oil level in the sight glass (A, **Figure 9**) on the alternator cover. Adjust the oil level if necessary.

12. Reinstall the front exhaust pipe and the mufflers as described in Chapter Eight.

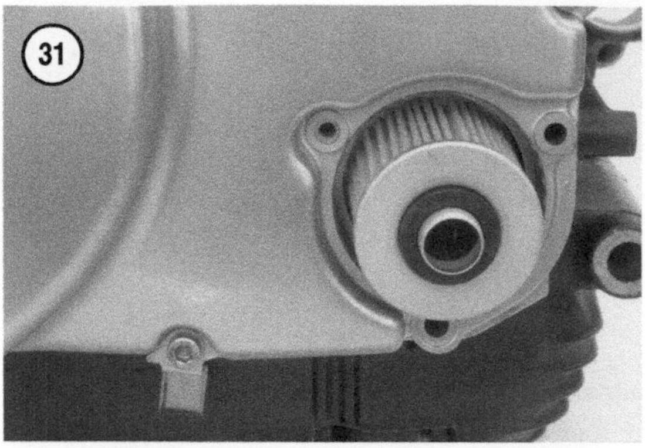
31

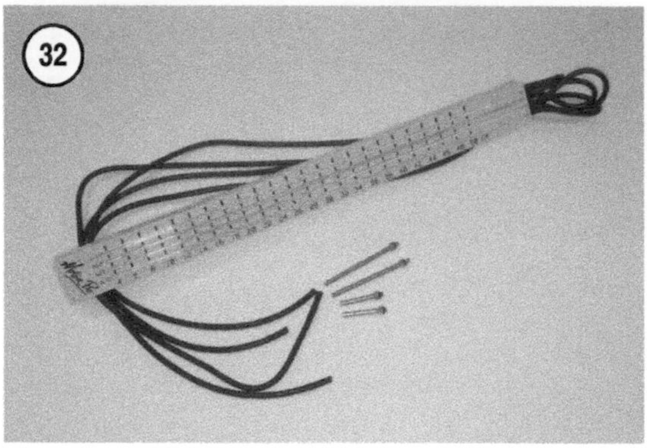
32

FUEL AND EXHAUST SYSTEMS

Fuel Line Inspection

Inspect the condition of all fuel lines for cracks or deterioration; replace them if necessary. Make sure the hose clamps are in place and holding securely.

Fuel Filter Replacement

Replace the fuel filter when it is dirty or at the interval specified in **Table 1**. Refer to the procedure in Chapter Eight.

Exhaust System Inspection

Check for leaks at all fittings. Tighten all bolts and nuts; replace any gaskets as necessary. Refer to Chapter Eight.

Emission Control System (California Models)

At the service intervals in **Table 1**, check all of the emission control lines and the EVAP canister for loose connections or damage. Refer to Chapter Eight.

Carburetor Synchronization

Synchronizing the carburetors makes sure one cylinder does not try to run faster than the other, cutting power and gas mileage. The only accurate way to synchronize the carburetors is to measure the intake vacuum of both cylinders at the same time with a set of gauges (**Figure 32**).

NOTE

Before synchronizing the carburetors, check the engine idle speed and valve clearances. Both must be within specification.

1. Securely support the motorcycle on a level surface.

2. Remove the rider and passenger seats as described in Chapter Fourteen.

3. Remove the fuel tank, air filter housing and surge tank (Chapter Eight).

4. Attach a portable tachometer to the No. 2 cylinder (front cylinder) spark plug lead following the manufacturer's instructions.

5. Remove the plug (**Figure 33**) from the vacuum fitting on each cylinder's intake manifold. On models with an AIS system, remove the AIS hose (**Figure 34**) from the vacuum fitting on the rear cylinder's intake manifold.

6. Connect the vacuum lines from the vacuum gauge to the vacuum fitting on each intake manifold following the manufacturer's instructions. Be sure to route the vacuum lines to the correct cylinder.

7. Attach an auxiliary fuel tank to the carburetor. Start the engine and let it warm up.

8. With the engine running at the idle speed listed in **Table 5**, check the gauge readings. The vacuum pressure at each cylinder must be within the range specified in **Table 5**.

9. If the difference between the vacuum readings in the two cylinders is greater than 1.33 kPa, (10 mm Hg [0.4 in. Hg]), synchronize the No. 1 carburetor to the No. 2 carburetor by adjusting the synchronizing screw (**Figure 35**). Turn the screw until the vacuum readings in the two cylinders are equal or as close as possible.

10. Check the idle speed. If necessary, adjust it as described in this section.

11. Stop the engine and detach the equipment.

12. Reinstall the plug (**Figure 33**) onto the intake manifold vacuum fittings. On models with an AIS system, reconnect the AIS hose (**Figure 34**) to the fitting on the rear cylinder intake manifold.

Idle Speed Adjustment

Before adjusting the idle speed, clean or replace the air filter, test the engine compression and synchronize the carburetors. Idle speed cannot be properly adjusted if these items are not within specification. Refer to the procedures described earlier in this chapter.

1. Remove the cylinder head cover from the rear cylinder as described in Chapter Four.

2. Attach a portable tachometer to the No. 1 cylinder (the rear cylinder) spark plug lead following the manufacturer's instructions.

3. Start the engine and warm it to normal operating temperature.

4. Sit on the seat while the engine is idling and adjust your weight to remove as much weight as possible from the front wheel. Turn the front wheel from side to side without touching the throttle grip. If the engine speed increases when the wheel is turned, the throttle cable may be damaged or incorrectly adjusted. Adjust the throttle cable as described in this chapter.

5. Turn the throttle stop screw (**Figure 36**) to set the idle speed to the specification in **Table 5**. When viewed from the left side, turning the throttle stop screw toward the rear increases idle speed; turning it toward the front decreases idle speed.

6. Rev the engine a couple of times to see if it settles down to the set speed. Readjust as necessary.
7. Shut off the engine and disconnect the portable tachometer.

Throttle Cable Adjustment

Always check the throttle cables before making any carburetor adjustments. Too much free play causes delayed throttle response; too little free play causes unstable idling.

Check the throttle cables from the throttle grip to the carburetors. Make sure they are not kinked or chafed. Replace the cables if necessary.

Make sure the throttle grip rotates smoothly from fully closed to fully open. Check free play with the handlebars at the center, full-left and full-right steering positions.

Check free play at the throttle grip flange (**Figure 37**). Free play should be within the range specified in **Table 5**. If adjustment is necessary, perform the following:

1. Remove the rider and passenger seats as described in Chapter Fourteen.
2. Remove the fuel tank, air filter housing and surge tank as described in Chapter Eight.
3. At the handlebar, make sure the throttle cable locknut and adjuster are tight.
4. At the carburetor assembly, loosen the pull cable locknut (A, **Figure 38**).
5. Rotate the pull cable adjuster (B, **Figure 38**) until the correct amount of free play is achieved.
6. Tighten the locknut securely.

NOTE
If the correct amount of free play cannot be achieved at the carburetor, additional adjustment can be obtained at the adjuster on the handlebar.

7. At the handlebar, loosen the throttle cable adjuster locknut (A, **Figure 39**).
8. Rotate the adjuster (B, **Figure 39**) in either direction until the correct amount of free play is achieved. Tighten the adjuster locknut.

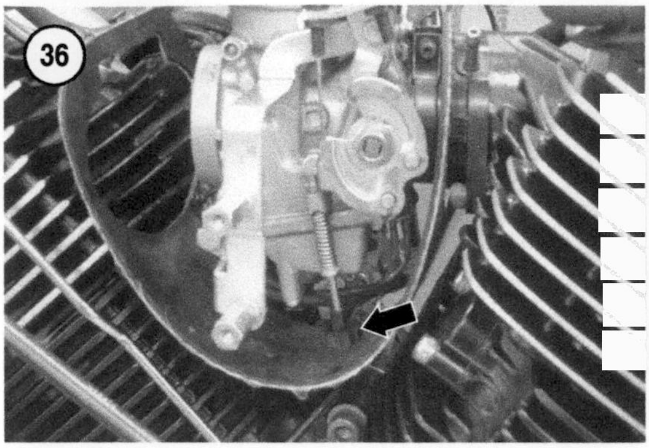

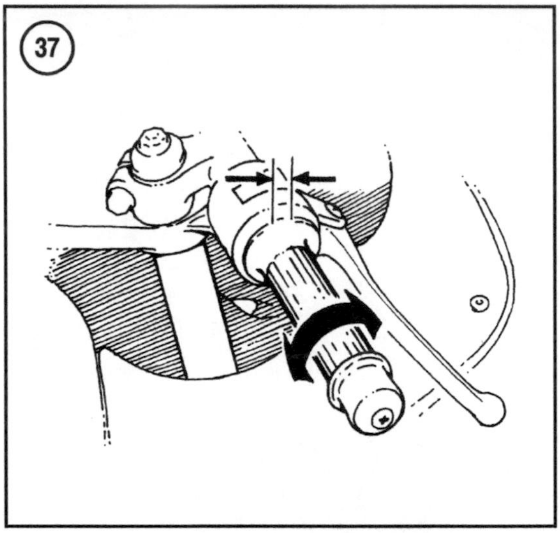

WARNING
If idle speed increases when the handlebar is turned to right or left, check the throttle cable routing. Do not ride the motorcycle in this unsafe condition.

9. Start the engine and let it idle in neutral. Listen to the engine speed while turning the handlebar from steering lock to steering lock. If idle speed changes as the handlebar is turned, the throttle cable is

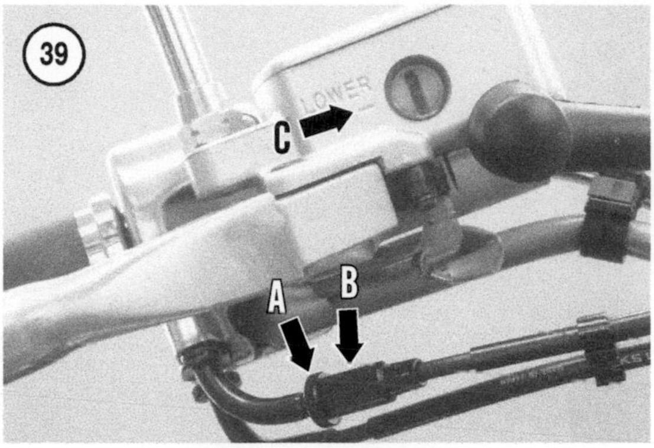

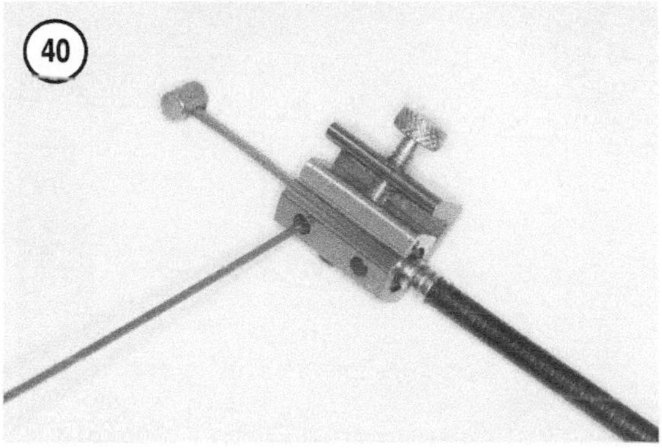

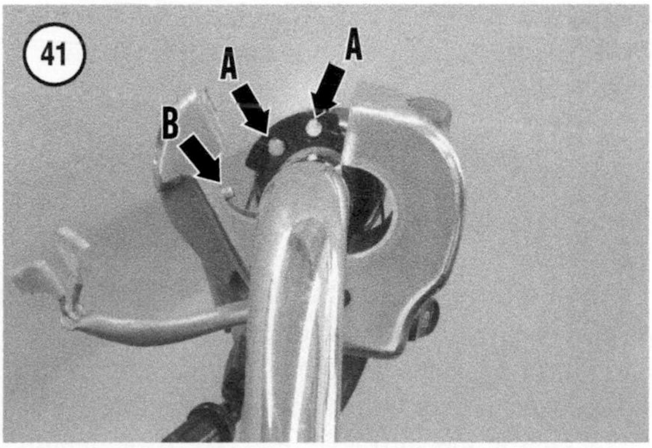

routed incorrectly or there is insufficient cable free play. Make the necessary corrections.

CONTROL CABLES

Lubricate and adjust the control cables at the intervals specified in **Table 1**. When lubricating a cable, also inspect it for fraying and check the cable sheath for chafing.

Cables can be lubricated with a cable lubricant and a cable lubricator as shown in **Figure 40**. Do not use chain lube to lubricate the cables.

If necessary, refer to *Throttle Cable Adjustment* in this chapter.

NOTE
The main cause of cable breaking or cable stiffness is improper lubrication. Maintaining the throttle cable as described in this section ensures a long service life.

Throttle Cable Lubrication

1. Remove the two mounting screws and separate the halves of the right handlebar switch assembly as described in Chapter Nine.
2. Disengage the ends (A, **Figure 41**) of both the pull and push cables from the throttle drum.
3. Attach a cable lubricator to the cable following the manufacturer's instructions.
4. Insert the nozzle of the lubricant can into the lubricator (**Figure 40**), press the button on the can and hold it down until the lubricant begins to flow out of the other end of the cable.

NOTE
Place a shop cloth at the carburetor end of the cable to catch all excess lubricant that flows out.

NOTE
If lubricant does not flow out the end of the cable, check the entire cable for fraying, bending or other damage.

5. Remove the lubricator, reconnect the cables and adjust the throttle cable as described in this chapter. When installing the right handlebar assembly, make sure the pin (B, **Figure 41**) on the switch assembly aligns with the hole in the handlebar.

Clutch Cable Lubrication

1. At the handlebar, slide the clutch lever boot (A, **Figure 42**) away from the adjuster.
2. Loosen the clutch cable locknut (B, **Figure 42**) and rotate the adjuster (C) to provide maximum slack in the cable.
3. Disconnect the cable end from the clutch hand lever.

4. Attach a cable lubricator to the cable following the manufacturer's instructions.
5. Insert the nozzle of the lubricant can into the lubricator (**Figure 40**), press the button on the can and hold it down until the lubricant begins to flow out of the other end of the cable.

NOTE
Place a shop cloth at the clutch lever end of the cable to catch all excess lubricant that flows out.

NOTE
If lubricant does not flow out the end of the cable, check the entire cable for fraying, bending or other damage.

6. Remove the lubricator, and reconnect and adjust the throttle cable as described in this chapter.

Clutch Cable Adjustment

Adjust the clutch cable free play at the interval indicated in **Table 1**. The clutch will not engage or disengage properly if the free play is not maintained within the range specified in **Table 5**. Please note that this clutch release mechanism loosens as the engine warms up. Therefore, set the free play to the minimum setting.

Measure clutch lever free play at the end of the hand lever. If the clutch lever free play is outside the range specified in **Table 5**, adjust free play as follows:

1. At the clutch hand lever, pull the rubber boot (A, **Figure 42**) back from the adjuster.
2. Loosen the clutch cable locknut (B, **Figure 42**).
3. Turn the adjuster (C, **Figure 42**) in or out to attain the amount of free play specified in **Table 5**.
4. Tighten the locknut and reinstall the rubber boot.
5. If the proper free play cannot be achieved at the hand lever adjuster, adjust the clutch release mechanism as follows:
   a. Remove the clutch adjuster cover (**Figure 43**) from the alternator cover.
   b. Loosen the locknut (A, **Figure 44**) on the lower clutch cable adjuster.
   c. Turn the adjuster screw (B, **Figure 44**) clockwise until it lightly bottoms, then back the adjuster out 1/4 turn.
   d. Torque the clutch adjuster locknut to 12 N•m (106 in.-lb.).

42
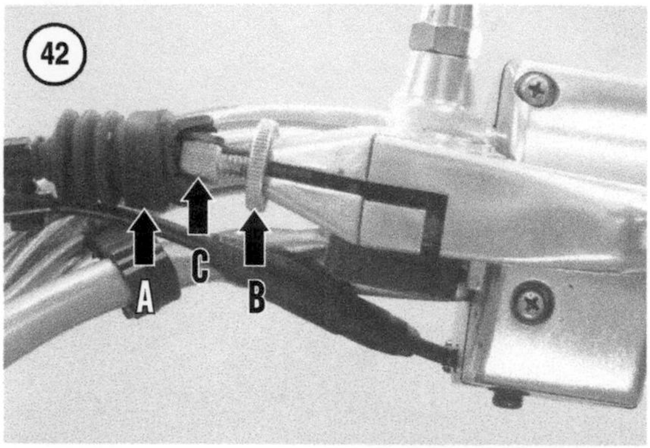

43
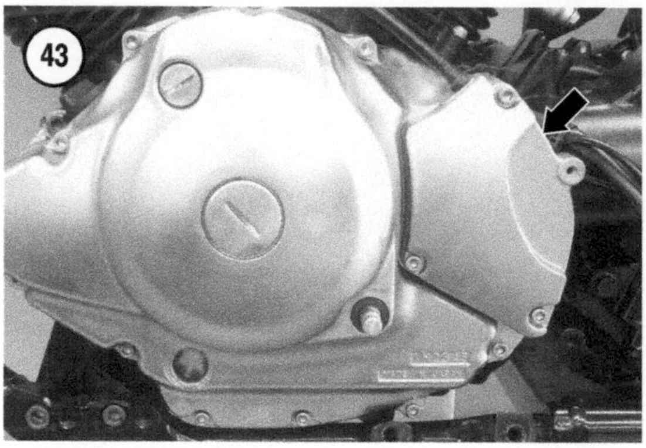

6. Recheck the amount of free play at the hand lever and perform any minor adjustments at the lever.

TRANSMISSION

Shift Pedal Adjustment

Measure the length of the shift rod on the shift pedal assembly. The shift rod length equals the distance from the outside edge of the locknut at the shift pedal end (A, **Figure 45**) to the outside edge of the locknut at the shift lever end (B). If the length does not equal the value specified in **Table 5**, adjust the length as follows:

1. Loosen the locknut at each end of the shift rod.
2. Turn the shift rod (C, **Figure 45**) to attain the desired length.
3. Tighten each locknut securely.

SIDESTAND

Check the operation of the sidestand and the sidestand switch at the service interval listed in **Table 1**.

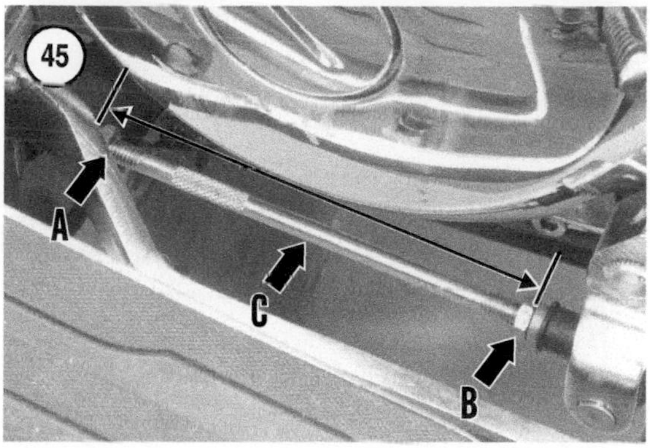

1. Securely support the motorcycle on a level surface.
2. Operate the sidestand, and check its movement and spring tension. Replace the spring if it is weak or damaged.
3. Lubricate the sidestand pivot surfaces with lithium soap grease.
4. Check the sidestand switch operation as follows:
   a. Park the motorcycle on a level surface. Both wheels must be on the ground.
   b. Sit on the motorcycle and raise the sidestand.
   c. Shift the transmission to neutral.
   d. Start the engine and let it idle.
   e. Pull in and hold the clutch lever. Shift the transmission into gear.
   f. While holding the clutch lever in, move the sidestand down. The engine should stop.
   g. If the engine does not stop when the sidestand is lowered, inspect the sidestand switch as described in Chapter Nine.
5. If the sidestand nut was loosened, torque it to 56 N•m (41 ft.-lb.).

GENERAL LUBRICATION

Swing Arm Bearings

Clean the swing arm bearings in solvent and pack them with molybdenum disulfide grease at the intervals specified in **Table 1**. The swing arm must be removed to service the bearings. Refer to Chapter Twelve.

Steering Stem Bearings

Remove, clean and lubricate the steering stem bearings with lithium soap grease at the interval specified in **Table 1**. Refer to the procedure in Chapter Eleven.

Wheel Bearings

Worn wheel bearings cause excessive wheel play that results in vibration and other steering troubles. Inspect the wheel bearings at the intervals specified in **Table 1**. Refer to the procedures in Chapter Ten.

Miscellaneous

Unless otherwise indicated, lubricate the following items with lithium soap grease: O-rings, oil seal lips, shift pedal shaft, brake pedal shaft, footrest pivots, clutch lever, front brake lever, control cable ends, and the sidestand bolt and sliding surfaces.

BATTERY

All the models covered in this manual are equipped with a maintenance free, sealed battery so the electrolyte level cannot be checked. Never attempt to remove the sealing bar cap from the top of the battery.

Since the negative side of the battery is grounded, always disconnect the negative cable, then the positive cable, when removing the battery. This minimizes the chance of a tool shorting to ground when the positive cable is disconnected.

Removal/Installation

1. Securely support the motorcycle on a level surface.
2. Remove the battery cover as described in Chapter Fourteen.
3. Disconnect the negative cable (A, **Figure 46**) from the battery.
4. Pull back the boot and disconnect the positive cable (B, **Figure 46**).
5. Remove the battery hold-down strap (C, **Figure 46**), and lift the battery from the battery box.
6. Set the battery on some newspapers or shop cloths to protect the workbench surface.
7. After the battery has been recharged or replaced, install it by reversing these removal steps. Note the following:
   a. Apply dielectric grease to each battery terminal.
   b. Connect the positive cable to the battery, then connect the negative cable.
   c. Tighten each screw securely.

Inspection

Check the state of charge in a maintenance free battery by measuring the voltage with the battery disconnected from the motorcycle.

If electrolyte spills onto your clothing or skin, immediately neutralize the electrolyte with a solution of baking soda and water.

WARNING

A damaged battery case could leak electrolyte. Electrolyte splashed into the eyes is extremely harmful. Always wear safety glasses when servicing a battery. If you get electrolyte in your eyes, call a physician immediately. Force your eyes open and flush them with cool, clean water for approximately 15 minutes or until medical help arrives.

1. Remove the battery as described in this section. Do not clean the battery while it is mounted in the frame.
2. Inspect the battery pads in the battery box for contamination or damage. Clean the pads and compartment with a solution of baking soda and water.
3. Set the battery on a stack of newspapers or shop cloths to protect the workbench surface.

4. Check the entire battery case for cracks or other damage. If the battery case is warped, discolored or has a raised top, the battery has been overcharging or overheating.
5. Check the battery terminals and bolts for corrosion or damage. Clean parts thoroughly with a solution of baking soda and water. Replace severely corroded or damaged parts.
6. If the top of the battery is corroded, clean it with a stiff bristle brush using the baking soda and water solution.
7. Check the battery cable terminals for corrosion and damage. If corrosion is minor, clean the battery cable terminals with a stiff wire brush. Replace severely worn or damaged cables.

NOTE

Measure the open circuit voltage when battery temperature is 20° C (68° F). Use a digital voltmeter when checking the battery's voltage. The precision of a digital meter helps to accurately determine the battery's state of charge.

8. Check the state of charge by connecting a digital voltmeter across the battery terminals. Connect the voltmeter negative test lead to the negative battery terminal and the positive test lead to the positive terminal (**Figure 47**).
   a. If the battery voltage is 12.8 volts or higher, the battery is fully charged.
   b. If the battery voltage is 12.0-12.8 volts, the battery is undercharged and requires charging.
   c. If battery voltage is less than 12.0 volts, replace the battery.

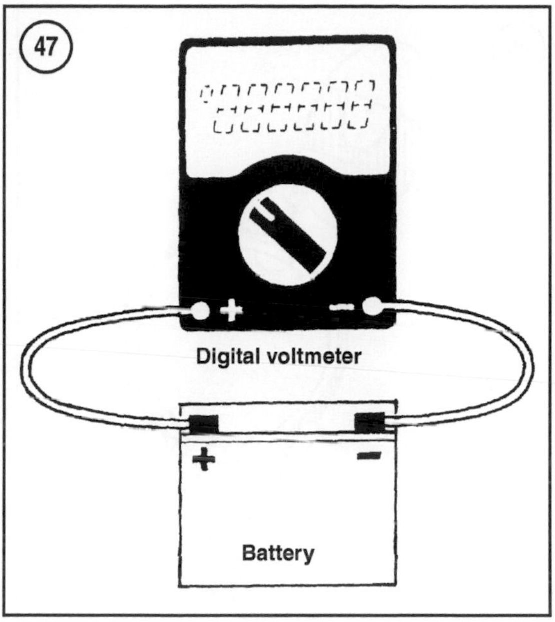

Charging

A digital voltmeter and a charger with an adjustable amperage output are required when charging a maintenance free battery. If this equipment is not available, have the battery charged by a shop with the proper equipment. Excessive voltage and amperage from an unregulated charger can damage the battery and shorten service life.

A battery self-discharges approximately one percent of its given capacity each day. If the battery is not in use (without any loads connected), and loses its charge within a week after charging, the battery is defective.

If the motorcycle is not used for long periods of time, an automatic battery charger with variable voltage and amperage outputs is recommended for optimum battery service life.

WARNING

During charging, highly explosive hydrogen gas is released from the battery. Only charge the battery in a well-ventilated area away from open flames, including appliance pilot lights. Do not allow smoking in the area. Never check the charge of the battery by arcing across the terminals; the resulting spark can ignite the hydrogen gas.

CAUTION

Always disconnect the battery cables from the battery. If the cables are left connected during the charging procedure, the charger may damage the diodes within the voltage regulator/rectifier.

NOTE

Some maintenance chargers can be used while the battery is connected to the motorcycle. These types of chargers are specifically designed for motorcycle batteries and will not damage the diodes in the regulator/rectifier.

1. Remove the battery from the motorcycle as described in this section.
2. Set the battery on a stack of newspapers or shop cloths to protect the surface of the workbench.
3. Connect the positive charger lead to the positive battery terminal and the negative charger lead to the negative battery terminal.
4. Set the charger to 12 volts. If the output of the charger is variable, it is best to select the low setting.

CAUTION

Never set the battery charger to more than 4 amps. The battery will be damaged if the charge rate exceeds 4 amps.

5. The charging time depends on the discharged condition of the battery. Use the charging amperage and length of time suggested on the battery label. Normally, a battery should be charged at a slow rate of 1/10 its rated capacity.
6. Turn the charger on.
7. After the battery has been charged for the pre-determined time, turn the charger off and disconnect the leads.
8. Wait 30 minutes, and then measure the battery voltage. Refer to the following:
   a. If the battery voltage is greater than 12.8 volts, the battery is fully charged.
   b. If the battery voltage is less than 12.8 volts, the battery is undercharged and requires charging.
9. If the battery remains stable for one hour, the battery is charged.

3

10. Install the battery into the motorcycle as described in this chapter.

New Battery Initialization

Always replace a maintenance free battery with another maintenance free battery. Also make sure the battery is charged completely before installing it. Failure to do so will reduce the life of the battery. Check with the dealership on the type of pre-service that the battery received.

NOTE

Recycle the old battery. *Most motorcycle dealerships will accept an old battery in trade with the purchase of a new one. Never place an old battery in the household trash. It is illegal in most states to place any acid or lead (heavy metal) contents in landfills.*

TIRES

Tire Pressure

Check and adjust tire pressure to accommodate the rider and cargo weight. A simple, accurate tire-pressure gauge can be purchased for a few dollars and should be carried in your motorcycle tool kit. The tire pressure specifications are shown in **Table 2**.

Tire Inspection

The likelihood of tire failure increases with tread wear. It is estimated that the majority of all tire failures occur during the last 10% of usable tread wear. Check tire tread for excessive wear, deep cuts, and embedded objects, such as stones, nails, etc. Also check for high spots that indicate internal tire damage. Replace tires that show high spots or swelling. If there is a nail in a tire, mark its location with a light crayon before pulling it out. This will help locate the hole for repair. Refer to Chapter Ten for tire changing procedures.

Measure tread wear (**Figure 48**) at the center of the tire tread with a tread depth gauge or small ruler. Because tires sometimes wear unevenly, measure wear at several points. Replace the original equipment tires if any tread depth measurement is less than the value specified in **Table 2**.

48

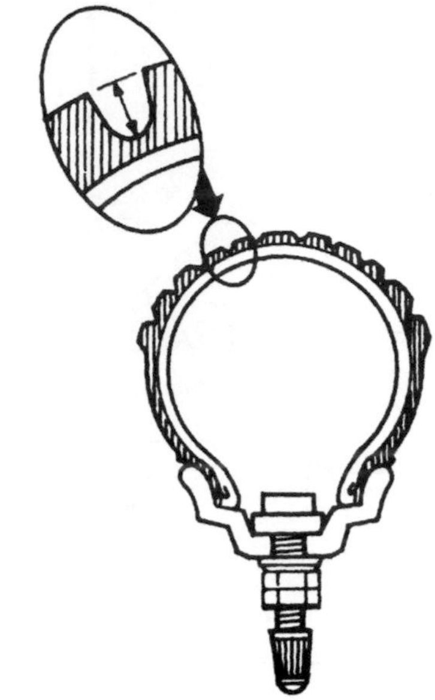

49

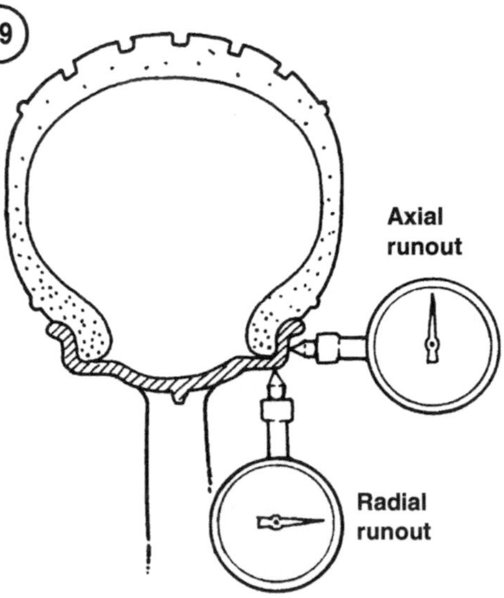

Rim Inspection and Runout

Frequently inspect the wheel rims for cracks, warping or dents. A damaged rim may cause an air leak.

Wheel rim runout is the amount of wobble a wheel shows as it rotates. To quickly check runout, simply support the bike with the wheel off the ground. Slowly turn the wheel while holding a pointer solidly against a fork leg or the swing arm with the other end against the rim. If either axial

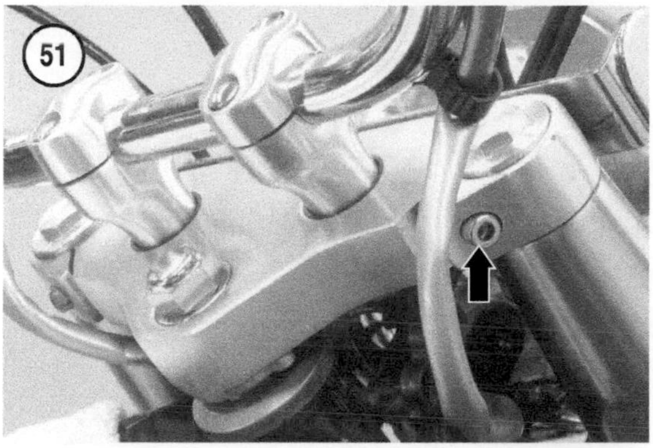

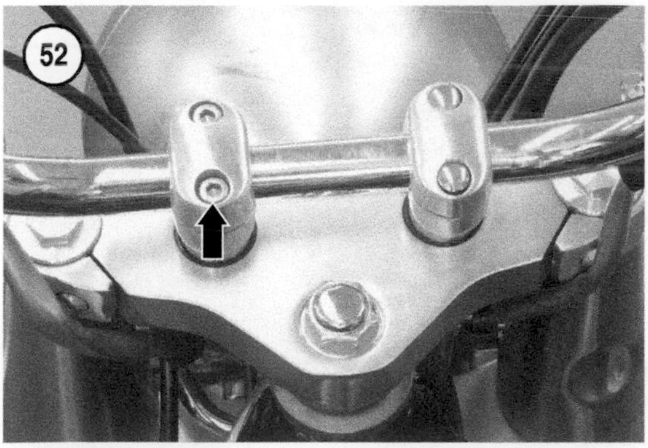

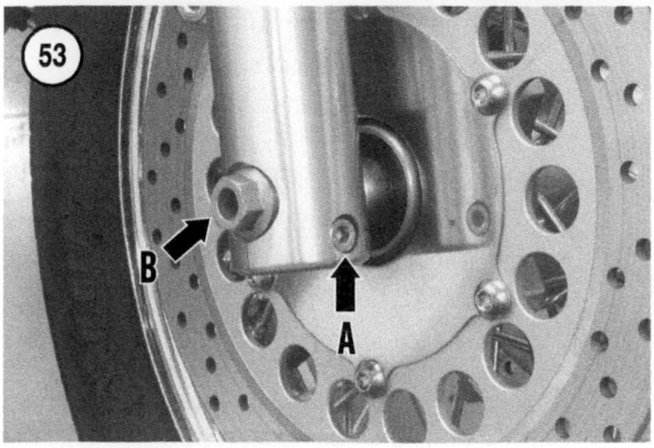

runout (side-to-side movement) or radial runout (up-and-down movement) exceeds the specification in **Table 5**, remeasure both axial and radial runout (**Figure 49**) as described in Chapter Ten.

FRONT SUSPENSION

Fork Oil Change

Yamaha does not provide an oil change interval for the front fork. Nonetheless, it is a good practice to change the fork oil once a year. If the fork oil becomes contaminated with dirt or water, change it immediately.

Changing the fork oil requires disassembling and assembling the fork leg. Refer to *Removal/Disassembly (Fork Leg Requires Service)* and *Fork Leg Assembly* in Chapter Eleven.

Inspection

1. Apply the front brake and pump the fork up and down as vigorously as possible. Check for smooth fork operation and check for any oil leaks.

NOTE

***Figure 50** shows the two lower fork bridge clamp bolts on an XVS1100A model. An XVS1100 model uses only one lower fork bridge clamp bolt.*

2. Make sure the upper (**Figure 51**) and lower (**Figure 50**) fork bridge clamp bolts are tight.
3. Make sure the handlebar holder bolts (**Figure 52**) are tight, and the handlebar is securely held in place.
4. Make sure the front axle clamp bolt (A, **Figure 53**) and axle (B) are tight.

WARNING

If any of the previously mentioned fasteners are loose, refer to Chapter Eleven for correct tightening procedures and torque specifications.

STEERING HEAD

Check the steering head for looseness at the intervals specified in **Table 1** or whenever the following conditions exist:

1. The handlebar vibrates more than normal.

2. The front fork makes a clicking or clunking noise when the front brake is applied.
3. The steering feels tight or slow.
4. The motorcycle does not want to steer straight on level road surfaces.

Inspection

1. Securely support the motorcycle so that the front tire clears the ground.
2. Check the bearing preload as follows:
   a. Center the front wheel. Push lightly against the left handlebar grip to start the wheel turning to the right, then let go. The wheel should continue turning under its own momentum until the fork legs hit their stop.
   b. Center the wheel and push lightly against the right handlebar grip.
   c. If the front wheel does not turn all the way to the stop when you lightly push a handlebar grip, the steering is too tight. Adjust the steering head bearings as described in *Steering Head Installation* in Chapter Eleven.
3. Check the bearing free play as follows:
   a. Center the front wheel. Grasp the bottom of the two fork sliders, and try to rock the fork legs back and forth. There should be little or no rocking in the steering head.
   b. If there is any play, the steering head is too loose. Adjust the steering head bearings as described in *Steering Head Installation* in Chapter Eleven.

REAR SUSPENSION

Inspection

1. Securely support the motorcycle on a level surface with the rear wheel off the ground.
2. Push the rear wheel sideways *hard* to check for side play in the swing arm bearings.
3. Remove the rider and passenger seats as described in Chapter Fourteen.
4. Remove the following items as described in Chapter Fourteen:
   a. Battery cover.
   b. Right side cover.
   c. Toolbox cover.
   d. Left side cover.
   f. Toolbox panel.

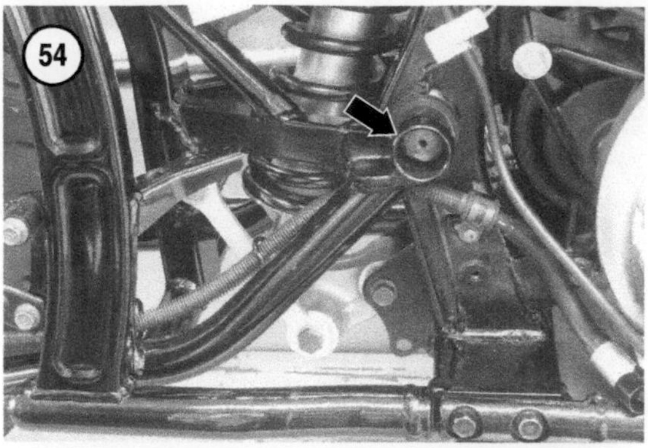

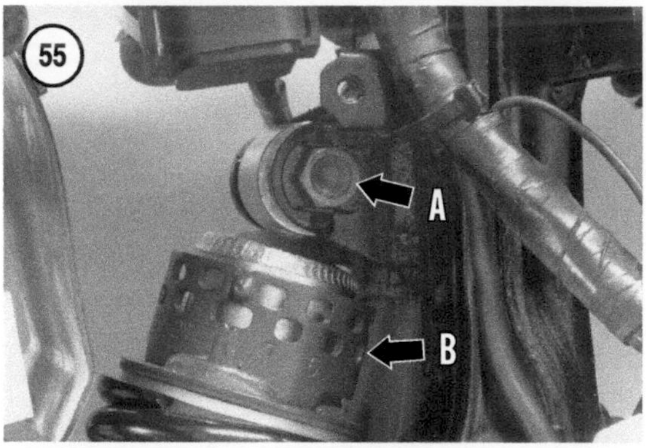

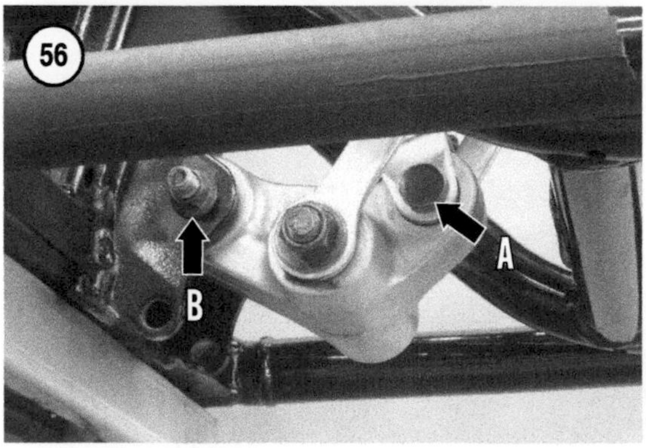

5. Check the tightness of the following hardware:
   a. The swing arm pivot bolt (**Figure 54**).
   b. The shock absorber's upper (A, **Figure 55**) and lower (A, **Figure 56**) mounts.
   c. The relay arm mount (B, **Figure 56**).
   d. The connecting arm mounts (**Figure 57**).
   e. The final gearcase bolts (A, **Figure 58**) and the axle nut (B).

WARNING
If any of the previously mentioned nuts or bolts are loose, refer to Chap-

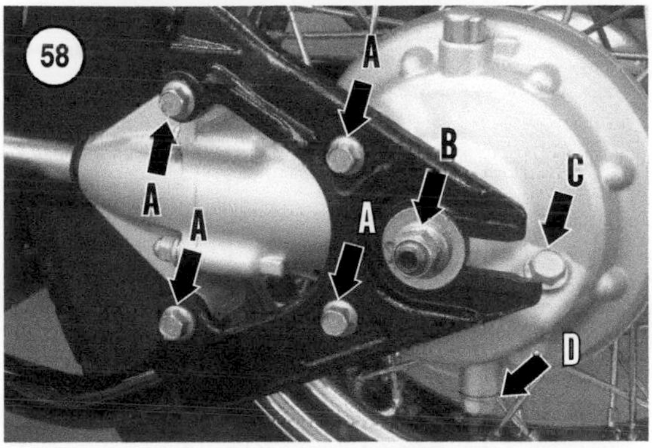

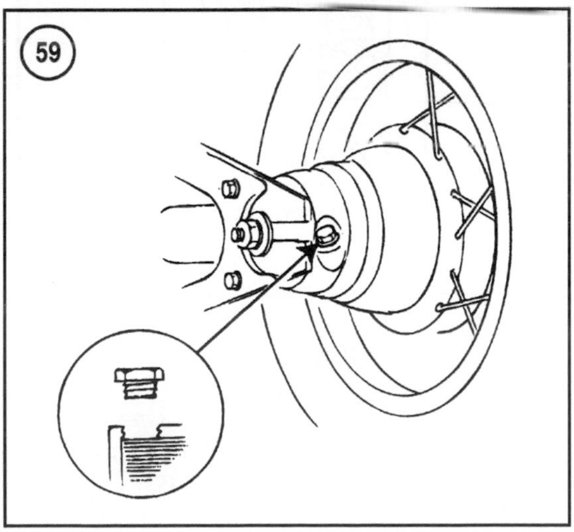

ter Twelve for correct tightening procedures and torque specifications.

Shock Absorber Spring Preload Adjustment

CAUTION
Never turn the cam ring beyond the maximum or minimum position.

The spring preload can be adjusted to seven different positions to suit riding, load and speed conditions. The third position is the default setting. The different adjuster ranges are: soft (No. 1 and No. 2), standard (No. 3), and hard (No. 4, 5, 6 and 7).

1. Remove the rider seat, passenger seat and ignitor panel as described in Chapter Fourteen. Remove the mud guard as described in *Rear Fender Removal* in Chapter Fourteen.
2. Using the spanner wrench and extension from the tool kit, adjust the preload by rotating the cam ring (B, **Figure 55**) on the shock absorber. Turning the cam ring to a higher numbered setting increases the preload. Turning the ring to a lower numbered setting decreases preload.

FINAL DRIVE

Final Gearcase Oil Check

1. Move the bike to a level surface. Use a suitable stand to support the bike in an upright position.
2. Remove the oil filler bolt (C, **Figure 58**) from the final gearcase.
3. The oil level should sit at the bottom on the oil filler brim as shown in **Figure 59**.
4. If the oil level is low, add final gear oil until the final gear case is full. Refer to **Table 4** for the recommended final gear oil.
5. Reinstall the oil filler bolt (C, **Figure 58**) and torque it to 23 N•m (17 ft.-lb.).

Final Gearcase Oil Change

1. Securely support the motorcycle on a level surface.
2. Place a drain pan under the drain bolt (D, **Figure 58**) on the final gearcase.
3. Remove the drain bolt and drain the oil from the final gearcase.
4. Reinstall the drain bolt and torque it to 23 N•m (17 ft.-lb.).
5. Refer to **Table 4** and add the recommended quantity of final gear oil.
6. Check the level of oil in the final gearcase. Add oil as necessary.
7. Reinstall the oil filler bolt (C, **Figure 58**) and torque it to 23 N•m (17 ft.-lb.).

BRAKES

Brake Hoses and Seals

Replace the brake hoses and piston seals every two years.

Check the brake hoses between the master cylinder and each brake caliper. If there is any leak, tighten the connections and bleed the brakes as described in Chapter Thirteen. If this does not stop the leak or if a line is obviously damaged, cracked, or chafed, replace the hose(s) and/or seals, then bleed the brake.

Brake Fluid Change

A small amount of dirt and moisture enters the brake fluid each time the reservoir cap is removed. The same thing happens if a leak occurs or when any part of the hydraulic system is loosened or disconnected. Dirt can clog the system and cause unnecessary wear. Water in the fluid vaporizes at high temperatures, impairing the hydraulic action and reducing brake performance.

Change the brake fluid at the intervals specified in **Table 1**. To do so, drain the fluid from the brakes as described in Chapter Thirteen. Add new fluid to the master cylinder, and bleed the brake at the caliper(s) until the fluid leaving the caliper is clean and free of contaminants and air bubbles. Refer to the brake bleeding procedure in Chapter Thirteen.

Checking Brake Fluid Level

If the brake fluid level reaches the lower level mark (front: C, **Figure 39**, rear: **Figure 60**), correct the fluid level by adding fresh brake fluid.

1. Securely support the motorcycle on level ground.

2A. When adding fluid to the front master cylinder, position the handlebar so the master cylinder reservoir is level.

2B. On the rear master cylinder, make sure the top of the reservoir is level.

3. Clean any dirt from the area around the top cover prior to removing the cover.

4. Remove the top cover, diaphragm plate and the diaphragm.

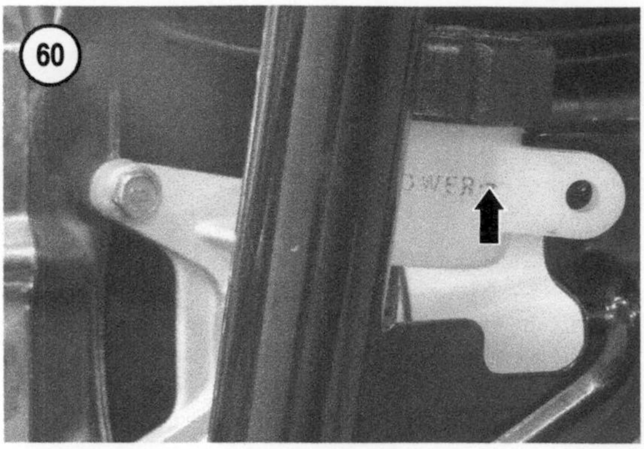

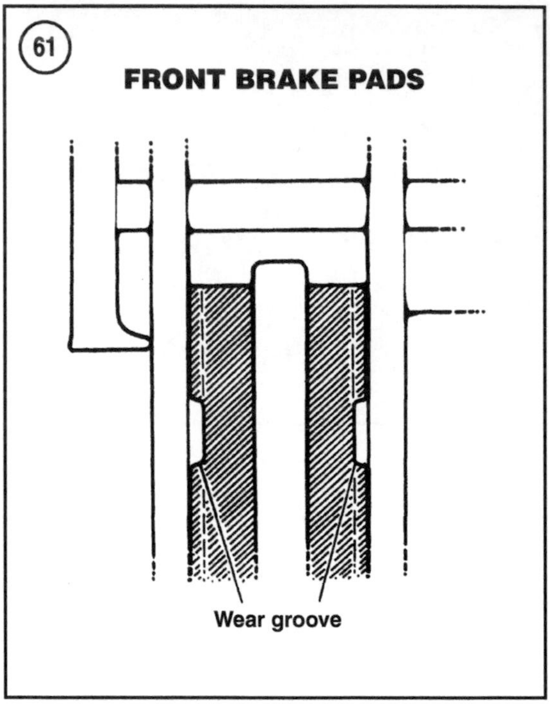

WARNING
Use brake fluid from a sealed container clearly marked DOT 4 (specified for disc brakes). Others may vaporize and cause brake failure. Do not mix different brands or types of brake fluid as they may not be compatible.

CAUTION
Be careful when handling brake fluid. Do not spill it on painted or plated surfaces or plastic parts as it will damge the surface. Wash the area immediately with soapy water and thoroughly rinse it off.

62

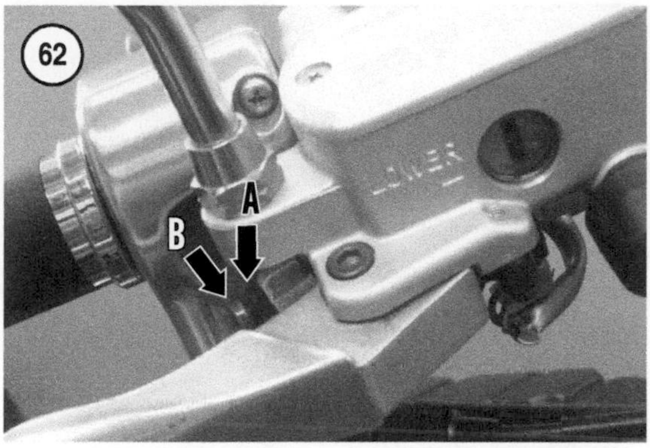

63

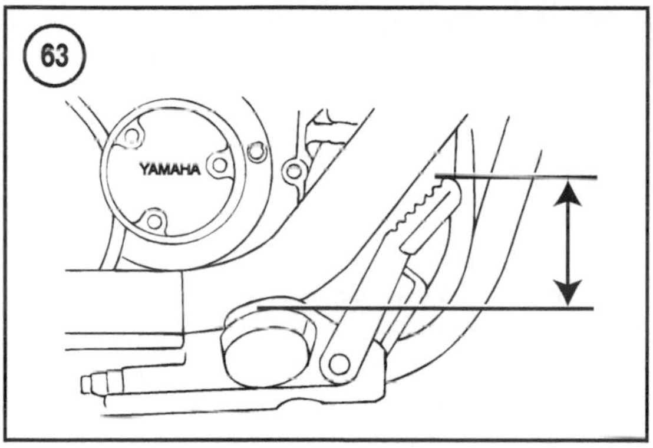

64

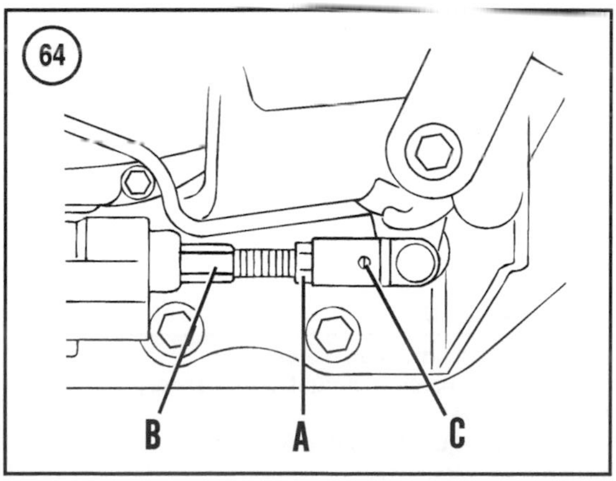

5. Add brake fluid from a sealed brake fluid container.
6. Reinstall the diaphragm, diaphragm plate and the top cover.

Brake Pad Inspection

Inspect the brake pads for excessive or uneven wear, scoring and oil or grease on the friction surface.

1. Securely support the motorcycle on level ground.
2. Look into the caliper assembly and inspect the wear indicators (**Figure 61**).
3. Replace both pads in the caliper if either pad is worn to the wear limit. On the front brakes, replace both pads in both front calipers if any pad is worn to the wear limit. Refer to Chapter Thirteen for brake pad replacement.

Front Brake Lever Adjustment

The front brake lever free play is the distance the brake lever moves before the master piston starts moving. Brake lever free play is measured at the end of the hand lever.

1. Push the brake lever forward, away from the handle grip.
2. Pull the lever, and measure free play.
3. If the free play is outside the range specified in **Table 5**, perform the following:
   a. Loosen the brake adjuster locknut (A, **Figure 62**).
   b. Turn the adjuster (B, **Figure 62**) in or out until free play is within the specification. Turning the adjuster in (clockwise) decreases free play; turning it out (counterclockwise) increases free play.
4. After adjusting free play, spin the wheel and check for any brake drag. Readjust free play as necessary.

Rear Brake Pedal Adjustment

Rear brake pedal height is the distance from the top of the brake pedal to the top of the footrest (**Figure 63**).

1. Securely support the motorcycle so it sits straight up.
2. Make sure the brake pedal is in the at-rest position.
3. Measure the distance from the top of the brake pedal to the top of the footrest. If the pedal height is not within the specification in **Table 5**, adjust the pedal height as follows:
   a. Loosen the adjuster locknut (A, **Figure 64**) on the rear master cylinder clevis. Turn the adjuster (B, **Figure 64**) until the pedal height is within the specification. Turn the adjuster

clockwise to raise the brake pedal and counterclockwise to lower it.

b. Tighten the locknut to 16 N•m (12 ft.-lb.).

c. Check the end of the brake pushrod. It must be visible through the hole (C, **Figure 64**) in the clevis.

4. After adjusting the brake pedal height, make sure the rear brake does not drag. Readjust the brake pedal height as necessary.

5. Adjust the rear brake light as described in this chapter.

Rear Brake Light Switch Adjustment

1. Turn the ignition switch on.

2. Depress the brake pedal. The brake light should come on when the brake pedal is depressed, and just before the rear brake is applied. If necessary, adjust as follows.

3. Hold the rear brake light switch body (A, **Figure 65**) and turn the adjuster nut (B) until the brake light operates properly. Turn the nut clockwise and the brake light comes on sooner; turn it counterclockwise and the light comes on later.

WARNING
Do not ride the motorcycle until the rear brake light operates properly.

65

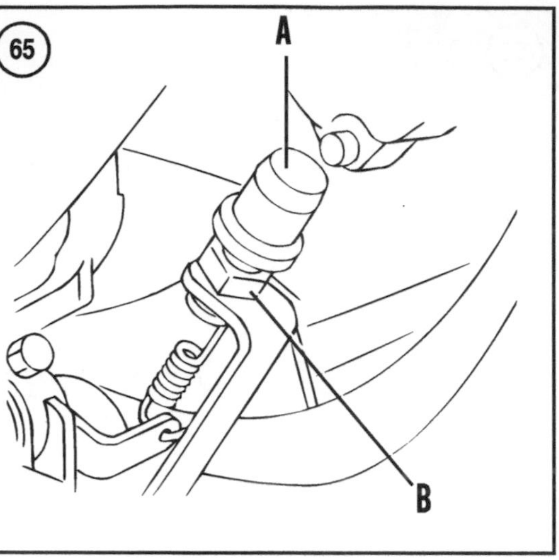

FASTENERS

Constant vibration can loosen many fasteners on a motorcycle. Check the tightness of all fasteners, especially those on:

1. Engine mounting hardware.
2. Engine crankcase covers.
3. Handlebar and front fork.
4. Gearshift lever.
5. Drive shaft components.
6. Brake pedal and lever.
7. Exhaust system.
8. Lighting equipment.

Table 1 MAINTENANCE SCHEDULE*

Initial 600 miles (1000 km) or 1 month	Check the valve clearance; adjust as necessary Check carburetor synchronization; adjust as necessary Change engine oil and filter Check clutch operation; adjust the cable free play or replace the cable as necessary Check the operation of the front and rear brakes; replace the pads if necessary Check brake fluid level in the master cylinders; adjust as necessary Check the operation of the sidestand Lightly lubricate the pivot and contact surfaces with lithium soap grease Check the operation of the sidestand switch; replace if necessary Lubricate all control and meter cables Check all fasteners; tighten them as necessary Replace the final gear oil

(continued)

Table 1 MAINTENANCE SCHEDULE* (continued)

4000 miles (7000 km) or 7 months	Check the valve clearance; adjust as necessary Check the condition of the spark plugs; clean and adjust the gap as necessary Check the crankcase ventilation hose for cracks or damage; replace as necessary Check the fuel lines for cracks or damage; replace as necessary Check carburetor synchronization; adjust as necessary Check the exhaust system for leaks Retighten hardware and/or replace gaskets as necessary Check the engine idle speed and throttle free play; adjust either as necessary Change engine oil and filter Clean the air filter; replace if damaged Check the operation of the front and rear brakes; replace the pads if necessary Check brake fluid level in the master cylinders; adjust as necessary Check clutch operation; adjust the cable free play or replace the cable as necessary. Lubricate all control and meter cables Lubricate the brake and clutch lever pivot shafts with lithium soap grease Lubricate the brake pedal and shift pedal pivot shafts with lithium soap grease Check the operation of the sidestand Lightly lubricate the sidestand pivot and contact surfaces with lithium soap grease Check the operation of the sidestand switch; replace if necessary Check the operation of the front fork and check for leaks Check the steering head bearings for smoothness or excessive play; adjust as necessary Check the operation of the shock absorber and check for leaks Check the wheel bearings for wear or damage; replace if necessary Check the wheels for runout and balance; adjust as necessary Check the spoke tightness; adjust as necessary Check tire tread for wear or damage; replace as necessary Check all fasteners; tighten them as necessary
8000 miles (13,000 km) or 13 months	Perform the 4000 mile (7000 km) checks Replace the spark plugs Check the final gear oil; adjust the level as necessary
12,000 miles (19,000 km) or 19 months	Perform the 4000 mile (7000 km) checks On California models, check the evaporative emission control system for damage; replace as necessary
16,000 miles (25,000 km) or 25 months	Perform the 4000 mile (7000 km) checks Replace the spark plugs Check the final gear oil; adjust the level as necessary Repack the swing arm pivot bearings with molybdenum disulfide grease Repack the steering head bearings with lithium soap grease Replace the final gear oil
20,000 miles (31,000 km) or 31 months:	Perform the 4000 mile (7000 km) checks Replace the fuel filter On California models, check the evaporative emission control system for damage; replace as necessary

*Consider this maintenance schedule a guide to general maintenance and lubrication intervals. Service these items more frequently if the motorcycle is exposed to mud, water, sand or high humidity, or if it is run harder than normal (stop-and-go traffic for example).

Table 2 TIRE SPECIFICATIONS

XVS1100 models	
Front tire size	110/90-18 61S
Manufacturer	Bridgestone Exedra L309, Dunlop K555F
Rear tire size	170/80-15M/C 77S
Manufacturer	
USA, California and Canada models	Bridgestone Exedra G546G, Dunlop K555
Europe and Australia models	Bridgestone Exedra G546, Dunlop K555
Tire Inflation Pressure[1]	
0-90 kg (0-198 lb.) load[2]	
Front	200 kPa (28.5 psi)
Rear	225 kPa (32 psi)
90-200 (198-441lb.) load[2]	
Front	225 kPa (32 psi)
Rear	250 kPa (36 psi)
XVS1100A models	
Front tire size	130/90-16 67S
Manufacturer	Dunlop D404F
Rear tire size	170/80-15M/C 77S
Manufacturer	Dunlop D404G
Tire Inflation Pressure[1]	
0-90 kg (0-198 lb.) load[2]	
Front	225 kPa (32 psi)
Rear	225 kPa (32 psi)
90-200 kg (198-441 lb.) load[2]	
Front	225 kPa (32 psi)
Rear	250 kPa (36 psi)
Minimum tread depth	1.6 mm (0.06 in.)

1. Tire inflation pressure for original equipment tires. Aftermarket tires may require different inflation pressures; refer to aftermarket manufacturer's specifications.
2. Load equals the total weight of the cargo, rider, passenger and accessories.

Table 3 BATTERY

Battery type	Maintenance free (sealed)
Capacity	12 V 12 AH
Charge rate	1.2 amp
Open circuit voltage @ 20° C (68° F)	
Fully-charged	12.8 volts or higher
Requires charging	12.0-12.8 volts
Replace the battery	Less than 12.0 volts

Table 4 RECOMMENDED LUBRICANTS AND FLUIDS

Fuel	Regular unleaded
Octane	86 [(R + M)/2 method] or research octane of 91 or higher
Capacity	17 liter (4.49 US gal.)
Reserve	4.5 liter (1.19 US gal.)

(continued)

Table 4 RECOMMENDED LUBRICANTS AND FLUIDS (continued)

Engine oil	
API classification	SE, SF or SG
Viscosity	
5° C (40° F) or above	SAE 20W40
15° C (60° F) or below	SAE 10W30
Capacity	
Oil change only	3.0 L (3.2 U.S. qt.)
Oil and filter change	3.1 L (3.3 U.S. qt.)
When engine completely dry	3.6 L (3.8 U.S. qt.)
Final gear oil	
Viscosity	
Single grade	SAE 80 hypoid gear oil
Multigrade	SAE 80W-90 hypoid gear oil
Grade	API GL-4, GL-5 or GL-6
Capacity	200 cc (6.8 U.S. oz.)
Brake fluid	DOT 4
Battery	Maintenance free
Fork oil	
Viscosity	SAE 10W fork oil
Capacity per leg	
1999-2003 all models	464 cc (15.7 U.S. oz.)
2004-on XVS1100 models	467-481 cc (15.8-16.3 U.S. oz.)
2004-on XVS1100A models	488 cc (16.5 U.S. oz.)
Oil level (measured from top of the fully compressed fork tube with the fork spring removed)	
1999-2003 all models	108 mm (4.25 in.)
2004-on XVS1100 models	99-105 mm (3.89-4.13 in.)
2004-on XVS1100A models	99 mm (3.89 in.)

Table 5 MAINTENANCE AND TUNE-UP SPECIFICATIONS

Item	Specification
Recommended spark plug	NGK BPR7ES, Denso W22EPR-U
Spark plug gap	0.7-0.8 mm (0.028-0.031 in.)
Idle Speed	950-1050 rpm
Pilot screw (Europe and Australia models)	3 turns out
Valve clearance	
Intake	0.07-0.12 mm (0.0028-0.0047 in.)
Exhaust	0.12-0.17 mm (0.0047-0.0067 in.)
Compression pressure (at sea level)	
Standard	1000 kPa (142 psi)
Minimum	900 kPa (128 psi)
Maximum	1100 kPa (156 psi)
Ignition timing	10° BTDC @ 1000 rpm
Vacuum pressure (at idle)	34.7-37.3 kPa (260-289 mm Hg [10.2-11.4 in. Hg])
Front brake pad wear limit	0.8 mm (0.03 in.)
Rear brake pad wear limit	0.5 mm (0.020 in.)
Brake pedal height	
XVS1100 models (above footpeg)	81.8 mm (3.22 in.)
XVS1100A models (above floorboard)	98.5 mm (3.88 in.)
Throttle cable free play (at the flange)	4-6 mm (0.16-0.24 in.)
Brake free play (at lever end)	5-8 mm (0.20-0.31 in.)
Clutch cable free play (at lever end)	5-10 mm (0.20-0.39 in.)
Shift rod length	
XVS1100 models	114.7 mm (4.52 in.)
XVS1100A models	114.9 mm (4.53 in.)

(continued)

Table 5 MAINTENANCE AND TUNE-UP SPECIFICATIONS (continued)

Item	Specification
Rim runout service limit	
Radial	1.0 mm (0.04 in.)
Axial	0.5 mm (0.02 in.)

Table 6 MAINTENANCE AND TUNE UP TORQUE SPECIFICATIONS

Item	N•m	in.-lb.	ft.-lb.
Air filter cover bolts	2	18	–
Air filter housing bolts	10	89	–
Brake pedal height adjuster locknut	16	–	12
Cam sprocket cover bolts	10	89	–
Clutch adjuster locknut	12	106	–
Final gearcase drain bolt	23	–	17
Final gearcase oil filler bolt	23	–	17
Fork bottom Allen bolt*	30		22
Fork cap bolt	23	–	17
Front axle	59	–	43
Front axle clamp bolt	20	–	15
Front brake caliper mounting bolt	40	–	30
Front caliper retaining bolt			
1999-on XVS1100 models	23	–	17
All models except 1999-on XVS1100 models	27	–	20
Lower fork bridge clamp bolt	30	–	22
Oil drain bolt	43	–	32
Oil filter outer cover bolt	10	89	–
Rear axle clamp nut	23	–	17
Rear axle nut	107	–	79
Rear brake adjuster locknut	16	–	12
Rear caliper bracket bolt	40	–	30
Rear caliper mounting bolt	40	–	30
Rocker arm bolt	38	–	28
Sidestand nut	56	–	41
Spark plugs	20	–	15
Steering head nut	110	–	81
Steering stem adjuster nut			
First stage	52	–	38
Second stage	18	–	13
Upper fork bridge clamp bolt	20	–	15
Valve adjuster locknut	27	–	20
Valve cover bolts	10	89	–

*Apply Loctite 242 (blue) or an equivalent medium strength threadlocking compound.

CHAPTER FOUR

ENGINE TOP END

This chapter provides complete service and overhaul procedures for the engine top end components. This includes the camshafts, valves, cylinder heads, pistons, piston rings and the cylinder blocks. Refer to Chapter Three for valve adjustment procedures. Refer to *Basic Service Methods* in Chapter One.

When inspecting components, compare any measurements to the specifications in **Table 2** at the end of this chapter. Replace any part that is damaged, worn or out of specification. During assembly, torque fasteners to the specified torque.

The V-Star 1100 engine is an air-cooled, four-stroke, single overhead camshaft (SOHC) V-twin.

Two main bearings support the crankshaft in a vertically split crankcase. The camshaft in each cylinder is chain-driven by a cam chain drive assembly that meshes with the timing gear on the crankshaft. Cam chain tension is maintained by an automatic, spring-loaded tensioner that bears against the rear run of the cam chain.

The engine and transmission share a common case and the same wet sump oil supply. The wet-plate clutch is located on the right side of the engine.

SERVICING ENGINE IN THE FRAME

The following components can be serviced while the engine is in the frame:

1. External shift mechanism.
2. Clutch.
3. Carburetors.
4. Starter.
5. Alternator and electrical system.
6. Oil pump.

CYLINDER HEAD COVERS

Two chrome covers crown the head on each cylinder: a large and a small cover. Mark each cover during removal so it can be easily identified and installed on the correct cylinder during assembly.

Removal/Installation

1. Securely support the motorcycle on level ground.
2. Remove the fuel tank, air filter housing, and surge tank and carburetor assembly as described in Chapter Eight.
3. The small cover snaps into place on the cylinder head. Press the arms that lock the small cover to the large cover, and pull the small cover (**Figure 1**) from the grommets on the bracket.
4. Unthread the cover screws (**Figure 2**) and remove the large cover from the cylinder head.
5. If necessary, repeat for the other cylinder head.
6. Installation is the reverse of removal.
   a. Install each cover in its original location.
   b. Make sure the posts on the small cover engage the grommets on the cylinder head bracket.
   c. Torque the cylinder head cover screws to 4 N•m (35 in.-lb.).

CYLINDER HEAD

Removal

The procedures for removing the rear (**Figure 3**) and front (**Figure 4**) cylinder heads are nearly identical. The few differences are noted. When removing both cylinder heads, remove the rear head first.

The Yamaha sheave holder (part No. YS-01880 or 90890-01701) or its equivalent is needed to perform this procedure.

1. Securely support the motorcycle on level ground.
2. Remove the engine from the frame as described in Chapter Five.
3. Remove the intake and exhaust valve covers (**Figure 5**) from the cylinder head. Watch for the cable clamp on one intake valve cover bolt on the front cylinder head.
4. Remove the two cam sprocket cover bolts (A, **Figure 6**), and remove the cam sprocket cover (B) and its O-ring from the cylinder head.

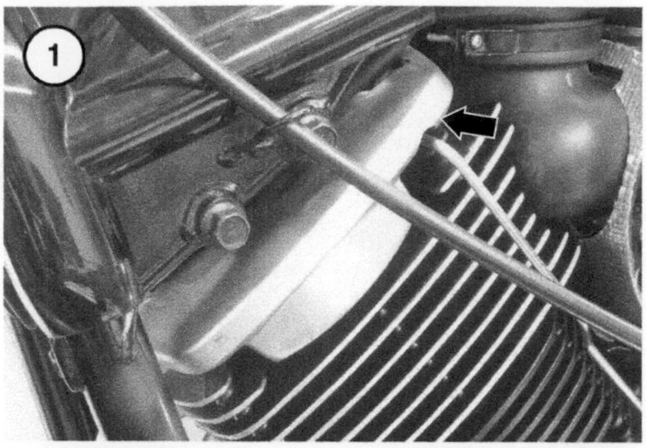

NOTE
A breather plate is not used in the rear cylinder head.

5. When working on the front cylinder, remove the breather plate (A, **Figure 7**) and its O-ring from the inside of the cam sprocket cover (B).
6. Remove both spark plugs. This makes it easier to rotate the engine.
7. Remove the timing cover (A, **Figure 8**) and flywheel nut cover (B) from the alternator cover.

CAUTION
When a cylinder is set to TDC on the compression stroke, the timing mark on the cam sprocket (rear cylinder) or cam sprocket plate (front cylinder) may not precisely align with the pointer on the cylinder head. On some models, the camshafts are slightly retarded from the factory. The mark could be off by as much as 1/2 tooth. Before removing a cylinder head, set that cylinder to top dead center on the compression stroke, and note the position of the cam sprocket timing

3

REAR CYLINDER HEAD

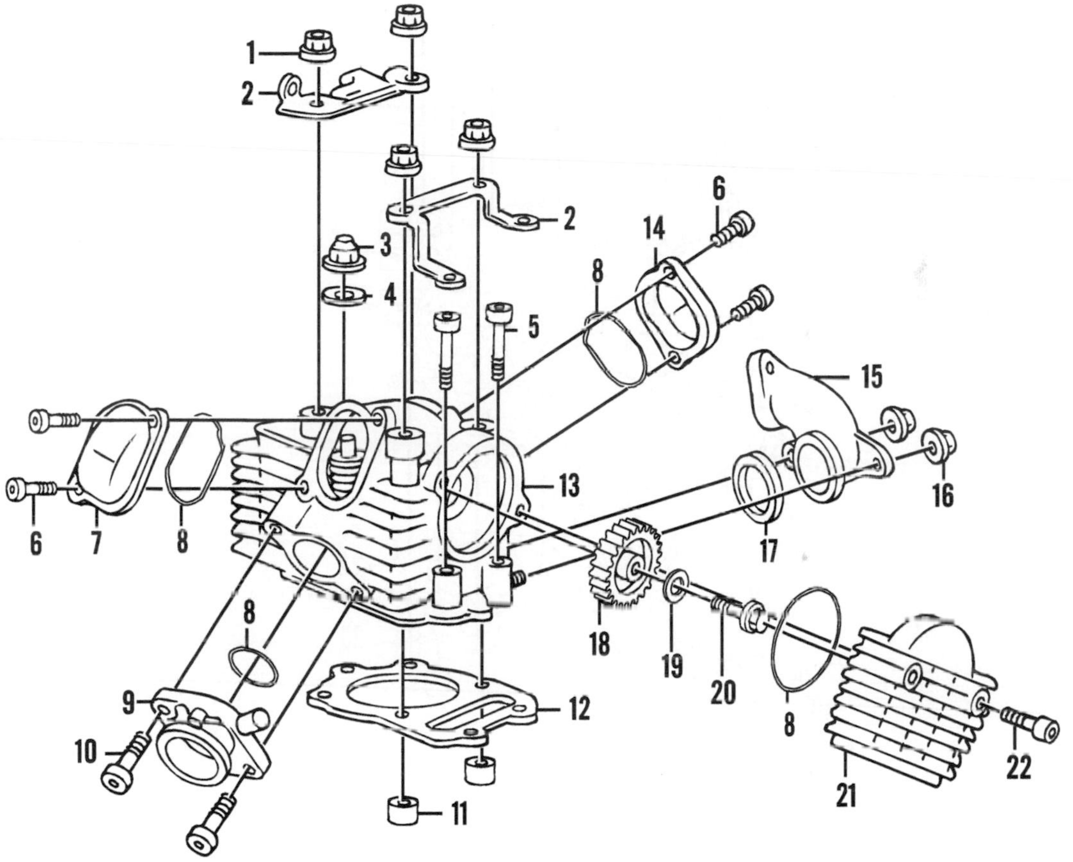

1. Cylinder head nut
2. Cylinder head cover bracket
3. Cylinder head cap nut
4. Washer
5. Cylinder bolt
6. Valve cover bolt
7. Intake valve cover
8. O-ring
9. Intake manifold
10. Intake manifold bolt
11. Dowel
12. Gasket
13. Rear cylinder head
14. Exhaust valve cover
15. Exhaust manifold
16. Manifold nut
17. Exhaust gasket
18. Cam sprocket
19. Washer
20. Cam sprocket bolt
21. Cam sprocket cover
22. Cam sprocket cover bolt

4

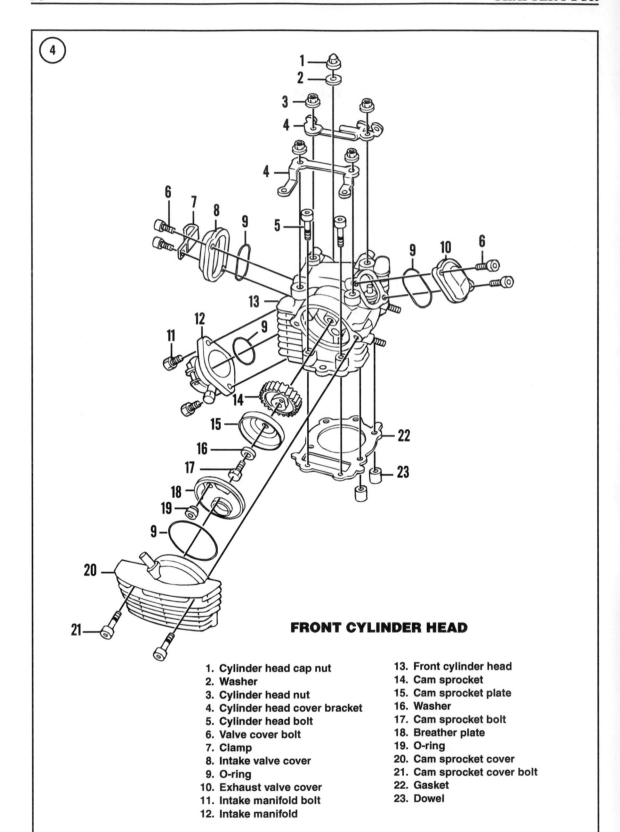

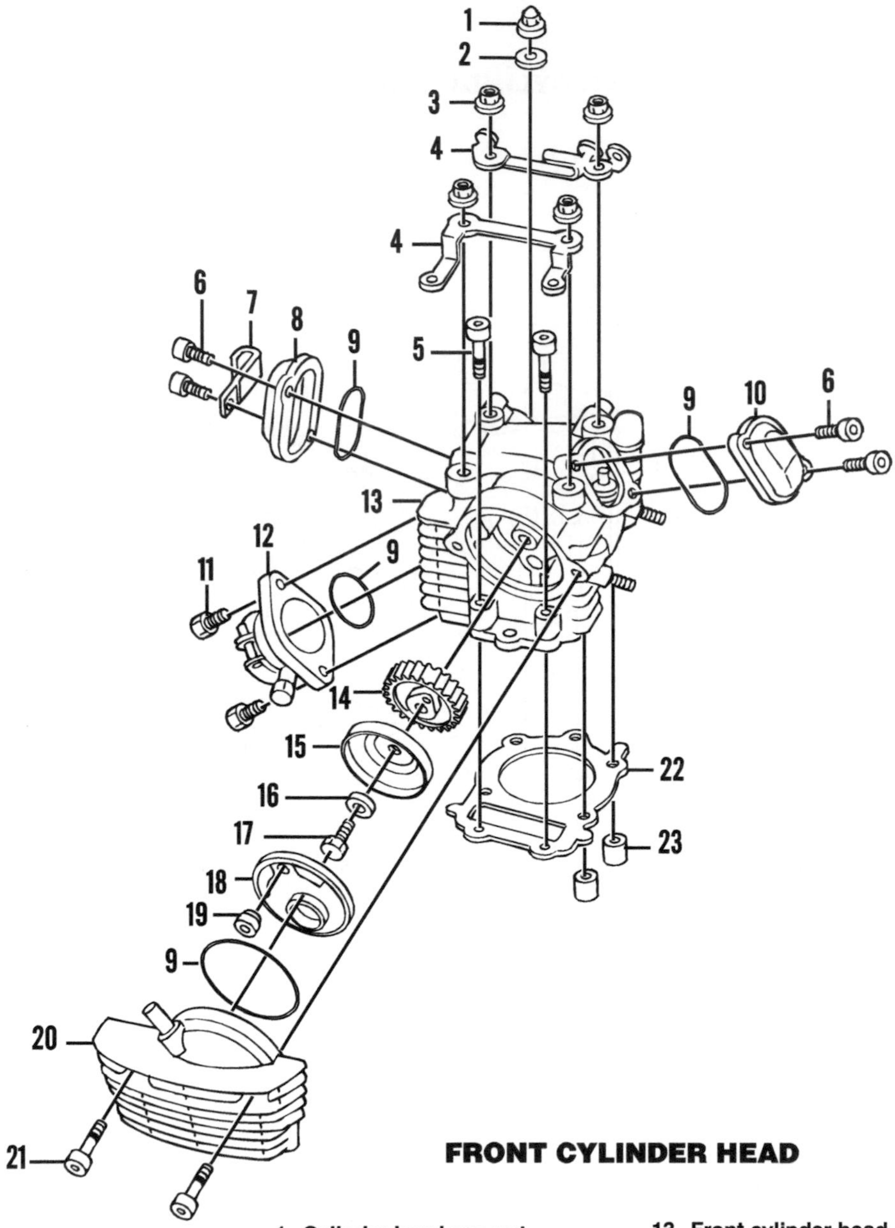

FRONT CYLINDER HEAD

1. Cylinder head cap nut
2. Washer
3. Cylinder head nut
4. Cylinder head cover bracket
5. Cylinder head bolt
6. Valve cover bolt
7. Clamp
8. Intake valve cover
9. O-ring
10. Exhaust valve cover
11. Intake manifold bolt
12. Intake manifold
13. Front cylinder head
14. Cam sprocket
15. Cam sprocket plate
16. Washer
17. Cam sprocket bolt
18. Breather plate
19. O-ring
20. Cam sprocket cover
21. Cam sprocket cover bolt
22. Gasket
23. Dowel

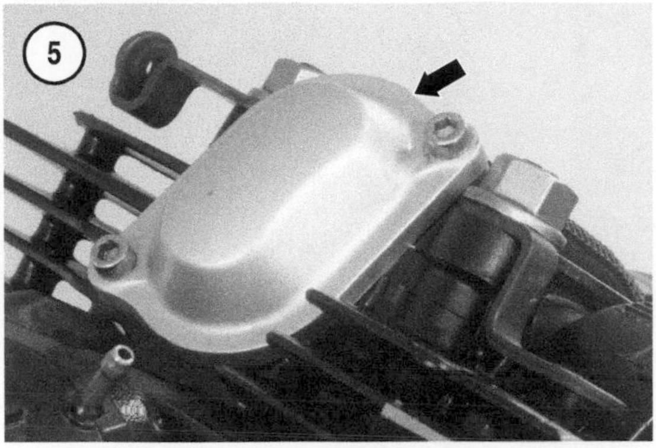

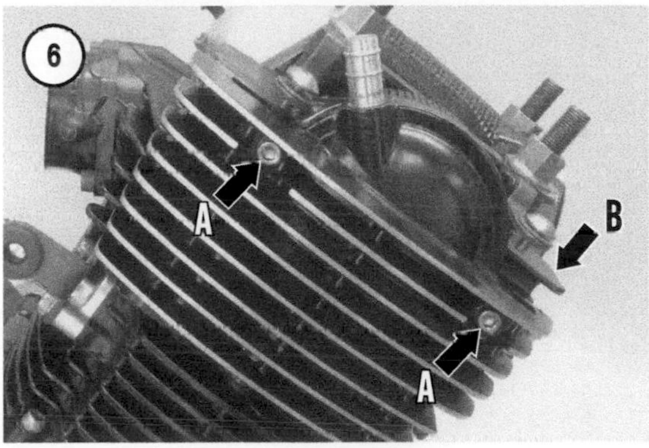

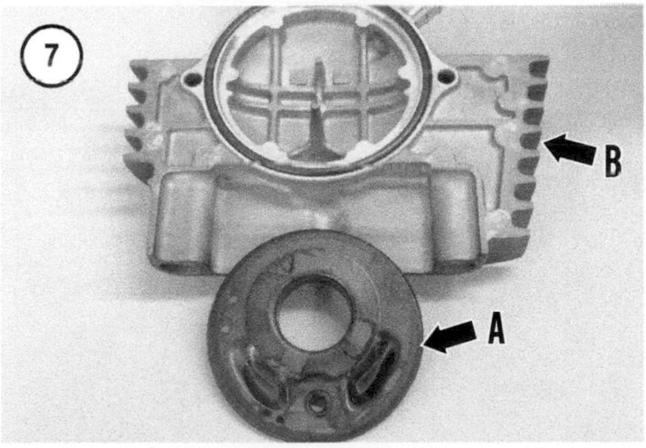

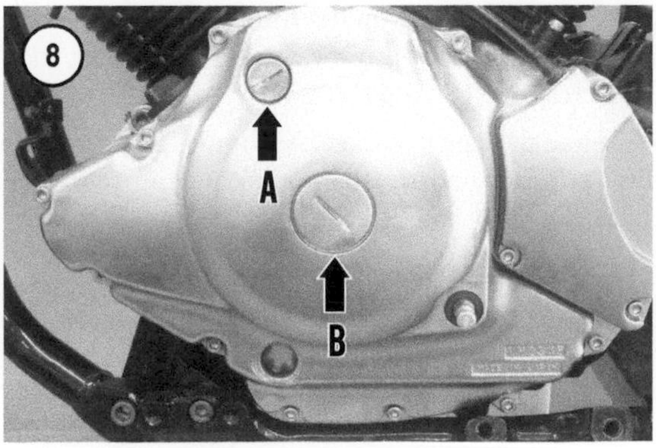

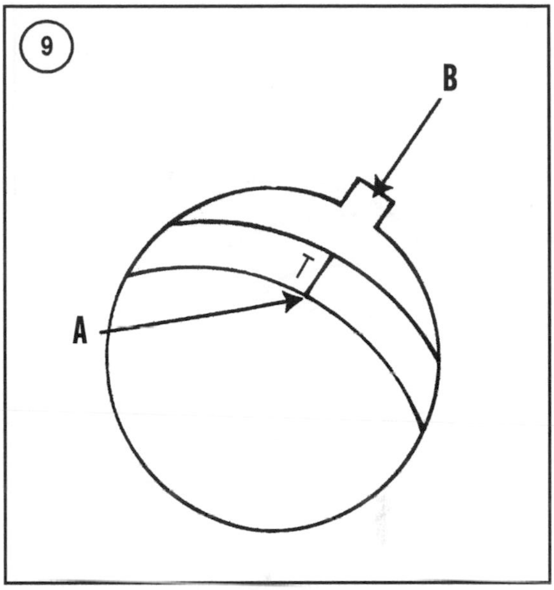

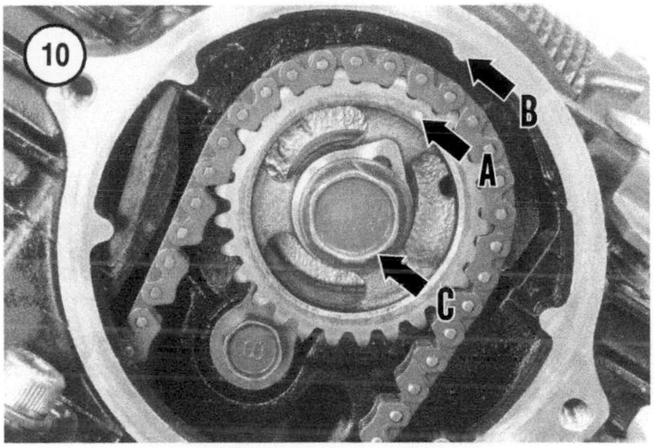

mark. Take a photograph or make a drawing so you can correctly time the camshaft during assembly.

8A. When servicing the rear cylinder, set it to top dead center on the compression stroke as follows:

a. Use the flywheel nut to turn the crankshaft clockwise until the T-mark on the flywheel (A, **Figure 9**) aligns with the cutout in the alternator cover (B).

b. Make sure the timing mark on the rear cam sprocket (A, **Figure 10**) aligns with the pointer on the rear cylinder head (B).

8B. When servicing the front cylinder, set it to TDC on the compression stroke as follows:

a. Set the rear cylinder to TDC by rotating the engine clockwise until the T-mark (A, **Figure**

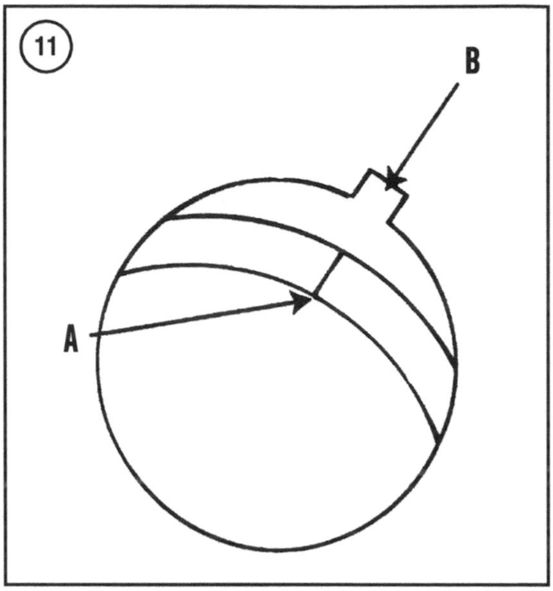

9) on the flywheel aligns with the alternator cover cutout (B).

b. Rotate the crankshaft another 290° clockwise until the I-mark on the flywheel (A, **Figure 11**) aligns with the cutout in the alternator cover (B).

c. Make sure the timing mark on the front cam sprocket plate (A, **Figure 12**) aligns with the pointer on the front cylinder head (B).

9. Make sure the appropriate cylinder is at TDC on the compression stroke by pressing each rocker arm. Both the intake and exhaust rocker arms should have free pay. If both rocker arms do not have free play, rotate the engine an additional 360° until they do.

10. Remove the two mounting bolts (A, **Figure 13**), then remove the cam chain tensioner (B) and gasket from the rear of the cylinder. The air filter housing bracket (C, **Figure 13**) comes out with the chain tensioner on the front cylinder.

11. Remove the alternator cover as described in Chapter Five.

12. Hold the flywheel with a sheave holder.

13A. When servicing the rear cylinder head, remove the cam sprocket bolt (C, **Figure 10**) along with its washer.

NOTE
A cam sprocket plate is not used on the rear cylinder.

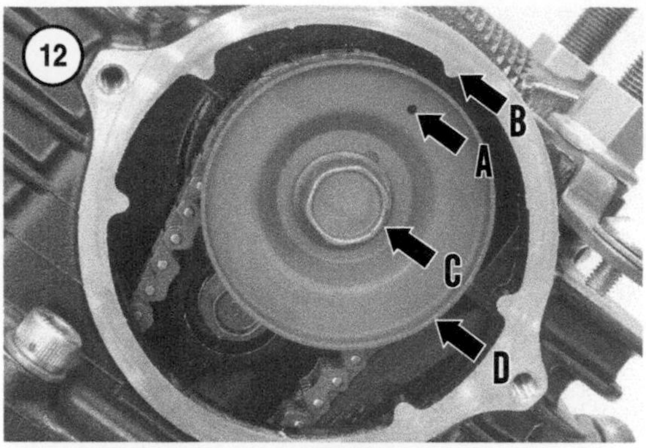

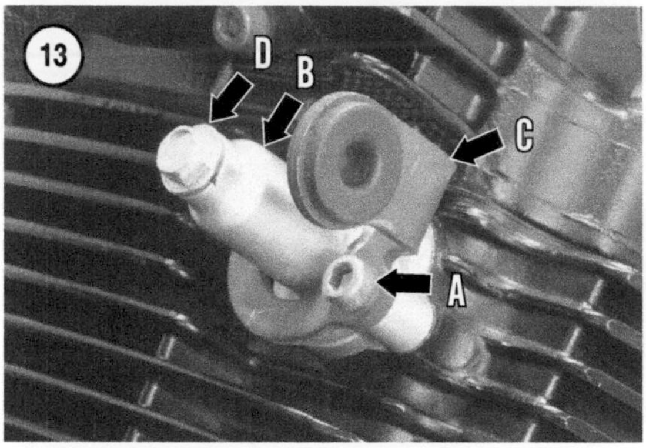

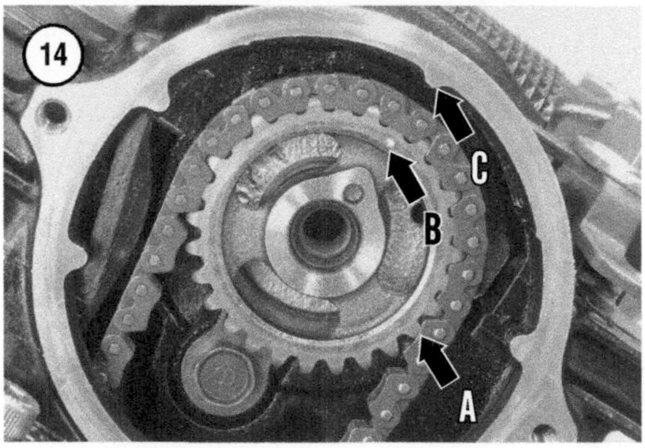

13B. When working on the front cylinder, remove the cam sprocket bolt (C, **Figure 12**), washer and the cam sprocket plate (D).

14. Slide the cam sprocket (A, **Figure 14**) from the camshaft and remove the sprocket from the cam chain.

CAUTION
If the crankshaft must be rotated with the cam sprocket removed, pull up on the cam chain and keep it taut so the chain remains meshed with the

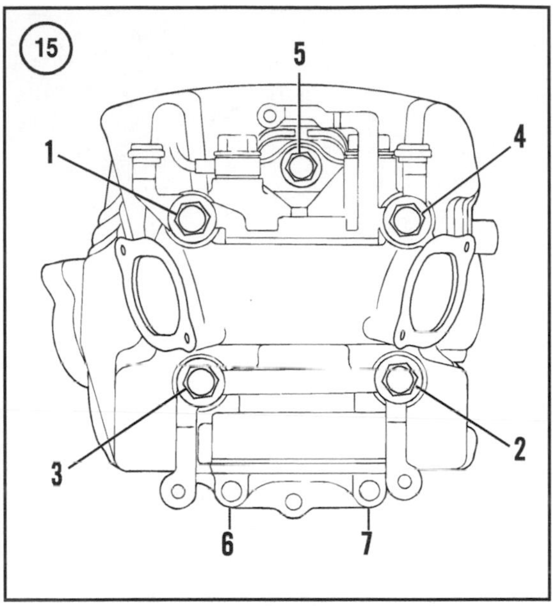

sprocket on the cam chain drive assembly. If the chain is not held taut, the chain may become kinked, and cause damage to the crankcase, the cam chain, or cam chain drive assembly.

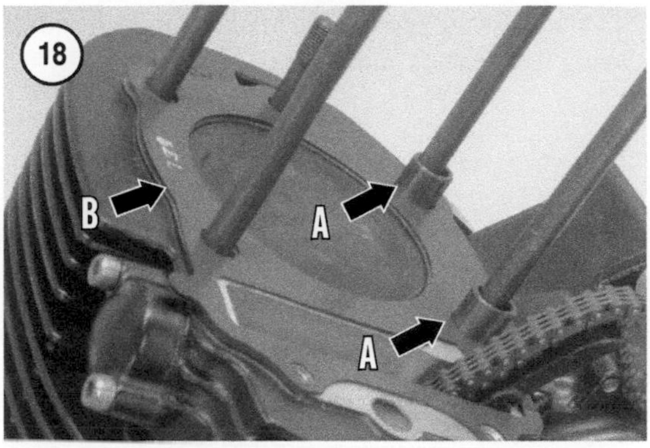

15. Tie a safety wire around the cam chain and secure the wire to the engine so the chain will not fall into the crankcase.
16. Evenly loosen all seven cylinder head fasteners 1/4-turn at a time. Loosen the fasteners in sequence by reversing the tightening sequence shown in **Figure 15**.
17. Once all the fasteners are loose, remove the two cylinder head bolts (6 and 7, **Figure 15**), remove the cylinder head nuts (1-4) from the cylinder studs, and remove the cylinder head cap nut (5). Make sure a washer comes out with the cap nut.
18. Lift the two cylinder head cover brackets (**Figure 16**) from the cylinder studs. Mark these brackets so they can be reinstalled in the correct locations during assembly.

CAUTION
The cooling fins are fragile and may be damaged if they are tapped too hard. Never use a metal hammer to loosen the cylinder head.

19. Loosen the head by tapping around its perimeter with a plastic mallet.
20. Remove the cylinder head by pulling it straight up and off the cylinder studs. Watch for the dowels on the two rear studs. Untie the cam chain from the engine while removing the head, then retie the chain.
21. Remove the front cam chain guide (**Figure 17**) from its boss in the crankcase.
22. Remove the two dowels (A, **Figure 18**) from the cylinder studs, and remove the cylinder head gasket (B).
23. Place a clean shop rag in the cam chain tunnel in the cylinder to keep foreign matter out of the crankcase.

24. If necessary, remove the intake manifold and its O-ring from the cylinder head.

NOTE
An exhaust manifold is not used on the front cylinder.

25. If necessary, remove the two manifold nuts (A, **Figure 19**) and remove the exhaust manifold (B) from the rear cylinder. Also remove and discard the exhaust gasket. A new gasket must be used during assembly.
26. Inspect the cylinder head as described in this section.

Installation

NOTE
If both cylinder heads have been removed, completely install the rear cylinder head, then install the front head.

1. Remove the shop cloth from the cylinder.
2. Install the two dowels into the cylinder (A, **Figure 18**).
3. Install a new cylinder head gasket onto the cylinder block (B, **Figure 18**) so the *5EL* mark faces up.

NOTE
The front chain guide is directional. The narrow end of the guide must sit in the seat in the crankcase.

4. Lower the narrow end of the front cam chain guide (A, **Figure 20**) through the cam chain tunnel in the cylinder block, and set the guide into the seat in the crankcase. See **Figure 17**. Make sure the guide's raised edge faces the cam chain.
5. Lower the cylinder head part way down the cylinder studs, and feed the cam chain and safety wire up through the cam chain tunnel in the head.
6. Carefully lower the cylinder head until it is seated on the cylinder block. While lowering the head into place, note the following:
   a. The top of the front chain guide must be captured by the cavity in the cylinder head.
   b. Make sure the two locating dowels engage the cylinder head.
7. Pull the cam chain taut and make sure it still properly engages the sprocket on the cam chain drive assembly. Secure the cam chain safety wire to the engine.
8. Install and finger-tighten the two cylinder head bolts (A, **Figure 21**).

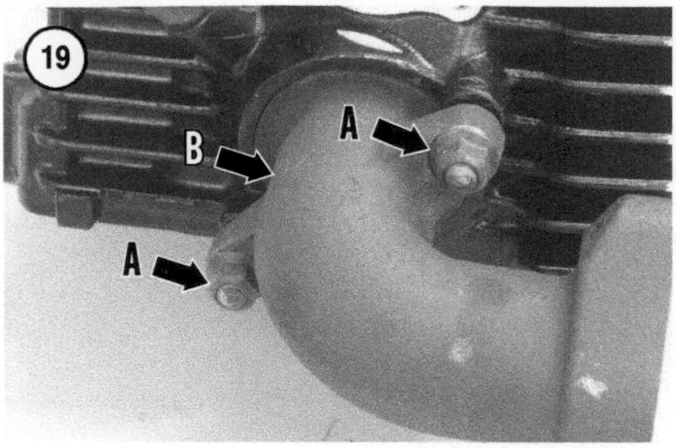

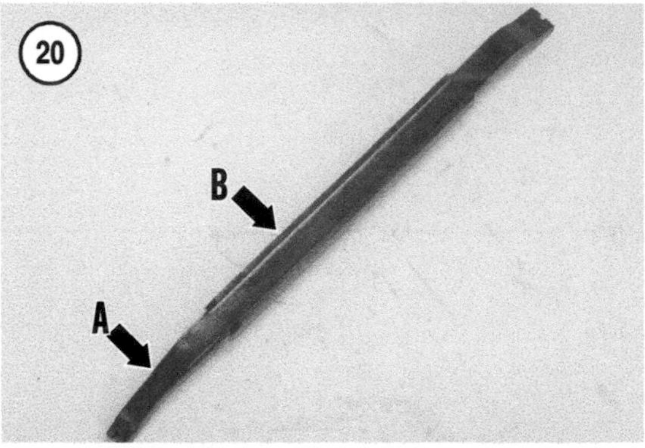

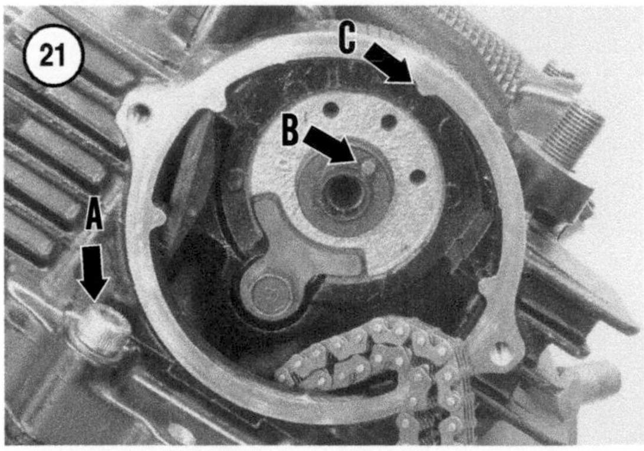

9. Lower the cylinder head cover brackets (**Figure 16**) over the cylinder studs. Check the marks made during removal and install each bracket in its original location.

10. Turn the cylinder head nuts (A, **Figure 22**) onto the cylinder studs and thread the cylinder head cap nut (B) into place in the spark plug well. Install a washer with the cap nut. Finger-tighten the nuts at this time.

CAUTION
The cylinder head fasteners must be evenly tightened in sequence. The head

could be damaged if it is not tightened properly.

11. Following the tightening sequence shown in **Figure 15**, torque the fasteners in 1/2 turn increments and in two stages to the specifications. Torque the fasteners to 1/2 the given torque, then torque them to their final specifications:
   a. Cylinder head nuts (1-4, **Figure 15**): 50 N•m (37 ft.-lb.).
   b. Cylinder head cap nut (5, **Figure 15**): 35 N•m (26 ft.-lb.).
   c. Cylinder head bolts (6 and 7, **Figure 15**): 20 N•m (15 ft.-lb.).

12. If the cylinder has also been reinstalled, torque the cylinder bolts (A and B, **Figure 23**) to 10 N•m (89 in.-lb.). One of the bolts sits in a recess (B, **Figure 23**) in the cylinder block.

13. If removed, install the locating pin (B, **Figure 21**) into the camshaft.

CAUTION
When rotating the crankshaft with the cam sprocket removed, pull up on the cam chain and keep it taut so the chain remains meshed with the sprocket on the cam chain drive assembly. If the chain is not held taut, the chain may become kinked, and cause damage to the crankcase, the cam chain and timing sprocket.

14. Temporarily install the alternator cover onto the crankcase.

15. Use the flywheel bolt to turn the crankshaft clockwise until the cylinder is a top dead center on the compression stroke.
   a. The *rear cylinder* is at TDC on the compression stroke when the T-mark (A, **Figure 9**) on the rotor aligns with the cutout in the alternator cover (B) and when the locating pin on the camshaft aligns (B, **Figure 21**) with the pointer on the cylinder head (C).
   b. The *front cylinder* is at TDC on the compression stroke when the I-mark (A, **Figure 11**) on the rotor aligns with the cutout in the alternator cover (B) and when the locating pin on the camshaft (B, **Figure 21**) aligns with the pointer on the cylinder head (C).

NOTE
A cylinder at TDC on the compression stroke has free play in both rocker arms, which indicates that both valves are closed.

16. Make sure the cylinder is at TDC by pressing each rocker arm. Both the intake and exhaust rocker arms should have free play. If both rocker arms do not have free play, rotate the engine an additional 360° until they do.

17. Install the cam sprocket (A, **Figure 14**). Install the sprocket onto the camshaft so the camshaft's locating pin fits into the cutout in the cam sprocket. Make sure the timing mark (B, **Figure 14**) on the cam sprocket faces out.

NOTE
When a cylinder is set to TDC on the compression stroke, the timing mark on the cam sprocket may not precisely align with pointer on the cylinder head. On some models, the mark could be off by as much as 1/2 tooth. Consult the notes made during removal.

18. Rotate the camshaft to remove any slack from the rear side of the cam chain. Insert a finger into the

cam chain tensioner hole and tension the chain. The timing mark on the cam sprocket (B, **Figure 14**) should align with the pointer on the cylinder head (C) as noted during removal.

19. If the timing marks do not align, remove the sprocket. Use a screwdriver to walk the cam chain one way or the other and reinstall the cam sprocket. Repeat this as necessary until the timing mark on the cam sprocket aligns with the pointer on the cylinder head while you press the rear chain guide inward.

NOTE
A cam sprocket plate is not used on the rear cylinder.

20. When working on the front cylinder, install the cam sprocket plate onto the camshaft. Make sure the dimple (A, **Figure 24**) on the sprocket plate engages the timing mark (B) on the camshaft.

21. Install the cam sprocket bolt and washer (rear cylinder: C, **Figure 10**; front cylinder: C, **Figure 12**). Finger-tighten the bolt at this time.

22. Install the cam chain tensioner and new gasket as follows:

   a. Apply a medium-grade threadlocking compound to the chain tensioner mounting bolts.
   b. Insert a screwdriver into the tensioner body.
   c. Turn the screwdriver clockwise until the plunger fully retracts into the chain tensioner body. Do not remove the screwdriver at this time. See **Figure 25**.
   d. Slide a new gasket onto the tensioner body.
   e. Insert the tensioner into the cylinder while holding the screwdriver. Make sure the UP mark on the tensioner faces up. Do not release the screwdriver until the tensioner is seated in the cylinder.
   f. Press the tensioner against the cylinder. Release the plunger by turning the screwdriver slightly counterclockwise, and remove the screwdriver. See **Figure 26**.
   g. When installing the chain tensioner into the front cylinder, slip the air filter bracket (C, **Figure 13**) over the tensioner.
   h. Install the mounting bolts (A, **Figure 13**). Evenly finger-tighten the bolts to hold the tensioner in place.
   i. Torque the cam chain tensioner mounting bolts to 10 N•m (89 in.-lb.).

24

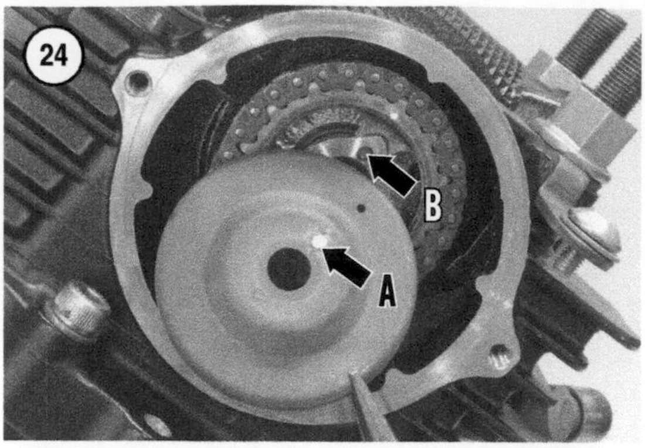

25

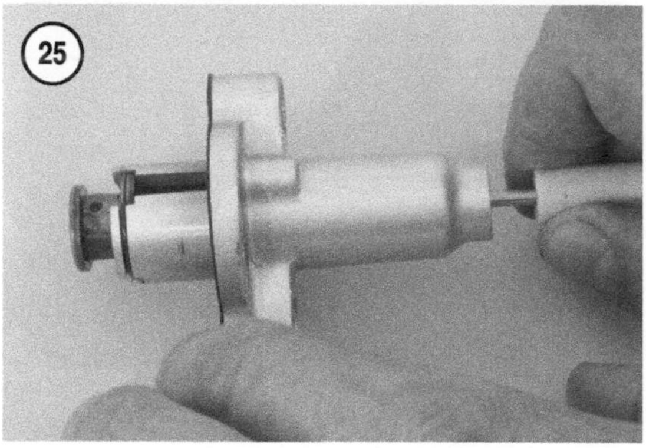

   j. Install the cap bolt (D, **Figure 13**) onto the cam chain tensioner. Torque the cap bolt to 8 N•m (71 in.-lb.).

23. Hold the flywheel with a sheave holder, and tighten the cam sprocket bolts (rear cylinder: C, **Figure 10**; front cylinder: C, **Figure 12**). Torque the cam sprocket bolts to 55 N•m (41 ft.-lb.).

NOTE
A breather plate is not used in the rear cylinder head.

23. When working on the front cylinder, install a new O-ring onto the breather plate (A, **Figure 7**) and set the breather plate in place in the cam sprocket cover (B).

24. Lubricate a new O-ring with lithium soap grease, and install it into the cam sprocket cover (**Figure 27**). Set the cam sprocket cover onto the cylinder head. Install the two cover bolts (A, **Figure 6**), and torque the cam sprocket cover bolts to 10 N•m (89 in.-lb.).

NOTE
The exhaust valve cover on the front cylinder is chrome plated.

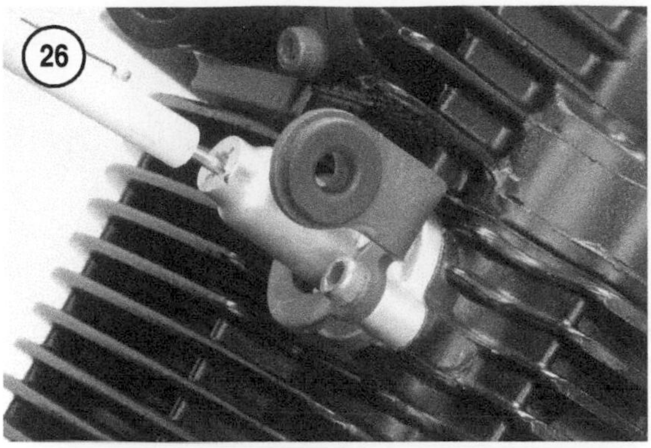

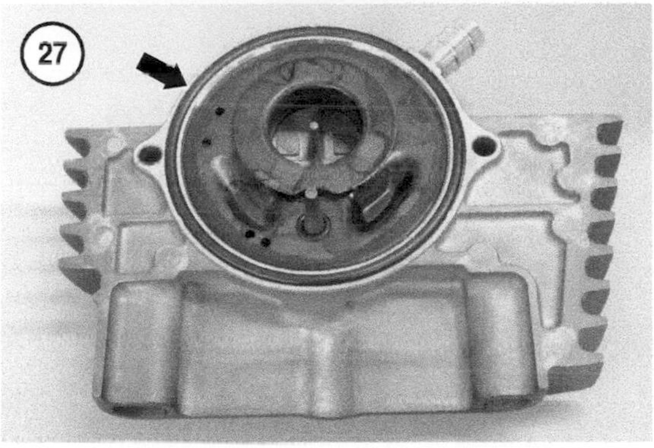

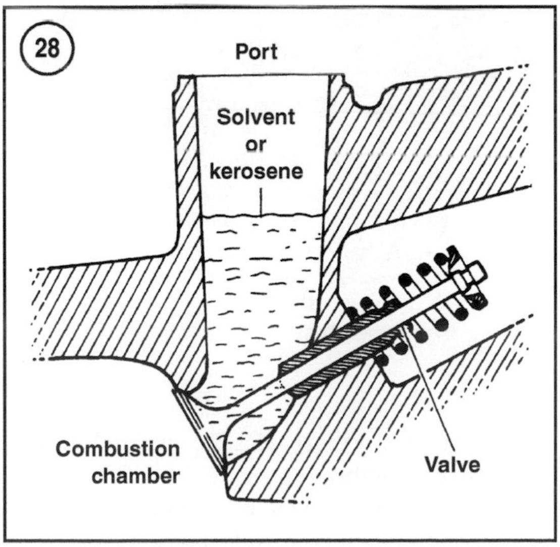

25. Install the valve covers (**Figure 5**) onto the cylinder head. Lubricate the new O-ring with lithium soap grease, and torque the valve cover bolts to 10 N•m (89 in.-lb.).

26. If the intake manifold was removed, install it. Lubricate a new O-ring with lithium soap grease, and torque the intake manifold bolts to 10 N•m (89 in.-lb.).

27. If removed, install the exhaust manifold (B, **Figure 19**) onto the rear cylinder. Install a new exhaust gasket, and torque the exhaust manifold nuts (A, **Figure 19**) to 20 N•m (15 ft.-lb.).

28. Install the spark plugs. Torque the plugs to 20 N•m (15 ft.-lb.).

29. Install the engine into the frame as described in Chapter Five.

30. Check and adjust the valves as described in Chapter Three.

Inspection

1. Before removing the valves or cleaning the cylinder head, perform the following leak test:
   a. Position the cylinder head so the exhaust port faces up. Pour solvent or kerosene into the exhaust port (**Figure 28**).
   b. Turn the head over slightly and check the exhaust valve area on the combustion chamber side. If the valve and seats are in good condition, there will be no leaks past the valve seats. If any area is wet, the valve seat is not sealing correctly. This can be caused by a damaged valve seat and/or valve face, or by a bent or damaged valve. Remove the valve, and inspect the valve and seat for wear or damage.
   c. Pour solvent into the intake port and check the intake valve.
2. Remove all traces of gasket material from the mating surfaces on the cylinder head and cylinder block.

CAUTION
Cleaning the combustion chamber with the valves removed can damage the valve seat surfaces. A damaged or even slightly scratched valve seat will cause poor valve seating.

3. Without removing the valves, remove all carbon deposits from the combustion chambers (A, **Figure 29**). Use a fine wire brush dipped in solvent or make a scraper from hardwood. Be careful not to damage the head, valves or spark plug threads.
4. Examine the spark plug threads (B, **Figure 29**) in the cylinder head for damage. If damage is minor or if the threads are dirty or clogged with carbon,

use a spark plug thread tap (**Figure 30**) to clean the threads. If the damage is severe, restore the threads by installing a steel thread insert. Thread insert kits can be purchased at automotive supply stores or they can be installed at a Yamaha dealership.

5. After all carbon is removed from the combustion chambers and valve ports, clean the entire head in solvent.

6. Clean away all carbon on the piston crowns. Do not remove the carbon ridge at the top of the cylinder bore.

NOTE

The intake manifolds are not interchangeable. Mark each manifold before removal so it will be reinstalled on the correct cylinder head during assembly.

7. Inspect the intake manifolds (**Figure 31**) for cracks or other damage that would allow unfiltered air into the engine. If necessary, remove the manifolds and discard the O-rings. Reinstall the manifolds with new O-rings. Install each manifold in its original location and tighten the mounting bolts to 10 N•m (89 in.-lb.).

8. Check for cracks in the combustion chambers and exhaust ports. If necessary, remove the exhaust manifold from the rear cylinder. A cracked head must be replaced.

9. Inspect the threads on the exhaust pipe mounting studs (A, **Figure 32**). Clean the threads with an appropriate size metric die. Replace a stud if the damage is severe.

10. After the head has been thoroughly cleaned, place a straightedge across the gasket surface at several points. Measure the warp by inserting a feeler gauge (B, **Figure 32**) between the straightedge and the cylinder head at each location. If warp exceeds the service limit listed in **Table 2**, the cylinder head must be replaced or resurfaced. Consult a Yamaha dealership or machine shop experienced in this type of work.

11. Visually inspect the cam chain tensioner assembly (**Figure 33**) for wear or damage. If any part is damaged, replace the chain tensioner assembly.

12. Check the chain tensioner operation as follows:

a. Insert a screwdriver in the tensioner body.

b. Turn the screwdriver clockwise until the plunger completely retracts and locks in place. See **Figure 25**. The pushrod should move smoothly.

c. Release the plunger by turning the screwdriver slightly counterclockwise.

d. The pushrod should fully extend from the tensioner body.

13. Inspect the cam sprockets (**Figure 34**) for wear or missing teeth. Replace the sprocket if necessary.

29

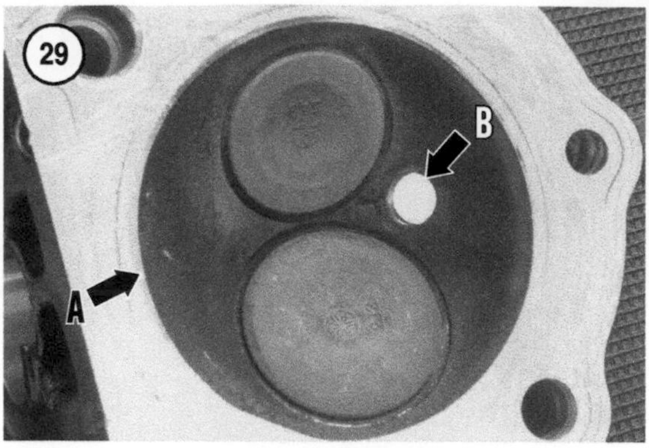

30

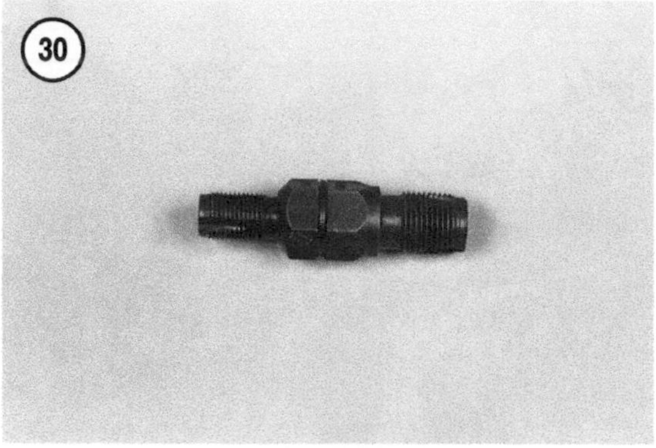

31

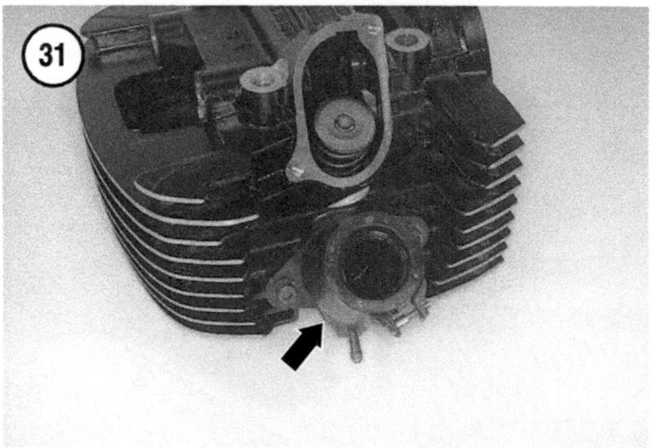

NOTE

If a cam sprocket is worn, also inspect the cam chain, chain guides and the

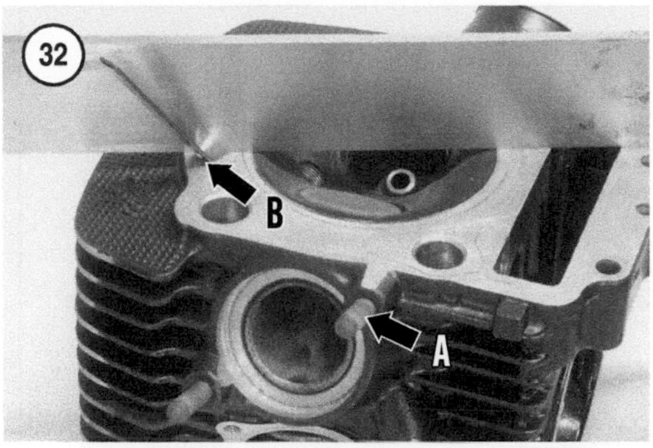

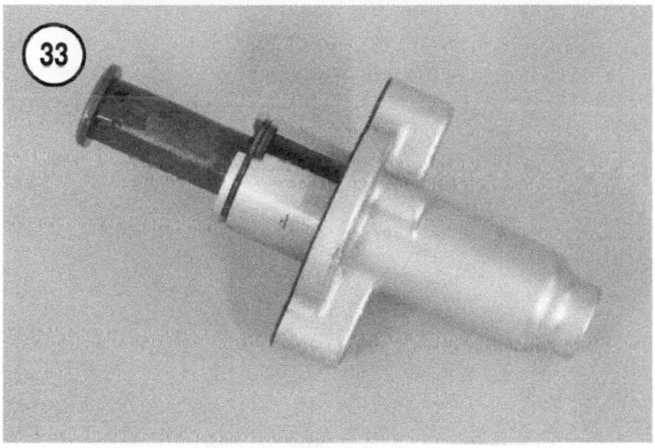

sprocket on the cam chain drive assembly.

14. Check the sliding surfaces (B, **Figure 20**) of the front cam chain guide and the rear chain guide (the bolted guide) for wear or damage. Replace the chain guide(s) as necessary.

15. If the exhaust manifold was removed, install it onto the rear cylinder. Install a new exhaust gasket into the exhaust port and torque the manifold nuts to 20 N•m (15 ft-lb.).

CAMSHAFT

Removal

Refer to **Figure 35**.

1. Remove the cylinder head as described in this chapter.
2. Unthread the camshaft bushing retainer bolt (A, **Figure 36**) and remove the retainer (B).
3. Turn a 10-mm bolt into the camshaft, and pull the camshaft and bushing (A, **Figure 37**) from the cylinder head. Watch for the locating pin (B, **Figure 37**) in the camshaft.

Installation

NOTE

*The camshafts in this engine are not interchangeable. Each camshaft is identified by a **1** or a **2** cast into the body (**Figure 38**). The No. 1 camshaft fits in the rear cylinder head; the No. 2 camshaft fits in the front. Make sure each camshaft is installed in the correct cylinder head.*

1. If removed, install the locating pin (A, **Figure 39**) into the end of the cylinder head.
2. Apply molybdenum disulfide oil to the surfaces of the camshaft journals.
3. Install the camshaft into the cylinder head so the locating pin (A, **Figure 39**) aligns with the pointer (B) on the top of the cylinder head.
4. Apply molybdenum disulfide oil to the inside of the camshaft bushing (A, **Figure 40**) and install the bushing (A, **Figure 37**) onto the camshaft. Position the bushing so the cutout faces down.
5. Install the bushing retainer (B, **Figure 36**) into the cutout in the bushing. Rotate the bushing as necessary so the retainer's mounting hole aligns with the hole in the cylinder head.
6. Install the camshaft bushing retainer bolt (A, **Figure 36**). Torque the bolt to 20 N•m (15 ft.-lb.).

Inspection

1. Clean all parts in solvent and blow them dry with compressed air.
2. Visually inspect the inside diameter (A, **Figure 40**) and the outside diameter (B) of the camshaft bushing. Replace it if necessary. If the camshaft

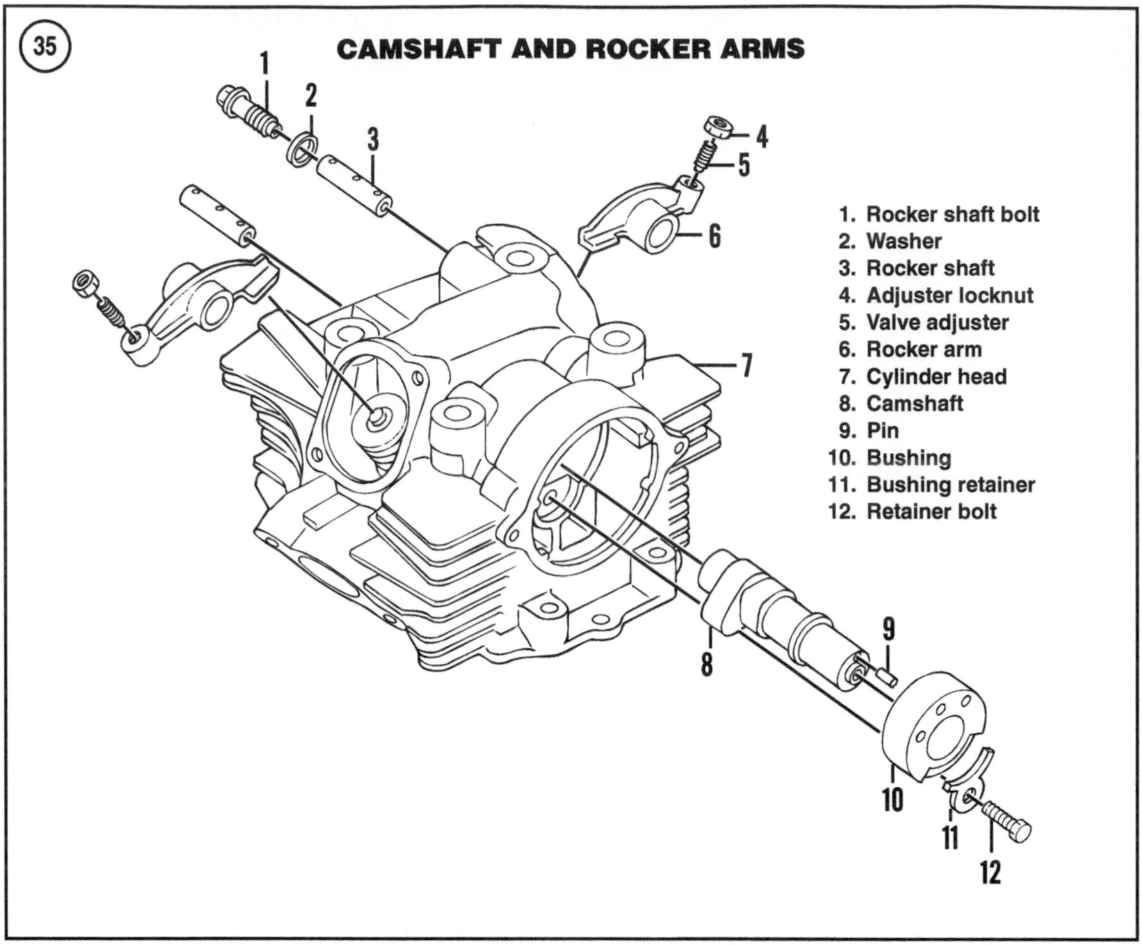

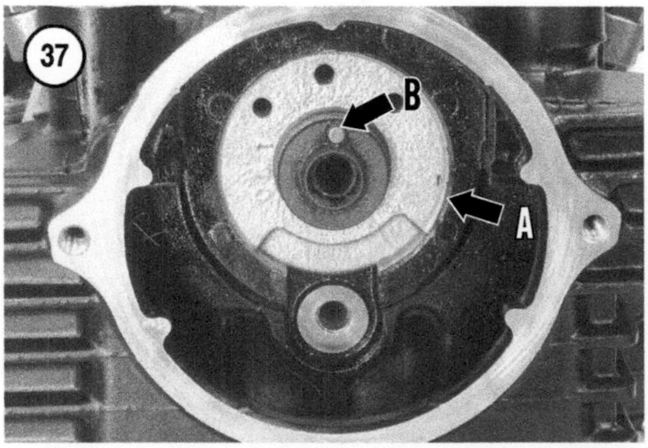

bushing is severely worn, inspect the bushing bore in the cylinder head. The surface should be smooth with no visible marks. Replace the cylinder head if this surface is worn.

3. Visually inspect the camshaft bearing journals (A, **Figure 38**) for wear and scoring. Replace as necessary.

4. Check the camshaft lobes (B, **Figure 38**) for wear. The lobes should not be scored and the edges should be square. Slight damage can be removed with silicon carbide oil stone. Use No. 100-120 grit initially, then polish the lobe with No. 280-320 grit.

5. Even if the cam lobe surfaces appears satisfactory, they must be measured with a micrometer.

38

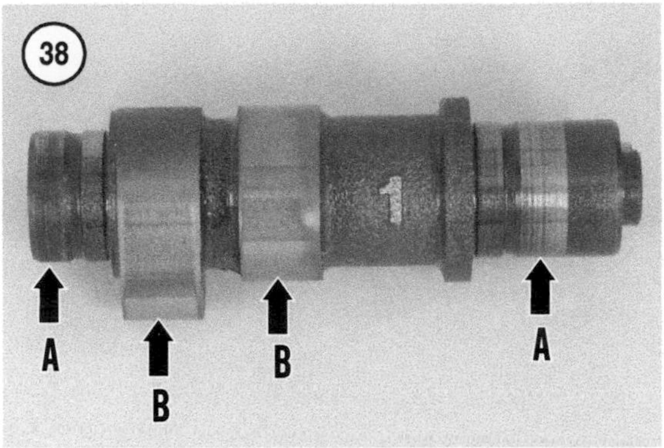

42

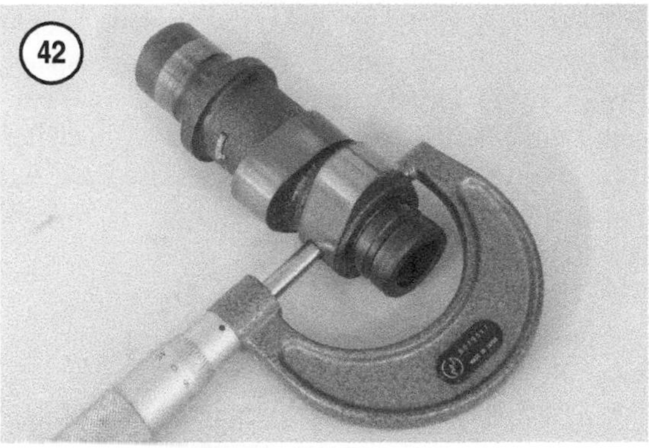

4

39

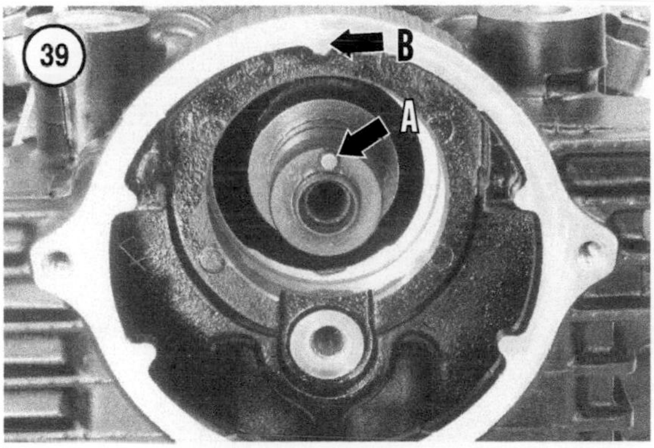

43

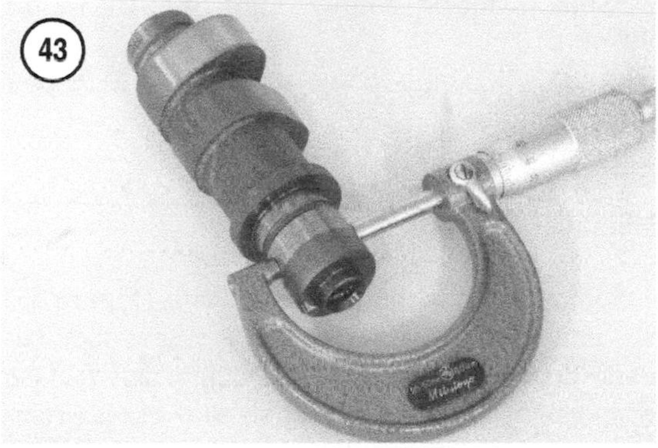

40

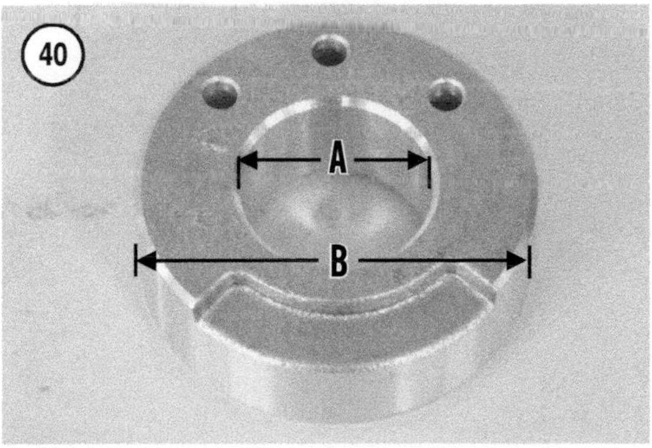

44

41

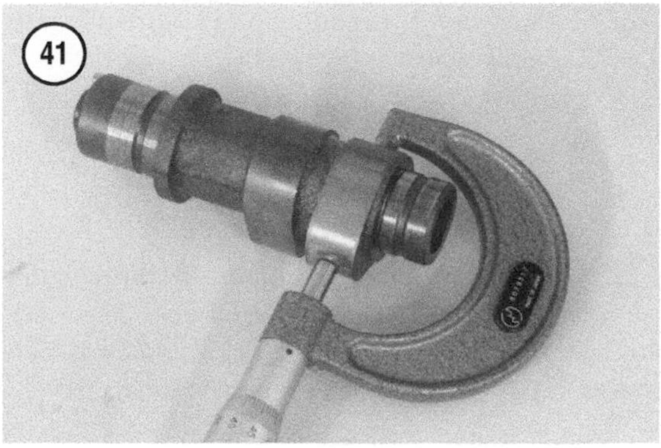

Measure the cam lobe height (**Figure 41**) and width (**Figure 42**) with a micrometer. Replace the camshaft if a lobe is worn beyond the service limits specified in **Table 2**.

6. Measure the camshaft runout with a dial indicator and V-blocks. Replace the camshaft if the runout exceeds the service limit.

7. Measure the camshaft journal outside diameter (**Figure 43**) and measure the inside diameter of the camshaft bushing (**Figure 44**). Calculate the camshaft-to-bushing clearance by subtracting the cam-

shaft outside diameter from the bushing inside diameter. If the clearance is outside the range specified in **Table 2**, compare the camshaft journal outside diameter and bushing inside diameter measurements to the specifications. Replace the camshaft and/or bushing, whichever is out of specification.

ROCKER ARMS

The following Yamaha special tools, or their equivalents, are required to remove the rocker arms:

1. Slide hammer bolt (8 mm): YU-1083-2 or 90890-01085.
2. Slide hammer weight: YU-1083-3 or 90890-0184.

Removal

Although the rocker arms and shafts (**Figure 35**) are interchangeable when they are new, they become mated through wear. Mark these parts (intake or exhaust) during removal, so they can be reinstalled in their original locations.

1. Remove the cylinder head as described earlier in this chapter, and remove the valve covers.
2. Remove the camshaft as described earlier in this chapter.

NOTE
*Two types of rocker arm bolts are used on each cylinder head. The bolt with the single oil hole (A, **Figure 45**) goes on the exhaust side; the bolt with the two oil holes (B) goes on the intake side. This bolt also secures the oil line to the head.*

3. If the rocker arm bolt is still installed, remove it from the exhaust side of the cylinder head.

4A. If a slide hammer is available, perform the following:

   a. Assemble the slide hammer per the manufacturer's instructions.
   b. Thread the slide hammer bolt (A, **Figure 46**) into the rocker arm shaft. Repeatedly slide the hammer weight (B, **Figure 46**) against the bolt head and remove the rocker arm shaft.

4B. If a slide hammer is not available, perform the following:

45

46
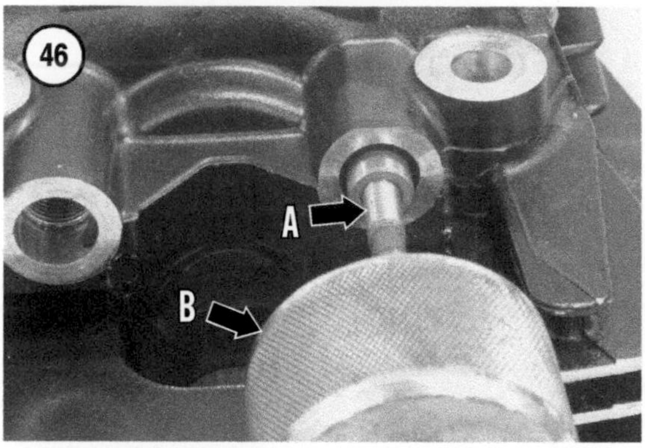

47
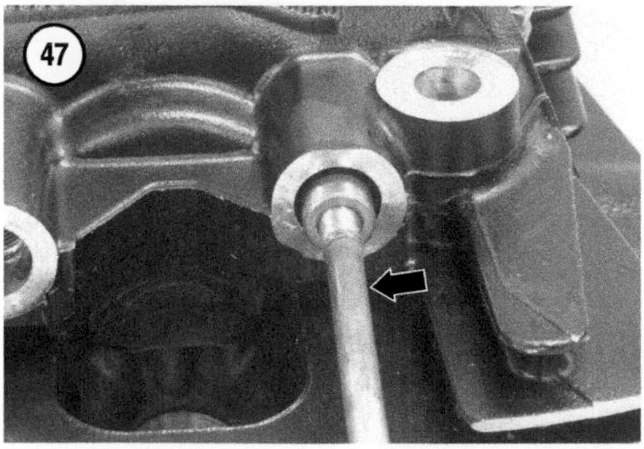

   a. Thread an 8 × 125 mm bolt (**Figure 47**) into the rocker arm.
   b. Grasp the bolt head with a pair of locking pliers.
   c. Tap the pliers with a hammer and pull the rocker shaft from the cylinder head.

5. Remove the rocker arm (**Figure 48**) through the camshaft port in the cylinder head. Label the rocker arm and its shaft (intake or exhaust) so they can be reinstalled in the same location in the head.
6. Repeat Steps 3-5 for the other rocker arm.

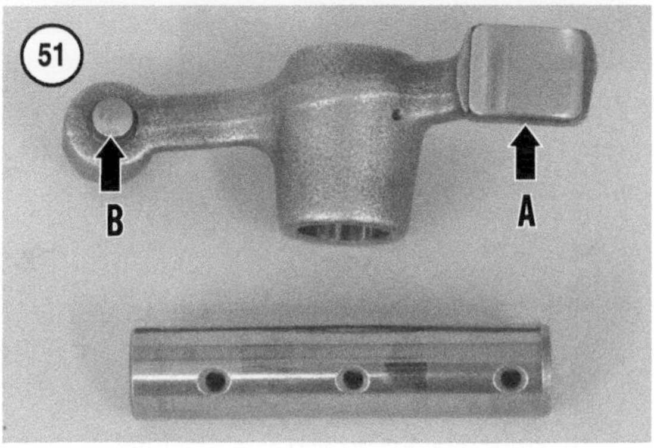

7. Inspect the rocker arms and shafts as described in this section.

Installation

NOTE
Each rocker arm and shaft have become mated through wear. Unless a part is being replaced, install each rocker arm with its original shaft, and install them in their original locations (intake or exhaust side) in the cylinder head.

1. Apply engine oil to the rocker arm bore and to the rocker shaft.

CAUTION
The rocker shaft is directional. The non-threaded end of the shaft is the inboard side.

2. Insert the rocker arm through the camshaft port (**Figure 48**) and set the rocker arm into its original place in the cylinder head (**Figure 49**).
3. Thread the slide hammer bolt, or the 8 × 125 mm bolt, into the threaded end of the rocker arm shaft. Use the bolt to install the shaft into the head (**Figure 50**). Make sure the shaft passes through the boss on the rocker arm and bottoms in the cylinder head.
4. When installing an exhaust rocker arm, slide a new copper washer on the rocker arm bolt (A, **Figure 45**) with a single hole. Turn the rocker arm bolt into the cylinder head and torque the bolt to 37 N•m (27 ft.-lb.).
5. Repeat Steps 1-4 for the other rocker arm. Do not install the intake rocker arm bolt at this time. It will be installed during oil line installation after the engine has been bolted into the frame (Chapter Five).
6. Reinstall the cylinder head as described in this chapter.

Inspection

1. Clean all parts in solvent and dry them thoroughly with compressed air.
2. Inspect the rocker arm pad (A, **Figure 51**) where it rides on the cam lobe. Replace the rocker arm if the pad surface is scratched, unevenly worn or shows signs of blue discoloration. If the rocker arm

pad is worn or damaged, also inspect the cam lobe for scoring, chipping or flat spots.

3. Inspect the valve adjuster (B, **Figure 51**) where it rides on the valve stem. Replace the adjuster if it is scratched, pitted or shows signs of blue discoloration.

4. Measure the inside diameter of the rocker arm bore (**Figure 52**). Replace the rocker arm if the bore exceeds the service limit specified in **Table 2**.

5. Measure the outside diameter of the rocker arm shaft (**Figure 53**). Replace the rocker arm shaft if its outside diameter is less than the service limit (**Table 2**).

6. Calculate the rocker arm-to-shaft clearance by subtracting the shaft outside diameter from the rocker arm bore inside diameter. If the clearance is less than the service limit (**Table 2**), replace the defective part.

7. Inspect the threads of each rocker arm bolt (**Figure 45**) for stretching or other signs of damage. Blow the passages clear with compressed air. Replace a bolt as necessary.

CAM CHAIN AND CAM CHAIN DRIVE ASSEMBLY

The camshaft for each cylinder is chain driven by the cam chain drive assembly, which consists of a spring-loaded gear and a sprocket. The cam chain drive assembly sits between the cam sprocket and the timing gear. The timing gear on the crankshaft drives the gear on the cam chain drive assembly. The cam chain runs from the sprocket on the cam chain drive assembly and turns the cam sprocket on the end of the camshaft.

NOTE

The front cylinder timing gear comes out during primary drive gear removal (Chapter Six); the rear cylinder timing gear comes out during flywheel removal (Chapter Five). Refer to the appropriate chapter when servicing a timing gear.

Removal (Rear Cylinder)

The Yamaha sheave holder (part No. YS-01880 or 90890-01701), or its equivalent, is required to perform this procedure.

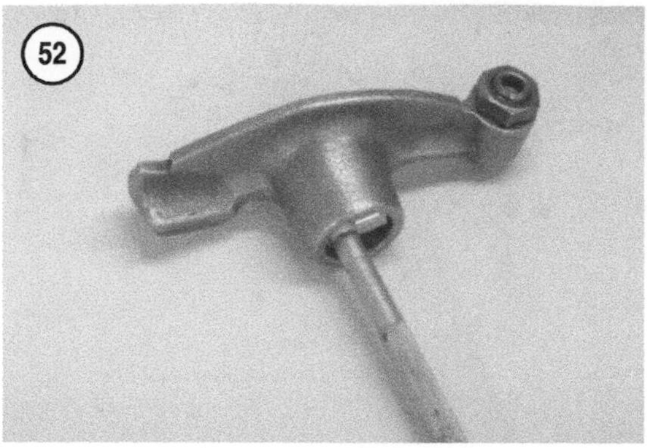

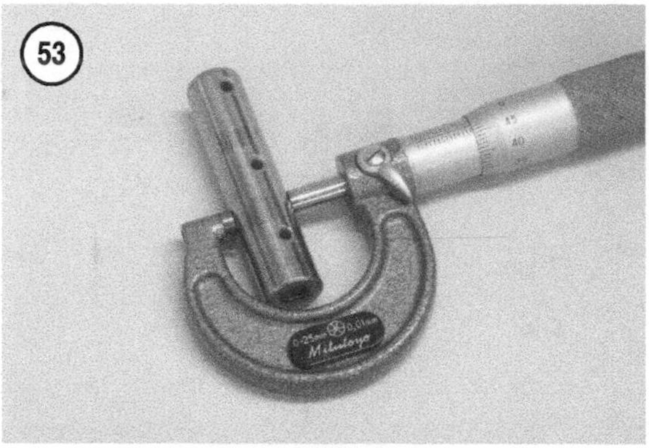

1. Remove the rear cylinder head as described in this chapter.

2. Remove the front chain guide by lifting it from its seat in the crankcase and pulling it from the cam chain tunnel in the cylinder.

NOTE

The cam chain and cam chain drive assembly can be removed with the flywheel installed. However, the flywheel must be removed for cam chain/chain drive assembly installa-

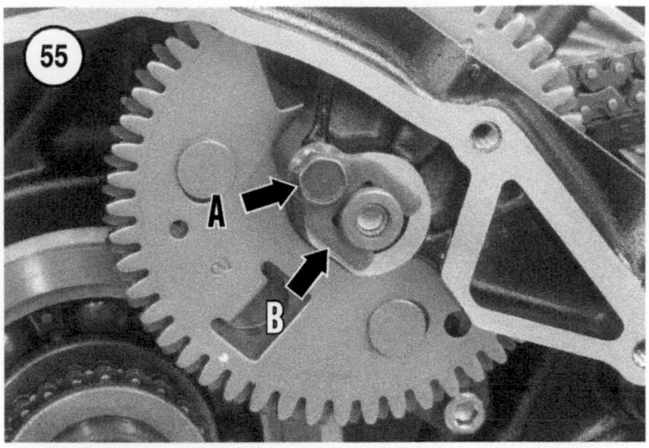

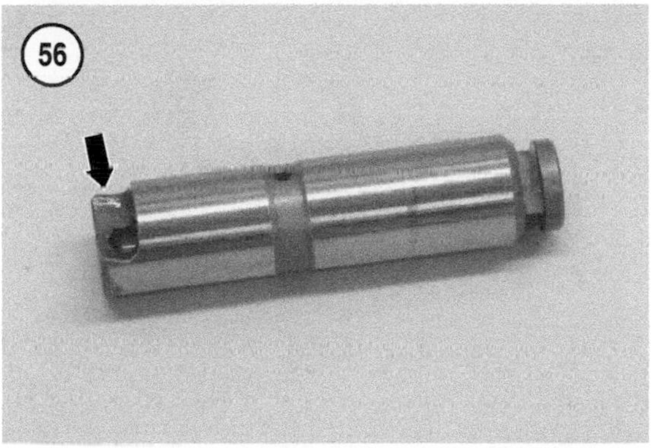

tion. The cam chain drive assembly cannot be properly timed unless the flywheel is removed. Remove the flywheel now in order to study the alignment of the timing marks.

3. Set the rear cylinder to top dead center on the compression stroke and remove the flywheel as described in Chapter Five. Note that the crankshaft keyway (A, **Figure 54**), the timing mark on the cam chain drive assembly (B), and the center of the shaft boss (C) for the cam chain drive assembly are aligned. These marks must be aligned during assembly.

4. Remove the retainer bolt (A, **Figure 55**) and slide the retainer (B) from the slots on the cam chain drive assembly shaft.

CAUTION
*An oil slot (**Figure 56**) in the end of the drive assembly shaft collects oil that lubricates the shaft and bushing in the cam chain drive assembly. The shaft is installed so this oil slot faces up.*

4. Remove the drive assembly shaft (**Figure 57**). Note that the oil slot in the shaft faces up. It must be reinstalled in this direction during assembly.

5. Lift the cam chain and cam chain drive assembly (**Figure 54**) from the cam chain tunnel.

Installation (Rear Cylinder)

1. Make sure the rear cylinder is still set to top dead center on the compression stroke.
   a. If necessary, install the Woodruff key into the crankshaft keyway and slide the flywheel onto the crankshaft. Temporarily install the alternator cover onto the crankcase. Make sure the T-mark (A, **Figure 9**) on the flywheel still aligns with the pointer (B) on the alternator cover. Remove the cover and flywheel.
   b. If the engine has been completely disassembled, set the rear cylinder to top dead center by pulling the rear connecting rod to the top of its stroke.

NOTE
*The cam chain drive assemblies are not interchangeable. The rear cylinder cam chain drive assembly is identified by a **2** stamped on the gear face; a **3** identifies the front cylinder cam chain drive assembly. Make sure each cam chain drive assembly is installed in the correct cylinder.*

2. Preload the cam chain drive assembly as follows:
   a. Cut a 6 × 15 mm pin from the shoulder (non-threaded portion) of a 6 mm bolt.

4

b. Use a screwdriver or similar tool to pry the drive teeth on the gear until one set of teeth aligns with the other.

c. Insert the 6-mm pin (**Figure 58**) into the aligned hole to lock the gear.

3. Position the timing gear so the side with the *2* stamp faces the outboard side of the engine. Seat the cam chain (A, **Figure 59**) on the sprocket of the cam chain drive assembly (B), and lower the cam chain/drive assembly into the cam chain tunnel.

4. Rotate the cam chain drive assembly as necessary and align the crankshaft keyway (A, **Figure 54**), the timing mark (B) on the cam chain drive assembly and the center of the shaft boss (C).

5. Install the shaft so the oil cutout faces up (**Figure 57**).

6. Slide the retainer (B, **Figure 55**) over the end of the shaft so the retainer fingers seat in the cutout in the shaft.

7. Install the retainer bolt (A, **Figure 55**). Apply a medium-strength threadlocking compound to the bolt threads, and torque the cam chain drive assembly retainer bolt to 10 N•m (89 in.-lb.).

8. If removed, set the starter idler gear assembly (A, **Figure 60**) into the crankcase, and secure it in place with the shaft (B).

9. If the timing gear was removed from the flywheel, install the timing gear as described in *Flywheel and Starter Gear* in Chapter Five.

10. Set the flywheel assembly face down on the bench so the timing gear is up. Mark the edge of the tooth (**Figure 61**) on either side of the timing mark so you can locate the mark as the flywheel is installed.

11. Position the flywheel so the two marked teeth align with the Woodruff key in the crankshaft (A, **Figure 62**). Slide the flywheel onto the crankshaft so the keyway engages the Woodruff key, the starter wheel gear (B, **Figure 62**) engages the starter idler gear assembly, and the timing gear (C) engages the cam chain drive assembly. See **Figure 63**.

12. Remove the 6 mm pin (**Figure 64**) from the gear of the cam chain drive assembly.

13. Install the washer (A, **Figure 65**) and the flywheel nut (B).

14. Hold the flywheel with the sheave holder, and torque the flywheel nut to 175 N•m (129 ft.-lb.). Make sure the sheave holder does not pass over a projection (C, **Figure 65**) on the flywheel.

Removal (Front Cylinder)

Refer to **Figure 66** when performing this procedure.

1. Remove the front cylinder head as described in this chapter.
2. Remove the front chain guide by lifting it from its seat in the crankcase and pulling it from the cam chain tunnel in the cylinder.

NOTE
*Note that the timing mark on the timing gear (A, **Figure 67**), the mark on the cam chain drive assembly (B) and the center of the drive assembly shaft (C) align when the front cylinder is at TDC. They must be aligned during assembly.*

3. Remove the retainer bolt (A, **Figure 68**), and slide the retainer (B) from the slots on the shaft of the cam chain drive assembly.

CAUTION
*An oil slot (**Figure 56**) in the end of the drive assembly shaft collects oil that lubricates the shaft and bushing in the cam chain drive assembly. The shaft must be installed so this oil slot faces up.*

4. Remove the drive assembly shaft. Note that the oil slot (A, **Figure 69**) in the shaft faces up. It must be reinstalled in this direction during assembly.
5. Lift the cam chain/cam chain drive assembly (**Figure 59**) from the cam chain tunnel.
6. Inspect the cam chain and its drive assembly as described in this section.

Installation (Front Cylinder)

1. Temporarily install the alternator cover onto the crankcase. Make sure the front cylinder is still set to top dead center on the compression stroke.

NOTE
*The cam chain drive assemblies are not interchangeable. The rear cylinder cam chain drive assembly is identified by a **2** stamped on the gear face; a **3** identifies the front cylinder cam chain drive assembly. Make sure each*

(66)

CAM CHAIN, CHAIN GUIDES AND TIMING GEARS

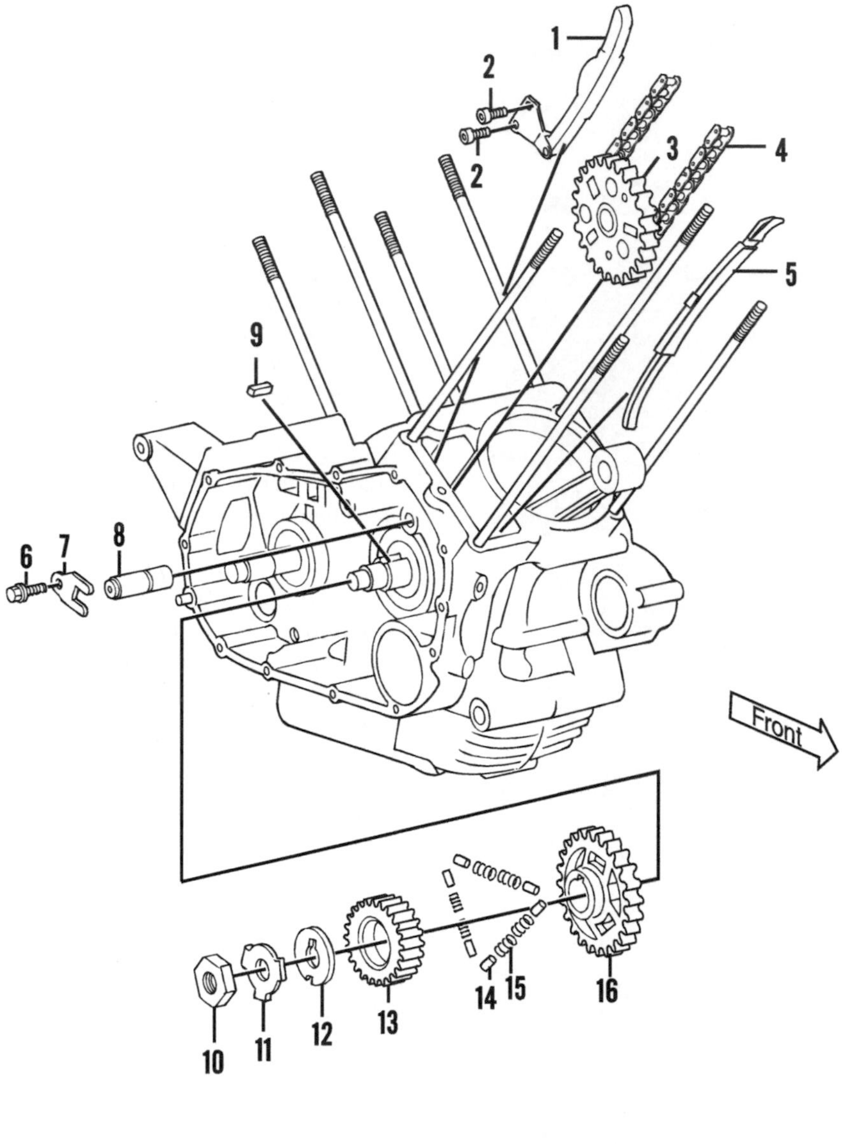

1. Rear chain guide
2. Chain guide bolt
3. Cam chain drive assembly
4. Cam chain
5. Front chain guide
6. Retainer bolt
7. Shaft retainer
8. Drive assembly shaft
9. Woodruff key
10. Primary drive nut
11. Lockwasher
12. Keyed washer
13. Timing gear
14. Pin
15. Spring
16. Primary drive gear

67

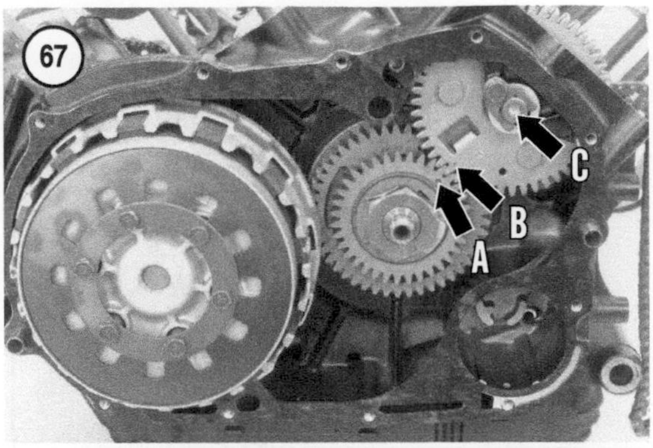

68

69

70

cam chain drive assembly is installed in the correct cylinder.

2. Preload the cam chain drive assembly as follows:
   a. Cut a 6 × 15 mm pin from the shoulder (non-threaded portion) of a 6-mm bolt.
   b. Use a screwdriver or similar tool to pry the drive teeth on the gear until one set of teeth aligns with the other.
   c. Insert the 6-mm pin (**Figure 58**) into the aligned hole to lock the gear.
3. Position the timing gear so the side with the *3* stamp faces the outboard side of the engine, and seat the cam chain (A, **Figure 59**) on the sprocket of the cam chain drive assembly (B).
4. Lower the cam chain drive assembly into the cam chain tunnel until the teeth of the drive assembly gear engage the teeth of the timing gear. Make sure the timing mark on the cam chain drive assembly (B, **Figure 69**) aligns with the mark on the front cylinder timing gear (C).
5. Install the shaft so the oil cutout faces up (A, **Figure 69**).
6. Slide the retainer (B, **Figure 68**) over the end of the shaft so the retainer fingers seat in the cutout in the shaft.
7. Install the retainer bolt (A, **Figure 68**). Apply a medium-strength threadlocking compound to the bolt threads, and torque the cam chain drive assembly retainer bolt to 10 N•m (89 in.-lb.).
8. Remove the 6-mm pin (**Figure 58**) from the cam chain drive assembly.

Inspection

NOTE

If a cam chain or cam chain drive assembly must be replaced, replace the chain, drive assembly and the cam sprocket as a set.

1. Inspect the cam chain (A, **Figure 59**) for wear, stretching or link damage.
2. Inspect the teeth of the cam sprocket (**Figure 70**) and the sprocket on the cam chain drive assembly (B, **Figure 59**) for worn or broken teeth.
3. Inspect the sliding surface (**Figure 71**) of each chain guide.
4. Inspect the pivot point on the rear chain guide (the bolted guide). Make sure the pivot moves freely. If necessary, replace the rear chain guide as follows:

4

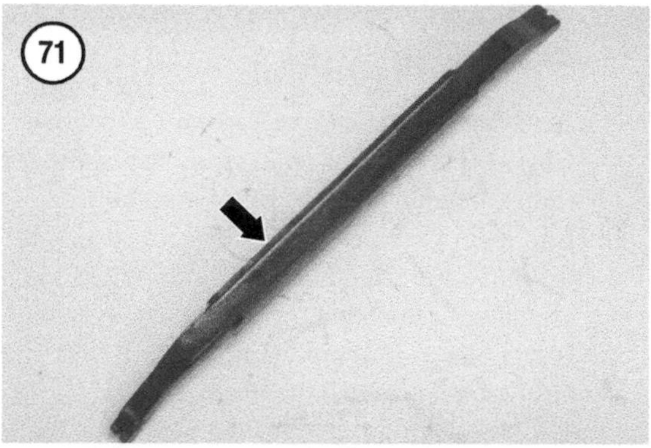

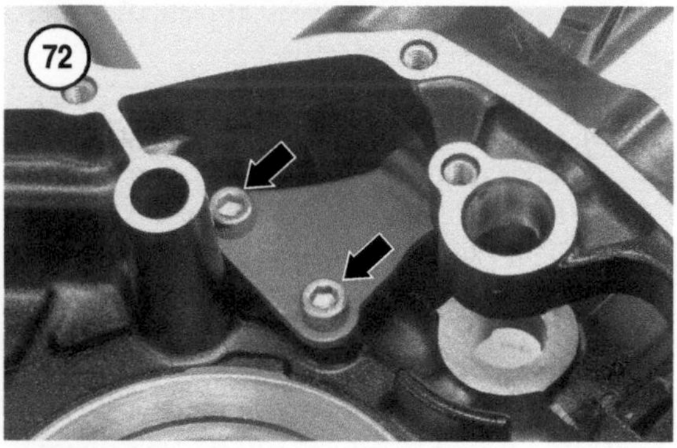

a. Remove the mounting bolts (**Figure 72**) from the bracket, and pull the assembly from the crankcase.

b. Fit the new cam chain assembly into place. Install the mounting bolts and torque them to 10 N•m (89 in.-lb.).

5. Replace any part that is worn or damaged.

VALVES AND VALVE COMPONENTS

Complete valve service requires a number of special tools. The following procedures describe how to check valve components and determine the needed service.

A valve spring compressor (Yamaha part No. YM-04019 or 90890-04019), or equivalent, is needed to remove and install the valves.

Valve Removal

Refer to **Figure 73**.

CAUTION

Keep the components of each particular valve assembly together. Do not mix components from different valve assemblies or excessive wear may occur.

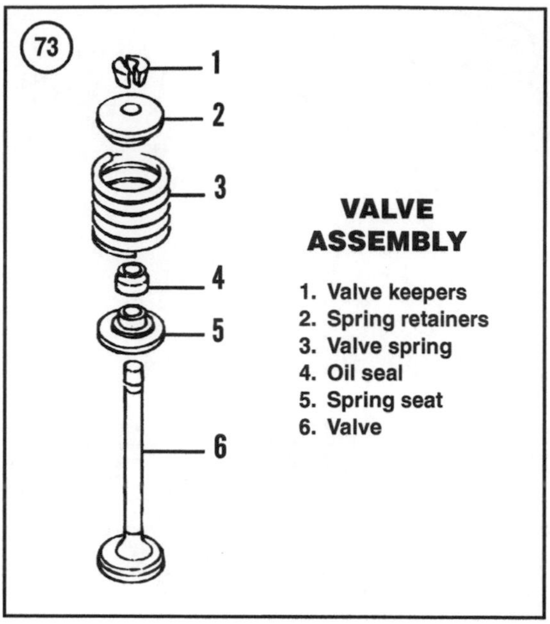

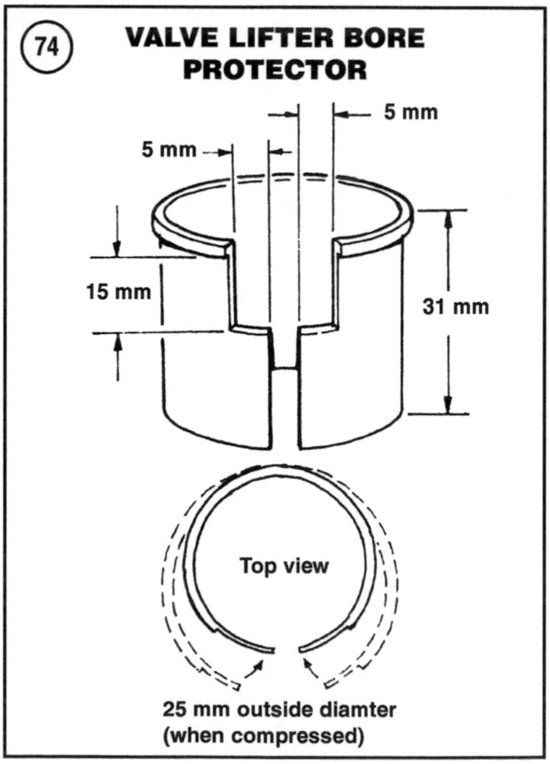

1. Remove the cylinder head as described in this chapter.

75

76

77

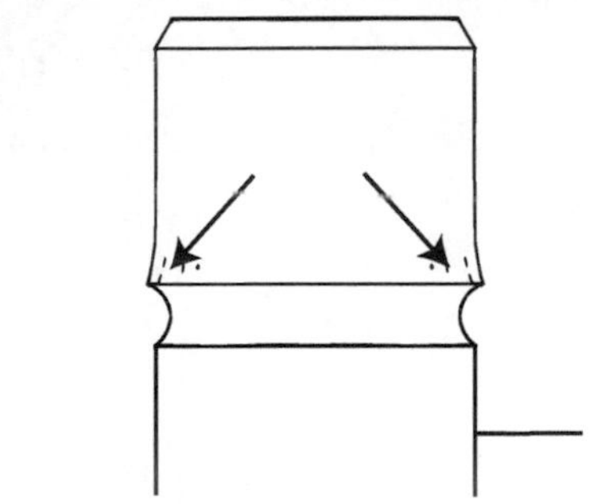

2. Perform the cylinder head leak test described in *Cylinder Head* in this chapter.

3. Remove the camshaft and rocker arms as described in this chapter.

NOTE
*A bore protector can be made from a plastic 35-mm film canister (**Figure 74**). Cut out the bottom of the canister and part of its side. Cut away enough material so the canister can slide between the valve assembly and the side of the bore. The plastic canister protects the bore from potential marring by the valve spring compressor.*

78

4. Insert a bore protector between the valve assembly and the bore.
5. Install a valve spring compressor squarely over the valve retainer. Make sure the opposite end of the compressor rests against the valve head. See **Figure 75**.

CAUTION
To avoid loss of spring tension, do not compress the valve spring any more than necessary to remove the valve keepers.

6. Tighten the compressor until the valve keepers separate from the valve stem. Remove both valve keepers with a magnet, tweezers or needlenose pliers (**Figure 76**).

CAUTION
Remove any burrs from the valve stem grooves before removing the valve. Burrs on the valve stem will damage the valve guide when the stem passes through it.

7. Inspect the valve stem grooves for burrs (**Figure 77**). Remove any burrs, then carefully remove the valve spring compressor.
8. Remove the spring retainer (**Figure 78**).
9. Remove the valve spring (**Figure 79**).
10. Remove the oil seal from the valve guide (**Figure 80**). Discard the oil seal.
11. Remove the spring seat (**Figure 81**).
12. Remove the valve (**Figure 82**) from the cylinder head while rotating it slightly.

CAUTION
*All the components of each valve assembly must be kept together (**Figure 83**). Place each set in a divided carton*

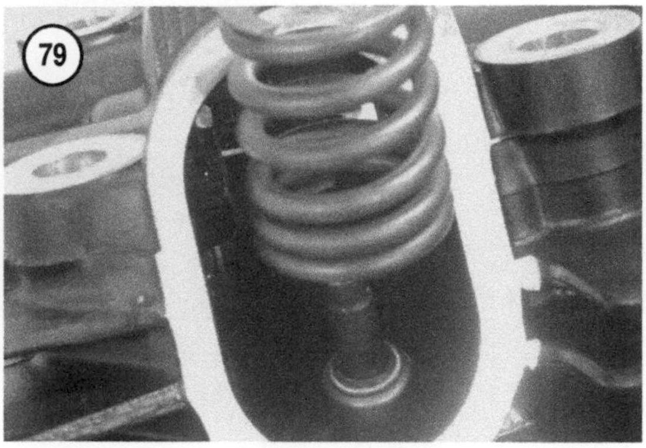
79

or into separate small boxes. Label the set so you will know what cylinder it came from and whether it is an intake or an exhaust valve. This keeps parts from getting mixed up and makes installation simpler. Do not mix components from different valve assemblies or excessive wear may occur.

13. Repeat Steps 3-12 for the remaining valve assembly. Keep the parts from each valve assembly separate.

Valve Installation

1. Clean the end of the valve guide.
2. Install the spring seat (**Figure 81**) and seat it in the cylinder head.
3. Apply molybdenum disulfide oil to a new oil seal and install the seal over the end of the valve guide (**Figure 80**). Push the seal straight down onto the valve guide until the seal bottoms.
4. Apply molybdenum disulfide oil to the valve stem. Install the valve partway into the guide (**Figure 82**). Slowly turn the valve as it enters the oil seal, and continue turning the valve until it is completely installed.
5. Install the valve spring (**Figure 79**) so the end with the closer wound coils faces in toward the combustion chamber.
6. Seat the spring retainer (**Figure 78**) on top of the spring.
7. Insert a bore protector between the valve assembly and the bore.
8. Install a valve spring compressor squarely over the valve retainer. Make sure the opposite end of the compressor sits against the valve head (**Figure 75**).

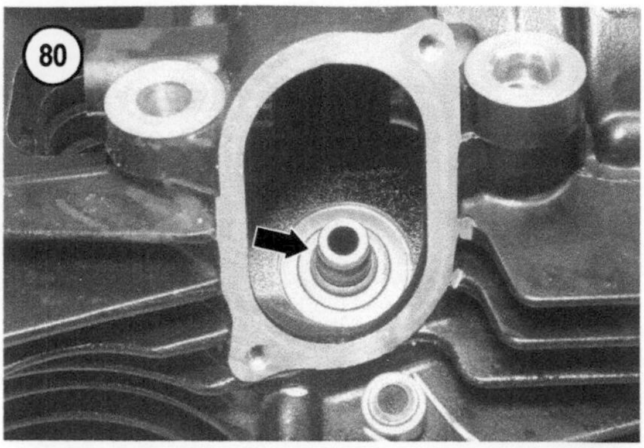
80

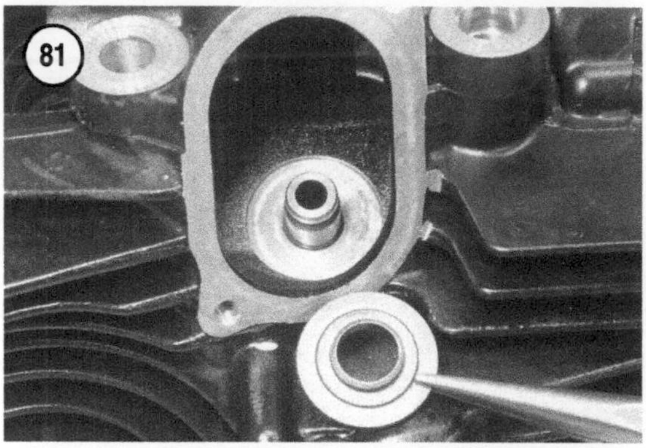
81

82

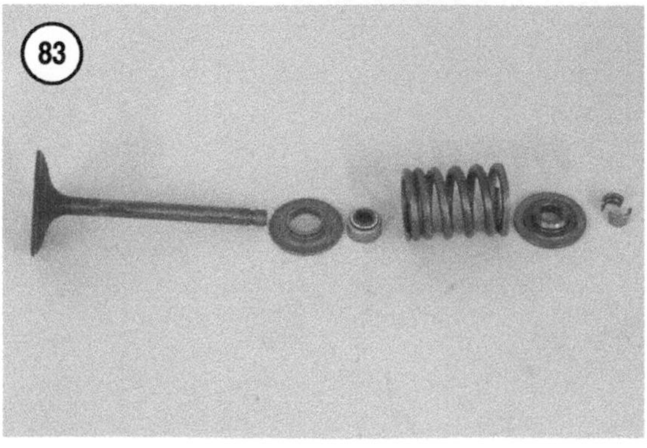
83

84

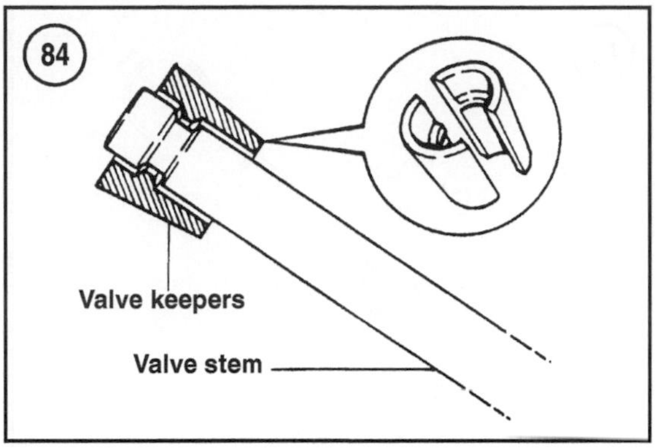

85

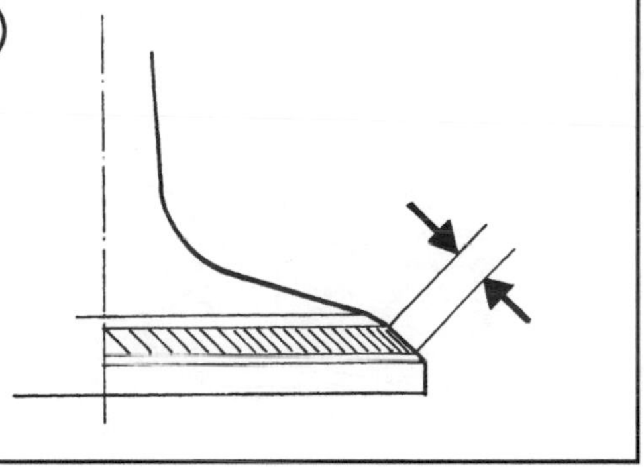

86

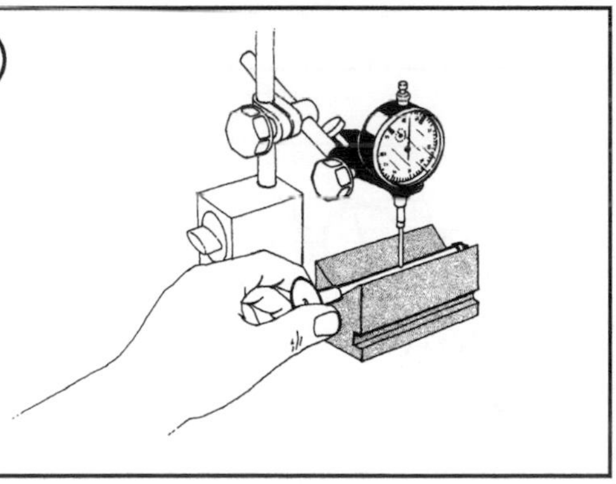

87

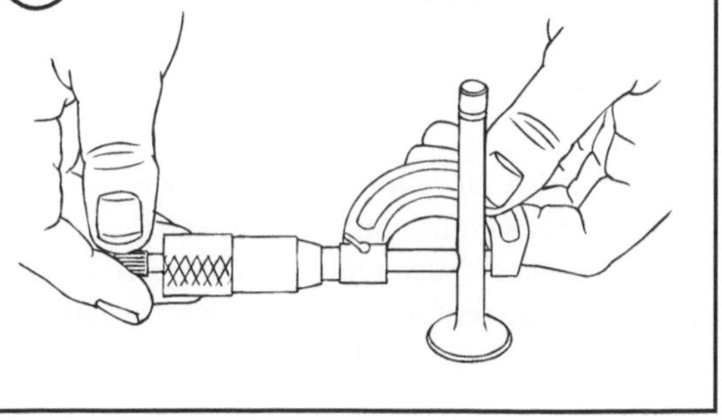

CAUTION
To avoid loss of spring tension, do not compress the springs any more than necessary to install the valve keepers.

9. Compress the valve springs with a valve spring compressor and install the valve keepers (**Figure 76**).
10. When both valve keepers are seated around the valve stem, slowly release the compressor. Remove the compressor and inspect the keepers (**Figure 84**). Tap the end of the valve stem with a soft-faced mallet to assure that the keepers are properly seated.
11. Repeat Steps 1-10 for the remaining valve.
12. Install the rocker arms and camshaft as described in this chapter.
13. Install the cylinder head as described in this chapter.
14. Adjust the valve clearance as described in Chapter Three.

Valve Inspection

NOTE
When a valve needs to be replaced also replace its valve guide. Do not install a new valve into an old guide or excessive wear will occur.

1. Clean the valve in solvent. Do not gouge or damage the valve seating surface.
2. Inspect the contact surface of each valve for burning (**Figure 85**). Minor roughness or pitting can be removed by lapping the valve as described in this chapter. Excessive unevenness indicates that the valve is not serviceable. Replace the valve and the valve guide.
3. Inspect the valve stem for wear and roughness. Measure the runout as shown in **Figure 86**. Replace the valve and valve guide if runout exceeds the service limit in **Table 2**.
4. Measure the diameter of the valve stem with a micrometer (**Figure 87**). Replace the valve and valve guide if the diameter of the valve stem is outside the specified range.
5. Remove all carbon and varnish from the valve guides with a stiff spiral wire brush.
6. Measure the inside diameter of the valve guide with a small bore gauge (**Figure 88**), then measure the gauge with a micrometer. Take a measurement at the top, middle and bottom of the guide. Replace

4

the valve guide if any measurement is outside the specified range.

7. Subtract the valve stem outside diameter (Step 4) from the valve guide inside diameter (Step 6). The difference is the valve stem-to-guide clearance. If the clearance exceeds the service limit, replace the valve and valve guide.

NOTE

If a small bore gauge is unavailable, perform Step 8 to check valve stem-to guide clearance. Measure the valve stem diameter before performing this test. This test is only accurate if the valve stem is within specification.

8. Insert the valve into its guide and attach a dial indicator as show in **Figure 89**. Hold the valve slightly off its seat and rock it sideways in two directions. Watch the dial indicator while rocking the valve. Compare the valve movement to the valve stem-to-guide clearance specification in **Table 2**. If the movement is outside the specified range, the valve guide should probably be replaced. As a final check, take the head to a dealership and have the valve guides measured.

9. Check the valve spring as follows:
   a. Visually inspect the valve spring for bends or other signs of distortion.
   b. Measure each valve spring free length with a vernier caliper **(Figure 90)**. Replace the spring if its free length is less than the service limit.
   c. Use a square to measure the tilt of each spring (**Figure 91**). Tilt should be within the specification in **Table 2**.
   d. Replace any defective or worn spring.

10. Measure the valve margin thickness (**Figure 92**) with a vernier caliper. Replace the valve and valve guide if the margin thickness is worn to the wear limit.

11. Check the spring seat, spring retainer and valve keepers for cracks or other damage.

12. Inspect each valve seat (**Figure 93**) in the cylinder head. If a seat is burned or worn, it must be reconditioned. This should be performed by a dealership or local machine shop. Seats and valves in near perfect condition can be reconditioned by lapping with fine carborendum paste, however, lapping is inferior to precision grinding.

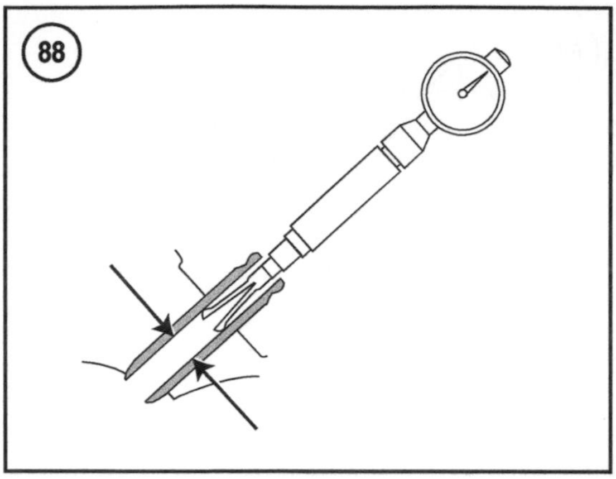

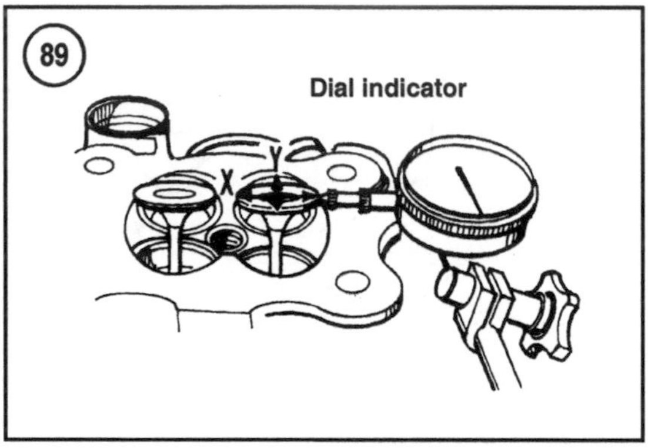

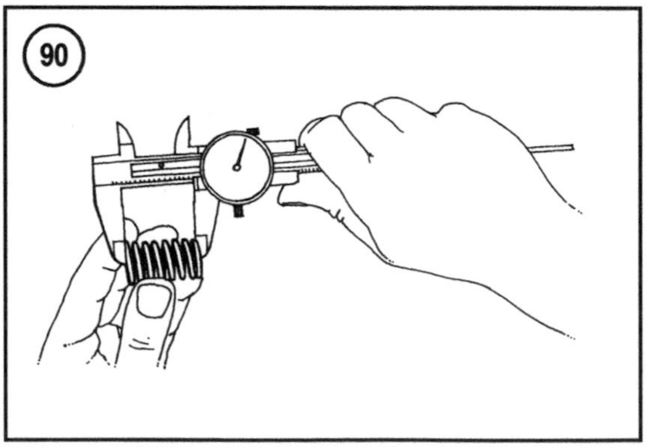

Valve Guide Replacement

Special tools

When valve stem-to-guide clearance is excessive, the valve guides must be replaced. If a valve guide is replaced, also replace its respective valve. This procedure requires the following special tools and should be entrusted to a Yamaha dealership or other qualified specialist.

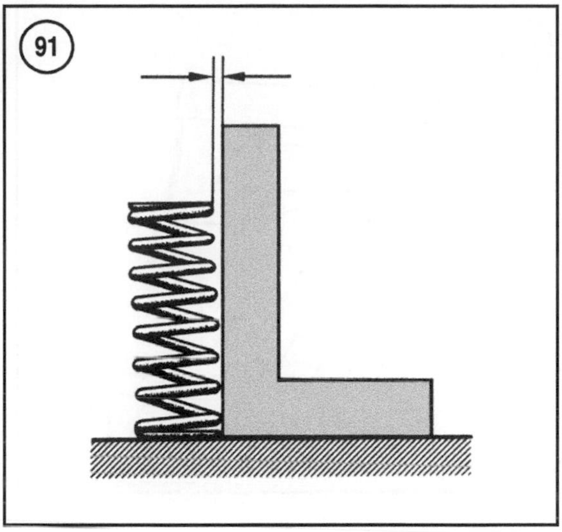

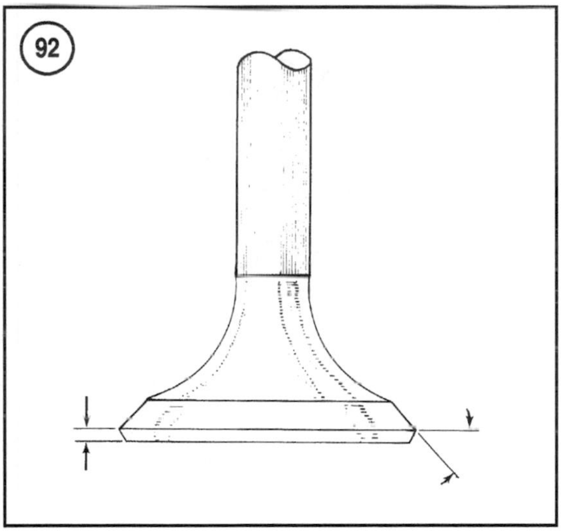

1. 8 mm (0.31 in.) valve guide remover (YM-01200 or 90890-01211).
2. 8 mm (0.31 in.) valve guide installer (YM-01200 or 90890-01200).
3. 8 mm (0.31 in.) valve guide reamer (YM-01201 or 90890-04013).

Procedure

NOTE

The valve guide contracts when it is cooled, which reduces the overall diameter of the guide. On the other hand, heating the cylinder head slightly increases the diameter of the guide bore due to expansion. Since

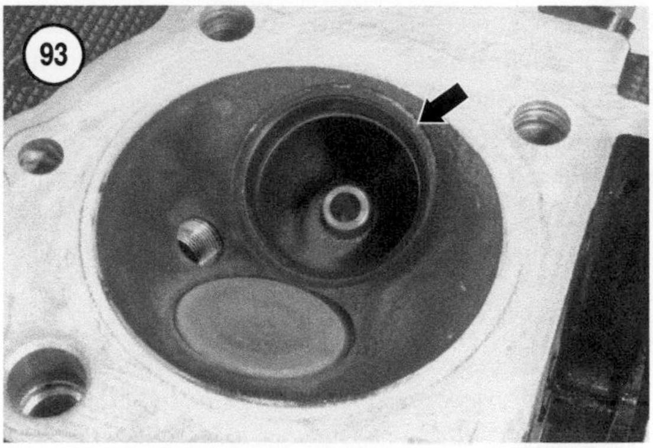

the valve guides have a slight interference fit, cooling the guides and heating the head makes installation easier.

1. Install new circlips onto the new valve guides, and place the valve guides in a freezer overnight.
2. Remove the intake manifold and O-ring from the cylinder head. Discard the O-ring.
3. If servicing the rear cylinder head, remove the exhaust manifold and exhaust gasket. Discard the gasket.

CAUTION

Do not heat the cylinder head with a torch. Never bring a flame into contact with the cylinder head. Direct flame can warp the cylinder head.

4. Place the cylinder head in a shop oven and warm it to 100° C (212° F). Check the temperature of the cylinder head by flicking tiny drops of water onto the head. The cylinder head is heated to the proper temperature if the drops sizzle and evaporate immediately.

WARNING

Wear heavy gloves when performing this procedure. The cylinder head will be very hot.

5. Using heavy gloves or kitchen pot holders, remove the cylinder head from the oven and place it on wooden blocks with the combustion chamber facing up.
6. From the combustion side of the head, drive the old valve guide (A, **Figure 94**) out of the cylinder head with the 8 mm valve guide remover (B) and a hammer.

7. Remove and discard the valve guide and circlip. Never reinstall a valve guide or circlip. They are no longer true and are not within tolerance.

8. After the cylinder head cools, check the guide bore for carbon or other contamination. Clean the bore thoroughly.

9. Reheat the cylinder head as described in Step 4.

WARNING
Wear heavy gloves when performing this procedure. The cylinder head will be very hot.

10. Using heavy gloves or kitchen pot holders, remove the cylinder head from the oven and place it on wooden blocks with the combustion chamber facing down.

11. Remove one valve guide from the freezer.

CAUTION
Failure to lubricate the new valve guide and guide bore will result in damage to the cylinder head and/or valve guide.

12. Apply clean engine oil to the new valve guide and to the valve guide bore in the cylinder head.

13. From the top side of the cylinder head (camshaft side), drive the new valve guide into the cylinder head with a hammer, the 8 mm (0.31 in.) valve guide installer (A, **Figure 95**) and the valve guide remover (B). Drive the valve guide into the bore until the circlip is completely seated against the cylinder head.

14. After the cylinder head has cooled down, ream the new valve guides as follows:

a. Apply cutting oil to both the new valve guide and to the valve guide reamer.

CAUTION
Always rotate the valve guide reamer clockwise. The valve guide will be damaged if the reamer is rotated counterclockwise.

b. Insert the 8 mm (0.31 in.) valve guide reamer from the top side (**Figure 96**) and rotate the reamer *clockwise*. Continue to rotate the reamer and work it down through the entire length of the new valve guide. Continue to apply additional cutting oil during this procedure.

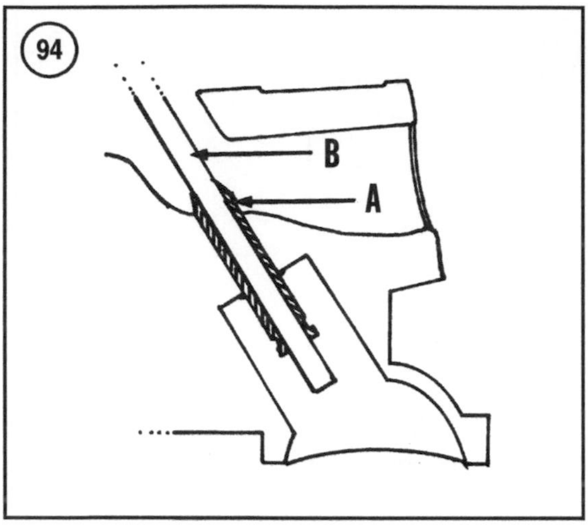

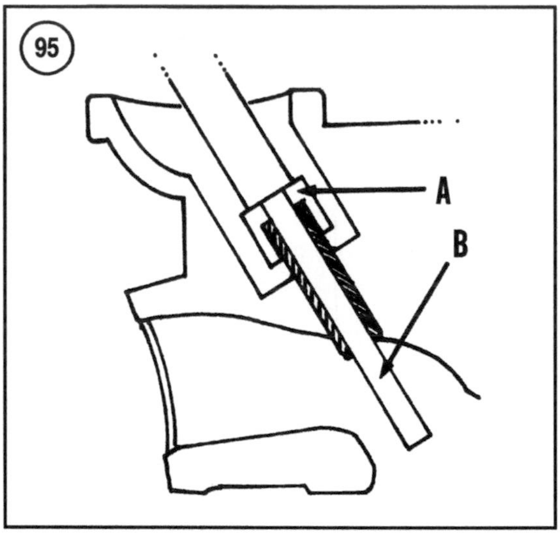

c. Rotate the reamer clockwise until it has traveled all the way through the new valve guide.

d. Rotate the reamer *clockwise* and completely withdraw the reamer from the valve guide.

e. Measure the inside diameter of the valve guide with a small bore gauge (**Figure 88**). Measure the gauge with a micrometer, and compare the measurement to the specification in **Table 2**. Replace the valve guide if it is not within specification.

15. If necessary, repeat Steps 1-14 for any other valve guide.

16. Thoroughly clean the cylinder head and valve guides with solvent to wash out all metal particles. Dry the head with compressed air.

17. Lightly oil the valve guides to prevent rust.

96

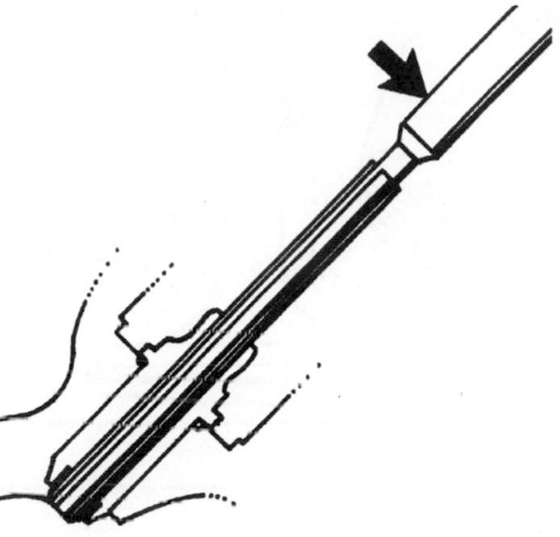

97

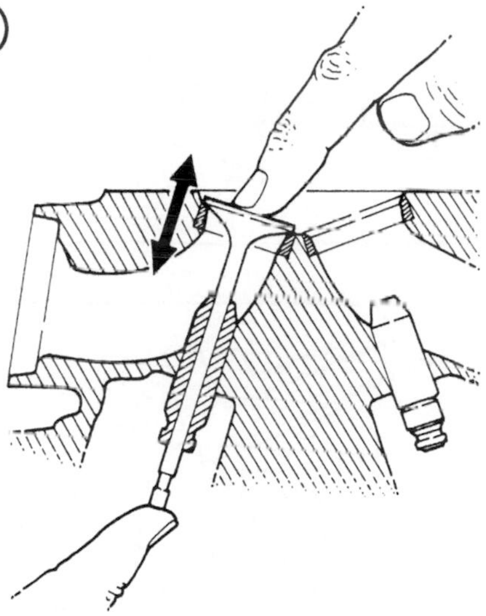

98

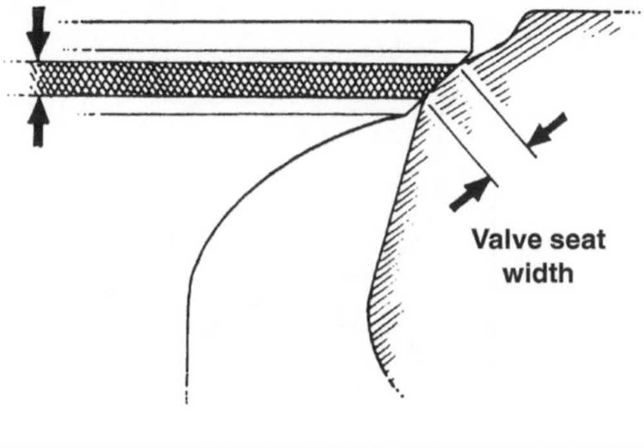

18. Reface the valve seats as described in this chapter.
19. Install the intake manifold. Use new O-rings, and torque the intake manifold bolts to 10 N•m (89 in.-lb.).
20. When servicing a rear cylinder head, install the exhaust manifold with a new exhaust gasket. Torque the exhaust manifold nuts to 20 N•m (15 ft.-lb.).

4

Valve Seat Inspection

1. Remove the valves as described in this chapter.
2. The most accurate means for checking the valve seal is to use marking compound, available from automotive parts stores or machine shops. Check the valve seal with marking compound as follows:
   a. Thoroughly clean all carbon deposits from the valve face with solvent or detergent. Completely dry the valve face.
   b. Spread a thin layer of marking compound evenly on the valve face.
   c. Insert the valve into its guide.
   d. Support the valve by hand (**Figure 97**), and tap the valve up and down in the cylinder head. Do not rotate the valve or the impression will not be accurate.
   e. Remove the valve and examine the impression left by the marking compound. If the impression left in the dye (on the valve or in the cylinder head) is not even and continuous, and if the valve seat width (**Figure 98**) is not within the tolerance specified in **Table 2**, the valve seat must be reconditioned.
3. Closely examine the valve seat in the cylinder head (**Figure 93**). It should be smooth and even, with a polished seating surface.
4. Measure the valve seat width (**Figure 98**) with a vernier caliper.
5. If the valve seat is within specification, install the valves as described in this section.
6. If the valve seat is not correct, recondition the valve seat in the cylinder head as described in this chapter.

Valve Seat Reconditioning

Special valve cutters and considerable expertise are required to properly recondition the valve seats in the cylinder head. You can save money by removing the cylinder head and taking it to a Yamaha dealership or machine shop to have the valve seats ground.

The following procedure is provided if you choose to perform this task yourself.

A valve seat cutter set (consisting of 30°, 45° and 60° cutters and the appropriate handle) is needed. These tool sets are available from a Yamaha dealership or from machine shop supply outlets. Follow the manufacturer's instruction when using the cutters.

The valve seats for both the intake valves and exhaust valves are machined to the same angles. The area below the contact surface (closest to the combustion chamber) is cut to a 30° angle (A, **Figure 99**). The valve contact surface is cut to a 45° angle (B, **Figure 99**). The area above the contact surface (closest to the valve guide) is cut to a 60° angle (C, **Figure 99**).

1. Using the 45° cutter, descale and clean the valve seat with one or two turns (**Figure 100**).

CAUTION

*Measure the valve seat contact area in the cylinder head (**Figure 101**) after each cut to make sure the contact area is correct and to prevent removing too much material. If too much material is removed, the cylinder head must be replaced.*

2. If the seat is still pitted or burned, turn the 45° cutter additional turns until the surface is clean. Avoid removing too much material from the cylinder head.

3. Remove the valve cutter and T-handle from the cylinder head.

4. Use marking compound to inspect the valve seat as described in *Valve Seat Inspection* in this section.

5. If the contact area is centered on the valve face but is too wide (**Figure 102**), use either the 30° or the 60° cutter and remove a portion of the valve seat material to narrow the contact area.

6. If the contact area is centered on the valve face but is too narrow (**Figure 103**), use the 45° cutter and remove a portion of the valve seat material to increase the contact area.

7. If the contact area is too narrow and up close to the valve head (**Figure 104**), first use the 30° cutter

99

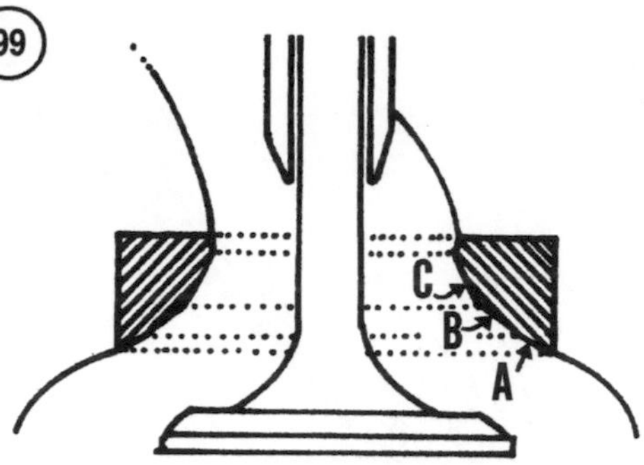

100

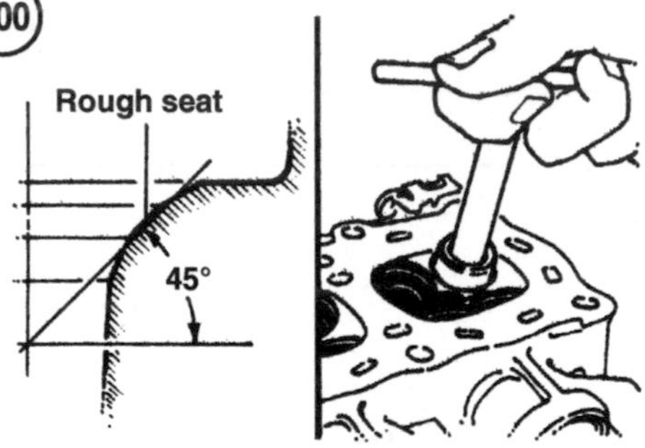

101

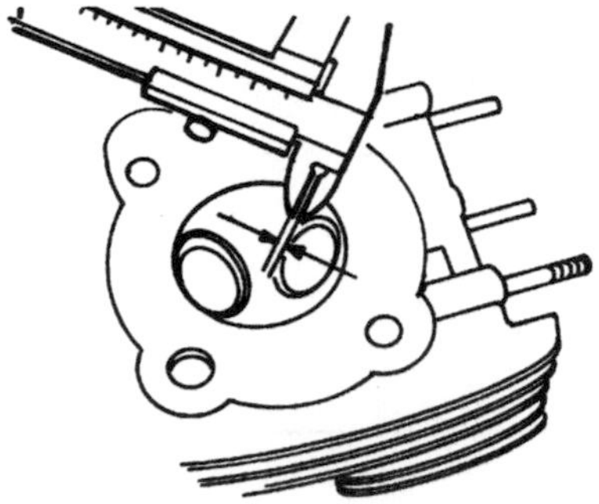

102

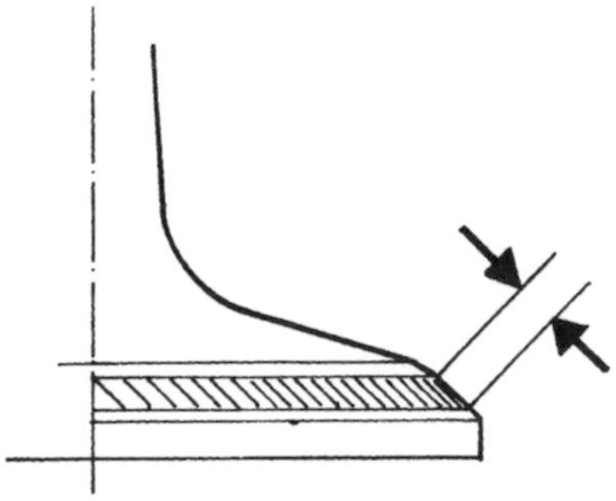

103

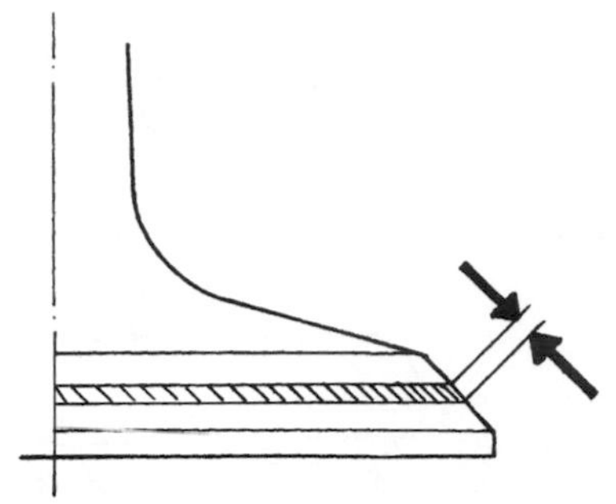

104

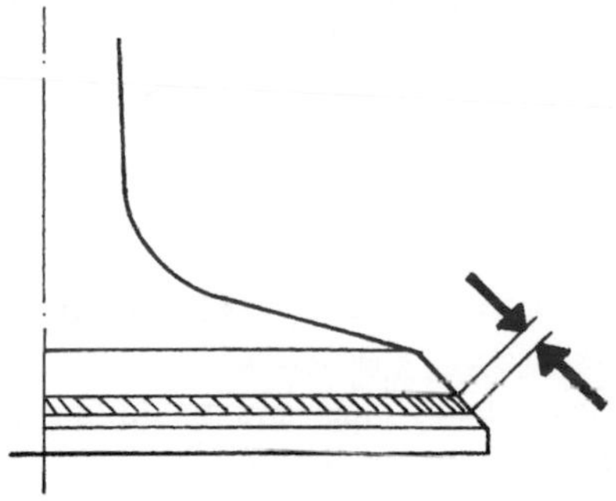

105

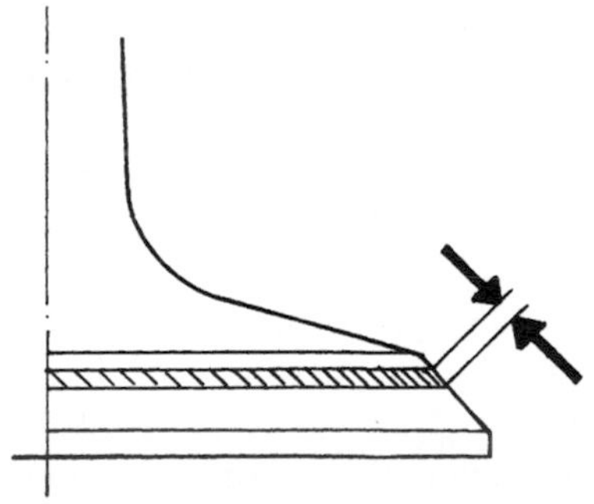

106

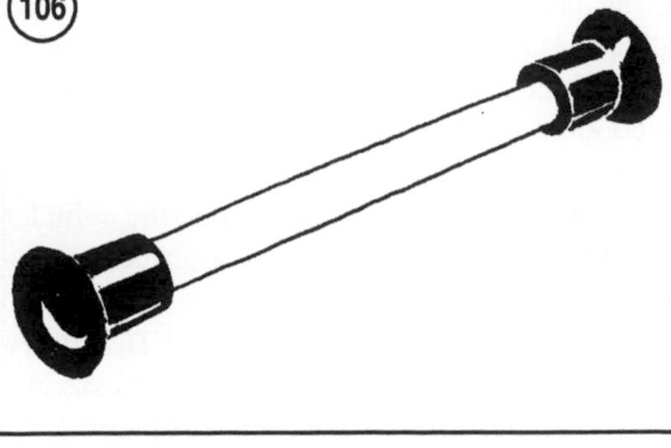

and then use the 45° cutter to center the contact area.

8. If the contact area is too narrow and down away from the valve head (**Figure 105**), first use the 60° cutter and then use the 45° cutter to center the contact area.

9. After the desired valve seat position and width is obtained, use the 45° cutter and T-handle and *very lightly* clean away any burrs that may have been caused by the previous cuts; remove only enough material as necessary.

10. Make sure the finish has a smooth and velvety surface, it should not be shiny or highly polished. The final seating will take place when the engine is first run.

11. Repeat Steps 1-10 for all remaining valve seats.

12. After the valve seat has been reconditioned, lap the seat and valve as described in this chapter.

Valve Seat Lapping

Valve lapping is a simple operation that can restore the valve seat without machining if the amount of wear or distortion is not too great. Lapping is also recommended after the valve seat has been refaced or when a new valve and valve guide have been installed.

1. Smear a light coat of fine grade valve lapping compound such as Carborendum or Clover Brand on the seating surface of the valve.

2. Apply molybdenum disulfide oil to the valve stem and insert the valve into the cylinder head.

3. Wet the suction cup of the valve lapping tool (**Figure 106**) and stick it onto the valve head.

4. Lap the valve to the valve seat (**Figure 107**) as follows:

   a. Lap the valve by rotating the lapping stick between your hands in both directions.

   b. Every 5 to 10 seconds, stop and rotate the valve 180° in the valve seat.

   c. Continue lapping until the contact surfaces of the valve and the valve seat in the cylinder head are a uniform gray. Stop as soon as they turn this color to avoid removing too much material.

5. Thoroughly clean the cylinder head and all valve components in solvent, followed by a wash with detergent and hot water.

6. After the lapping has been completed and the valve assemblies have been reinstalled into the cyl-

inder head, the valve seat should be tested. Check the seat by performing the leakage test described in the *Cylinder Head* section earlier in this chapter. If fluid leaks past any of the seats, disassemble that valve assembly and repeat the lapping procedure until there are no leaks.

7. After the cylinder head and valve components are cleaned in detergent and hot water, apply a light coat of engine oil to all bare metal surfaces to prevent rust.

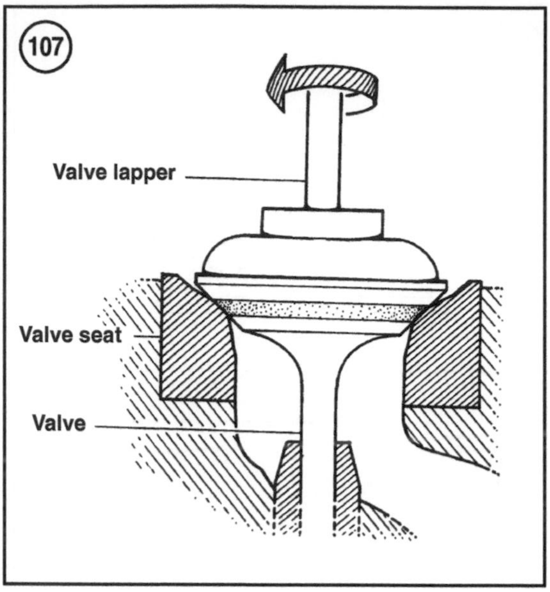

CYLINDER

Refer to **Figure 108**.

Removal

1. Remove the cylinder head and head gasket as described in this chapter.
2. If the front cam chain guide is still installed, remove it from the cylinder.
3. Remove the cylinder bolts (A and B, **Figure 109**) from the cam chain side of the cylinder. One bolt sits in a recess (B, **Figure 109**) in the top of the cylinder.
4. Loosen the cylinder by tapping around the perimeter with a rubber or plastic mallet.
5. Pull the cylinder straight up, and lift it off the pistons and cylinder studs.

NOTE
Be sure to keep the cam chain wired up to prevent it from falling into the crankcase.

6. Remove the two dowels (A, **Figure 110**) from the exhaust side of the cylinder.
7. Remove and discard the base gasket (B, **Figure 110**).
8. Stuff clean shop rags into the crankcase opening to prevent objects from falling into the crankcase. Also place a length of hose (**Figure 111**) over each crankcase stud the piston leans against so the rings will not be marred if the piston is accidentally struck.
9. Inspect the cylinder as described in this section.

Installation

1. Make sure the top and bottom cylinder surfaces are clean of all gasket residue.
2. Install the two dowels (A, **Figure 110**) and new base gasket (B).
3. Lubricate the cylinder and piston liberally with engine oil prior to installation.
4. Rotate the crankshaft so the piston is at top dead center.
5. Carefully install the cylinder onto the crankcase studs. Make sure the cam chain and rear cam chain guide pass through the cam chain tunnel in the cylinder.
6. Slowly lower the cylinder until the piston rings are within the cylinder sleeve (**Figure 112**). Compress each ring as it enters the cylinder with your fingers or with a piston ring compressor. Take the time to carefully compress each ring individually if necessary.
7. Run the cam chain and safety wire up through the cam chain tunnel in the cylinder, and secure the safety wire to the outside of the engine.
8. Carefully lower the cylinder all the way down onto the crankcase and install the cylinder bolts (**Figure 113**) into the cam chain side of the cylinder. Finger-tighten the bolts at this time. They will be torqued after the cylinder head hardware has been torqued.
9. Install the cylinder head and torque the cylinder head hardware to specification as described in this chapter.
11. Torque the cylinder bolts (A and B, **Figure 109**) to 10 N•m (89 in.-lb.). One of the bolts sits in a recess (B, **Figure 109**) in the cylinder block.

(108)

CYLINDER AND PISTON

1. Cylinder bolt
2. Cylinder
3. Compression ring
4 Oil ring
5. Piston pin circlip
6. Piston
7. Base gasket
8. Dowel
9. Front cam chain guide
10. Piston pin

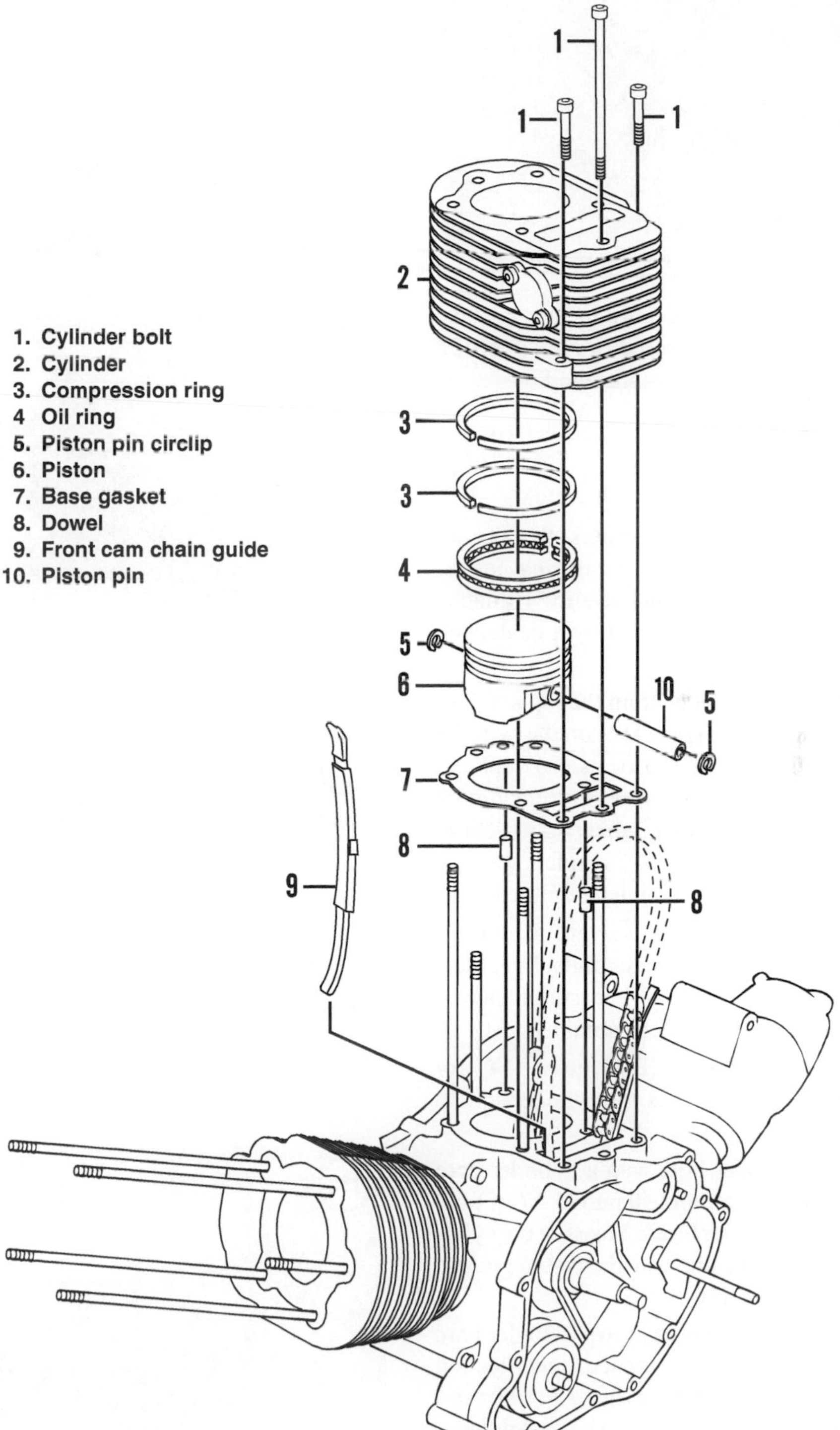

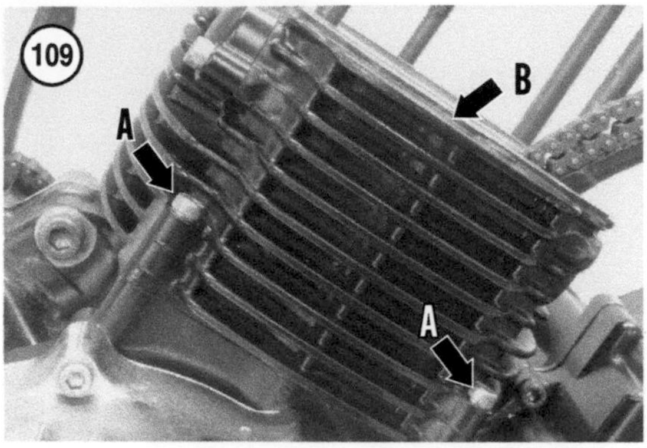

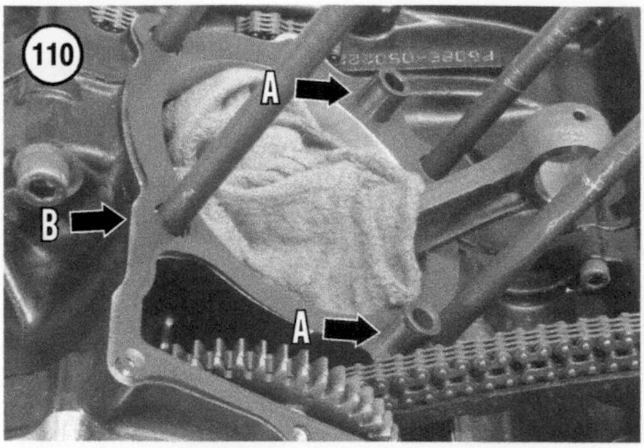

12. Follow the break-in procedure in Chapter Five if the cylinder block was rebored or honed, or if a new piston or piston rings were installed.

Inspection

The following procedure requires the use of highly specialized and costly measuring instruments. If such equipment is not readily available, have the measurements performed by a dealership or qualified machine shop.

1. Remove all gasket residue from the top and bottom gasket surfaces (A, **Figure 114**) on the cylinder block. Apply a gasket remover or use solvent and soak any old gasket material stuck to the cylinder block. If necessary, use a *dull*, broad-tipped chisel and gently scrape off all gasket residue. Do not gouge a sealing surface or an oil leak will result.
2. Wash the cylinder block in solvent to remove any oil and carbon particles. The cylinder bore must be cleaned thoroughly before any measurement otherwise incorrect readings may be obtained.
3. Check the cylinder wall (B, **Figure 114**) for scratches. If there are scratches, the cylinder should be rebored.

5. Measure the cylinder bore with a cylinder bore gauge (**Figure 115**) at a point 40 mm (1.57 in.) below the top of the cylinder block. Measure the cylinder bore in two axes: in-line with the piston pin and 90° to the pin (**Figure 116**).
6. Calculate the average of these two cylinder bore measurements. If this average is out of specification (**Table 2**), replace the cylinder, piston and rings as a set. Oversized pistons and rings are not available.
7. Calculate the piston-to-cylinder clearance as described in *Piston Clearance* later in this chapter.

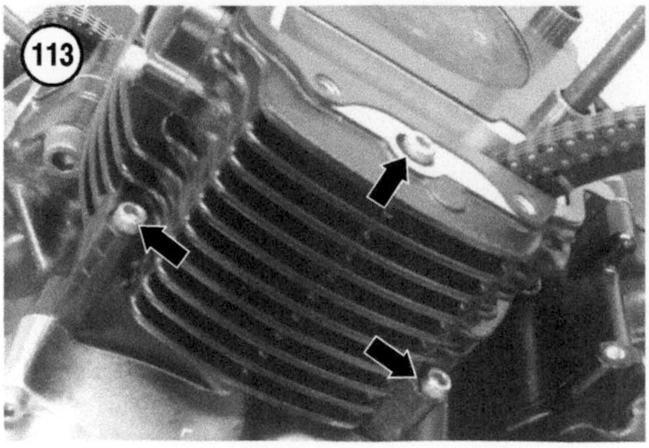

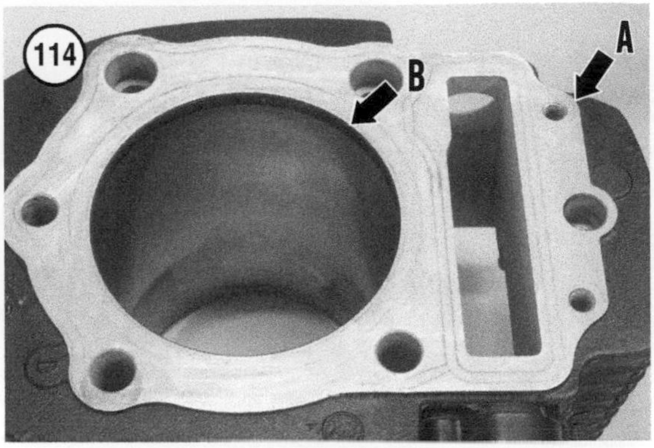

114

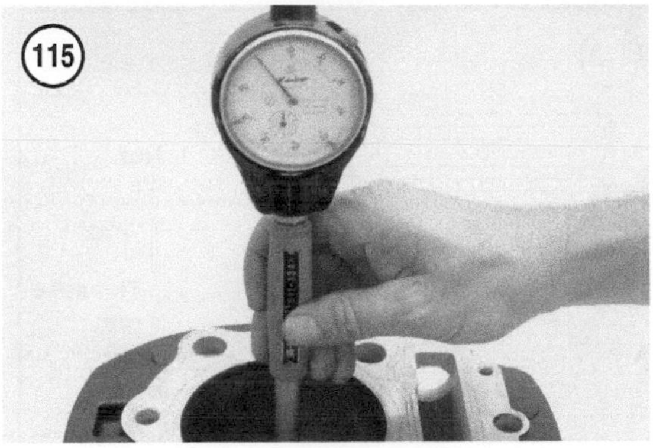

115

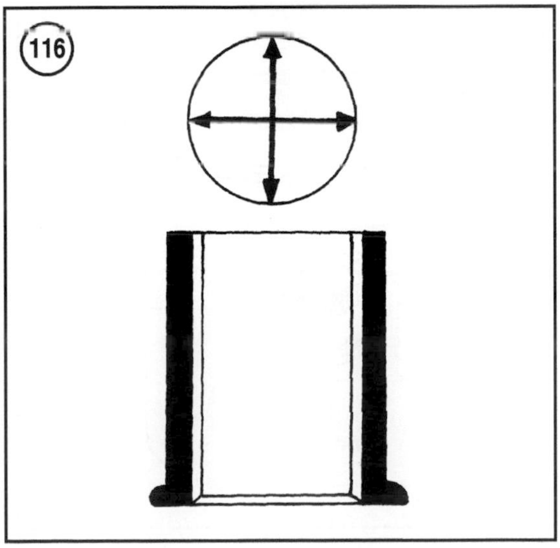

116

PISTONS AND PISTON RINGS

The pistons are made of an aluminum alloy. The piston pins are made of steel and are a precision fit. The piston pin is held in place by a clip at each end.

Refer to **Figure 108** when servicing the piston and rings.

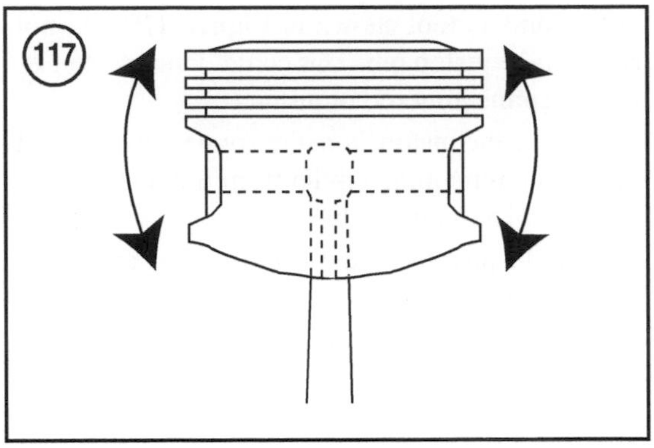

117

Piston Removal

1. Remove the cylinder head and cylinder as described in this chapter.
2. Lightly mark the top of the piston (front or rear) so it can be installed in the correct cylinder during installation.

WARNING
The edges of all piston rings are very sharp. Be careful when handling them to avoid cutting fingers.

3. Before removing the piston, hold the connecting rod tightly and rock the piston (**Figure 117**). Any rocking motion (do not confuse with the normal sliding motion) indicates wear on the piston pin, piston pin bore or connecting rod small-end bore (more likely a combination of these). If necessary, replace the piston and piston pin as a set.

NOTE
Wrap a clean shop cloth under the piston so the piston pin clip will not fall into the crankcase.

4. Remove a circlip (**Figure 118**) from one side of the piston pin bore with a small screwdriver or scribe. Hold your thumb over one edge of the clip when removing it to prevent the clip from springing out. Deburr the piston pin and the circlip groove as necessary.

NOTE
Discard the piston pin circlip. New circlips must be installed during assembly.

5. From the other side, push the piston pin out of the piston by hand. If the pin is tight, remove it with

4

the homemade tool shown in **Figure 119**. Do not drive out the piston pin. This could damage the piston pin, connecting rod or piston.

6. Remove the piston from the connecting rod and remove the remaining circlip from the piston. Discard both piston pin circlips.

7. Mark the piston pin and piston so they can be reassembled as a set.

8. If the piston is going to be left off for some time, place a piece of foam insulation tube over the end of the connecting rod to protect it.

118

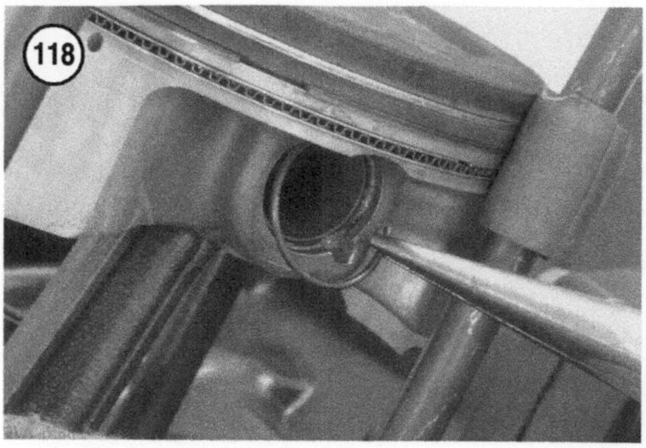

119

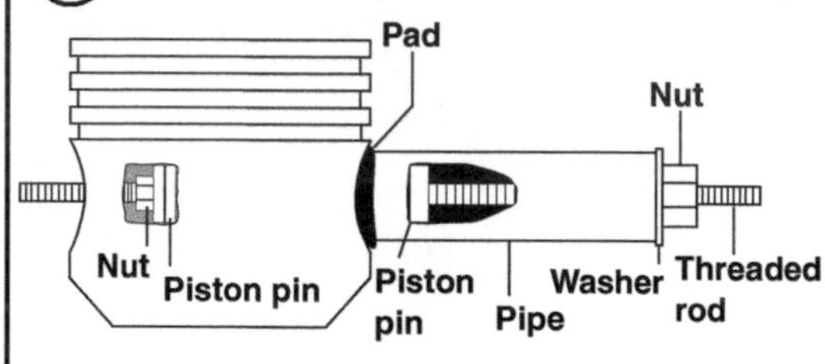

Piston Installation

1. Apply fresh engine oil to the inside surface of the connecting rods.

CAUTION

Install new piston pin circlips during assembly. Install the circlips with the gap away from the cutout in the piston.

2. Install a new piston pin circlip into one side of the piston. Make sure the circlip end gap does **not** align with the notch in the piston (**Figure 120**).

3. Apply fresh engine oil to the piston pin, and install the pin into the piston until its is flush with the inside of the piston pin boss (**Figure 121**).

4. Place the piston over the connecting rod (**Figure 111**). Make sure the EX on the piston crown (A, **Figure 122**) faces the exhaust side of the cylinder. This is the front side on the front cylinder and the rear side on the rear cylinder.

CAUTION

When installing the piston pin in Step 5, do not push the pin in too far. The piston pin circlip installed in Step 2 will be forced into the piston metal, destroying the clip groove and loosening the clip.

5. Line up the piston pin with the hole in the connecting rod. Push the piston pin through the connecting rod and into the piston boss on the other side of the piston. It may be necessary to move the piston until the piston pin enters the connecting rod. Do not use force during installation or damage may occur. Push the piston pin in until it bottoms against the pin clip on the other side of the piston.

120

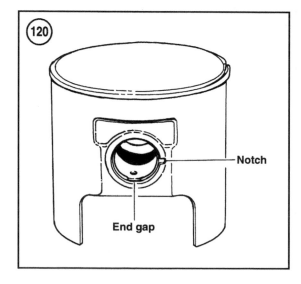

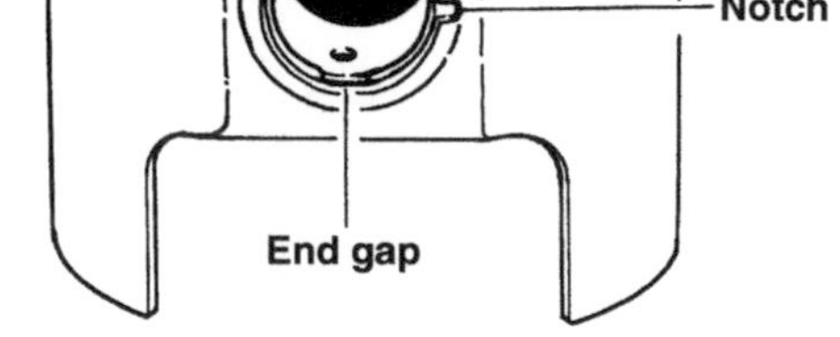

6. If the piston pin does not slide easily, use the homemade tool (**Figure 119**) used during removal but eliminate the piece of pipe. Pull the piston pin in until it stops.

7. After the piston is installed, recheck and make sure the EX on the piston crown faces the exhaust side of the cylinder.

(121)

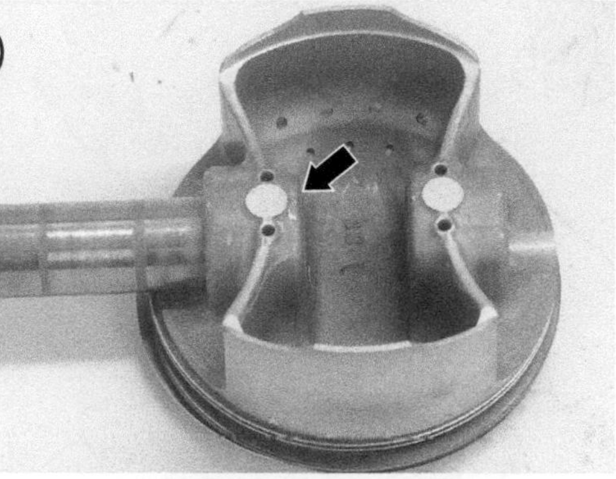

(122)

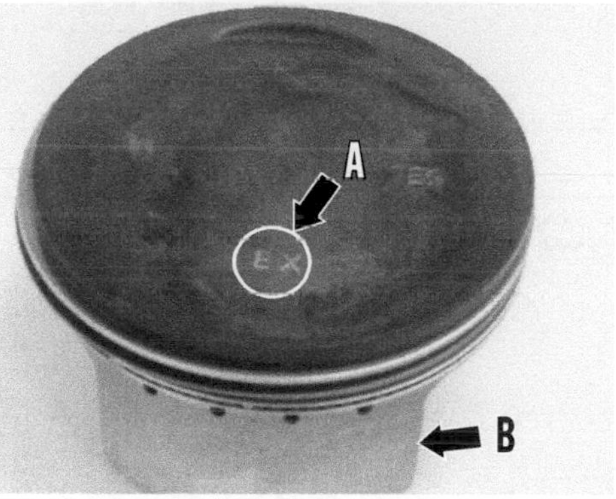

(123)

(124)

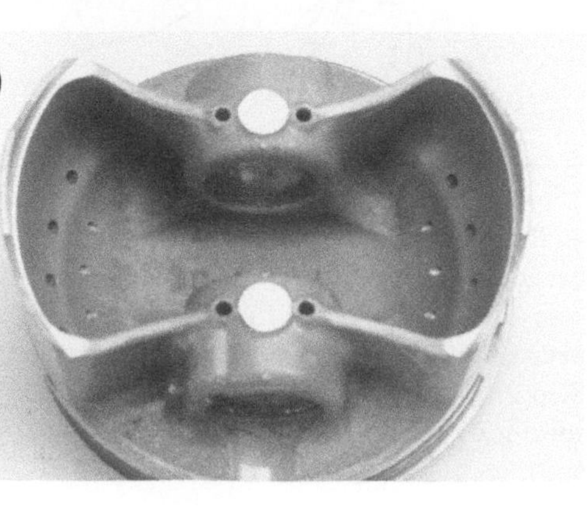

NOTE
In the next step, install the second circlip with the gap away from the cutout in the piston.

8. Install the second piston pin circlip (**Figure 118**) into the groove in the piston. Make sure the circlip's end gap does **not** align with the notch in the piston (**Figure 120**). Also, make sure both piston pin circlips are seated in their grooves in the piston.
9. Check the installation by rocking the piston back and forth around the pin axis and from side to side along the axis. It should rotate freely back and forth, but not from side to side.
10. If necessary, install the piston rings as described in this chapter.
12. Install the cylinder and cylinder head as described in this chapter.

Piston Inspection

1. Carefully clean the carbon from the piston crown (**Figure 123**) with a chemical remover or with a soft scraper. Re-mark the piston as soon as it is cleaned. Do not remove or damage the carbon ring around the circumference of the piston above the top ring. If the piston, rings and cylinder are dimensionally correct and can be reused, removal of the carbon ring from the top of the piston or removal of the carbon ridge from the top of the cylinder block wall will promote excessive oil consumption in this cylinder.

CAUTION
Do not use a wire brush on the piston skirts.

2. After cleaning the piston, examine the crown. It should show no signs of wear or damage.
3. Examine each ring groove for burrs, dented edges and wide wear. Pay particular attention to the top compression ring groove. It usually wears more than the other grooves. Since the oil rings are constantly bathed in oil, these rings and grooves wear little compared to compression rings and their grooves. If the oil ring groove shows signs of wear, or if the oil ring assembly is tight and difficult to remove, the piston skirt may have collapsed. If so, replace the piston.
4. Check the oil control holes (**Figure 124**) in the piston for carbon or oil sludge build-up. If neces-

4

sary, clean the holes and blow them out with compressed air.

5. Check the piston skirt (B, **Figure 122**) for galling and abrasion, which may have been caused by piston seizure. If a piston shows signs of partial seizure (bits of aluminum build-up on the piston skirt), replace the piston to reduce the possibility of engine noise and further piston seizure. When replacing the piston, lightly hone the cylinder with a bottlebrush hone.

NOTE

If the piston skirt is worn or scuffed unevenly from side to side, the connecting rod may be bent or twisted.

6. Check the circlip groove on each side of the piston for wear or other damage. Install a new circlip into each piston circlip groove and try to move the clip from side to side. If the circlip has any side play, the groove is worn and the piston must be replaced.

7. Measure the outside diameter of the piston across the skirt at right angles to the piston pin. Measure the piston at a point 5 mm (0.20 in.) up from the bottom of the piston skirt (**Figure 125**). If the piston diameter is out of specification, replace the piston and rings as a set. When installing the new piston, lightly hone the cylinder with a bottlebrush hone.

8. Measure the piston-to-cylinder clearance as described in *Piston Clearance* in this section. If clearance is out of specification (**Table 2**), replace the cylinder, piston and rings as a set. Oversized pistons are not available.

Piston Clearance

1. Make sure the pistons and cylinder walls are clean and dry.

2. Measure the cylinder bore with a cylinder bore gauge (**Figure 115**) at a point 40 mm (1.57 in.) below the top of the cylinder block. Measure the cylinder bore in two axes: in-line with the piston pin and 90° to the pin. Calculate the average of the two measurements. This average is the cylinder bore diameter.

3. Measure the outside diameter of the piston across the skirt at right angles to the piston pin. Measure the piston at a point 5 mm (0.20 in.) up from the bottom of the piston skirt (**Figure 125**).

125

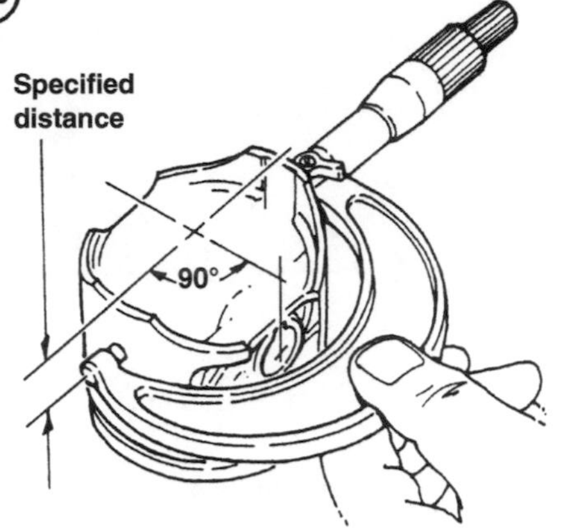

126

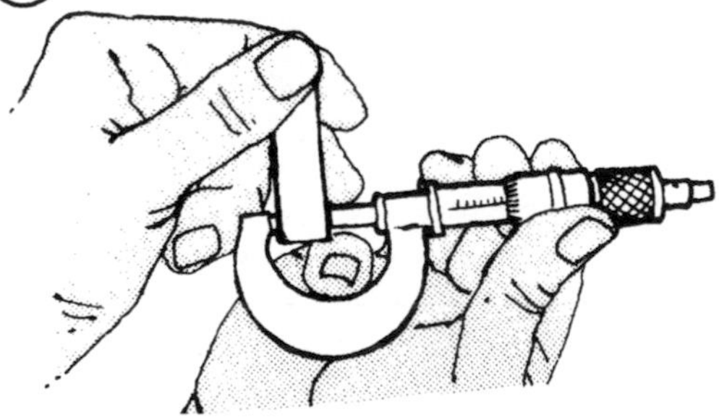

127

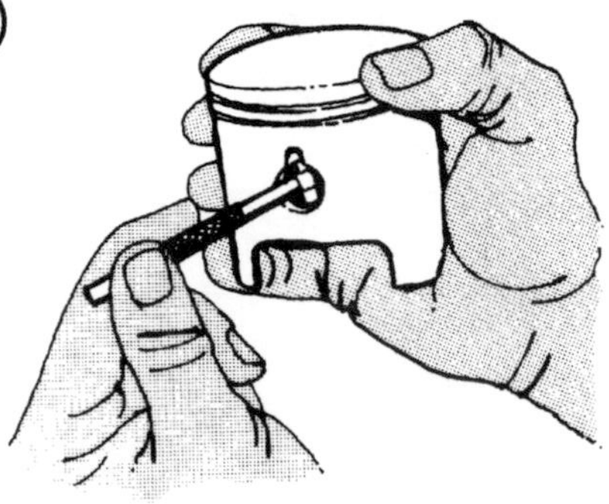

4. Piston-to-cylinder clearance is the difference between the piston outside diameter and the cylinder bore diameter. Subtract the outside diameter of the piston from the cylinder bore diameter calculated in Step 2. If the piston-to-cylinder clearance is out of specification (**Table 2**), replace the cylinder, piston and rings as a set. Oversized pistons are not available.

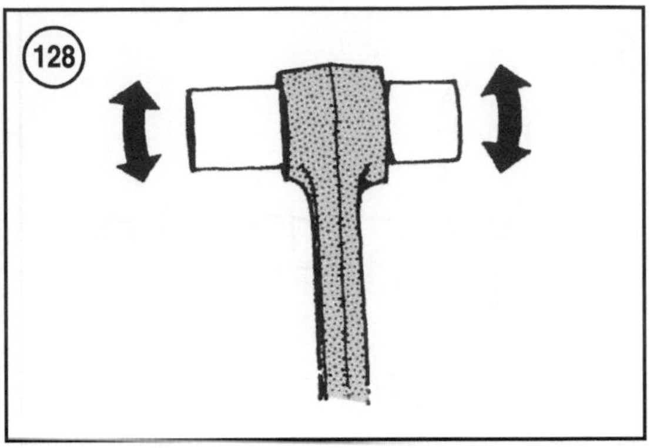

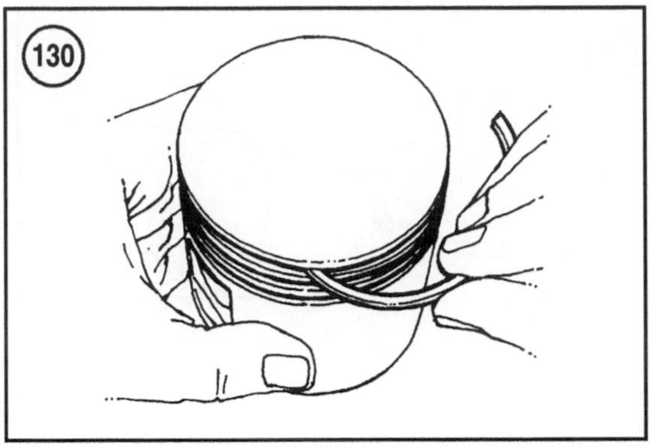

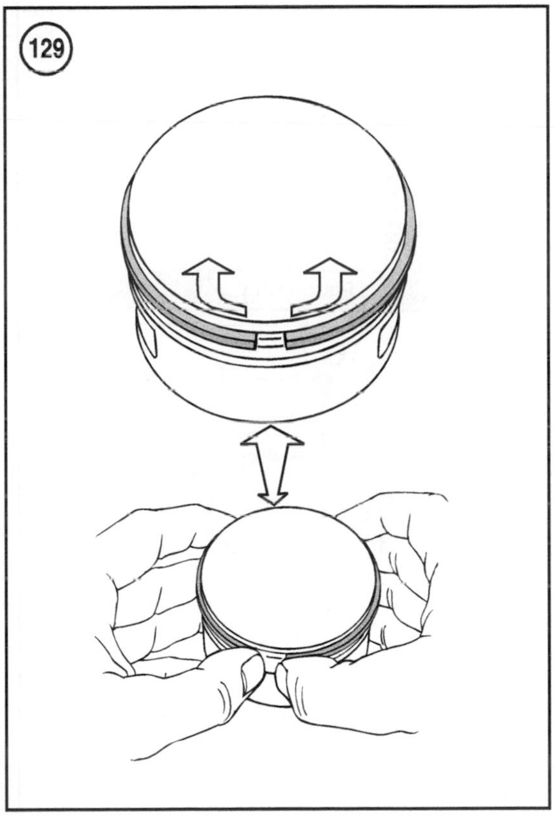

Piston Pin Inspection

1. Clean the piston pin in solvent and dry it thoroughly.

2. Inspect the piston pin for chrome flaking or cracks. Replace the pin if necessary.

3. Measure the outside diameter of the piston pin with a micrometer (**Figure 126**). If the measurement is outside the range specified in **Table 2**, replace the piston pin.

4. Measure the inside diameter of the piston pin bore in the piston with a small bore gauge (**Figure 127**).

5. Calculate piston pin-to-piston clearance by subtracting the piston pin outside diameter from the piston pin bore inside diameter. If the clearance is outside the range specified in **Table 2**, replace the piston pin (if the piston bore inside diameter is within specification).

6. Oil the piston pin and install it in the connecting rod. Slowly rotate the piston pin, and check for radial and lateral play (**Figure 128**). If there is play, the connecting rod should be replaced (if the piston pin outside diameter is within specification). Inspect the connecting rod as described in Chapter Five.

Piston Ring Removal and Inspection

A three-ring assembly is used with each piston. The top and second rings are compression rings. The lower ring is an oil control ring assembly (consisting of two ring rails and an expander spacer).

WARNING
The edges of all piston rings are very sharp. Be careful when handling them to avoid cutting fingers.

1. Remove the compression rings with a ring expander tool or by spreading the ends with your thumbs just enough to slide the ring up over the piston (**Figure 129**). Repeat for the remaining rings.

2. Carefully remove all carbon build-up from the ring grooves with a broken piston ring (**Figure 130**). Do not remove aluminum material from the ring grooves. This will increase ring side clearance.

131

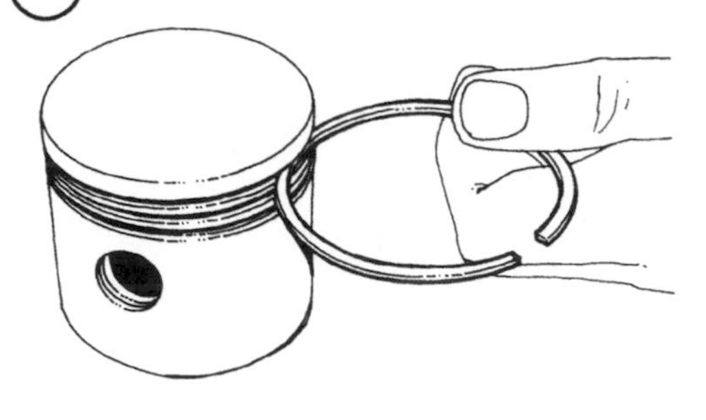

132

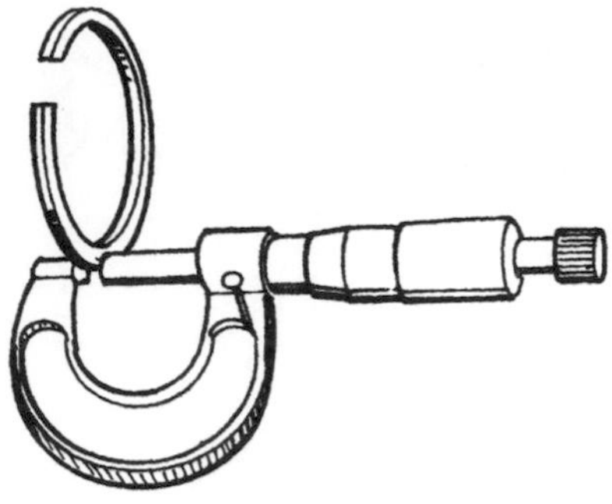

3. Inspect the grooves carefully for burrs, nicks or broken and cracked lands. Recondition or replace the piston if necessary.

4. Roll each ring around its piston groove as shown in **Figure 131** to check for binding. Minor binding may be cleaned up with a fine-cut file.

5. Measure the thickness of each ring with a micrometer (**Figure 132**). If the thickness is less than the value specified in **Table 2**, replace the ring(s).

NOTE

When checking the oil control ring assembly, just measure the end gap of each ring rail. The end gap of the expander spacer cannot be measured. If either ring rail has excessive end gap, replace the entire oil ring assembly.

6. Place each ring, one at a time, into the cylinder, and push the ring to a point 40 mm (1.57 in.) below the top of the cylinder. Push the ring with the crown of the piston to ensure the ring is square in the cylinder bore. Measure the ring end gap with a flat feeler gauge (**Figure 133**). If the gap is out of specification (**Table 2**), replace the rings.

7. Install the piston rings as described below and measure the side clearance of each ring in its groove with a flat feeler gauge (**Figure 134**). If the clearance is greater than specified, replace the piston and rings as a set.

8. When installing new rings, measure their end gaps as described in Step 5, and compare the measurements to the dimensions given in **Table 2**. If the end gap is greater than specified, return the rings for another set(s). If the end gap is smaller than specified, secure a small file in a vise, grip the ring with your fingers, and enlarge the gap (**Figure 135**).

133

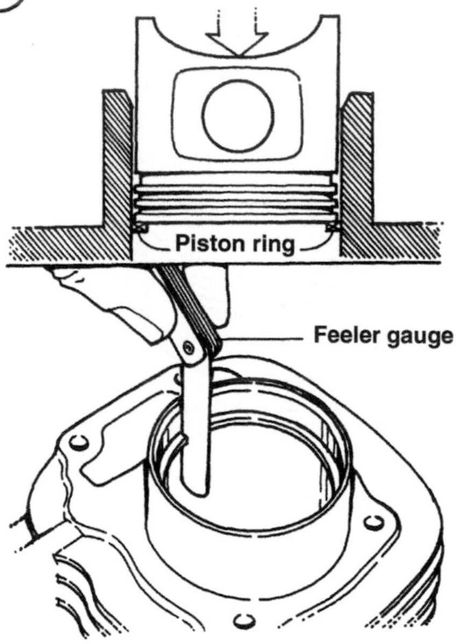

134

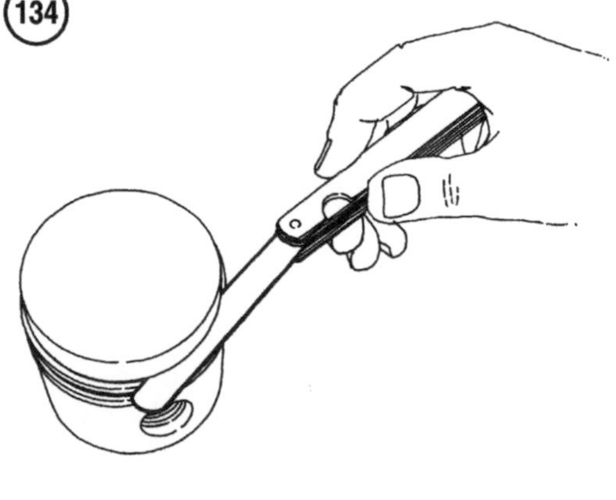

135

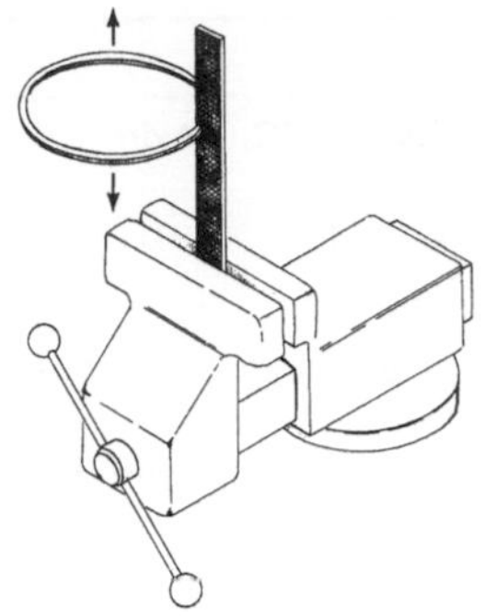

136

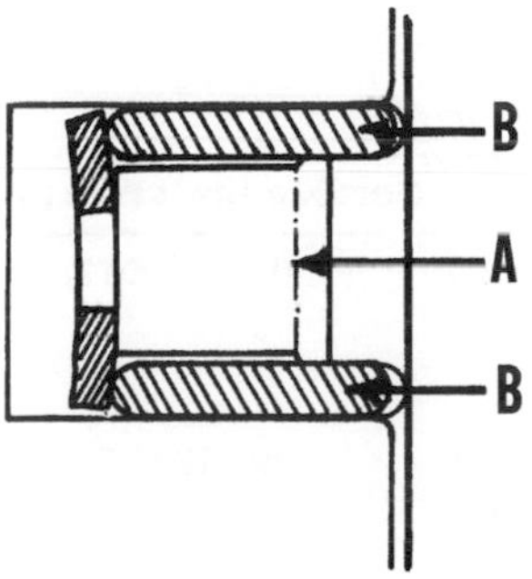

137

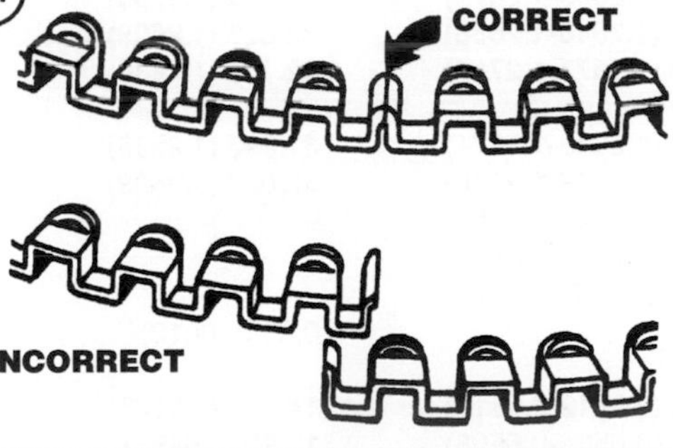

138

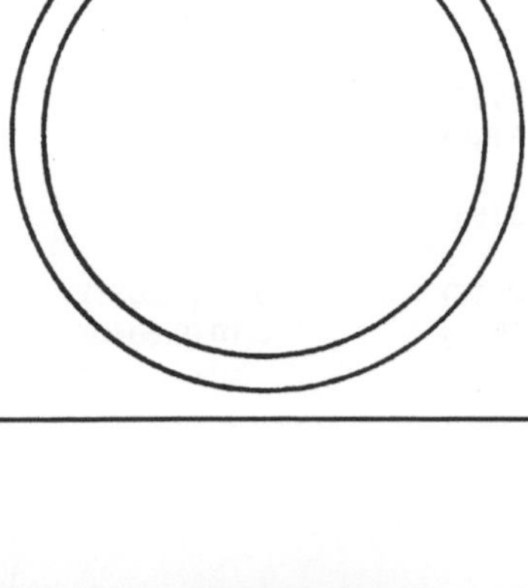

139

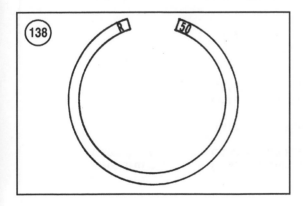

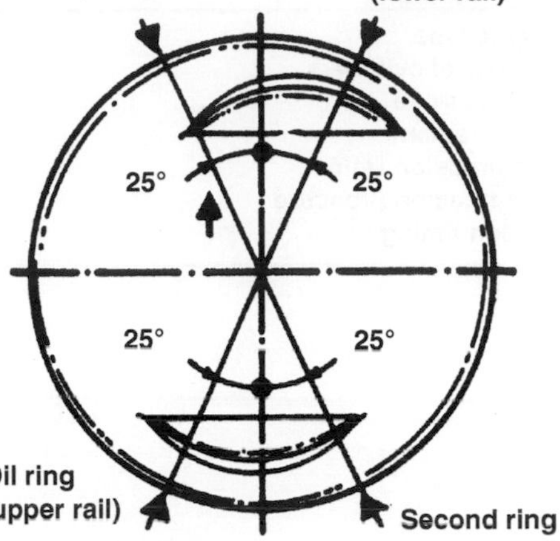

Piston Ring Installation

NOTE
When installing any ring, liberally lubricate the ring and piston groove with clean engine oil.

1. Install the oil control ring assembly into the bottom ring groove. Install the oil ring expander spacer first (A, **Figure 136**), then install each ring rail (B). Make sure the ends of the expander spacer butt together (**Figure 137**). They should not overlap. When reassembling used parts, install the ring rails as they were removed.

2. Install the second compression ring, then install the top ring. Carefully spread the ends of each ring with your thumbs and slip the ring over the top of the piston (**Figure 129**). Install each compression ring with its manufacturing marks (**Figure 138**) facing up.

3. Make sure the rings are seated completely in their grooves all the way around the piston and the ends are distributed around the piston.

4. Check the side clearance of each ring as shown in **Figure 134**. If the side clearance is not within the specification shown in **Table 2**, re-examine the condition of the piston and rings.

5. Distribute the ring gaps around the piston as shown in **Figure 139**.

6. Follow the break-in procedure in Chapter Five if a new piston or new piston rings have been installed, or if the cylinder was honed.

Table 1 GENERAL ENGINE SPECIFICATIONS

Item	Specification
Engine type	Four-stroke, air-cooled, SOHC, V-twin
Number of cylinders	2
Bore × stroke	95 × 75 mm (3.74 × 2.95 in.)
Displacement	1063 cc (64.87 cu. in.)
Compression ratio	8.3 : 1
Compression pressure	1000 kPa (142 psi) @ 400 rpm
Ignition timing	10° B.T.D.C. @ 1000 rpm

Table 2 ENGINE TOP END SPECIFICATIONS

Item	New mm (in.)	Service limit mm (in.)
Cylinder head warp	–	0.03 (0.0012)
Camshaft		
Cam lobe height		
All models except 2001-on XVS1100		
Intake	39.112-39.212 (1.5398-1.5438)	39.012 (1.5359)
Exhaust	39.145-39.245 (1.5411-1.5451)	39.045 (1.5372)
2001-on XVZ1100		
Intake and exhaust	39.112-39.212 (1.5398-1.5438)	39.012 (1.5359)
Cam lobe width		
All models except 2001-on XVS1100		
Intake		
No. 1	32.093-32.193 (1.2635-1.2674)	31.993 (1.2596)
No. 2	32.127-32.227 (1.2648-1.2688)	32.027 (1.2609)
Exhaust	32.200-32.300 (1.2677-1.2717)	32.100 (1.2638)
2001-on XVS1100		
Intake	32.093-32.193 (1.2635-1.2674)	31.993 (1.2596)
Exhaust	32.127-32.227 (1.2648-1.2688)	32.027 (1.2609)
Camshaft bushing inside diameter	25.000-25.021 (0.9843-0.9851)	–
Camshaft journal outside diameter	24.96-24.98 (0.9827-0.9835)	–
Camshaft-to-bushing clearance	0.020-0.061 (0.0008-0.0024)	–
Camshaft runout	–	0.03 (0.0012)
Rocker arm		
Rocker arm bore inside diameter	14.000-14.018 (0.5512-0.5519)	14.036 (0.5526)
Rocker arm shaft outside diameter	13.985-13.991 (0.5506-0.5508)	13.95 (0.5492)
Rocker arm-to-rocker shaft clearance	0.009-0.033 (0.00035-0.0013)	0.086 (0.0034)
Valves and valve springs		
Valve clearance		
Intake	0.07-0.12 (0.0028-0.0047)	–
Exhaust	0.12-0.17 (0.0047-0.0067)	–
Valve stem runout	–	0.03 (0.0012)
Valve stem outside diameter		
Intake	7.975-7.990 (0.3140-0.3146)	–
Exhaust	7.960-7.975 (0.3134-0.3140)	–
Valve guide inside diameter (intake and exhaust)	8.000-8.012 (0.3150-0.3154)	–
Valve stem-to-guide clearance		
Intake	0.010-0.037 (0.0004-0.0015)	0.08 (0.0031)
Exhaust	0.025-0.052 (0.0010-0.0020)	0.10 (0.0040)

(continued)

Table 2 ENGINE TOP END SPECIFICATIONS (continued)

Item	New mm (in.)	Service limit mm (in.)
Valve head diameter		
Intake	47.0-47.2 (1.850-1.858)	–
Exhaust	39.0-39.2 (1.535-1.543)	–
Valve face width (intake and exhaust)	2.1 (0.083)	–
Valve seat width (intake and exhaust)	1.2-1.4 (0.047-0.055)	1.8 (0.071)
Valve margin thickness (intake and exhaust)	1.1-1.5 (0.043-0.060)	0.8 (0.031)
Valve seat cutting angle	30°, 45°, 60°	
Valve spring free length	44.6 (1.76)	43.5 (1.71)
Valve spring tilt	2.5°/1.9 (0.075)	
Cylinder		
Bore[1]	95.00-95.01 (3.7402-3.7405)	95.1 (3.7441)
Pistons		
Outside diameter[2]	94.960-94.975 (3.7386-3.7392)	–
Piston-to-cylinder clearance	0.025-0.050 (0.0010-0.0020)	0.15 (0.0060)
Piston off-set	0	–
Piston pin bore inside diameter	22.004-22.015 (0.8663-0.8667)	–
Piston pin outside diameter	21.991-22.000 (0.8658-0.8661)	–
Piston pin-to-piston clearance	0.004-0.024 (0.0001-0.0009)	–
Piston rings		
Side clearance (ring-to groove clearance)		
Top	0.04-0.08 (0.0015-0.0031)	0.1 (0.004)
Second	0.03-0.07 (0.0012-0.0027)	0.1 (0.004)
Ring thickness		
Top	1.5 (0.059)	–
Second	1.2 (0.047)	–
Oil ring	2.5 (0.098)	–
Ring end gap		
Top	0.30-0.50 (0.012-0.020)	0.8 (0.031)
Second	0.30-0.45 (0.012-0.018)	0.8 (0.031)
Oil ring	0.2-0.7 (0.008-0.028)	–

1. Measured 40 mm (1.54 in.) from top of cylinder.
2. Measured 5 mm (0.197 in.) from bottom of piston skirt.

Table 3 ENGINE TOP END TORQUE SPECIFICATIONS

Item	N•m	in.-lb.	ft.-lb.
Cam sprocket bolts	55	–	41
Cam sprocket cover bolts	10	89	–
Cam chain drive assembly retainer bolt	10	89	–
Cam chain guide bolt	10	89	–
Cam chain tensioner cap bolt	8	71	–
Cam chain tensioner mounting bolts	10	89	–
Camshaft bushing retainer bolt	20	–	15
Cylinder bolts	10	89	–
Cylinder head cap nut	35	–	26
Cylinder head cover screws	4	35	–
Cylinder head bolts	20	–	15
Cylinder head nuts	50	–	37
Exhaust manifold-to-cylinder head nuts (rear cylinder)	20	–	15
Exhaust pipe-to-cylinder head nuts (front cylinder)	20	–	15

(continued)

Table 3 ENGINE TOP END TORQUE SPECIFICATIONS (continued)

Item	N•m	in.-lb.	ft.-lb.
Exhaust pipe-to-manifold bolts (rear cylinder)	20	–	15
Flywheel nut	175	–	129
Intake manifold bolts	10	89	–
Oil delivery pipe banjo bolt	20	–	15
Primary drive nut	110	–	81
Rear chain guide bolt	10	89	–
Rocker arm bolts			
Exhaust	37	–	27
Intake	20	–	15
Spark plugs	20	–	15
Timing sprocket shaft retainer bolt	10	89	–
Valve adjuster locknut	27	–	20
Valve cover bolts	10	89	–

CHAPTER FIVE

ENGINE LOWER END

This chapter describes the service procedures for the following lower end components:

1. Crankcase assembly.
2. Crankshaft.
3. Connecting rods.
4. Flywheel and starter clutch.
5. Oil pump.
6. Transmission shaft assemblies (removal and installation only).

When inspecting lower end components, compare measurements to the specification in **Table 1** at the end of this chapter. Replace any part that is worn, damaged or out of specification. During assembly, torque fasteners to the given specification.

The text makes frequent references to the left and right side of the engine. This refers to the engine as it sits in the frame not how it may sit on the workbench.

Refer to *Basic Service Methods* in Chapter One.

SERVICING THE ENGINE IN THE FRAME

The following components can be serviced while the engine is in the frame:

1. External gearshift mechanism.
2. Clutch.
3. Carburetors.
4. Starter.
5. Alternator and starter clutch.
6. Oil pump.

ENGINE

CAUTION

*Examine the position of the cam timing marks for each cylinder before removing the engine. When a cylinder is set to TDC on the compression stroke, the timing mark on the cam sprocket (rear cylinder) or cam sprocket plate (front cylinder) may not precisely align with the pointer on the cylinder head. Make a drawing of each cam sprocket (or cam sprocket plate) and cylinder head pointer so the camshafts can be correctly timed during assembly. Refer to **Cylinder Head Removal** in Chapter Four for checking the cam timing.*

1

ENGINE REMOVAL

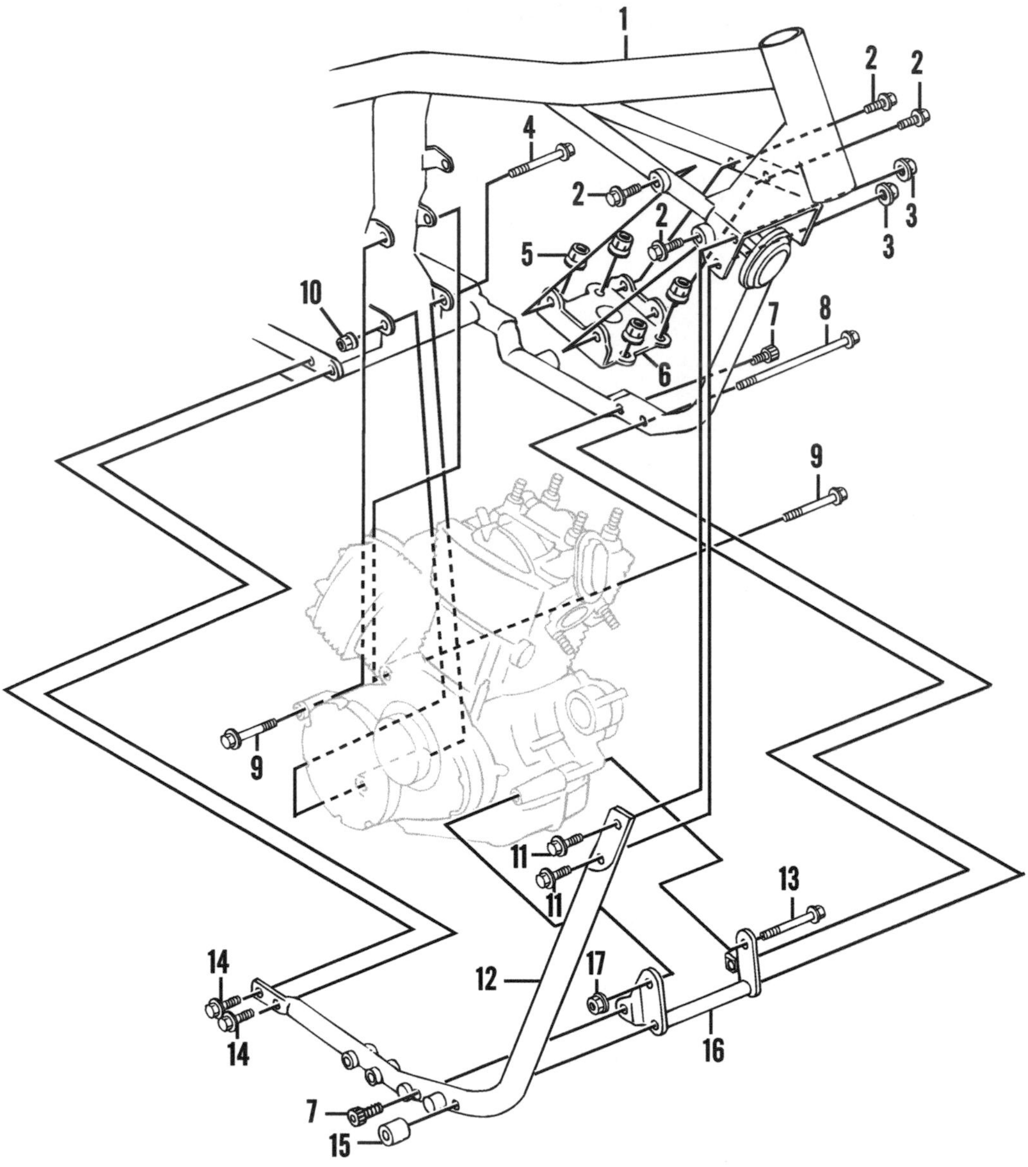

1. Frame
2. Cylinder head bracket bolt
3. Front frame member nut
4. Rear engine through bolt
5. Cylinder head bracket nut
6. Cylinder head bracket
7. Engine bracket bolt
8. Engine bracket lower through bolt
9. Upper rear engine mounting bolt
10. Engine mounting nut
11. Front frame member bolt
12. Removable frame member
13. Engine bracket upper through bolt
14. Rear frame member bolt
15. Engine bracket lower nut
16. Engine bracket
17. Engine bracket upper nut

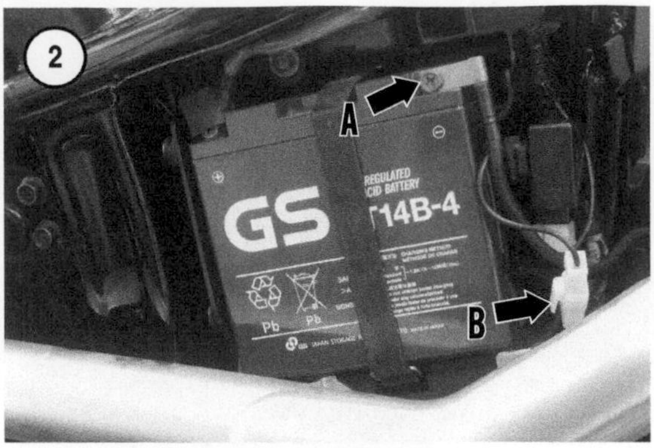

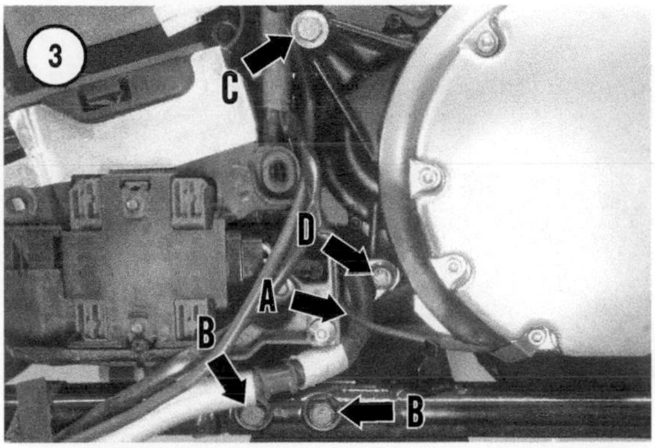

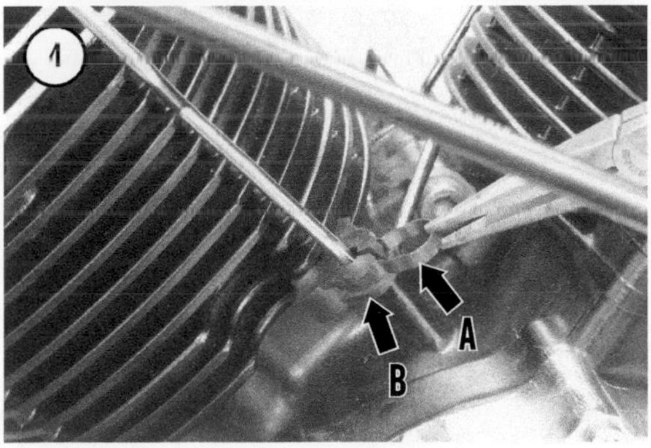

Removal

CAUTION
Yamaha recommends removing the alternator cover before removing the engine from the frame. Though this is necessary, it also exposes the alternator cover mating surface on the crankcase, which could be damaged during engine removal. If this surface is marred and leaks oil, the crankcases will have to be replaced. Use an old coolant hose to protect this mating surface as described in Step 15.

Refer to **Figure 1**.

1. Securely support the motorcycle in an upright position with the rear wheel off the ground.
2. Drain the engine oil and remove the oil filter as described in Chapter Three.
3. Remove the battery cover (Chapter Fourteen). Disconnect the negative lead (A, **Figure 2**) from the battery and disconnect the ground connector (B) from its harness mate.

CAUTION
Stuff clean shops rags into the intake manifold openings to prevent the entry of foreign matter into the cylinder head.

4. Refer to Chapter Eight and remove the fuel tank, air filter housing, surge tank, carburetor assembly and exhaust system.
5. Remove the brake pedal/footrest assembly (Chapter Thirteen).
6. Remove the right and left side covers, toolbox panel, frame neck covers and the sidestand as described in Chapter Fourteen.
7. Follow the oil level switch wire (A, **Figure 3**) and disconnect the bullet connector from its harness mate.
8. Remove the starter, speed sensor and horn (Chapter Nine).
9. Remove the shift pedal/footrest assembly (Chapter Seven).

NOTE
Before removing the spark plug caps, twist the caps from side to side to break the mating seal. Also label each spark plug wire if the original labels are no longer in place.

10. Disconnect the spark plug cap from each spark plug.
11. Remove the cylinder head covers from each cylinder head (Chapter Four).
12. Remove the external oil pipes, so they will not be damaged during engine removal, as follows:
    a. Pull the clip (A, **Figure 4**) from the oil pipe damper (B) on the front cylinder. Remove the damper from the cooling fins.

NOTE
*Two types of rocker arm bolts are used in the cylinder head: one with a single oil hole (A, **Figure 5**) and one with two oil holes (B). Note that the bolt with two oil holes (B, **Figure 5**) secures the oil pipe to the cylinder head.*

b. Remove the rocker arm bolt (**Figure 6**) that secures the oil pipe to the intake side of each cylinder head. Discard the two copper washers installed with each bolt. New washers must be used during assembly.
c. Remove the oil pipe banjo bolt (**Figure 7**) and the three copper washers from the crankcase. Discard the three copper washers.
d. Remove the oil pipes from the engine.

13. Remove the alternator cover as described in this chapter.
14. Place a hydraulic jack under the crankcase to support the engine once the mounting bolts are removed (**Figure 8**).
15. Split a few lengths of coolant hose lengthwise. Install the hoses over the left frame member (**Figure 9**) so the alternator cover mating surface on the crankcase will not be damaged.
16. Loosen each of the four cylinder head bracket nuts (A, **Figure 10**).
17. Remove the front engine bracket as follows:
a. Hold the nuts (A, **Figure 11** and A, **Figure 12**) and remove the upper through bolt (B, **Figure 11**) and lower through bolt (B, **Figure 12**).
b. Remove the engine bracket bolt from each side (C, **Figure 11** and C, **Figure 12**) and lower the engine bracket from the frame. The engine bracket bolts thread into weld nuts on the bracket.

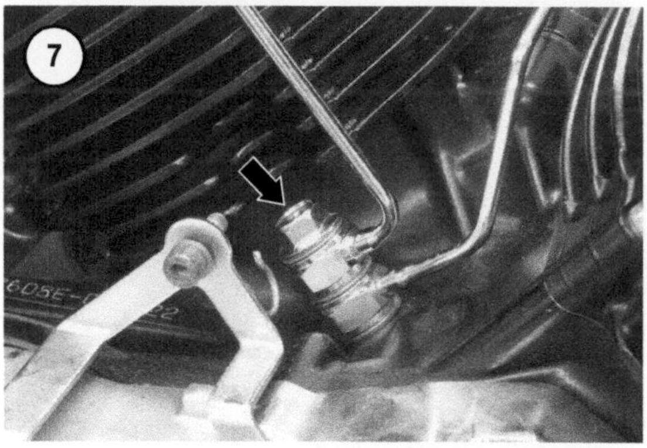

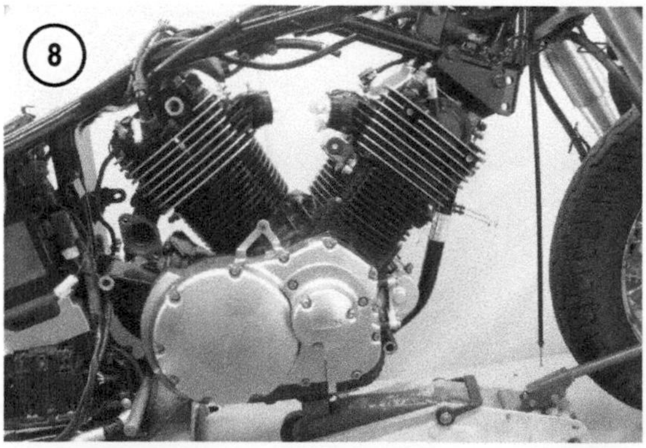

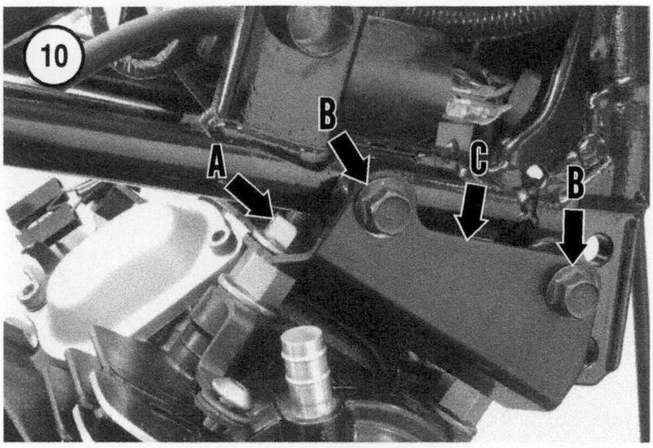

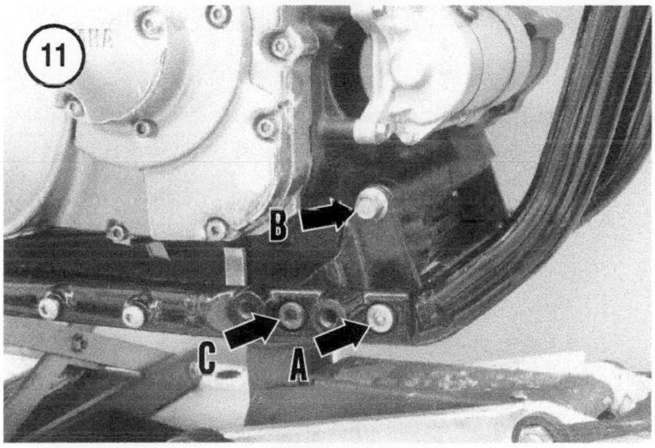

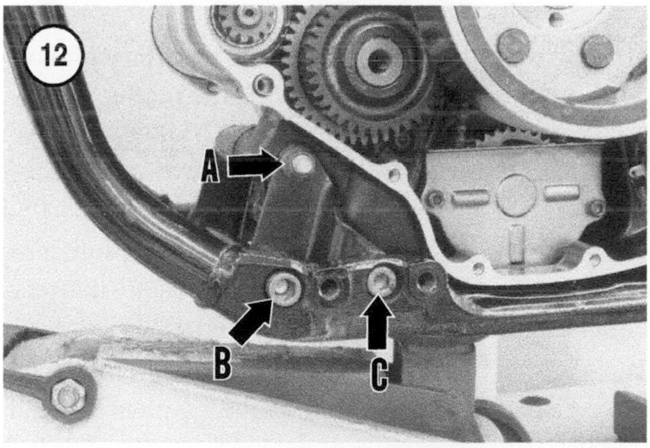

18. Remove the removable frame member as follows:
   a. Remove the nuts from the upper frame member bolts (**Figure 13**) and remove the bolts.
   b. Remove the lower frame member bolts (B, **Figure 3**) and lift the removable frame member from the frame.

19. Remove the rear upper engine mounting bolt (C, **Figure 3**) from the right side and remove the nut (D, **Figure 3**) from the through bolt.

20. Remove the rear upper engine mounting bolt (A, **Figure 14**) and pull the through bolt (B) from the right side.

21. Remove the engine bracket bolts (C and D, **Figure 14**). Lower the upper and lower engine brackets from the frame.

22. Make sure all cables, wires and hoses are disconnected from the engine and safely moved out of the way.

23. Remove the cylinder head bracket bolts (B, **Figure 10**) from each side of the frame.

24. Lower the engine slightly and tilt it to the right.

25. Turn the cylinder head nuts (A, **Figure 10**) off the studs and remove the cylinder head bracket (C) from the head studs. The engine cannot be removed with this bracket in place.

WARNING

The engine is very heavy and has many sharp edges. It may shift or drop suddenly once the mounting bolts are removed. Never place your hands or any other part of your body where the engine could drop and crush your hands or arms. One or more assistants will be required to remove the engine from the frame. Do not attempt engine removal by yourself.

26. Pull back the universal joint boot. Roll the engine forward and disengage the universal joint from the driveshaft.
27. With an assistant, lift the engine off the jack and remove it from the right side of the frame.
28. While the engine is removed for service, check all of the frame engine mounts for cracks or other damage. If any cracks are detected, take the chassis assembly to a Yamaha dealership for further examination.

Installation

1. If they are not already installed, secure lengths of coolant hose onto the left frame member (**Figure 15**). The hoses help protect the crankcase mating surface during engine installation.

WARNING
The engine is very heavy and has many sharp edges. Never place your hands or any other part of your body where the engine could drop and crush your hands or arms. One or more assistants will be required to install the engine into the frame. Do not attempt engine removal by yourself.

2. With the aid of an assistant, lift the engine and set it into the frame from the right side (**Figure 16**).
3. While your assistant holds the engine, place a hydraulic jack under the crankcase to support the engine (**Figure 8**).
4. Raise the engine and roll the jack rearward until the splines of the universal joint engage those of the driveshaft.
5. Lower the engine slightly and tilt it to the right. Slip the cylinder head bracket (C, **Figure 10**) onto the cylinder head studs. Install and finger-tighten the cylinder head bracket nuts (A, **Figure 10**).
6. Roll the engine rearward so the universal joint completely engages the driveshaft and the lower engine mount aligns with the frame mount on the right side.
7. Install and finger-tighten the cylinder head bracket bolts (B, **Figure 10**).
8. Set the removable frame member into place on the right side.
9. Loosely install the upper frame member bolts (**Figure 13**) and the lower frame member bolts (B, **Figure 3**). Finger-tighten them at this point.

15

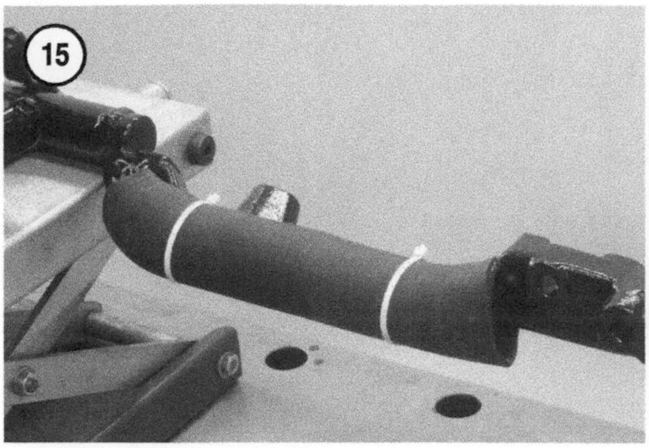

10. Install the front engine bracket as follows:
    a. Position the engine bracket between the frame members.
    b. Loosely install the engine bracket bolt (C, **Figure 11** and C, **Figure 12**) onto each side.
    c. Install the upper through bolt (B, **Figure 11**) and lower through bolt (B, **Figure 12**).
    d. Turn the nuts (A, **Figure 11** and A, **Figure 12**) onto the through bolts.
    e. Finger tighten the engine bracket bolts.
11. Install the upper and lower rear brackets into place on the frame and loosely install the frame bracket bolts (C and D, **Figure 14**).
12. From the left side, install the rear through bolt (B, **Figure 14**) and loosely install the rear upper engine mounting bolt (A, **Figure 14**). Make sure these bolts pass through the upper and lower rear brackets.
13. On the right side, install the nut (D, **Figure 3**) onto the through bolt, and loosely install the rear upper engine mounting bolt (C, **Figure 3**).
14. Tighten the bolts and nuts to the indicated torque specifications. Follow the sequence listed below.
    a. Torque the rear engine through bolt (B, **Figure 14**) and the lower engine bracket bolts (D) to 48 N•m (35 ft.-lb.).
    b. Torque the right upper rear engine bolt (C, **Figure 3**) to 48 N•m (35 ft.-lb.).
    c. Torque the left upper rear engine bolt (A, **Figure 14**) and the upper engine bracket bolts (C) to 48 N•m (35 ft.-lb.).
    d. Torque the cylinder head bracket bolts (B, **Figure 10**) to 48 N•m (35 ft.-lb.).
    e. Torque the cylinder head bracket nuts (A, **Figure 10**) to 74 N•m (55 ft.-lb.).

f. Torque the front frame member bolts (**Figure 13**) and rear frame member bolts (B, **Figure 3**) to 48 N•m (35 ft.-lb.).
g. Torque the front engine bracket bolts (A, B and C, **Figure 12** and **Figure 11**) to 48 N•m (35 ft.-lb.).

15. Pull the coolant hose from the left side of the frame and install the alternator cover as described in this chapter.
16. Connect the clutch cable as described in *Clutch Release Mechanism* in Chapter Six.
17. Install the external oil pipes as follows:
    a. Set the oil pipes into place on the engine.
    b. Loosely install the oil pipe banjo bolt (**Figure 7**) that secures the oil pipe to the crankcase. Use three new copper washers.

NOTE

*Two types of rocker arm bolts are used in the cylinder head: one with a single oil hole (A, **Figure 5**) and one with two oil holes (B). The bolt with two oil holes (B, **Figure 5**) secures the oil pipe to the cylinder head.*

    c. Loosely install the rocker arm bolt (**Figure 6**) that secures an oil pipe to each cylinder head. Use two new copper washers on each bolt; one washer on either side of the pipe fitting.
    d. Hand tighten the bolts, and torque the oil pipe banjo bolt (**Figure 7**) and the intake rocker arm bolts (**Figure 6**) to 20 N•m (15 ft.-lb.).
18. Install the cylinder head covers (Chapter Four).
19. Connect each spark plug cap to the spark plug in its respective cylinder.
20. Install the shift pedal/footrest assembly (Chapter Seven).
21. Install the starter, speed sensor and horn (Chapter Nine).
22. Route the oil level switch wire (A, **Figure 3**) along the path noted during removal and connect the bullet connector to its harness mate.
23. Install the right and left side covers, toolbox panel, frame neck covers and the sidestand (Chapter Fourteen).
24. Install the brake pedal/footrest assembly (Chapter Thirteen).
25. Install the carburetor assembly, exhaust system, surge tank, air filter housing and fuel tank (Chapter Eight).
26. Install the battery (Chapter Three), connect the ground lead connector (B, **Figure 2**) to its harness mate, and install the battery cover (Chapter Fourteen).
27. Fill the crankcase with the recommended type and quantity of oil as described in Chapter Three.
28. Adjust throttle cable and clutch cable free play as described in Chapter Three.
29. Start the engine and check for oil and exhaust leaks.

ALTERNATOR COVER

Removal/Installation

Refer to **Figure 17**.

1. Drain the engine oil as described in Chapter Three.
2. Remove the battery cover and disconnect the electrical lead from the negative battery terminal (A, **Figure 2**).
3. Remove the shift pedal/footrest assembly (Chapter Seven).
4. Disconnect the clutch cable from the clutch release lever as follows:
    a. Remove the mounting bolts (A, **Figure 18**), and pull the clutch adjuster cover (B) from the alternator cover.
    b. At the handlebar, slide the clutch lever boot (A, **Figure 19**) away from the adjuster. Loosen the clutch cable locknut (B, **Figure 19**) and rotate the adjuster (C) to provide maximum slack in the cable.
    c. Disconnect the clutch cable (**Figure 20**) from the clutch lever. Pull the clutch cable from the fitting on the alternator cover.

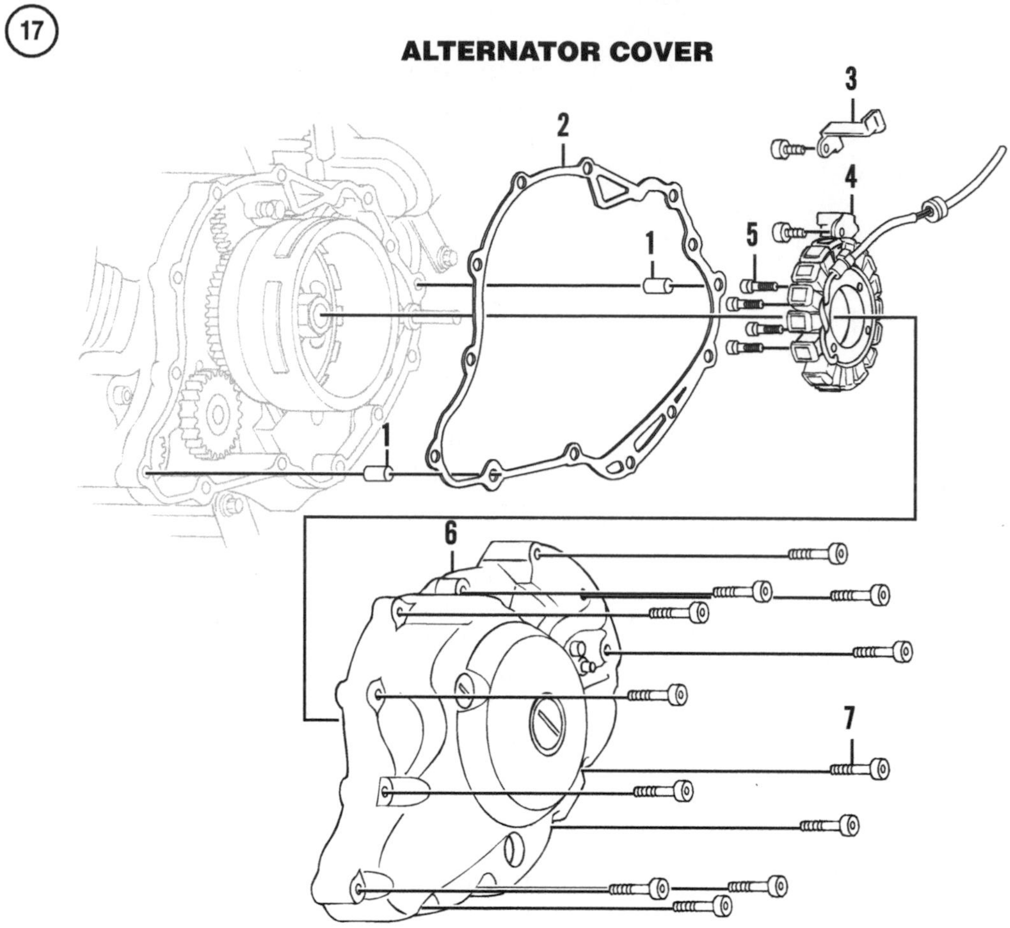

1. Dowel
2. Gasket
3. Wire clamp
4. Stator/pickup coil assembly
5. Stator mounting bolt
6. Alternator cover
7. Alternator cover bolt

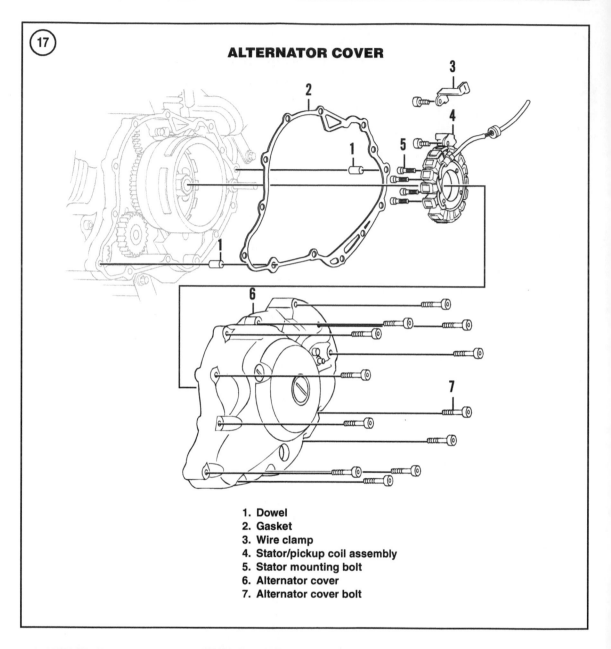

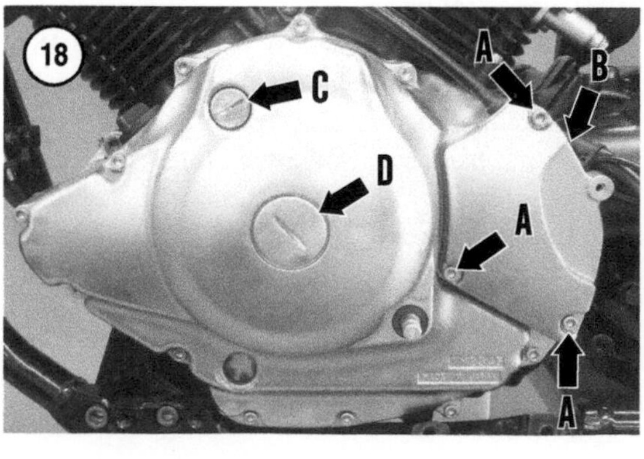

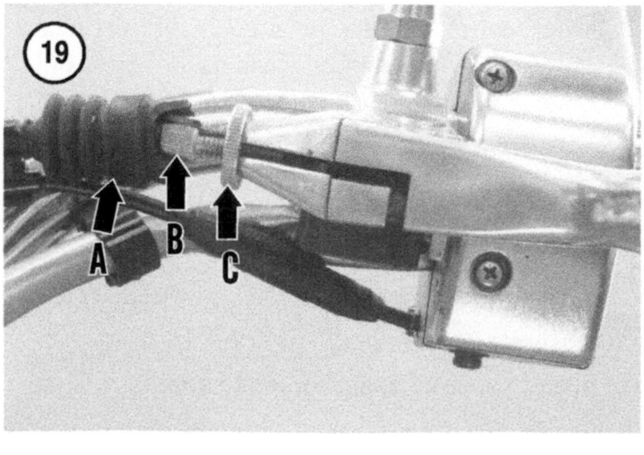

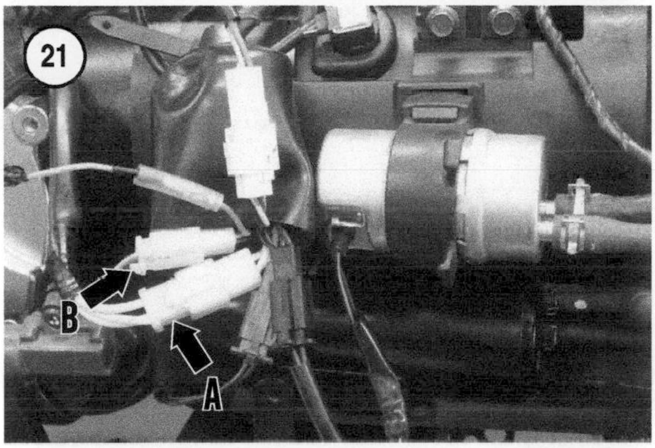

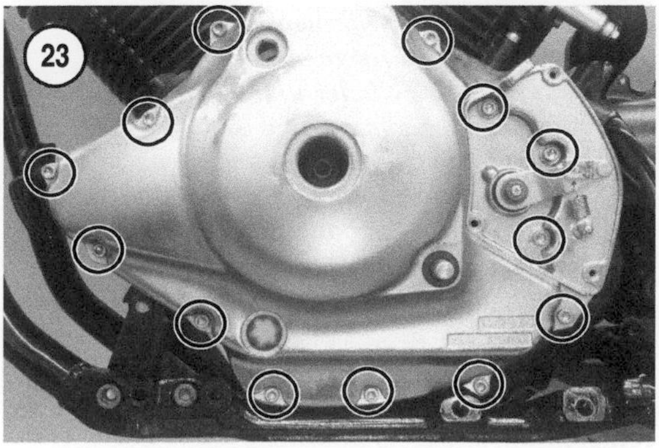

5. Remove the sidestand, toolbox cover and right side cover (Chapter Fourteen).

6. Roll the boot on the toolbox panel from the electrical connectors. Disconnect the three-pin stator connector (A, **Figure 21**) and the two-pin pickup coil connector (B) from their respective mates on the wiring harness.

7. Release any cable ties that secure the stator and pickup coil wires to the frame. Note how the wires are routed along the frame. Reroute these wires along the same path during installation.

8. Before removing the alternator bolts, note how the clutch cable/AIS bracket is secured behind two bolts (**Figure 22**). The bracket must be installed behind these bolts during installation.

9. Evenly loosen the alternator cover bolts (**Figure 23**). Remove the bolts and pull the alternator cover from the crankcase. Watch for the starter idler gear assembly (A, **Figure 24**). It may come out with the cover. If it does, reinstall the assembly into the crankcase so the idler assembly engages the starter wheel gear and the starter drive gear.

10. Remove the alternator cover gasket (B, **Figure 24**) and the two dowels (C) from the crankcase.

11. Inspect the shift shaft oil seal for damage or signs of leaking. If necessary, replace the seal as follows:

   a. Pry the oil seal from the alternator cover (**Figure 25**).
   b. Lubricate a new seal with lithium soap grease.
   c. Set the new seal in place so the manufacturer's marks face up.
   d. Drive the oil seal into the cover with a driver or socket that matches the outside diameter of the seal.

12. Installation is the reverse of removal. Note the following:

a. Make sure the two dowels (C, **Figure 24**) are in place in the crankcase.
b. Install a new alternator cover gasket (B, **Figure 24**).
c. Lubricate the shift shaft oil seal so it will not be damaged during installation.
d. Install the clutch cable/AIS bracket in its original location on the alternator cover (**Figure 22**).
e. Seal the threads of the three indicated alternator cover bolts (**Figure 26**) with silicone sealant.
f. Torque the alternator cover bolts and the clutch release mechanism cover bolts to 10 N•m (89 in.-lb.).
g. Install the shift pedal/footrest assembly so the indexing mark on the shift shaft (A, **Figure 27**) aligns with the slot in the shift lever. Tighten the shift lever clamp bolt (B, **Figure 27**) to 10 N•m (89 in.-lb.).
h. Add engine oil and adjust the clutch free play as described in Chapter Three.

STATOR AND PICKUP COILS

See Chapter Nine for stator and pickup coil removal and inspection procedures.

FLYWHEEL AND STARTER CLUTCH

The following Yamaha special tools, or their equivalents, are needed to remove or install the flywheel and starter clutch:

1. Sheave holder: part No. YS-01880 or 90890-01701.
2. Flywheel puller: part No. YU-33270 or 90890-01362.
3. Flywheel puller adapter: part No. YM-38145 or 90890-04131.

Refer to **Figure 28**.

Removal

NOTE

When the rear cylinder is set to TDC on the compression stroke, the timing mark on the cam sprocket may not precisely align with the pointer on the cylinder head. On some models, the camshafts are slightly retarded. The mark could be off by as much as 1/2 tooth. Before removing the flywheel, set the rear cylinder to top dead center on the compression stroke, and note the position of the timing marks on the cam sprocket and cylinder head. Take a photograph or make a drawing so the camshaft can be correctly timed during assembly.

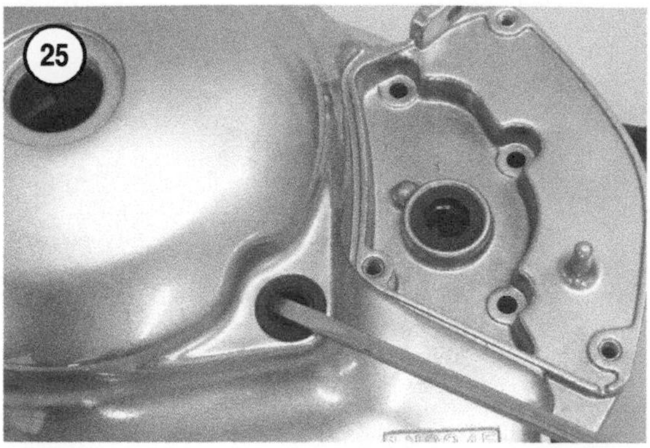

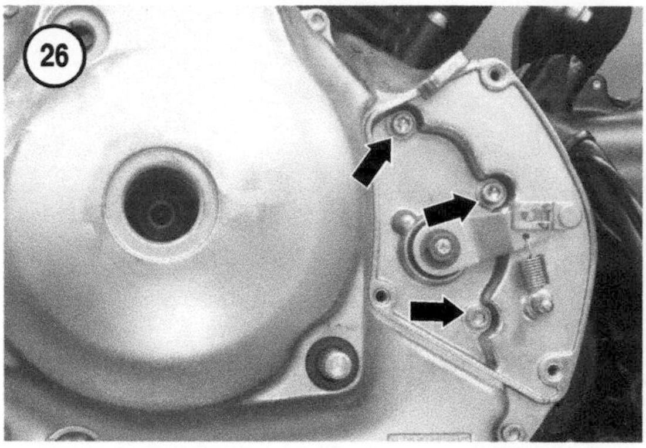

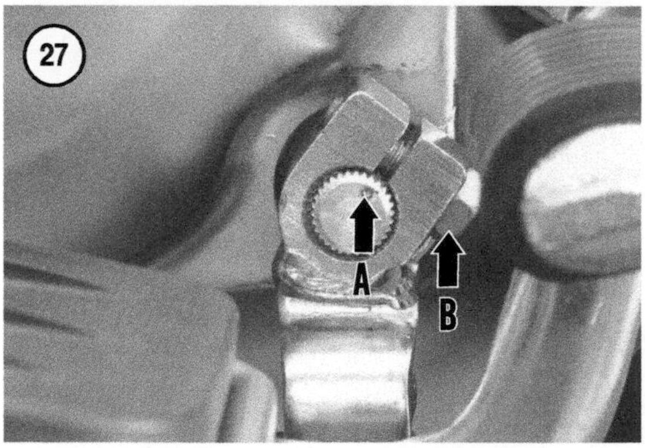

1. Set the rear cylinder to top dead center on the compression stroke as follows:

28

FLYWHEEL AND STARTER CLUTCH

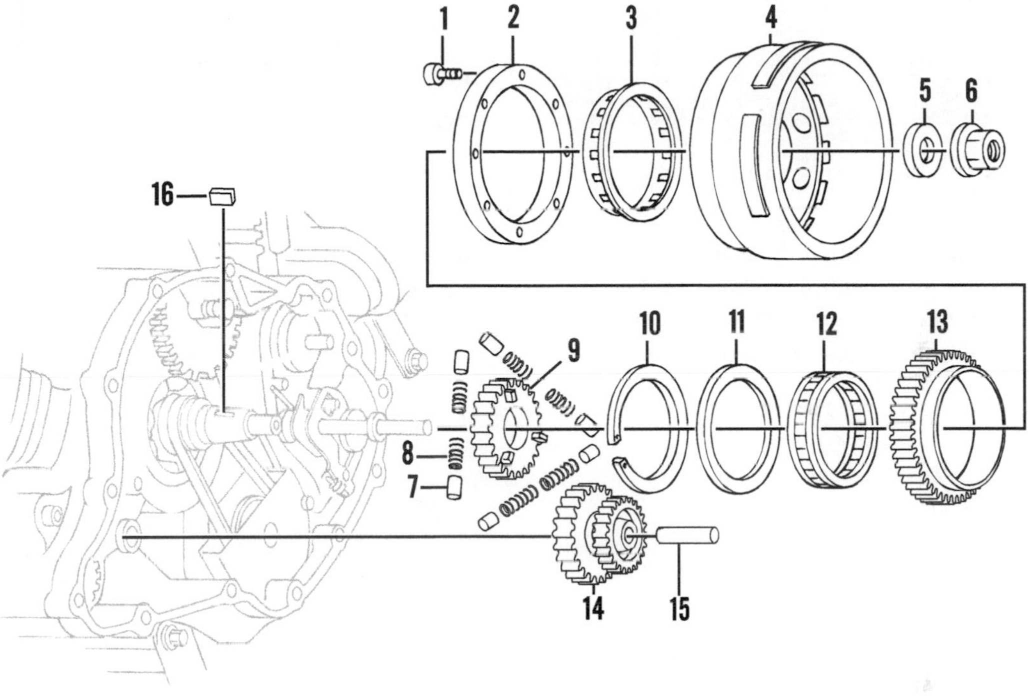

1. Bolt
2. One-way clutch housing
3. One-way clutch
4. Flywheel
5. Washer
6. Nut
7. Pin
8. Spring
9. Timing gear
10. Snap ring
11. Washer
12. Bearing
13. Starter wheel gear
14. Starter idler gear assembly
15. Idler gear shaft
16. Woodruff key

29

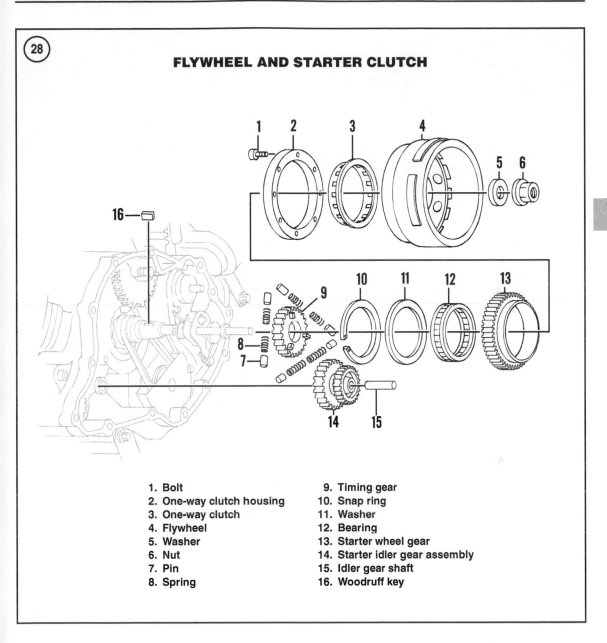

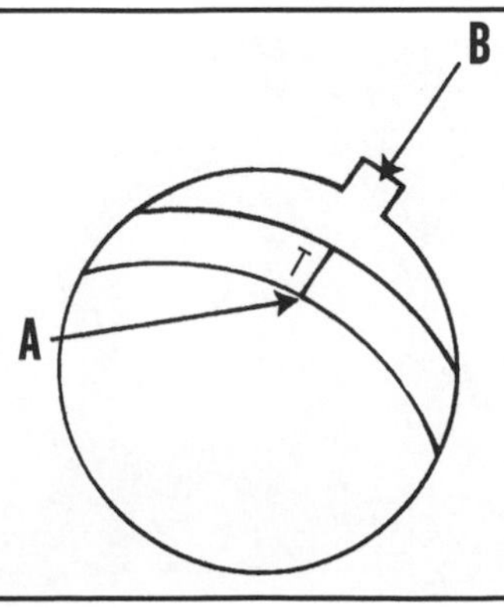

a. Remove the timing cover (C, **Figure 18**) and the flywheel bolt cover (D) from the alternator cover.
b. Remove the cam sprocket cover from the rear cylinder head.
c. Use the flywheel nut to rotate the crankshaft clockwise until the T-mark on the flywheel (A, **Figure 29**) aligns with the cutout in the alternator cover (B).
d. Check the timing mark on the rear cam sprocket (A, **Figure 30**). It should align with the pointer on the rear cylinder head (B, **Figure 30**).

e. If the timing mark on the rear cam sprocket does not align with the pointer on the rear head, rotate the engine one turn clockwise.

2. Remove the alternator cover as described in this chapter.

NOTE
*Install the sheave holder so it sits completely flat against the flywheel. Do not let the sheave holder sit across any raised portion (A, **Figure 31**) of the flywheel.*

3. Hold the flywheel with the sheave holder, and remove the flywheel nut (B, **Figure 31**) and its washer (C).

CAUTION
The timing gear normally comes out with the flywheel. However, if it remains behind on the crankshaft, the six springs and six pins may get scattered. Be prepared to catch loose parts. Stuff rags into any openings in the crankcase so nothing falls into the case.

4. Install the flywheel puller and adapter onto the flywheel. (**Figure 32**). Make sure the puller is parallel to the flywheel.

5. Turn the puller's center screw and drive the rotor off the crankshaft. If necessary, adjust a puller screw to keep the puller parallel to the flywheel.

6. Remove the Woodruff key (A, **Figure 33**) from the crankshaft. Note that the timing mark (B, **Figure 33**) on the cam chain drive assembly aligns with the Woodruff key.

7. Remove the starter idler gear assembly (A, **Figure 34**) and the idler shaft (B) from the crankcase.

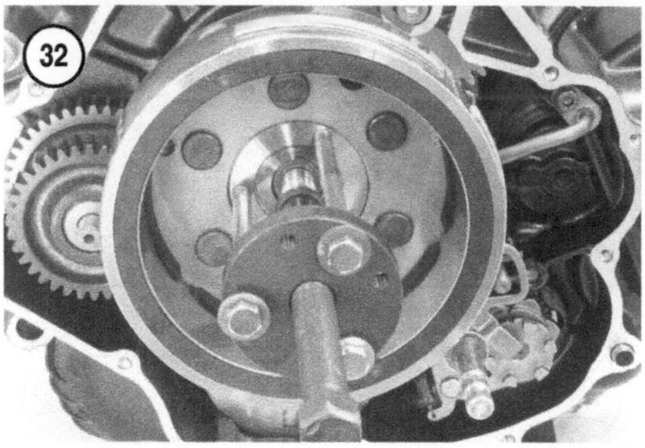

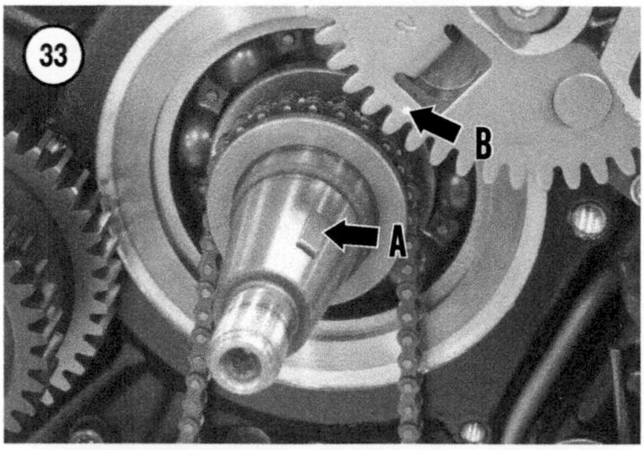

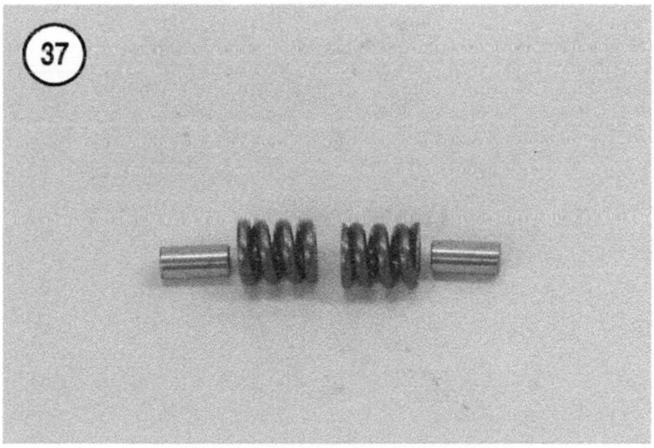

8. If necessary, lift the timing gear (A, **Figure 35**) from the back of the flywheel. Remove each pin and spring (**Figure 36**) from the slots in the flywheel.

Installation

1. Insert a pin into each spring (**Figure 37**).
2. Set the flywheel face down on the bench and install two sets of pins/springs into each slot on the back of the flywheel (**Figure 36**). Push the springs as far apart as possible.
3. Apply grease to the dogs (A, **Figure 38**) on the timing gear.
4. Position the timing gear so its index mark (B, **Figure 35**) aligns with the keyway on the flywheel, set the dogs between the springs in each flywheel slot and press the timing gear into place (A, **Figure 35**).
5. Mark the edge of the tooth (**Figure 39**) on either side of the timing mark so you can locate the mark as the flywheel is installed.
6. If the Woodruff key was removed, install it into the keyway in the crankshaft.
7. Install the starter idler gear assembly (A, **Figure 34**) into the crankcase. Secure it in place with the idler shaft (B, **Figure 34**).
8. Preload the cam chain drive assembly as follows:
   a. Cut a 6 × 15 mm pin from the shoulder (non-threaded portion) of a 6 mm bolt.
   b. Use a screwdriver or similar tool to pry the drive teeth on the gear until one set of teeth aligns with the other.
   c. Insert the 6 mm pin (**Figure 40**) into the aligned hole to lock the gear.
9. Make sure the rear cylinder is still set to top dead center. The timing mark on the cam chain drive as-

sembly (B, **Figure 33**) aligns with the Woodruff key (or the crankshaft keyway) when the cylinder is at top dead center.

10. Position the flywheel so the two marked teeth (A, **Figure 41**) align with the Woodruff key in the crankshaft. Slide the flywheel onto the crankshaft so the keyway engages the Woodruff key (B, **Figure 41**), the starter wheel gear (C) engages the starter idler gear, and the timing gear (D) engages the cam chain drive assembly.

11. Install the washer (C, **Figure 31**) and flywheel nut (B). Hold the flywheel with the sheave holder and torque the flywheel nut to 175 N•m (129 ft.-lb.). Make sure the sheave holder does not cross any raised portion (A, **Figure 31**) of the flywheel.

12. Install the alternator cover as described earlier in this chapter.

Flywheel Inspection

1. Clean the parts in solvent and dry them with compressed air.

WARNING
Replace a cracked or chipped flywheel. A damaged flywheel can fly apart at high speed, throwing metal fragments into the engine. Do not attempt to repair a damaged flywheel.

2. Inspect the flywheel (**Figure 42**) for cracks or breaks. Make sure the magnet is free of all metal parts.

3. Check the flywheel tapered bore and the crankshaft taper for damage. Replace damaged parts as necessary.

4. Inspect the threads of the flywheel nut bolt. Replace the nut if the threads are stretched or damaged.

5. Inspect the dogs (A, **Figure 38**) and teeth (B) on the timing gear for chips, cracks or other signs of damage. Replace the timing gear if there is any damage. If the teeth on the timing gear are damaged, also check the teeth on the gear of the cam chain drive assembly.

6. Check the operation of the starter clutch as described in this section.

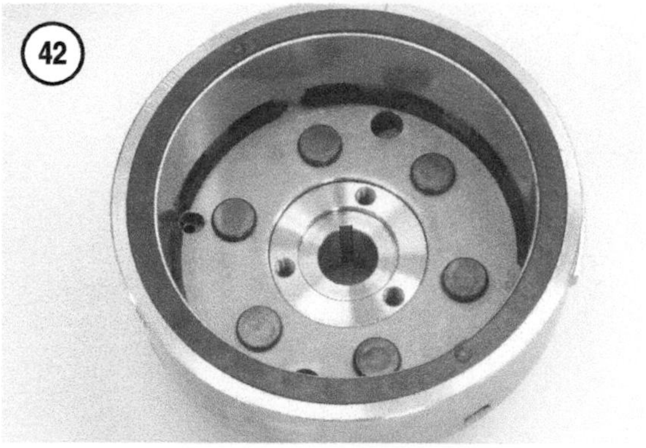

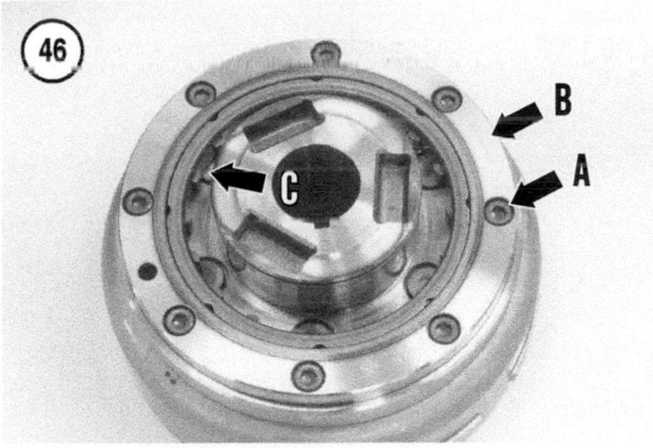

Starter Clutch Operational Test

1. Set the flywheel on the bench so the starter clutch faces up as shown in **Figure 36**.

2. Hold the flywheel and turn the starter wheel gear clockwise. The wheel gear should turn freely within the starter clutch.

3. Hold the flywheel and try to turn the starter wheel gear counterclockwise. It should not turn in this direction.

NOTE

The one-way clutch and one-way clutch housing are not available separately. If either part is worn or damaged, replace the starter clutch assembly.

4. The one-way clutch is faulty if it fails either test. Replace the starter clutch assembly.

Starter Clutch Disassembly/Assembly

Refer to **Figure 28**.

1. Remove the flywheel and remove the timing gear as described in this section.
2. Remove the snap ring (**Figure 43**) and washer (**Figure 44**).
3. Remove the bearing (A, **Figure 45**), then remove the starter wheel gear (B).
4. If necessary, remove the starter clutch bolts (A, **Figure 46**) and lift the starter clutch assembly (B) from the back of the flywheel.
5. Installation is the reverse of removal. Note the following:
   a. Apply a medium-strength threadlocking compound to the threads of each starter clutch bolt and install the bolts.
   b. Hold the rotor with the sheave holder and torque the starter clutch bolts to 12 N•m (106 in.-lb.).
6. Install the starter wheel gear into the starter clutch as follows:
   a. Set the rotor assembly face down on the bench.
   b. Set the bearing surface of the starter wheel gear between rollers (C, **Figure 46**) in the one-way clutch.
   c. Press the starter wheel gear down while rotating it clockwise. Gently press the gear until it bottoms within the one-way clutch.

Starter Clutch Inspection

1. Inspect the teeth of the starter idler gears (A, **Figure 47**). Replace a gear if any teeth are worn, broken or missing.
2. Inspect the idler gear shaft and the bearing surface (B, **Figure 47**) of the starter idler gear assembly for nicks or other signs of damage. Replace the idler gear shaft or starter gear assembly if worn or damaged.

5

3. Inspect the teeth (A, **Figure 48**) of the starter wheel gear. Replace the gear if any teeth are worn, broken or missing.
4. Inspect the bearing surface (B, **Figure 48**) of the starter wheel gear for nicks or scratches. Replace the wheel gear if it shows signs of wear.
5. Inspect the starter wheel gear bearing (**Figure 49**) for wear of other signs of damage. Replace the bearing as necessary.

NOTE

The one-way clutch and one-way clutch housing are not available separately. If either part is worn or damaged, replace the starter clutch assembly.

6. Inspect the rollers in the one-way clutch (C, **Figure 46**) for wear or damage. All the rollers should rotate freely. If there is damage or wear, replace the starter clutch assembly.

OIL PUMP

The V-Star 1100 uses a two-stage oil pump to lubricate the engine and the transmission. The primary side of the pump delivers oil to the oil filter, through the crankshaft and onto the upper end. The secondary side pumps oil to the transmission shafts.

In general, this oil pump is not serviceable. The oil pump sprocket, sprocket cover and chain are the only parts that can be replaced; other parts are not available separately. If inspection reveals any faulty part(s), replace the oil pump.

Removal/Installation

1. Drain the oil and remove the flywheel as described in this chapter.
2. Remove the cover bolts (A, **Figure 50**) and remove the sprocket cover (B) from the oil pump.
3. Hold the oil pump driven sprocket with a Grabbit or similar tool, and remove the oil pump driven sprocket bolt (A, **Figure 51**).
4. Pull the oil pump driven sprocket (B, **Figure 51**) from the shaft and remove the sprocket from the drive chain.
5. Remove the oil pump bolts (A, **Figure 52**) and pull the pump (B) from the crankcase. Watch for the dowel (A, **Figure 53**) and two O-rings (B) behind the pump.

47

48

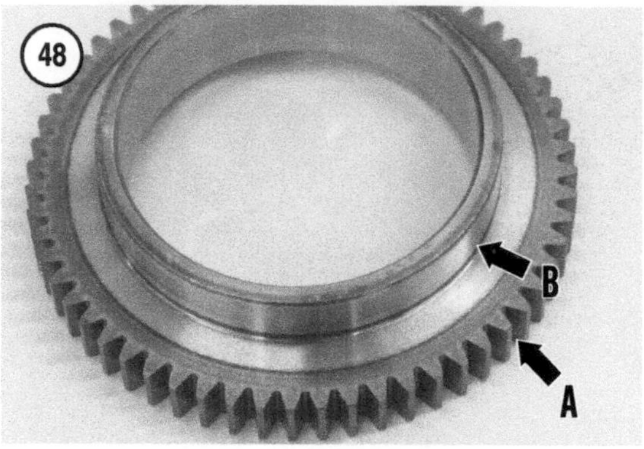

49

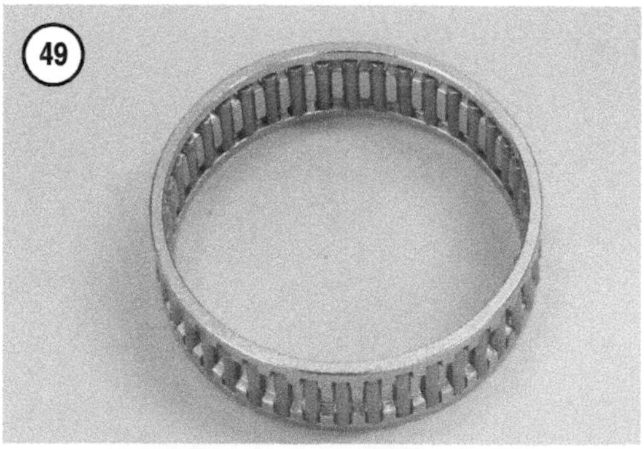

50

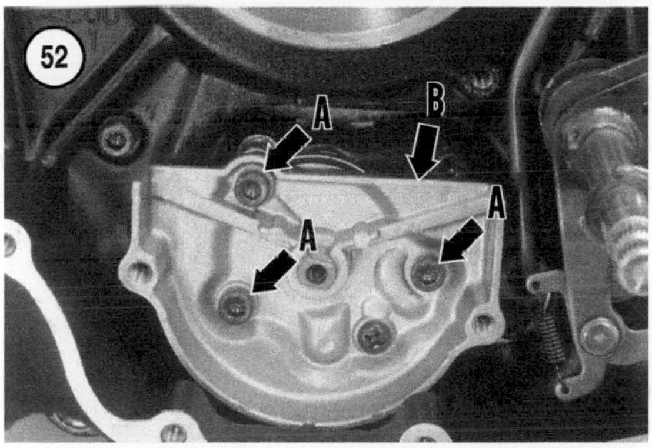

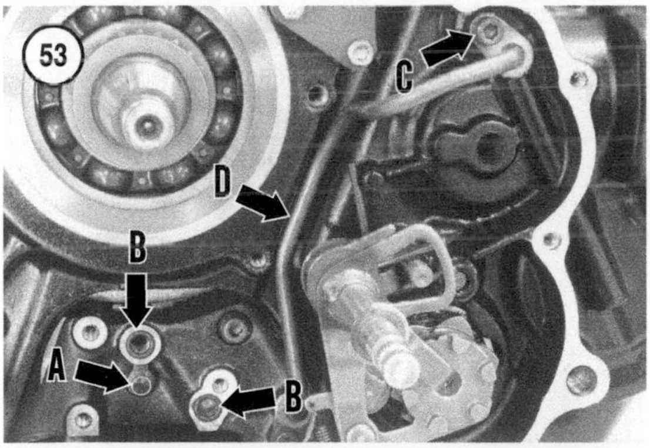

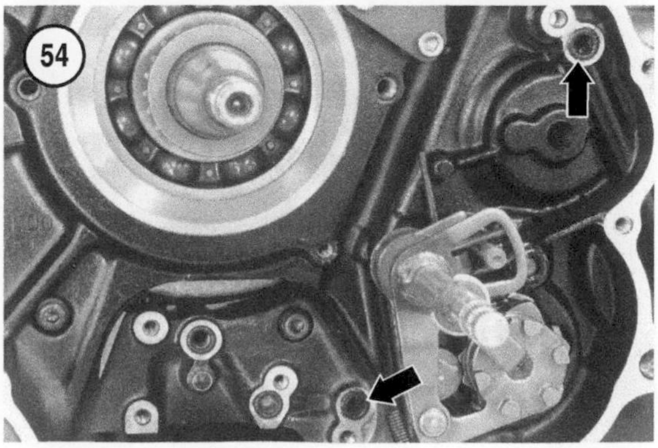

6. If necessary, remove the oil delivery pipe mounting bolt (C, **Figure 53**) at each end of the pipe, and remove the oil delivery pipe (D) from the crankcase. Watch for the O-ring (**Figure 54**) behind the fitting on each end of the pipe.

NOTE
The oil pump drive sprocket is damaged during removal. If removed, it must be replaced.

7. If necessary, use a universal bearing puller to remove the oil pump drive sprocket from the crankshaft.

8. Installation is the reverse of removal. Note the following:
   a. If removed, install a new oil pump drive sprocket onto the crankshaft. Apply oil to the crankshaft and drive the new sprocket into place with a pipe that matches the inside diameter of the sprocket.
   b. Install new oil delivery pipe O-rings (**Figure 54**) and new oil pump O-rings (B, **Figure 53**). Lubricate each O-ring with lithium soap grease.
   c. Make sure the oil pump properly engages the dowel (A, **Figure 53**) in the crankcase.
   d. Torque the oil delivery pipe mounting bolt (C, **Figure 53**) to 10 N•m (89 in.-lb.). Torque the oil pump mounting bolts (A, **Figure 52**) to 10 N•m (89 in.-lb.). Torque the oil pump driven sprocket bolt (A, **Figure 51**) to 12 N•m (106 in.-lb.). Torque the oil pump sprocket cover bolts (A, **Figure 50**) to 10 N•m (89 in.-lb.).

Disassembly/Inspection/Assembly

Refer to **Figure 55**.

CAUTION
Be careful when removing the oil strainer. A replacement strainer is not available. If the strainer is damaged, the oil pump must be replaced.

1. If necessary, carefully remove the oil strainer (**Figure 56**) by prying at one of the four pry points.
2. Remove the oil pump guide (**Figure 57**) from the primary housing.
3. Remove the Phillips screw (A, **Figure 58**) from the oil pump cover (B).

55

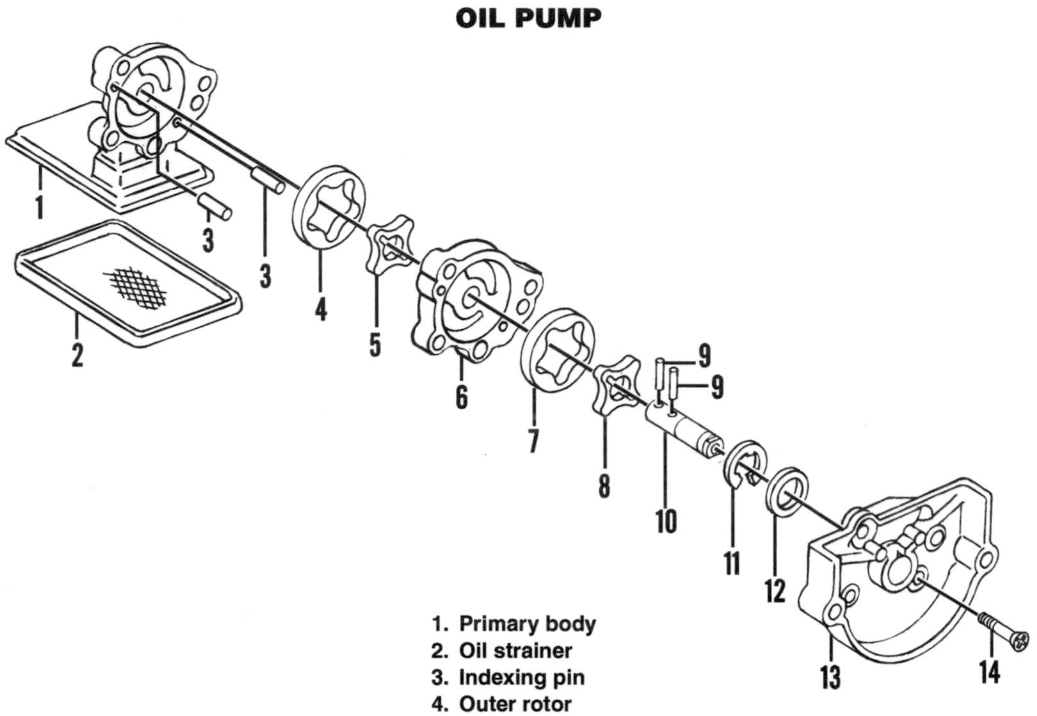

1. Primary body
2. Oil strainer
3. Indexing pin
4. Outer rotor
5. Inner rotor
6. Secondary body
7. Outer rotor
8. Inner rotor
9. Drive pin
10. Pump shaft
11. Snap ring
12. Washer
13. Oil pump cover
14. Screw

56

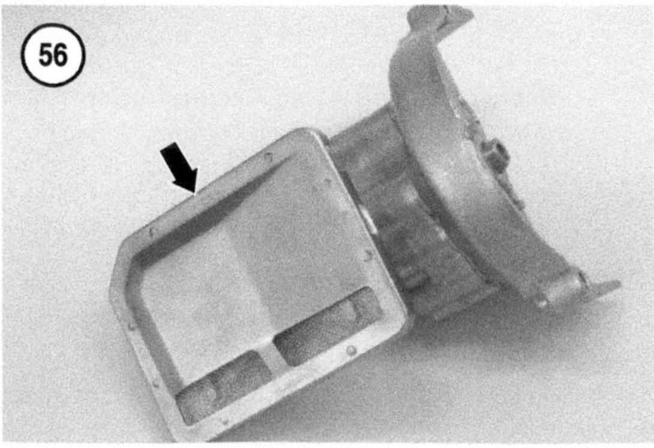

57

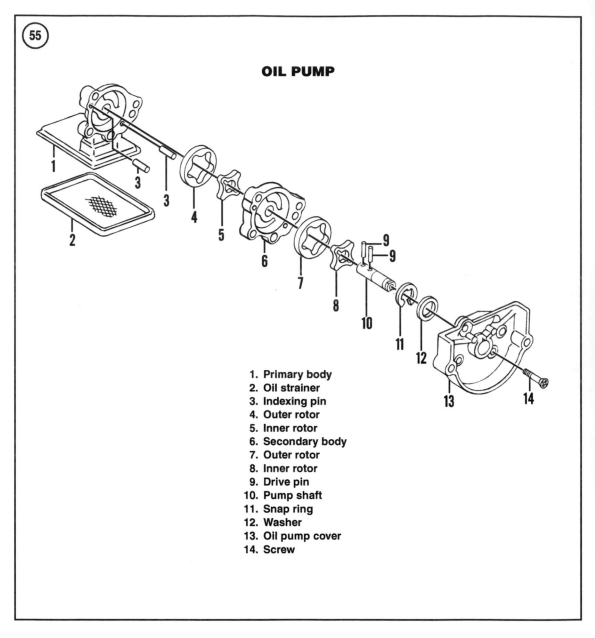

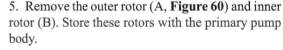

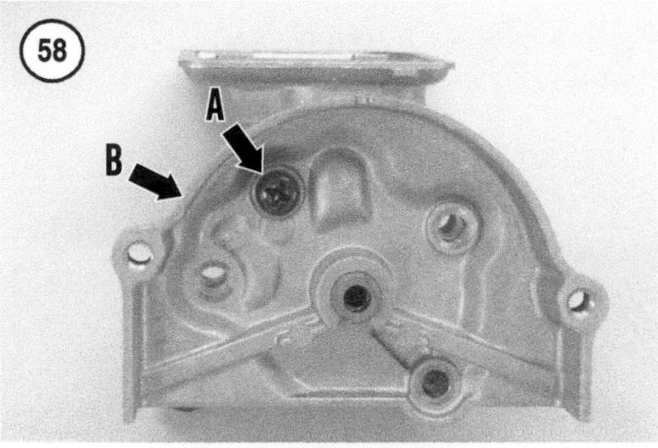

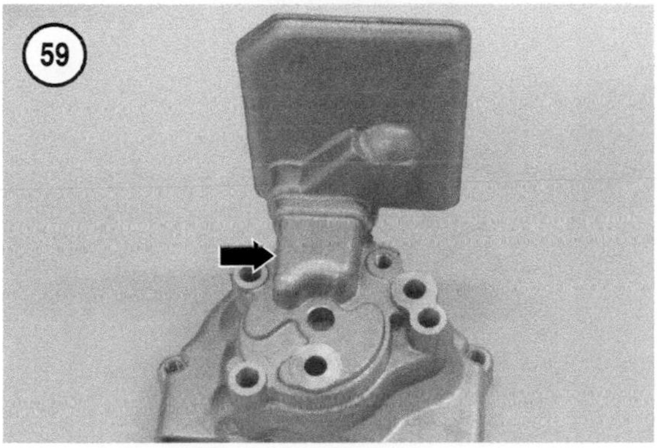

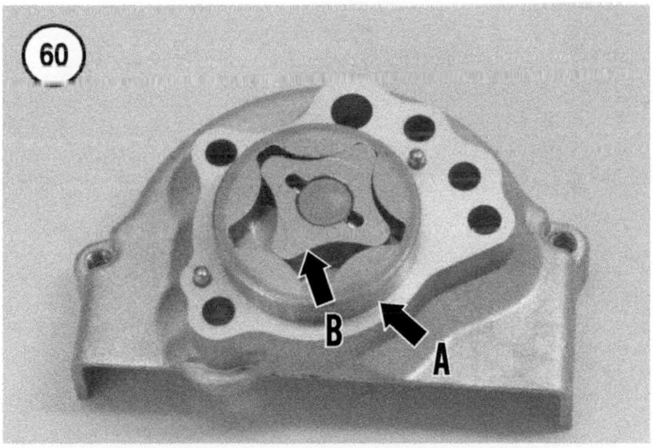

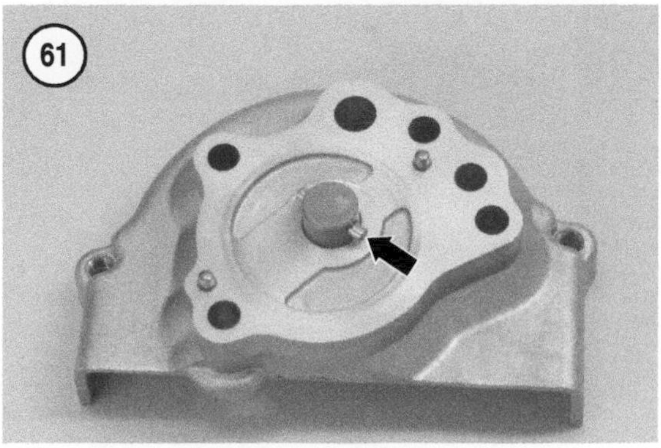

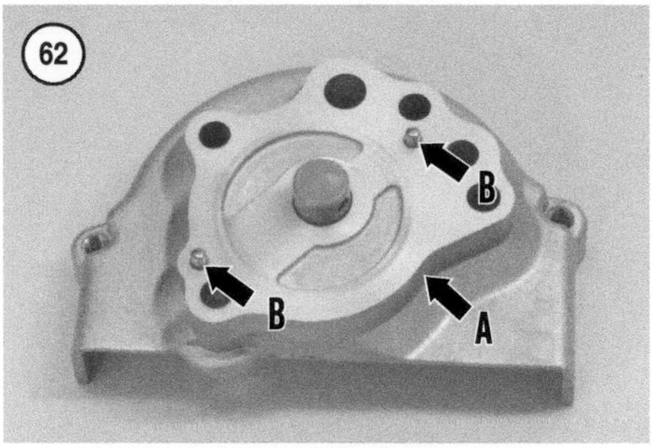

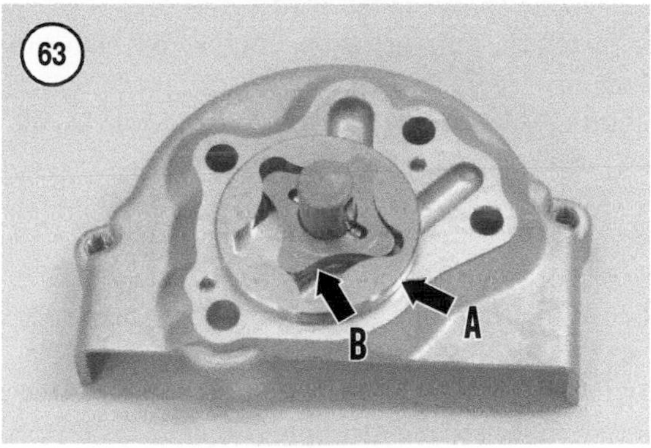

4. Turn the pump over and lift the primary pump body (**Figure 59**) from the pump.

NOTE
This pump has two sets of rotors: a thick set that fits in the primary body and a thin set for the secondary body. Keep each set of rotors together with its respective body.

5. Remove the outer rotor (A, **Figure 60**) and inner rotor (B). Store these rotors with the primary pump body.

6. Remove the drive pin (**Figure 61**) from the pump shaft.

7. Remove the secondary body (A, **Figure 62**) and the two indexing pins (B).

8. Remove the outer rotor (A, **Figure 63**) and inner rotor from the pump cover. Store these rotors with the secondary body.

9. Remove the drive pin (**Figure 64**) from the shaft.

10. Remove the snap ring (**Figure 65**) and the washer (**Figure 66**) from the shaft.

11. Remove the shaft (**Figure 67**) from the pump cover.

Assembly

1. Coat all parts with fresh oil prior to assembly.
2. Install the pump shaft (**Figure 67**) into the pump cover. The end with the holes enters from the outboard side of the cover.
3. Turn the cover over and install the washer (**Figure 66**).
4. Install the snap ring (**Figure 65**) so it completely seats in the groove in the shaft.
5. Install the first drive pin (**Figure 64**) into the shaft.
6. Install the thin set of rotors onto the cover. Install the inner rotor (B, **Figure 63**) onto the shaft so the rotor engages the drive pin, then install the outer rotor (A).
7. Install the secondary body (A, **Figure 62**) over the rotors. Align the holes in the secondary body with those in the pump cover, then install the two indexing pins (B, **Figure 62**).
8. Install the second drive pin (**Figure 61**) into the shaft.
9. Install the thick set of rotors. Install the inner rotor (B, **Figure 60**) onto the shaft so the rotor engages the drive pin, then install the outer rotor (A).
10. Install the primary body (**Figure 59**) over the rotors. Make sure the body engages the indexing pins.
11. Turn the assembly over and install the Phillips screw (A, **Figure 58**). Apply medium-strength threadlocking compound to the threads of the screw, and tighten it securely.
12. If removed, install the oil pump guide (**Figure 57**) into the primary body, then carefully install the oil strainer (**Figure 56**). Crimp the oil strainer to hold it in place.
13. Check the pump operation by manually rotating the pump shaft. The pump should turn smoothly without binding or excessive noise.

Inspection

NOTE
Except for the oil pump driven sprocket, drive chain and sprocket cover, replacement parts are not available.

64

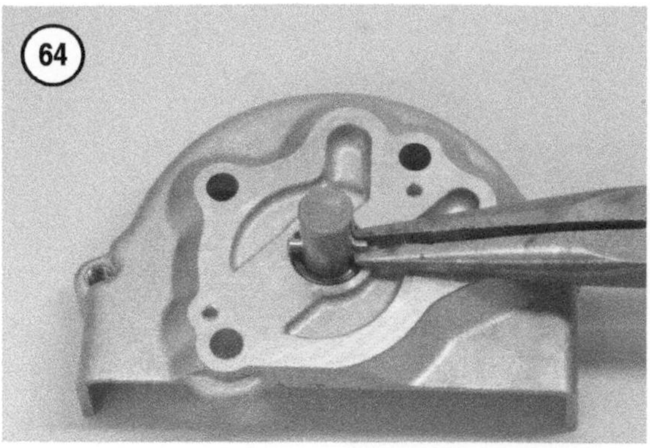

65

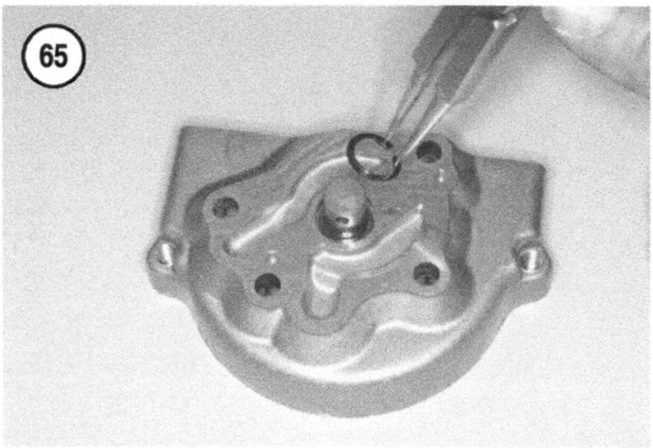

66

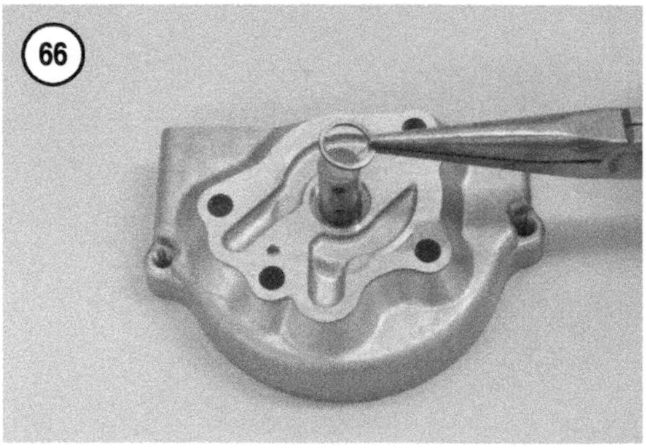

67

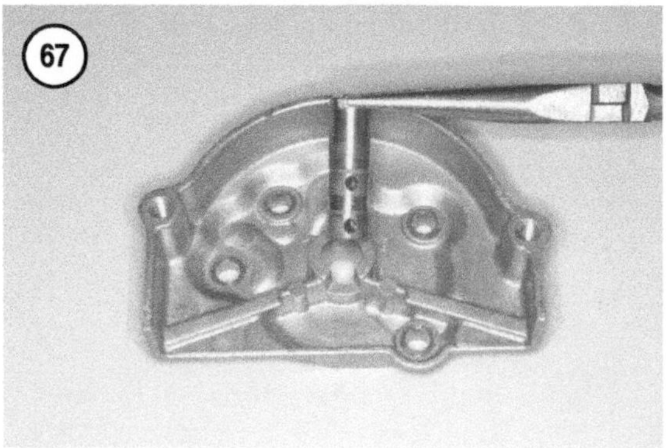

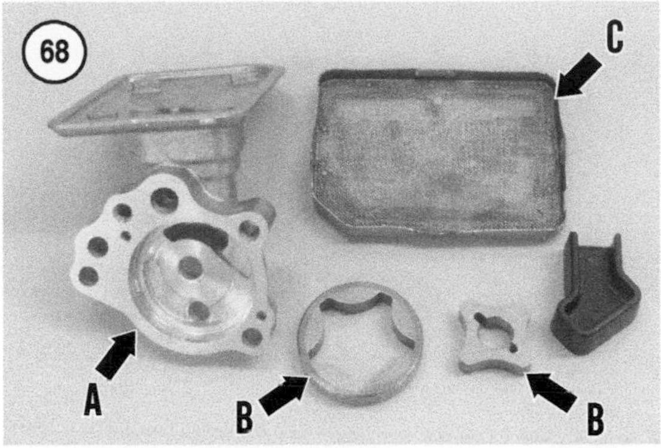

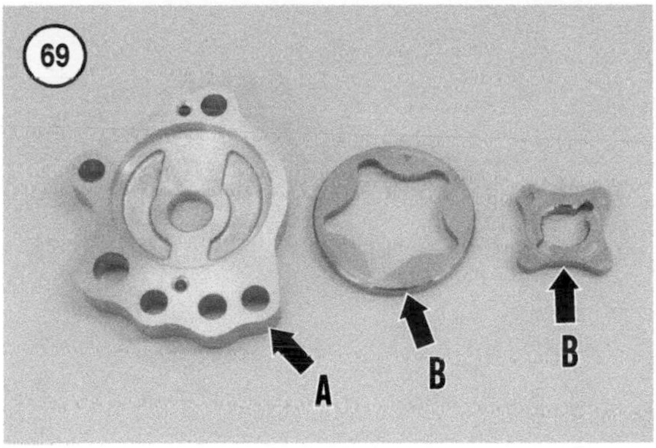

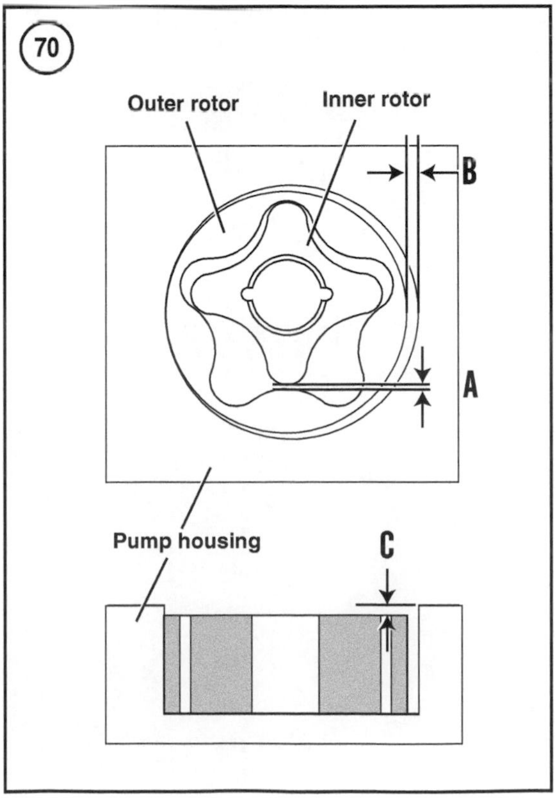

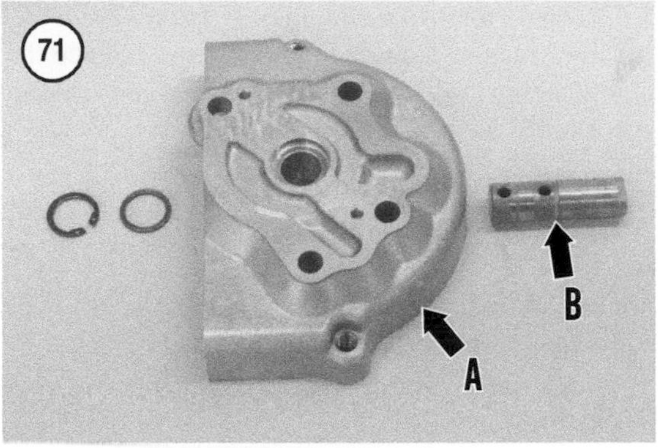

1. Clean all parts in solvent and dry them thoroughly with compressed air.

2. Inspect the primary body (A, **Figure 68**) for cracks and inspect the primary body rotors (B) for wear or abrasion.

3. Inspect the strainer (C, **Figure 68**) for tears or other damage that will allow contaminants into the pump.

4. Inspect the secondary body (A, **Figure 69**) for cracks and inspect the secondary body rotors (B) for wear or abrasion.

5. Install each outer and inner rotor into its respective body.

6. Use a flat gauge to measure the clearance between the tip of the inner rotor and the outer rotor (A, **Figure 70**).

7. Use a flat feeler gauge to measure the clearance between the outer rotor and the oil pump body (B, **Figure 70**).

8. Measure the pump rotor depth, which is the distance from the top of the pump body to the top of the rotors (C, **Figure 70**).

9. Replace the oil pump if any measurement exceeds the service limit in **Table 1**.

10. Inspect the oil pump cover (A, **Figure 71**) for cracks, and inspect the oil pump shaft (B) for wear or abrasion.

11. Replace the pump if any of the above parts are worn or damaged. Replacement parts are not available.

12. Inspect the teeth on the oil pump driven sprocket (A, **Figure 72**). Replace the sprocket if any teeth are worn or broken. Also inspect the oil pump drive sprocket on the crankshaft. Replace the driven sprocket, drive chain and drive sprocket as a set.

13. Inspect the drive chain (B, **Figure 72**) for excessive wear or damage. If necessary, replace the chain along with both the drive and driven sprocket.

CRANKCASE

The following Yamaha special tools, or their equivalents, are needed to service the crankcase:
1. Crankcase separating tool (part No. YU-01135-A or 90890-01135).
2. Crankshaft installer pot (part No. YU-90058 or 90890-01274).
3. Crankshaft installer bolt (part No. YU-90060 or 90890-01275).
4. Adapter (part No. YM-4059 or 90890-04130).
5. Spacer (part No. YU-90070 or 90890-04060).

Crankcase Disassembly

This procedure describes the separation of the crankcase halves, and the removal of the crankshaft, transmission shaft assemblies and the internal shift mechanism. Disassembly and inspection procedures for the transmission shaft assemblies, middle driven gear and internal shift mechanism are in Chapter Seven. Removal/installation, disassembly/assembly and inspection procedures for the middle drive assembly and the middle driven assembly also appear in Chapter Seven.
1. Remove the engine as described in this chapter.
2. Remove the following exterior assemblies from the crankcase as described in the respective chapters:
   a. Oil filter (Chapter Three).
   b. Cylinder head (Chapter Four).
   c. Cam chain and chain guides (Chapter Four).
   d. Cylinder and piston (Chapter Four).
   e. Clutch (Chapter Six).
   f. Primary drive gear (Chapter Six).
   g. Flywheel (this chapter).
   h. Oil pump (this chapter).
   i. External shift mechanism (Chapter Seven).
3. Remove the mounting screws (A, **Figure 73**) and lower the oil level switch cover (B) from the crankcase.
4. Pull the oil level switch (**Figure 74**) from the crankcase.

72

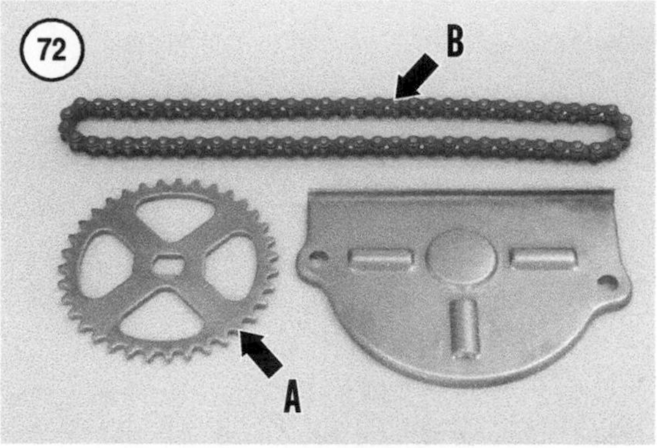

73

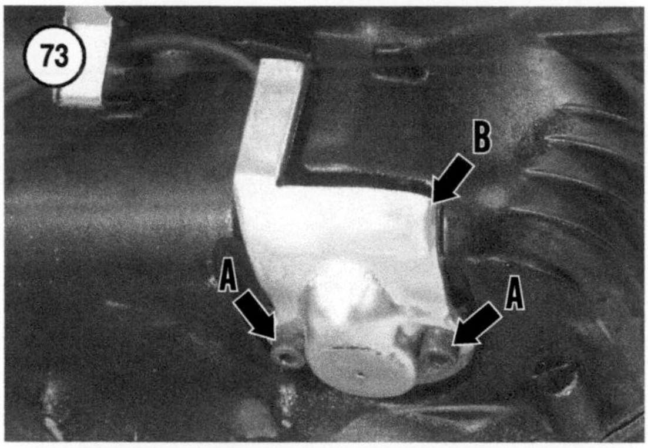

74

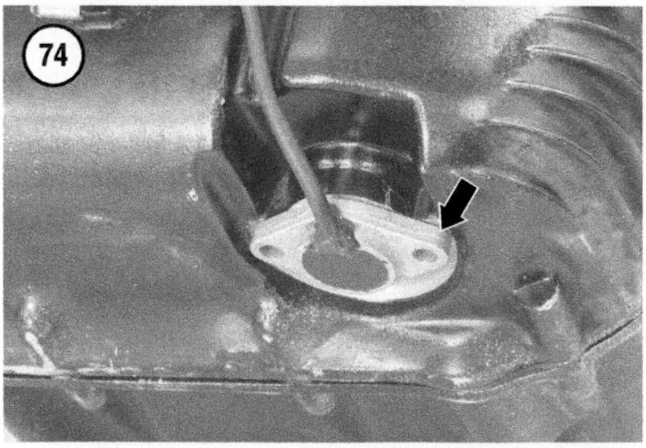

75

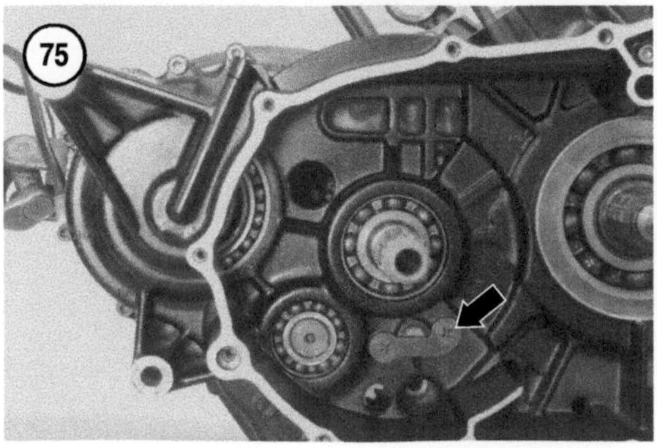

76

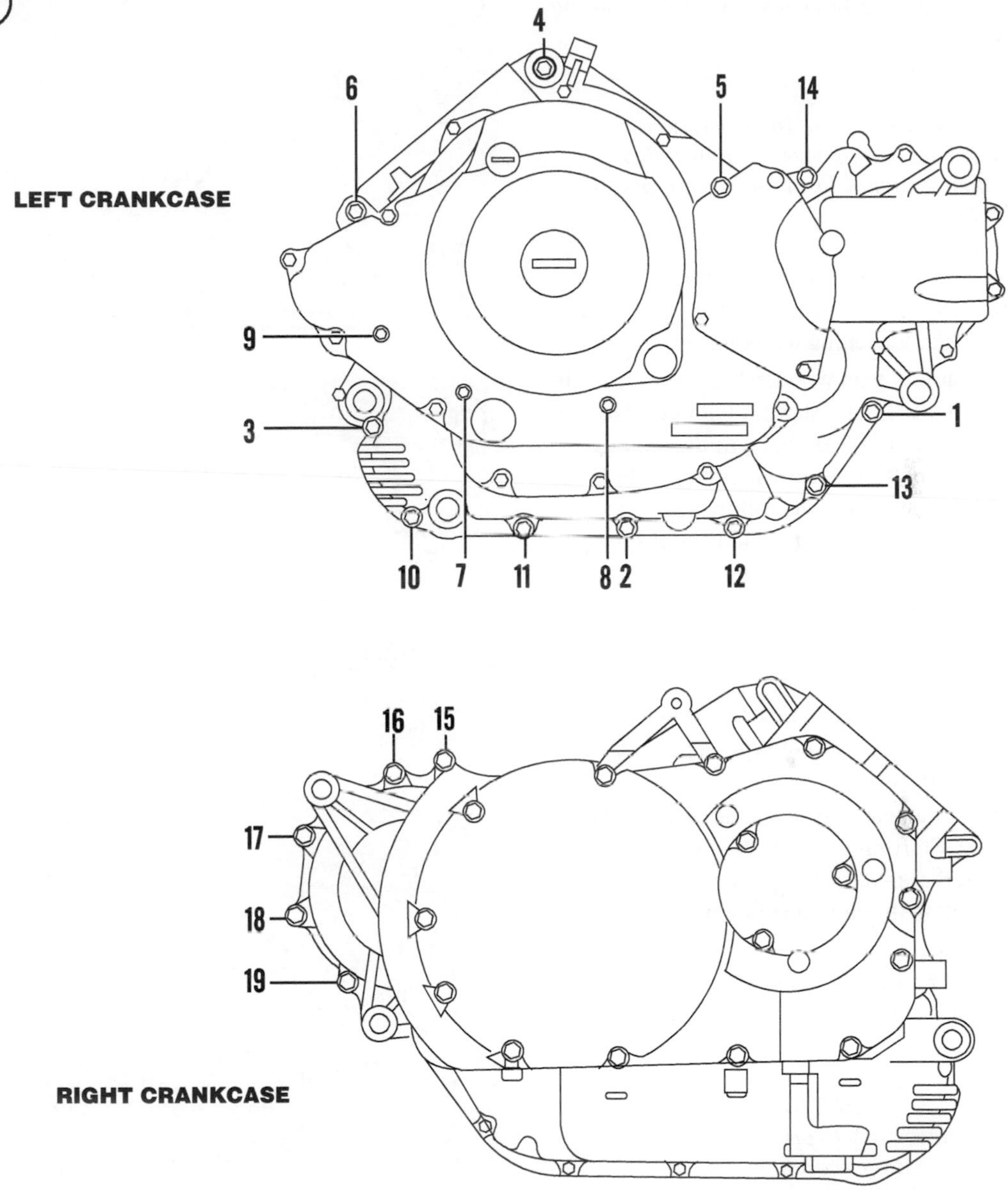

5. Remove the mounting screw, and lift the shift fork shaft stopper plate (**Figure 75**) from the right crankcase half.

NOTE

*Crankcase bolts are identified by a number embossed on the crankcase near each bolt hole. These numbers also indicate the crankcase bolt tightening sequence. See **Figure 76**.*

6. Before removing the crankcase bolts, draw an outline of each case half on a piece of cardboard. Punch holes along the drawing outline corresponding to the bolt location in each crankcase half shown in **Figure 76**. As each bolt is removed, insert it and

any washer into its respective hole in the cardboard template so bolts can be quickly identified during assembly.

NOTE

Bolts No. 15 and No. 16 are silver colored. These bolts must be reinstalled in their original locations during assembly.

7. Evenly loosen all the crankcase bolts 1/4 turn at a time. Reverse the tightening sequence shown in **Figure 76** when loosening the bolts.

8. Set the crankcase on wooden blocks so the right crankcase half faces up. Starting with bolt No. 19, remove the right crankcase bolts in descending order. Place each bolt in its corresponding hole in the cardboard template. Note the following:

   a. Discard the copper washer behind bolt No. 15. A new copper washer must be installed during assembly.
   b. Watch for the ground wire installed beneath bolt No. 18.
   c. Leave one right crankcase bolt finger-tight so the crankcases will not separate as the cases are turned over.

9. Turn the crankcase over so the left crankcase half faces up. Starting with bolt No. 14, loosen and remove the left crankcase bolts in descending order. Place each bolt in its corresponding hole in the cardboard template.

10. Turn the crankcase over so the right crankcase half faces up. Remove the one remaining right crankcase bolt.

CAUTION

If it is necessary to pry apart the case halves, do it very carefully. The crankcase mating surfaces must not be damaged. If they are, the cases will leak and must be replaced as a set. They cannot be repaired.

11. Separate the crankcase halves by carefully tapping around the crankcase perimeter with a plastic mallet. Do not use a metal hammer. Separate the crankcase halves by lifting the right case half off the left half. The transmission and crankshaft assemblies should remain in the left crankcase half.

12. Watch for the following items when splitting the cases. Remove them once the cases are apart.

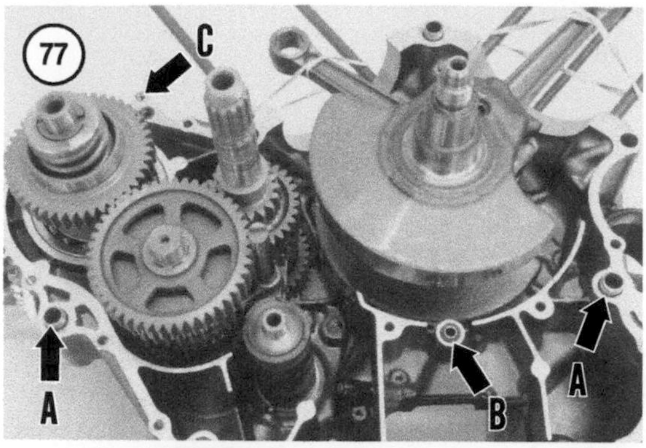

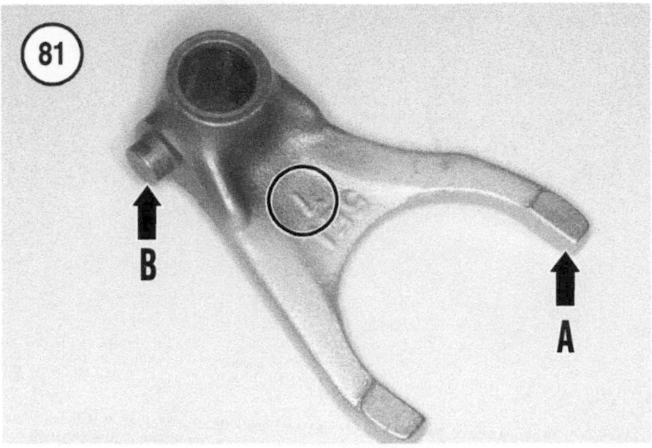

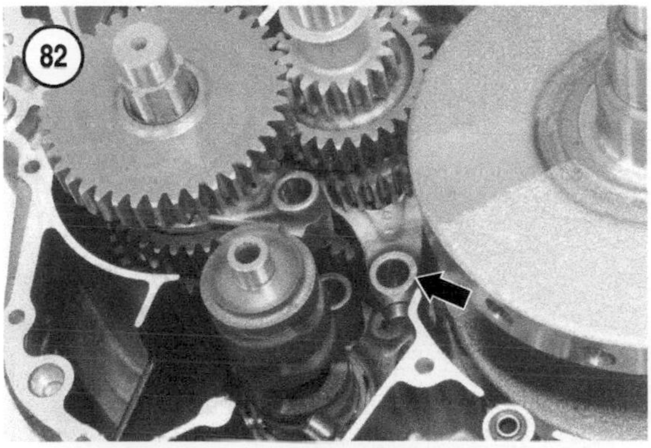

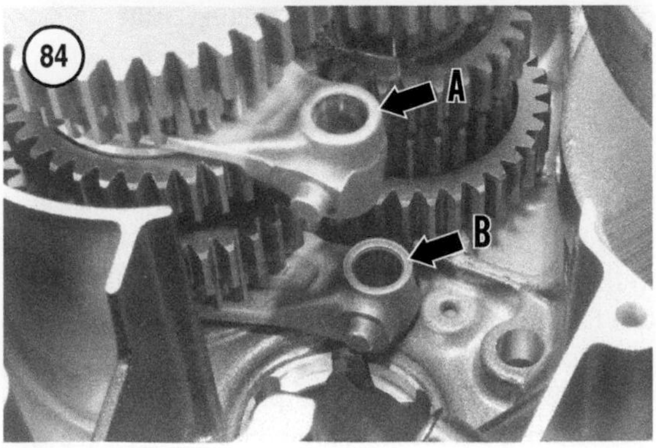

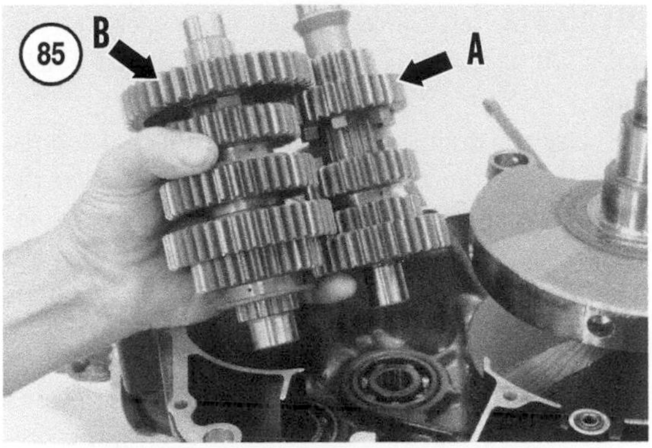

a. Remove the two dowels (A **Figure 77**) from the crankcase mating surface.

b. Remove the O-ring and its short collar (B, **Figure 77**) from the boss beneath the crankshaft.

c. Remove the O-ring from the crankcase mating surface above the middle drive shaft.

13. Remove the oil pressure relief valve (**Figure 78**) and its O-ring from the right crankcase half. Store the valve in a recloseable plastic bag.

14. Remove the middle drive gear (A, **Figure 79**) from the countershaft and remove the middle driven gear (B) from the middle drive shaft.

15. Remove the shift fork shaft (**Figure 80**). Note that the cutout on the shaft faces the countershaft. The shift fork shaft must be installed in this orientation during assembly.

NOTE

*The left, center and right shift forks are identified by a letter (L, C or R) embossed on their faces (**Figure 81**). Note the location and orientation of the shift forks. Each fork must be reinstalled into its original location with the identification mark facing up out of the crankcase.*

16. Rotate the center shift fork (**Figure 82**) away from the shift drum and remove the fork from the mainshaft second-third combination gear.

17. Rotate the left (A, **Figure 83**) and right (B) shift forks away from the shift drum (C), and lift the drum from the crankcase.

18. Remove the right shift fork (A, **Figure 84**) from the countershaft fourth gear and remove the left shift fork (B) from countershaft fifth gear.

19. Simultaneously lift both the mainshaft assembly (A, **Figure 85**) and the countershaft assembly

(B) from the left crankcase half. Store each individual shaft assembly in a sealed and labeled plastic bag.

20. Use the crankcase separating tool to drive the crankshaft from the left crankcase half. Keep the tool centered over the crankshaft and follow the instructions from the tool's manufacturer. The left main bearing should come out with the crankshaft.

Crankcase Assembly

This procedure describes the installation of the crankshaft, transmission shaft assemblies, and internal shift mechanism as well as the assembly of the crankcase halves.

NOTE
Coat all parts with engine oil prior to assembly.

1. Install the oil pressure relief valve (**Figure 78**), with a new O-ring, into the right crankcase half. Lubricate the O-ring with lithium soap grease.
2. Place the left crankcase half on wooden blocks.
3. If removed, install the neutral switch and its washer (**Figure 86**) into the crankcase. Apply a medium-strength threadlocking compound to the threads of the switch and torque the neutral switch to 20 N•m (15 ft.-lb.).

CAUTION
Do not attempt to install the crankshaft without the special tools. If these tools are not available, have a Yamaha dealership or other qualified service shop install the crankshaft.

4. Following the tool manufacturer's instructions, assemble the crankshaft installer pot (A, **Figure 87**), crankshaft installer bolt (B), adapter (C) and spacer (D) onto the crankshaft, and pull the crankshaft into the left crankcase half. Make sure the connecting rods align within their respective cylinder openings. Continually monitor this alignment until the crankshaft is completely seated in the left crankcase half.
5. Mesh the mainshaft (A, **Figure 85**) and countershaft (B) assemblies together and simultaneously lower both assemblies into place in the left crankcase half. Make sure each shaft slides into its respective bearing.

86

87

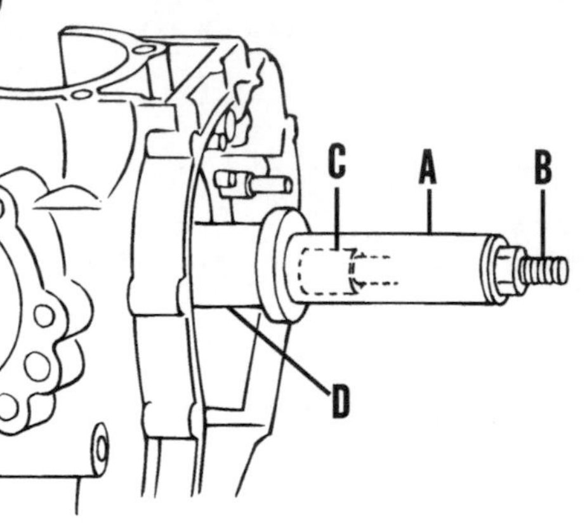

NOTE
*Each shift fork is identified by a letter embossed on its face. See **Figure 81**. Install each shift fork with the letter facing up toward the right side of the engine.*

6. Install the left shift fork into the countershaft fifth gear (**Figure 88**) so its identification mark faces up. Rotate the countershaft so the fifth gear dogs can slide up and engage second gear.
7. Install the right shift fork (A, **Figure 84**) into the countershaft fourth gear. The identification mark must face up.
8. Rotate the shift forks outward as necessary so the shift drum can be lowered into place.
9. Install the shift drum (C, **Figure 83**). Use a pick to lift the left shift fork (B, **Figure 83**) so the drum can fall into place in the case.
10. Rotate each shift fork (A and B, **Figure 83**) so its guide pin engages the respective groove in the shift drum. Turn the shift drum into the neutral posi-

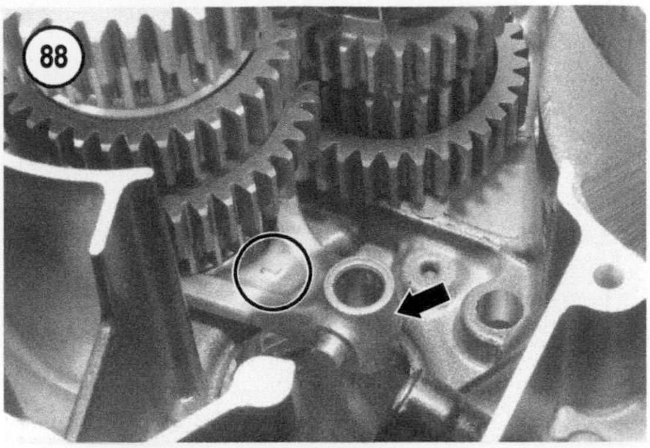

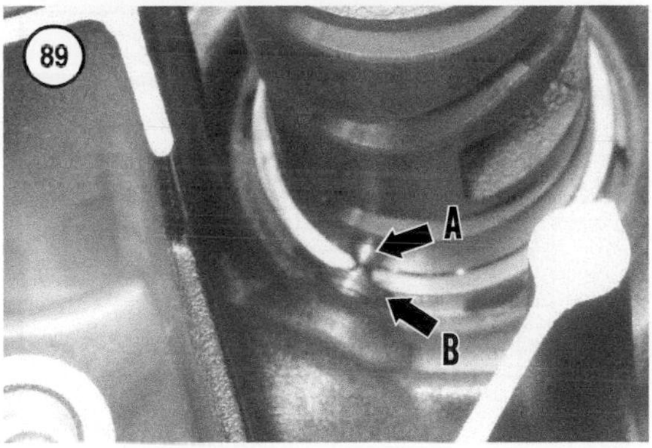

tion so the neutral knob (A, **Figure 89**) on the drum aligns with the neutral switch (B) in the crankcase.

11. Install the center shift fork (**Figure 82**) into the mainshaft second-third combination gear. Rotate the shift fork so its guide pin engages the groove in the shift drum.

12. Install the shift fork shaft (**Figure 80**) through the shift forks. Rotate the shaft as necessary so the cutout in the shaft faces the countershaft, and seat the shaft into the crankcase boss.

13. Spin the transmission shafts and use the shift drum to shift through the gears. Make sure you can shift into all gears. This is the time to find a problem with the transmission, not after the engine is assembled and installed in the frame.

14. Once it is confirmed that the transmission shifts normally, shift it into neutral.

15. Install the middle driven gear (B, **Figure 79**). Align the internal splines of the gear with those of the middle drive shaft and lower the middle driven gear into place so its bearing bottoms in the boss.

16. Install the middle drive gear (A, **Figure 79**) onto the countershaft. The splines of the gear must engage those of the countershaft, and the middle drive gear teeth must mesh with the teeth of the middle driven gear.

17. Install the two dowels (A, **Figure 77**) into the crankcase mating surface.

18. Lubricate a new O-ring with lithium soap grease. Install the O-ring onto the short collar and seat the assembly (B, **Figure 77**) in the boss beneath the crankshaft web.

19. Lubricate a new O-ring with lithium soap grease and install the O-ring (C, **Figure 77**) into the crankcase mating surface.

20. Check the left and right crankcase mating surfaces for old sealant material or other residue. Clean them as necessary.

21. Apply engine oil to the transmission shafts and journals in the left case half. Also lubricate the bearings, including the inner races, in the right crankcase half. Do not add so much oil that it will drip from the bearing when the case half is turned over.

22. Apply a light coat of Yamaha Bond No. 1215 (part No. ACC-1100-15-01 or 90890-85505) or on equivalent sealant to the mating surface of the left crankcase half. Make the coating as thin as possible while also completely covering the mating surface.

23. Lower the right crankcase half onto the left case half until it is completely seated. If the right case will not seat on the left, remove the right case and determine the source of the obstruction. Do not use the crankcase bolts to pull the cases together. Pay attention to the following:

   a. Align the bearings in the right case half with their respective assemblies in the left case half.
   b. Rotate the shift drum so the ramps on the shift cam align with the cutouts in the right crankcase half.

24. Install the right crankcase bolts (**Figure 76**). Tighten the bolts enough to hold the cases together while they are turned over. These bolts will be torqued after the left crankcase bolts are tightened. Note the following:

   a. The silver colored bolts must be installed in the No. 15 and 16 positions.
   b. Install a new copper washer with bolt No. 15.
   c. Install the engine ground wire with bolt No. 18. Position the wire so its free end extends up above the top of the crankcase.

25. Install the shift fork shaft retainer (**Figure 75**). Apply a high-strength threadlocking compound to the retainer screws and tighten the screws securely.

5

26. Turn the crankcase over and install the left crankcase bolts.

CAUTION
Rotate the crankshaft frequently during the tightening process. If there is any binding, stop and correct the cause before proceeding.

27. Evenly tighten the left crankcase bolts in 1/4-turn increments in the sequence shown in **Figure 76**. Once all the bolts are snug, torque them in sequence to the following specifications:
   a. Bolts Nos. 4-6 (10 mm): 39 N•m (28 ft.-lb.).
   b. bolts Nos. 1-3 and 7-19 (6 mm): 10 N•m (89 in.-lb.).
28. Turn the crankcase over and torque the right crankcase bolts in sequence (**Figure 76**).
29. Check the operation of the crankshaft and transmission. Rotate the universal joint yoke and shift the transmission through the gears. If necessary, disassemble the crankcase and correct any problem now.
30. Install all external assemblies that were removed.
31. Install the engine as described in this chapter.

Crankcase Inspection

1. Using a scraper, remove all sealer residue from all crankcase mating surfaces.

WARNING
When drying the crankcase bearings in Step 2, do not allow the inner bearing race to spin. The bearings lack lubrication and damage will occur. When drying the bearings with compressed air, hold both races with your hand so the bearing will not rotate. The air jet will force the bearings to turn at speeds that exceed their design limit. The likelihood of a bearing disintegrating and causing serious injury and damage is very great.

2. Clean both crankcase halves and all crankcase bearings with cleaning solvent. Thoroughly dry them with compressed air.
3. Clean all crankcase oil passages with compressed air.
4. Lightly oil the crankcase bearings with engine oil before checking the bearings in Step 5.

90

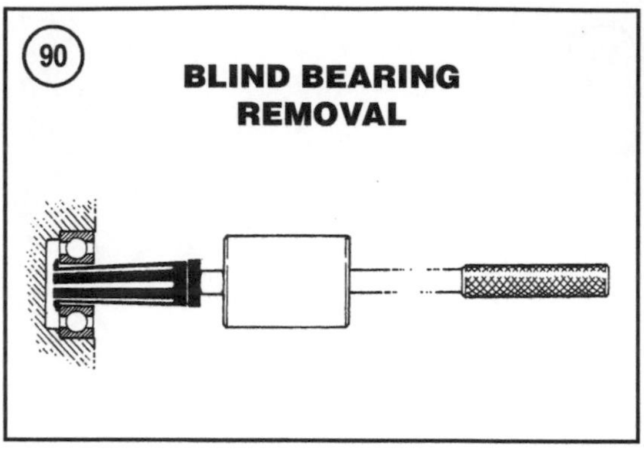

91

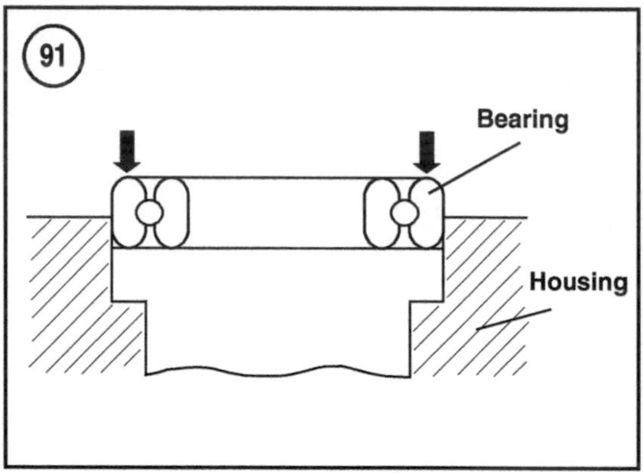

92

NOTE
When replacing a bearing, also replace its mate in the opposite case half. Crankcase bearings should be replaced as a set.

5. Rotate the bearings slowly by hand, and check them for roughness, pitting, galling and play. Replace any bearing that turns roughly or shows excessive play.

93

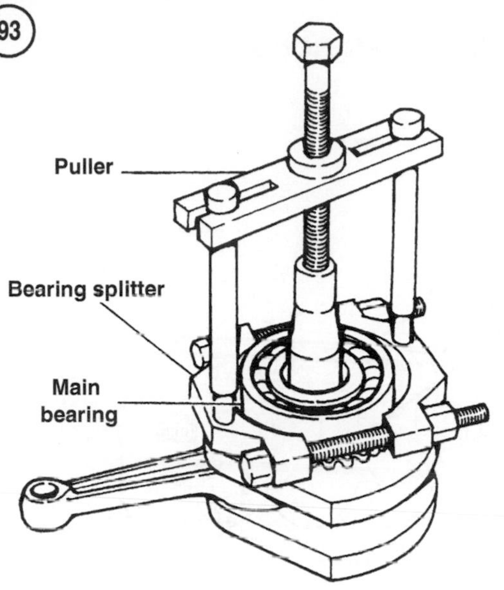

6. Inspect the crankcase studs. Make sure they are straight and their threads are in good condition. Make sure they are tightly screwed into the crankcase.
7. Inspect the mating surfaces of both crankcase halves. They must be free of gouges, burrs or any damage that could cause an oil leak.
8. Inspect the cases for cracks and fractures, especially in the lower areas where they are vulnerable to rock damage.
9. Check the areas around the stiffening ribs and around bearing bosses for damage. Repair or replaced damaged cases.
10. Check the threaded holes in both crankcase halves for thread damage, dirt or oil buildup. If necessary, clean or repair the threads with a suitable size metric tap. Coat the tap threads with kerosene or an aluminum tap fluid before use.

Crankcase Bearing Replacement

Before replacing the crankcase bearings and bushings, note the following:
1. Because of the number of bearings used in the left and right crankcase halves, make sure to identify a bearing and note its location before removing it. Use the size code markings to identify a bearing.
2. Refer to *Bearing Replacement* in Chapter One for general information on bearing removal and installation.
3. Before heating the crankcase, remove the middle drive shaft assembly and middle driven shaft assembly (Chapter Seven).
4. Heat the crankcase to approximately 95-125° C (205-257° F) in an shop oven or on a hot plate. Do not heat the crankcase with a torch. This type of localized heating may warp the cases.
5. Drive the bearing out with a suitable size bearing driver, socket or a drift.
6. After removing bearings and bushings, clean the crankcase half in solvent and dry it thoroughly.
7. A blind bearing remover (**Figure 90**) is required to remove some of the blind bearings in the following procedures.
8. When installing a new bearing into the crankcase, press the outer bearing race only (**Figure 91**). Use a bearing driver or socket that matches the outside diameter of the bearing.

Right crankcase half

1. Press the main bearing (A, **Figure 92**), mainshaft bearing (B) and countershaft bearing (C) from the case.
2. Use a blind bearing puller to remove the small middle drive shaft bearing (D, **Figure 92**).
3. To remove the large middle drive shaft bearing (E, **Figure 92**), evenly heat the crankcase and lift the bearing out.

Left crankcase half

1. Use a blind bearing puller to remove the mainshaft and countershaft bearings.
2. Replace the main bearing as follows:
   a. Remove the crankshaft from the left crankcase as described earlier in this chapter. The left main bearing comes out with crankshaft.
   b. Use a bearing puller to remove the oil pump drive gear from the crankshaft.
   c. Use a larger bearing puller (**Figure 93**) to remove the main bearing.
3. Install the main bearing as follows:
   a. Place support between the crankshaft webs and set the crankshaft in a press.
   b. Press the bearing onto the crankshaft.
   c. Press the oil pump gear to the shoulder on the crankshaft.

94

CRANKSHAFT AND CONNECTING RODS

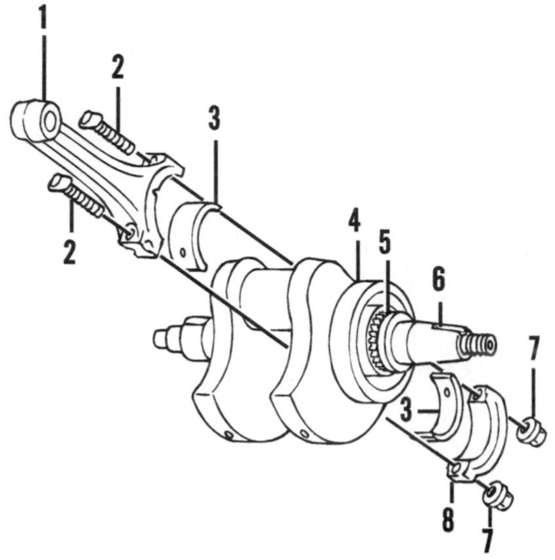

1. Connecting rod
2. Connecting rod bolt
3. Connecting rod bearing
4. Main bearing
5. Oil pump drive sprocket
6. Crankshaft
7. Nut
8. Cap

CRANKSHAFT

Refer to **Figure 94**.

Removal/Installation

Remove and install the crankshaft as described in *Crankcase* in this chapter.

Inspection

1. Clean the crankshaft thoroughly with solvent and dry it with compressed air. Lightly oil the journal surfaces immediately to prevent rust.
2. Blow the oil passages clear with compressed air.
3. Visually inspect the main bearing journals for scratches, heat discoloration or other defects.
4. Check the flywheel taper, threads and keyway for damage.
5. Check the oil pump drive sprocket for excessive wear or tooth damage. Replace the sprocket as necessary.
6. Check the connecting rod big end for signs of seizure, bearing or thrust washer damage or for connecting rod damage.
7. Check the connecting rod small end for signs of excessive heat (blue coloration) or other damage.

95

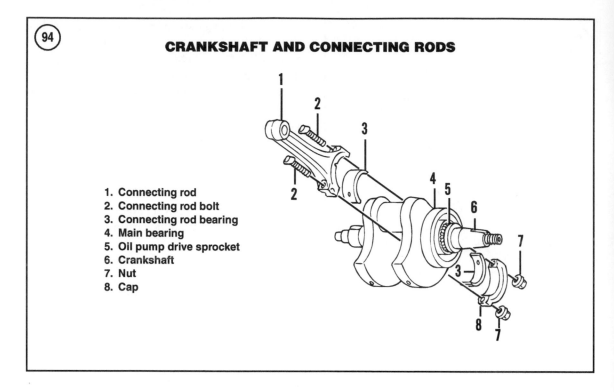

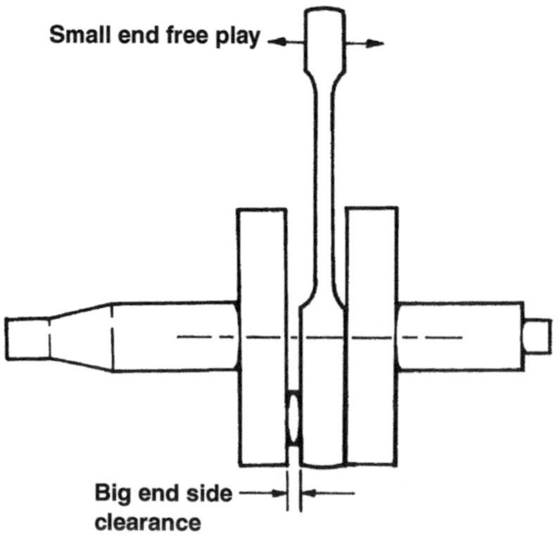

8. Check the connecting rod big end side clearance. Slide the connecting rods to one side. Measure the big end side clearance (**Figure 95**) with a flat feeler gauge. If the clearance is outside the range specified in **Table 1**, replace the connecting rods and recheck

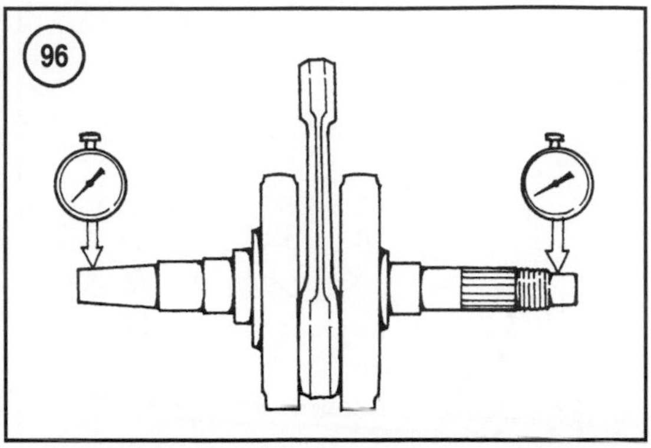

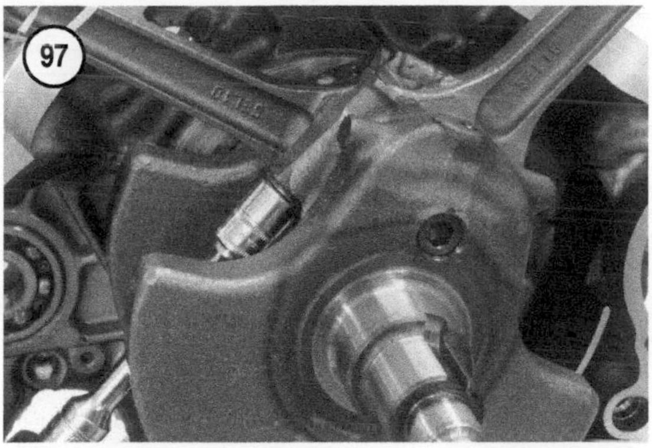

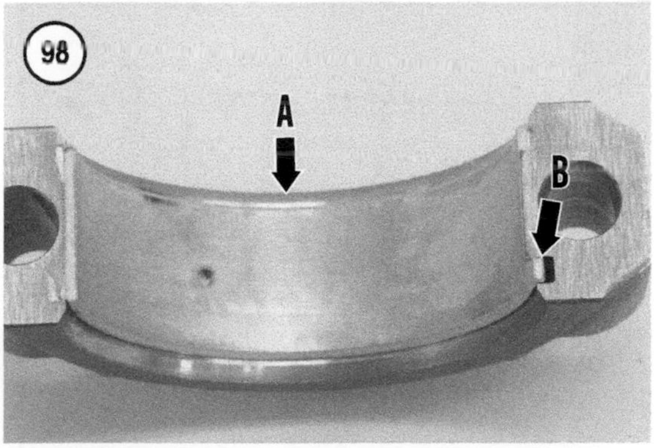

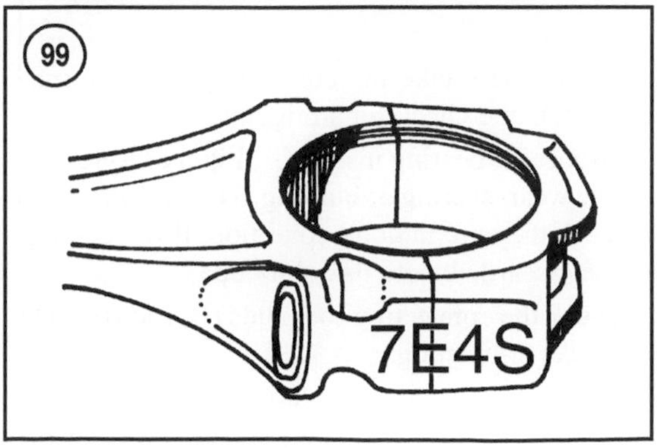

the clearance. If the side clearance is still outside the specified range, replace the crankshaft.

9. Use V-blocks and a dial gauge to check the crankshaft runout (**Figure 96**). If the runout exceeds the specification in **Table 1**, replace the crankshaft.

CONNECTING RODS

Removal/Installation

NOTE

The connecting rods can be serviced while the crankshaft is installed in the left case half. Crankshaft removal is not necessary.

1. Split the crankcase as described in this chapter.
2. Check the big end slide clearance if it has not been checked. Slide the connecting rods to one side. Measure the connecting rod big end side clearance (**Figure 95**) with a flat feeler gauge. If the clearance is outside the range specified in **Table 1**, replace the connecting rods and recheck the clearance. If the side clearance is still outside the specified range, replace the crankshaft.

CAUTION

*Prior to disassembly, mark each connecting rod and cap with an **F** (Front) or **R** (rear) so the rods can be installed in their original locations.*

3. Remove the connecting rod cap nuts (**Figure 97**) and separate the rods from the crankshaft.

CAUTION

*Keep each bearing insert (A, **Figure 98**) in its original place in the connecting rod or rod cap. If bearing inserts are going to be reused, they must be installed in their original locations, or rapid wear will occur.*

4. Mark each rod cap and bearing so they can be reinstalled in their original locations. Make sure the weight mark on the end of the cap matches the mark on the rod (**Figure 99**).
5. Installation is the reverse of removal. Note the following:
   a. Install the bearing insert (A, **Figure 98**) into each connecting rod and cap. Make sure the

tab (B) on the bearing insert locks into the cutout in the rod cap or connecting rod.

CAUTION
*Each connecting rod has a **Y** embossed on one side (**Figure 100**). The connecting rod must be installed so this side faces the left crankshaft end (the tapered end).*

b. Apply engine oil to the connecting rod bearing inserts.
c. If new bearings are going being installed, check the connecting rod oil clearance as described in this chapter.
d. Install the bearing caps so the numbers on the rod and cap align with each other (**Figure 99**).
e. Apply a light-grade molybdenum disulfide grease to the threads of the connecting rod studs.
f. Tighten the connecting rod nuts evenly.

CAUTION
When torquing a connecting rod nut to specification, once 43 N•m (32 ft.-lb.) of torque is applied,tightening cannot be stopped until the final torque value is achieved. If tightening is interrupted between 43-48 N•m (32-35 ft.-lb.), loosen the nut to less than 43 N•m (32 ft.-lb.), then tighten it to 48 N•m (35 ft.-lb.) in one continuous motion.

g. Using a beam-type torque wrench, tighten each connecting rod nut to 48 N•m (35 ft.-lb.). Apply continuous torque when tightening the nut from 43 N•m (32 ft-lb.) to 48 N•m (35 ft.-lb.). Do not stop tightening the nut until the final torque specification is reached.

Connecting Rod Inspection

When inspecting the connecting rods, compare any measurements to the specifications in **Table 1**. Replace any connecting rods that are damaged or out of specification.

1. Check each connecting rod for obvious damage.
2. Check the small end bore (A, **Figure 101**) for wear or scoring.

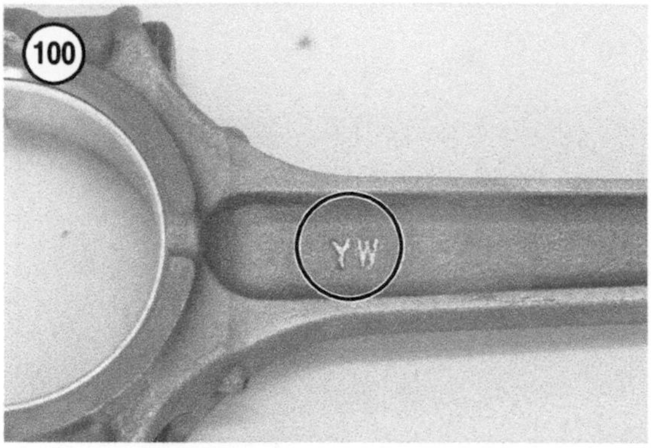

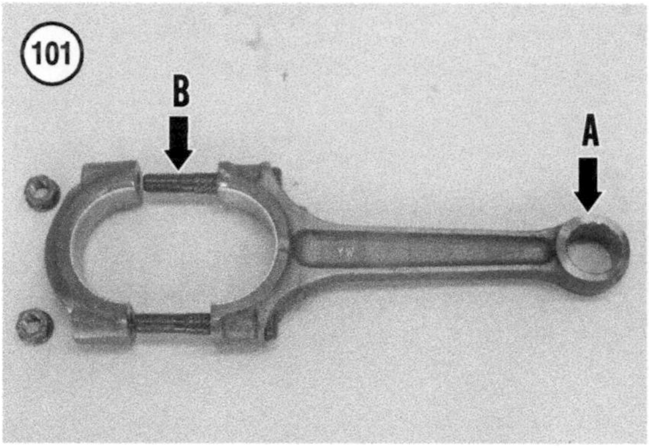

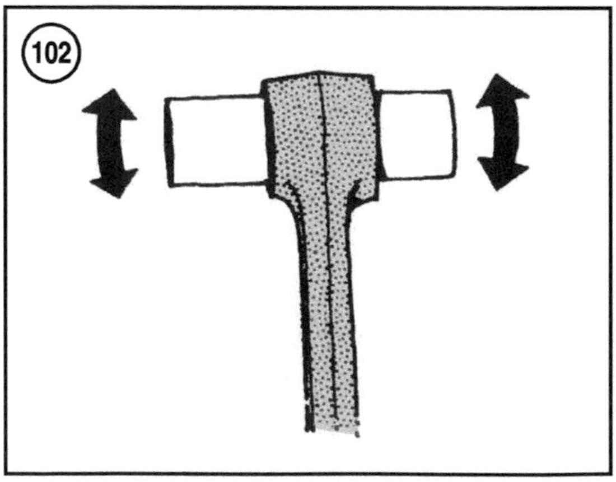

3. If necessary, take the connecting rod to a machine shop and check for straightness.
4. Inspect the bearing inserts (A, **Figure 98**) for excessive wear, scoring or burning. The inserts can be reused if they are in good condition. If the bearing will be discarded, note its color code.
5. Check the connecting rod studs (B, **Figure 101**) for cracks or twisting.
6. Oil the piston pin and install it in the connecting rod. Slowly rotate the piston pin and check for ra-

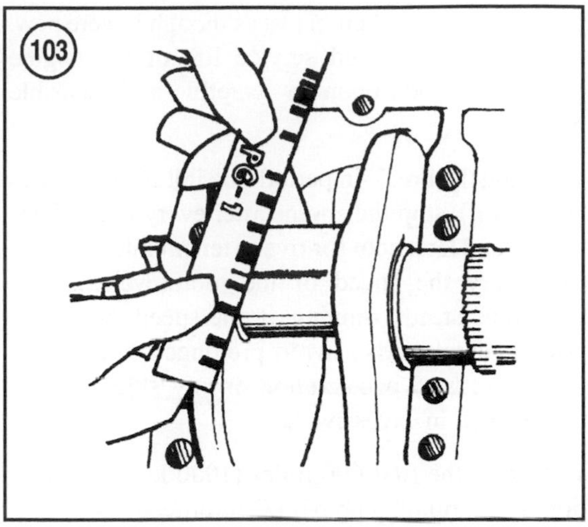

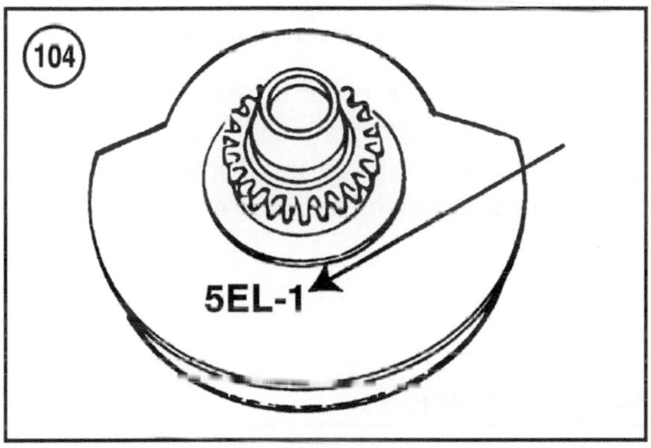

dial and lateral play (**Figure 102**). If any play exists, replace the connecting rod (if piston pin outside diameter is within specification).

7. Check the connecting rod oil clearance as described in this chapter.

Connecting Rod Oil Clearance

1. Wipe the bearing inserts and crankpins clean. Install the inserts (A, **Figure 98**) into their original connecting rod or rod cap.

2. Place a piece of Plastigage on the crankpin parallel to the crankshaft (**Figure 103**).

NOTE

*Each connecting rod has a **Y** embossed on one side. (**Figure 100**). The connecting rod must be installed so this side faces the left crankshaft end (the tapered end).*

3. Install rod and cap. Be sure the weight marks (**Figure 99**) on the side of the rod and cap align.

CAUTION

When torquing a connecting rod nut to specification, once 43 N•m (32 ft.-lb.) of torque is applied, tightening cannot be stopped until the final torque value is achieved. If tightening is interrupted between 43-48 N•m (32-35 ft.-lb.), loosen the nut to less than 43 N•m (32 ft.-lb.), then tighten it to 48 N•m (35 ft.-lb.) in one continuous motion.

4. Using a beam-type torque wrench, tighten each connecting rod nut to 48 N•m (35 ft.-lb.). Apply continuous torque when tightening the nut from 43 N•m (32 ft-lb.) to 48 N•m (35 ft.-lb.). Do not stop tightening the nut until the final torque specification is reached.

CAUTION

Do not rotate the crankshaft while Plastigage is in place.

5. Remove the rod cap.

6. Determine the oil clearance by measuring the width of flattened Plastigage according to the manufacturer's instructions (**Figure 103**). Replace the bearings if the connecting rod oil clearance is outside the range specified in **Table 1**.

7. Clean all Plastigage from the crankshaft and bearing inserts. Install the connecting rods as described in this chapter.

Connecting Rod Bearing Selection

Bearing inserts are identified by color. To determine the proper bearing inserts, calculate the bearing number as described below.

1. The connecting rods and caps are marked with a No. 4 or No. 5 (**Figure 99**).

2. The crankweb is marked with a number (**Figure 104**) that relates to the connecting rod crankpin.

3. To select the proper bearing insert number, subtract the crankpin number from the number on the connecting rod and cap. For example, if the connecting rod is marked with a 4 and the matching crankpin number is a 1, 4 - 1 = 3. The new bearing insert is a No. 3.

4. Refer to **Table 2** and use the bearing number to determine the color code for the bearing insert. In the above example, brown bearing inserts would be installed in the connecting rod.

NOTE
Determine the bearing insert number for both connecting rods. Then take insert numbers and colors to a Yamaha dealership for bearing purchase.

5. Repeat Steps 1-4 for the other connecting rod.
6. After new bearings have been selected, recheck the oil clearance as described in this section. If clearance is still out of specification, take the crankshaft and connecting rods to a Yamaha dealership for further service. Yamaha does not provide connecting rod or crankpin wear specifications.

BREAK-IN

Following cylinder servicing (boring, honing, new rings, etc.) and major lower end work, the engine should be broken in just as though it were new. The performance and service life of the engine greatly depend upon a careful and sensible break-in.

During the break-in period (the initial 1000 miles [1600 km]), stop the engine after every hour of operation and let it cool for five to ten minute. Periodically vary the speed of the motorcycle. Avoid prolonged steady running at one speed, no matter how moderate. Also avoid prolonged full-throttle operation, hard acceleration or any situation that could result in excessive heat.

1. During the first 600 miles (1000 km) of operation, avoid running above 1/3 throttle.

2. Between 600 to 1000 miles (1000-1600 km) of operation, avoid running above 1/2 throttle.

3. At the end of the break-in period (1000 miles [1600 km]), replace the engine oil, oil filter and final gear oil. This helps ensure that all of the particles produced during break-in are removed from the engine.

Table 1 ENGINE LOWER END SPECIFICATIONS

Item	New mm (in.)	Service limit mm (in.)
Connecting rods		
Oil clearance	0.044-0.073 (0.0017-0.0029)	–
Big end side clearance	0.032-0.474 (0.0126-0.0187)	–
Crankshaft		
Runout	–	0.02 (0.0008)
Crank web-to-web width	101.95-102.00 (4.014-4.016)	–
Oil pump		
Inner rotor-to-outer rotor tip clearance	0.03-0.09 (0.001-0.004)	0.15 (0.006)
Outer rotor to housing clearance	0.03-0.08 (0.001-0.003)	0.15 (0.006)
Rotor depth	0.03-0.08 (0.001-0.003)	0.15 (0.006)
Relief valve operating pressure	450-550 kPa (64.0-78.2 psi)	–
Engine oil temperature	75-85° C (167-185° F)	–

Table 2 CONNECTING ROD BEARING INSERT SELECTION*

Connecting rod bearing number	Bearing insert color
1	Blue
2	Black
3	Brown
4	Green
5	Yellow

*Refer to text procedure.

Table 3 ENGINE LOWER END TORQUE SPECIFICATIONS

Item	N•m	in.-lb.	ft.-lb.
Alternator cover bolts	10	89	–
Connecting rod nuts	48	–	35
Crankcase stud			
10 mm	20	–	15
12 mm	24	–	17
Crankcase bolts			
6 mm	10	89	–
10 mm	39	–	29
Engine mounting hardware			
Cylinder head bracket bolts	48	–	35
Cylinder head bracket nuts	74	–	55
Engine bracket bolts	48	–	35
Engine bracket through bolt	48	–	35
Lower frame member bolt	48	–	35
Lower front engine mounting bolt/nut	48	–	35
Lower rear engine mounting bolt/nut	48	–	35
Frame member bolts	48	–	35
Rear engine through bolt	48	–	35
Upper rear engine mounting bolt	48	–	35
Flywheel nut	175	–	129
Neutral switch	20	–	15
Oil pipe banjo bolt	20	–	15
Oil delivery pipe mounting bolt	10	89	–
Oil drain bolt	43	–	32
Oil filter cover bolt	10	89	–
Oil pump sprocket cover bolts	10	89	–
Oil pump gear cover bolts	10	89	–
Oil pump mounting bolts	10	89	–
Oil pump driven sprocket bolt	12	106	–
Oil pump sprocket cover bolt	10	89	–
Pickup coil mounting screw	7	62	–
Primary drive gear nut	110	–	81
Rocker arm bolts			
Exhaust	37	–	27
Intake	20	–	15
Starter clutch bolts	12	106	–
Stator mounting bolt	10	89	–
Shift-lever clamp bolt	10	89	–

5

CHAPTER SIX

CLUTCH AND PRIMARY DRIVE GEAR

This chapter includes service procedures for the clutch, clutch release mechanism and primary drive gear. When inspecting components, compare any measurement to the specifications in **Table 1** at the end of the chapter. Replace any part that is damaged, worn or out of specification. During assembly, tighten fasteners to the specified torque.

The wet, multi-plate clutch is mounted on the right side of the transmission mainshaft.

The clutch release mechanism is mounted in the alternator cover on the left side of the crankcase. It is cable operated by the clutch lever on the left side of the handlebar.

NOTE

In the interest of clarity, the clutch and clutch release mechanism procedures were photographed with the engine removed from the frame. These components can be serviced while the engine is in the frame.

CLUTCH COVER

Removal/Installation

Refer to **Figure 1**.

1. Securely support the motorcycle on level ground.
2. Remove the brake pedal/footrest assembly as described in Chapter Thirteen.
3. Remove the exhaust pipe and air filter housing (Chapter Eight).
4. Drain the engine oil and remove the oil filter (Chapter Three).
5. Open the wire clamps and release the wires running under the cover.
6. Remove the clutch cover bolts and remove the clutch cover. Note the placement of the bolts that secure the wire clamps (A, **Figure 2**) and air filter housing bracket (B). The clamps and bracket must be installed in the same locations during assembly.

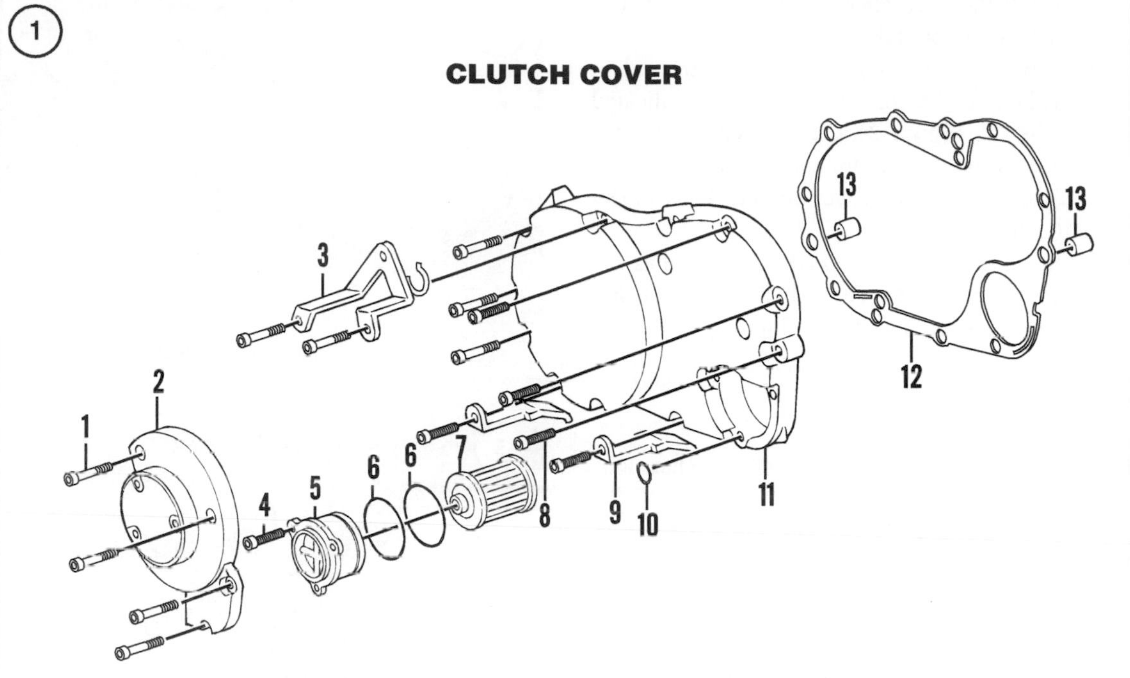

1. Oil filter cover bolt
2. Oil filter outer cover
3. Air filter housing bracket
4. Cap bolt
5. Cap
6. O-ring
7. Oil filter
8. Clutch cover bolt
9. Wire clamp
10. O-ring
11. Clutch cover
12. Gasket
13. Dowel

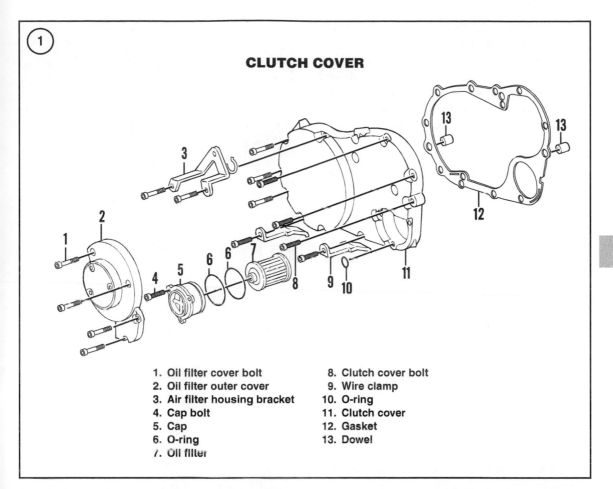

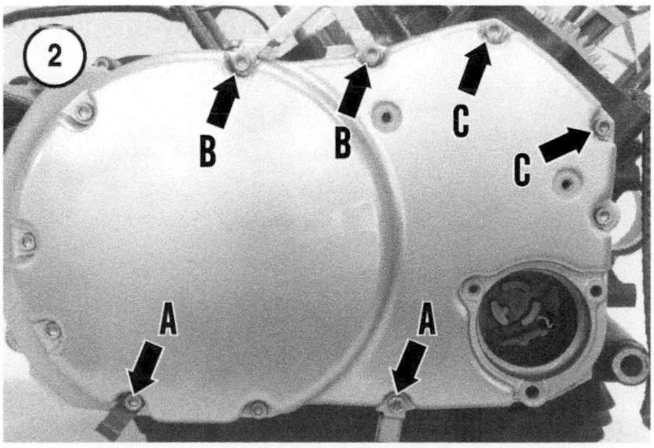

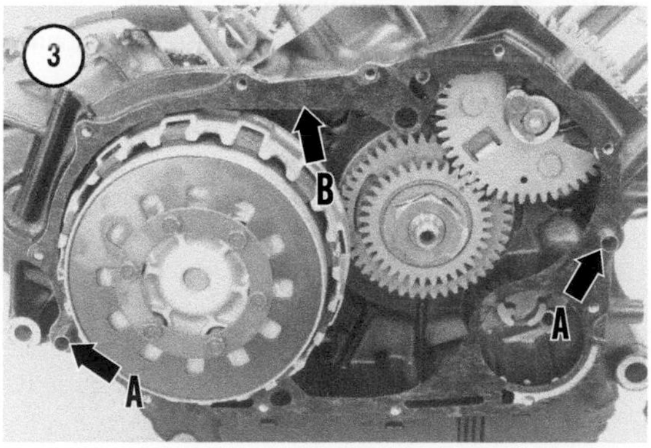

7. Remove the clutch cover from the crankcase. Watch for the dowels (A, **Figure 3**) behind the cover. Remove the dowels.

8. Remove and discard the clutch cover gasket (B, **Figure 3**).

9. Inspect the clutch cover for cracks and other signs of wear. Replace the cover if necessary.

10. Installation is the reverse of removal. Note the following:

a. Install the two dowels (A, **Figure 3**) and a new gasket (B) into the crankcase.

b. Install the air filter housing bracket behind the two bolts (B, **Figure 2**) noted during removal.

c. Install the wire clamps behind the bolts (A, **Figure 2**) noted during removal.
d. Apply Yamaha Bond 1215 or an equivalent sealant to the threads of the two indicated bolts (C, **Figure 2**). Oil will leak past these bolts if the threads are not sealed.
e. Torque the clutch cover bolts to 10 N•m (89 in.-lb.).
f. Install the oil filter and add engine oil as described in Chapter Three. Install a new O-ring (**Figure 4**) behind the oil filter cap.

Inspection

1. Inspect the clutch cover for cracks or other signs of damage.
2. Closely inspect the crankshaft oil seal (A, **Figure 5**) in the clutch cover.

CAUTION

The crankshaft oil seal is a critical part of the oil system. It seals the crankshaft and facilitates main oil delivery. When in doubt, replace this seal.

3. If necessary, remove the oil seal retainer (B, **Figure 5**) for a closer inspection. Replace the oil seal (**Figure 6**) if there is any damage or if the seal is becoming brittle.
4. Replace the seal as follows:
   a. If still installed, remove the oil seal retainer (B, **Figure 5**).
   b. Pry the crankshaft oil seal (**Figure 6**) from the clutch cover.
   c. Lubricate a new seal with lithium soap grease. Position the seal so the manufacturer's marks face out, and drive the seal into the cover with a driver or socket that matches the outside diameter of the seal.
   d. Install the oil seal retainer (B, **Figure 5**) so its concave side faces the oil seal.
   e. Apply a medium-strength threadlocking compound to the threads of the retainer bolts, and tighten them securely.

CLUTCH

The Yamaha clutch holding tool (part No. YM-91042 or 90890-04086), a Grabbit, or an equivalent tool is needed during clutch service.

Removal

Refer to **Figure 7**.

1. Securely support the motorcycle on a level surface.
2. Remove the clutch cover as described in this chapter.
3. Evenly loosen the clutch spring bolts (A, **Figure 8**) in a crisscross pattern. Remove the bolts and the spring plate (B, **Figure 8**).
4. Remove the clutch spring (**Figure 9**) and spring seat (A, **Figure 10**).

7

CLUTCH

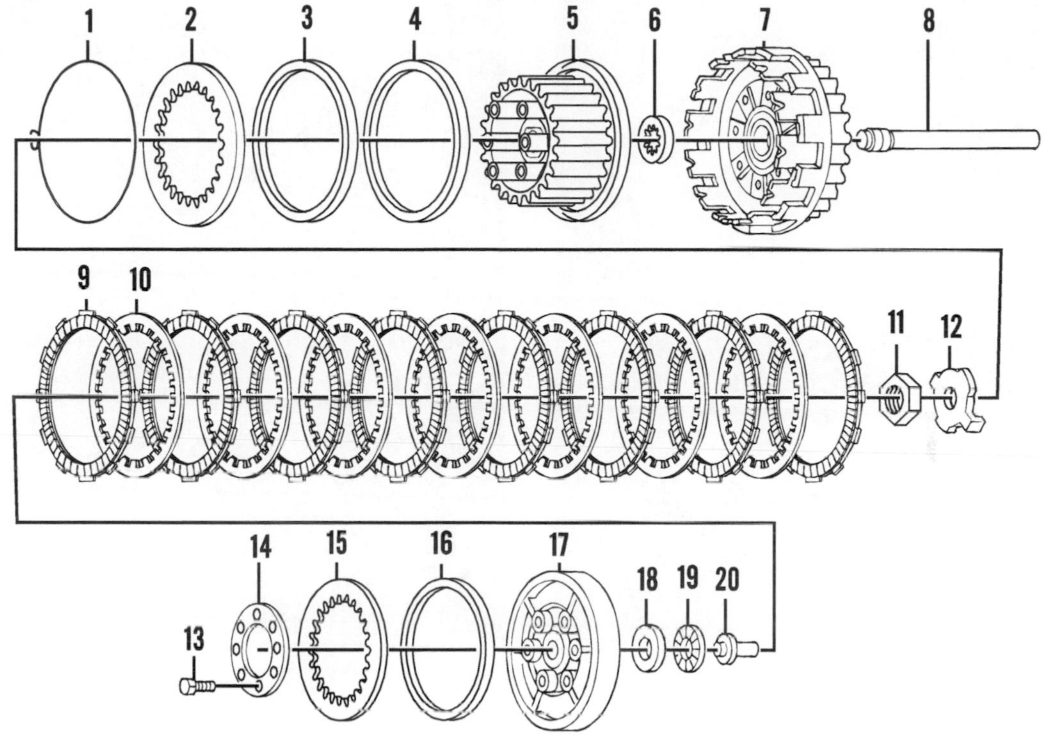

1. Retaining wire
2. Clutch plate No. 1
3. Clutch boss spring
4. Plate seat
5. Clutch boss
6. Thrust washer
7. Clutch housing
8. Pushrod No. 2
9. Friction disc
10. Clutch plate
11. Clutch nut
12. Lock plate
13. Clutch spring bolt
14. Spring plate
15. Clutch spring
16. Spring seat
17. Pressure plate
18. Washer
19. Release bearing
20. Pushrod No. 1

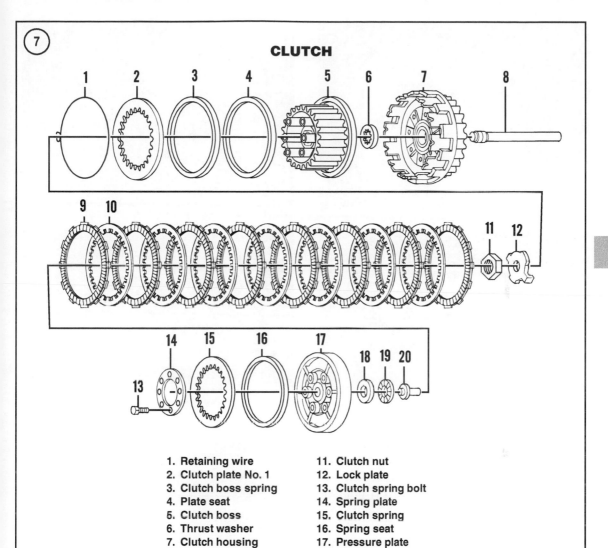

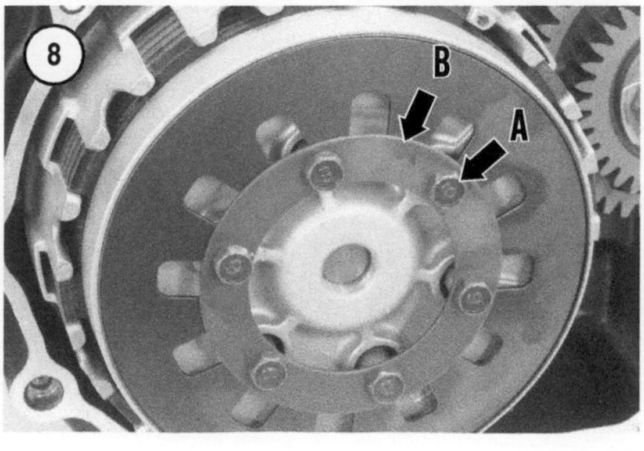

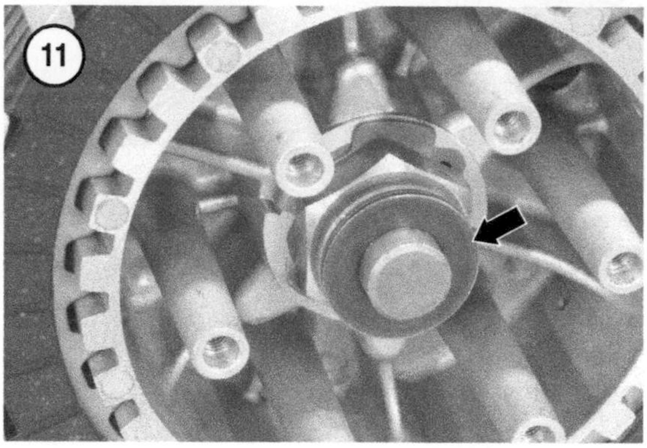

5. Remove the pressure plate (B, **Figure 10**).
6. Remove the washer (**Figure 11**) and the release bearing (**Figure 12**) from pushrod No. 1.
7. Remove pushrod No. 1 (**Figure 13**) from the mainshaft, then remove pushrod No. 2 (**Figure 14**).
8. Note that each friction disc is installed so the tab with the two semi-circular cutouts fits into the clutch basket slot with the two marks. See **Figure 15**. The marked tab on each friction disc must be installed in this marked slot during assembly.
9. Remove all of the friction discs and clutch plates. Stack the disc and plates in the order of removal.
10. Straighten the ears (A, **Figure 16**) of the lockwasher away from the flats on the clutch nut (B).

CAUTION
Do not clamp the special tool too tightly. The tool could damage the grooves in the clutch hub.

11A. Hold the clutch boss with a clutch holding tool (**Figure 17**) and loosen the clutch nut.

CAUTION
Any soft-metal washer, such as copper, brass, or aluminum, will work in

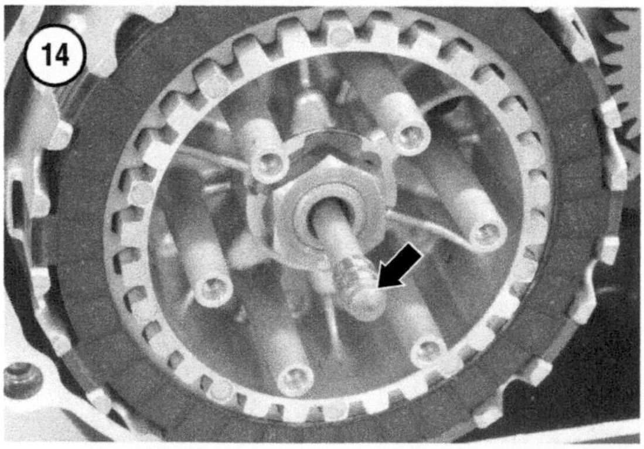

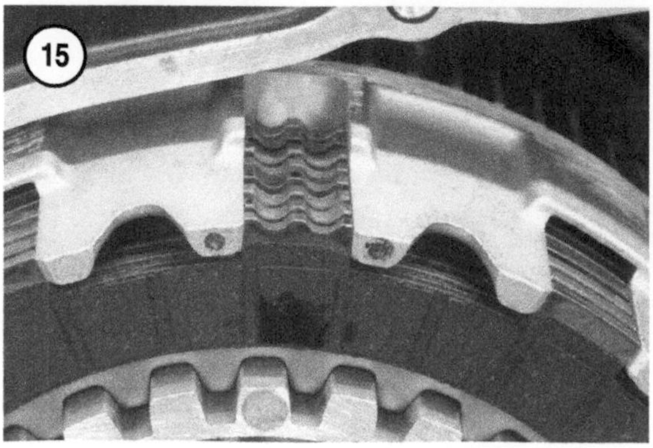

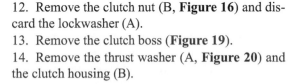

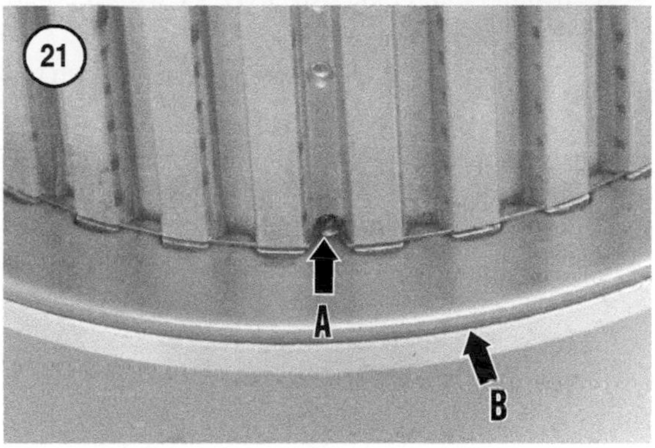

the next step. However, ***do not*** *use a steel washer. Steel will damage the gear teeth.*

11B. If a clutch holder is not available, stuff a shop cloth, copper penny, or brass washer (**Figure 18**) between the primary drive gear and the primary driven gear on the clutch housing. Remove the nut. The washer holds the clutch during nut removal.

12. Remove the clutch nut (B, **Figure 16**) and discard the lockwasher (A).

13. Remove the clutch boss (**Figure 19**).

14. Remove the thrust washer (A, **Figure 20**) and the clutch housing (B).

NOTE

Clutch plate No. 1 is part of a damper assembly on the clutch boss. This assembly does not have to be disassembled unless the clutch experiences severe clutch chatter.

15. If necessary, remove the clutch hub damper assembly as follows:

a. Pull one end of the retaining wire (A, **Figure 21**) from its hole in the clutch boss.

b. Lift the other end from the hole and remove the retaining wire from the clutch hub.
c. Remove clutch plate No. 1 (B, **Figure 21**), the clutch boss spring and the plate seat from the clutch hub.

Installation

1. Assemble the clutch boss damper assembly, if it was removed, as follows:
   a. Install the plate seat onto the clutch hub and then install the clutch boss spring.
   b. Install clutch plate No. 1 (B, **Figure 21**) onto the clutch boss.
   c. Seat one end of the retaining wire through the hole in the clutch boss.
   d. Press the retaining wire into the grooves around the circumference of the clutch boss, then insert the remaining end of the retaining wire into the hole (A, **Figure 21**).
   e. Make sure both ends of the retaining wire are locked in the clutch boss hole (**Figure 22**).
2. Install the clutch housing onto the mainshaft as follows:
   a. Apply oil to the bushing in the clutch housing hub.
   b. Slide the clutch housing (B, **Figure 20**) onto the mainshaft until the teeth of the primary driven gear (A, **Figure 23**) on the housing engage the teeth of the primary drive gear (C, **Figure 20**). Gently push the clutch housing onto the mainshaft until it bottoms.
3. Install the thrust washer (A, **Figure 20**) onto the mainshaft.
4. Slide the clutch boss (**Figure 19**) onto the mainshaft.
5. Install a new lockwasher so its arms (**Figure 24**) engage the cutouts in the clutch boss.
6. Turn the clutch nut (B, **Figure 16**) onto the mainshaft. The concave side of the nut must face in toward the lockwasher.
7. Use the same tool set-up used during removal to hold the clutch boss (**Figure 17**) in place, and torque the clutch nut to 70 N•m (52 ft.-lb.).
8. Bend each lockwasher ear (A, **Figure 16**) flat against the nut.

NOTE
Apply fresh engine oil to each friction disc to avoid clutch lock up. Install each friction disc so the tab with the

22
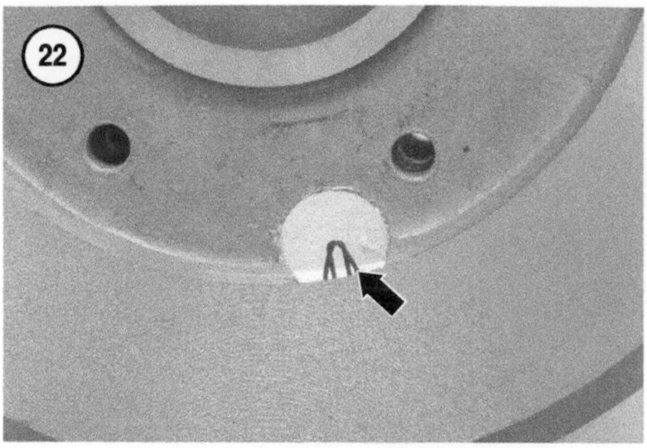

23

24
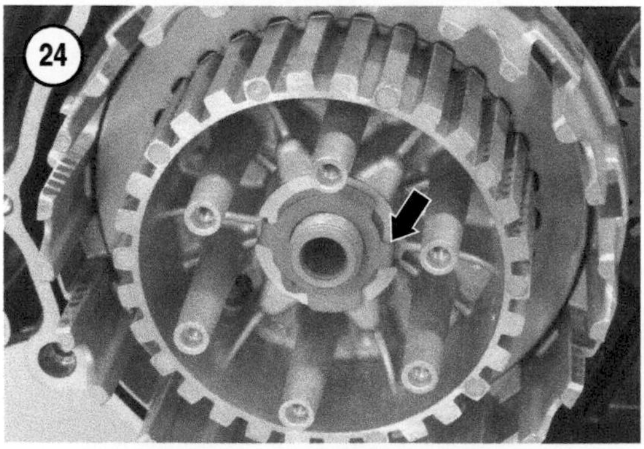

25

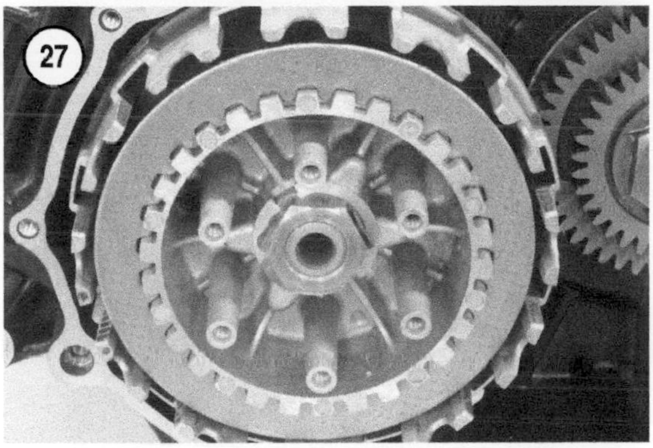

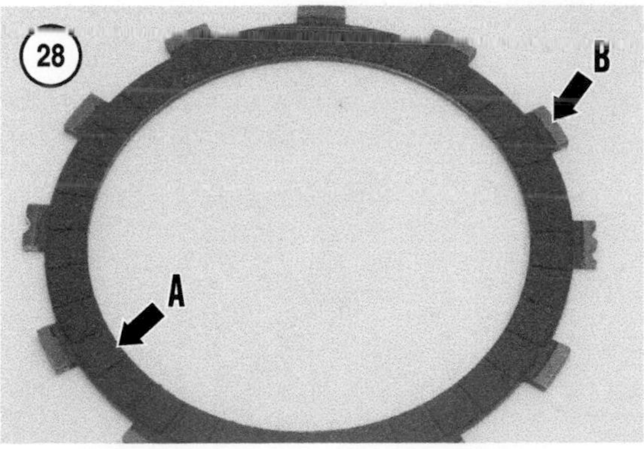

*two semi-circular cutouts fits into the clutch housing slot with the two marks. See **Figure 15**. This slot (A, **Figure 25**) is cut down to the bottom of the clutch housing.*

9. Install the first friction disc (**Figure 26**) so the tab with the two semi-circular cutouts fit into the marked slot on the clutch housing.
10. Install the first clutch plate (**Figure 27**), then install the next friction disc. Continue alternately installing a clutch plate, then a friction disc, until all plates and discs are installed. The last item installed should be friction disc.
11. Make sure the marked tab on each friction disc sits in the marked slots on the clutch housing as shown in **Figure 15**. If necessary, remove the friction discs and clutch plates. Reinstall them correctly.
12. Lubricate pushrod No. 2 with lithium soap grease and install it into the mainshaft so its round end faces out (**Figure 14**).
13. Lubricate pushrod No. 1 with lithium-soap grease and install it into the mainshaft (**Figure 13**).
14. Install the release bearing (**Figure 12**) onto pushrod No. 1, then install the washer (**Figure 11**).
15. Install the pressure plate (B, **Figure 10**) into place so its splines engage those of the clutch boss.
16. Install the spring seat (A, **Figure 10**) onto the pressure plate and set the clutch spring (**Figure 9**) into the spring seat. Make sure the convex side of the spring faces out.
17. Install the spring plate (B, **Figure 8**) into place and loosely install the clutch spring bolts (A). Apply a medium-strength threadlocking compound to the threads of the bolts.
18. Evenly tighten the bolts in a crisscross pattern. Torque the clutch spring bolts 8 N•m (71 in.-lb.).
19. Adjust the clutch lever free play as described in Chapter Three.

Inspection

1. Clean all clutch parts in a petroleum-based solvent, such as a commercial solvent or kerosene. Thoroughly dry the parts with compressed air.
2. Inspect the friction discs as described below. If any disc must be replaced, replace all the friction discs as a set.
   a. Check the friction material (A, **Figure 28**) for excessive wear, cracks and other damage.
   b. Inspect the friction disc tangs (B, **Figure 28**) for surface damage. Pay particular attention to the sides of the tangs where they slide along the clutch housing slots. If these tangs are not smooth, the clutch will not disengage and engage correctly.
   c. Measure the thickness of each friction disc (**Figure 29**) at four places around the disc. If any disc is worn to the service limit (**Table 1**), replace all the friction discs as a set.
3. Inspect the clutch plates as follows:

6

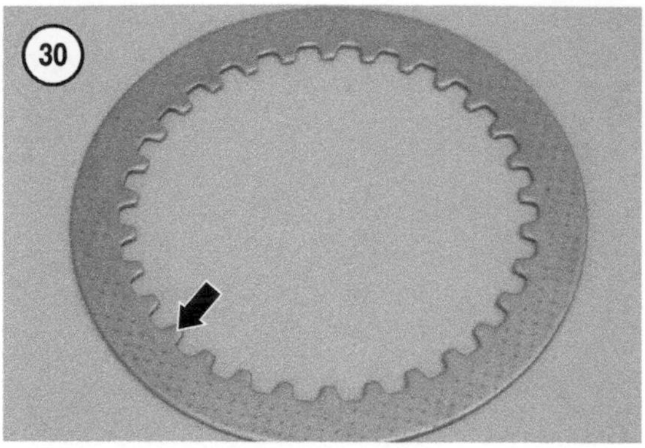

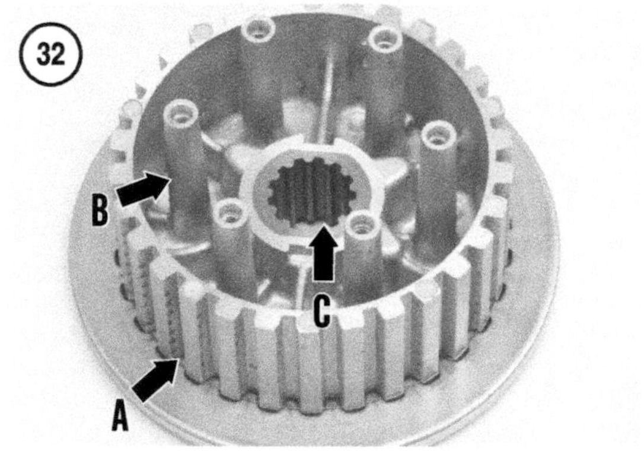

a. Check the inner splines (**Figure 30**) on the clutch plates. Minor roughness can be cleaned with an oilstone or fine file. If any one plate has excessive roughness or wear, replace all the clutch plates as a set.

NOTE
The clutch plate thickness does not apply to clutch plate No. 1. Clutch plate No. 1 is much thicker than the seven standard clutch plates.

b. Measure the thickness of each clutch plate at several places around the plate. If any plate is worn to the service limit, replace all the clutch plates as a set.
c. Check the clutch plates for warp on a surface plate or a piece of plate glass (**Figure 31**). Replace all the clutch plates as a set if the warp in any plate equals or exceeds the service limit in **Table 1**.

4. If clutch plate No. 1 was disassembled from the clutch boss, perform the following:
   a. Measure the thickness of clutch plate No. 1 at four places around its circumference. Replace it if any measurement exceeds the service limit.
   b. Check the warp on a surface plate or piece of plate glass (**Figure 31**). Replace clutch plate No. 1 if its warp equals or exceeds the service limit.
5. Inspect the clutch housing as follows:
   a. Check the slots (A, **Figure 25**) for cracks, nicks or galling where they come in contact with the friction disc tabs. They must be smooth for chatter-free operation. If there is any excessive damage, replace the clutch housing and inspect the friction disc tabs for excessive wear.
   b. Inspect the bushing in the housing for discoloration or other signs of heat damage.
   c. Check the primary driven gear (A, **Figure 23**) on the clutch housing for tooth wear, damage

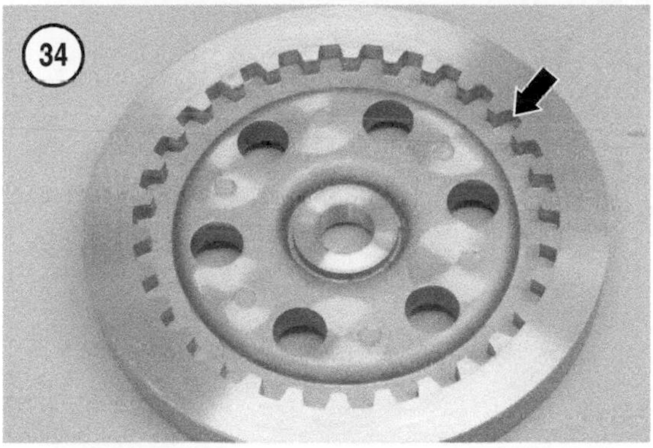

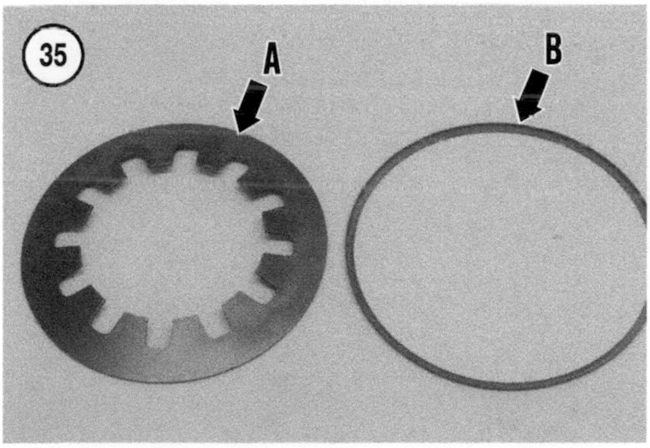

or cracks. Replace the clutch housing if necessary.

d. If the primary driven gear is worn or damaged, also inspect the primary drive gear as described in this chapter.

e. Check the damper springs (B, **Figure 23**) in the clutch housing. Replace the housing if any spring is damaged.

6. Inspect the clutch boss for the following:

a. Check the grooves for cracks, nicks or galling (A, **Figure 32**) where they come in contact

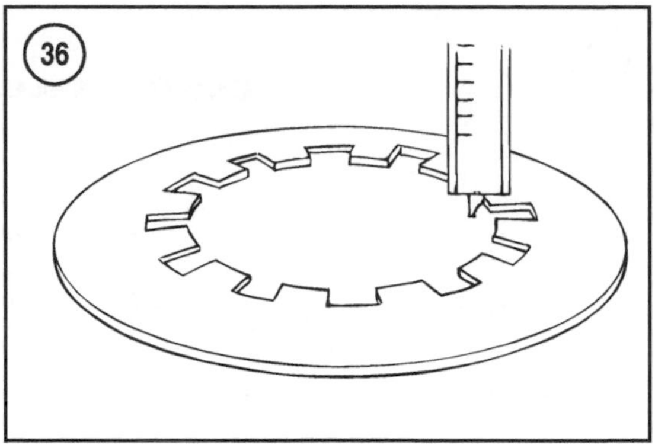

with the clutch plate tabs. They must be smooth for chatter-free operation. If there is any excessive damage, replace the components.

b. Inspect the posts (B, **Figure 32**) for wear or galling. If there is any excessive damage, replace the clutch boss.

c. Inspect the inner splines (C, **Figure 32**) in the hub for damage. Remove any small nicks with an oilstone. If damage is excessive, the clutch boss must be replaced.

8. Check the pressure plate (**Figure 33**) for cracks, wear or galling. Inspect the splines (**Figure 34**) on the inside face of the pressure plate for roughness or excessive wear. Replace the pressure plate as necessary.

9. Visually inspect the clutch spring (A, **Figure 35**) and spring seat (B) for cracks, wear and other signs of fatigue.

10. Measure the free height of the clutch spring with a vernier caliper (**Figure 36**). Replace the spring if its height is less than the service limit.

11. Inspect the ends of pushrods No. 1 and No. 2. Replace either pushrod as necessary.

12. Use a feeler gauge and surface plate to measure the warp of pushrod No. 2. Replace the pushrod if warp exceeds the service limit.

13. Inspect the clutch nut for wear or damage. Maybe sure the threads are in good condition. Replace the nut as necessary.

CLUTCH RELEASE MECHANISM

Disassembly/Inspection

Refer to **Figure 37**.

37

CLUTCH RELEASE MECHANISM

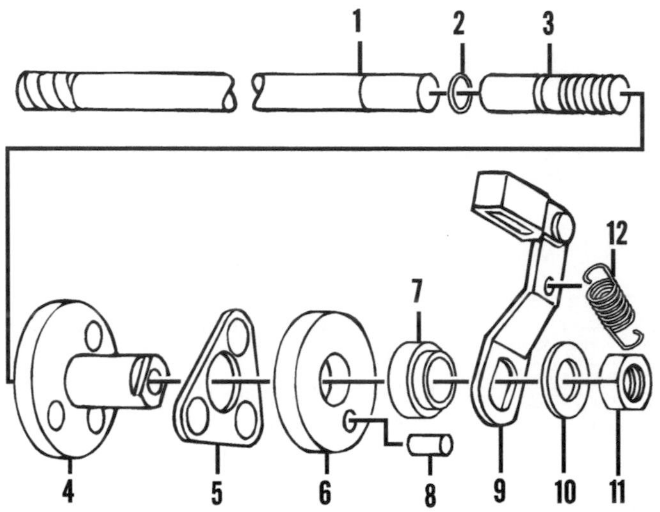

1. Pushrod No. 2
2. O-ring
3. Adjuster
4. Push screw
5. Ball assembly
6. Housing
7. Oil seal
8. Pin
9. Clutch lever
10. Washer
11. Nut
12. Spring

1. Remove the alternator cover as described in Chapter Five.

2. Remove the clutch adjuster locknut (A, **Figure 38**) and washer (B).

3. Disconnect the return spring (A, **Figure 39**) from the post on the alternator cover, and remove the clutch lever (B).

4. Turn the alternator cover over and remove the push screw assembly (**Figure 40**) from the housing in the cover.

5. Remove the ball assembly (A, **Figure 41**) from the push screw (B).

6. Inspect the clutch lever (A, **Figure 42**) for signs of wear or damage.

7. Inspect the balls in the ball assembly (B, **Figure 42**) for scratches, heat discoloration or other signs of damage.

8. Inspect the detents (C, **Figure 42**) in the push screw for scratches, wear or heat discoloration.

9. Inspect the clutch release mechanism oil seal (**Figure 43**) for damage or brittleness.

10. Replace any part that is worn or damaged.

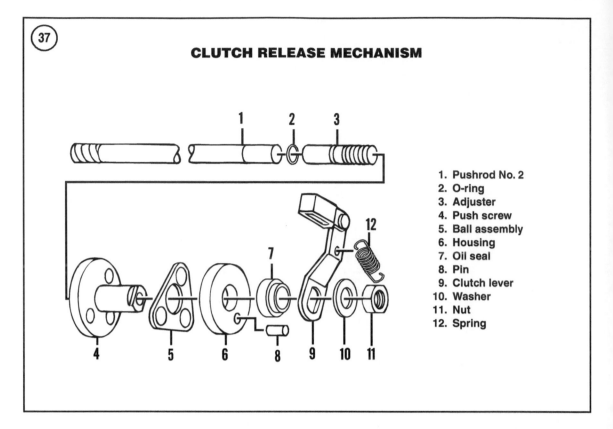

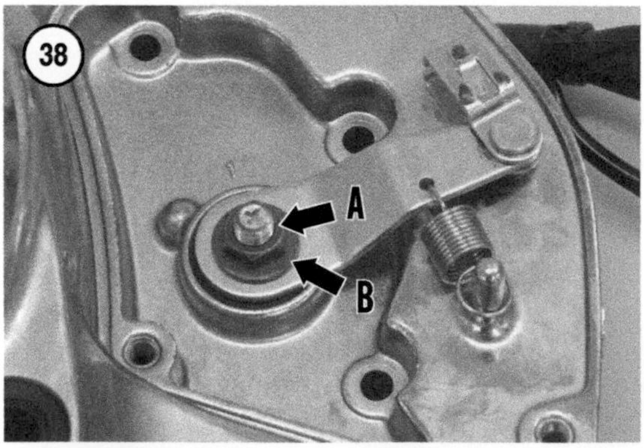

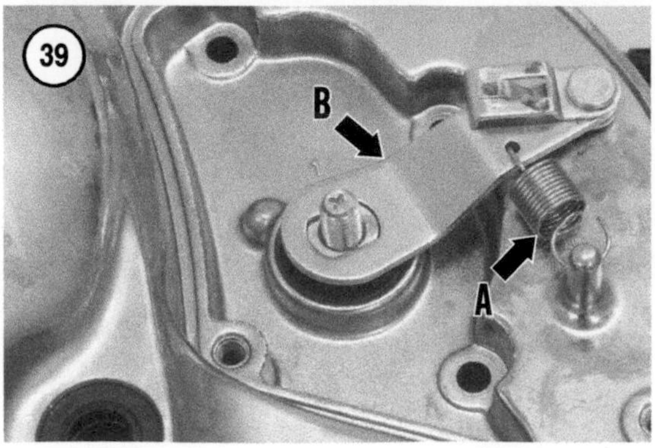

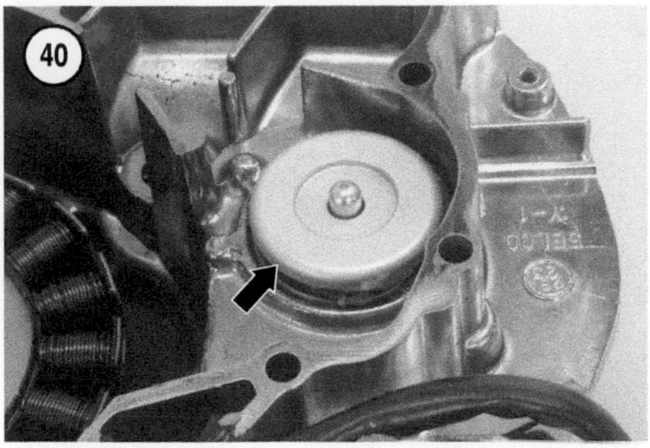
40

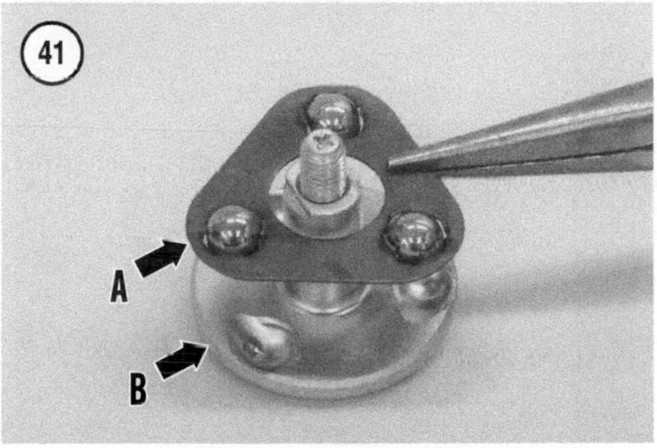

41

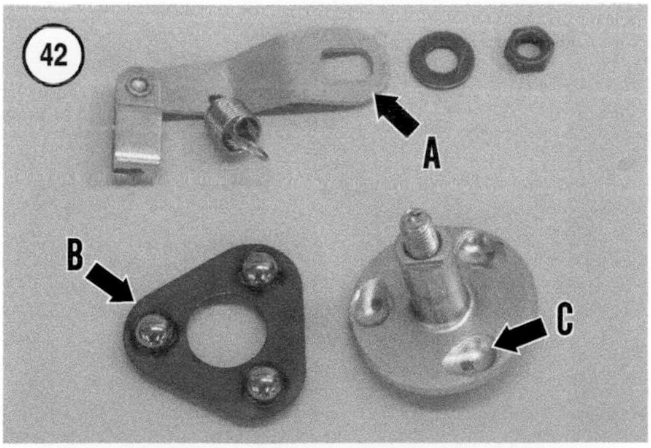

42

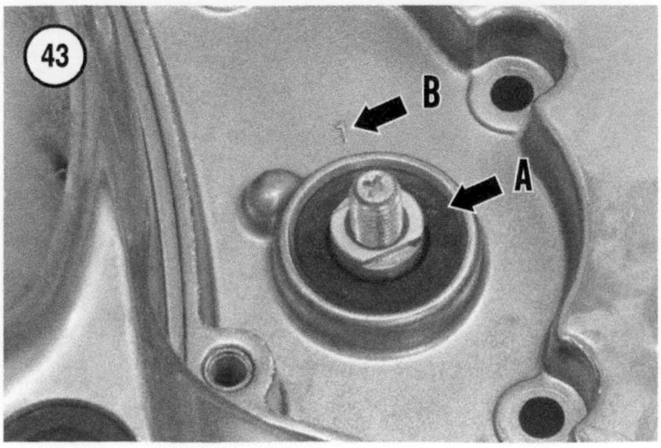

43

Assembly

1. Apply grease to the balls in the ball assembly (A, **Figure 41**).
2. Lower the ball assembly (A, **Figure 41**) onto the push screw (B). Seat the balls in the detents.
3. Invert the assembly and install it into the housing on the inboard side of the alternator cover (**Figure 40**).
4. Turn the alternator cover over.

NOTE
*The clutch will not operate smoothly if a flat on the push screw shaft is not opposite the **1** on the alternator cover (B, **Figure 43**) as described in the next step.*

5. Rotate the push screw shaft so a flat is opposite the *1* on the alternator cover (B, **Figure 43**).
6. Install the clutch lever (B, **Figure 39**) so it engages the flats on the push screw shaft.
7. Hook the return spring (A, **Figure 39**) over the post on the alternator cover.
8. Install the washer (B, **Figure 38**) and turn the locknut (A) onto the shaft. Finger-tighten the locknut. It will be tightened once the clutch cable free play is adjusted.
9. Install the alternator cover as described in Chapter Five.
10. Adjust the clutch cable free play (Chapter Three).

CLUTCH CABLE REPLACEMENT

In time, the clutch cable will stretch to the point where it can no longer function and must be replaced.

1. Remove the fuel tank as described in Chapter Eight.
2. Disconnect the clutch cable from the hand lever as follows:
   a. At the handlebar, slide the clutch cable boot (A, **Figure 44**) away from the adjuster.
   b. Loosen the clutch cable locknut (B, **Figure 44**) and rotate the adjuster (C) to provide maximum slack in the cable.
   c. Disconnect the clutch cable end from the hand lever.
3. Disconnect the cable end from the clutch lever as follows:

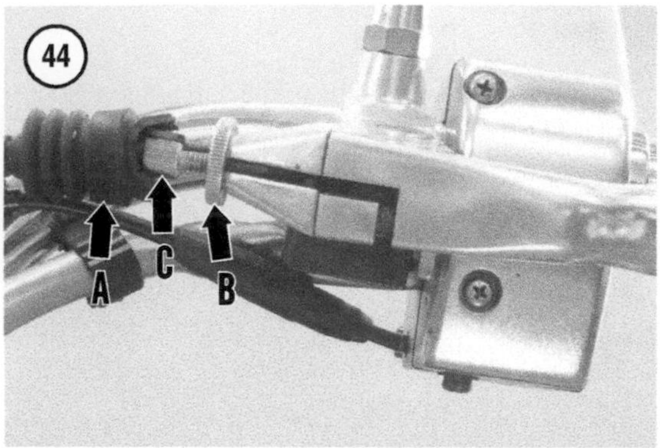

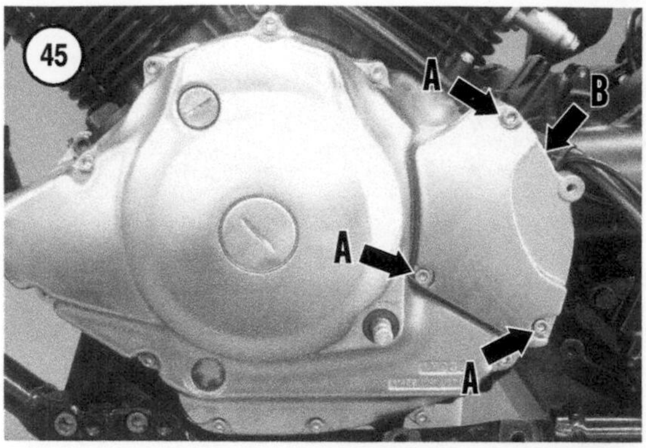

a. Remove the mounting bolts (A, **Figure 45**), and pull the clutch adjuster cover (B) from the alternator cover.

b. Disconnect the clutch cable (**Figure 46**) from the clutch lever.

c. Pull the clutch cable end from the port in the alternator cover.

4. Remove any cable ties that secure the clutch cable to the motorcycle.

5. Release the clutch cable from the AIS/clutch cable bracket on the alternator cover.

NOTE

Before removing the cable, make a drawing of the cable routing through the frame. It is very easy to forget how it was routed once it has been removed. The new cable must be rerouted along the same path as the old cable. Avoid any sharp turns.

6. Pull the clutch cable out from behind the steering head area, through the retaining loop (**Figure 47**) on the left side of the frame and from any clips on the frame.

7. Remove the cable and replace it with a new cable.

8. Installation is the reverse of removal.

a. Route the new cable along the same path as the old.

b. Secure the new cable to the motorcycle at the same points noted during removal.

c. Adjust the clutch cable free play as described in Chapter Three.

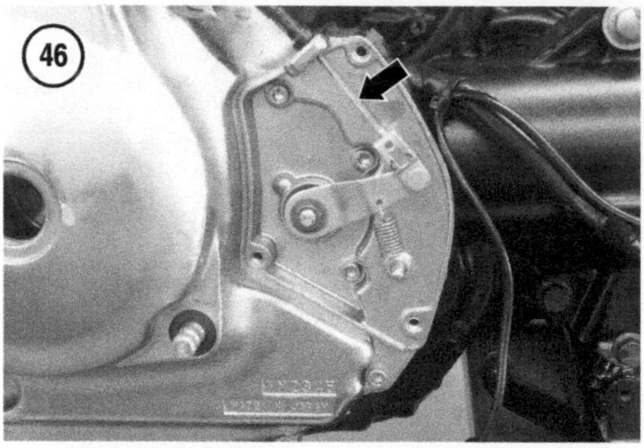

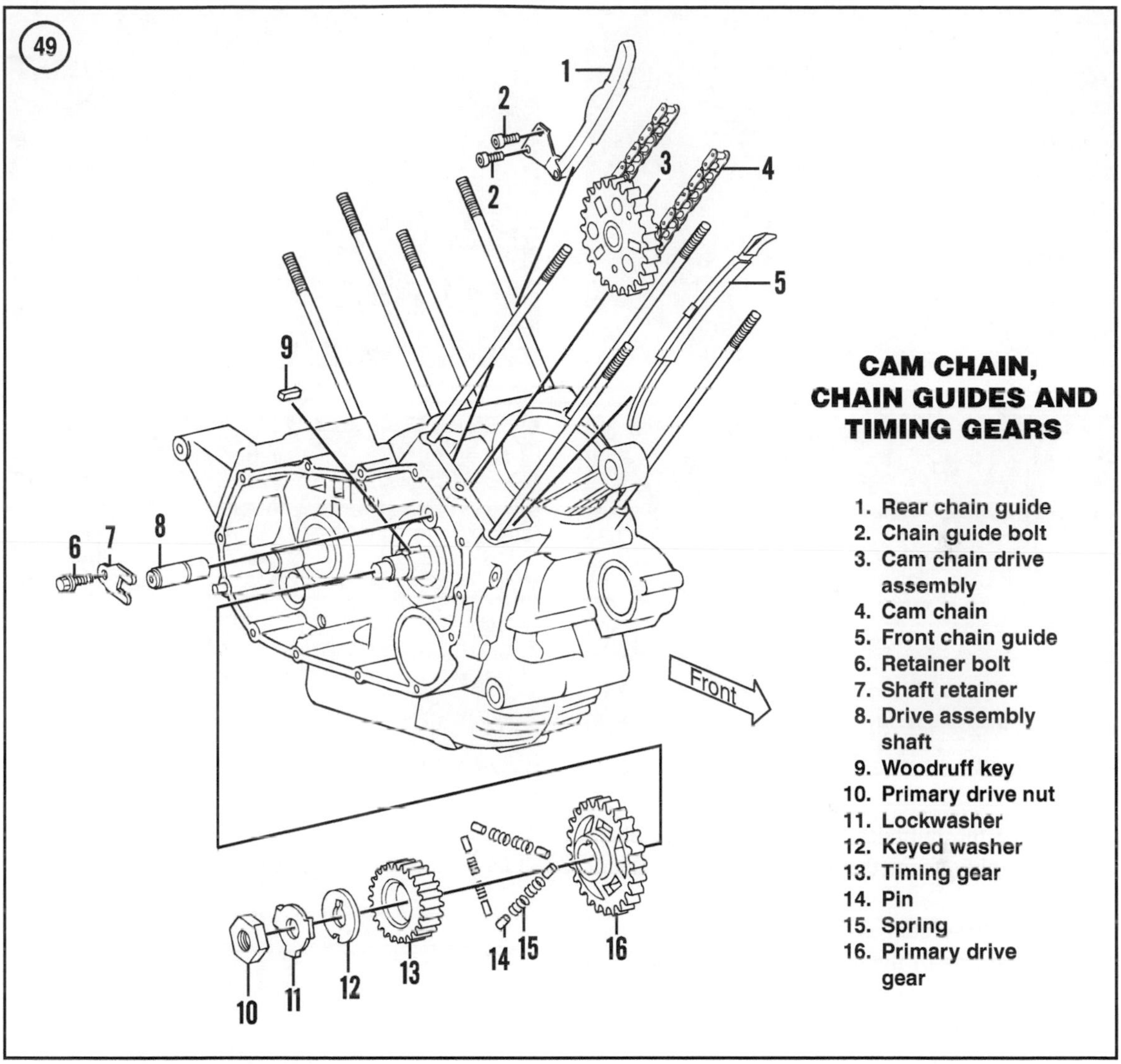
49
1
2
2
3
4
5
9
6
7
8
Front
10
11
12
13
14
15
16
CAM CHAIN, CHAIN GUIDES AND TIMING GEARS
1. Rear chain guide
2. Chain guide bolt
3. Cam chain drive assembly
4. Cam chain
5. Front chain guide
6. Retainer bolt
7. Shaft retainer
8. Drive assembly shaft
9. Woodruff key
10. Primary drive nut
11. Lockwasher
12. Keyed washer
13. Timing gear
14. Pin
15. Spring
16. Primary drive gear

PRIMARY DRIVE GEAR

The Yamaha sheave holder (part No. YU-01880 or 90890-04131), or an equivalent flywheel holder, is needed to service the primary drive gear.

Removal

NOTE

When the front cylinder is set to TDC on the compression stroke, the timing mark on the cam sprocket plate may not precisely align with the pointer on the cylinder head. On some models, the camshafts are slightly retarded. The mark could be off by as much as 1/2 tooth. Before removing the primary drive gear, set the front cylinder to top dead center on the compression stroke, and note the position of the timing marks on the cam sprocket plate and cylinder head. Take a photograph or make a drawing so the camshaft can be correctly timed during assembly. Also note that the timing mark on the timing gear (A, ***Figure 48****), the mark on the cam chain drive assembly (B) and the center of the drive assembly shaft (C) align when the front cylinder is at TDC.*

Refer to **Figure 49**.

1. Remove the engine as described in Chapter Five.

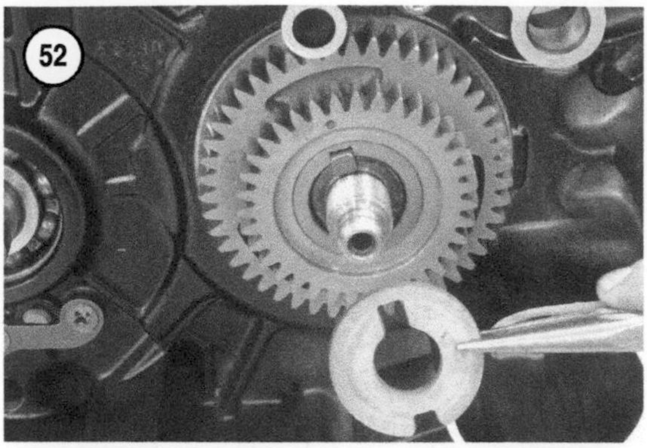

2. Remove the cylinder head and the cam chain drive assembly from the front cylinder (Chapter Four).
3. Remove the clutch as described in this chapter.
4. Bend the ears of the lockwasher (A, **Figure 50**) away from the primary gear nut.
5. Hold the flywheel with the sheave holder and remove the primary drive nut (B, **Figure 50**).
6. Remove and discard the lockwasher (**Figure 51**).
7. Remove the slotted washer (**Figure 52**) from the crankshaft.

NOTE
*The primary drive gear (A, **Figure 53**) and the timing gear (B) come out as an assembly. However, if the timing gear slips from the primary drive gear during removal, the six springs and six pins may be scattered. Stuff rags into the opening in the crankcase so nothing falls into the case.*

8. Grasp the primary drive gear (A, **Figure 53**) and slide the primary drive gear/timing gear (B) assembly from the crankshaft.

56

57

58

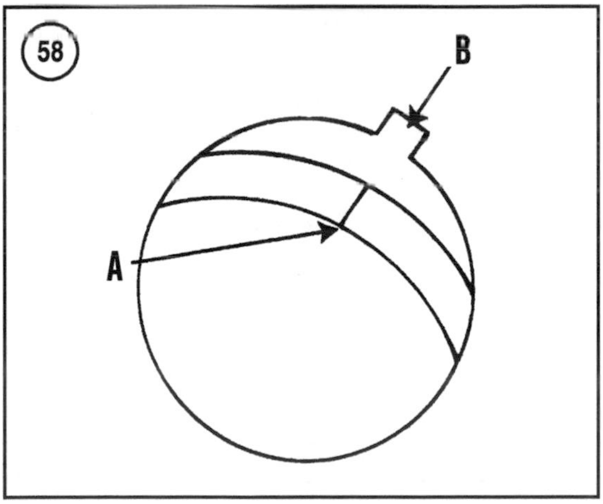

9. Remove the Woodruff key (C, **Figure 53**) from the crankshaft.

10. Set the assembly down on the bench. If necessary, remove the timing gear (A, **Figure 54**) from the primary drive gear (B).

11. Remove each pin/spring assembly (**Figure 55**) from the slots in the primary drive gear. A total of six pins and six springs should be found. Store them in a recloseable plastic bag.

59

Installation

6

1. Insert a pin into each spring (**Figure 56**) and install two sets of pins/springs into each slot on the primary drive gear (**Figure 55**). Push the springs as far apart as possible.
2. Apply grease to the dogs (A, **Figure 57**) on the timing gear.
3. Position the timing gear so its index mark (C, **Figure 54**) aligns with the keyway (D) on the primary drive gear. Set the dogs between the springs in each gear slot and press the timing gear into place.
4. Install the primary drive gear/timing gear assembly onto the crankshaft and install the Woodruff key (C, **Figure 53**) into the keyway.
5. Install the slotted washer (**Figure 52**) so its slot engages the Woodruff key.
6. Install a new lockwasher (**Figure 51**) and the primary drive gear nut (B, **Figure 50**).
7. Hold the flywheel with the sheave holder and torque the primary drive gear nut to 110 N•m (81 ft.-lb.).
8. Bend the ears of the lockwasher (B, **Figure 50**) flat against the nut.
9. Make sure the front cylinder is still set to top dead center on the compression stroke. If necessary, set it as follows:
   a. Set the rear cylinder to TDC by rotating the engine clockwise until the T-mark aligns with the alternator cover cutout.
   b. Rotate the crankshaft another 290° clockwise until the I-mark on the flywheel (A, **Figure 58**) aligns with the cutout in the alternator cover (B).
   c. The timing mark (A, **Figure 59**) on the timing gear and cam chain drive assembly (B) should align with the center of the mounting

boss for the cam chain drive assembly shaft (C). See **Figure 48**.

10. Install the cam chain drive assembly and the cylinder head as described in Chapter Four.

11. Install the clutch as described in this chapter.

Inspection

1. Inspect the primary drive gear (**Figure 60**) for broken or missing teeth. If it is worn, replace the primary drive gear and inspect the primary driven gear on the clutch housing.

2. Inspect the dogs (A, **Figure 57**) and teeth (B) on the timing gear for chips, cracks or other signs of damage. If there is wear or damage, replace the timing drive gear and inspect the front cylinder cam chain drive assembly (Chapter Four).

60

3. Inspect the Woodruff key for nicks or other signs of wear. If it is worn, replace the key and inspect the keyway in the primary drive gear, timing gear, slotted washer and in the crankshaft.

4. Inspect the threads of the primary gear nut. Replace the nut as necessary.

Table 1 CLUTCH SPECIFICATIONS

Item	New mm (in.)	Service limit mm (in.)
Friction disc		
Thickness	2.9-3.1 (0.114-0.122)	2.8 (0.11)
Quantity	8	–
Clutch plate		
Thickness	1.9-2.1 (0.075-0.083)	0.1 (0.004)
Quantity	7	
Warp	–	Less than 0.1 mm (0.004 in.)
Clutch plate No. 1		
Thickness	2.5-2.7 (0.098-0.106)	0.1 (0.004)
Quantity	1	
Warp	–	Less than 0.1 mm (0.004 in.)
Clutch lever free play		
At lever end	5-10 (0.20-0.39)	–
Clutch spring free height	7.2 (0.283)	6.5 (0.256)
Clutch housing thrust clearance	0.05-0.40 (0.002-0.016)	–
Clutch housing radial clearance	0.010-0.044 (0.0004-0.0017)	–
Push rod No. 2 warp	–	0.5 (0.02)

Table 2 CLUTCH AND PRIMARY DRIVE TORQUE SPECIFICATIONS

Item	N•m	in.-lb.	ft.-lb.
Clutch adjuster locknut	12	106	–
Clutch cover bolts	10	89	–
Clutch nut	70	–	52
Clutch spring bolts	8	71	–
Primary drive gear nut	110	–	81
Shift-lever clamp bolt	10	89	–

CHAPTER SEVEN

TRANSMISSION, SHIFT MECHANISM AND MIDDLE GEAR

This chapter describes service procedures for the transmission, internal shift mechanism, external shift mechanism, middle drive shaft assembly and the middle driven shaft assembly.

When inspecting these components, compare any measurement to the specifications listed in **Table 1** at the end of the chapter. Replace any components that are worn, damaged or out of specification. During assembly, tighten fasteners to the specified torque.

SHIFT PEDAL/FOOTREST ASSEMBLY

Removal/Installation

Refer to **Figure 1**.

1. Note that the index mark (A, **Figure 2**) on the shift shaft aligns with the slot in the shift lever. If the mark is worn or missing, make a new one so the shift lever can be properly installed on the shift shaft during assembly.
2. Loosen the shift lever clamp bolt (B, **Figure 2**).
3. Remove the footrest bracket bolts (**Figure 3**).
4. Slide the shift lever (C, **Figure 2**) from the shift shaft and remove the shift pedal/footrest. Watch for the washer behind the shift lever.
5. If necessary, remove the shift pedal from the footrest bracket as follows:
   a. Remove the snap ring from the pivot boss on the footpeg bracket.
   b. Remove the washer and slide the shift pedal from the pivot boss.
   c. On XVS1100A models, watch for the wave washer on the outboard side of the shift pedal.
6. Installation is the reverse of assembly. Note the following:
   a. Lubricate the pivot boss on the footrest bracket with lithium soap grease.
   b. Make sure the snap is properly seated in the groove on the footpeg bracket pivot boss.

(1)

EXTERNAL SHIFT MECHANISM

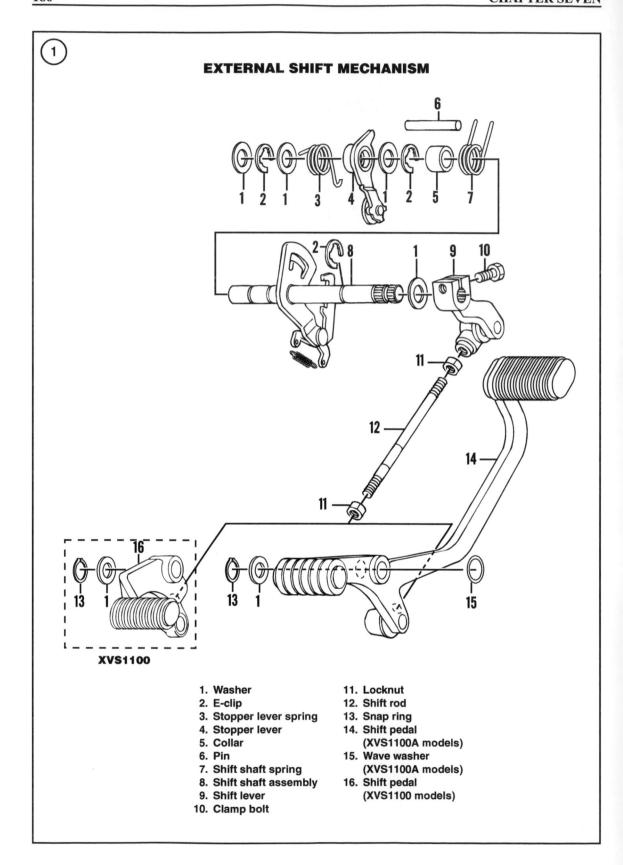

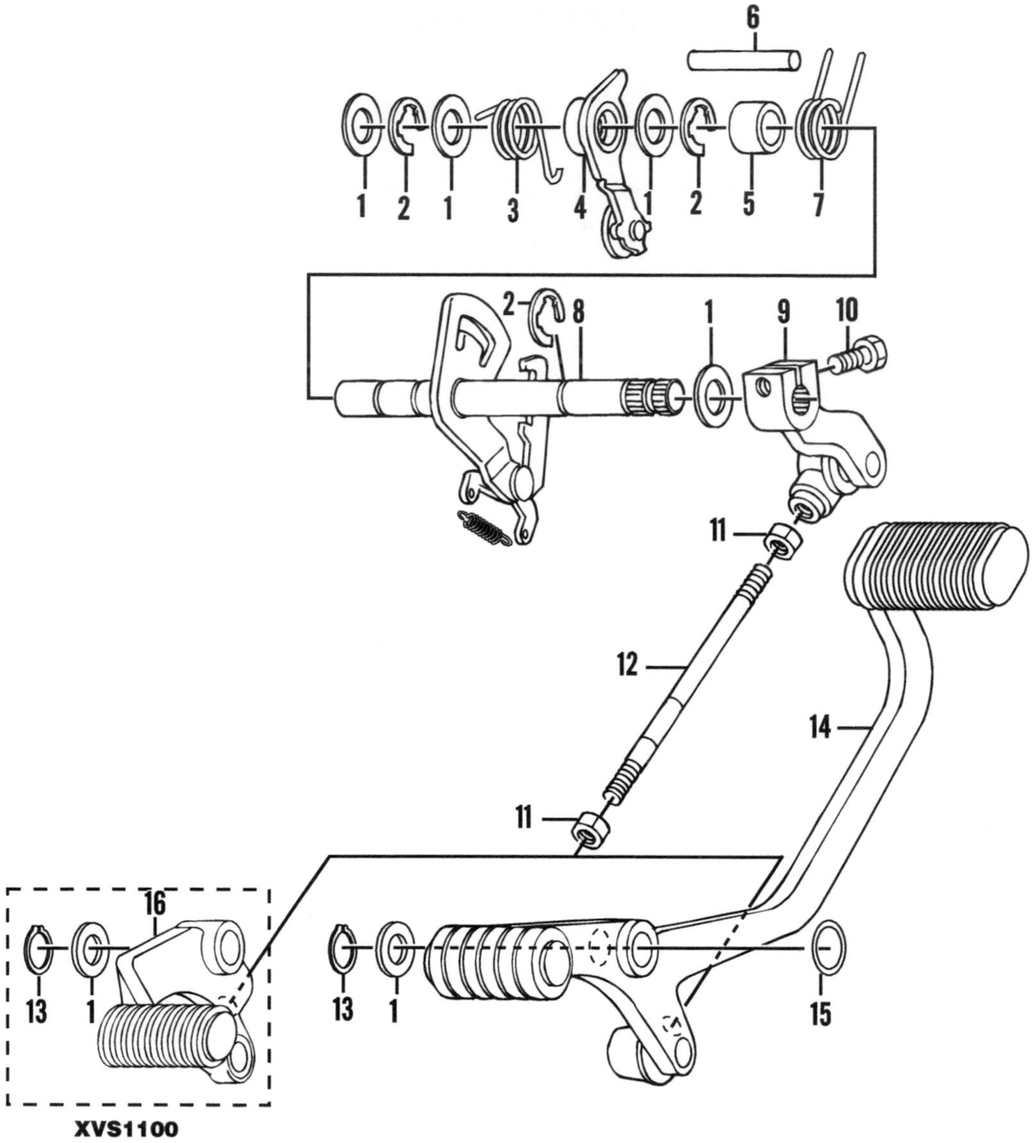

1. Washer
2. E-clip
3. Stopper lever spring
4. Stopper lever
5. Collar
6. Pin
7. Shift shaft spring
8. Shift shaft assembly
9. Shift lever
10. Clamp bolt
11. Locknut
12. Shift rod
13. Snap ring
14. Shift pedal (XVS1100A models)
15. Wave washer (XVS1100A models)
16. Shift pedal (XVS1100 models)

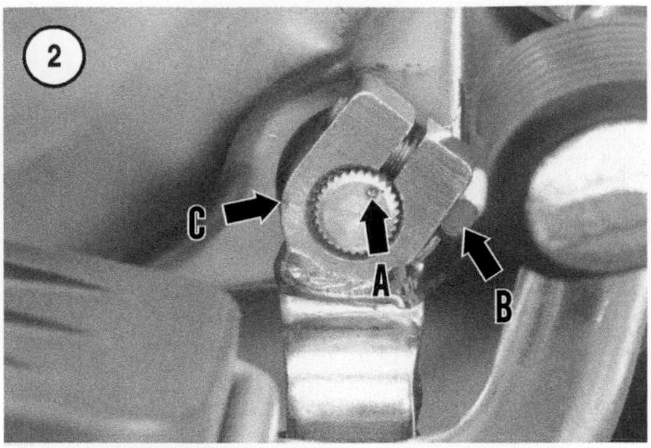

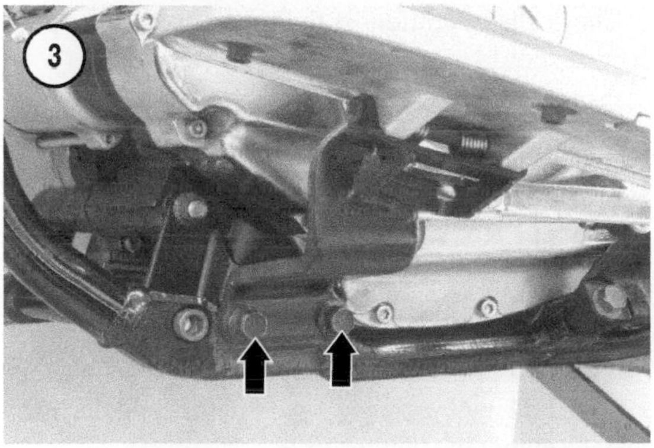

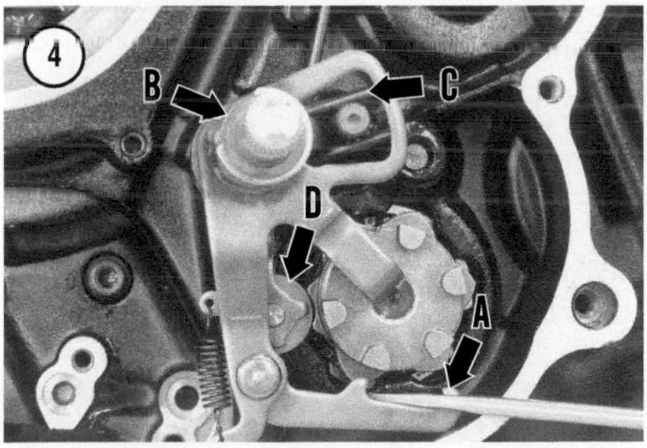

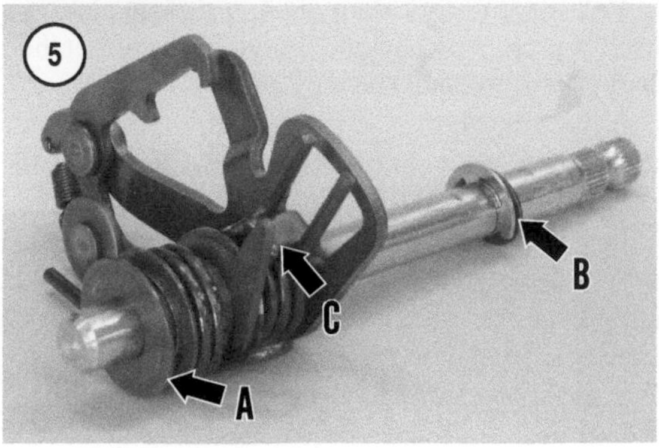

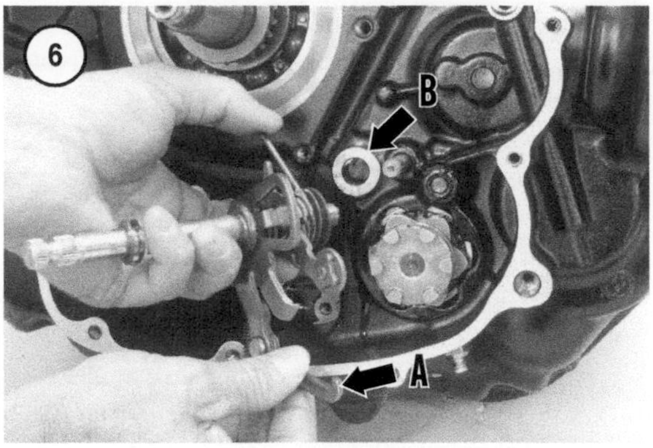

c. Install the shift lever so the index mark (A, **Figure 2**) on the shift shaft aligns with the slot in the shift lever.

d. Torque the shift lever clamp bolt (B, **Figure 2**) to 10 N•m (89 in.-lb.).

e. Torque the footrest bracket bolts (**Figure 3**) to 64 N•m (47 ft.-lb.).

f. Adjust the shift pedal as described in Chapter Three.

EXTERNAL SHIFT MECHANISM

Removal

Refer to **Figure 1**.

1. Securely support the motorcycle on level ground, and shift the transmission into neutral.
2. Remove the shift pedal/footrest assembly as described earlier in this chapter.
3. Drain the engine oil as described in Chapter Three.
4. Remove the alternator cover and pull it from the flywheel (Chapter Five).
5. Press the shift arm away from the shift cam so the shift pawls (A, **Figure 4**) clear the shift pins.
6. Grasp the shift shaft (B, **Figure 4**) and pull the assembly from the crankcase. Watch for the washer (A, **Figure 5**) on the inboard end of the shift shaft.

Installation

1. Make sure a washer is in place on each end of the shift shaft (A and B, **Figure 5**).
2. Spread the shift arm (A, **Figure 6**) so the pawls can clear the pins in the shift drum.

3. Slide the end of the shift shaft into the crankcase boss (B, **Figure 6**) until the assembly bottoms. Note the following:
   a. The arms of the shift shaft spring (C, **Figure 4**) must straddle the post in the crankcase.
   b. When the shift arm is released, the shift pawls must engage the pins on the shift drum as shown in A, **Figure 7**.
   c. The roller (D, **Figure 4**) on the stopper lever must engage the detent on the shift cam.
4. Pull the arm of the stopper lever spring (B, **Figure 7**) and seat it on the forward side of the crankcase boss (C).

Disassembly/Assembly

NOTE
Replace any removed E-clip. An E-clip should not be reused.

1. Remove the washer from each end of the shift shaft.
2. Remove the inboard E-clip and the washer. Discard the E-clip.
3. Slide the stopper lever and stopper lever spring from the shift shaft.
4. Remove the washer and the middle E-clip.
6. Slide the shift shaft spring and its collar from the assembly.
7. If necessary, remove and discard the remaining E-clip.
8. Assembly is the reverse of removal. Note the following:
   a. Use new E-clips. Make sure each clip completely engages its groove in the shift shaft.
   b. The arms of the shift shaft spring must straddle the tab (C, **Figure 5**) on the shift shaft.
   c. The hook on the stopper lever spring must sit in the cutout on the stopper lever

Inspection

Replace any part that is worn or damaged.
1. Inspect the shift shaft for bends or damage.
2. Inspect the shift arm for bends or damage. Pay particular attention to the shift pawls.
3. Check the shift arm spring, the shift shaft spring and the stopper lever spring for cracks or signs of fatigue. Replace any spring as necessary.

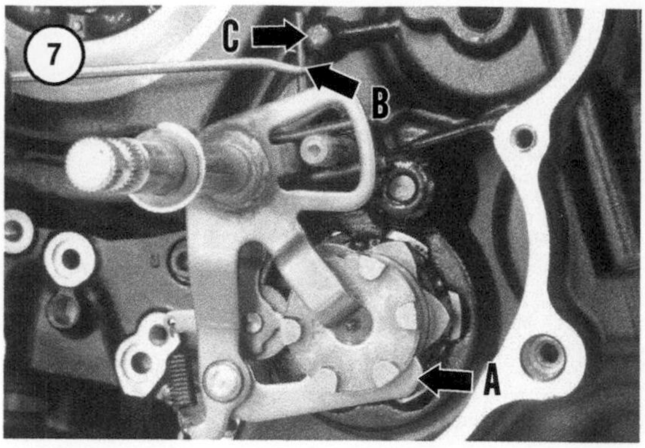

4. Check the operation of the roller on the stopper lever. It should move smoothly.
5. Inspect the stopper lever for bends or other signs of damage.

INTERNAL SHIFT MECHANISM

Removal/Installation

Remove and install the internal shift mechanism as described in *Crankcase Disassembly* and *Crankcase Assembly* in Chapter Five.

Inspection

Refer to **Figure 8**.
1. Inspect each shift fork for signs of wear or cracking. Examine each fork at the points where their fingers (A, **Figure 9**) contact the gears and where the guide post (B) contacts the shift drum. These surfaces should be smooth with no signs of wear or damage.
2. Make sure each fork slides smoothly on the shaft (**Figure 10**). If there is any binding, replace the fork shaft and related shift fork(s).
3. Roll the shift fork shaft along a surface plate or piece of glass. Any clicking sounds indicate that the shaft is bent and must be replaced.
4. Inspect the grooves in the shift drum (A, **Figure 11**) for wear or roughness. Replace the shift drum if any groove is worn.
5. Inspect the shift drum journal (B, **Figure 11**) for wear or discoloration.
6. Spin the drum bearing (C, **Figure 11**), and check for excessive play or roughness.
7. Inspect the neutral post (A, **Figure 12**) for wear or damage.

(8)

INTERNAL SHIFT MECHANISM

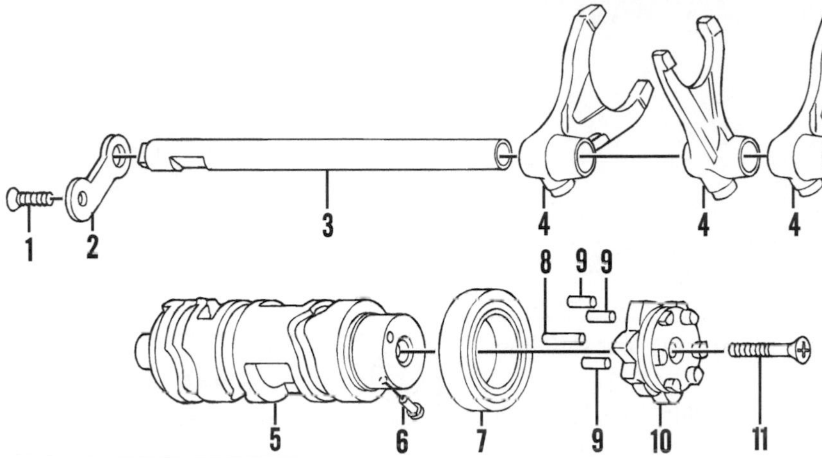

1. Retainer screw
2. Shift fork shaft retainer
3. Shift fork shaft
4. Shift fork
5. Shift drum
6. Neutral pin
7. Bearing
8. Pin (20 mm)
9. Pin (16 mm)
10. Shift cam
11. Shift drum screw

(9)

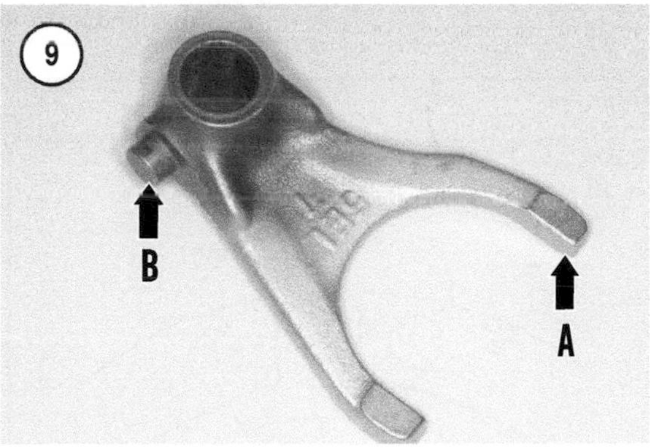

(11)

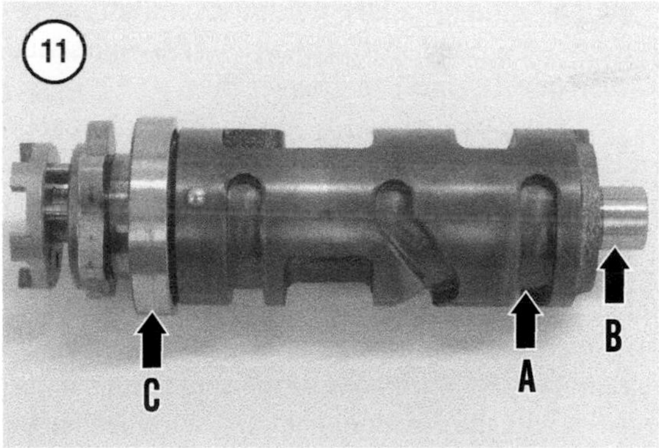

(10)

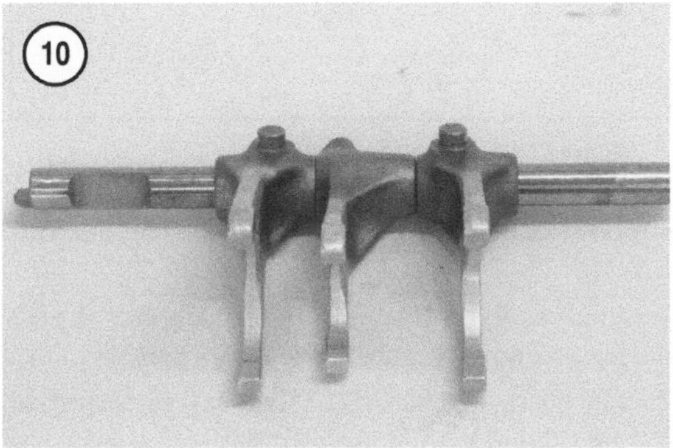

(12)

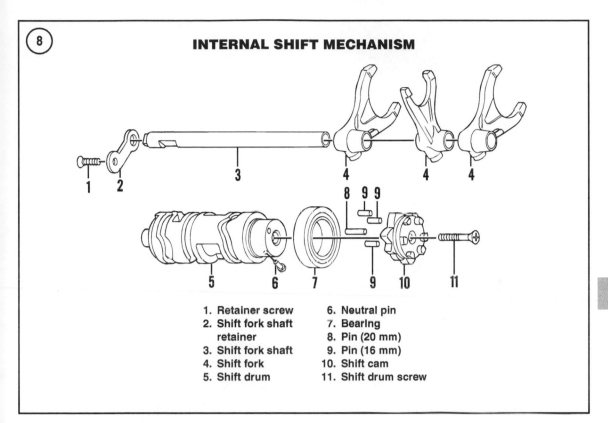

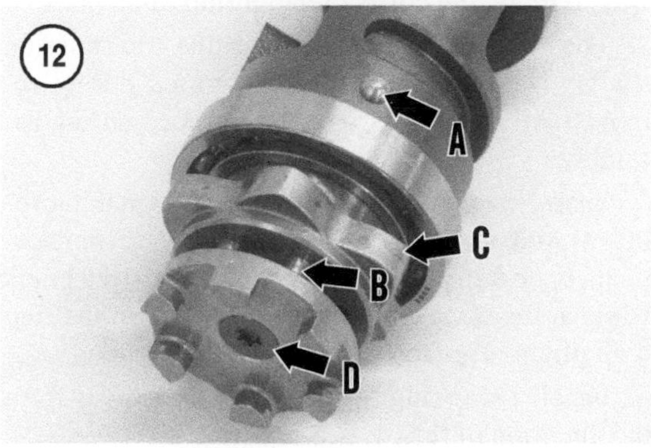

8. Check the pins (B, **Figure 12**) and shift cam ramps (C) for wear.
   a. If necessary, remove the shift drum screw (D, **Figure 12**) to replace the shift cam, pins or bearing.
   b. During assembly, torque the shift drum screw to 4 N•m (35 in.-lb.).

TRANSMISSION

Removal/Installation

Remove and install the transmission shaft assemblies as described in *Crankcase Disassembly* and *Crankcase Assembly* in Chapter Five.

Service Notes

1. A large egg flat (the type that restaurants get their eggs in) can be used to help maintain correct alignment and positioning of the parts. As each part is removed, set it into one of the depressions in the egg flat with the same orientation it had when on the transmission shaft. See **Figure 13**. This is an easy way to retain the correct relationship of all parts.
2. The snap rings fit tightly on the transmission shafts. They usually become distorted during removal. All snap rings must be replaced during assembly.
3. Snap rings will turn and fold over, making removal and installation difficult. To ease replacement, open a snap ring with a pair of snap ring pliers while at the same time holding the back of the ring with pliers (**Figure 14**).
4. Install a snap ring so its flat side faces away from the direction of thrust. See **Figure 15**.

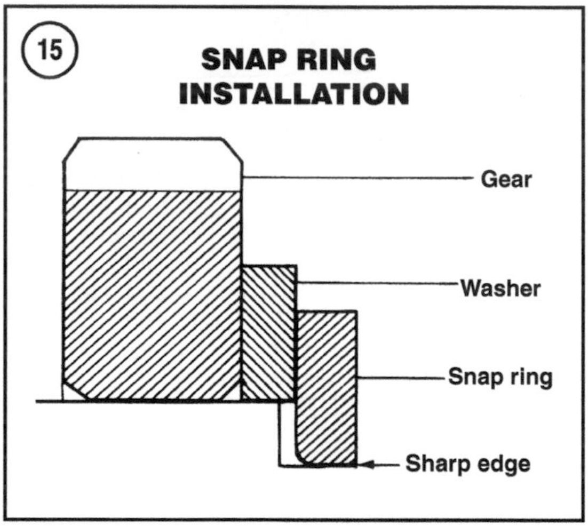

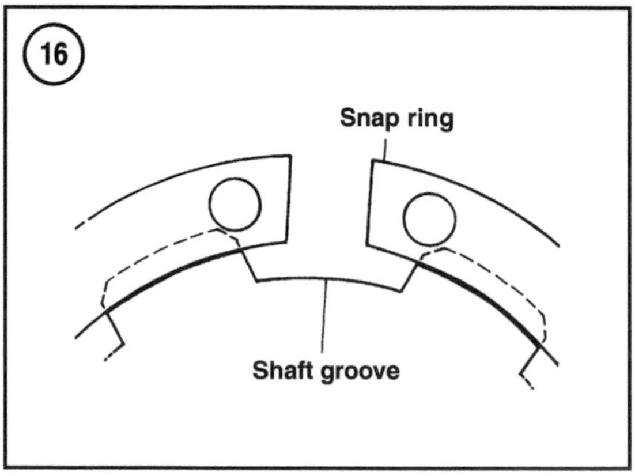

5. Position each snap ring so its end gap sits above a groove in the transmission shaft as shown in **Figure 16**.
6. Apply molybdenum disulfide oil to the each gear during assembly.
7. The gears on a particular shaft can be identified by their diameter. On the mainshaft, first gear has

17

TRANSMISSION

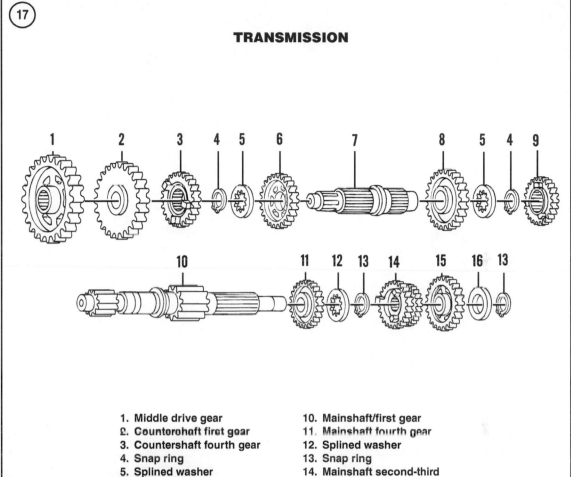

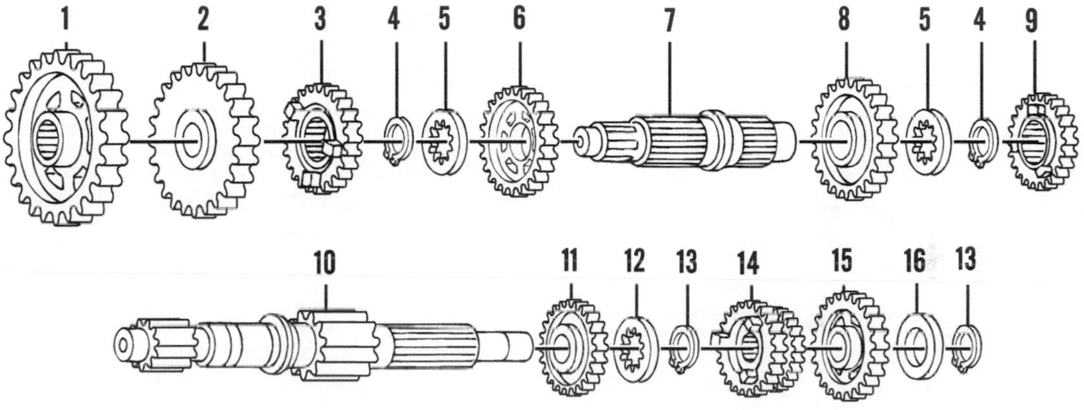

1. Middle drive gear
2. Countershaft first gear
3. Countershaft fourth gear
4. Snap ring
5. Splined washer
6. Countershaft third gear
7. Countershaft
8. Countershaft second gear
9. Countershaft fifth gear
10. Mainshaft/first gear
11. Mainshaft fourth gear
12. Splined washer
13. Snap ring
14. Mainshaft second-third combination gear
15. Mainshaft fifth gear
16. Washer

the smallest diameter and fifth gear has the largest. On the countershaft, fifth gear has the smallest diameter and first gear has the largest.

Mainshaft Disassembly

Refer to **Figure 17**.

1. Clean the assembled shaft in solvent. Dry all components with compressed air or let the assembly sit on rags to drip dry.
2. Remove the snap ring and washer from the mainshaft.
3. Slide fifth gear from the mainshaft, then slide off the second-third combination gear.
4. Remove the snap ring and the washer.
5. Remove fourth gear from the mainshaft.
6. Inspect the mainshaft (**Figure 18**) and gears as described in this chapter.

Mainshaft Assembly

NOTE
Before installing any component, coat all surfaces with molybdenum disulfide oil.

18
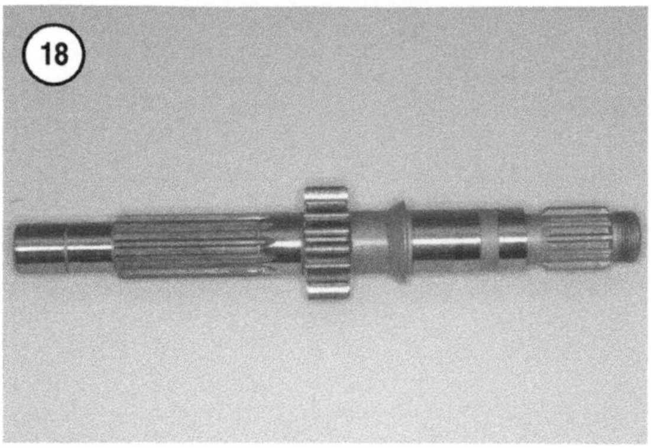

20

19
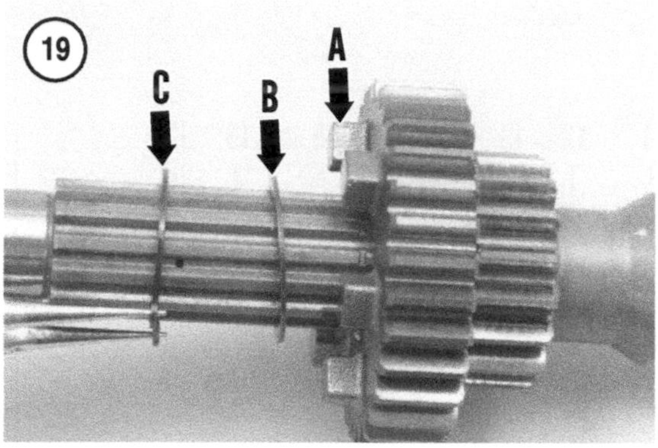

21
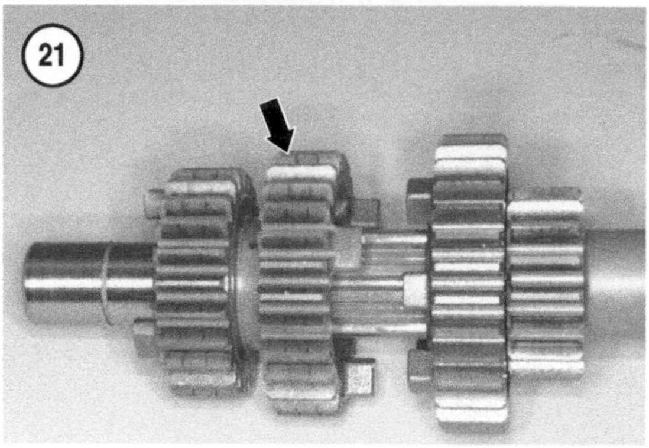

1. Install fourth gear (A, **Figure 19**) onto the mainshaft, and slide it up against first gear. Make sure the engagement dogs on the gear face away from first gear.
2. Install a splined washer (B, **Figure 19**) and a new snap ring (C). The flat side of the snap ring must face away from the splined washer, and the snap ring must be completely seated in the shaft groove (**Figure 20**). Position the snap ring so its end gap sits within a groove in the mainshaft (**Figure 16**).
4. Install the second-third combination gear so third gear (**Figure 21**) faces toward fourth gear.
5. Install fifth gear (A, **Figure 22**) so its flat side faces away from the second-third combination gear.
6. Install the washer (B, **Figure 22**) and a new snap ring (C). The flat side of the snap ring should face out away from the washer.

Countershaft Disassembly

Refer to **Figure 17**.

1. Clean the assembled shaft in solvent. Dry all components with compressed air or let the assembly sit on rags to drip dry.

22

23

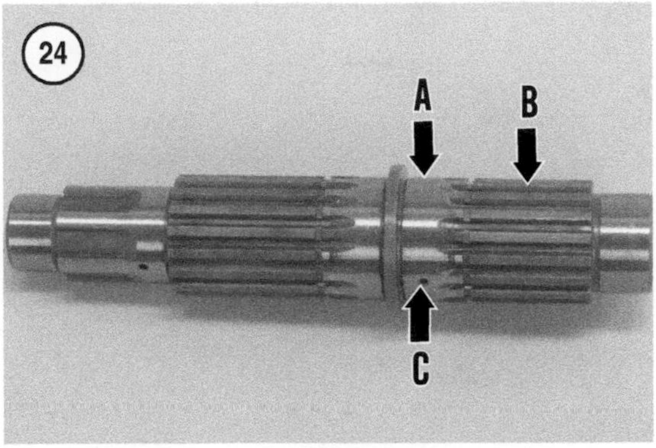

24

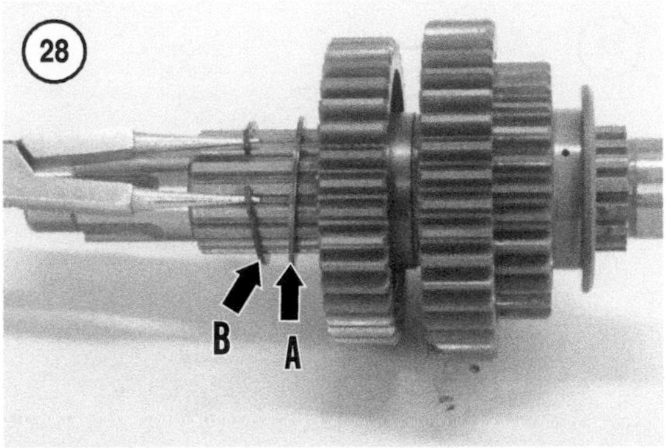

28

25

26

27

2. If it is still installed, remove the middle drive gear (**Figure 23**) from the countershaft.
3. Remove first gear, then remove fourth gear.
4. Remove the snap ring and the splined washer.
5. Remove the countershaft third gear.
6. At the opposite end of the countershaft (the short end), remove fifth gear.
7. Remove the snap ring and splined washer.
8. Remove second gear.
9. Inspect the countershaft (**Figure 24**) and gears as described in this chapter.

Countershaft Assembly

NOTE
Before installing any component, coat all surfaces with molybdenum disulfide oil.

1. Install second gear (A, **Figure 25**) onto the short end of the countershaft and slide the gear up against the stop. The gear's engagement slots must face out.
2. Install a splined washer (B, **Figure 25**) and a new snap ring (C). The snap ring's flat side must face away from splined washer and the snap ring must be completely seated in its groove. Position the end gap (**Figure 16**) so it sits above a groove in the countershaft.
4. Install fifth gear (**Figure 26**) so its fork groove faces away from second gear.
5. On the opposite end of the shaft, install third gear so its shouldered side (**Figure 27**) faces toward the stop on the countershaft.
6. Install the splined washer (A, **Figure 28**) and a new snap ring (B). The flat side of the ring must face away from the splined washer.
7. Install fourth gear (**Figure 29**) so the side with the fork groove faces in.

(29)

(30)

8. Install first gear (**Figure 30**) so its flat side faces out.
9. Install the middle drive gear (**Figure 23**) so its splines engage those on the countershaft.
10. Mesh both assembled transmission shafts together with the middle driven gear (**Figure 31**). Make sure all gears mate properly.

(31)

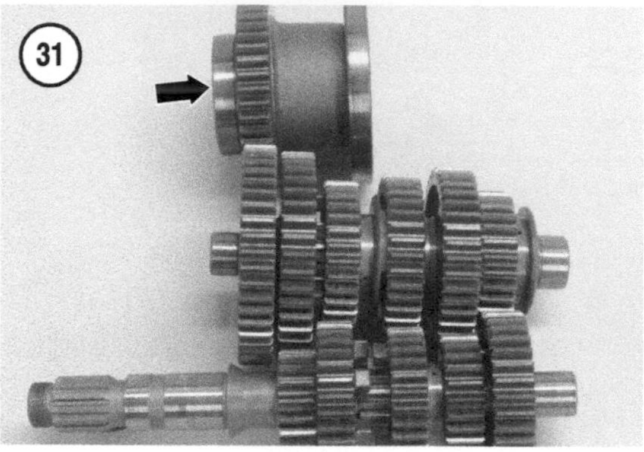

Transmission Inspection

1. Clean all parts in cleaning solvent and dry them thoroughly.

NOTE

Any defective gear should be replaced. It is also a good idea to replace the gear's mate from the opposite shaft, even though the mate may not show as much wear or damage. Worn parts usually cause accelerated wear on new parts. Replace gears in sets to ensure proper mating and wear.

2. Visually inspect the gears for cracks or chips as well as for broken or burnt teeth (**Figure 32**).
3. Check the engagement dogs (A, **Figure 33**) and engagement slots (**Figure 34**). Replace any gear(s) with rounded or damaged edges on the dogs or within the slots.
4. Inspect all free-wheeling gear bearing surfaces (B, **Figure 33**) for wear, discoloration and galling. Also inspect the respective shaft's bearing surfaces (A, **Figure 24**). If there is any metal flaking or visual damage, replace both parts.
5. Inspect the splines (B, **Figure 24**) on each shaft for wear or discoloration. Also check the internal splines (**Figure 32**) on the sliding gears and the middle driven gear (A, **Figure 35**). If no visual damage is apparent, install each gear onto its respective shaft and work the gear back and forth to make sure it moves smoothly.
6. Make sure any oil hole (A, **Figure 36**) in a gear or shaft (C, **Figure 24**) is clear.

(32)

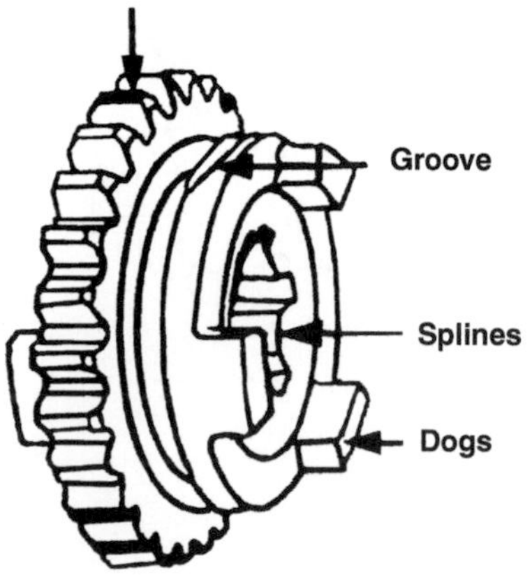

33

34

35

36

7. Inspect each shift fork groove (B, **Figure 36**) for wear or damage. Replace the gear(s) if necessary.
8. Inspect the bearing (B, **Figure 35**) in the middle driven gear. It should turn smoothly without roughness or binding.
9. Replace any washers that are worn.
10. Discard all snap rings and replace them during assembly.
11. If any transmission parts are worn or damaged, inspect the shift drum and shift forks as described in this chapter.

MIDDLE DRIVE SHAFT ASSEMBLY

The following Yamaha special tools, or their equivalents, are needed to remove and service the middle drive shaft assembly. Have these tools on hand before beginning this procedure.

1. Bearing retainer wrench (part No. YM-04137 or 90890-04137).
2. Middle gear backlash band (part No. YM-01231 or 90890-01231)
3. Damper spring compressor (part No. YM-33286 or 90890-04090).
4. Middle drive shaft holder (part No. YM-04055 or 90890-04055).
5. Middle drive shaft nut wrench (part No. YM-045054 or 90890-04138).

Refer to **Figure 37** when servicing the middle drive shaft assembly

Removal

1. Remove the engine, split the crankcase and disassemble the cases as described in Chapter Five.
2. The bearing retainer was staked (**Figure 38**) during assembly. Unstake it now.
3. Use the bearing retainer wrench (**Figure 39**) to remove the bearing retainer (A, **Figure 40**) from the left crankcase half.
4. Pull the middle drive shaft assembly (B, **Figure 40**) from the crankcase.
5. If necessary, disassemble the middle drive shaft assembly as described in this section.

Installation

1. Install the middle driven shaft assembly if it was removed.

7

37

MIDDLE DRIVE SHAFT ASSEMBLY

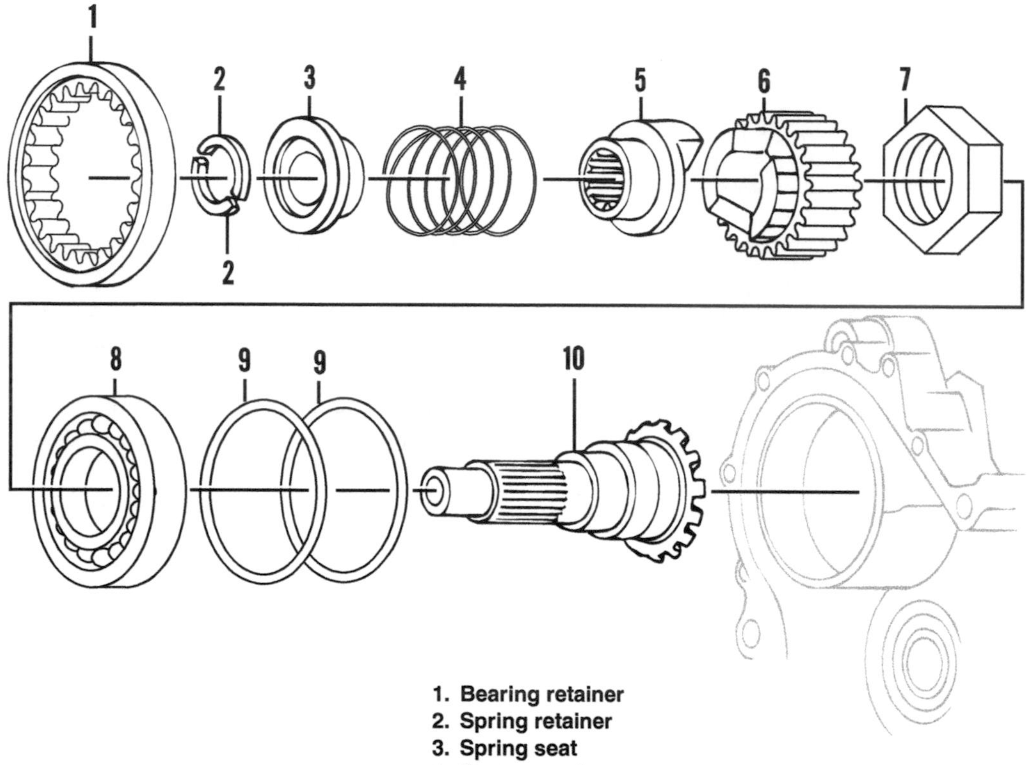

1. Bearing retainer
2. Spring retainer
3. Spring seat
4. Damper spring
5. Damper cam (male)
6. Damper cam (female)
7. Middle drive shaft nut
8. Bearing
9. Middle drive shaft shims
10. Middle drive shaft

38

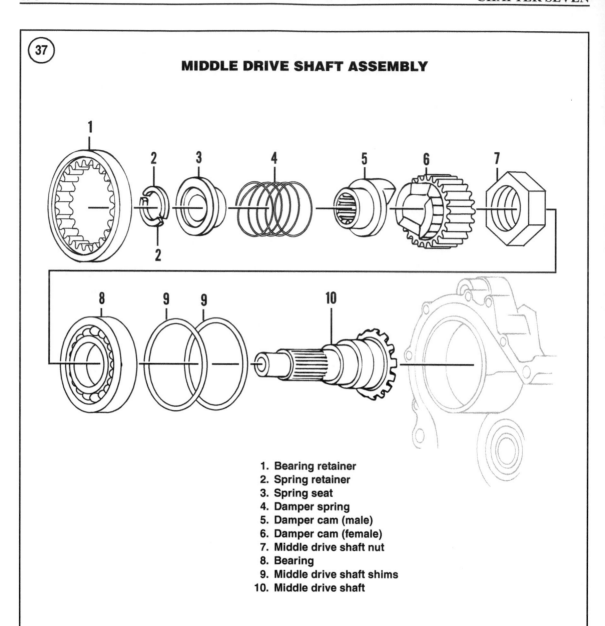

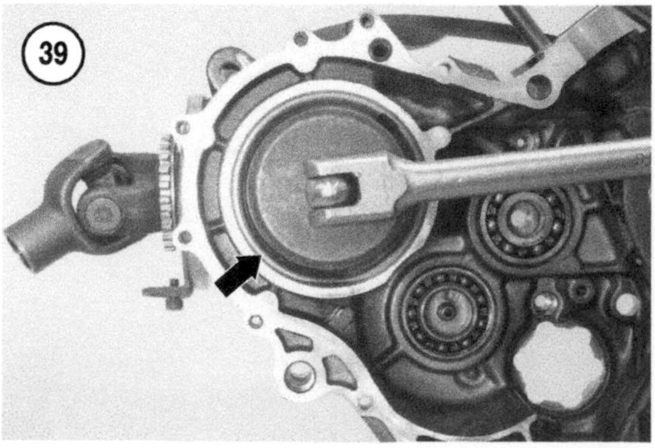

39

40

2. Apply engine oil to the teeth of the middle drive shaft and install the middle drive shaft assembly into the crankcase.
3. Turn the bearing retainer (A, **Figure 40**) until it is in place.
4. Using the bearing retainer wrench (**Figure 39**), torque the bearing retainer to 110 N•m (81 ft.-lb.).
6. Stake the bearing retainer threads (**Figure 38**) to lock it in place.
7. Check the middle gear backlash as described later in this chapter.

41

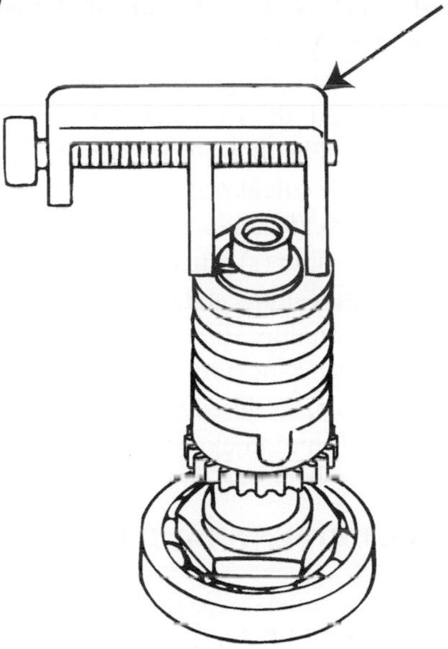

Disassembly

1. Compress the damper spring with the damper spring compressor (**Figure 41**) and remove the spring retainers.
2. Remove the spring seat, damper spring and both spring cams.
3. Install the middle drive shaft holder (A, **Figure 42**) onto the middle drive shaft and secure the holder in a vise.
4. Unstake the threads of the middle drive shaft nut.
5. Install the middle drive shaft nut wrench (B, **Figure 42**). Loosen and remove the nut from the middle drive shaft.
6. Install a bearing splitter onto the middle drive shaft. Support the splitter in a press and press the shaft from the bearing.
7. Remove the bearing and shims.
8. Inspect the middle drive shaft assembly as described in this section.

42

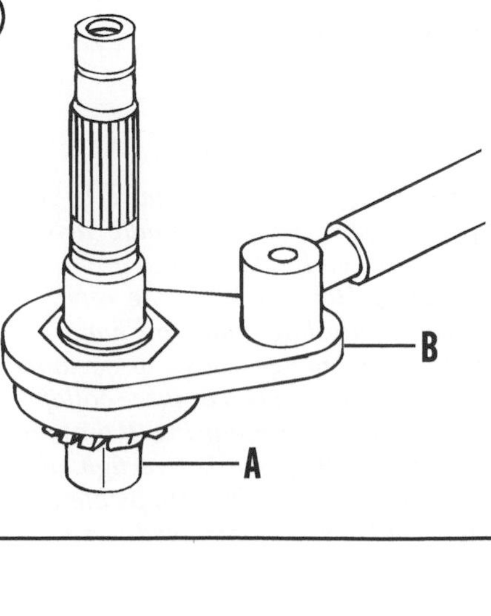

Assembly

NOTE
New middle drive shaft shims must be installed whenever the middle drive shaft or the crankcases have been replaced. If necessary, select new shims as described in this section.

1. If necessary, select new middle drive shaft shims.
2. Install the shims and install the bearing onto the middle drive shaft. Apply oil to the bearing and use a hydraulic press to press the shaft into the bearing.
3. Screw a new middle drive shaft nut onto the shaft.

7

4. Install the middle drive shaft holder (A, **Figure 42**) onto the middle drive shaft and secure the holder in a vise.
5. Install the middle drive shaft nut wrench (B, **Figure 42**) onto the nut.
6. Install a torque wrench at 90° to the middle drive shaft nut wrench and torque the nut to 110 N•m (81 ft-lb.).
7. Stake the threads of the nut to lock it into place.
8. Install the female damper cam and the male damper cam.
9. Install the damper spring and the spring seat.
10. Compress the spring with the damper spring compressor (**Figure 41**) and install the spring retainers.

Inspection

1. Inspect the gear teeth on the middle drive shaft for galling, pitting or chipped teeth. If there is any damage, replace the middle drive shaft and the middle driven shaft as a set.
2. Inspect the splines on the middle drive shaft. If it is damaged, replace both the middle drive shaft and the middle driven shaft. Also inspect the splines on the middle driven gear (A, **Figure 35**) for damage.
3. Inspect the mating surfaces of the male and female sides of the damper cam. Replace both parts as necessary.
4. Inspect the internal splines on the male and female sides of the damper cam.
5. Inspect the damper spring for wear, cracks or damage. Replace the spring if necessary.

Middle Drive Shaft Shim Selection

1. Calculate shim thickness for the middle drive shaft assembly using the following formula:
   a. Shim thickness = 43.00 – (42.00 + A)
   b. *A* equals the number on the top of the left crankcase half.
   c. For example, if the number on the upper crankcase is 46 (**Figure 43**), then shim thickness = 43.00 – (42.00 + 0.46) or 0.54.
2. Middle drive shaft shims are available in three sizes (0.10, 0.15 and 0.20 mm), round off shim thickness to the nearest five hundredths of a millimeter as follows:
   a. If the hundredths digit in the calculated value is 0-2, round shim thickness down to the existing tenths digit. For example, if the calculated value is 0.52, round shim thickness down to 0.50 mm.
   b. If the hundredths digit in the calculated value is 3-6, round shim thickness to the nearest five hundredths of a millimeter. In the above example, the calculated value is 0.54. Round this up to 0.55, and install the appropriate number of shim(s). In this instance, install two 0.20 mm shims and one 0.15 mm shim.
   c. If the hundredths digit is 7-9, round shim thickness up to the next tenth of a millimeter. For example, if the shim thickness is 0.58 mm, round shim thickness up to 0.60 mm.

43

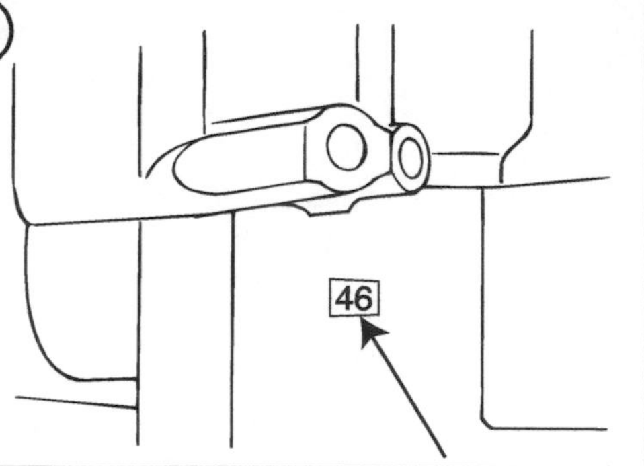

MIDDLE DRIVEN SHAFT ASSEMBLY

The Yamaha universal joint holder (part No. YM-04062 or 90890-04062), or its equivalent, is needed to remove and service the middle driven shaft assembly.

Removal/Installation

Refer to **Figure 44**.

NOTE

The collapsible collar must be replaced whenever the middle driven shaft assembly is removed from the housing. Have at least one replacement collapsible collar on hand before beginning this procedure. Consider having two or three collapsible collars before starting. If the middle driven shaft nut is overtightened during assembly, the collapsible collar will have to be replaced.

44

MIDDLE DRIVEN SHAFT ASSEMBLY

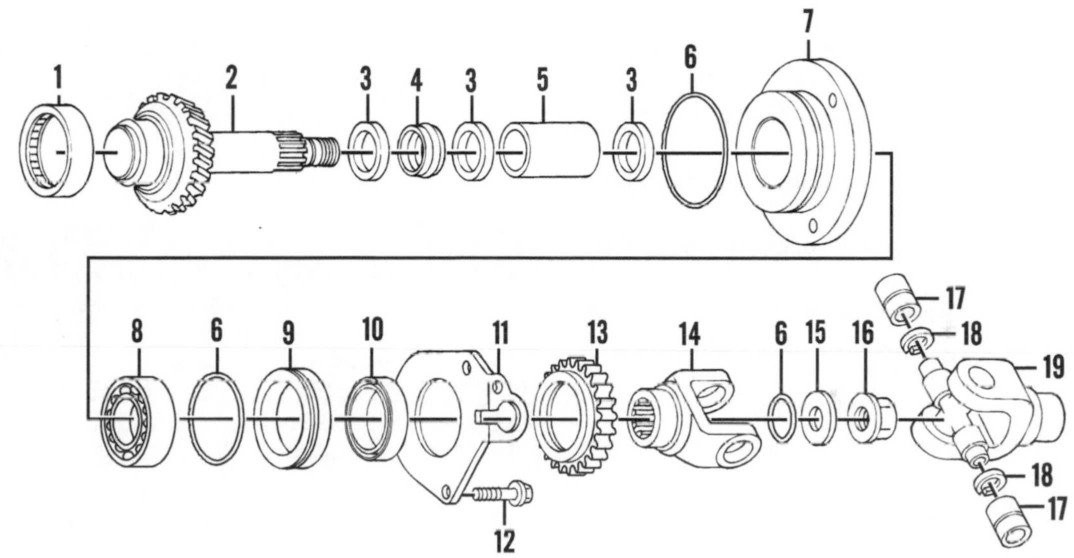

1. Bearing
2. Middle driven shaft
3. Washer
4. Collapsible collar
5. Collar
6. O-ring
7. Bearing housing
8. Bearing
9. Bearing retainer
10. Oil seal
11. Speed sensor bracket
12. Bolt
13. Rotor
14. Drive yoke
15. Washer
16. Middle driven shaft nut
17. Bearing
18. Circlip
19. Driven yoke

45

1. Remove the rear wheel and swing arm as described in Chapter Twelve.

2. Remove the bearing housing bolts (A, **Figure 45**) from the middle driven shaft assembly.

3. Separate the universal joint as follows:

   a. Use a small screwdriver to drive the circlip (A, **Figure 46**) from each side of the drive yoke (B).

   b. Using a front end ball joint remover (**Figure 47**) or a hydraulic press with the proper size

socket, press the bearing from each side of the drive yoke.
   c. Remove the driven yoke (C, **Figure 46**) from the drive yoke (B).

4. Hold the drive yoke with the universal joint holder (A, **Figure 48**) and loosen the middle driven shaft nut (B).
5. Remove the universal joint holder, and remove the middle driven shaft nut and its washer.
6. Remove the drive yoke, rotor (B, **Figure 45**) and the speed sensor bracket (C).
7. Remove the bearing housing and O-ring. Discard the O-ring.
8. Remove the washer and collar from the middle driven shaft.
9. Remove the washer, the collapsible collar and the washer. Discard the collapsible collar. Install a new one during assembly.
10. Remove the middle driven shaft.
11. If necessary, disassemble the bearing housing as follows:
   a. Pry the oil seal from the housing.
   b. Remove the bearing retainer and the O-ring. Discard the O-ring.
   c. Use a drift to drive the bearing from the bearing housing.
12. If necessary, remove the bearing from the end of the middle driven shaft.

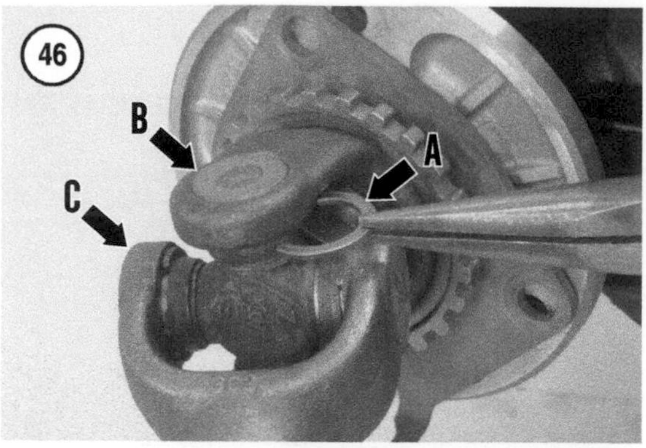

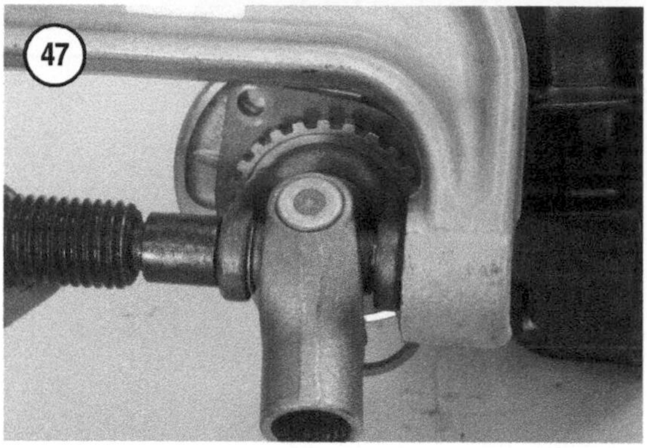

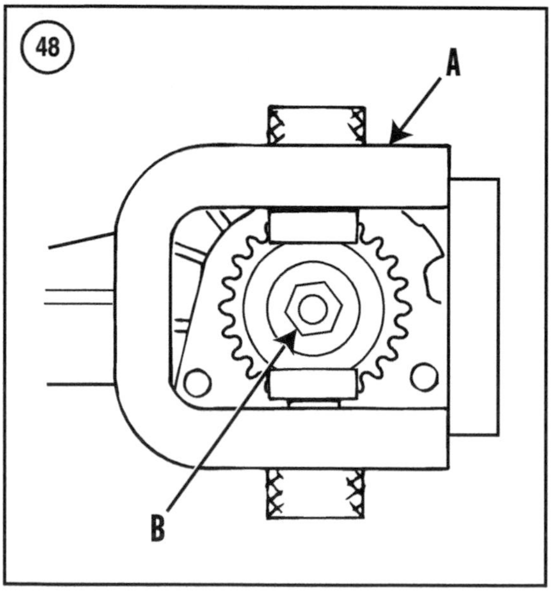

Assembly

1. Install the bearing onto the inboard end of the middle driven shaft if it was removed. Thoroughly lubricate the bearing with engine oil.
2. If the bearing housing was disassembled, assemble it as follows:
   a. Lubricate the bearing with engine oil and set the bearing into the bearing housing.
   b. Use a driver or socket that matches the diameter of the outer race to drive the bearing in until it bottoms in the housing.
   c. Lubricate a new O-ring with lithium soap grease and install the O-ring into the bearing housing.
   d. Install the bearing retainer.
   e. Pack the lips of a new oil seal with lithium soap grease and drive the oil seal in until it bottoms in the housing. Use the same driver used to install the bearing.
3. Install the middle driven shaft into the crankcase.
4. Install a washer, a new collapsible collar and another washer onto the middle driven shaft.
5. Slide the collar and washer onto the shaft.

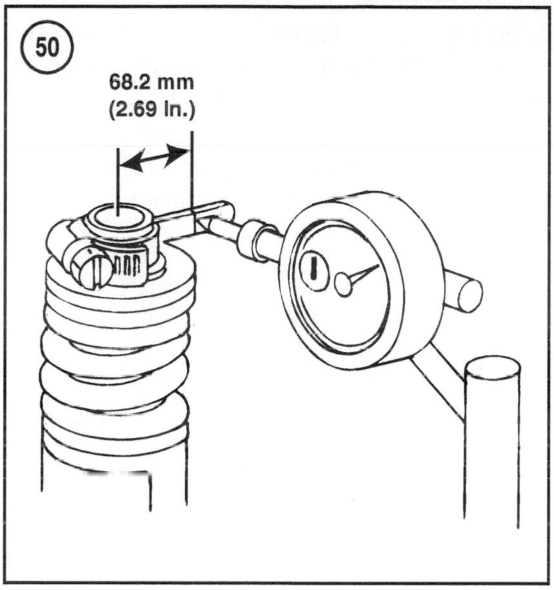

6. Lubricate the splines of the middle driven shaft with lithium soap grease and slide the bearing housing onto the shaft.

7. Seat the speed sensor bracket (C, **Figure 45**), rotor (B) and the drive yoke on the bearing housing.

8. Lubricate a new O-ring with lithium soap grease and carefully roll the O-ring onto the middle drive shaft. Make sure the shaft threads do not tear the O-ring.

NOTE
A specified torque is not provided for the middle driven shaft nut. Instead, the nut is tightened until the middle gear backlash is within specification.

9. Install the washer and the middle driven shaft nut (B, **Figure 48**). Apply a medium-strength threadlocking compound to the nut threads and finger-tighten the nut at this time.

10. Install the universal joint holder (A, **Figure 48**) onto the drive yoke and install the middle gear backlash band (**Figure 49**) onto the middle drive shaft assembly.

11. Set a dial gauge against the backlash band so the plunger sits 68.2 mm (2.69 in.) from the center of the middle drive shaft. See **Figure 50**.

CAUTION
Gradually tighten the nut in small increments and check the middle gear backlash. If the nut is overtightened, the collapsible collar will be crushed and will have to be replaced.

12. Gradually tighten the middle driven gear nut (B, **Figure 48**) while measuring the middle gear backlash.
   a. Tighten the nut in a small increment and stop.
   b. Turn the universal joint holder and read the backlash on the dial gauge.
   c. Repeat substeps a and b until the backlash is within the range specified in **Table 1**.

13. Stake the middle driven shaft nut to lock it in place.

14. Assemble the universal joint as follows:
   a. Install the driven yoke into place in the drive yoke.
   b. When using a hydraulic press, support the assembly in a vise.
   c. Lubricate each bearing with engine oil.

CAUTION
The needles can easily fall out of their races. Check each bearing before installation.

   d. Set a bearing into the drive yoke so the side with the circlip groove faces down into the yoke.

CAUTION
Do not use a hammer to install the bearing. Hammering could distort the collapsible collar in the middle driven shaft assembly. If this occurs, the collar must be replaced.

   e. Install a new circlip (A, **Figure 46**) so it is completely seated in the bearing's circlip groove.

7

f. Turn the universal joint over and repeat substeps d-f for the opposite bearing.

15. Install the bearing housing bolts (A, **Figure 45**) and secure the speed sensor bracket and bearing housing to the crankcase. Apply Yamaha Bond 1215 or an equivalent sealant to the threads of the bolts and torque the speed sensor bracket bolts to 25 N•m (18 ft.-lb.).

Inspection

1. Inspect the gear on the end of the middle driven shaft for galling, pitting or chipped teeth. If there is any damage replace the middle driven shaft and the middle drive shaft as a set.

2. Inspect the splines on the middle driven shaft. If damaged, replace both the middle drive and middle driven shafts.

3. Inspect the inner splines of the universal joint. If the splines are damaged, replace the universal joint. Also inspect the outer splines on the drive shaft. They may also be damaged.

CHECKING MIDDLE GEAR BACKLASH

1. Remove the engine, separate the crankcase, and disassemble the cases as described in Chapter Five.

2. Install the middle gear backlash band (**Figure 49**) onto the middle drive shaft assembly.

3. Set a dial gauge against the backlash band so the plunger sits 68.2 mm (2.69 in.) from the center of the middle drive shaft. See **Figure 50**.

4. Rotate the universal joint and read the backlash on the dial gauge. The middle gear backlash should be within the range specified in **Table 1**.

Table 1 TRANSMISSION AND GEARSHIFT SPECIFICATIONS

Item	New mm (in.)	Service limit mm (in.)
Mainshaft runout	–	0.08 (0.003)
Countershaft runout	–	0.08 (0.003)
Transmission gear ratios		
First gear	40/17 (2.353)	–
Second gear	40/24 (1.667)	–
Third gear	36/28 (1.286)	–
Fourth gear	32/31 (1.032)	–
Fifth gear	29/34 (0.853)	–
Primary reduction ratio	78/47 (1.660)	–
Secondary reduction ratio	44/47 × 19/18 × 32/11 (2.875)	–
Middle gear backlash	0.1-0.2 (0.004-0.008)	–

Table 2 TRANSMISSION, SHIFT MECHANISM AND MIDDLE GEAR TORQUE SPECIFICATIONS

Item	N•m	in.-lb.	ft.-lb.
Footrest bracket bolts	64	–	47
Middle drive assembly bearing retainer nut	110	–	81
Middle drive shaft nut	110	–	81
Shift drum screw	4	35	–
Shift lever clamp bolt	10	89	–
Shift pedal adjuster locknut*	10	89	–
Shift shaft stopper bolt	22	–	16
Speed sensor bracket bolts	25	–	18

*Left-hand threads on rear nut.

CHAPTER EIGHT

AIR/FUEL, EXHAUST AND EMISSION CONTROL SYSTEMS

This chapter describes service procedures for the air/fuel, exhaust and emission control systems. During inspection, compare measurements to the specifications in the tables at the end of this chapter. Replace any components that are worn, damaged or out of specification. Tighten fasteners to the specified torque during assembly.

FUEL TANK

Refer to **Figure 1**.

WARNING
Some fuel may spill during the following procedure. Work in a well-ventilated area at least 50 feet from any sparks or flames, including gas appliance pilot lights. Do not allow any smoking in the area. Keep a B:C rated fire extinguisher on hand.

Removal/Installation

NOTE
During removal, label each hose and its fitting. Also note how a hose is routed through the frame. This assures that each hose will be connected to the correct fitting and properly routed during assembly.

1. Securely support the motorcycle on a level surface.
2. Make sure the main switch is off.
3. Disconnect the electrical lead from the negative battery terminal (**Figure 2**).
4. Turn the fuel valve (A, **Figure 3**) off.
5. Disconnect the fuel hose (A, **Figure 4**) from the fuel valve fitting. Plug the fuel hose to keep out dirt and other contaminants.
6. Remove the passenger and rider seats as described in Chapter Fourteen.
7. Remove the fuel tank bolts (A, **Figure 5**).

1

FUEL TANK AND SEATS

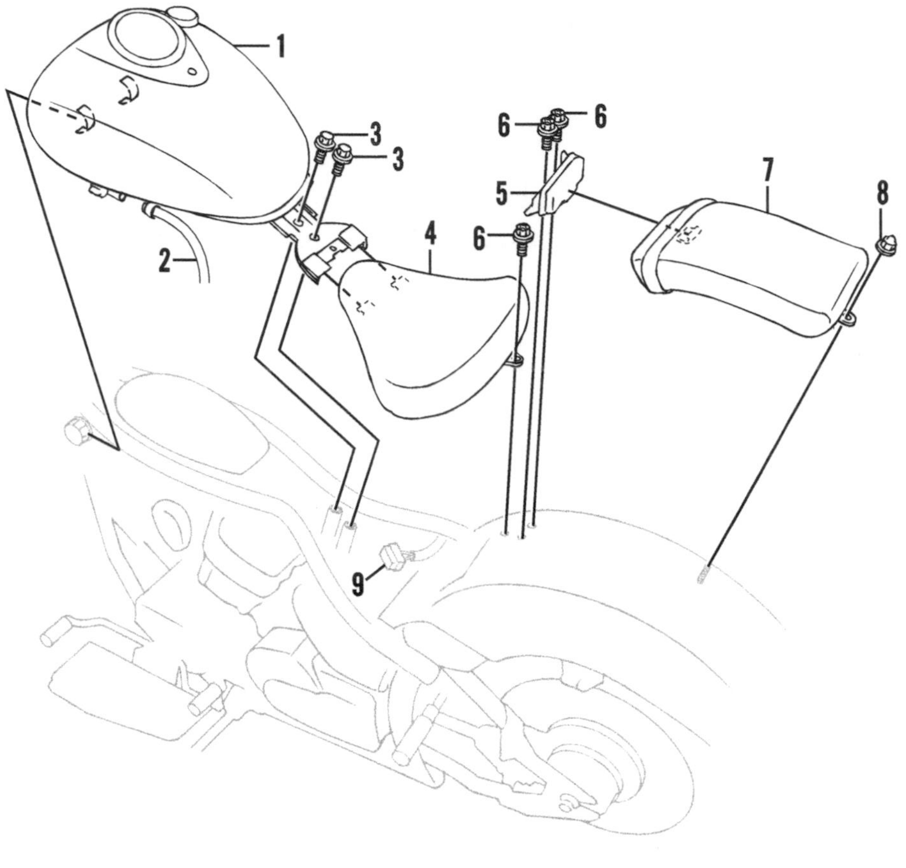

1. Fuel tank
2. Fuel hose
3. Fuel tank bolt
4. Rider's seat
5. Seat bracket
6. Seat bracket bolt
7. Passenger seat
8. Nut
9. Meter connector

8. Raise the fuel tank bracket and support the tank. Pull the meter connector (B, **Figure 5**) from beneath the ignitor panel and disconnect the 14-pin meter connector (A, **Figure 6**) from its harness mate.

9. On California models, disconnect the EVAP hose (B, **Figure 6**) from the fitting on the fuel tank.

10. If necessary, release any cable holder that secures a hose or wire to the motorcycle. Note the location of these cable holders.

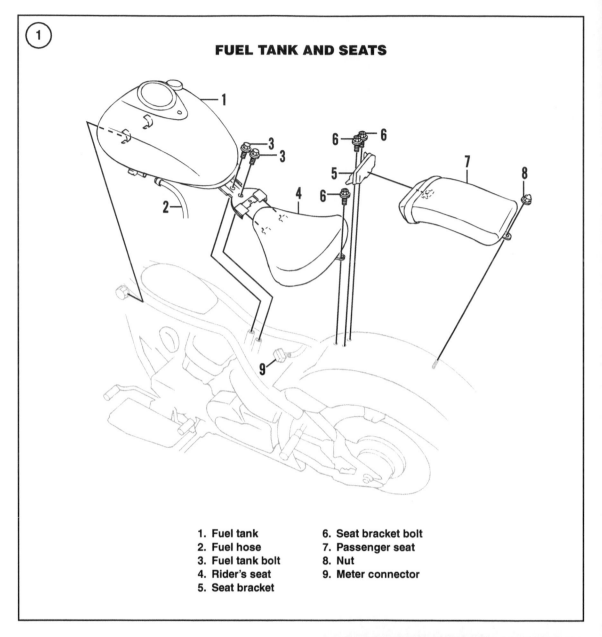

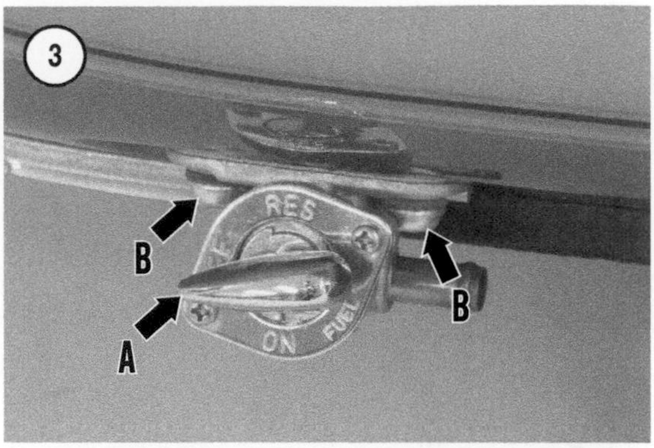

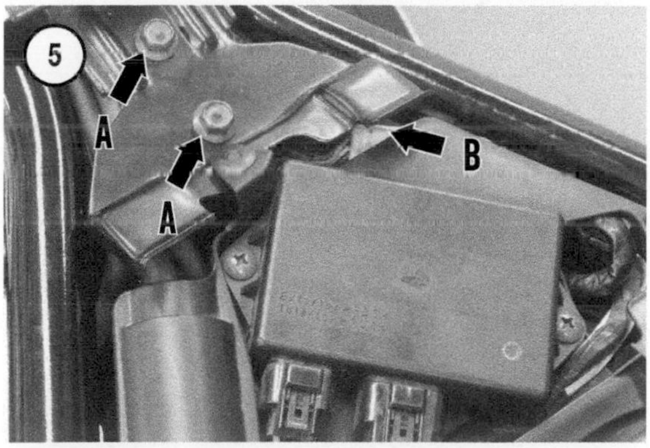

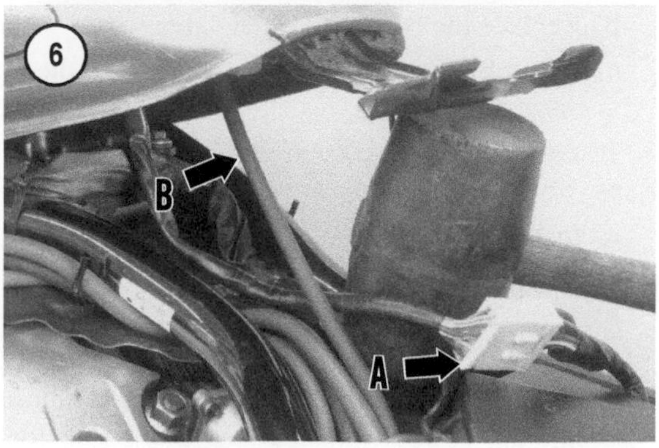

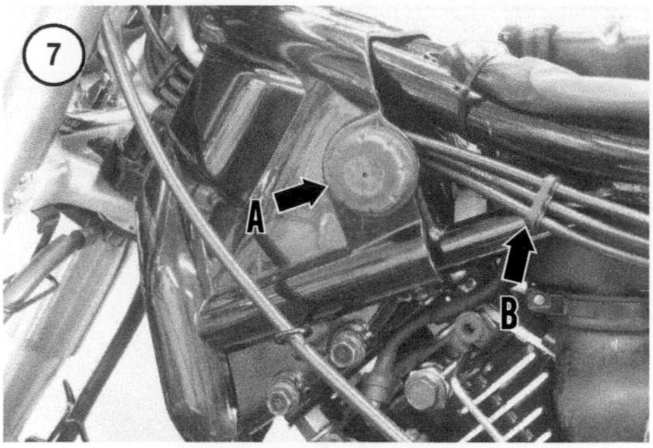

11. Pull the tank bracket rearward until the tank is free of the damper (A, **Figure 7**) on each side of the frame. Remove the fuel tank.

12. Installation is the reverse of removal. Note the following:

   a. Set the tank on the frame and route the meter assembly cable and hoses along the paths noted during removal.
   b. Slide the tank forward until the tank engages the damper (A, **Figure 7**) on each side of the frame.
   c. Route the meter connector beneath the ignitor panel.
   d. Torque the fuel tank bolts to 23 N•m (17 ft.-lb.).
   e. Insert the meter connector (B, **Figure 5**) beneath the ignitor panel.
   d. Secure the harness and hoses with any cable holders released during removal.

FUEL VALVE

Removal/Installation

WARNING
Some fuel may spill during the following procedure. Work in a well-ventilated area at least 50 feet from any sparks or flames, including gas appliance pilot lights. Do not allow any smoking in the area. Keep a B:C rated fire extinguisher on hand.

Refer to **Figure 8**.

1. Remove the fuel tank as described in this chapter.
2. Set the fuel tank on a protective pad or blanket.

8

FUEL VALVE

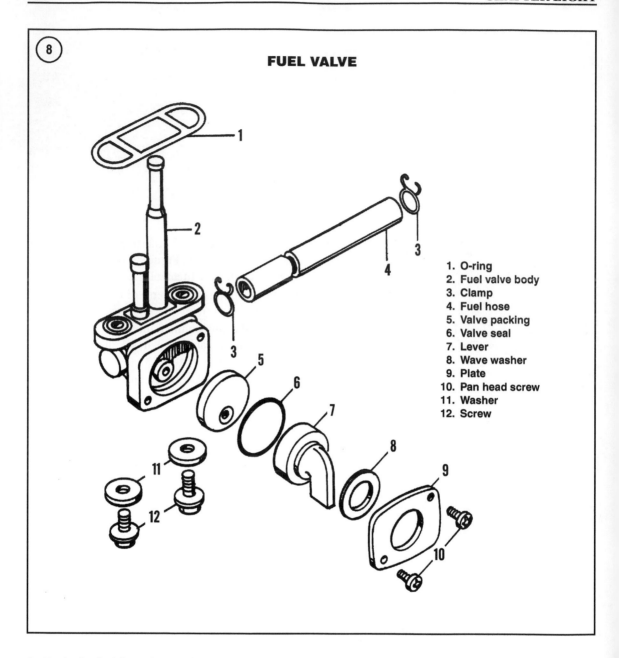

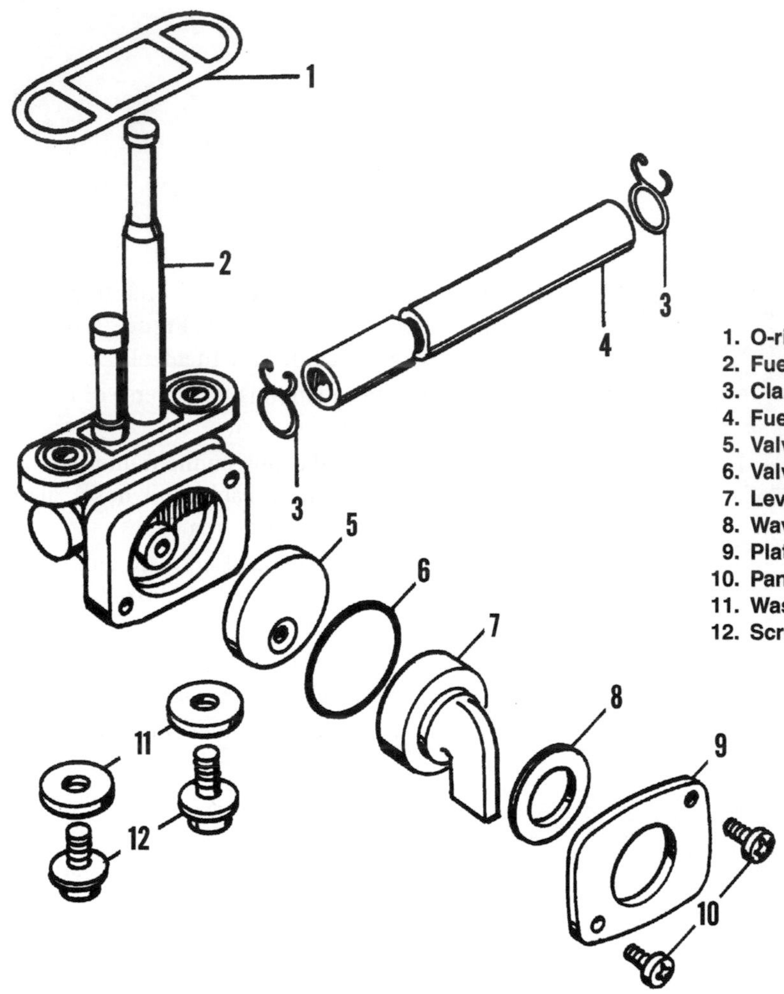

1. O-ring
2. Fuel valve body
3. Clamp
4. Fuel hose
5. Valve packing
6. Valve seal
7. Lever
8. Wave washer
9. Plate
10. Pan head screw
11. Washer
12. Screw

3. Drain the fuel from the tank into a clean, sealable container.

4. Position the tank so residual fuel will not spill from the tank when the fuel valve is removed.

5. Remove each fuel valve screw and washer (B, **Figure 3**), and remove the valve and O-ring.

6. The packing and fuel valve seal are the only replaceable parts in the fuel valve. Replace the seals as follows:

   a. Remove the two pan head screws and disassemble the valve as shown in **Figure 8**.
   b. Replace the packing or fuel valve seal.
   c. Assemble the fuel valve by reversing these disassembly steps.

7. Install the fuel valve by reversing these removal steps. Note the following:

   a. Install a new O-ring (1, **Figure 8**) during installation.
   b. Tighten the fuel valve bolts (B, **Figure 3**) to 7 N•m (62 in.-lb.).
   c. After installing the fuel valve, pour a small amount of fuel into the tank and check for

9

AIR INTAKE

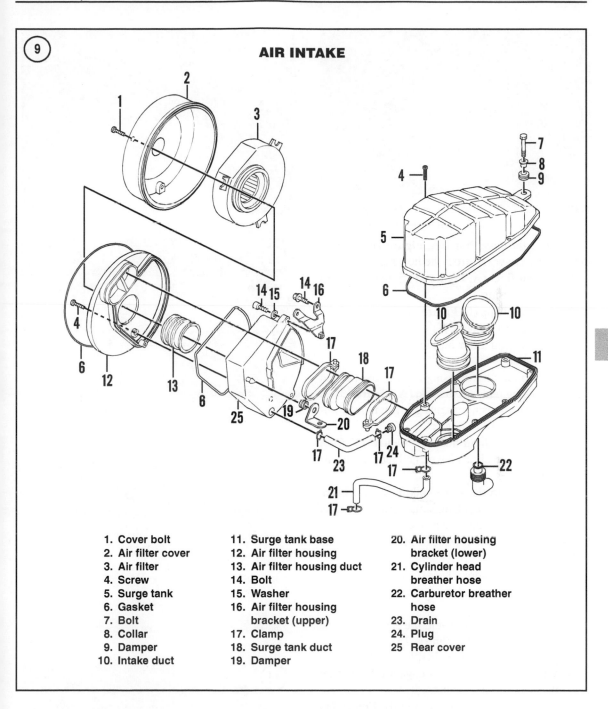

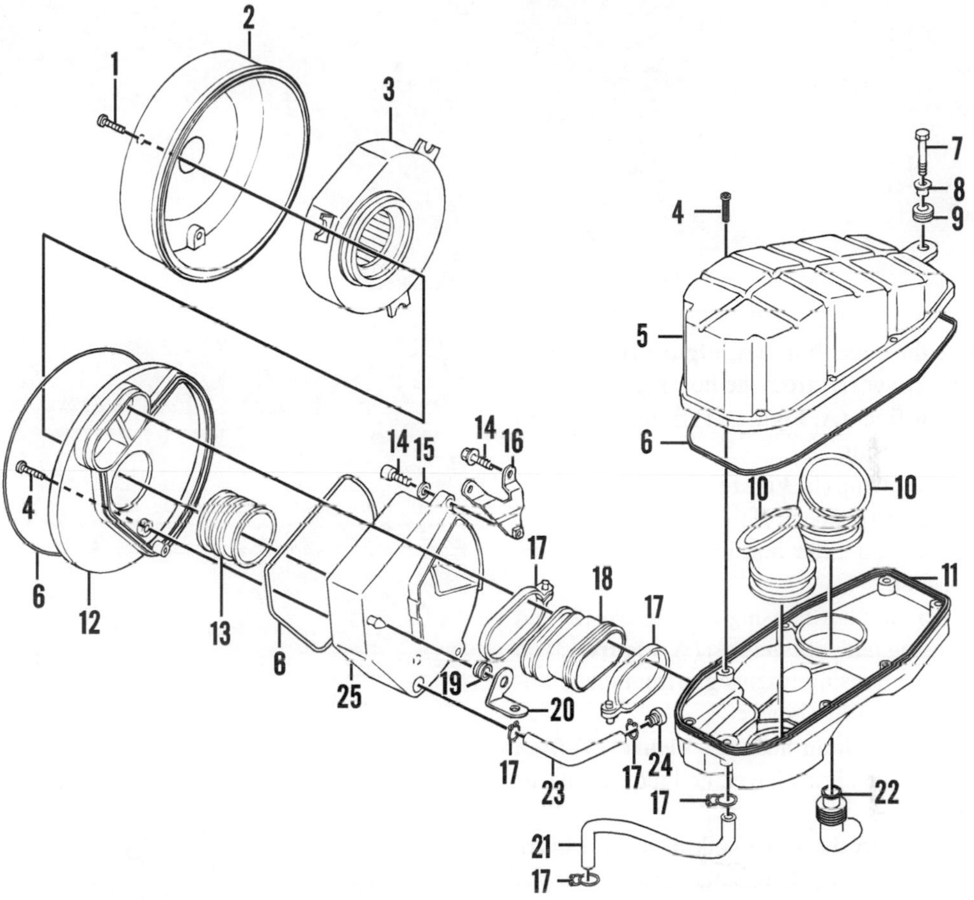

1. Cover bolt
2. Air filter cover
3. Air filter
4. Screw
5. Surge tank
6. Gasket
7. Bolt
8. Collar
9. Damper
10. Intake duct
11. Surge tank base
12. Air filter housing
13. Air filter housing duct
14. Bolt
15. Washer
16. Air filter housing bracket (upper)
17. Clamp
18. Surge tank duct
19. Damper
20. Air filter housing bracket (lower)
21. Cylinder head breather hose
22. Carburetor breather hose
23. Drain
24. Plug
25 Rear cover

leaks. If there is a leak, solve the problem before installing the fuel tank.

AIR FILTER HOUSING

NOTE
Air filter service is described in Chapter Three.

Removal/Installation

Refer to **Figure 9**.

1. Securely support the motorcycle on level ground.

2. Remove the fuel tank as described in this chapter.

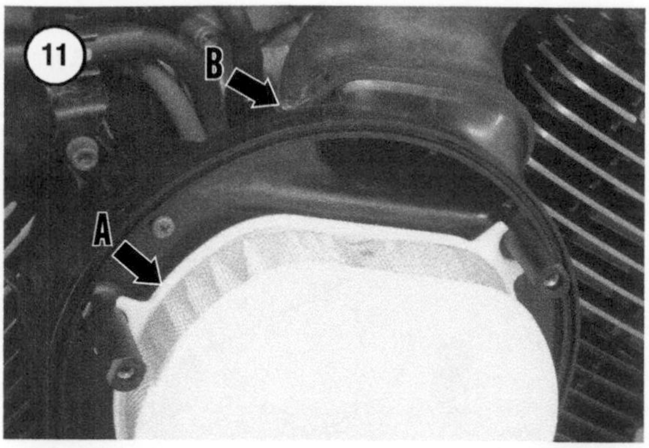

3. Remove the cover bolts (A, **Figure 10**) and pull the air filter cover (B) from the housing.
4. Lift the air filter (A, **Figure 11**) from the posts in the air filter housing.
5. Loosen the clamp (B, **Figure 11**) that secures the surge tank duct to the back of the air filter housing.
6. Remove the air filter housing bolts (A, **Figure 12**) and washers.
7. Grasp the air filter housing (B, **Figure 12**) and pull the housing out until the post (A, **Figure 13**) on the back of the housing disengages from the grommet (B) in the air filter bracket.
8. Disengage the drain hose (C, **Figure 13**) from the bracket on the clutch cover, and remove the air filter housing.
9. Stuff a rag into the surge tank duct (A, **Figure 14**) so dirt or other contaminants cannot enter the duct while the housing is removed.
10. Installation is the reverse of removal. Note the following:
   a. Make sure the drain hose (C, **Figure 13**) is secured to the holder of the bracket on the clutch cover.
   b. Press the housing toward the engine until the housing post (A, **Figure 13**) engages the grommet (B) on the air filter bracket. Make sure the surge tank duct engages the port on the back of the housing.
   c. Tighten the surge tank clamp (B, **Figure 11**) securely.
   d. Tighten the air filter housing bolts (A, **Figure 12**) to 10 N•m (89 in.-lb.).
   e. Install the air filter (A, **Figure 11**) so the notches on the filter engage the posts on the air filter housing.
   f. Tighten the air filter cover bolts (A, **Figure 10**) to 2 N•m (18 in.-lb.).

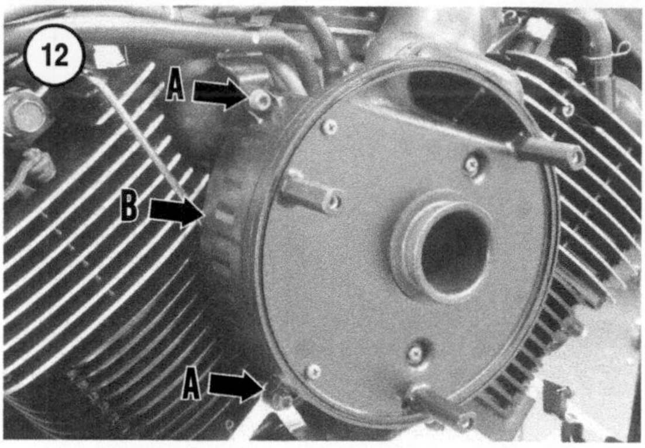

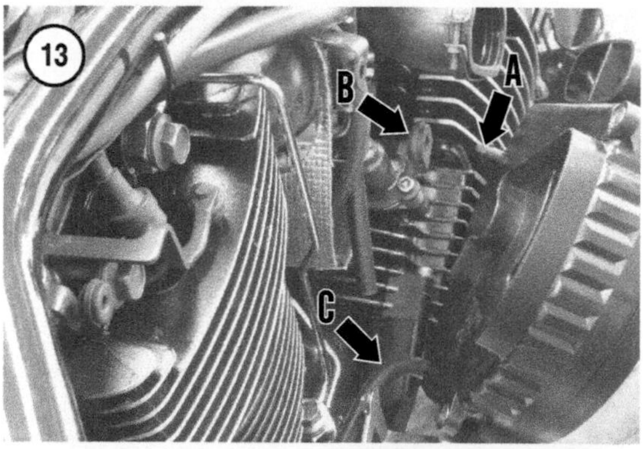

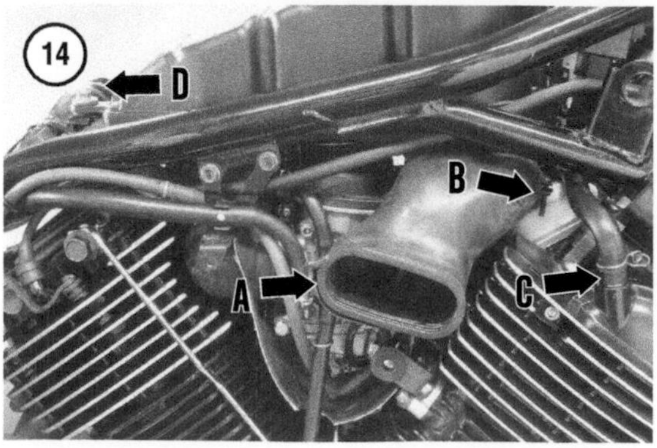

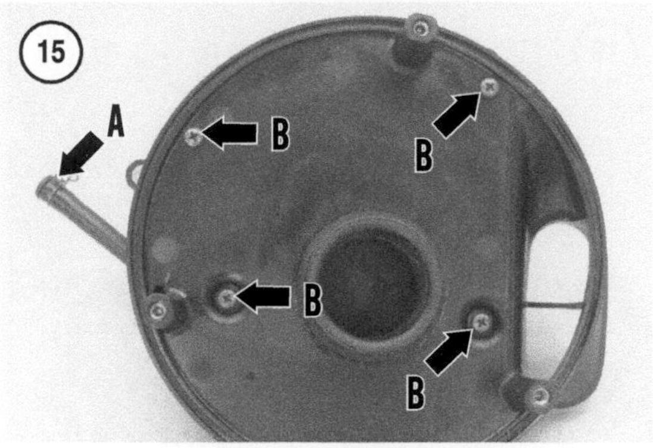

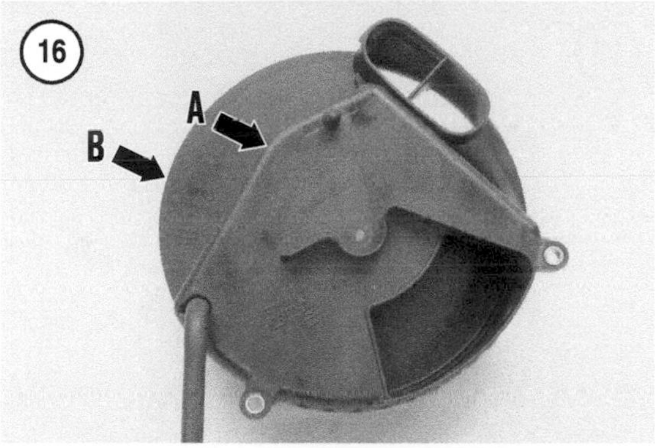

Inspection

1. Remove the plug (A, **Figure 15**) from the drain hose on the air filter housing. Clean out accumulated moisture or dirt. If necessary, blow the hose clear with compressed air.

2. Inspect all components of the air filter housing assembly. Look closely for cracks or other damage that would allow unfiltered air into the engine. Replace any part that is damaged or starting to deteriorate.

3. Wipe out the inside of the air filter housing with a clean rag.

4. If necessary, remove the self-taping screws (B, **Figure 15**) and separate the rear cover (A, **Figure 16**) from the housing (B).

SURGE TANK

Removal

Refer to **Figure 9**.

1. Remove the air filter housing as described in this chapter.

2. Release the clamp (B, **Figure 14**) and remove the surge tank duct (A) from the surge tank.

3. Disconnect the cylinder head breather hose (C, **Figure 14**) from the cam sprocket cover on the front cylinder.

4. Remove the surge tank bolt (D, **Figure 14**) and collar.

5. Remove the air filter housing bracket bolts (A, **Figure 17**). On California models, note that the EVAP purge hose (B, **Figure 17**) runs behind the bracket. Release the fuel hose (C, **Figure 17**) from the holder on the bracket and lower the air filter housing bracket from the frame.

6. Loosen the clamp (A, **Figure 18**) that secures each carburetor duct (B) to the surge tank. Also refer to A, **Figure 19**.

7. Lift the surge tank from the frame, separate each male fitting on the bottom of the surge tank from its carburetor duct, and remove the tank. Watch for the carburetor breather hose (B, **Figure 19**). It may have to be manually pulled from the recess on the bottom of the surge tank.

8

Installation

1. If removed, install the cylinder head breather hose (**Figure 20**) onto the surge tank fitting so the dots on the hose face forward.

2. Make sure the clamp (A, **Figure 19**) on each carburetor duct is loose and the carburetor breather hose (B) faces up.

3. Apply soap to the male fittings on the bottom of the surge tank.

4. Lower the surge tank into the frame so the surge tank fittings engage the carburetor ducts. Make sure the carburetor breather hose (B, **Figure 19**) fits into the recess in the surge tank.

5. Connect the cylinder head breather hose (**Figure 20**) to the fitting on the front cylinder cam sprocket cover (C, **Figure 14**).

6. Install the surge tank bolt (D, **Figure 14**) and collar. Make sure the damper is in place in the surge tank mount. Tighten the bolt securely.

7. Tighten the clamps (A, **Figure 18**) that secure the surge tank to the carburetor ducts (B). Also refer to A, **Figure 19**.

8. Secure the air filter housing bracket to the frame with its mounting bolts (A, **Figure 17**). On California models route the EVAP purge hose (B, **Figure 17**) behind the bracket. Secure the fuel hose (C, **Figure 17**) in the holder on the bracket.

9. Install the surge tank duct (A, **Figure 14**) onto the port (D, **Figure 17**) on the surge tank. Install the end with the dot into the surge tank port and tighten the clamp (B, **Figure 14**) securely.

10. Install the air filter housing as described in this chapter.

Inspection

1. Remove the surge tank screws (**Figure 21**) and separate the base from the surge tank.

2. Wipe the inside of the base and surge tank (**Figure 22**).

3. Visually inspect the components for cracks or other damage that would allow unfiltered air into the engine. Replace any part that is damaged or starting to deteriorate.

19

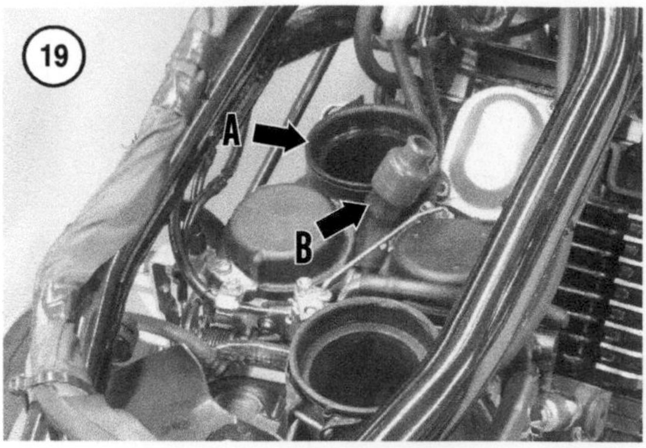

20

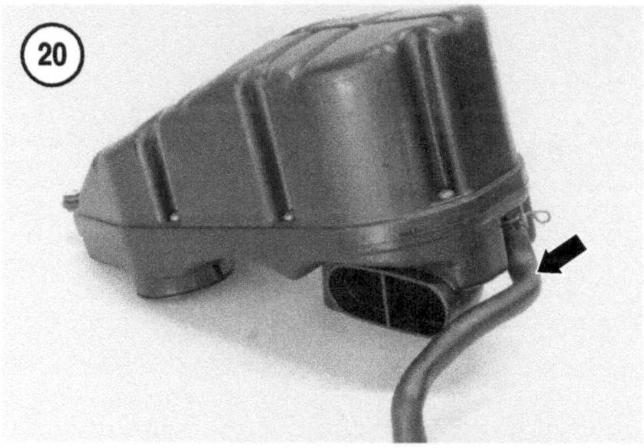

21

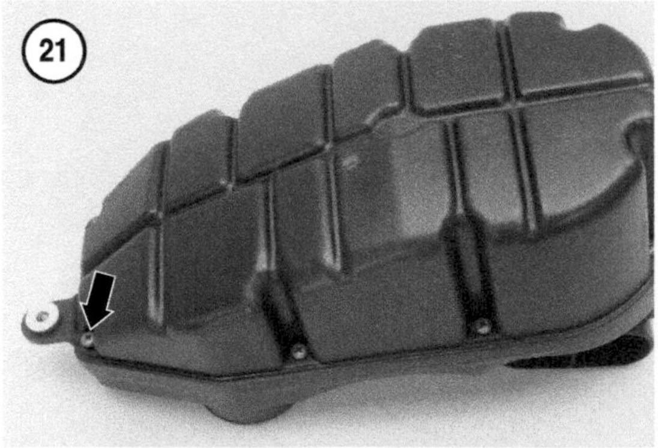

22

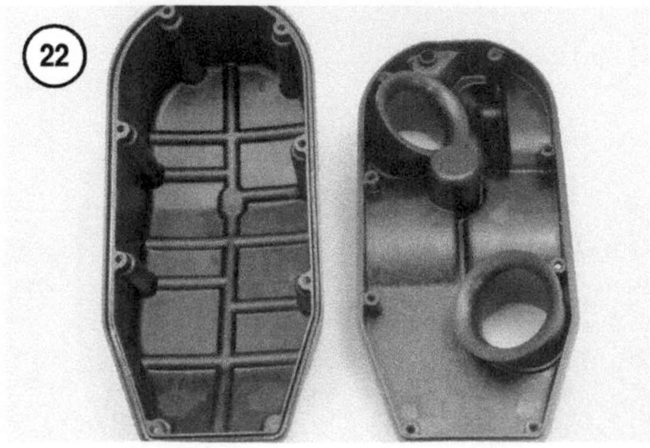

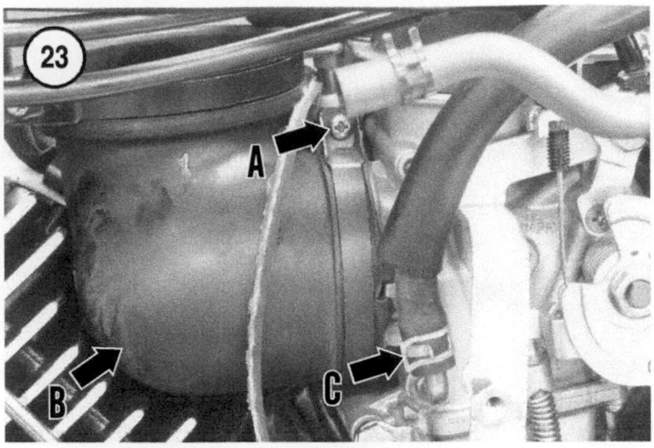

CARBURETOR

Operation

The carburetor atomizes fuel and mixes it in correct proportions with air that is drawn in through the air intake. At the primary throttle opening (idle), a small amount of fuel is siphoned through the pilot jet by the incoming air. When the throttle is opened further, the air stream begins to siphon fuel through the main jet and needle jet. As the tapered needle is lifted, it occupies progressively less area within the needle jet, which increases the effective flow capacity of the jet. At full throttle, the carburetor venturi is fully open and the needle is lifted far enough to permit the main jet to flow at full capacity.

The choke circuit is a bystarter system. The choke lever in this system opens a starter valve rather than closing the butterfly in the venturi. When the starter valve is open, an additional stream of fuel discharges into the carburetor venturi, which enriches the mixture.

Removal/Installation

NOTE

The carburetors have several hoses connected to various fittings. During removal, mark each hose and its fitting so the hoses can be installed on the correct fittings during assembly. Also note how each hose is routed along the motorcycle. Hoses must be rerouted along their original paths.

1. Securely support the motorcycle on a level surface and disconnect the electrical lead from the negative battery terminal (**Figure 2**).
2. Remove the fuel tank as described in this chapter.
3. Remove the air filter housing and the surge tank as described in this chapter.
4. Loosen the clamp (A, **Figure 23**) and remove the carburetor duct (B) from each carburetor.
5. On California models, disconnect the EVAP purge hose (A, **Figure 24**) from the fitting on the front of the front carburetor.
6. Pull the carburetor slide breather hose from the holder (B, **Figure 24**) on the side of the front carburetor, and disconnect the fuel line (C) from the front carburetor fitting.
7. Remove the mounting screws and lower the carburetor cover (B, **Figure 4**) from its mounting bracket.
8. Remove the AIS vacuum line (**Figure 25**) from the fitting on the rear carburetor intake manifold.
9. Disconnect the fuel line (C, **Figure 23**) from the rear carburetor fitting.
10. Note how the throttle and choke cables are routed along the left side of the motorcycle. If necessary, release the cables from the holder (B, **Figure 7**).
11. Loosen the pull cable locknut (A, **Figure 26**) and the return cable locknut (B). Release each cable

from the carburetor bracket and disconnect the cable ends from the throttle wheel. If necessary, turn the pull cable adjuster (C, **Figure 26**) to create additional slack.

12. Loosen the cable clamp screw (A, **Figure 27**) and disconnect the choke cable end (B) from the choke linkage.

13. Move the throttle and choke cables out of the way.

NOTE

*Do not remove the throttle position sensor from the carburetor body. Refer to **Throttle Position Sensor** in this chapter for TPS inspection and adjustment procedures.*

14. Disconnect the carburetor heater connector (A, **Figure 28**) and the throttle position sensor (B) from the harness.

15. On carburetors equipped with fuel cut solenoid valves, disconnect the leads. Label the leads so they can be installed on their respective solenoid valve.

16. Loosen the clamp screw (**Figure 29**) that secures each carburetor to its intake manifold.

17. Pull the carburetors from the intake manifolds and remove the assembly from the left side of the motorcycle.

18. On California models, remove the heat shield (**Figure 30**).

19. While the carburetor assembly is removed, examine the intake manifold on each cylinder head. Look for cracks or other damage that would allow unfiltered air into the engine. Replace any damaged part.

20. Stuff clean shops rag into each intake manifold to keep foreign matter out of the cylinder heads.

21. Install the carburetor assembly by reversing these removal steps. Note the following:

CAUTION

The carburetors form an airtight seal with the manifolds. Air leaks can cause engine damage due to a lean air/fuel mixture or dirt entering with the incoming air.

a. On California models, hook the heat shield (**Figure 30**) over each intake.
b. Lubricate each intake manifold with a little soap to ease installation.
c. Install the carburetor assembly from the left side and fully seat each carburetor in its intake

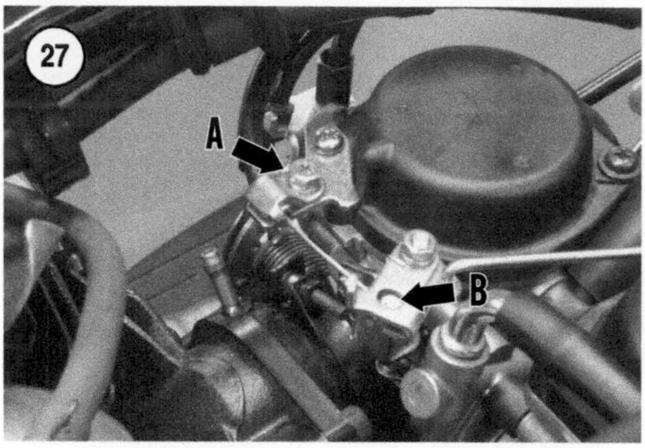

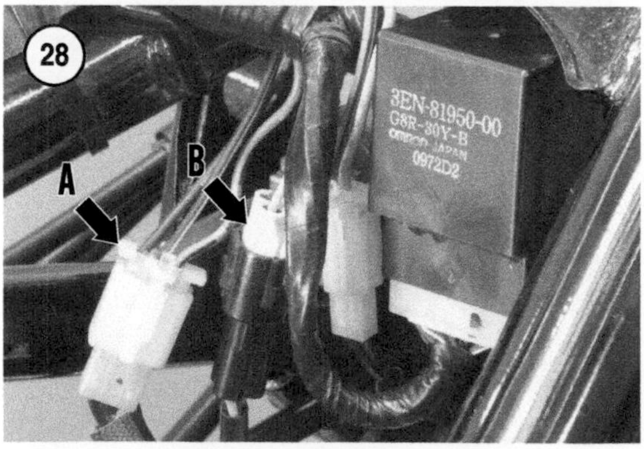

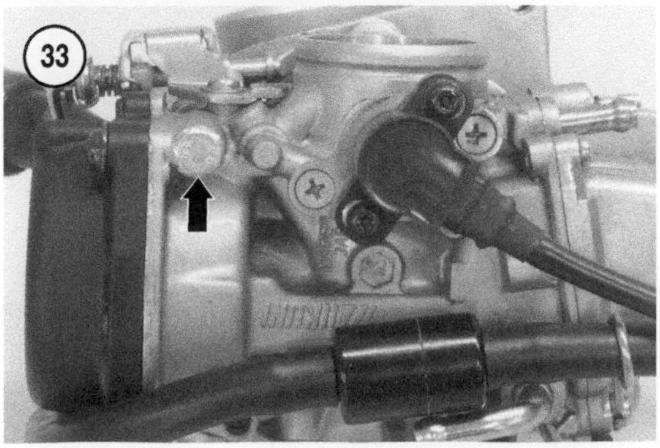

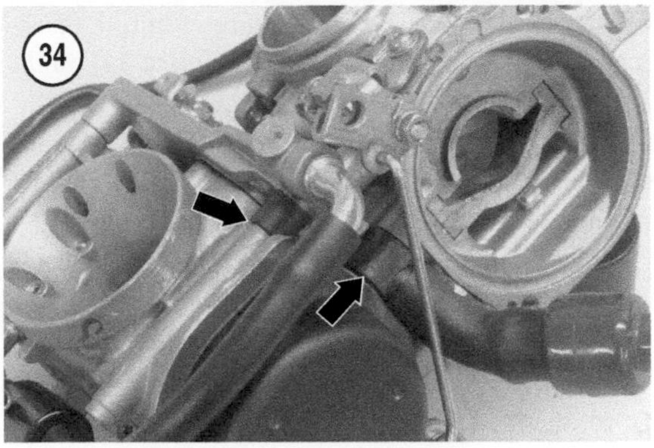

manifold. A solid bottoming will be felt when a carburetor is correctly installed. Tighten each carburetor clamp (**Figure 29**) securely.

d. Route the throttle cables and choke cable along their original paths. These cables must not be twisted, kinked or pinched. Secure the cables in the holder (B, **Figure 7**).
e. Connect each hose to the fitting noted during removal and route the hose along its original path.
f. Adjust the throttle cable as described in Chapter Three.

22. Reinstall the surge tank, air filter housing and fuel tank as described in this chapter.

Separation

Except for the coasting enricher and the starter plunger, the components that require cleaning can be removed from each carburetor body without separating the carburetors. *Do not* separate the carburetors unless the coasting enricher or starter plunger requires service.

The carburetors mount to a bracket that sits between them. The following procedure describes removing the front carburetor from this bracket.

1. Disconnect the connector (A, **Figure 31**) from the carburetor heater on the front carburetor.
2. Release the cable ties (B, **Figure 31**) that secure the wires to the bracket.
3. Disconnect the connectors from the carburetor heater(s) (A, **Figure 32**) and the heater ground (B) on the rear carburetor.
4. Release the wires from the clamp (C, **Figure 32**) on the float bowl.
5. Remove the bolts (**Figure 33**) that secure the front carburetor to the carburetor bracket.

6. Carefully pull the front carburetor until it disengages from the connector tubes (**Figure 34**) on the rear carburetor.

7. Remove the cotter pin (A, **Figure 35**) and washer. Remove the throttle arm (B, **Figure 35**) from the throttle shaft on the front carburetor.

8. Remove the mounting screw (A, **Figure 36**) and lift the choke linkage (B) from the starter plunger on the rear carburetor.

9. Remove the front carburetor (C, **Figure 36**) from the bracket/rear carburetor assembly.

Joining

1. Set the rear carburetor/bracket assembly (**Figure 37**) on the bench.

2. Position the front carburetor (C, **Figure 36**) next to the bracket.

3. Install the choke linkage (B, **Figure 36**) onto the rear carburetor so the linkage fingers engage the starter plunger. Secure the linkage to the carburetor with the mounting screw (A, **Figure 36**).

4. Slide the throttle arm (B, **Figure 35**) onto the front carburetor throttle shaft. Install the washer and secure the arm in place with a new cotter pin (A, **Figure 35**).

5. Press the front carburetor toward the carburetor bracket until the front carburetor seats within the connector tubes (**Figure 34**).

6. Install the bracket bolts (**Figure 33**) and secure the front carburetor to the bracket.

7. Connect the connectors to the carburetor heater(s) (A, **Figure 32**) and the heater ground (B) on the rear carburetor.

8. Secure the wires in the clamp (C, **Figure 32**) on the float bowl.

9. Connect the connector (A, **Figure 31**) to the carburetor heater(s) on the front carburetor.

10. Secure the wires to the bracket with new the cable ties (B, **Figure 31**).

Disassembly

Refer to **Figure 38**.

Disassemble, clean and reassemble one carburetor at a time. This will prevent the accidental mixing of parts.

CAUTION

*When cleaning a carburetor, do not adjust the pilot screw. Refer to **Pilot Screw** in this chapter.*

1. If the coasting enricher (A, **Figure 39**) or starter plunger (B) requires service, separate the carburetors as described in this chapter.

2. Unscrew and remove the starter plunger (B, **Figure 39**).

3. Disassemble the coasting enricher as follows:

   a. Remove the two mounting screws (C, **Figure 39**) and lift the coasting enricher cover (A) from the carburetor body.

38

CARBURETOR

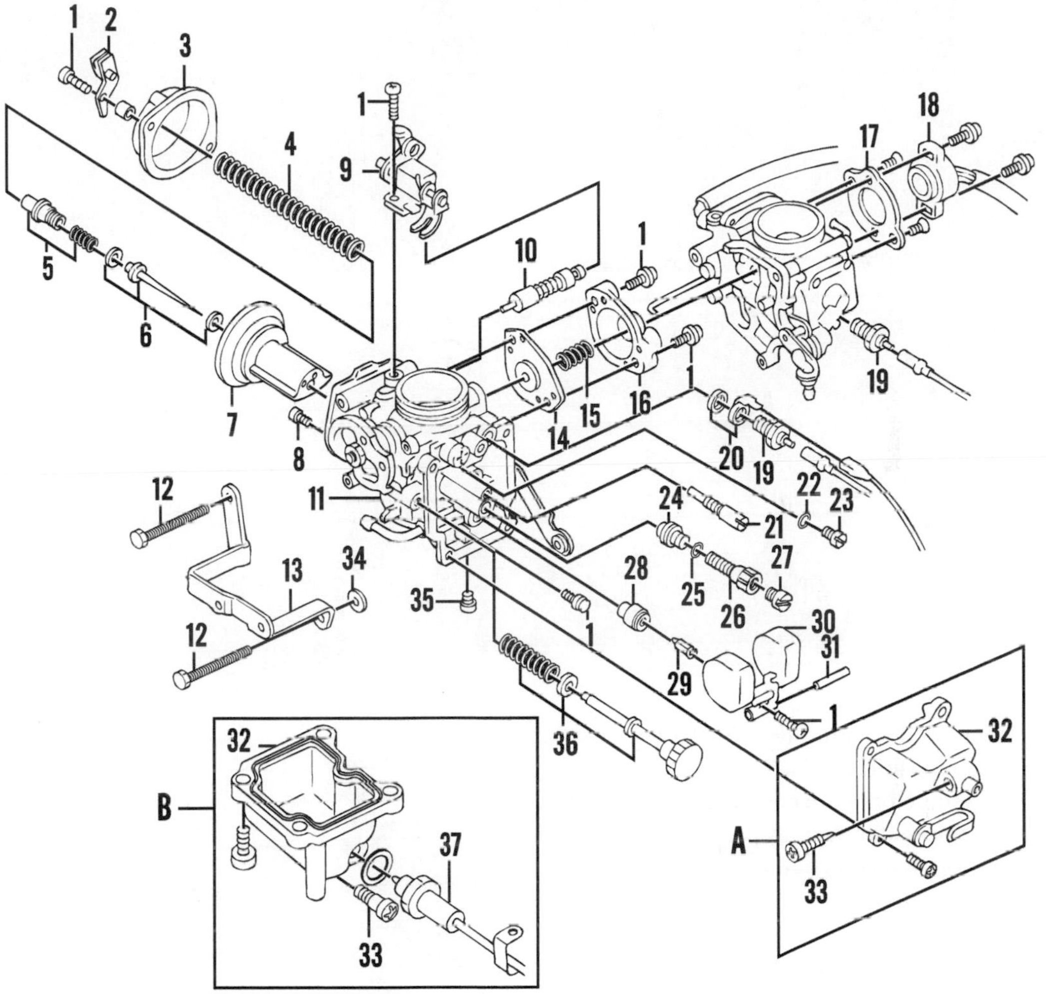

1. Screw
2. Choke cable bracket
3. Top cover
4. Spring
5. Needle retainer
6. Jet needle
7. Vacuum piston
8. Pilot air jet No. 2
9. Choke linkage
10. Starter plunger
11. Carburetor body
12. Bracket bolt
13. Cover bracket
14. Coasting enricher diaphragm
15. Spring
16. Coasting enricher cover
17. Gasket
18. Throttle position sensor (TPS)
19. Carburetor heater
20. Ground terminal
21. Pilot jet
22. Spacer
23. Starter jet
24. Needle jet
25. O-ring
26. Jet holder
27. Main jet
28. Valve seat
29. Float valve
30. Float
31. Pin
32. Float bowl
33. Drain screw
34. Washer
35. Pilot air jet No. 1
36. Throttle stop screw
37. Fuel cut solenoid valve

A. 1998-2003: All models
2004 and 2005: All models except California models

B. 2004 and 2005: California models
2006-on: all models

8

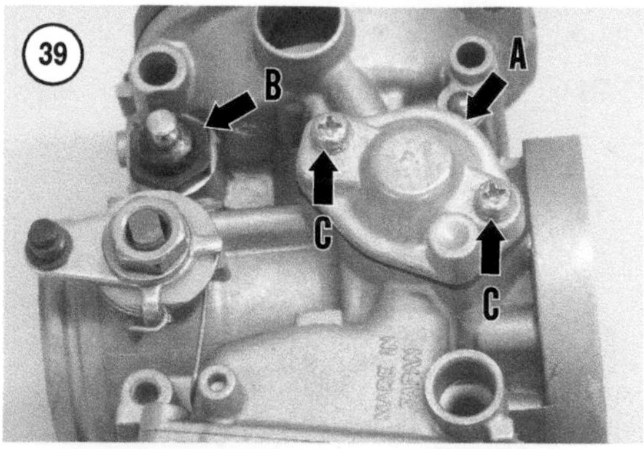

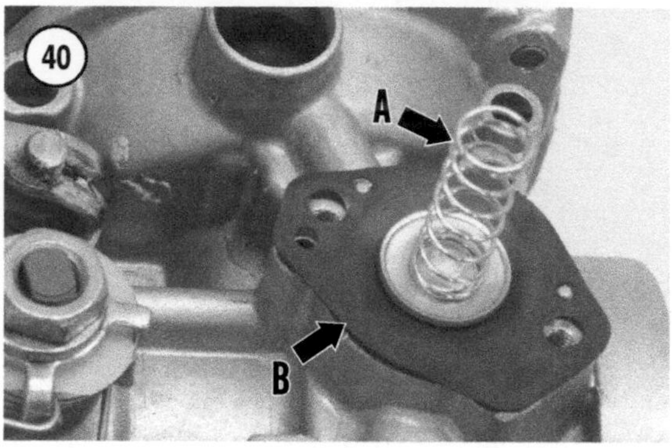

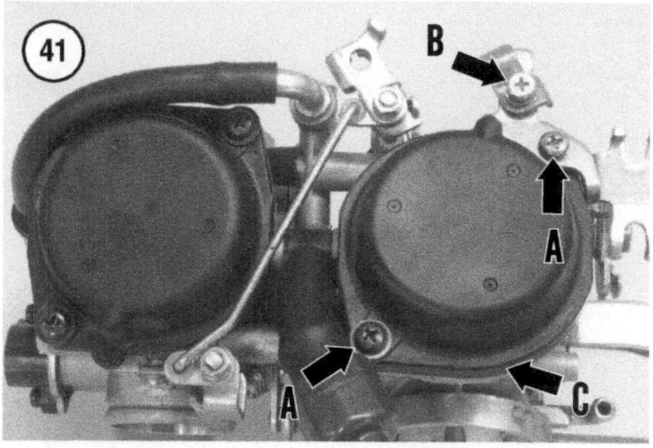

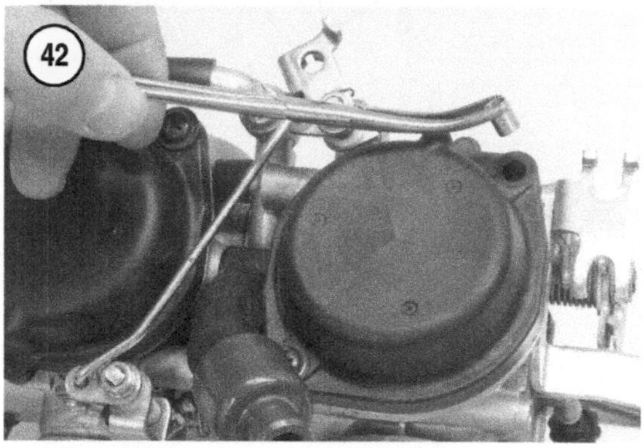

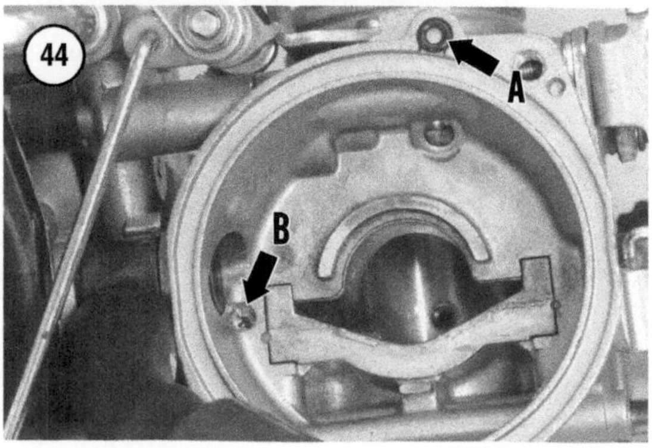

   b. Remove the spring (A, **Figure 40**) and the coasting enricher diaphragm (B).

4. If the carburetors were not separated, disconnect the carburetor heater wires by performing Steps 1-4 in *Separation* in this section.

5. Remove the top cover as follows:

   a. Remove the top cover screws (A, **Figure 41**).

   b. When servicing the rear carburetor, remove the choke cable bracket (B, **Figure 41**) from the top cover.

   c. Watch for the collar (**Figure 42**) beneath the outboard screw on the rear carburetor.

   d. Remove the top cover (C, **Figure 41**) from the carburetor.

6. Remove the spring (A, **Figure 43**) and vacuum piston (B).

7. Remove and discard the small O-ring (A, **Figure 44**).

8. Remove the needle retainer (**Figure 45**) from the vacuum piston and remove the jet needle (**Figure 46**).

9. Remove pilot air jet No. 2 (B, **Figure 44**) from the vacuum piston bore.

10. Remove pilot air jet No. 1 (**Figure 47**) from the carburetor intake horn.

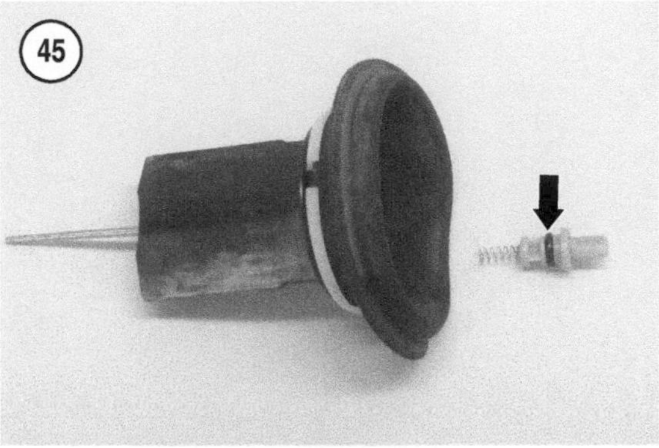

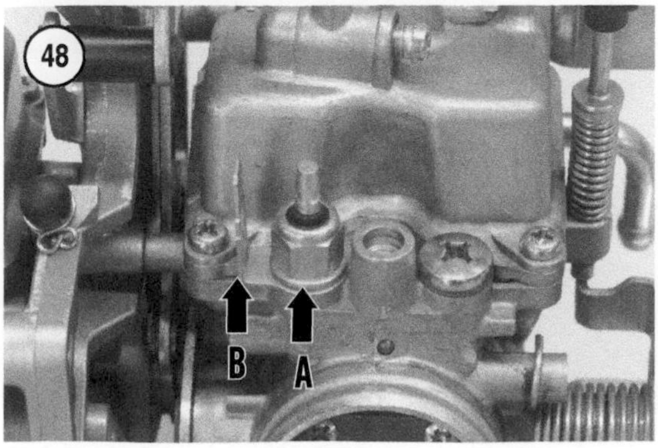

11. Remove the carburetor heater(s) and washer(s) (A, **Figure 48**). When servicing the rear carburetor, also remove the heater ground terminal (B, **Figure 48**).

12. On carburetors equipped with a fuel cut solenoid, remove the solenoid valve.

13. Remove the mounting screws (A, **Figure 49**) and lift the float bowl (B) from the carburetor body. Watch for the float bowl O-ring (**Figure 50**).

14. Remove the screw (A, **Figure 51**), and pull the float pin (B) from the float.

15. Lift out the float and float valve (A, **Figure 52**).

8

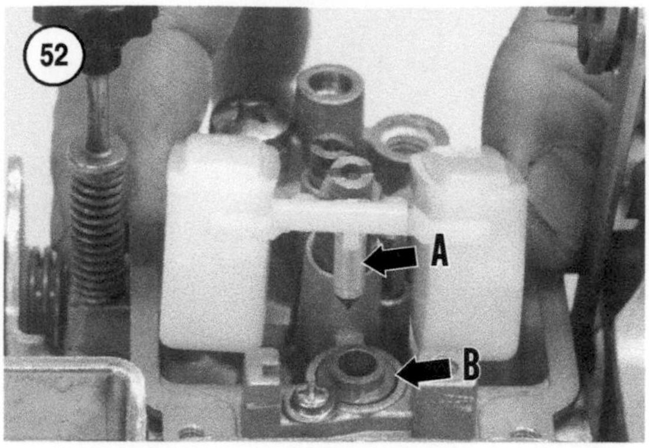

16. Remove the screw (A, **Figure 53**) and pull the valve seat (B) from the carburetor body.
17. Remove the pilot jet (**Figure 54**).
18. Remove the starter jet (**Figure 55**) and spacer.
19. Remove the main jet (**Figure 56**).
20. Remove the jet holder (**Figure 57**) and O-ring.
21. Insert a finger into the carburetor venturi, press the needle jet toward the float bowl and remove it (**Figure 58**).

CAUTION
When cleaning a carburetor, do not adjust the pilot screw setting. Changing this setting will decrease engine performance.

22. If necessary, remove the pilot screw as described in *Pilot Screw* in this chapter.
23. Clean and inspect the components as described in this section.

Assembly

1. If the pilot screw was removed, install it as described in *Pilot Screw* in this chapter.
2. Install the needle jet (**Figure 58**) so the concave side goes into the carburetor body.
3. Install the needle jet holder (**Figure 57**) with a new O-ring and install the main jet (**Figure 56**).
4. Install the starter jet (**Figure 55**) and spacer.
5. Install the pilot jet (**Figure 54**).
6. Install the valve seat into the carburetor body (**Figure 59**). Secure it in place with the screw (A, **Figure 53**).
7. Hook the needle valve onto the tang on the float. Install the float so the needle valve (A, **Figure 52**) seats in the valve seat (B).

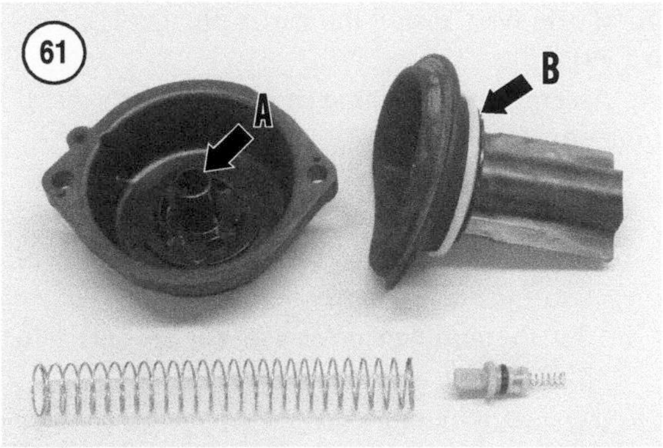

8. Install the float pin (B, **Figure 51**) and secure it with the screw (A).
9. Install a new float bowl O-ring (**Figure 50**) and set the float bowl onto the bottom of the carburetor (B, **Figure 49**). Secure the float bowl in place with the mounting screws (A, **Figure 49**).
10. Install the carburetor heater(s) (A, **Figure 48**) and its washer. If working on the rear carburetor, install the washer and ground terminal (B, **Figure 48**) over the carburetor heater threads, then install the carburetor heater.
11. Install pilot air jet No. 1 (**Figure 47**) into the carburetor horn.
12. Install pilot air jet No. 2 (B, **Figure 44**) into the vacuum piston bore.
13. Install a new O-ring (A, **Figure 44**) onto the top of the vacuum piston bore.
14. Assemble the vacuum piston, if it was disassembled, as follows:
   a. Lower the jet needle (**Figure 46**) into the vacuum piston.
   b. Secure the needle in place with the needle retainer (**Figure 45**).
15. Lower the vacuum piston (**Figure 60**) into the slide bore so the piston diaphragm sits within the lip on top of the bore.
16. Install the spring (A, **Figure 43**) into the vacuum piston.
17. Install the top cover (C, **Figure 41**) and secure it in place with the mounting screws (A).
   a. Make sure the post (A, **Figure 61**) inside the top cover sits inside the spring.
   b. When working on the rear carburetor, install a collar into the outboard mounting position (**Figure 42**), and secure the choke cable bracket (B, **Figure 41**) with the top cover mounting screw (A).

8

18. If removed, install the starter plunger (B, **Figure 39**).

19. Assemble the coasting enricher, if it was disassembled, as follows:

  a. Install the coasting enricher diaphragm (B, **Figure 40**) onto the carburetor (**Figure 62**).
  b. Install the spring (A, **Figure 40**) into the center of the diaphragm.
  c. Set the enricher cover (A, **Figure 39**) into place and install the mounting screws (C).

20. If the carburetors were not separated, connect the carburetor heaters as follows:

  a. Connect the leads to the carburetor heater (A, **Figure 63**) and the heater ground (B) on the rear carburetor.
  b. Secure the wires in the clamp (C, **Figure 63**) on the float bowl.
  c. Connect the leads (A, **Figure 64**) to the carburetor heater(s) on the front carburetor.
  d. Secure the wires to the bracket with new cable ties (B, **Figure 64**).

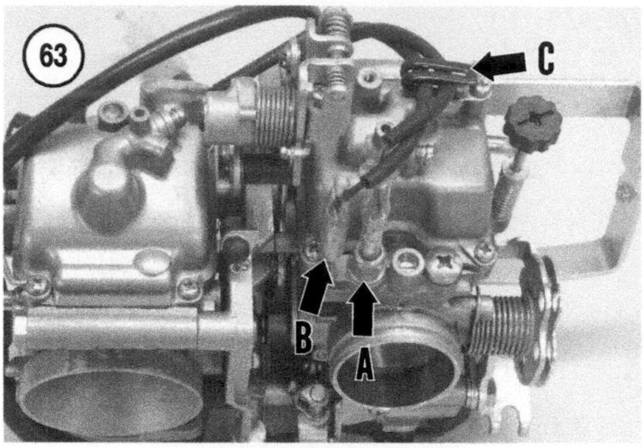

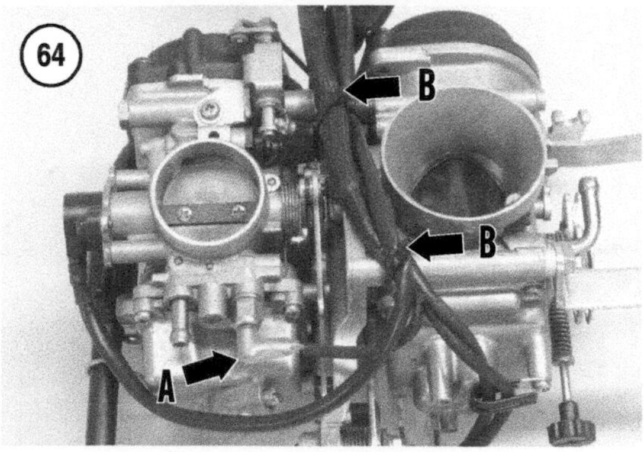

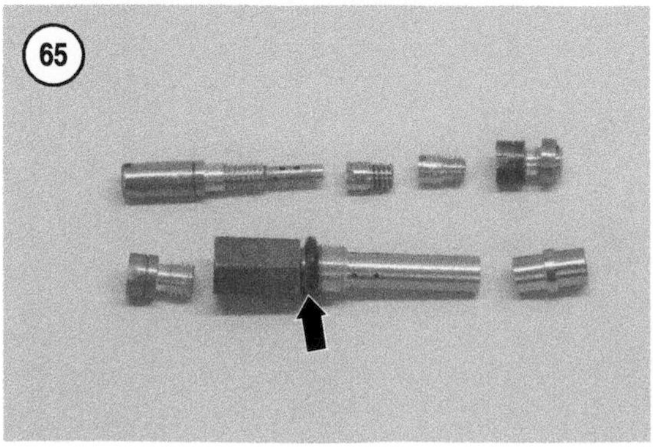

Cleaning and Inspection

1. Throughly clean and dry all parts. Use a commercial cleaner that is specifically for carburetors.

CAUTION

Do not submerge the carburetor body in solvent if the throttle positon sensor, heater(s) or fuel cut solenoid are installed.

CAUTION

***Do not** use wire to clean the jets. Scratches in the jet can alter the air/fuel mixture.*

2. Allow the carburetor to dry thoroughly before assembly or dry parts with compressed air. Blow out the jets and jet holder with compressed air.
3. Replace the jet holder O-ring (**Figure 65**) during assembly.
4. Make sure the holes in all jets and holders are clear. Clean them out if they are plugged in any way. Replace any jet or holder if its holes cannot be unplugged.
5. Make sure all openings in the carburetor body are clear. Clean them out if they are plugged in any way.

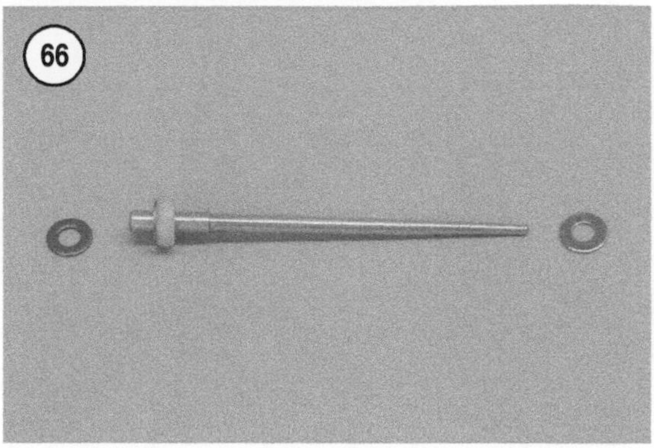

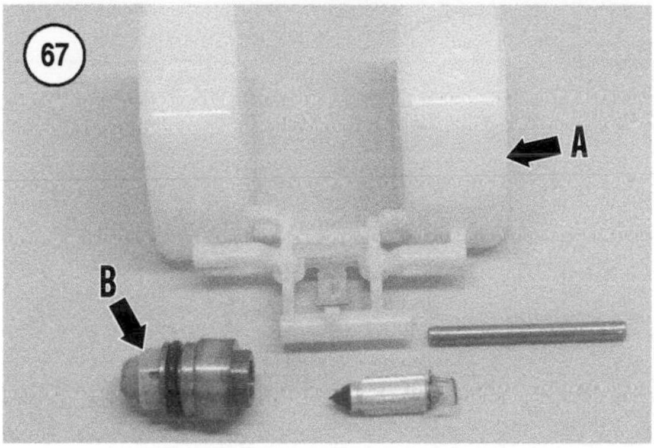

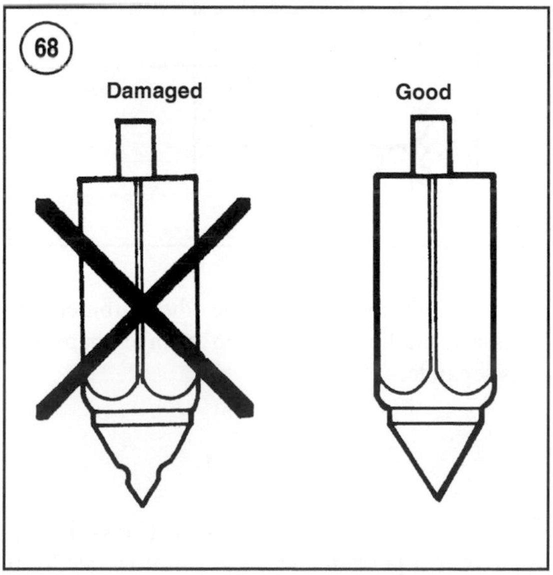

6. Inspect the slide bore in the carburetor body. Make sure it is clean and free of any burrs or obstructions that may cause the vacuum piston to hang up during normal operation.
7. Inspect the vacuum piston (B, **Figure 61**) for scoring and wear. Replace it if necessary.

8. Inspect the diaphragm on the vacuum piston for tears, cracks or other damage. Replace the vacuum piston if the diaphragm is damaged.
9. Inspect the jet needle (**Figure 66**) for excessive wear at the tip or other damage.
10. Inspect the float (A, **Figure 67**) for deterioration or damage. Check for leaks by placing the float in a container of water. If the float sinks or if bubbles appear, replace the float.
11. Inspect the needle valve assembly as follows:
   a. Inspect the end of the needle valve for wear or damage. See **Figure 68**.
   b. Check the inside of the valve seat (B, **Figure 67**). A damaged needle valve or a particle of dirt or grit in the float valve assembly will cause the carburetor to flood and fuel to overflow.
   c. Inspect the filter on the end of the valve seat. If there are any holes or if the filter is starting to deteriorate, replace the needle valve assembly.
   d. Inspect the O-ring on the valve seat. If it is torn or beginning to harden, replace the needle valve assembly.
   e. If any part is worn or damaged, replace the entire assembly, the valve seat and the needle valve as a set.
12. If the coasting enricher diaphragm was removed, inspect it for tears, cracks or other damage. Replace the diaphragm if there is any damage.
13. If removed, inspect the starter plunger (**Figure 69**) for wear or damage.
14. If removed, inspect the pilot screw (**Figure 70**) for wear or damage that may have occurred during removal. Replace the pilot screw in both carburetors even if only one requires replacement. This is necessary for correct pilot screw adjustment as described in this chapter.

15. Replace all O-rings during assembly. O-rings tend to harden after prolonged use and exposure to heat, which reduces their ability to seal properly.

PILOT SCREW

Removal/Installation

The pilot screw is sealed. A plug has been installed at the top of the pilot screw bore. A pilot screw does not require adjustment unless the carburetor is overhauled, the pilot screw has been incorrectly adjusted, or the pilot screw was replaced.

NOTE
An exhaust gas analyzer is needed to precisely set the air/fuel mixture with the pilot screws. If a pilot screw must be removed, perform the procedure as described, then have a Yamaha dealership perfom the final adjustment.

1. Set a stop 6 mm from the end of a 1/8 in. drill bit. See **Figure 71**.
2. Carefully drill a hole in the plug at the top of the pilot screw bore on the carburetor body as shown in **Figure 71**. Do not drill too deeply. The pilot screw will be difficult to remove if the head is damaged.
3. Screw a sheet metal screw into the plug and pull the plug from the bore.
4. Screw the pilot screw clockwise until it *lightly* seats in the bore while counting and recording the number of turns. The pilot screw must be reinstalled to this same position during assembly.
5. Remove the pilot screw, spring, washer and O-ring from the carburetor body (**Figure 72**).
6. Inspect the O-ring and the end the pilot screw. Replace the screw and/or O-ring if damaged or worn (grooved).
7. Install the pilot screw assembly and turn the pilot screw until it *lightly* seats in the bore. Turn the pilot screw counterclockwise and back it out the number of turns noted during removal.
8. Install a new plug by tapping it into place with a punch.

70

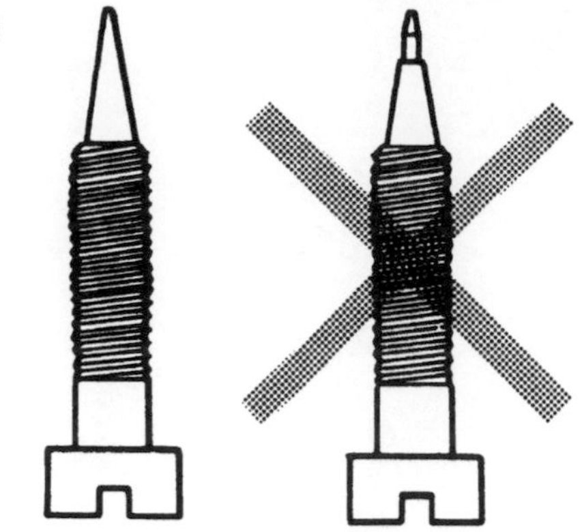

71

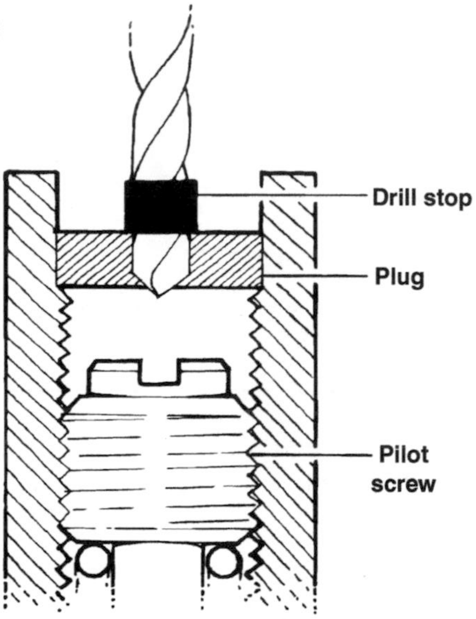

9. Repeat this procedure for the other carburetor, if necessary. Make sure to keep each carburetor's parts separate.

FUEL LEVEL

The fuel level in the carburetor float bowls is critical to proper performance. The fuel flow rate from the bowl up to the carburetor bore depends not only on the vacuum in the throttle bore and the size of the jets, but also on the fuel level in the float bowl. **Table 1** provides the specification for fuel level, measured from (**Figure 73**) the upper edge of the float bowl with the carburetors mounted on the motorcycle.

72

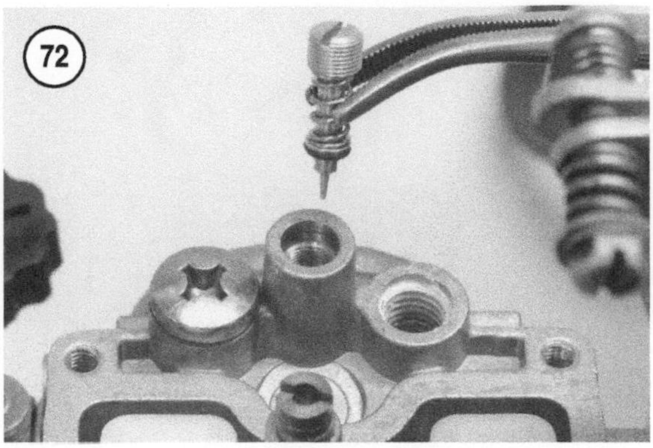

73

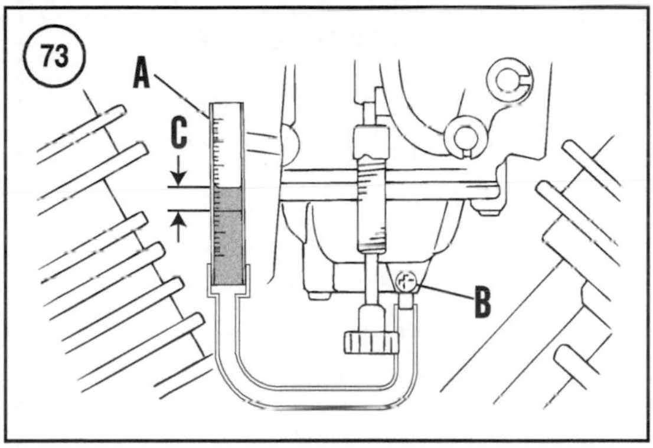

74

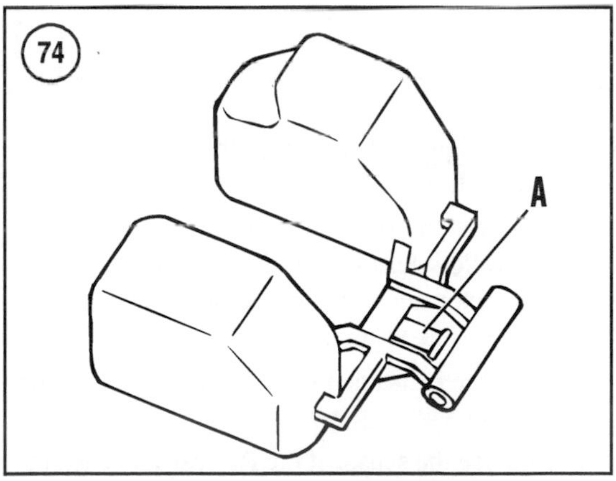

This measurement is more accurate than a float height measurement because the actual fuel level can vary between carburetors, even when the floats are set at the same height. Fuel level inspection requires a special fuel level gauge (Yamaha part No. YM-01312-A or 90890-01312).

Inspection/Adjustment

Inspect or adjust the fuel level as follows:

WARNING
Some gasoline will drain from the carburetors during this procedure. Work in a well-ventilated area at least 50 feet from any open flame. Do not allow anyone to smoke. Wipe up spills immediately.

1. Remove the fuel tank, air filter housing and surge tank as described in this chapter.
2. Use a jack and/or wood blocks to level the motorcycle. Make sure the carburctors are level.
3. Connect the fuel level gauge (A, **Figure 73**) to the drain on the carburetor. Secure the gauge so it sits vertically against the float bowl.
4. Loosen the carburetor drain screw (B, **Figure 73**).
5. Wait until the fuel in the gauge settles.
6. Fuel level (C, **Figure 73**) equals the distance between the fuel level in the gauge and the upper edge of the float bowl. Record the fuel level and compare the reading to the specification in **Table 1**.
7. Repeat this procedure for the other carburetor. The fuel level of one carburetor should equal the fuel level of the other.

NOTE
If the float bowls empty during this procedure, temporarily install the fuel tank and refill the float bowls. Remove the fuel tank, then proceed to the next step.

8. If the fuel level in a particular carburetor requires adjustment, adjust the float height as follows:
   a. Remove the carburetor assembly as described in this chapter.
   b. Remove the float bowl from the appropriate carburetor.
   c. Remove the float pin screw (A, **Figure 51**), pull pin (B), and lift out the float.
   d. Remove the needle valve from the float tang.
   e. Bend the float tang (**Figure 74**) as required to attain the correct fuel level. Install the float bowl.
   f. Reinstall the carburetor assembly and re-check the fuel level.

IDLE SPEED ADJUSTMENT

Refer to Chapter Three.

75

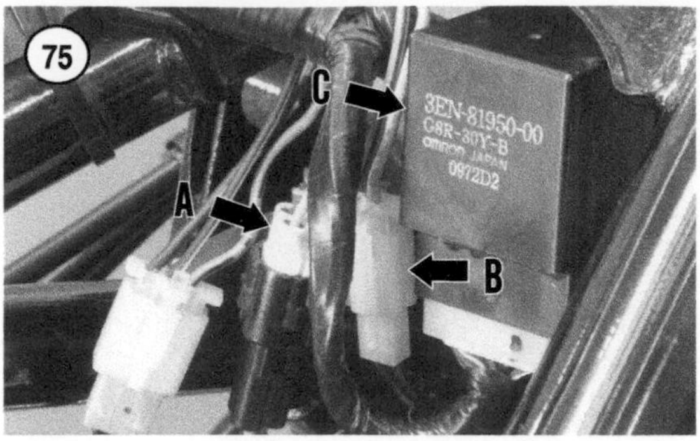

76

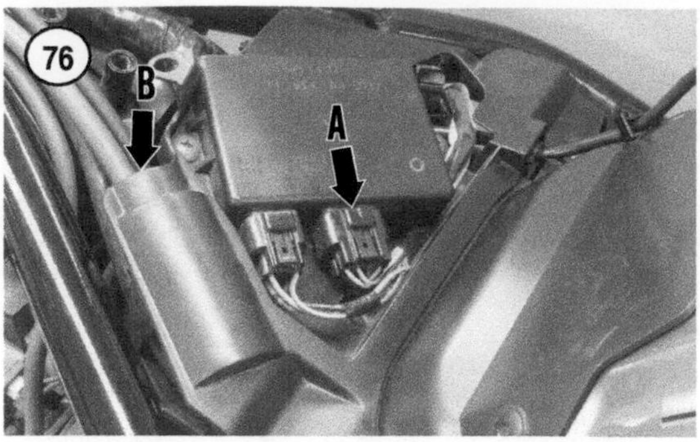

THROTTLE CABLE ADJUSTMENT

Refer to Chapter Three.

THROTTLE POSITION SENSOR

Testing

Perform the following test procedure whenever the self-diagnostic system (see Chapter Nine) flashes a TPS trouble code. Perform the test in the listed sequence. Each step presumes that the components tested in the earlier steps are working properly. The tests can yield invalid results if they are performed out of sequence. If a test indicates that a component is working properly, reconnect the electrical connections and proceed to the next step.

1. Remove the fuel tank, air filter housing and surge tank as described in this chapter.
2. Refer to the wiring diagram at the back of the manual and check the continuity of the throttle position sensor wiring as follows:
   a. Disconnect the throttle position sensor connector (A, **Figure 75**).
   b. Disconnect the connector (**Figure 76**) from the ignitor unit.
   c. Check the continuity on the blue, yellow and black/blue wires between the ignitor connector and throttle position sensor connector.
   d. Make any necessary repairs to the wiring.
3. Refer to **Figure 77** and check the sensor's R1 resistance as follows:
   a. Connect the ohmmeter positive test lead to the blue terminal in the sensor side of the connector and connect the negative test lead to the black terminal.
   b. Replace the sensor if the R1 resistance is outside the range specified in **Table 2**.

77

TPS CONNECTOR

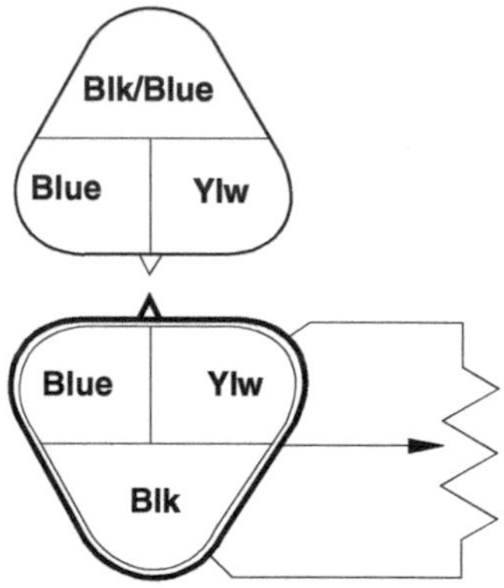

4. Refer to **Figure 77** and check the sensor's R2 resistance as follows:
   a. Connect the ohmmeter positive test lead to the yellow terminal in the sensor side of the connector and connect the negative test lead to the black terminal.
   b. Note the resistance while slowly opening the throttle.
   c. Replace the sensor if the R2 resistance is outside the range specified in **Table 2**.

Adjustment

Adjust the throttle position sensor by turning the sensor until its resistance is within the specified range.

NOTE

The throttle position sensor is mounted on the side of the rear carburetor.

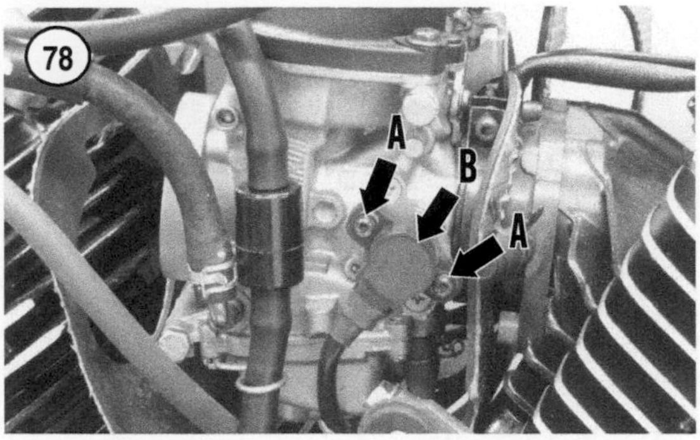

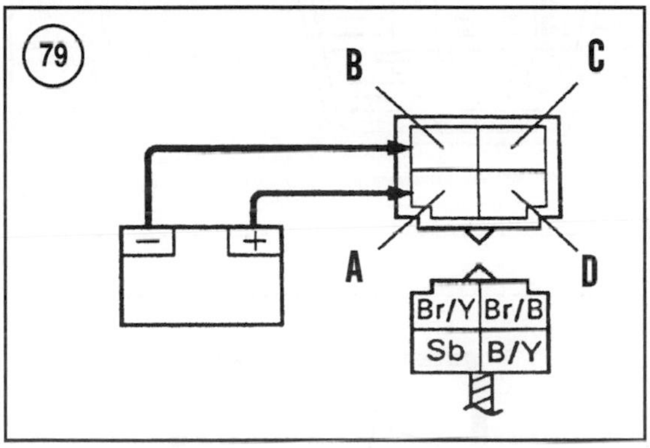

1. Properly adjust the idle speed as described in Chapter Three.
2. Remove the fuel tank, air filter housing and the surge tank as described in Chapter Eight.
3. Disconnect the throttle position sensor (A, **Figure 75**) from its harness mate.
4. Loosen the throttle position sensor mounting screws (A, **Figure 78**).
5. Connect the positive lead of an ohmmeter to the yellow terminal (**Figure 77**) on the sensor side of the connector and connect the negative lead to the black terminal.
6. With the throttle fully closed, rotate the sensor body (B, **Figure 78**) until the resistance is within the fully closed range specified in **Table 2**.
7. Tighten the throttle position sensor screws securely.

FUEL CUT SOLENOID

On 2004-on California models and all other models from 2006-on, each carburetor is equipped with a fuel cut solenoid valve. The valves stop fuel flow into the carburetors whenever the main switch or engine stop switch is turned off when the engine is running.

Perform the following whenever the self-diagnostic system (Chapter Nine) flashes a fuel cut solenoid valve trouble code.

Fuel Cut Solenoid Valve Test

1. Remove the fuel cut solenoid valve from the carburetor.
2. Use an ohmmeter and check the resistance of the valve as follows:
   a. Connect an ohmmeter lead to the wire connector and ground the other lead to the valve body.
   b. Check the resistance.
   c. Refer to **Table 2** for specifications. Replace the valve if it is not within the specification.

CARBURETOR HEATER

Troubleshooting

Refer to Chapter Two.

Carburetor Heater Relay Test (1999-2003 Models)

1. Remove the carburetor heater relay as described below.
2. Use jumpers to connect the battery positive terminal to the brown/yellow terminal (A, **Figure 79**) on the relay and connect the battery negative terminal to the relay's sky blue terminal (B).
3. Connect the ohmmeter positive test lead to the brown/black terminal (D, **Figure 79**) in the relay, and connect the negative test lead to the relay's black/yellow terminal (C).
4. If the ohmmeter does not show continuity, replace the carburetor heater relay.

Carburetor Heater Relay Removal/Installation (1999-2003 Models)

1. Remove the right-side cover as described in Chapter Fourteen.
2. The carburetor heater relay is in the bottom of the battery box, near the voltage regulator.

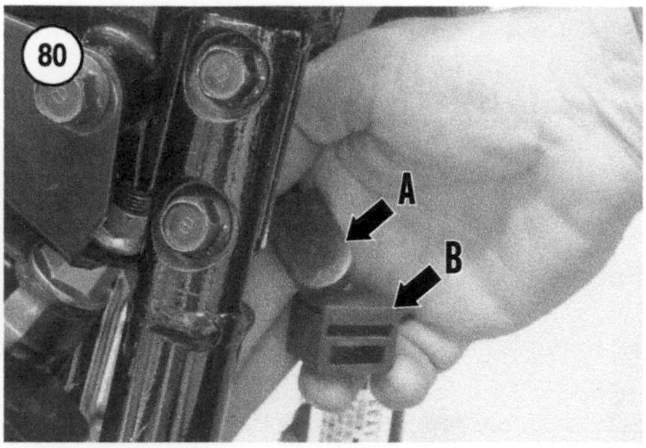

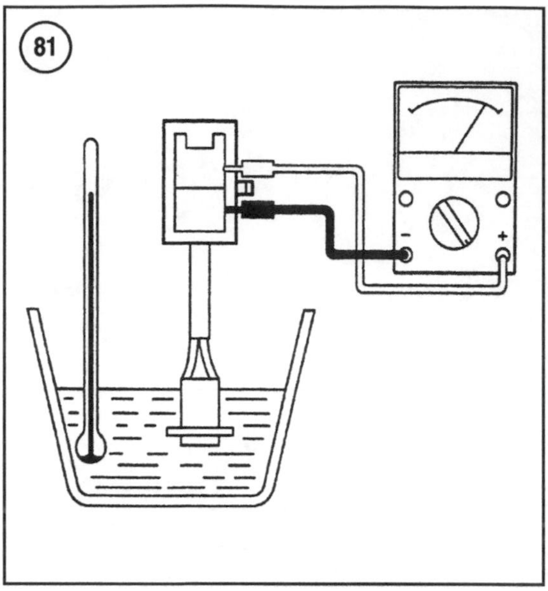

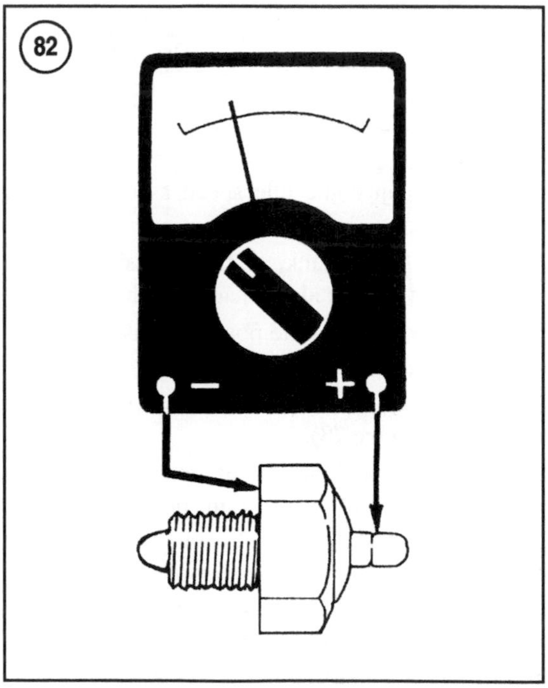

3. Disconnect the harness from the carburetor heater relay and remove the relay.
4. Installation is the reverse of removal.

Thermo Switch Test

For models with four carburetors, the thermo switch passes voltage to the No. 1 carburetor heater and to the carburetor heater relay when the engine is cold. For models with two carburetor heaters, the thermo switch passes voltage directly to the heaters when the engine is cold. The switch turns off power when the engine warms. Because the switch is sensitive to very low temperature, accurately test the switch to determine if it operates within the required temperature range.

1. Disconnect the thermo switch connector (B, **Figure 75**) from the harness.

CAUTION
The switch is sensitve to shock. Do not drop or mishandle the switch.

2. Remove the thermo switch dampter (B, **Figure 80**) from the tang below the frame neck and remove the switch (A).
3. Partially fill a container with water and suspend the thermo switch and a thermometer in the water as shown in **Figure 81**. Use a container that is large enough to have additional water added during the test.
   a. For 1999-2002 models, the water temperature must be below 68°F (20°C)
   b. For 2003-on models, the water temperature must be below 55°F (13°C)
4. Connect an ohmmeter to the switch connector and allow the switch to stablize to the water temperature. When the swtich is stablized, the switch should be on. There should be continuity.

5. Add hot water to the container, keeping the switch wires above the water. Observe the thermometer and ohmmeter. Note the following:

   a. For 1999-2002 models, when the water temperature is 68°-78°F (20-26°C), the switch

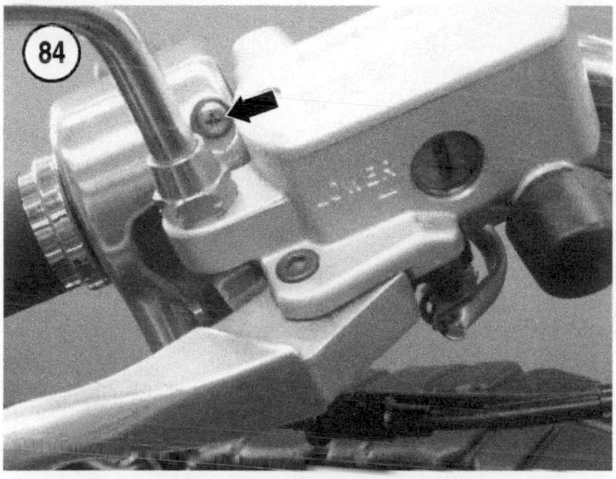

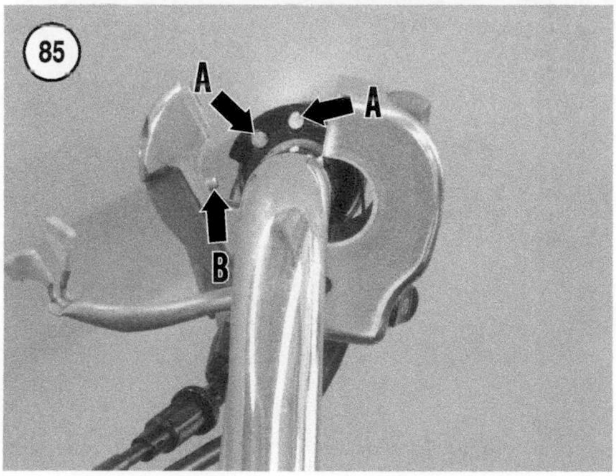

should turn off. There should be no continuity.

b. For 2003-on models, when the water temperature is 55°-66°F (13°-19° C), the switch should turn off. There should be no continuity.

6. Allow the water to cool. If necessary, add ice to lower the temperature below ambient temperature.

For all models, when the water temperature is 46°-60°F (8°-16° C), the switch should turn on. There should continuity.

7. Replace the switch if the test results are not within the test specifications.

Carburetor Heater Test

1. Disconnect the electrical lead from the carburetor heater and remove the heater from the carburetor.

2. Connect an ohmmeter to the carburetor heater as shown in **Figure 82** and check the heater's resistance.

3. Replace the carburetor heater if the resistance is not within the range specified in **Table 2**.

THROTTLE CABLE

Replacement

1. Securely support the motorcycle on a level surface.

2. Remove the fuel tank, air filter housing and surge tank as described in this chapter.

3. Remove the frame neck covers (Chapter Fourteen).

4. Note how the throttle cables are routed along the left side and through the fork legs. The new cable must be rerouted along the same path.

5. Loosen the pull cable locknut (A, **Figure 83**) and the return cable locknut (B). Release each cable from the carburetor bracket and disconnect the cable ends from the throttle wheel. If necessary, turn the pull cable adjuster (C, **Figure 83**) to create additional slack.

6. Remove the front brake master cylinder (Chapter Thirteen).

7. Remove the mounting screws (**Figure 84**) and separate the halves of the right handlebar switch assembly.

8. Disengage the ends (A, **Figure 85**) of both the pull and return cables from the throttle grip. Remove each cable from the handlebar switch housing.

9. Release the cables from any ties or holders (**Figure 86**) that secure them in place. The ties or holder must be reinstalled in their original locations.

NOTE
The piece of string in the next step is used to pull the new throttle cable through the frame so the cable will be properly routed.

10. Tie a piece of heavy string or cord to the ends of the throttle cable at the carburetor. Wrap this end with tape. Do not use an excessive amount of tape. Too much tape could interfere with the cable's passage through the frame. Tie the other end of the string to the frame.
11. Starting at the handlebar, carefully pull the throttle cable along the frame and through the fork legs. Make sure the attached string follows the same path as the throttle cable through the frame.
12. Untie the string from the old cable and tie it to the carburetor ends of the new cable.
13. Carefully pull the string back through the frame, routing the new cable through the same path as the old.
14. Remove the string and lubricate the ends of the new cable with lithium soap grease.
15. Connect the return cable end to the carburetor wheel. Fit the cable into the bracket and secure it in place with the locknut (B, **Figure 83**).
16. Connect the pull cable to the carburetor throttle wheel and secure it to the bracket with its locknut (A, **Figure 83**).
17. At the handlebar, lubricate the ends of the new cable with lithium soap grease. Connect the cable ends (A, **Figure 85**) to the throttle grip. Make sure the cables rests in the grooves of the throttle drum.
18. Reinstall the halves of the right handlebar switch assembly. Make sure the index pin (B, **Figure 85**) on the switch assembly aligns with the hole in the handlebar. Secure the switch with the screws (**Figure 84**).
19. Operate the throttle grip. Make sure the throttle linkage operates correctly without binding. If necessary, make sure the cable is correctly attached and there are no tight bends in the cable.
20. Adjust throttle cable free play as described in Chapter Three.
21. Reinstall the surge tank, air filter housing and the fuel tank as described in this chapter.
22. Start the engine. Turn the handlebar from side to side without operating the throttle. If the engine speed increases as the handlebar is turned, the throttle cable is routed incorrectly. Recheck the cable routing.

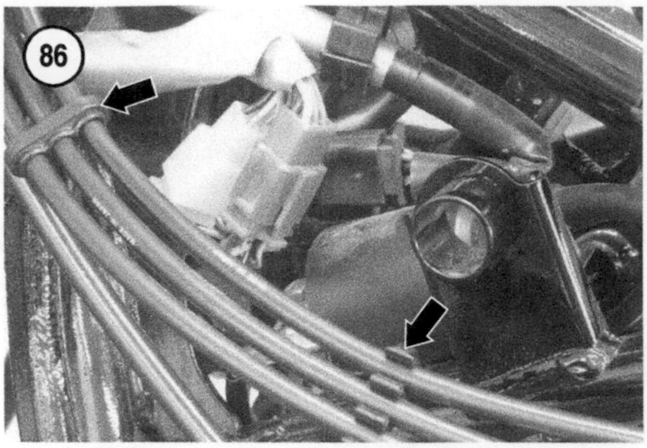

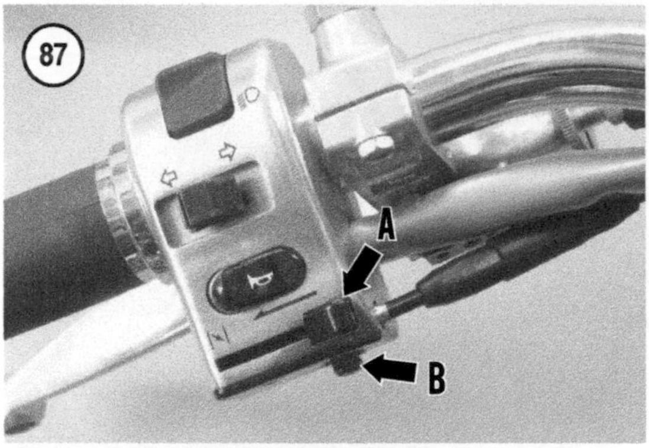

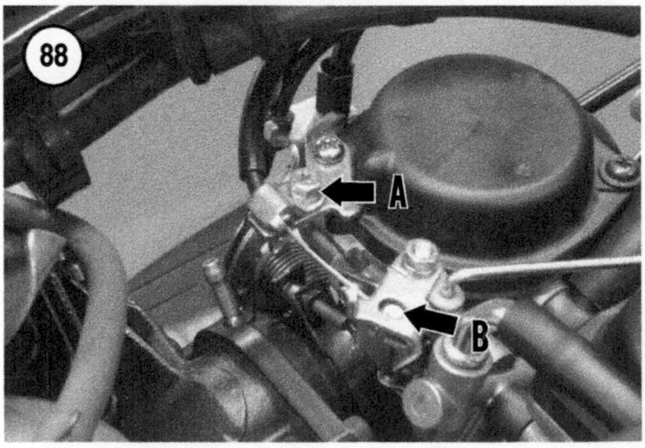

CHOKE CABLE

Replacement

NOTE
*The carburetor starter valve is controlled by the choke lever (A, **Figure 87**) on the left handlebar switch.*

1. Securely support the motorcycle on a level surface.

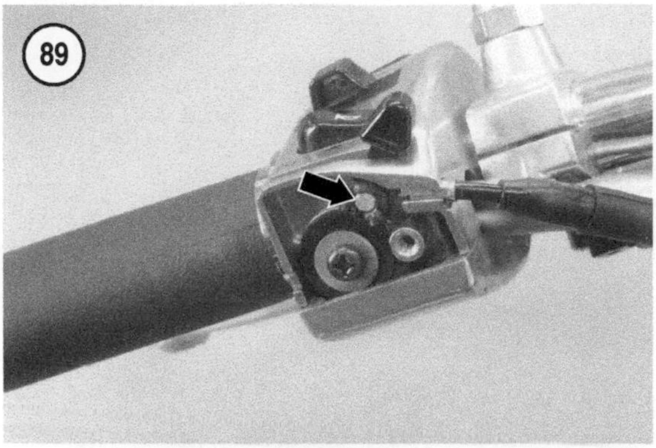

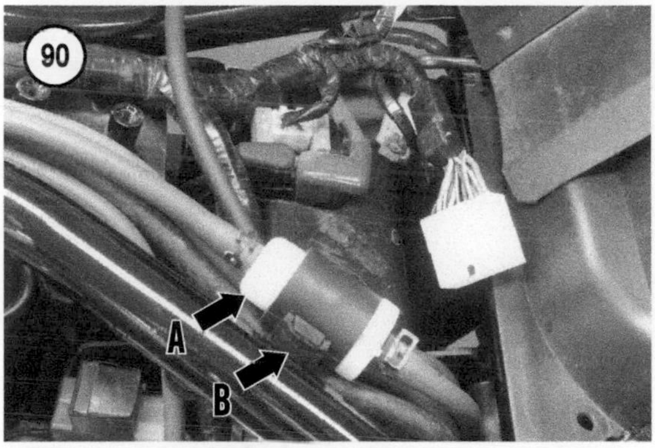

2. Remove the fuel tank, air filter housing and the surge tank as described in this chapter.
3. Remove the neck covers (Chapter Fourteen).
4. Note how the choke cable is routed along the frame and through the fork legs. The new cable will have to follow the same path.
5. At the carburetor, loosen the cable clamp screw (A, **Figure 88**) and disconnect the choke cable end (B) from the choke linkage.
6. At the left handlebar switch, remove the screw (B, **Figure 87**) and lower the cover from the bottom of the switch.
7. Disconnect the cable end from the choke lever fitting (**Figure 89**) and remove the cable from the switch housing.
8. Release the choke cable from any ties or holders (**Figure 86**) that secure the choke cable in place. The ties or holders must be reinstalled in their original locations.
9. Tie a piece of heavy string or cord to the carburetor end of the choke cable. Wrap this end with tape. Do not use an excessive amount of tape. Too much tape could interfere with the cable's passage through the frame.
10. Starting at the choke lever, carefully pull the cable from the motorcycle, following the path of the original cable.
11. Lubricate the new cable as described in Chapter Three.
12. Untie the string from the old cable and tie it to the carburetor end of the new cable.
13. Carefully pull the string back through the frame, routing the new cable through the same path as the old.
14. Remove the string and lubricate the cable end with lithium soap grease.
15. Connect the cable end (B, **Figure 88**) to the choke linkage and secure the cable with the clamp screw (A).
16. At the handlebar, lubricate the cable end with lithium soap grease and connect the cable to the choke lever linkage (**Figure 89**).
17. Operate the choke lever and check the operation of the carburetor starter valve assembly. If it does not operate properly or if it is binding, make sure the cable is correctly attached, and check for tight bends in the cable.
18. Secure the handlebar switch cover in place with its mounting screw (B, **Figure 87**).

FUEL FILTER

All models are equipped with a separate fuel filter that cannot be cleaned. The fuel filter must be replaced whenever it is dirty or at the interval specified in **Table 1** in Chapter Three.

Removal/Installation

1. Remove the rider and passenger seats as described in Chapter Fourteen.
2. Disconnect the electrical lead (**Figure 2**) from the negative battery terminal.
3. Remove the fuel tank as described in this chapter.
4. Remove the ignitor panel (B, **Figure 76**) as described in Chapter Fourteen.
5. Disconnect the inlet and outlet hoses from the fuel filter (A, **Figure 90**).
6. Remove the filter from the rubber grommet (B, **Figure 90**).
7. Installation is the reverse of removal. Note the following:
   a. Securely seat the filter in its grommet.

b. Position the filter so its arrow points toward the fuel pump.
c. Check the hose clamps for damage; replace them if necessary.
d. After installation is complete, thoroughly check for leaks.

FUEL PUMP

Removal/Installation

1. Disconnect the electrical lead from the negative battery terminal.
2. Remove the toolbox cover and the left side cover as described in Chapter Fourteen.
3. Roll the boot back from the electrical connectors.
4. Disconnect the fuel pump connector (A, **Figure 91**).

NOTE
During removal, label each hose and its fitting, and note how a hose is routed through the frame. This assures that each hose will be connected to the correct fitting and properly routed during assembly.

5. Disconnect the inlet hose (B, **Figure 91**) and outlet hose (C) from their fittings on the fuel pump.
6. Slide the fuel pump from its rubber grommet, and remove it.
7. Installation is the reverse of removal. Note the following:
   a. Securely seat the fuel pump in its mounting grommet.
   b. Connect the inlet (B, **Figure 91**) and outlet hoses (C) to the proper fittings on the pump.

Troubleshooting

Refer to Chapter Two.

Operational Test

WARNING
Perform the test in a safe, well-ventilated location away from the shop area, Use caution when handling the fuel tanks, fuel lines and jumper wires. Immediately clean up fuel spills. Connect the power source to the fuel pump after the test equipment is secure.

Check the operation of the fuel pump by applying battery power directly to the fuel pump.

91
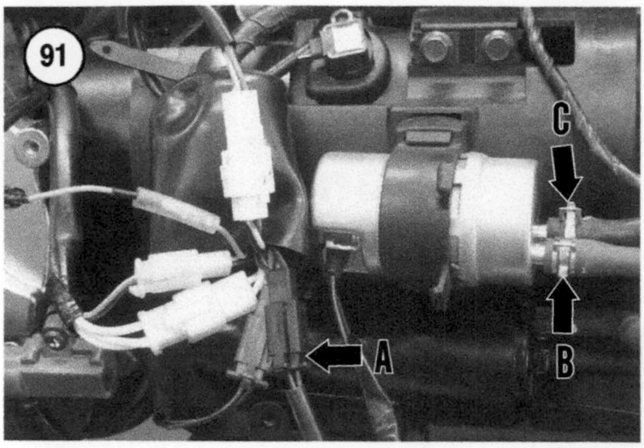

92
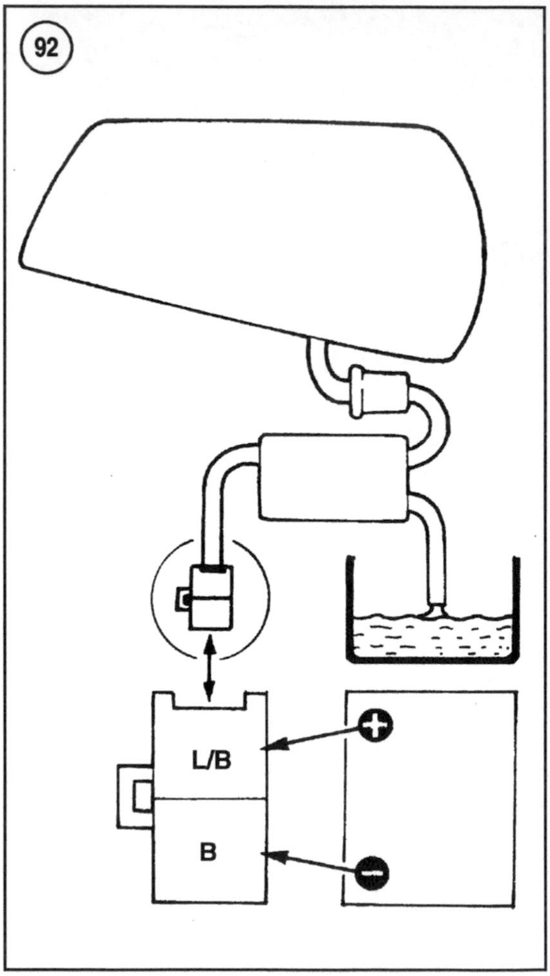

93

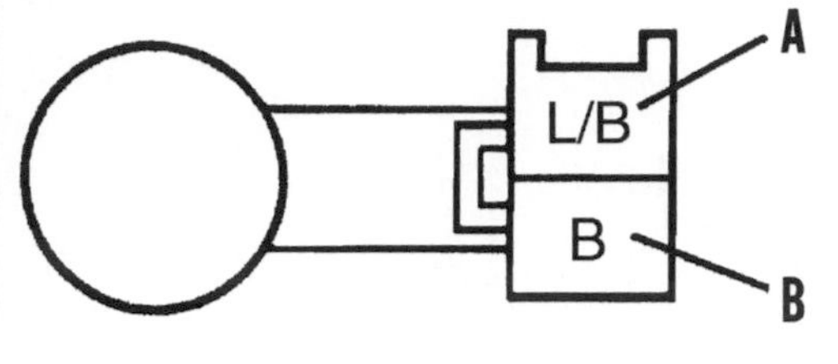

1. Remove the toolbox cover and the side cover from the left side as described in Chapter Fourteen.
2. Roll the boot from the connectors on the toolbox panel.
3. Disconnect the fuel pump connector (A, **Figure 91**).
4. Disconnect the fuel pump outlet hose (C, **Figure 91**) from the fitting on the pump. Connect a test hose to the outlet fitting on the pump and feed the hose into a container.
5. Disconnect the input hose (B, **Figure 91**) from the fitting on the fuel pump and connect an auxiliary fuel tank to the pump's input fitting.
6. Use jumpers to connect the battery to the fuel pump side of the connector as shown in **Figure 92**. Fuel should flow from the fuel pump outlet hose.
7. The fuel pump is faulty if fuel does not flow from the fuel pump outlet hose. Replace the fuel pump.

Resistance Test

1. Remove the toolbox cover and the left side cover as described in Chapter Fourteen.
2. Roll the boot from the connectors on the toolbox panel.
3. Disconnect the fuel pump connector (A, **Figure 91**).
5. Connect the ohmmeter positive test lead to the blue/black terminal (A, **Figure 93**) in the pump side of the connector. Connect the negative test lead to the black terminal (B, **Figure 93**).
5. Measure the fuel pump resistance. Replace the fuel pump if its resistance is outside the range specified in **Table 2**.

Relay Continuity Test

1. Remove the fuel tank and surge tank as described in this chapter.
2. Remove the frame neck cover (Chapter Fourteen).
3. Remove the starting circuit cutoff relay (C, **Figure 75**) from the mounting tang and disconnect the harness from the relay.
4. Use an ohmmeter and check the continuity of the fuel pump relay as follows:
   a. Use a jumper wire to connect the positive battery terminal to the red/black terminalon the relay and connect the negative battery terminal to the blue/red terminal.
   b. Connect the ohmmeter positive test lead to the red/black terminal on the relay and connect the negative test lead to the blue/black terminal.
   c. The fuel pump relay should have continuity during this test. Replace the starting circuit cutoff relay if there is no continuity.

AIR INDUCTION SYSTEM

The air induction system (AIS) reduces hydrocarbon emissions by promoting more complete combustion. The system injects fresh air into each exhaust port so any unburned fuel in the exhaust gasses are burned instead of being released into the atmosphere.

The system consists of an air cut valve, AIS filter, a reed valve, as well as air and vacuum lines. During normal operation, the air cut valve is open so secondary air flows from the AIS filter, through the air cut valve, to the reed valve, then into AIS fitting at the exhaust port in each cylinder.

The air cut valve, however, closes during deceleration to prevent backfiring. When the throttle is closed, vacuum from the intake port closes the air cut valve so secondary air cannot flow to the reed valve.

Removal/Installation

Refer to **Figure 94**.

1. Remove the toolbox cover and the side cover from the left side as described in Chapter Fourteen.

8

(94)

AIR INDUCTION SYSTEM

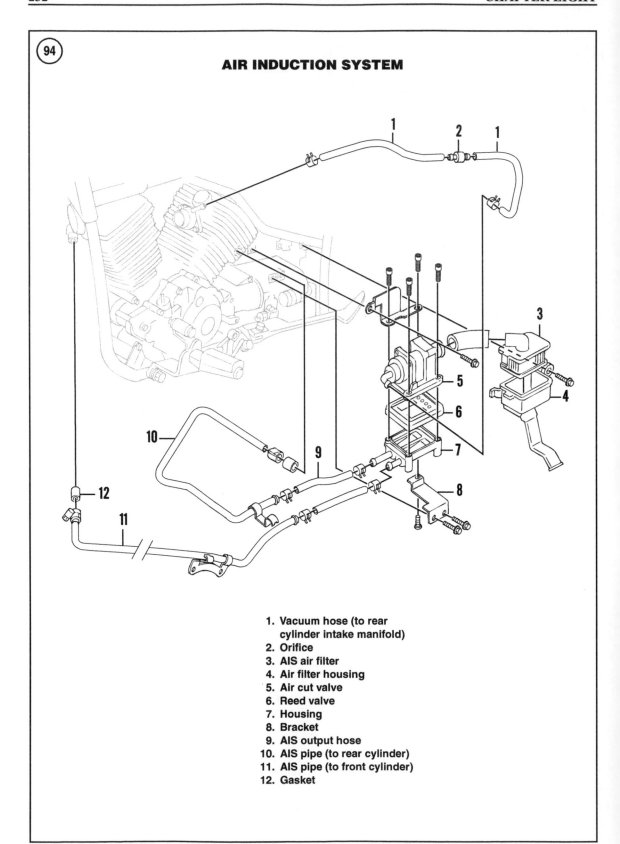

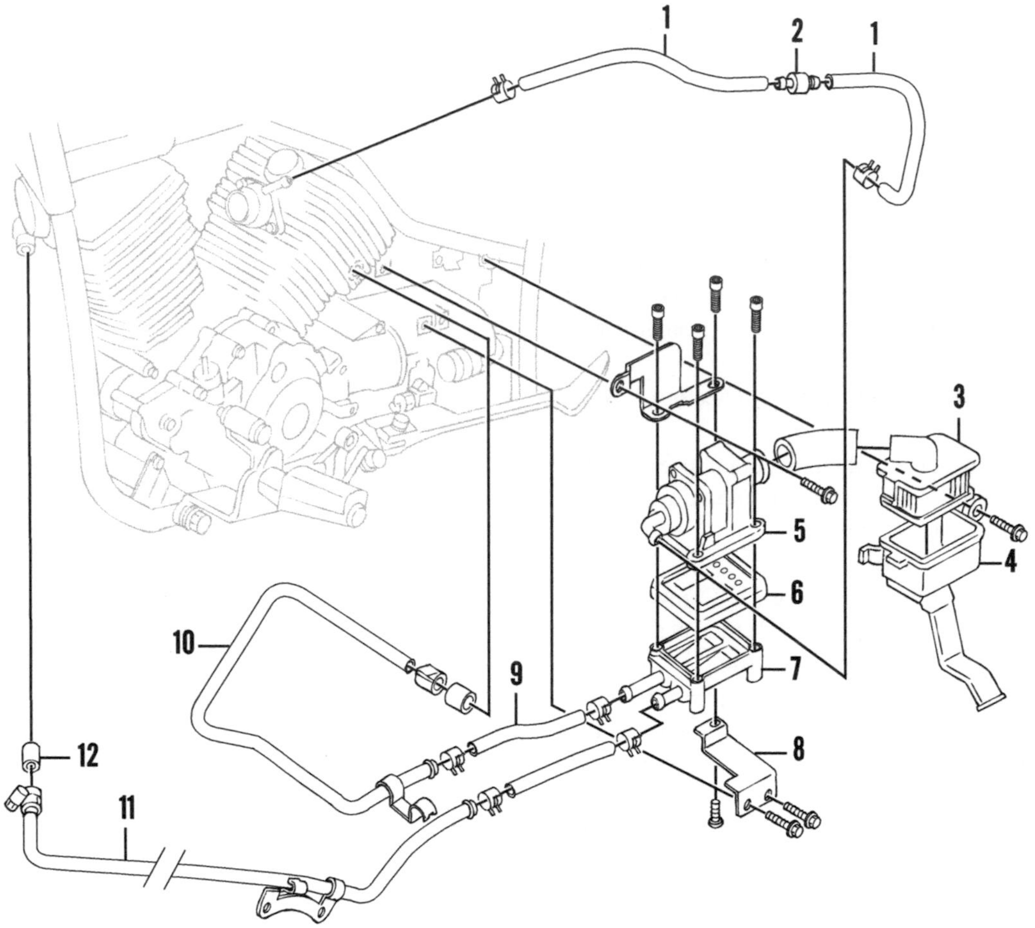

1. Vacuum hose (to rear cylinder intake manifold)
2. Orifice
3. AIS air filter
4. Air filter housing
5. Air cut valve
6. Reed valve
7. Housing
8. Bracket
9. AIS output hose
10. AIS pipe (to rear cylinder)
11. AIS pipe (to front cylinder)
12. Gasket

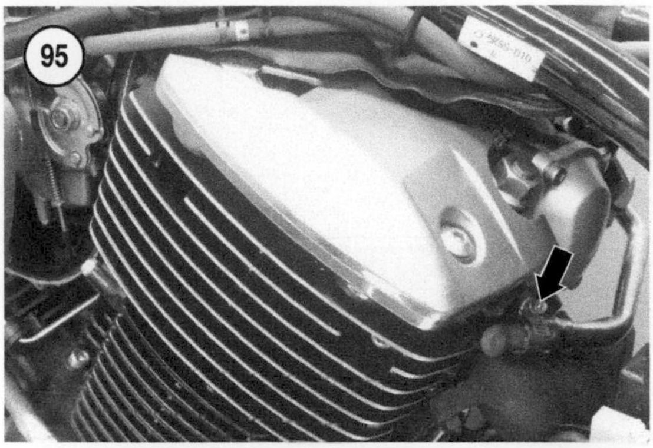

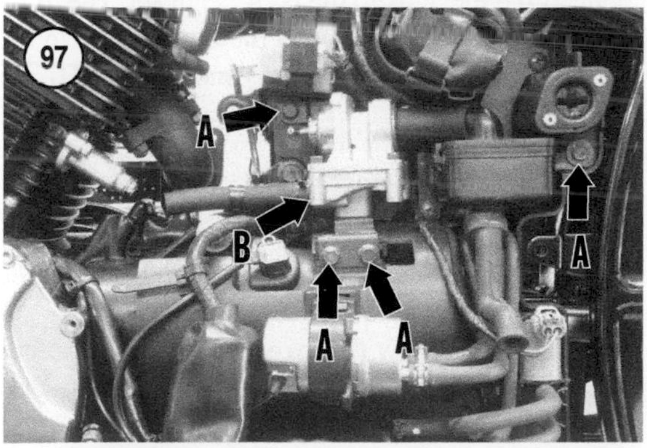

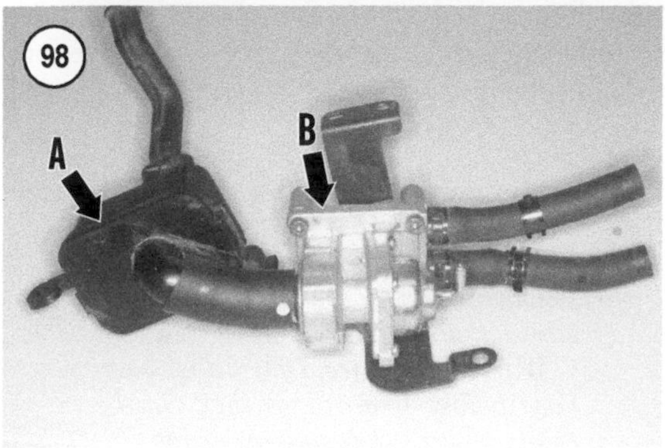

2. Release the hose clamp (**Figure 95**) and disconnect the AIS pipe from its fitting on each cylinder head.
3. Release the clamp and disconnect the output hose (A, **Figure 96**) on the AIS valve from each AIS pipe. Mark each pipe and hose. They must be reinstalled in their original positions during assembly.
4. Disconnect the vacuum line (B, **Figure 96**) from the fitting on the air cut valve.
5. Remove the two alternator cover bolts (C and D, **Figure 96**) and release the bracket from the alternator cover. Lift the AIS pipes from the engine.
6. Remove the mounting screws (A, **Figure 97**) and lower the AIS assembly (B) from the toolbox panel.
7. Installation is the reverse of removal. Note the following:
   a. Make sure the lines are correctly routed as shown in **Figure 94**.
   b. Properly secure each hose to its respective fitting.
   c. Apply silicone sealant to the threads of the rear alternator cover bolt (D, **Figure 96**). Torque the two alternator cover bolts to 10 N•m (89 in.-lb.).

Inspection

1. Release the lock tabs and lift the AIS air filter (A, **Figure 98**) from the air filter housing.
2. Inspect the air filter and housing for cracks or damage.
3. Remove the mounting screws and lift the air cut valve (B, **Figure 98**) and the reed valve from the AIS valve housing.
4. Inspect the reed valve and the air cut valve. Replace any component that is damaged.
5. Make sure the ports in the reed valve, air cut valve and AIS filter are unobstructed. Clean them if possible or replace the clogged component.
6. Inspect the air and vacuum lines for cracks or signs of leaks. Replace any hose or pipe as necessary.
7. Inspect the hose connections for signs of leaks.
8. Measure the reed valve height as shown in **Figure 99**. Replace the reed valve if its height exceeds the specification in **Table 3**.
9. Replace any component that is worn, damaged or out of specification.

EVAPORATIVE EMISSION CONTROL (CALIFORNIA MODELS)

All models sold in California are equipped with an evaporative emission control (EVAP) system, which reduces the amount of fuel vapors released into the atmosphere. The system consists of a charcoal canister, rollover valve, solenoid valve, assorted hoses, and modified carburetors and fuel tank. A schematic of the emission control system is on a special label on the right frame downtube.

The EVAP system captures fumes that are created in the fuel tank and stores them in a charcoal canister. See **Figure 100**. While the motorcycle is parked or when it is operated at low engine speeds, the fuel vapors remain in the charcoal canister. When the motorcycle is ridden at high speed, the vapors pass through a hose to the carburetor and are burned.

The rollover valve, which is installed in line between the fuel tank and charcoal canister, assures that the fumes remain in the canister until they can be safely burned. The gravity-operated rollover valve is opened and closed by an internal weight. During normal riding (when the motorcycle is upright), the weight keeps the valve open so fuel vapors can flow from the tank to the charcoal canister. When the motorcycle is leaned over (like when it is parked on the sidestand), the weight closes the valve so vapors stored in the canister cannot flow back into the fuel tank and escape into the atmosphere.

Inspection

Maintenance to the evaporative emission control system consists of inspecting the condition and routing of the hoses, making sure the canister is securely mounted to the engine mounting bracket, and testing the EVAP solenoid valve. No attempt should be made to modify or remove the emission control system.

WARNING
Because the evaporative emission control system stores fuel vapors, make sure the work area is free of flames or sparks before working on the EVAP system.

1. When servicing the evaporative system, make sure the ignition switch is off.
2. Make sure all hoses are attached and are not damaged or pinched.

99

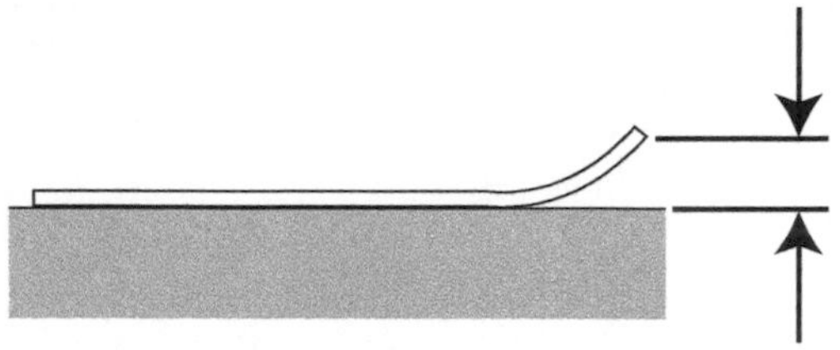

3. Replace any worn or damaged parts immediately. Replacement parts must be specific to California models.
4. The canister is capable of working through the motorcycle's life without maintenance provided that it is not damaged or contaminated.

Rollover Valve Removal/Installation

1. Remove the toolbox cover and the side cover from the left side of the motorcycle.
2. The rollover valve sits in-line with the hose that connects the fuel tank to the EVAP canister. The rollover valve (A, **Figure 101**) is mounted on the toolbox panel, just forward of the solenoid valve (B).

NOTE
The toolbox panel has been removed for photographic clarity. The rollover valve can be serviced while the panel is installed on the motorcycle.

3. Remove the screw (**Figure 102**) and release the rollover valve from the clamp that secures it to the toolbox panel.
4. Release the hose clamps, pull the hoses from the valve fittings, and remove the rollover valve.
5. Installation is the reverse of removal. Make sure the hose clamps are tight.

Canister Removal/Installation

NOTE
*The two ports on the top of the EVAP canister are identified as **TANK** and **CARB**. See **Figure 103**. Label each hose before removal so they can be easily identified during assembly.*

(100)

EVAPORATIVE EMISSIONS CONTROL SYSTEM (CALIFORNIA MODELS)

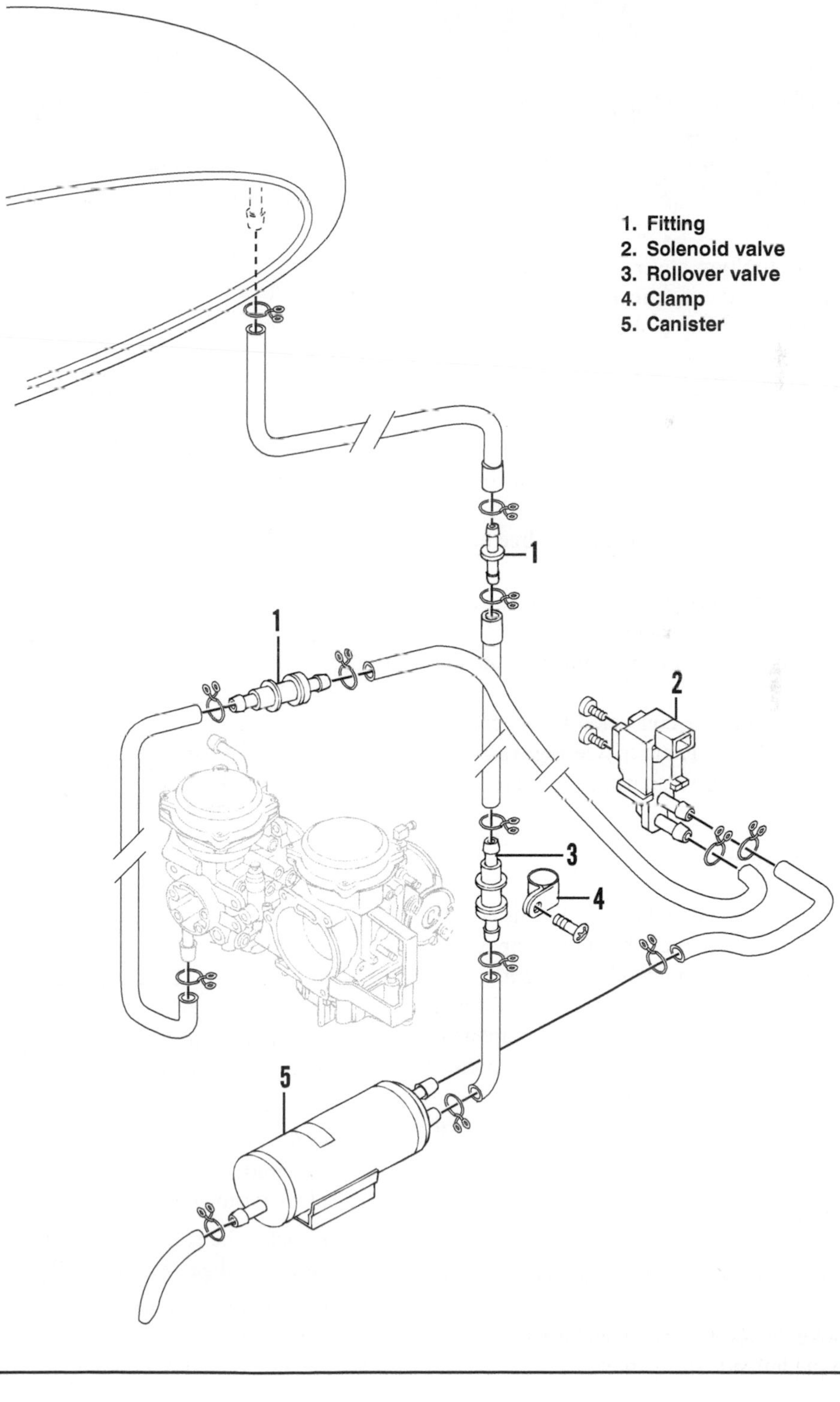

1. Fitting
2. Solenoid valve
3. Rollover valve
4. Clamp
5. Canister

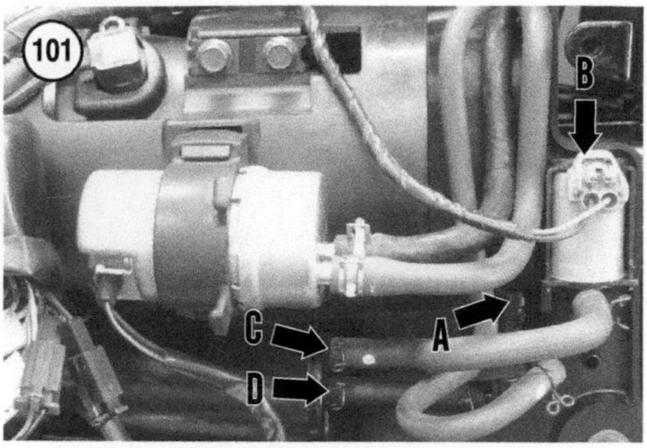

1. Remove the toolbox cover and the side cover from the left side of the motorcycle.
2. Label each hose on top of the canister.
3. Release the hose clamp and disconnect the solenoid valve hose (C, **Figure 101**) from the CARB port on the EVAP canister.
4. Release the clamp and disconnect the fuel tank hose (D, **Figure 101**) from the TANK fitting on the canister.
5. Make sure the canister's vent hose is free of the frame member.
6. Remove the two bolts that secure the canister to the frame, and remove the canister.
7. Installation is the reverse of removal. Note the following:
   a. Each hose must be connected to the correct port on the EVAP canister. See **Figure 100**.
   b. Make sure the hose clamps and bolts are tight.

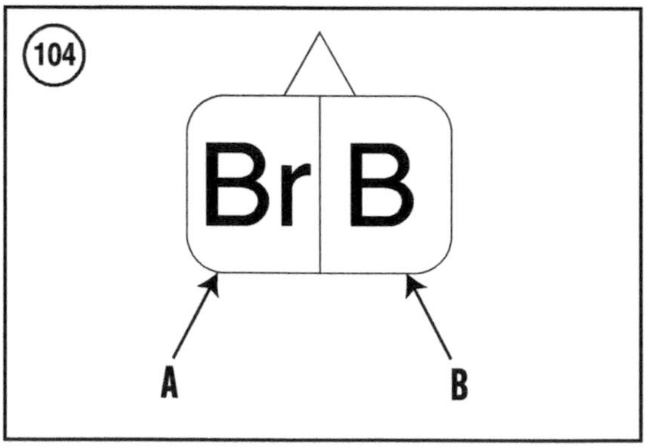

Solenoid Valve Testing

Perform these test procedures in the listed sequence. Each test presumes that the components tested in the earlier steps are working properly. The tests can yield invalid results if they are performed out of sequence. If a test indicates that a component is working properly, reconnect the electrical connections and proceed to the next step.

1. Check the main and ignition fuses as described in Chapter Nine.
2. Check the condition of the battery (Chapter Three).
3. Check the continuity of the main switch (Chapter Nine).
4. Check the resistance of the solenoid valve as follows:
   a. Remove the toolbox cover and the side cover from the left side of the motorcycle.
   b. Disconnect the connector from the solenoid valve (B, **Figure 101**).
   c. Connect the positive test lead of an ohmmeter to the brown terminal (A, **Figure 104**) in the solenoid. Connect the negative test lead to the black terminal (B, **Figure 104**).
   d. Replace the solenoid if the resistance is outside the range specified in **Table 2**.
5. Check the circuit wiring and grounds from the battery, through the main fuse, main switch, ignition fuse and to the EVAP solenoid valve.

(105)

EXHAUST SYSTEM

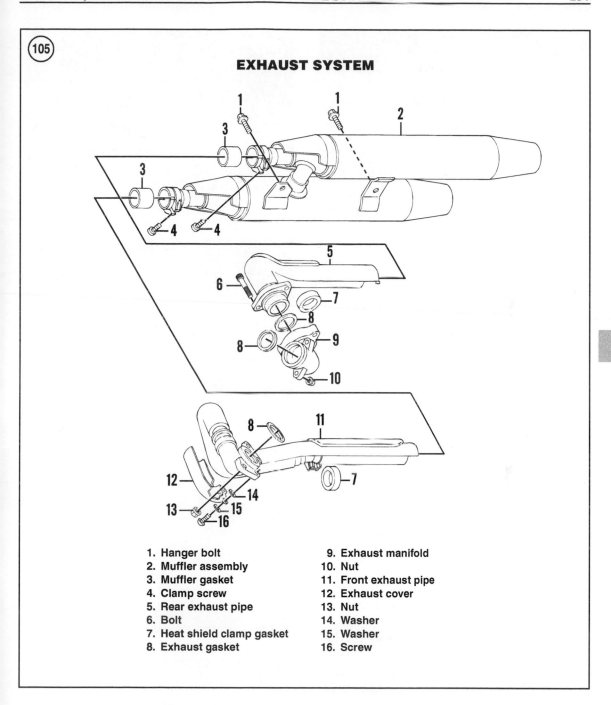

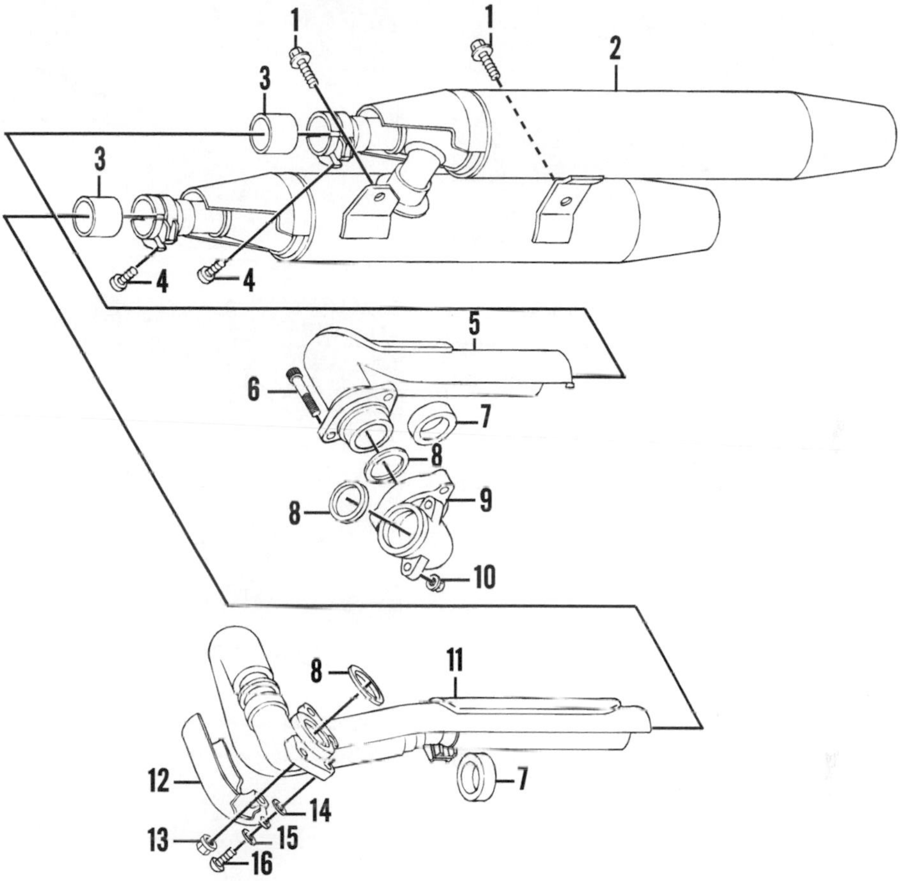

1. Hanger bolt
2. Muffler assembly
3. Muffler gasket
4. Clamp screw
5. Rear exhaust pipe
6. Bolt
7. Heat shield clamp gasket
8. Exhaust gasket
9. Exhaust manifold
10. Nut
11. Front exhaust pipe
12. Exhaust cover
13. Nut
14. Washer
15. Washer
16. Screw

EXHAUST SYSTEM

Removal/Installation

Refer to **Figure 105**.

1. Securely support the motorcycle on level ground.

2. Loosen the muffler clamp screw (A, **Figure 106**) on each muffler.

3. Remove the muffler hanger bolts (B, **Figure 106**).

4. Pull the muffler assembly rearward. Separate each muffler from its respective exhaust pipe and remove the muffler assembly. Discard each muffler gasket.

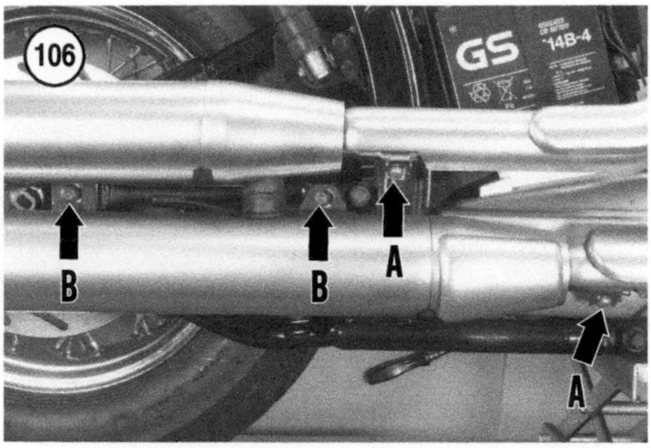

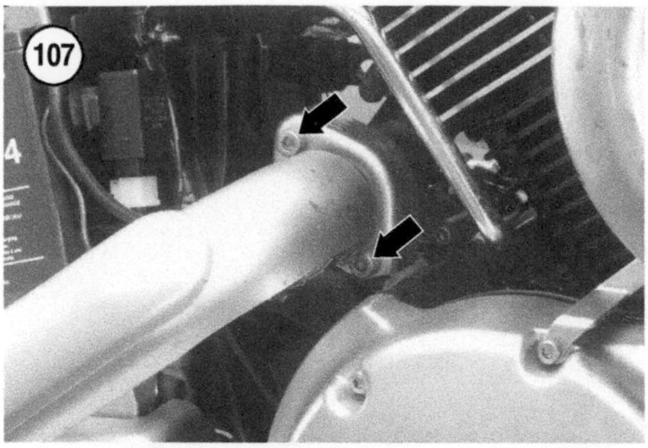

5. Remove the rear exhaust pipe as follows:
   a. Remove the exhaust pipe-to-manifold bolts (**Figure 107**) from the exhaust manifold on the rear cylinder.
   b. Pull the exhaust pipe from the manifold.
   c. Remove and discard the exhaust gasket (**Figure 108**).
6. Remove the front exhaust pipe as follows:

NOTE
*The front exhaust pipe can be removed with the cover (A, **Figure 109**) installed on the pipe.*

   a. Remove the nuts (B, **Figure 109**), and pull the exhaust pipe from the exhaust port.
   b. Remove and discard the exhaust gasket (**Figure 108**).
7. Inspect the system as described below.
8. Installation is the reverse of removal. Note the following:
   a. Install a new exhaust gasket into the rear exhaust manifold (**Figure 108**) and into the exhaust port on the front cylinder.
   b. Install a new muffler gasket (**Figure 110**) into each muffler.
   c. Make sure each exhaust pipe is correctly seated in the exhaust port or exhaust manifold.
   d. Loosely install the entire exhaust system and finger-tighten the hardware.
   e. First, torque the exhaust pipe-to-cylinder head nuts (B, **Figure 109**) and the exhaust manifold bolts (**Figure 107**) to 20 N•m (15 ft-lb.).
   f. Then, torque the muffler clamp screws (A, **Figure 106**) to 20 N•m (15 ft.-lb.).

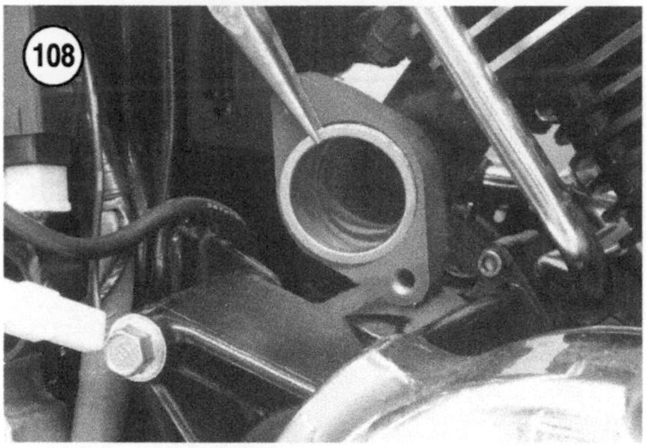

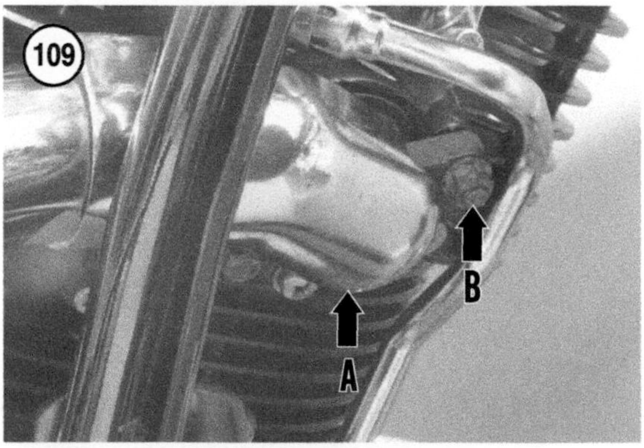

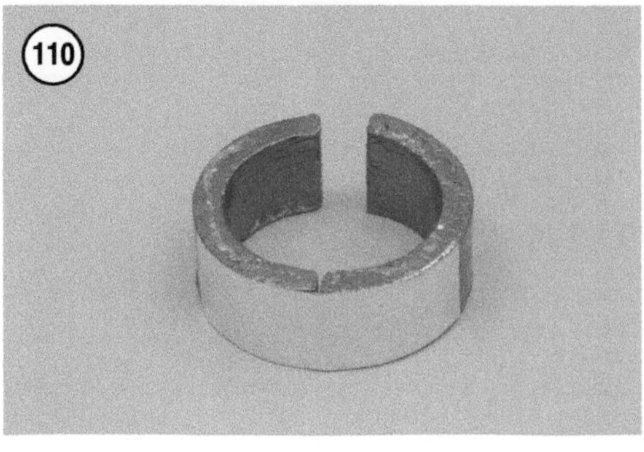

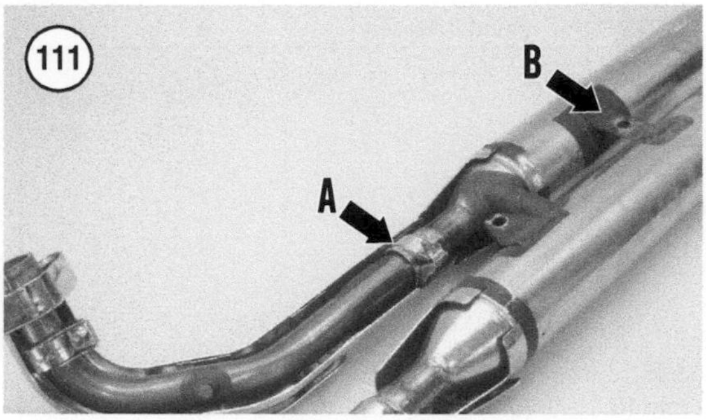

g. Then, torque the muffler hanger bolts (B, **Figure 106**) to 25 N•m (18 ft.-lb.).

h. After installation is complete, start the engine and make sure there are no exhaust leaks. Correct any leak prior to riding the bike.

Inspection

The exhaust system is vital to the motorcycle's operation and performance. Periodically inspect the exhaust system.

Replace parts that have excessive dents that cause flow restrictions.

To prevent internal rust buildup, periodically remove the drain bolt from each muffler to drain away trapped moisture.

1. Check for leaks where each muffler joins an exhaust pipe (A, **Figure 111**).
2. Inspect the drain bolts for corrosion or exhaust leaks. Replace the bolts and washers if necessary.
3. Inspect the muffler mounting brackets (B, **Figure 111**) for wear or damage. Replace the muffler as necessary.

Table 1 CARBURETOR SPECIFICATIONS

Carburetor type	Mikuni BSR37
Carburetor identification	
Canada models	5EL5 20
California models	
1999-2003	5EL6 30
2004-on	5YS1 00
USA models	
1999-2005	5EL5 20
2006-on	5YS1 00
Main jet	
Canada models	
Carburetor 1	No. 110
Carburetor 2	No. 112.5
California models	
1999-2003	
Carburetor 1	No. 110
Carburetor 2	No. 112.5
2004-on	
Carburetor 1	No. 112.5
Carburetor 2	No. 115
USA models	
1998-2005	
Carburetor 1	No. 110
Carburetor 2	No. 112.5
2006-on	
Carburetor 1	No. 112.5
Carburetor 2	No. 115
Main air jet	No. 55
Needle jet	P-0M
Jet needle	
USA, California and Canada models	
(Carburetors No. 1 and No. 2)	5DL43-53-1
Pilot jet	No. 17.5
Pilot outlet size	1.0
Pilot air jet	
P.A.J. 1	No. 63.8
P.A.J. 2	No. 145

(continued)

Table 1 CARBURETOR SPECIFICATIONS (continued)

Pilot screw	
Europe and Australia models	3 turns out
Valve seat size	1.2
Starter jet	
G.S. 1	No. 42.5
G.S. 2	0.8
Bypass 1	0.8
Bypass 2	0.8
Bypass 3	0.8
Throttle valve size	No. 125
Fuel level	4-5 mm (0.16-0.20 in.) above float bowl mark
Idle speed	950-1050 rpm
Intake vacuum	34.7-37.3 kPa (260-289 mm Hg [10.2-11.4 in. Hg])

Table 2 FUEL AND EMISSION SYSTEM ELECTRICAL SPECIFICATIONS*

Carburetor heater	
Resistance	
1999-2001	6-10 ohms
2002-on	3.2-5.8 ohms
Voltage, wattage	12 volts, 30 watts
EVAP solenoid valve resistance	28-34 ohms
Fuel cut solenoid valve resistance	12 ohms at 68° F
Fuel pump	
Amperage	0.8 amps
Resistance	1.6-2.2 ohms
Throttle position sensor	
R1 resistance	4.0-6.0 k ohms
R2 resistance	0.56-0.84 k ohms to 3.01-4.51 k ohms
Fully closed resistance	0.56-0.84 k ohms

*Resistance specification @ 20° C (68° F)

Table 3 AIR/FUEL AND EXHAUST SYSTEM SPECIFICATIONS

AIS reed valve height	0.4 mm (0.016 in.)
Fuel pump output pressure	
1998-2003	12 kPa (1.7 psi)
2004-on	8.3-12.3 kPa (1.2-1.8 psi)
Fuel tank capacity	
Total	17 liter (4.49 U.S. gal.)
Reserve	4.5 liter (1.19 U.S. gal.)
Fuel	Regular unleaded
Octane	86 ([R+M] / 2 method) or research octane 91 or higher 91 or higher

Table 4 AIR/FUEL AND EXHAUST SYSTEM TORQUE SPECIFICATIONS

Item	N•m	in.-lb.	ft.-lb.
Air filter housing bolts	10	89	–
Air filter cover bolts	2	18	–
Alternator cover bolts	10	89	–
Exhaust manifold-to-cylinder head nuts (rear cylinder)	20	–	15
Exhaust pipe-to-cylinder head nuts (front cylinder)	20	–	15
Exhaust pipe-to-manifold bolts (rear cylinder)	20	–	15
Fuel tank bolts	23	–	17
Fuel valve bolts	7	62	–
Muffler clamp screws	20	–	15
Muffler hanger bolts	25	–	18

CHAPTER NINE

ELECTRICAL SYSTEM

This chapter describes service procedures for the following electrical sub-systems and components:

1. Charging system.
2. Ignition system.
3. Starting system.
4. Lighting system.
5. Signal system.
6. Switches.
7. Self-diagnostic system.

Battery and spark plug information is in Chapter Three.

All electrical connections must be completely coupled to each other. During electrical service, apply dielectric compound to all electrical connectors when they are reconnected. This helps seal out moisture and prevents corrosion of the electrical terminals.

When testing electrical components, compare any measurements to the specifications in the **Table 1** at the end of this chapter. Replace parts that are damaged, worn or out of specification. During assembly, tighten fasteners to the specified torque. Refer to *Electrical Component Replacement* in Chapter Two.

Refer to the appropriate wiring diagram at the back of the manual for specific wire colors and connection points.

CHARGING SYSTEM

The charging system consists of the battery, main fuse, main switch, alternator and the regulator/rectifier assembly.

Alternating current generated by the alternator is rectified to direct current. The voltage regulator maintains the voltage to the battery and additional electrical loads (lights, ignition, etc.) at a constant voltage regardless of variations in engine speed and load.

Troubleshooting

Refer to Chapter Two.

Current Draw Test

1. Turn the main switch off.
2. Remove the battery cover as described in Chapter Fourteen.
3. Disconnect the lead from the battery negative terminal (A, **Figure 1**).
4. Connect an ammeter between the battery negative lead and the negative terminal of the battery (**Figure 2**).

5. The ammeter should read less than 0.1 mA. If the amperage is greater, there is a current draw on the system that will discharge the battery.

Charging Voltage Test

1. Connect an engine tachometer to the spark plug lead on the No. 1 (rear) cylinder.
2. Connect a 0-20 DC voltmeter to the battery terminals as shown in **Figure 3**.
3. Start the engine and increase engine speed to approximately 5000 rpm. The measured voltage should equal the charging voltage specified in **Table 1**.
4. If the charging voltage is out of specification, check the stator coil resistance as described in this chapter.

STATOR

Resistance Test

Test the stator assembly at a minimum temperature of 20° C (68° F).

1. Remove the toolbox cover and the left side cover as described in Chapter Fourteen.
2. Roll back the boot from the electrical cables.
3. Disconnect the three-pin stator connector (A, **Figure 4**) from its mate on the wiring harness.
4. Measure the stator coil resistance.

NOTE
In each of the following tests, connect the ohmmeter's positive lead to the connector's center terminal (A, ***Figure 5****) and the ohmmeter's negative lead to the other terminal (B or C).*

5. Measure the resistance between the center terminal (A, **Figure 5**) and the left terminal (B) in the stator side of the connector.
6. Measure the resistance between the center terminal (A, **Figure 5**) and the right terminal (C) in the stator side of the connector.
7. Replace the stator assembly if either resistance is not within the range specified in **Table 1**.
8. Check the continuity between each terminal and ground. There should be no continuity (infinite resistance). Continuity between any stator wire and ground indicates that either the stator or one of the

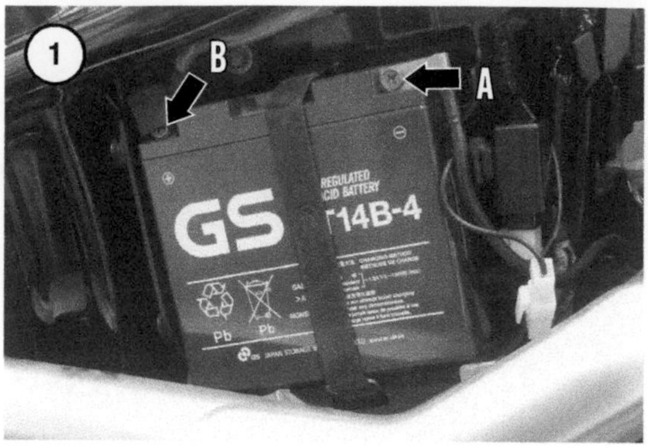

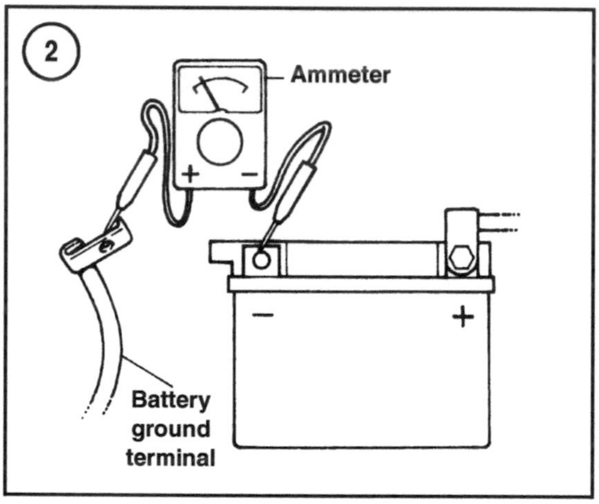

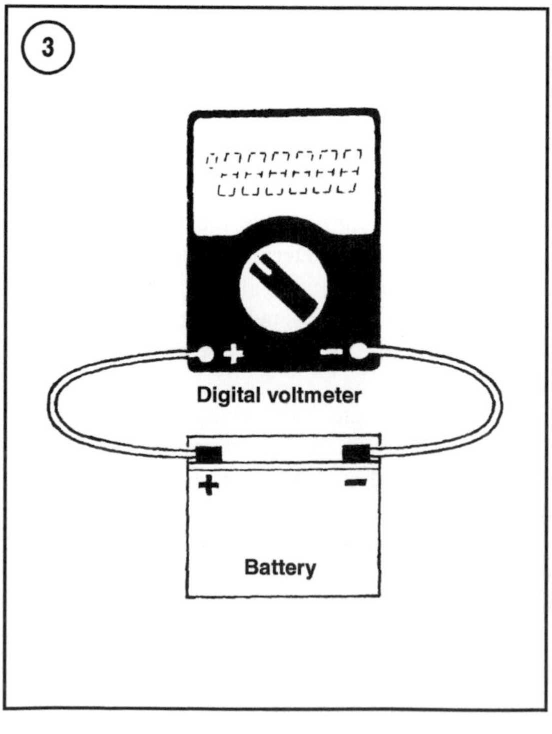

4

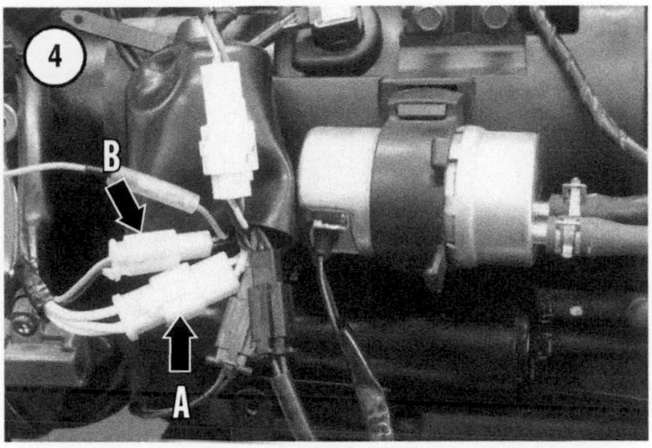

5

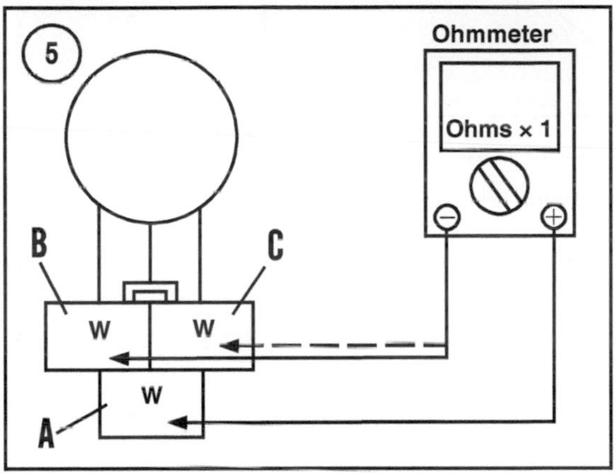

stator wires is shorted to ground. Replace the stator/pickup coil assembly.

Stator/Pickup Coil Assembly Removal/Installation

Refer to **Figure 6**.

The stator/pickup coil is an assembly. The stator and pickup coil are not available separately.

1. Remove the alternator cover as described in Chapter Five.
2. Disconnect the three-pin stator connector (A, **Figure 4**) and the two-pin pickup coil connector (B) from their harness mates.
3. Remove the mounting screw and lift the wire clamp (A, **Figure 7**) from the alternator cover.
4. Remove the pickup coil mounting screws (B, **Figure 7**) and the stator mounting screws (C).
5. Pry the two grommets (D, **Figure 7**) from the cover.
6. Note how the coil assembly cable is routed through the cover, and remove the coil assembly and its cable.
7. Installation is the reverse of removal.
   a. Apply a medium-strength threadlocking compound to the threads of the stator mounting screws (C, **Figure 7**) and the pickup coil mounting screws (B). Torque the stator screws to 10 N•m (89 in.-lb.) and the pickup coil screws to 7 N•m (62 in.-lb.).
   b. Securely seat the grommets (D, **Figure 7**) in the cover.
   c. Route the cable along the same path noted during removal.
   d. Apply a dielectric compound to the electrical connectors prior to reconnecting them.
   e. Make sure the electrical connector is free of corrosion and both sides are completely coupled to each other.
   f. Reinstall the alternator cover as described in Chapter Five.

VOLTAGE REGULATOR/RECTIFIER

Removal/Installation

The voltage regulator/rectifier mounts to the inside of the battery box, which must be removed for access to the regulator/rectifier.

1. Securely support the motorcycle on a level surface.
2. Disconnect the electrical lead from the negative battery terminal (A, **Figure 1**).
3. Remove the battery box as described in Chapter Fourteen.
4. Press the release tab and disconnect the connector (A, **Figure 8**) from the voltage regulator/rectifier.
5. Remove the regulator/rectifier bolts (B, **Figure 8**) and remove the regulator/rectifier from the battery box.
6. Installation is the reverse of removal. Note the following:
   a. Apply a dielectric compound to the electrical connector prior to reconnecting it.
   b. Make sure the electrical connector is free of corrosion and is completely coupled to its mate.

IGNITION SYSTEM

All models are equipped with a fully transistorized ignition system. This solid-state system pro-

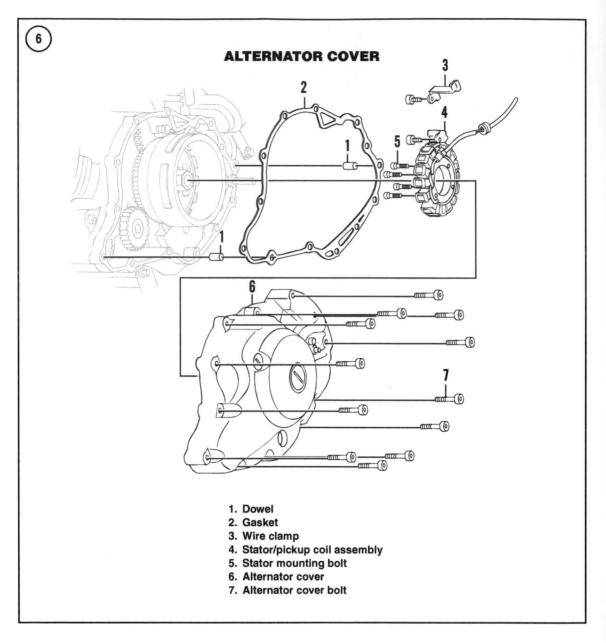

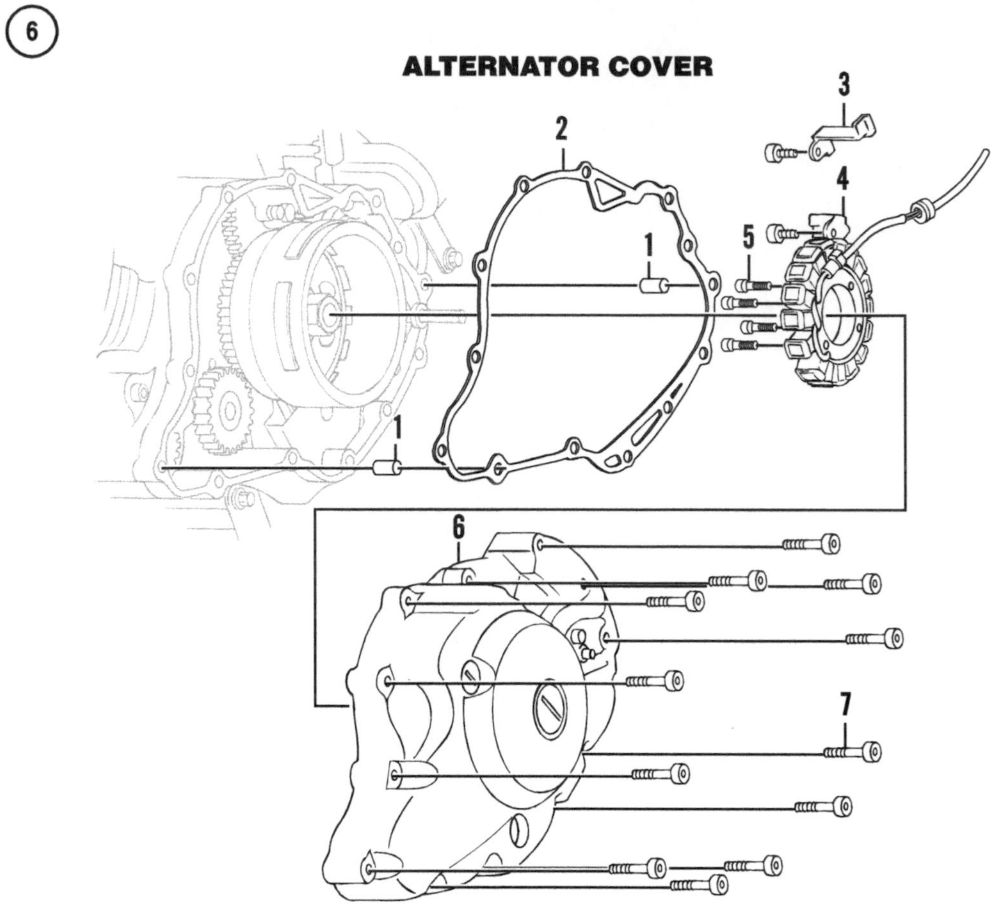

1. Dowel
2. Gasket
3. Wire clamp
4. Stator/pickup coil assembly
5. Stator mounting bolt
6. Alternator cover
7. Alternator cover bolt

vides longer component life and delivers a strong, reliable spark throughout the entire range of engine speed. Ignition timing and advance are maintained without adjustment. Ignition timing procedures given in Chapter Three can be used to determine if the ignition system is operating properly.

When the crankshaft driven flywheel passes the pickup coil, an electrical pulse is generated within the pickup coil. This pulse flows to the switching and distribution circuits in the ignitor unit. The ignitor unit interrupts current flow through the ignition coil and the magnetic field within the coil collapses. This induces a very high voltage in the secondary windings of the ignition coil. This voltage is sufficient to jump the gap at the spark plugs.

Troubleshooting

Refer to Chapter Two.

The ignition system is designed to operate only when the sidestand is up or when the transmission is in neutral. If there is a no-spark condition, check the

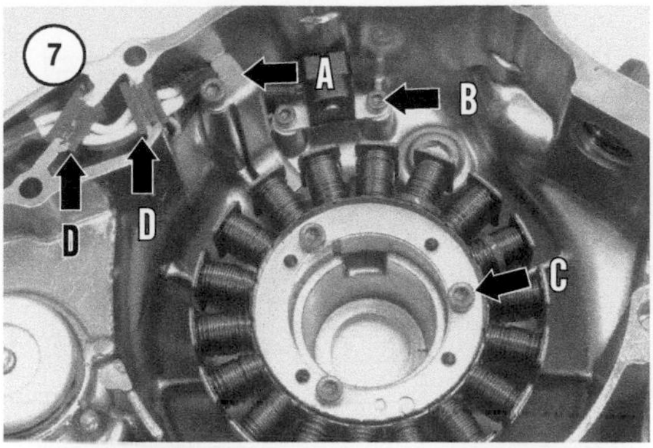

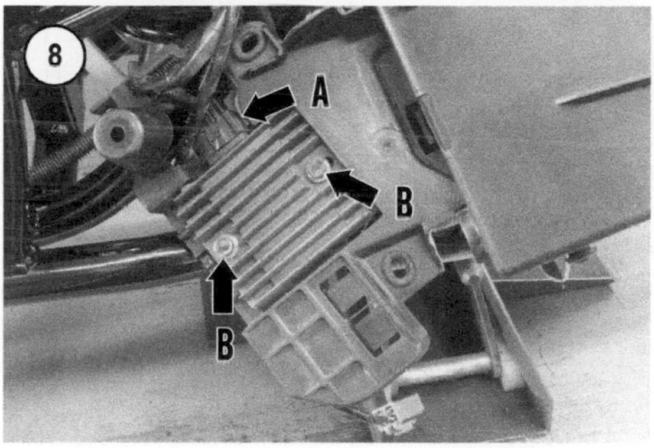

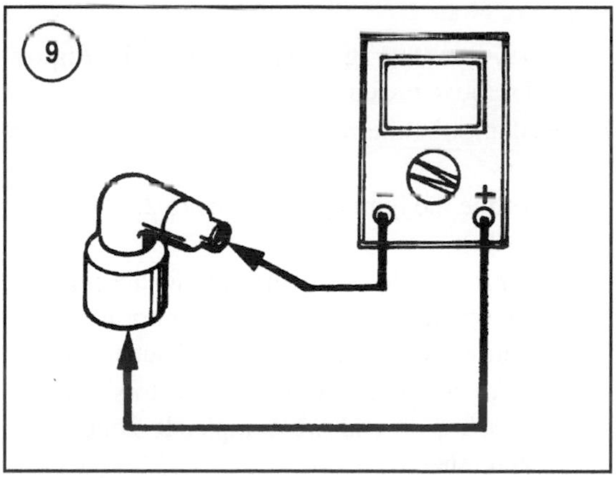

neutral switch and sidestand switch as described in this chapter.

Precautions

Damage to the semiconductors in the system may occur if the following precautions are not observed.

1. Never connect the battery backwards. If the battery polarity is incorrect, the voltage regulator, alternator and ignitor unit will be damaged.

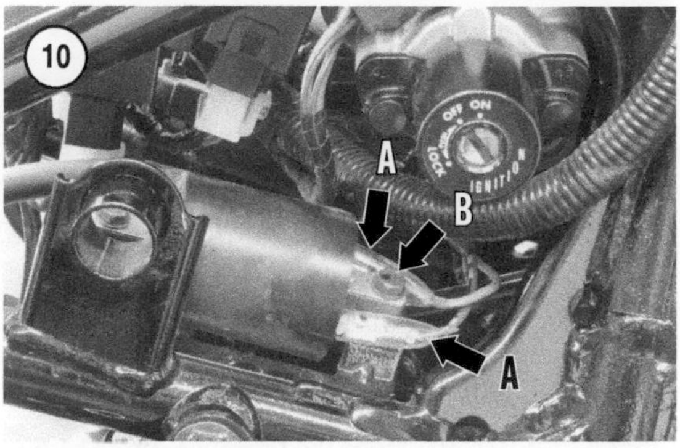

2. Do not disconnect the battery while the engine is running.
3. Keep all connections between the various units clean and tight. Make sure the wiring connectors are firmly pushed together.
4. Do not substitute another type of ignition coil or battery.
5. Each solid-state unit is mounted on a rubber vibration isolator. Make sure the isolators are in place when replacing any units.

SPARK PLUG CAP

Resistance

1. Disconnect the spark plug cap from the spark plug.
2. Measure the resistance between each end of the cap as shown in **Figure 9**.
3. Replace the spark plug cap if the resistance exceeds the specification in **Table 1**.
4. Repeat this test for the other spark plug cap.

IGNITION COIL

The front cylinder ignition coil is mounted on the left side of the frame; the rear cylinder ignition coil sits on the right side. Occasionally check to see that the coils are mounted securely.

Perform the following test whenver the self-diagnostic system flashes an ignition coil trouble code.

Resistance Test

Test the ignition coils at a minimum temperature of 20° C (68° F).

1. Securely support the motorcycle on level ground.
2. Remove the fuel tank, air filter housing and surge tank as described in Chapter Eight, and the remove the frame neck cover (Chapter Fifteen).
3. Disconnect the electrical lead from the negative battery terminal (A, **Figure 1**).
4. Disconnect the spark plug cap from the respective spark plug.
5. Disconnect the two primary coil spade connectors (A, **Figure 10**) from the terminals on the ignition coil.
6. Measure the primary coil resistance as follows:
   a. Connect the ohmmeter positive test probe to the red/black terminal on the ignition coil; connect the negative test probe to the orange (or gray) terminal. See **Figure 11**.
   b. Replace the ignition coil if the primary resistance is not within the specification listed in **Table 1**.
7. Measure the secondary coil resistance as follows:
   a. Connect the ohmmeter positive test probe to the spark plug lead; connect the negative test probe to the orange (or gray) terminal. See **Figure 12**.
   b. Replace the ignition coil if the secondary coil resistance is not within the specification listed in **Table 1**.
8. Repeat this test for the other ignition coil.

Removal/Installation

1. Use jack stands or a scissors jack to securely support the motorcycle on level ground.
2. Remove the fuel tank and surge tank as described in Chapter Eight, and remove the frame neck cover (Chapter Fourteen).
3. Disconnect the electrical lead from the negative battery terminal (A, **Figure 1**).
4. Disconnect the spark plug cap from the respective spark plug.

NOTE
Note how the ignition coil's secondary lead is routed through the frame. It will have to be rerouted along the same path during installation.

11

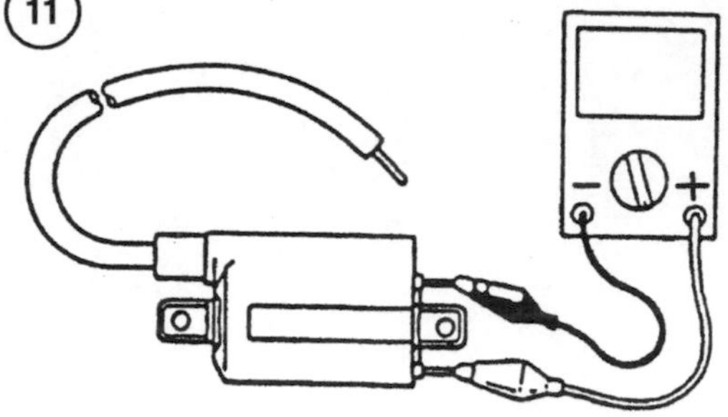

12

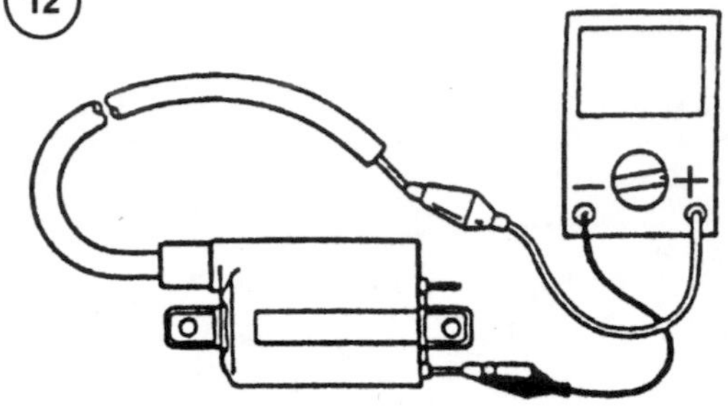

5. Disconnect the two primary coil spade connectors (A, **Figure 10**) from the terminals on the ignition coil.
6. Remove the two ignition coil bolts (B, **Figure 10**) and remove the coil.
7. Installation is the reverse of removal. Note the following:
   a. Torque the ignition coil bolts to 2.5 N•m (22 in.-lb.).
   b. Apply a dielectric compound to the primary coil connectors (A, **Figure 10**) prior to reconnecting them. This will help seal out moisture.
   c. Make sure the electrical connectors are free of corrosion and are completely coupled to each other.
   d. Install all removed items.

PICKUP COIL

Resistance Test

Test the pickup coil at a minimum temperature of 20° C (68° F).

1. Remove the toolbox cover and the side cover from the left side as described in Chapter Fourteen.
2. Roll back the boot from the electrical cables.

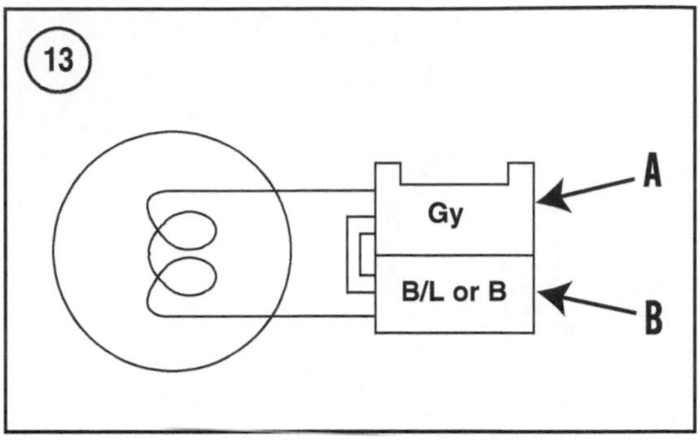

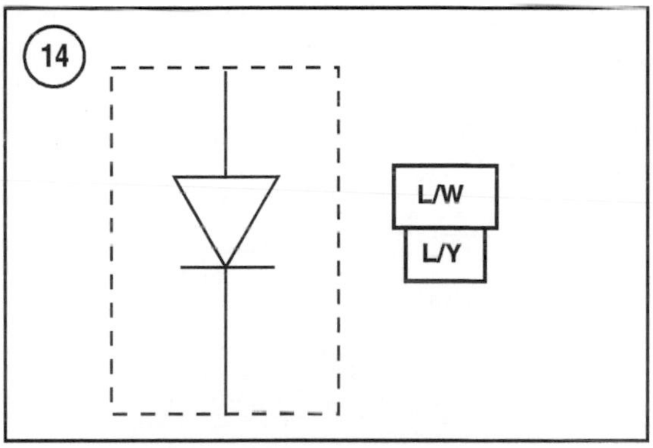

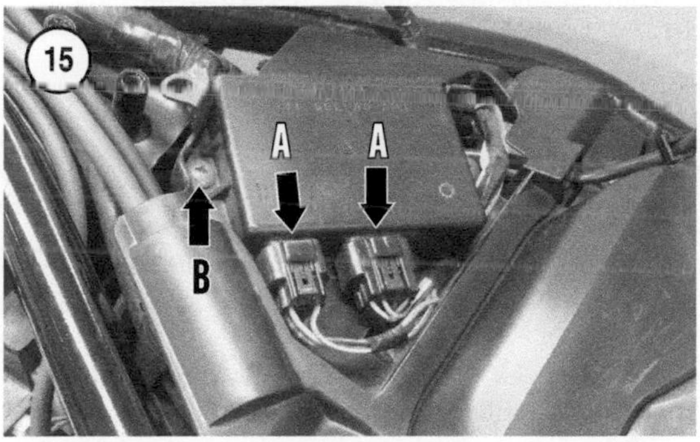

3. Disconnect the two-pin pickup coil connector (B, **Figure 4**) from its mate on the wiring harness.

4. If using an analog ohmmeter, set it to the R × 100 scale, and measure the stator coil resistance.

5. Connect the ohmmeter positive test probe to the gray terminal (A, **Figure 13**) in the pickup coil side of the connector; connect the negative test probe to the black/blue (or black) terminal (B).

6. Replace the stator/pickup coil assembly, as described in *Charging System* of this chapter, if the pickup coil resistance is outside the range specified in **Table 1**.

DIODE

Continuity Test

1A. For 1999-2003 models, remove the diode from the wiring harness and use an ohmmeter to test the diode as follows:

a. Connect the ohmmeter leads to the blue/yellow wire terminal and the blue/white wire terminal (**Figure 14**). Check for continuity.
b. Reverse the ohmmeter lead connections. Check for continuity.
c. The diode should have continuity in only one direction. If there is no continuity, or continuity in both directions, replace the relay unit.

1B. For 2004-on models, the diode is part of the starting circuit cutoff relay. Remove the relay as described in this chapter and use an ohmmeter to test the diode as follows:

a. Connect the ohmmeter leads to the blue/yellow wire terminal and the blue/green wire terminal. Check for continuity.
b. Reverse the ohmmeter lead connections. Check for continuity.
c. The diode should have continuity in only one direction. If there is no continuity, or continuity in both directions, replace the relay unit.

IGNITOR UNIT

Testing

The ignitor unit cannot be tested. If all other ignition system components perform within test specifications, consider the ignitor unit defective by a process of elimination.

Before purchasing a new ignitor unit, have the system checked by a Yamaha dealership or another qualified shop. Most motorcycle dealerships will *not* accept returns on electrical components.

Removal/Installation

1. Remove the rider and passenger seats as described in Chapter Fourteen.
2. Carefully disconnect the two electrical connectors (A, **Figure 15**) from the ignitor unit.
3. Remove the two ignitor mounting screws (B, **Figure 15**), and lift the ignitor unit from the ignitor panel.

4. Installation is the reverse of removal. Note the following:
   a. Apply a dielectric compound to the electrical connectors prior to reconnecting them.
   b. Make sure the electrical connections are free of corrosion and are completely coupled to each other.
   c. Tighten the screws securely.
   d. Install all removed items.

STARTING SYSTEM

When the starter button is pressed under the correct conditions, control current flows through the starter relay coil, which energizes the relay. The starter relay contacts close and load current flows directly from the battery to the starter.

The starter will only operate when the transmission is in neutral or when the clutch lever is pulled in while the sidestand is up. The starting circuit cutoff relay prevents the flow of control current to the starter relay unless one of these conditions has been met. The starting circuit cutoff relay is energized and the contacts close only when the neutral switch is closed (the transmission is in neutral) or when both the clutch switch and sidestand switch are closed (when the clutch lever is pulled in and the sidestand is up).

Refer to Chapter Five for starter clutch.

CAUTION
Do not operate the starter for more than five seconds at a time. Let it cool approximately ten seconds, then use it again.

Troubleshooting

If the meter on the instrument panel is flashing, refer to *Self-Diagnostic System* in this chapter to begin troubleshooting. If there is no self-diagnostic trouble code, refer to Chapter Two.

STARTER

Operational Test

1. Securely support the motorcycle on a level surface and shift it into neutral.
2. Make sure the main switch is off.

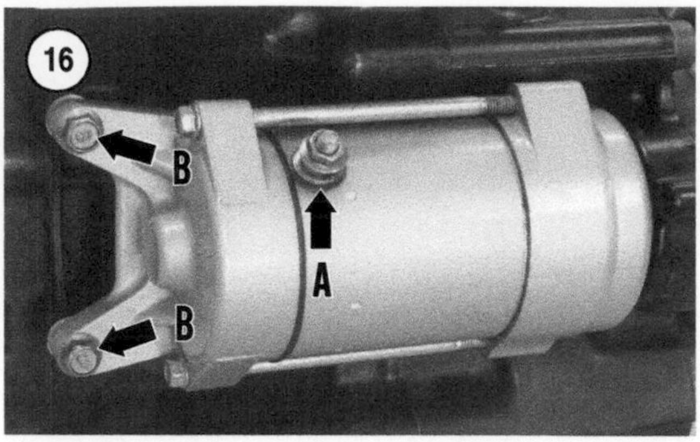

3. Remove the fuel tank as described in Chapter Eight.

WARNING
The negative lead is still connected to the battery in the following steps. While working with the battery positive lead, make sure it cannot touch any metal on the motorcycle.

4. Disconnect the electrical lead (B, **Figure 1**) from the positive battery terminal. Secure the lead safely out of the way.
5. Pull back the rubber boot and disconnect the cable from the starter terminal (A, **Figure 16**).

WARNING
The jumper wire must be large enough to handle the current flow from the battery. If the wire is too small, it could melt.

WARNING
The following test may produce sparks. Make sure no flammable gas or fluid is in the vicinity.

6. Apply battery voltage directly to the starter by connecting a jumper from the battery positive ter-

(18)

STARTER

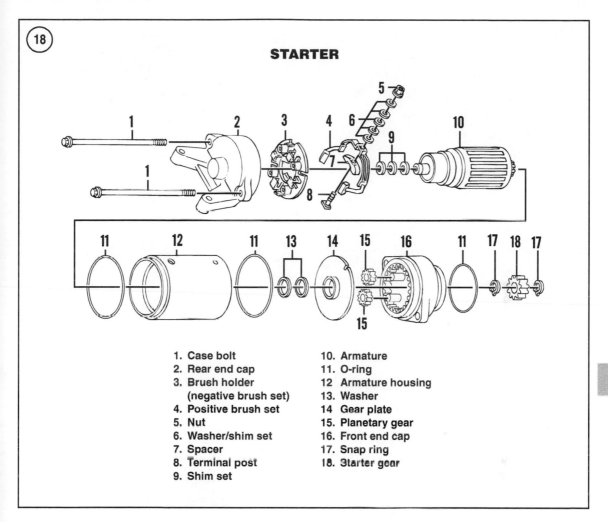

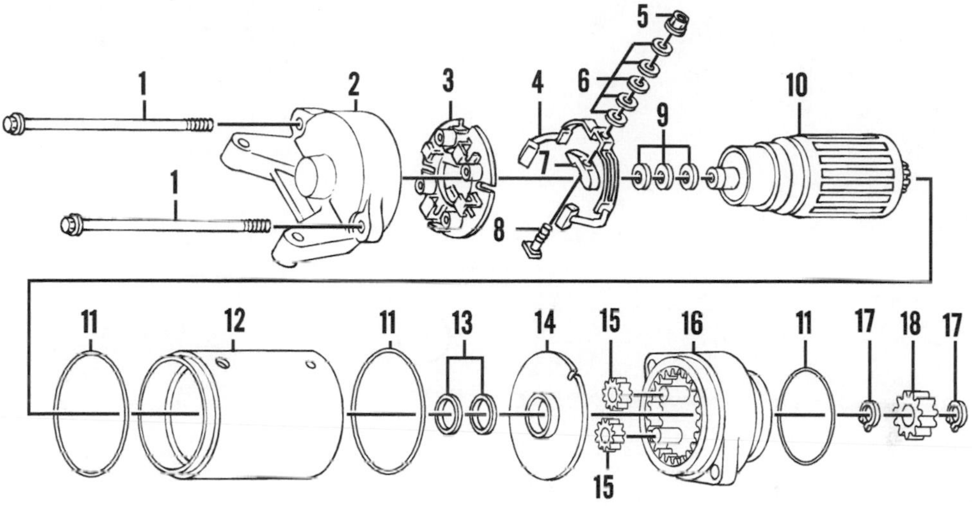

1. Case bolt
2. Rear end cap
3. Brush holder (negative brush set)
4. Positive brush set
5. Nut
6. Washer/shim set
7. Spacer
8. Terminal post
9. Shim set
10. Armature
11. O-ring
12. Armature housing
13. Washer
14. Gear plate
15. Planetary gear
16. Front end cap
17. Snap ring
18. Starter gear

minal (B, **Figure 1**) to the starter terminal (A, **Figure 16**). The starter should operate.

7. If the starter does not operate when battery voltage is applied, repair or replace the starter.

Removal/Installation

1. Securely support the motorcycle on a level surface.
2. Make sure the main switch is off.
3. Disconnect the electrical lead from the negative battery terminal (A, **Figure 1**).
4. Pull back the rubber boot and disconnect the cable from the starter terminal (A, **Figure 16**).
5. Remove the two starter mounting bolts (B, **Figure 16**). Pull the starter toward the right and remove it.
6. Remove and discard the O-ring (A, **Figure 17**).
7. Installation is the reverse of removal. Note the following:
   a. Install a new O-ring (A, **Figure 17**). Lubricate the O-ring with lithium soap grease.
   b. Tighten the starter mounting bolts (B, **Figure 16**) to 10 N•m (89 in.-lb.).
   c. Make sure the electrical connector is free of corrosion and is tight.

Disassembly

Refer to **Figure 18**.

1. If still installed, remove and discard the O-ring (A, **Figure 17**) from the front cover.
2. Remove the outer snap ring (B, **Figure 17**), the starter gear (C) and the inner snap ring.
3. Remove the case bolts (**Figure 19**) from the starter.

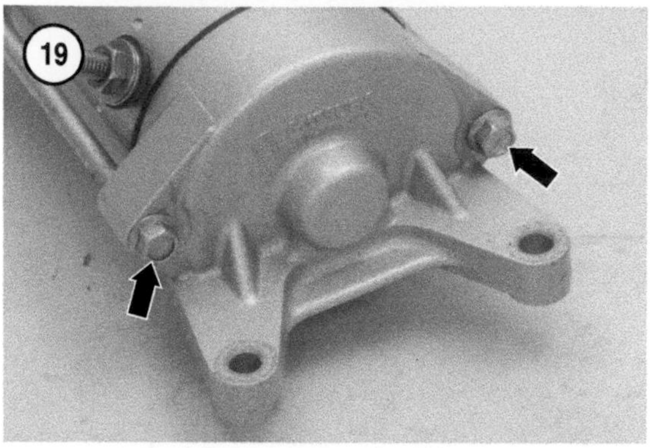

19

21

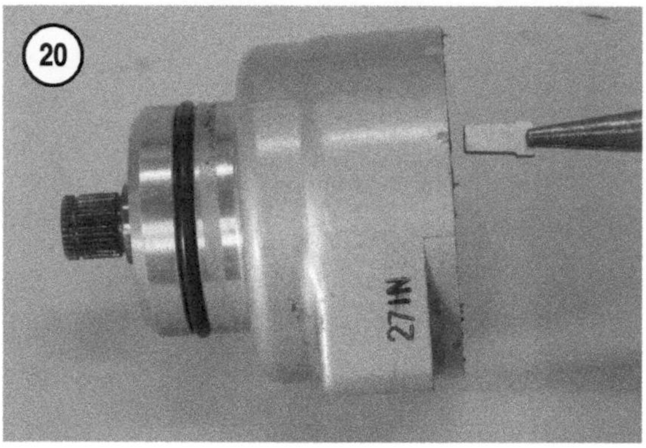

20

22

4. Remove the front end cap (D, **Figure 17**) from the housing and remove the locating key (**Figure 20**). It could come out with the end cap or it could remain in the housing.

5. Lift each planetary gear (**Figure 21**) from its post in the front end cap.

6. Remove the ring gear (**Figure 22**). Note that its male index faces down and engages the recess in the ring gear holder.

7. Remove the ring gear holder (**Figure 23**) and washer (**Figure 24**) from the front end cap.

8. Remove the gear plate (**Figure 25**) from the armature housing.

9. Slide the washers (**Figure 26**) from the armature shaft.

10. Remove the rear end cap (A, **Figure 27**) from the armature housing.

NOTE

Count and label the number of shims removed in Step 11. These shims must be reinstalled in the order in which they are removed. Do not mix these shims with other washers or shims.

23

24

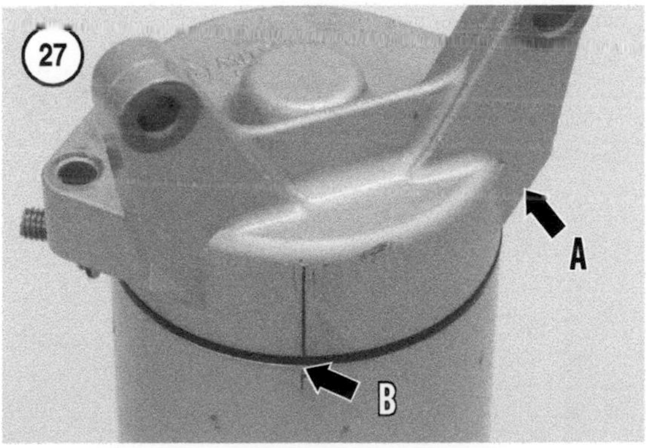

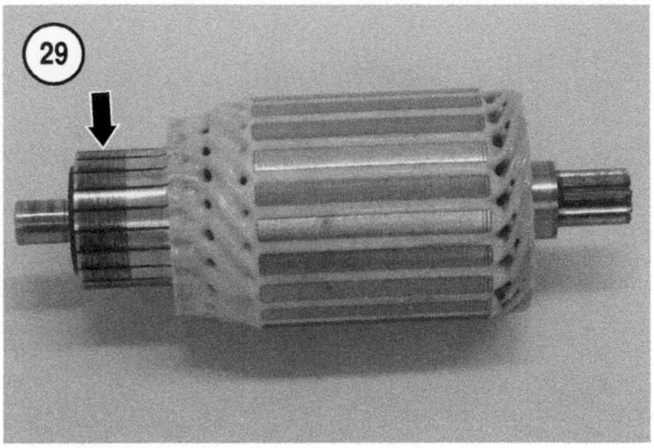

11. Remove the shims (A, **Figure 28**) from the commutator end of the armature shaft. Record the number of shims and their locations. Store the shims in a marked plastic bag.

12. Pull the armature (**Figure 29**) from the armature housing.

13. Remove and discard the O-ring on each end of the armature housing. New O-rings must be installed during assembly.

9

CAUTION

Do not immerse the wire windings in the housing or the armature coil in solvent. Doing so could damage the insulation. Wipe the windings with a cloth lightly moistened with solvent and thoroughly dry them.

14. Clean all grease, dirt and carbon from the armature, housing and end caps.

NOTE

Further disassembly is not necessary unless the brush holder assembly must be replaced. The brushes can be removed and inspected at this stage.

CAUTION

The cable terminal assembly consists of several washers, shims and a nut. Label each component as it is removed. They must be reinstalled in the same order to insulate the positive brush assembly from the housing.

15. Remove the nut (**Figure 30**), washers and shims from the terminal post. Note their order so they can be reinstalled in the same order during assembly. In this instance, the assembly consists of two small fiber washers, a large fiber washer, a steel

washer and the nut (**Figure 31**). The number and order of the washers and shims varies.

16. Remove the brush holder assembly (**Figure 32**) from the armature housing.

17. Remove the insulator (A, **Figure 33**) from the brush holder assembly.

18. Separate the positive brush assembly (A, **Figure 34**), including the spacer (B) and terminal bolt (C), from the brush holder (D).

Assembly

1. Seat the washer (**Figure 24**) in the front end cap.

2. Install the ring gear holder (**Figure 23**) so its tabs fit into the slots in the end cap. Install the holder so its female recess faces up out of the end cap.

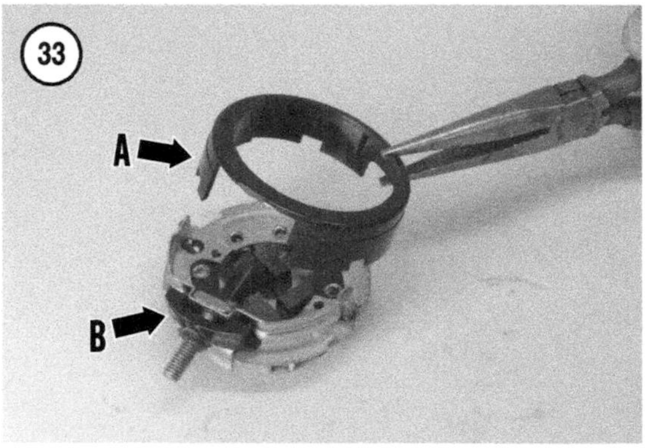

3. Install the ring gear (**Figure 22**) so its male index engages the female recess in the ring gear holder.

4. Install a planetary gear (**Figure 21**) onto each post in the front end cap.

CAUTION

*When installing the positive brush assembly, make sure an insulated portion of each positive brush wire (E, **Figure 34**) sits within a cutout in the brush holder. If bare wire touches the holder, the starter will short to ground.*

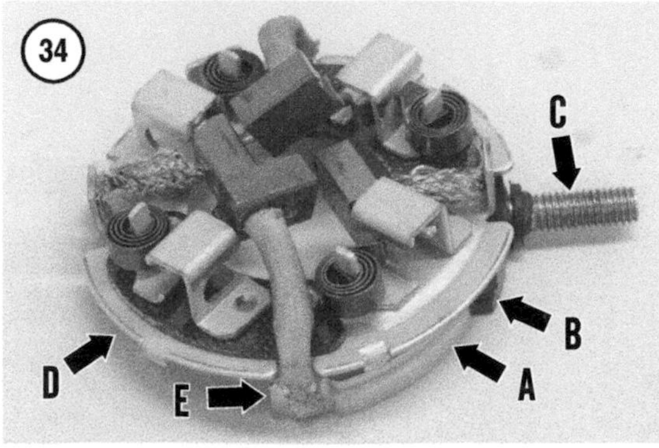

5. Install the positive brush assembly (A, **Figure 34**) onto the brush holder so an insulated portion of each positive brush wire (E) sits within a cutout in the brush holder.

6. Lower the assembly into the insulator (A, **Figure 33**) so the cutout in the insulator engages the spacer (B, **Figure 33**) on the terminal bolt.

35

36

37

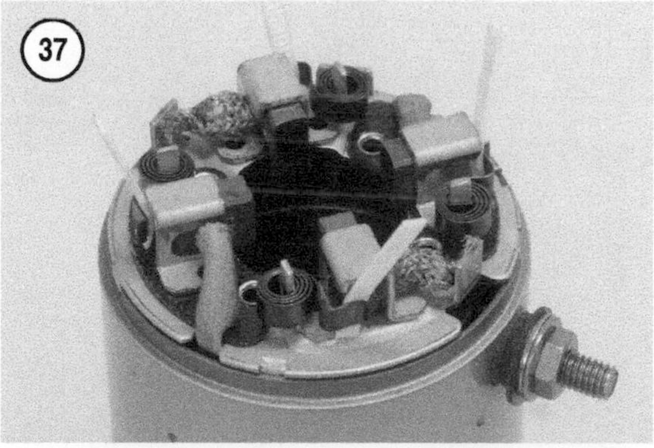

38

39

a. Make sure an insulated portion of each positive brush wire (**Figure 35**) touches the brush holder.

b. If bare wire from the positive brush touches the brush holder, disassemble the brush holder assembly and reassemble it properly.

7. Make sure the O-ring is installed on the terminal bolt, and install the brush holder assembly into the armature housing (**Figure 36**).

8. Install the washer and shims (**Figure 31**) onto the terminal post in the same order noted during removal, and install the nut (**Figure 30**).

9. Install each brush into its holder as follows:

a. Cut a cable tie into four 1.5 in. strips.

b. Pull back the brush spring, and insert a strip between the brush holder and the spring.

c. Press the brush into the holder. See **Figure 37**. Each cable tie will hold the spring away from the brush during armature installation.

10. Insert the armature into the housing (**Figure 38**) until the armature bottoms against the brush holder. Exercise caution so the commutator (**Figure 29**) is not damaged.

11. Slip the cable tie (**Figure 39**) from each brush holder. The springs can now press the brushes against the commutator.

12. Install the shims (A, **Figure 28**) onto the armature shaft in the same order noted during removal.

13. Lubricate a new O-ring (B, **Figure 28**) with lithium soap grease and install it onto the end of the armature housing.

14. Install the rear end cap (A, **Figure 27**) over the brush holder and seat it on the armature housing. Make sure the index mark on the end cap aligns with the mark on the armature housing (B, **Figure 27**).

15. Install the washers onto the armature shaft in the same order noted during removal.

16. Install the gear plate (**Figure 25**) so its cutout engages the notch in the armature housing. Make sure the flat side faces out and seat the plate (A, **Figure 40**) in the armature housing.
17. Lubricate a new O-ring with lithium soap grease, and install it onto the armature housing (B, **Figure 40**).
18. The front end cap has four machined slots (A, **Figure 41**) around its inner circumference. The locating key (**Figure 20**) must be installed in the one slot that sits opposite the cutout (C, **Figure 40**) in the armature housing when the front end cap is installed. Identify that particular slot and install the key as follows:
   a. Temporarily fit the front end cap onto the armature housing.
   b. Align the index mark on the end cap (**Figure 42**) with the mark on the housing.
   c. While keeping the index marks directly opposite one another, carefully slide the front end cap straight off the housing until the cutout in the housing is exposed (C, **Figure 40**).
   d. The slot in the end cap that is aligned with the housing cutout receives the locating key.
   e. Remove the front end cap and install the locating key (**Figure 20**) in the slot identified in substep d.
19. Install the front end cap onto the housing so the index mark on the cap aligns with the mark on the housing (**Figure 42**).
20. Install the starter case bolts (**Figure 19**). Apply a medium-strength threadlocking compound to the bolt threads and tighten the bolts securely.
21. Lubricate a new O-ring (A, **Figure 17**) with lithium soap grease and install it onto the front end cap.
22. Install the inner snap ring onto the armature shaft.
23. Install the starter gear (C, **Figure 17**) onto the armature shaft and install the outer snap ring (B).

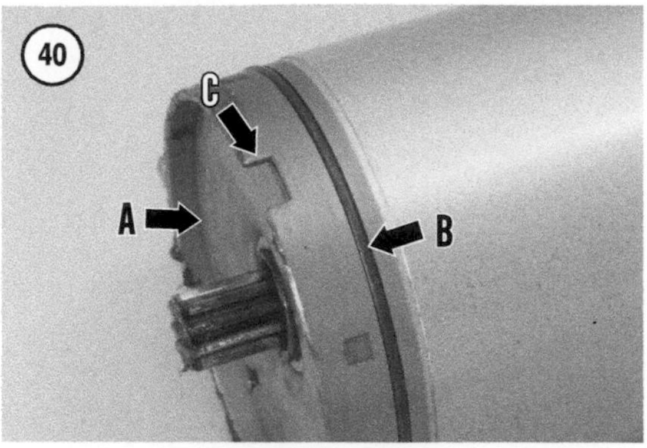

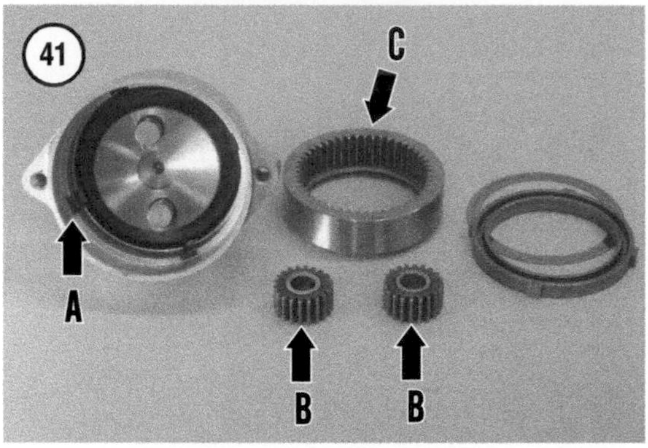

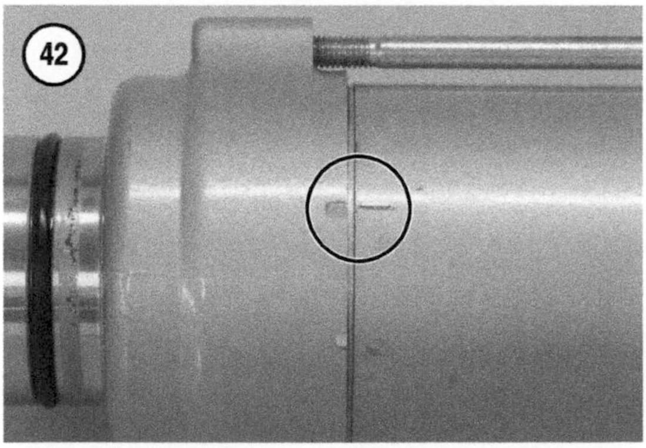

Inspection

NOTE
The O-rings, case bolts and brush holder assembly are the only starter components available.

1. Pull the spring away from each brush and pull the brush out of its holder. Measure the length of each brush with a vernier caliper. If the length of any brush is less than the service limit in **Table 1**, the brush holder assembly must be replaced. The brushes cannot be replaced individually.

2. Inspect the commutator (**Figure 29**). The mica in a good commutator sits below the surface of the copper bars. Measure the mica undercut, which is the distance between the top of the mica and the top of the adjacent copper bars (**Figure 43**). If the mica undercut is less than specification (**Table 1**), have the commutator serviced by a Yamaha dealership or electrical repair shop.

43

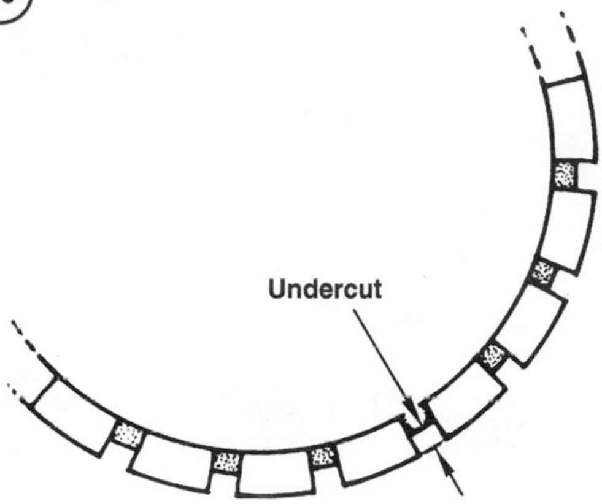

44

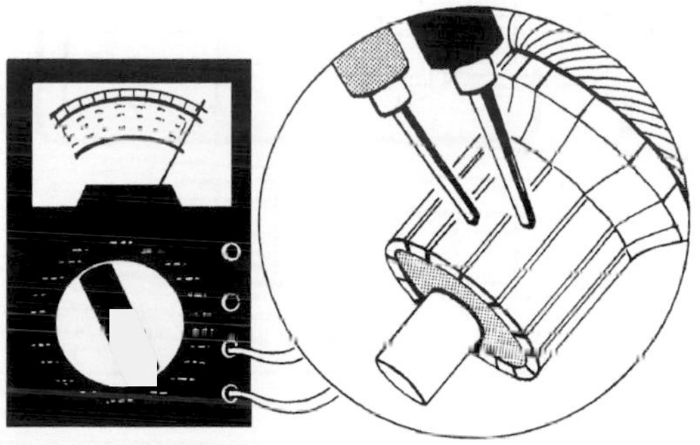

45

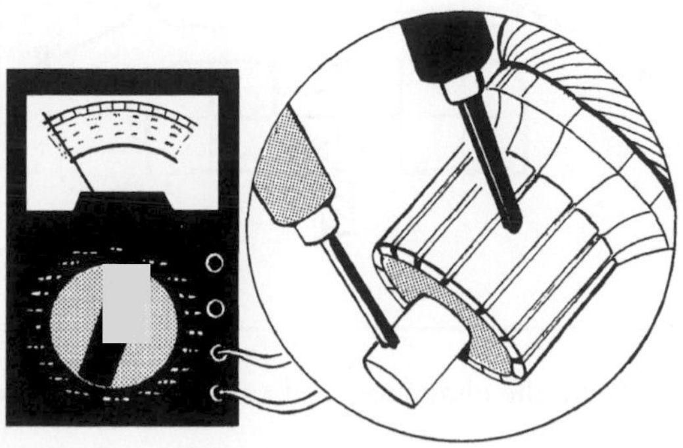

46

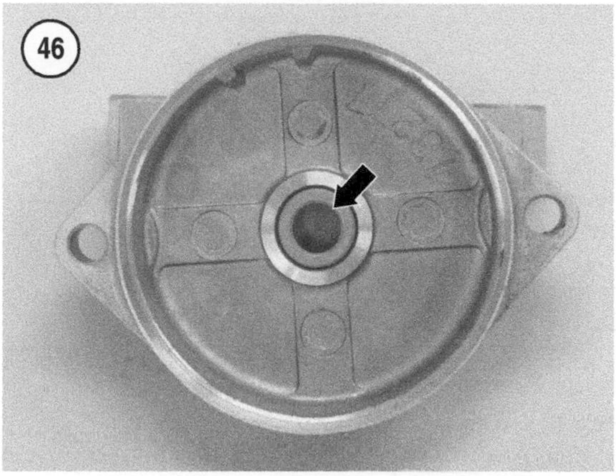

3. Inspect the commutator copper bars for discoloration. If a pair of bars is discolored, armature coils are grounded. Replace the starter.

4. Measure the diameter of the commutator with a vernier caliper. Replace the starter if the commutator diameter is less than the service limit specified in **Table 1**.

5. Use an ohmmeter to perform the following:

   a. Check for continuity between the commutator bars (**Figure 44**). There should be continuity (zero or low resistance) between pairs of bars.

   b. Check for continuity between the commutator bars and the armature shaft (**Figure 45**). There should be *no* continuity (infinite resistance).

   c. If the unit fails any of these tests, replace the starter.

6. Inspect the bushing (**Figure 46**) in the rear end cap for wear or damage. If it is damaged, replace the starter.

7. Inspect the bearing and oil seal in the front end cover (A, **Figure 41**). If either is worn or damaged, replace the starter.

8. Inspect the teeth of the planetary gears (B, **Figure 41**) and ring gear (C) for chipped, cracked or broken teeth.

9. Clean all dirt and contaminants from the mounting lugs of the rear end cover. The lugs ground the starter so there must be good contact between the starter lugs and the crankcase.

9

STARTING CIRCUIT CUTOFF RELAY

The starting circuit cutoff relay (SCCR) interrupts the flow of current to the starter relay unless the transmission is in neutral or unless the clutch lever is pulled in and the sidestand is up.

SCCR Removal/Installation

1. Securely support the motorcycle on a level surface.
2. Remove the fuel tank, air filter housing and surge tank as described in Chapter Eight.
3. Remove the frame neck covers (Chapter Fourteen).
4. Lift the starting circuit cutoff relay (**Figure 47**) from its tang and disconnect the 12-pin connector from the relay.
5. Installation is the reverse of removal. Make sure the electrical terminals are clean and pack the connector with dielectric grease.

SCCR Continuity Test

1. Remove the starting circuit cutoff relay as described in this section.
2. Disconnect the harness connector from the relay.
3. Check the continuity of the starting circuit cutoff relay as follows:
   a. Use jumpers to connect the positive battery terminal to the red/black terminal in the relay (A, **Figure 48**); connect the negative battery terminal to the black/yellow terminal (B).
   b. Connect the ohmmeter positive test lead to the blue terminal (C, **Figure 48**) in the relay and connect the negative test lead to the blue/white terminal (D, **Figure 48**).
   c. The unit should have continuity during this test.
4. Replace the starting circuit cutoff relay if it has no continuity.

SCCR Diode Test (Starting Circuit)

1. Remove the starting circuit cutoff relay as described in this chapter.
2. Connect the ohmmeter leads to the sky blue terminal (B, **Figure 49**) and the black/yellow wire terminal (B). Check for continuity.

47

48

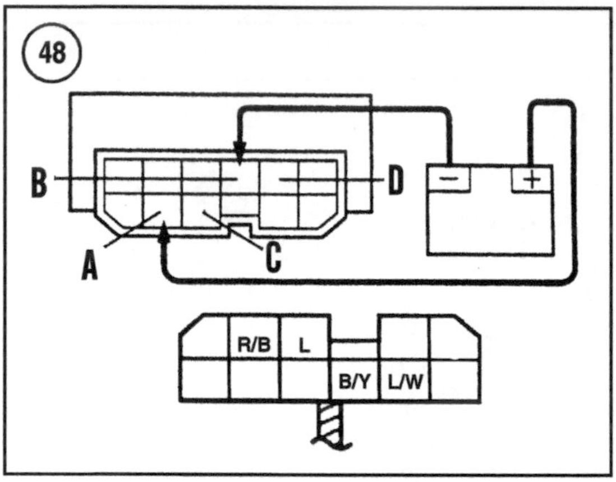

49

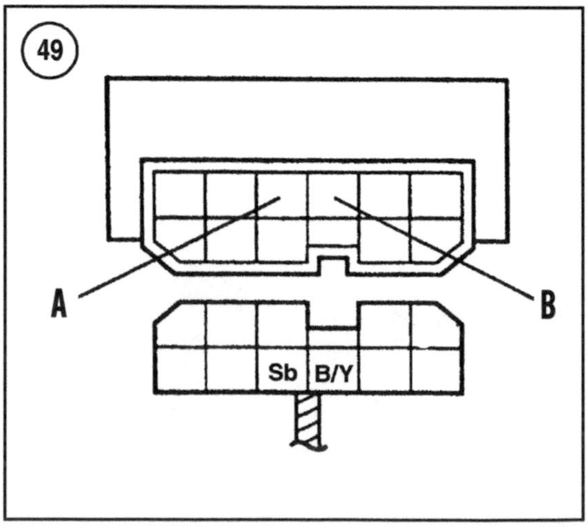

3. Reverse the ohmmeter lead connections. Check for continuity.

4. The diode should have continuity in only one direction. If there is no continuity, or continuity in both directions, replace the relay unit.

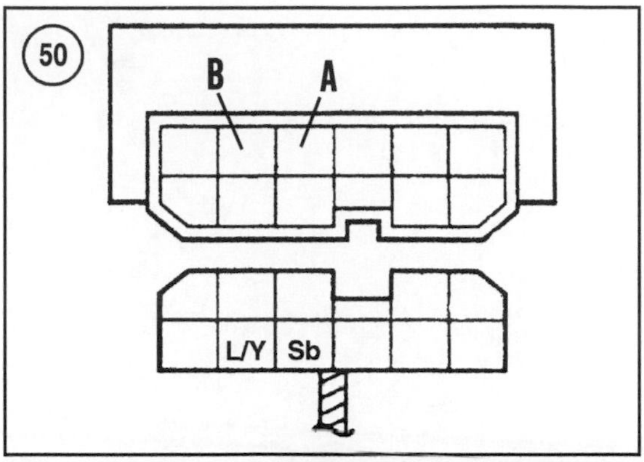

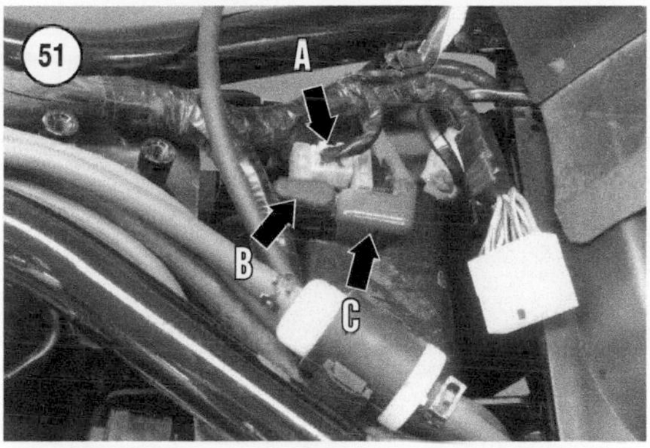

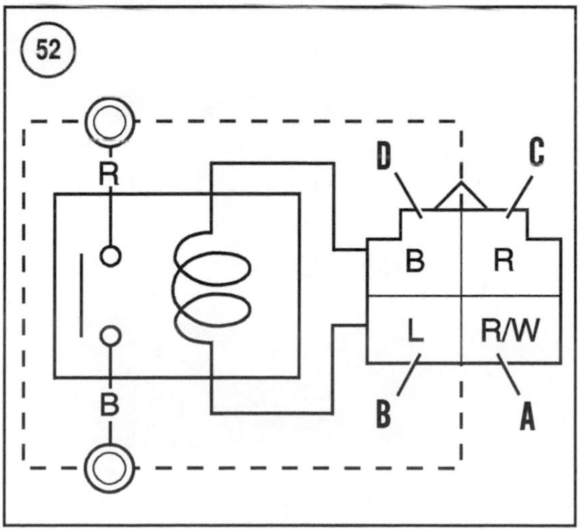

SCCR Diode Test (Ignition Circuit)

1. Remove the starting circuit cutoff relay as described in this chapter.
2. Connect the ohmmeter leads to the skly blue wire terminal (A, **Figure 50**) and the blue/yellow wire terminal (B). Check for continuity.
3. Reverse the ohmmeter lead connections. Check for continuity.
4. The diode should have continuity in only one direction. If there is no continuity, or continuity in both directions, replace the relay unit.

STARTER RELAY

Removal/Installation

1. Securely support the motorcycle on a level surface.
2. Make sure the main switch is turned off. Disconnect the negative terminal from the battery.
3. Remove the seats and ignitor panel as described in Chapter Fourteen.
4. Press the release tab on the four-pin connector (A, **Figure 51**) and disconnect the connector from the starter relay.
5. Pull back the boot on each starter relay cable connector and remove the terminal nuts.
6. Disconnect the red battery lead (B, **Figure 51**) from the relay's positive terminal and disconnect the black starter lead (C) from the relay's negative terminal.
7. Remove the relay from the rubber mount.
8. Installation is the reverse of removal. Note the following:
   a. Install the relay in the rubber mount.
   b. Make sure all connections are free of corrosion and are tight.
   c. Torque the relay terminal nuts to 7 N•m (62 in.-lb.).

Continuity Test

1. Securely support the motorcycle on a level surface.
2. Turn the main switch off. Disconnect the negative terminal from the battery.
3. Remove the seats and ignitor panel as described in Chapter Fourteen.
4. Press the release tab on the four-pin connector (A, **Figure 51**) and disconnect the connector from the starter relay.
5. Use jumper wires to connect the positive battery terminal to the red/white terminal (A, **Figure 52**) in the relay; connect the negative battery terminal to the blue terminal (B).

9

6. Connect the ohmmeter positive test lead to the red (C, **Figure 52**) terminal in the relay; connect the negative test lead to the black terminal (D). The relay should have continuity.
7. Replace the starter relay if it fails this test.

LIGHTING SYSTEM

The lighting system consists of the headlight, taillight, high beam indicator light, meter illumination lights, and front turn signal/position light. If there is trouble with any of these lights, check the affected bulb first. Replacement bulbs are listed in **Table 2**. If the bulb is good, follow the lighting system troubleshooting procedures listed in Chapter Two.

HEADLIGHT

Headlight Bulb Replacement

WARNING
If the headlight has just burned out or just has just been turned off, the bulb will be ***hot****. Do not touch the bulb until it cools off.*

CAUTION
All models are equipped with a quartz-halogen bulb. Do not touch the bulb glass with your fingers because traces of oil on the bulb will drastically reduce the life of the bulb. Clean any traces of oil from the bulb with a cloth moistened in alcohol or lacquer thinner.

1. Remove the mounting screw (**Figure 53**) on either side of the headlight housing and lower the lens assembly from the housing.
2. Disconnect the headlight connector (A, **Figure 54**) from the bulb and remove the lens assembly.
3. Remove the boot (**Figure 55**) from the lens assembly.
4A. On 1999-2000 XVS1100 models and 2000-on XVS1100A models, turn the headlight bulb holder counterclockwise, remove the bulb holder and pull the bulb from the lens assembly.
4B. On 2001-on XVS1100 models, release the securing clip (**Figure 56**), and remove the bulb (**Figure 57**) from the lens assembly.

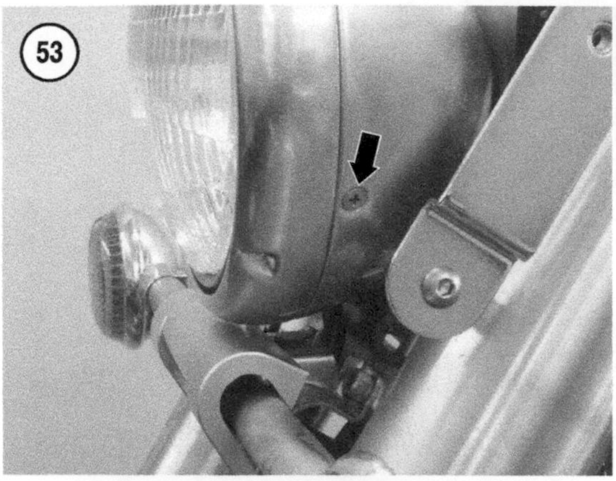

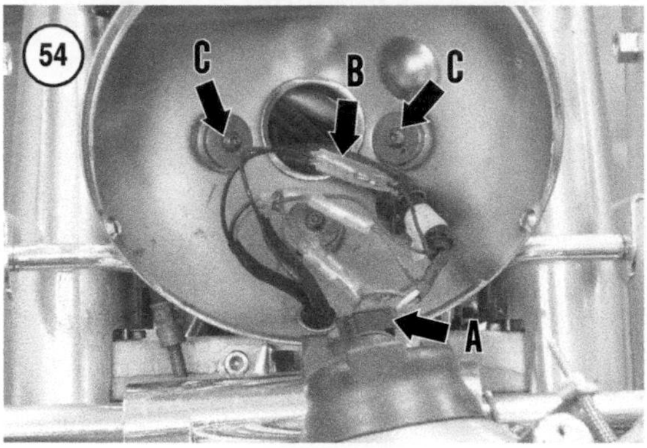

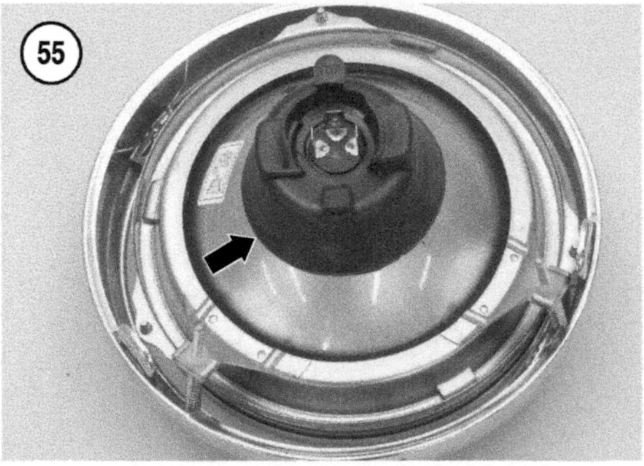

5. Installation is the reverse of removal. Note the following:

a. Install the bulb and make sure the projections on the bulb engage the slots (**Figure 58**) in the lens assembly.

b. Make sure the electrical connector is free of corrosion and is tightly pushed onto the bulb terminals.

56

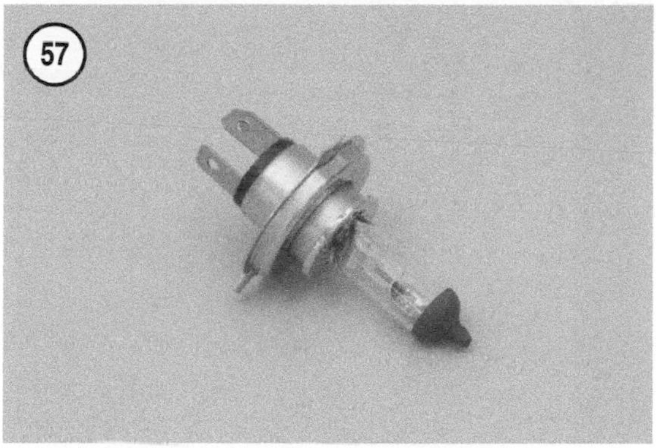

57

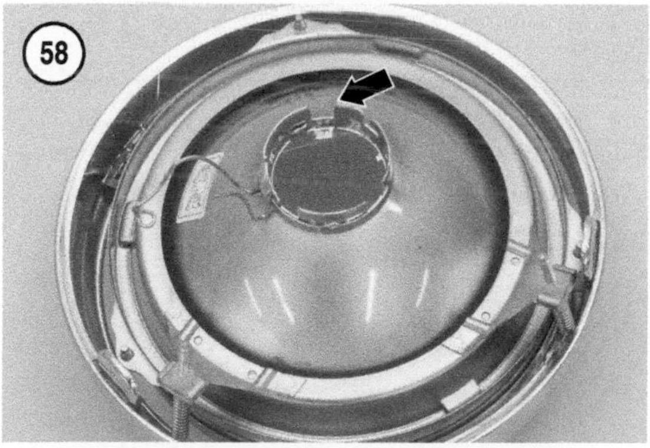

58

Headlight Housing Removal/Installation

Refer to **Figures 59-61**.

1. Remove the mounting screw (**Figure 53**) on either side of the headlight housing and lower the lens assembly from the housing.
2. Disconnect the headlight connector (A, **Figure 54**) from the bulb and remove the lens assembly.
3. Disconnect the front turn signal bullet connectors (B, **Figure 54**) from their harness mates.
4. Feed the wiring harness lead and the turn signal leads through the opening in the headlight housing.

5A. On 1999-2000 XVS1100 models, remove the two headlight housing bolts and remove the housing from the headlight bracket.

5B. On 2001-on XVS1100 models, remove the headlight housing bolts and remove the housing from the headlight bracket.

5C. On XVS1100A models, remove the two headlight housing bolts (C, **Figure 54**) and remove the housing from the bracket on the lower fork bridge. Watch for the washer and collar on each bolt.

6. Installation is the reverse of removal. Note the following:
   a. Make sure the electrical connectors are free of corrosion and are pushed tightly together.
   b. Adjust the headlight as described in this chapter.
   c. Torque the headlight housing bolts to 8 N•m (71 in.-lb.).

Headlight Adjustment

Adjust the headlight horizontally and vertically according to the Department of Motor Vehicles regulations in your area.

1999-2000 XVS1100 models

The horizontal adjuster on these models is at 2 o'clock (A, **Figure 62**) on the front of the headlight rim and the vertical adjuster is at 8 o'clock (B).

1. To adjust the beam to the right, turn the horizontal adjuster clockwise. To adjust the beam to the left, turn the adjuster counterclockwise.
2. To raise the beam, turn the vertical adjuster clockwise. To lower the beam, turn the adjuster counterclockwise.

2001-on XVS1100 models

The horizontal adjuster on these models is at 4 o'clock (A, **Figure 63**) on the side of the headlight housing and the vertical adjuster is at 8 o'clock on the housing (B, **Figure 63**).

1. To adjust the beam to the right, turn the horizontal adjuster counterclockwise. To adjust the beam to the left, turn the adjuster clockwise.
2. To raise the beam, turn the vertical adjuster clockwise. To lower the bear, turn the adjuster counterclockwise.

59

HEADLIGHT (1999-2000 XVS1100 MODELS)

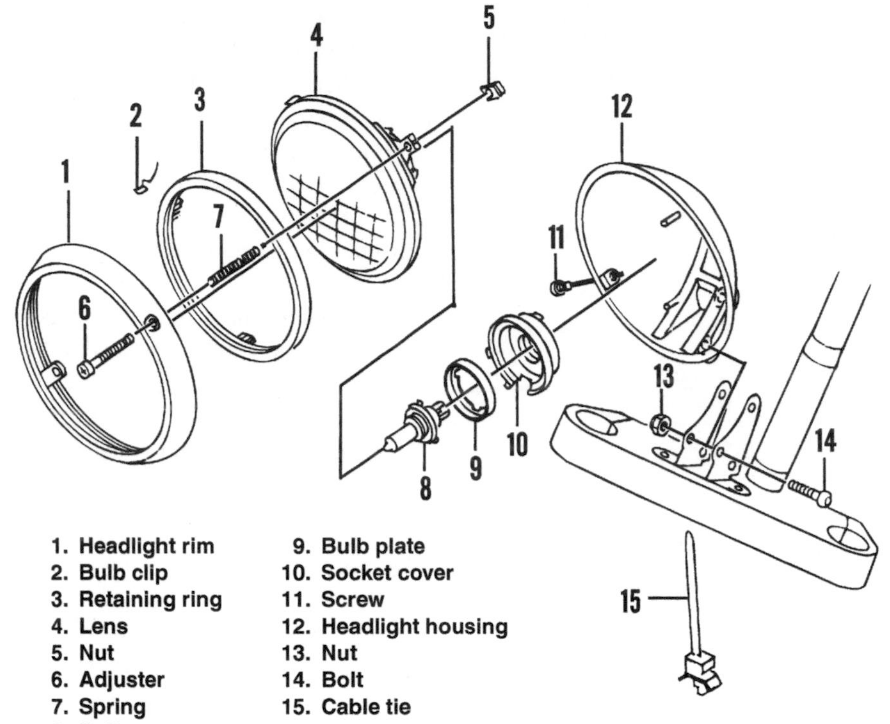

1. Headlight rim
2. Bulb clip
3. Retaining ring
4. Lens
5. Nut
6. Adjuster
7. Spring
8. Bulb
9. Bulb plate
10. Socket cover
11. Screw
12. Headlight housing
13. Nut
14. Bolt
15. Cable tie

XVS1100A models

The horizontal adjuster is at 8 o'clock (B, **Figure 62**) on the front of the headlight rim and the vertical adjuster is at 4 o'clock on the rim (C).

1. To adjust the beam to the right, turn the horizontal adjuster clockwise. To adjust the beam to the left, turn the adjuster counterclockwise.
2. To raise the beam, turn the vertical adjuster clockwise. To lower the bear, turn the adjuster counterclockwise.

Headlight Voltage Test

If the headlight does not turn on but its bulb is in good working order, test the headlight circuit voltage as follows:

1. Remove the mounting screw (**Figure 53**) on either side of the headlight housing and lower the lens assembly from the housing.
2. Disconnect the headlight connector (A, **Figure 54**) from the bulb and remove the lens assembly.
3. Set a voltmeter to the DC 20 volt range.
4. Turn the dimmer switch to LO, and check the voltage as follows:
   a. Connect the voltmeter negative test lead to the black terminal (A, **Figure 64**) in the headlight connector; connect the voltmeter positive test lead to the green terminal (B).
   b. Turn the main switch on and check the voltmeter. It should read battery voltage.
5. Turn the dimmer switch to HI, and check the voltage as follows:
   a. Connect the voltmeter negative test lead to the black terminal (A, **Figure 64**) in the headlight connector; connect the voltmeter positive lead to the yellow terminal (C).
   b. Turn the main switch on, and check the voltmeter. It should read battery voltage.

(60)

HEADLIGHT (2001-ON XVS1100 MODELS)

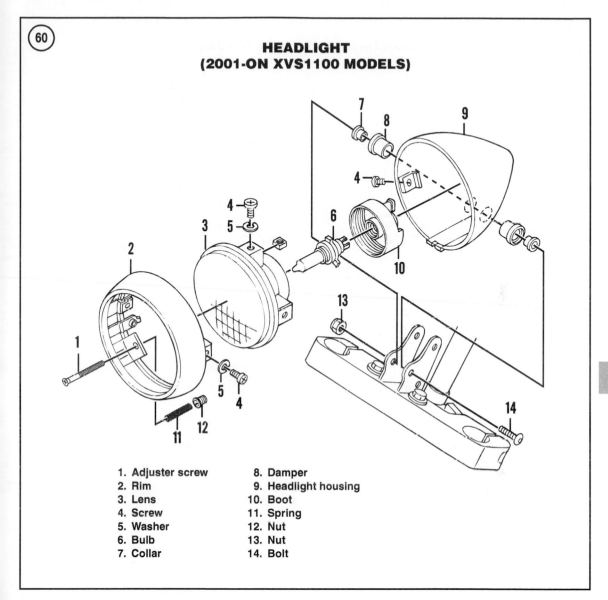

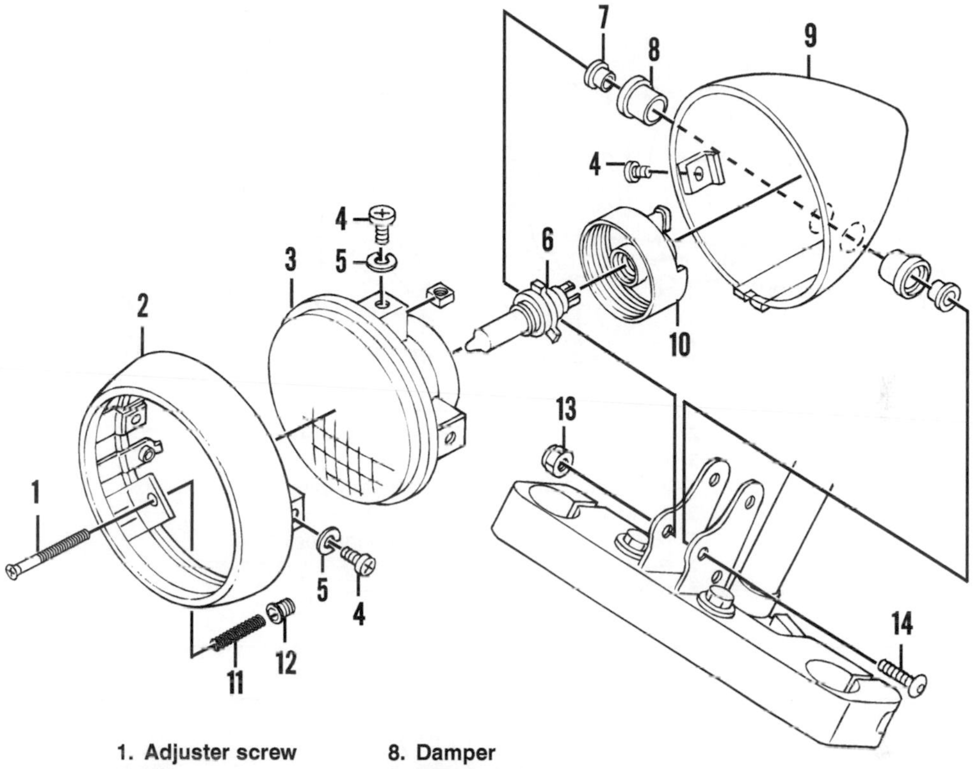

1. Adjuster screw
2. Rim
3. Lens
4. Screw
5. Washer
6. Bulb
7. Collar
8. Damper
9. Headlight housing
10. Boot
11. Spring
12. Nut
13. Nut
14. Bolt

6. Remove the fuel tank (Chapter Eight) and disconnect the meter assembly connector (**Figure 65**) from its harness mate.

7. Connect the voltmeter positive test lead to the yellow terminal (A, **Figure 66**) in the harness side of the connector; connect the voltmeter negative test probe to the black terminal (B).

8. Turn the main switch on and turn the dimmer switch to HI. The meter should read battery voltage.

9. If any voltage reading does not equal battery voltage, the wiring between the main switch and the headlight connector is faulty.

METER ASSEMBLY

Removal/Installation

Refer to **Figure 67**.

1. Remove the fuel tank as described in Chapter Eight.
2. Remove the mounting screws (A, **Figure 68**).
3. Raise the meter cover (B, **Figure 68**), and disconnect the two-pin, trip meter button connector (A, **Figure 69**) from it mate. Remove the cover.
4. Remove the mounting nuts (B, **Figure 69**) and lift the meter assembly from the mounting studs in the fuel tank.

61

HEADLIGHT (XVS1100A MODELS)

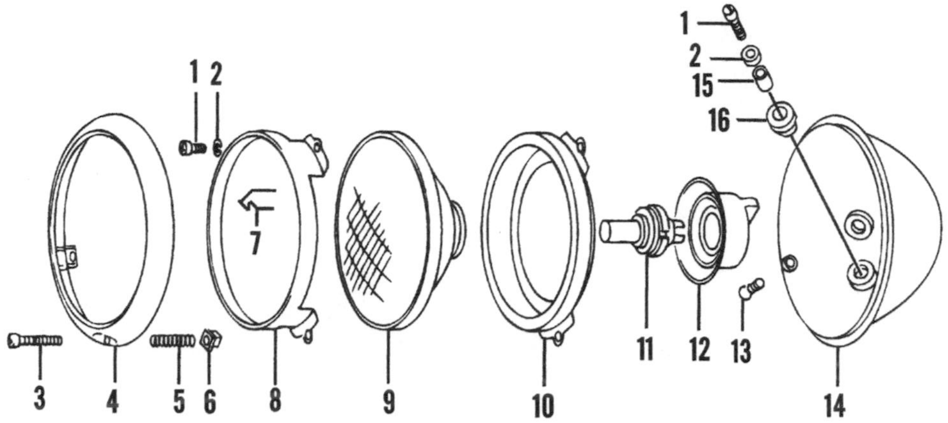

1. Screw
2. Washer
3. Adjuster
4. Headlight rim
5. Spring
6. Nut
7. Bulb clip
8. Retaining clip
9. Lens
10. Mounting ring
11. Bulb
12. Socket cover
13. Screw
14. Headlight housing
15. Collar
16. Damper

62

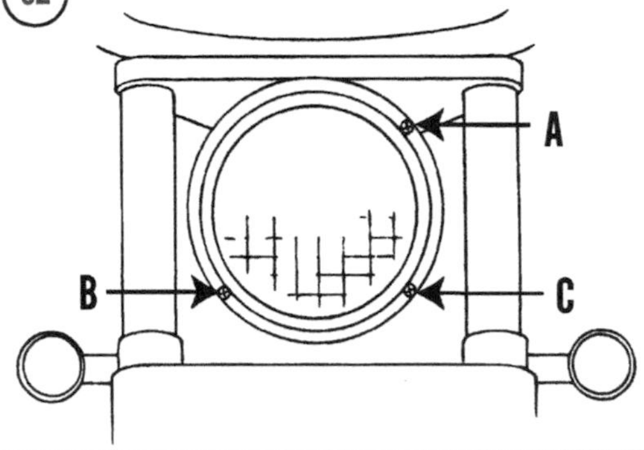

63

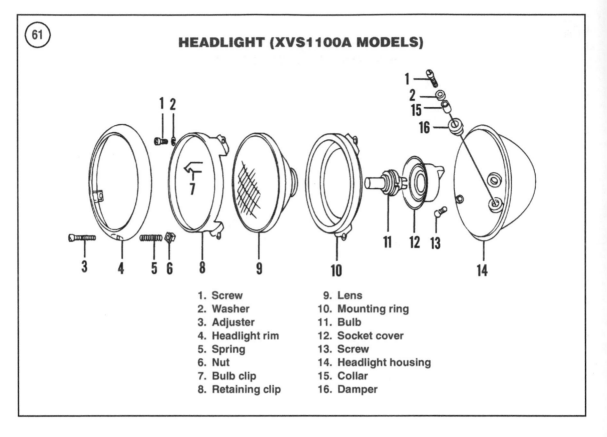

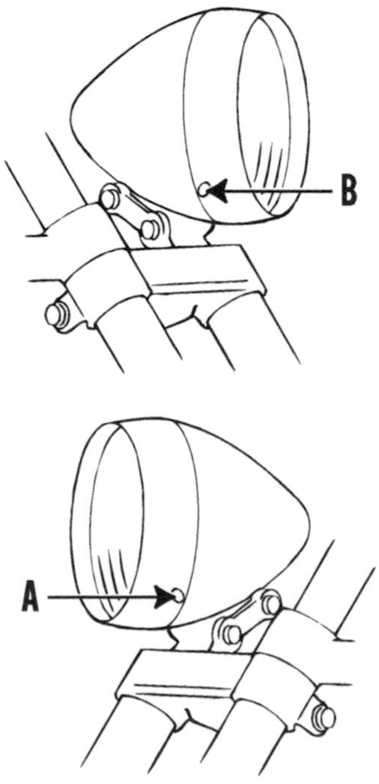

5. Installation is the reverse of removal.
   a. Make sure the dampers are in place on each mount in the meter bracket.
   b. Do not overtighten the meter cover screws as it may damage the cover.

Meter Indicator Light Test

If the meter illumination lights do not operate, perform the following test.

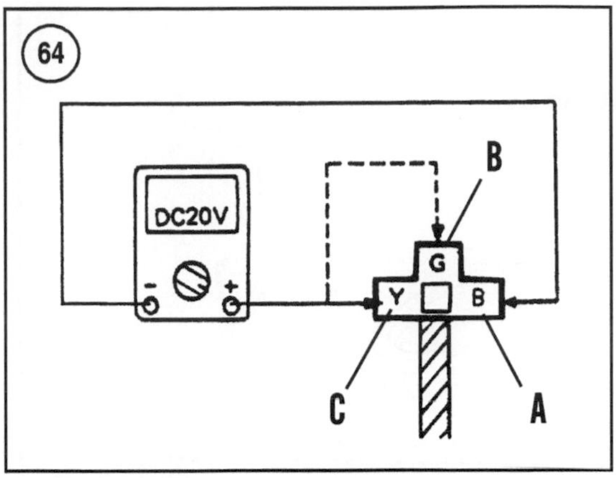

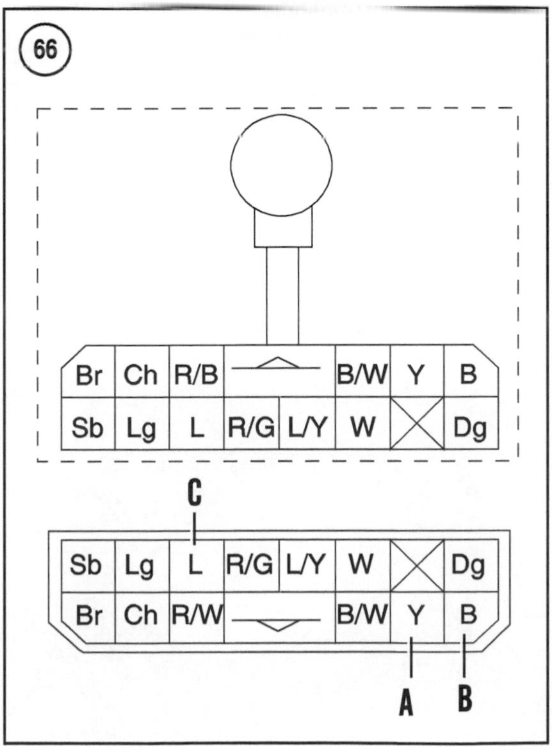

1. Remove the meter assembly as described in this chapter.
2. Check the voltage in the illumination light circuit as follows:
   a. Set a voltmeter to the 20 DC volt range.
   b. Connect the voltmeter positive test lead to the blue terminal (C, **Figure 66**) in the harness side of the meter connector; connect the voltmeter negative test lead to the black terminal (B).
   c. Turn the main switch on.
   d. If the voltmeter does not read battery voltage, the wiring between the main switch and the meter connector is faulty. Make the necessary repairs.

TAILLIGHT/BRAKE LIGHT

Taillight/Brake Light Replacement

1. Remove the screws (A, **Figure 70**) and pull the taillight lens (B) from the taillight assembly.
2. Turn the bulb (A, **Figure 71**) counterclockwise and remove it.
3. Push the new bulb into the socket and turn it clockwise to lock it in position.
4. Make sure the gasket (B, **Figure 71**) is in place and reinstall the lens.

Taillight/Brake Light Assembly Removal/Installation

Refer to **Figure 72** and **Figure 73**.

1. Remove the taillight lens (B, **Figure 70**) and gasket (B, **Figure 71**).

2A. On XVS1100 models, remove the taillight housing as follows:
   a. Remove the nut and washer from each housing bolt.
   b. Pull the housing from the taillight bracket.
   c. Disconnect the taillight harness connector and remove the taillight bracket.

2B. On XVS1100A models, remove the taillight housing as follows:
   a. Remove the license plate from the bracket.
   b. Disconnect the black, blue and yellow taillight bullet connectors inside the housing on the license plate bracket (**Figure 74**).
   c. Remove the lens (B, **Figure 70**) and gasket (B, **Figure 71**) from the taillight housing.

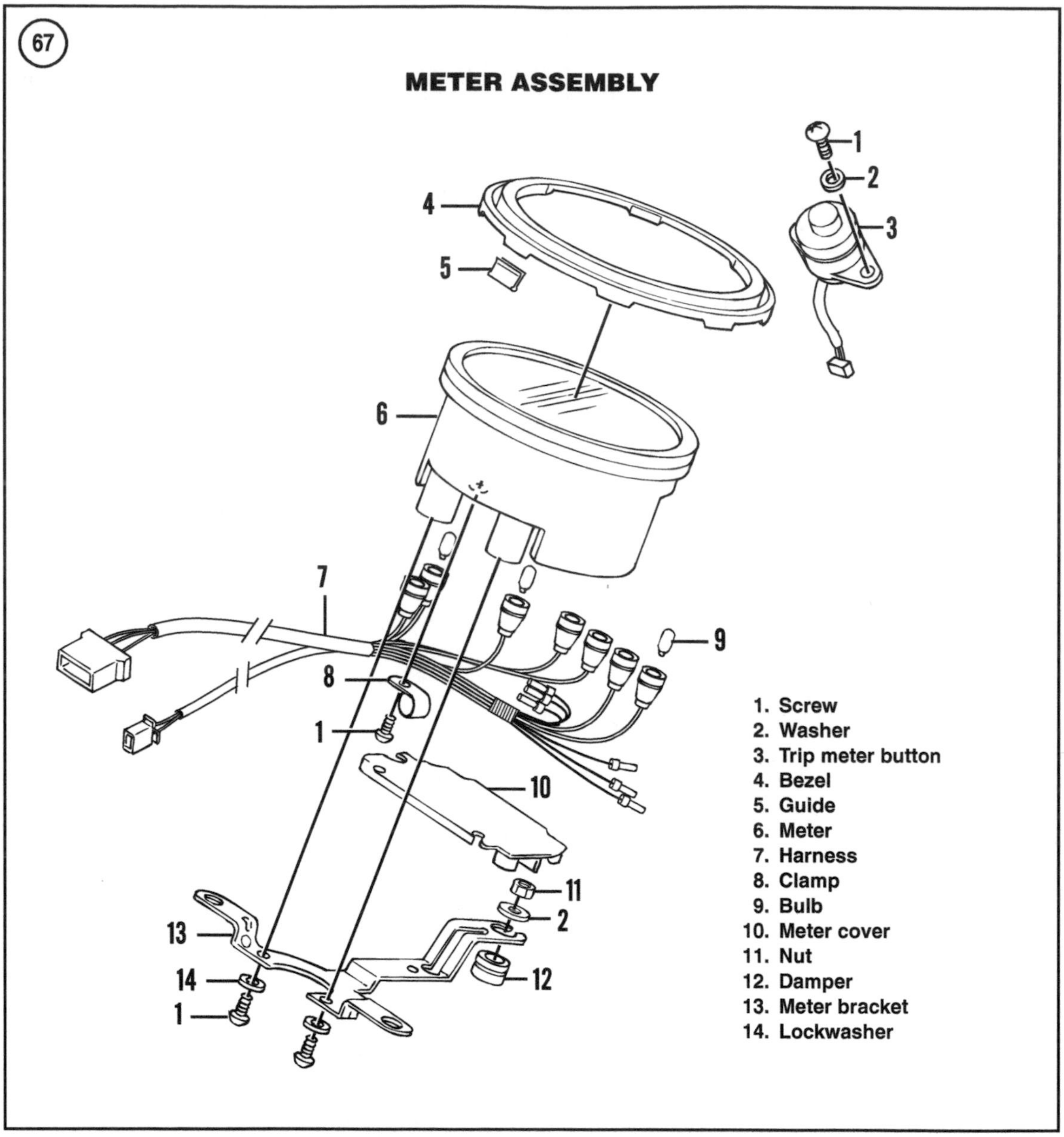
67
METER ASSEMBLY
1
2
3
4
5
6
7
8
9
1
10
11
2
12
13
14
1
1. Screw
2. Washer
3. Trip meter button
4. Bezel
5. Guide
6. Meter
7. Harness
8. Clamp
9. Bulb
10. Meter cover
11. Nut
12. Damper
13. Meter bracket
14. Lockwasher

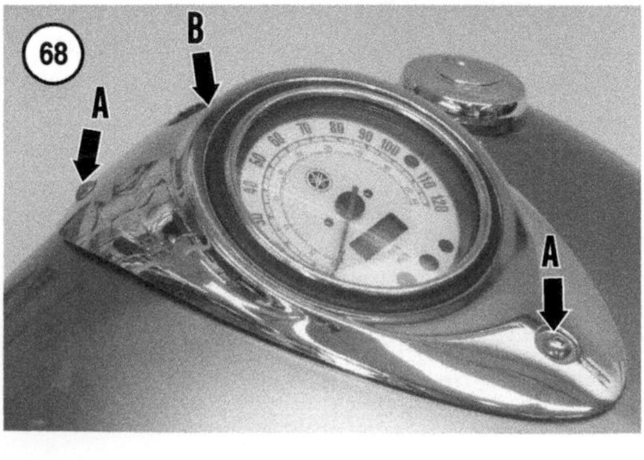
68
B
A
A

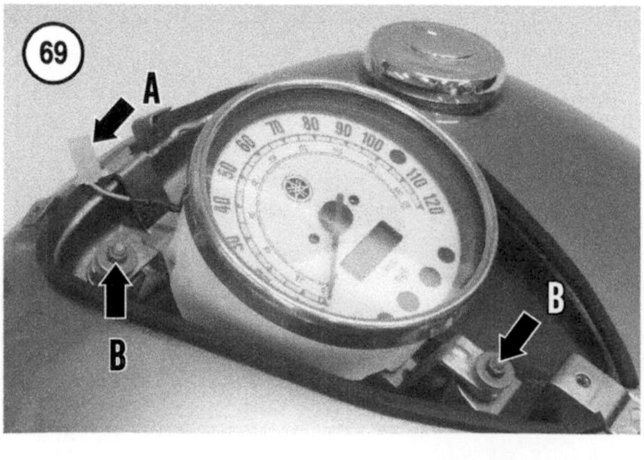
69
A
B
B

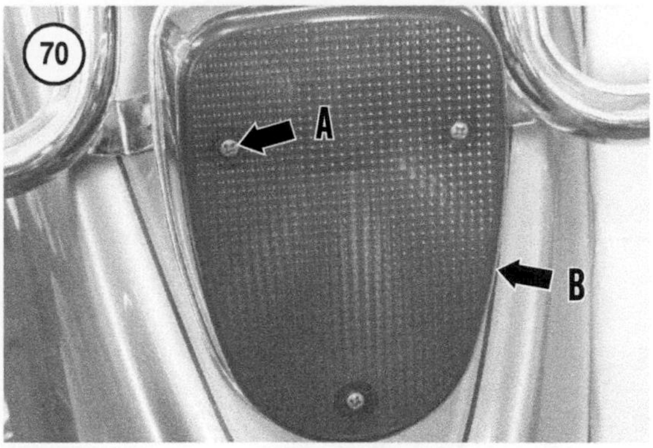
70
A
B

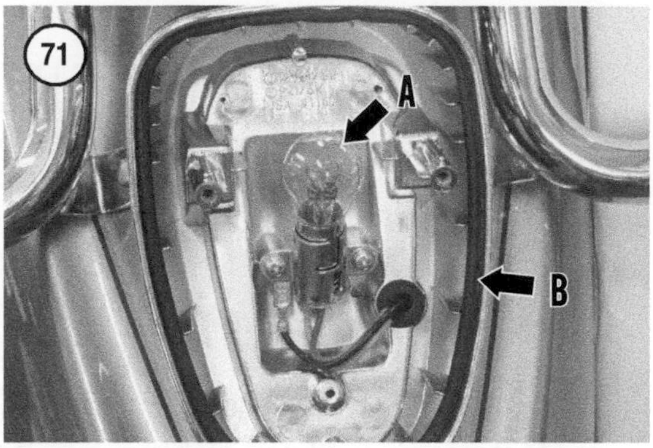
71
A
B

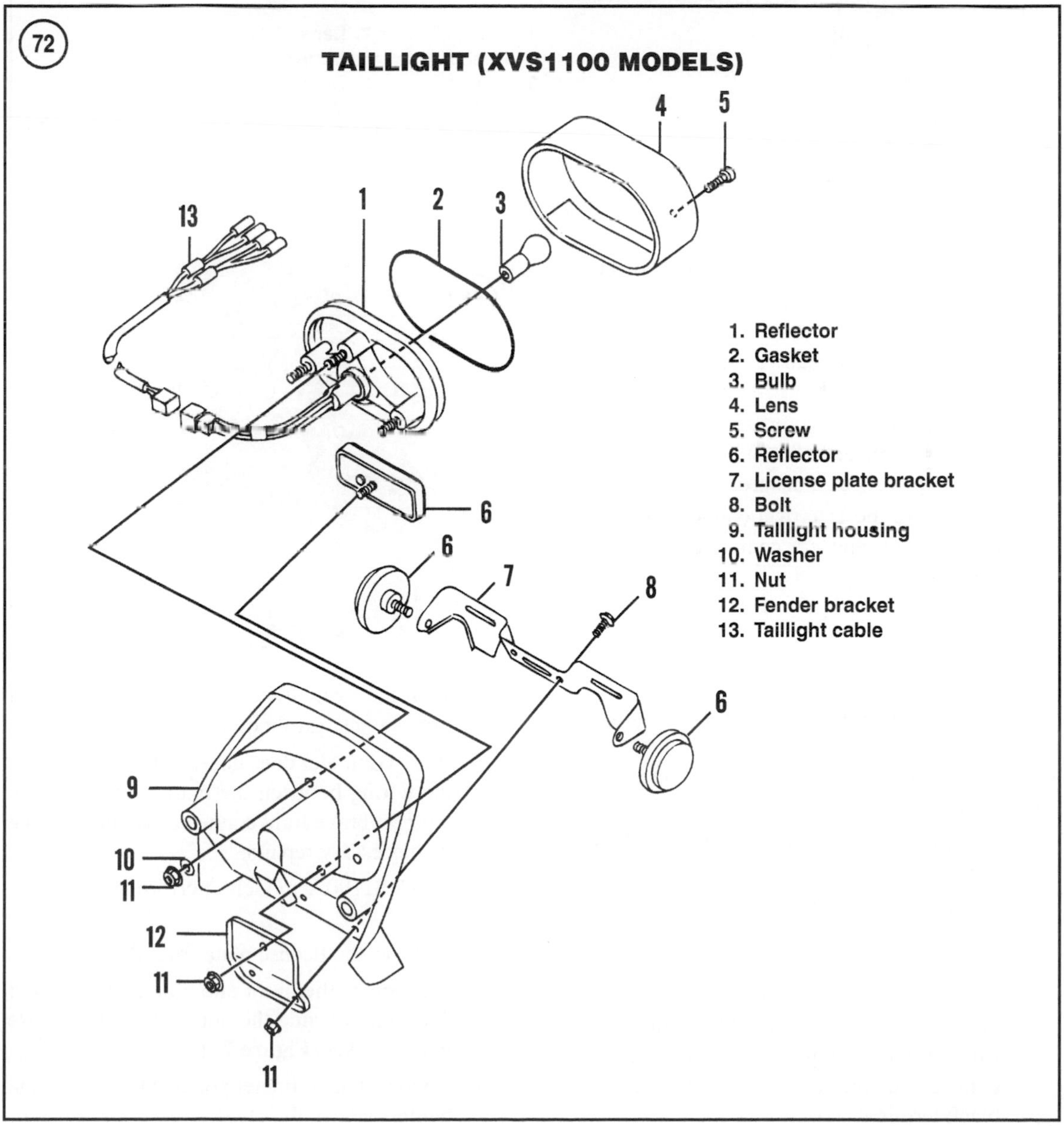
72
TAILLIGHT (XVS1100 MODELS)
1. Reflector
2. Gasket
3. Bulb
4. Lens
5. Screw
6. Reflector
7. License plate bracket
8. Bolt
9. Taillight housing
10. Washer
11. Nut
12. Fender bracket
13. Taillight cable

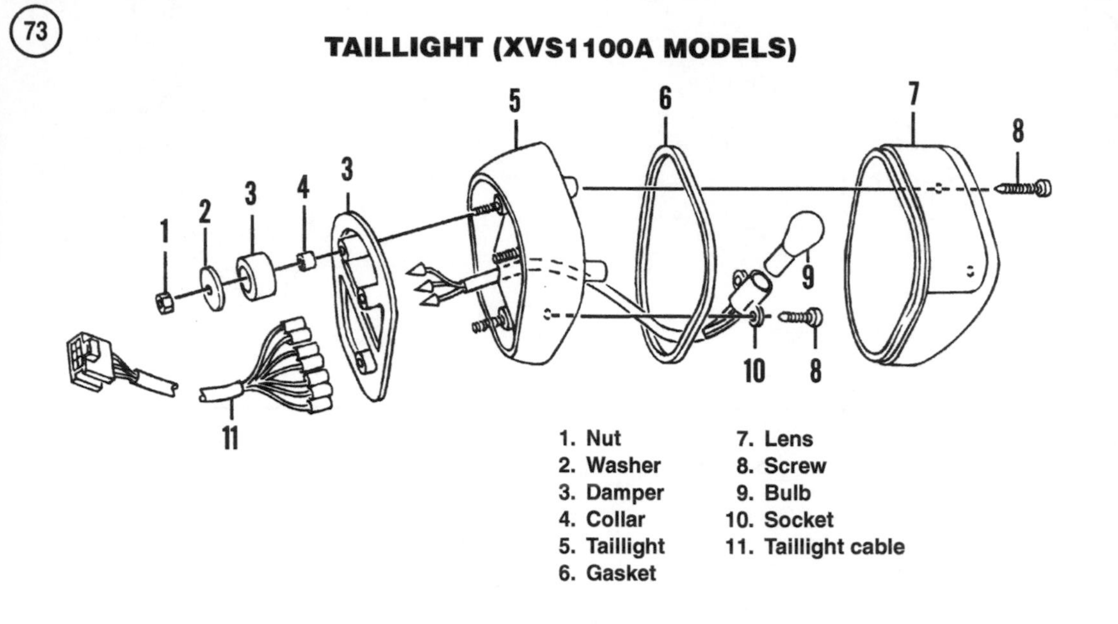

d. Remove the nut and washer, and lower the housing and its wiring from the fender. Watch for the collar on each stud.

3. Installation is the reverse of removal. Note the following:

a. Apply a dielectric compound to the electrical connectors prior to reconnecting them. This will help seal out moisture.

b. Make sure the electrical connectors are free of corrosion and are completely coupled to each other.

Taillight Test

If a taillight does not operate, test the circuit as follows:

1. Set a voltmeter to the 20 DC volt range.

2A. On XVS1100 models, perform the following:

a. Remove the nut and washer from each housing bolt.

b. Pull the housing from the taillight bracket.

c. Disconnect the taillight harness connector and remove the taillight bracket.

d. Connect the voltmeter positive test lead to the blue terminal (A, **Figure 75**) in the harness side of the taillight/brake light connector; connect the voltmeter negative test lead to the black terminal (B).

e. Turn the main switch on. On models with a light switch, turn it on also.

f. If the voltmeter does not read battery voltage, the wiring between the main switch and the taillight/brake light connector is faulty. Make the necessary repairs.

2B. On XVS1100A models, perform the following:

a. Remove the license plate from the bracket.

b. Disconnect the black and blue taillight bullet connectors inside the housing on the license plate bracket (**Figure 74**).

c. Connect the voltmeter positive test lead to the harness side of the blue bullet connector; con-

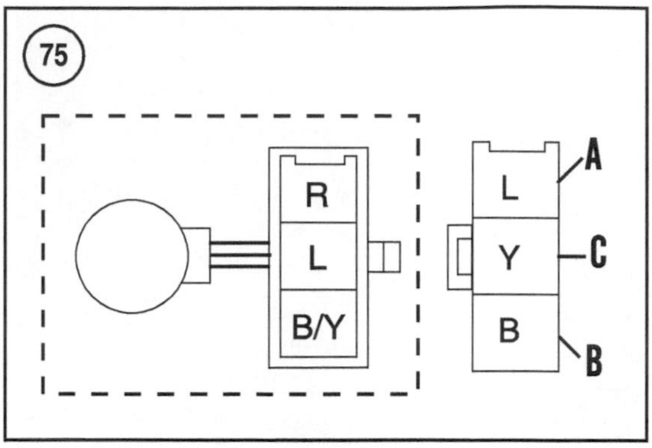

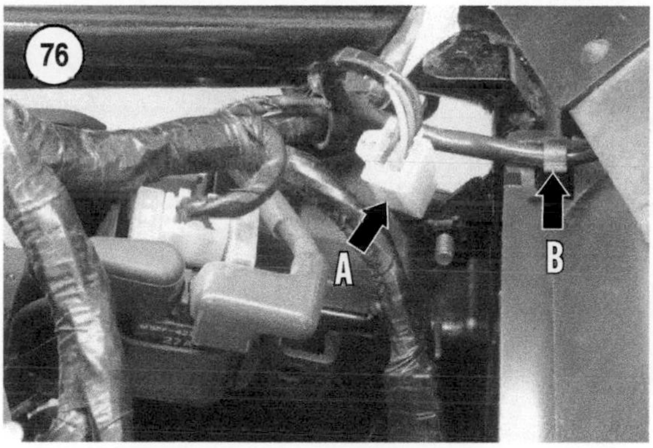

nect the voltmeter negative test lead to the harness side of the black bullet connector.

e. Turn the main switch on. On models with a light switch, turn it on also.
f. If the voltmeter does not read battery voltage, the wiring between the main switch and the taillight/brake light connector is faulty. Make the necessary repairs.

Brake Light Test

1. Set a voltmeter to the 20 DC volt range.

2A. On XVS1100 models, perform the following:
   a. Remove the nut and washer from each housing bolt.
   b. Pull the housing from the taillight bracket.
   c. Disconnect the taillight harness connector and remove the taillight bracket.
   d. Connect the voltmeter positive test lead to the yellow terminal (C, **Figure 75**) in the harness side of the taillight/brake light connector; connect the voltmeter negative test lead to the black terminal (B).
   e. Turn the main switch on. On models with a light switch, turn it on also.
   f. Apply the brake lever or pedal.
   g. If the voltmeter does not read battery voltage, the wiring between the main switch and the taillight/brake light connector is faulty. Make the necessary repairs.

2B. On XVS1100A models, remove the taillight housing as follows:
   a. Remove the license plate from the bracket.
   b. Disconnect the black and yellow taillight bullet connectors inside the housing on the license plate bracket (**Figure 74**).
   c. Connect the voltmeter positive test lead to the harness side of the yellow bullet connector; connect the voltmeter negative test lead to the harness side of the black bullet connector.
   e. Turn the main switch on. On models with a light switch, turn it on also.
   f. Apply the brake lever or pedal.
   g. If the voltmeter does not read battery voltage, the wiring between the main switch and the taillight/brake light connector is faulty. Make the necessary repairs.

Taillight Sub-Harness Connector

The taillight sub-harness connector is beneath the ignition plate. Perform the following to access this connector during taillight/brake light troubleshooting.

1. Remove the seats and ignitor plate as described in Chapter Fourteen.
2. Disconnect the five-pin taillight sub-harness connector (A, **Figure 76**) from its harness mate.
3. If necessary, release the wire from the clamp (B, **Figure 76**) on the mud guard.

Auxiliary Light Test (Europe and Australia Models)

1. Check the continuity of the auxiliary light bulb and socket.
2. Set a voltmeter to the 20 DC volt range.
3. Back probe the auxiliary light bullet connectors. Connect the positive test probe to the harness side of the blue/red connector; connect the negative test probe to the harness side of the black connector.
4. Turn on the main switch and the light switch. The voltmeter should read battery voltage.

5. If the reading is less than battery voltage, the wiring between the main switch and the auxiliary light connectors is faulty and must be repaired.

SIGNAL SYSTEM

The signal system includes the horn, turn signal lights, brake light and indicator lights (except the high beam indicator, which is part of the lighting system). In the event of trouble with any of these lights, check the affected bulb/component first. Replacement bulbs are listed in **Table 2**. If the bulb is good, follow the signal system troubleshooting procedures listed in Chapter Two.

HORN

Removal/Installation

The horn sits on the front of the left frame member just below the frame neck.

1. Remove the horn mounting bolt (**Figure 77**) and turn the horn over.
2. Disconnect the connectors (**Figure 78**) from the two spade terminals on the back of the horn and remove the horn.
3. Installation is the reverse of removal.

Circuit Test

Perform the following test if the horn does not sound.

1. Check the continuity of the horn switch as described in this chapter. Replace the left handlebar switch if the horn switch is faulty.
2. Check the voltage on the battery side of the horn circuit as follows:
   a. Set a voltmeter to the 20 DC volt range.
   b. Back probe the connector. Connect the voltmeter positive test lead to the brown horn terminal and connect the negative test lead to a good frame ground. See **Figure 79**.
   c. Turn the main switch to on and check the voltage on the meter. It should read battery voltage.
   d. If the reading is less than battery voltage, the wiring between the main switch and the horn is faulty.
3. Check the voltage on the ground side of the horn circuit as follows:

77

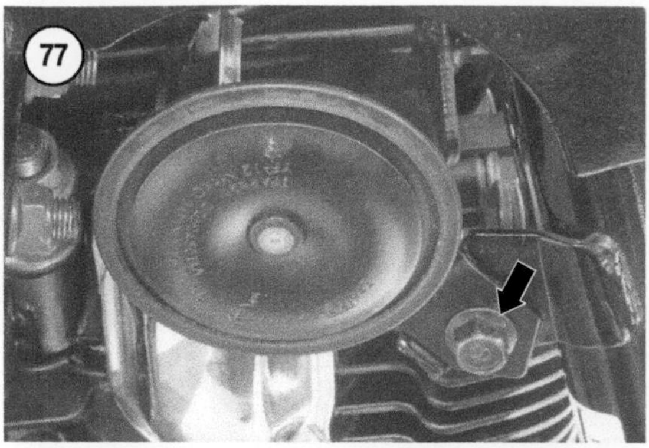

78

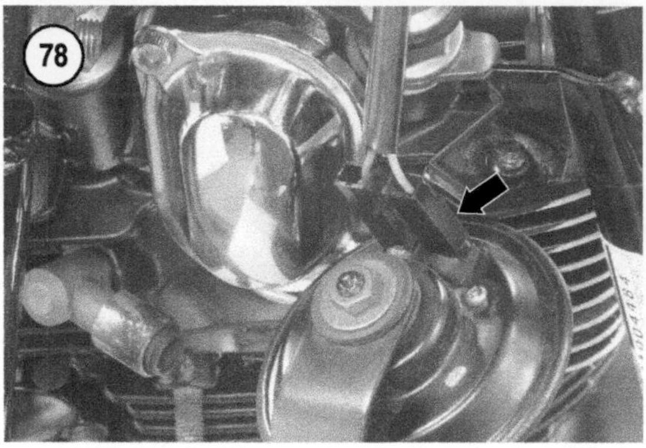

79

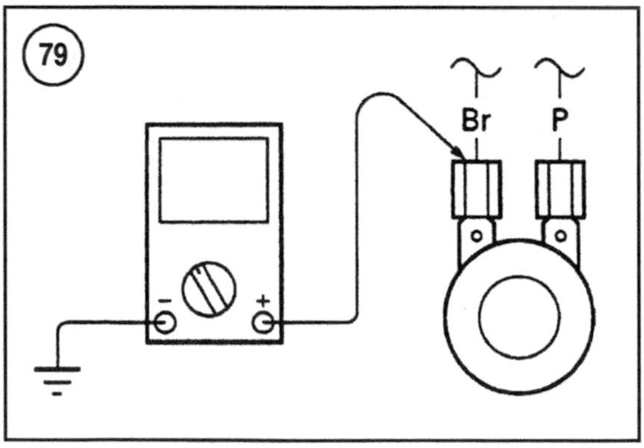

80

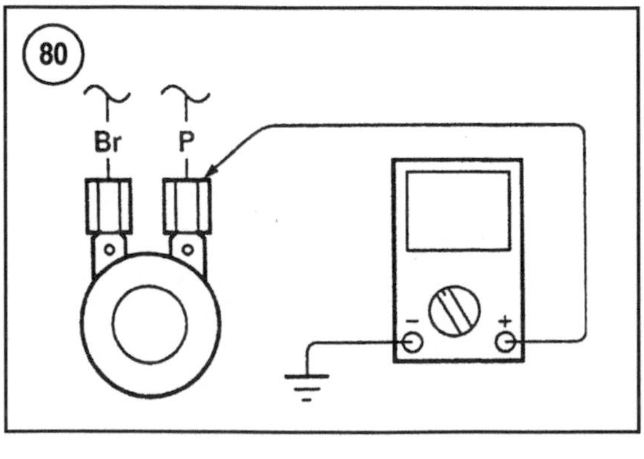

81

TURN SIGNALS (XVS1100 MODELS)

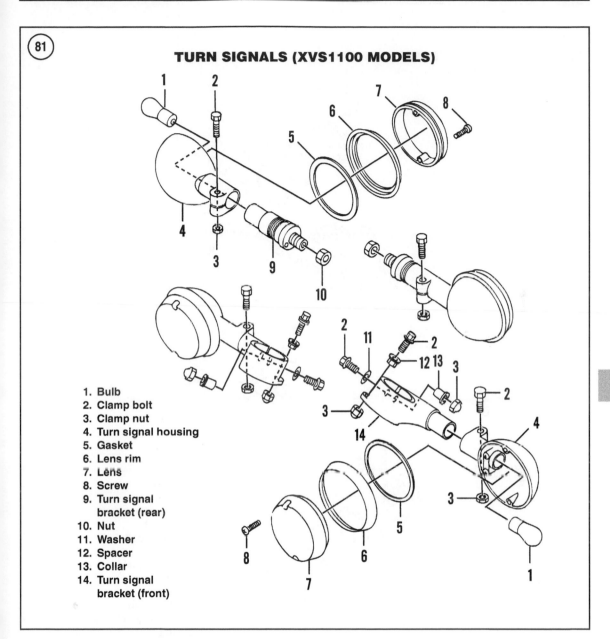

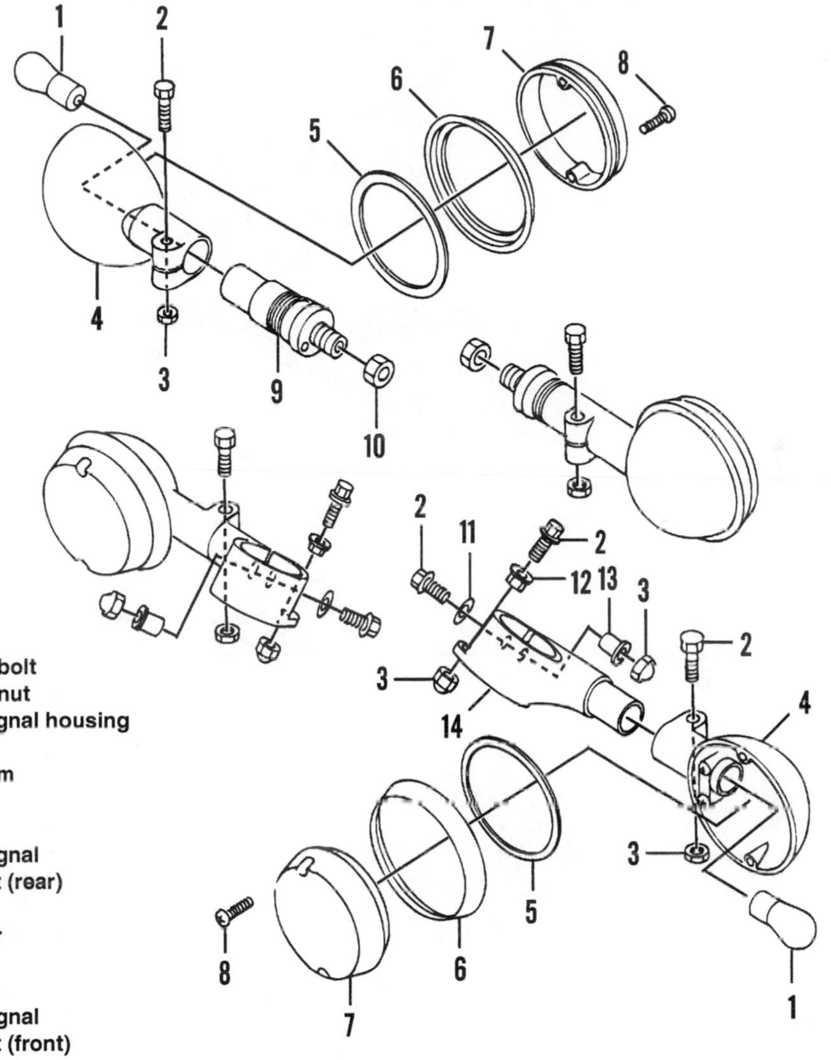

1. Bulb
2. Clamp bolt
3. Clamp nut
4. Turn signal housing
5. Gasket
6. Lens rim
7. Lens
8. Screw
9. Turn signal bracket (rear)
10. Nut
11. Washer
12. Spacer
13. Collar
14. Turn signal bracket (front)

a. Set a voltmeter to the DC 20 volt range.

b. Back probe the connector. Connect the voltmeter positive test lead to the pink horn terminal and connect the negative test lead to a good frame ground. See **Figure 80**.

c. Turn the main switch on and check the voltage on the meter. It should read battery voltage.

d. If the voltage is less than battery voltage, replace the horn.

TURN SIGNALS

Refer to **Figure 81** and **Figure 82**.

Turn Signal Bulb Replacement

1. Remove the two screws (A, **Figure 83**) and the lens ring from the housing.

2. Wash the lens (inside and outside) with a mild detergent.

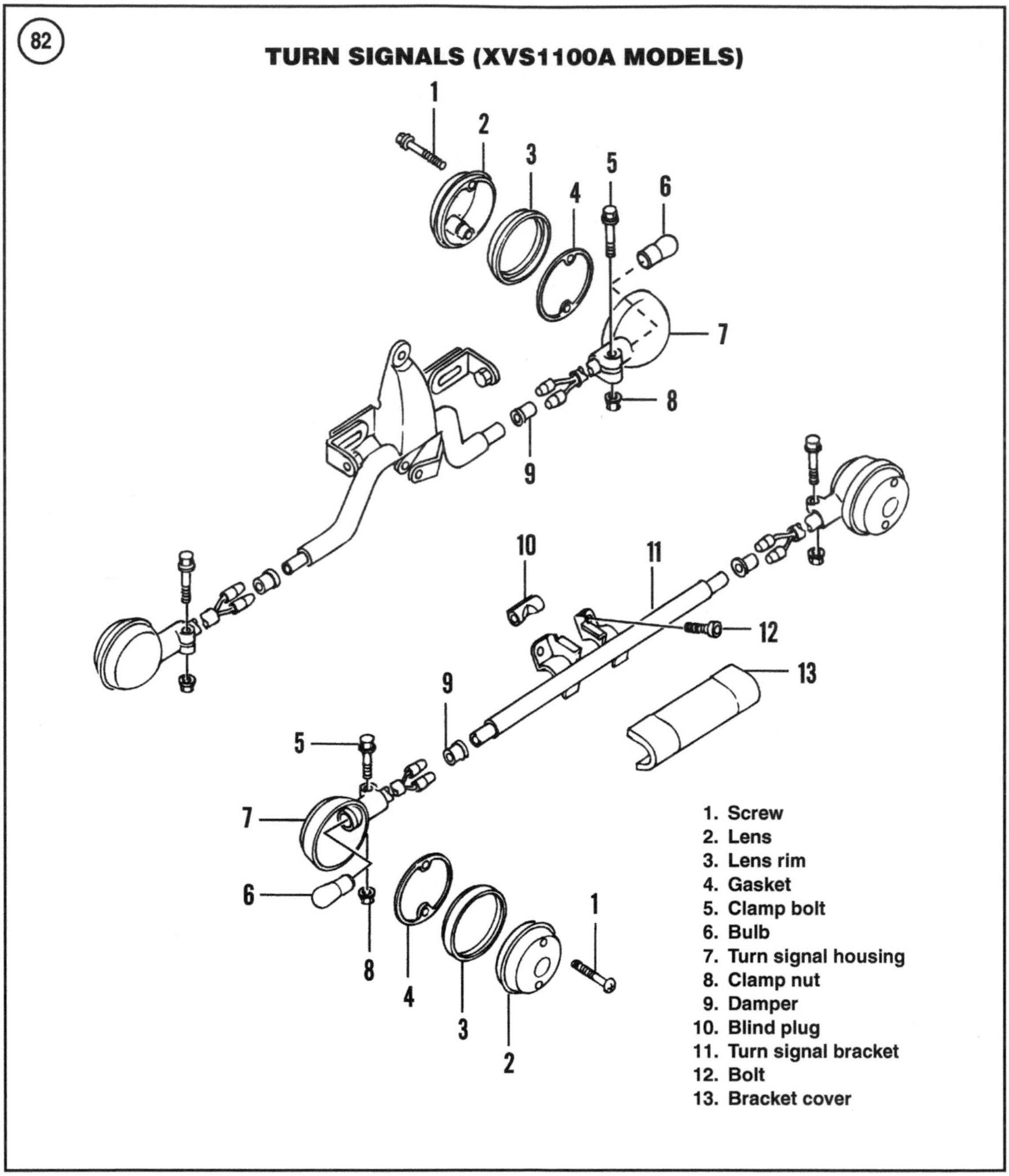

3. Turn the bulb (A, **Figure 84**) counterclockwise and remove it. Install the new bulb and turn it clockwise to lock it in place.

4. Make sure the gasket (B, **Figure 84**) is in place on the housing.

5. Install the lens ring and lens. Do not overtighten the lens screws (A, **Figure 83**), as this will crack the lens.

Turn Signal Assembly Removal/Installation

Front turn signal

1. Remove the mounting screw (**Figure 53**) on either side of the headlight housing and lower the lens assembly from the housing.

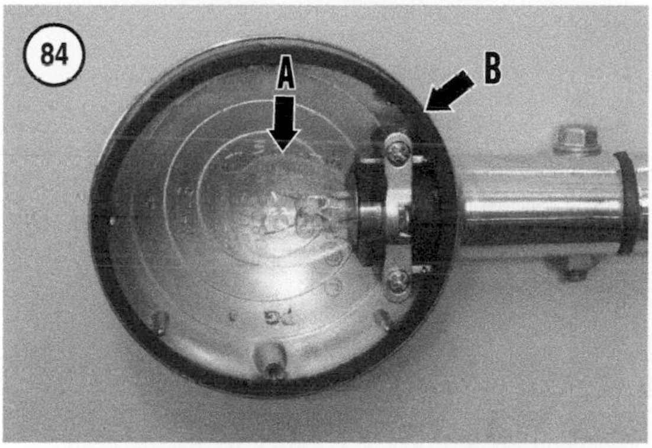

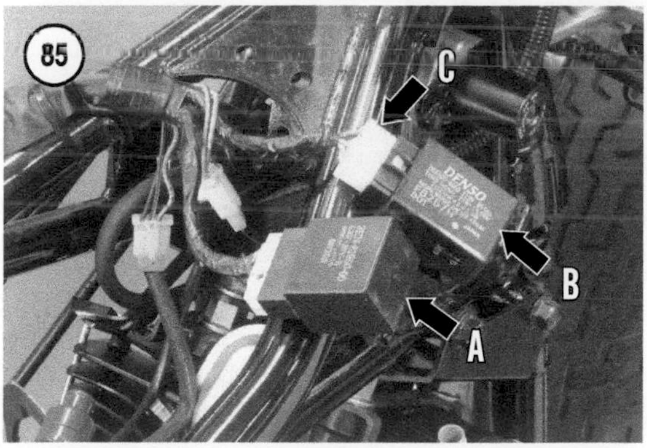

2. Disconnect the headlight connector (A, **Figure 54**) from the bulb and remove the lens assembly.

3. Disconnect the front turn signal bullet connectors (B, **Figure 54**) for the affected turn signal.

4. Feed the turn signal wires out through the hole in the headlight housing.

5. Remove the nut from the clamp bolt (C, **Figure 83**, typical), and pull the turn signal assembly and its wires from the mounting stalk.

6. Installation is the reverse of removal.

   a. On XVS1100A models, make sure the damper is in place on the mounting stalk.
   b. Apply a dielectric compound to the electrical connectors prior to reconnecting them.
   c. Make sure the electrical connectors are free of corrosion and are completely coupled to each other.

Rear turn signal

1. Locate and disconnect the bullet connectors for the affected turn signal assembly. On XVS1100A models, these are located in the housing behind the license plate (**Figure 74**).
2. Loosen and remove the nut from the turn signal clamp bolt (C, **Figure 83**, typical).
3. Pull the turn signal assembly and its wiring from the mounting stalk. Watch for the damper.
4. Installation is the reverse of removal. Note the following:
   a. On XVS1100A models, make sure the damper is in place.
   b. Apply a dielectric compound to the electrical connectors prior to reconnecting them. This will help seal out moisture.
   c. Make sure the electrical connectors are free of corrosion and are completely coupled to each other.

Turn Signal Flash Test

Perform the following check if a turn signal light or the turn signal indicator light does not flash.

1. Remove the flasher relay as described later in this chapter.

NOTE

On USA, California and Canada models, the flasher relay connector is a five-pin connector. It is a three-pin connector on Europe and Australia models.

2. Check the input voltage into the relay as follows:
   a. Set a voltmeter to the DV 20 volt range.
   b. Disconnect the five-pin connector (C, **Figure 85**) from the relay.
   c. Connect the voltmeter positive test lead to the brown terminal in the harness side of the connector (A, **Figure 86**). Connect the voltmeter negative test lead to a good ground.

d. Turn the main switch on and check the voltage on the meter. If it does not read battery voltage, the wiring between the main switch and the flasher relay is faulty. Repair the wiring if input voltage is less than battery voltage.

e. Reconnect the connector to the relay.

3. Check the output voltage from the relay as follows:

a. Set a voltmeter to the 20 DV volt range.

b. Back probe the flasher relay and connect the voltmeter positive test lead to the relay's brown/white terminal (B, **Figure 86**). Connect the voltmeter negative test lead to a good ground.

c. Turn the main switch on, and turn the flasher switch on (either left or right).

d. Check the voltage on the meter. It should read battery voltage.

e. Replace the relay if output voltage is less than battery voltage.

f. Seal the wiring with silicon sealant.

5A. If a turn signal is not flashing, check the wiring for the affected turn signal light as follows:

a. Set a voltmeter to the 20 DC volt range.

b. Locate the bullet connector for the affected flasher, and disconnect the connector.

c. Connect the voltmeter positive test lead to the harness side of the bullet connector; connect the voltmeter negative test lead to a good ground.

d. Turn the main switch on, turn on the turn signal switch for that side, and check the voltage on the voltmeter. It should read battery voltage.

e. The wiring between the turn signal switch and the connector is faulty if voltage is less than battery voltage.

5B. If the turn signal indicator lamp is not flashing, check the indicator lamp wiring as follows:

a. Remove the fuel tank and disconnect the meter assembly connector (**Figure 87**) from its harness mate.

b. Set the voltmeter to the 20 DC volt range.

c. Connect the voltmeter positive test lead to the affected terminal in the harness side of the connector. Connect the voltmeter negative test lead to a good frame ground.

86

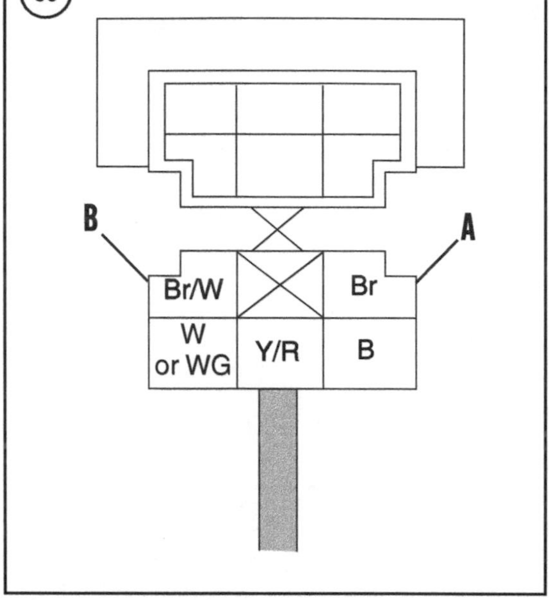

87

d. Turn the main switch on, turn on the turn signal switch for that side, and check the voltage on the meter. It should read battery voltage.

e. If the reading is less than battery voltage, the wiring between the turn signal switch and the connector is faulty.

Flasher Relay Removal/Installation

The flasher relay sits directly in front of the starting circuit cutoff relay (**Figure 88**) in the frame neck. Removing the SCCR provides access the flasher relay.

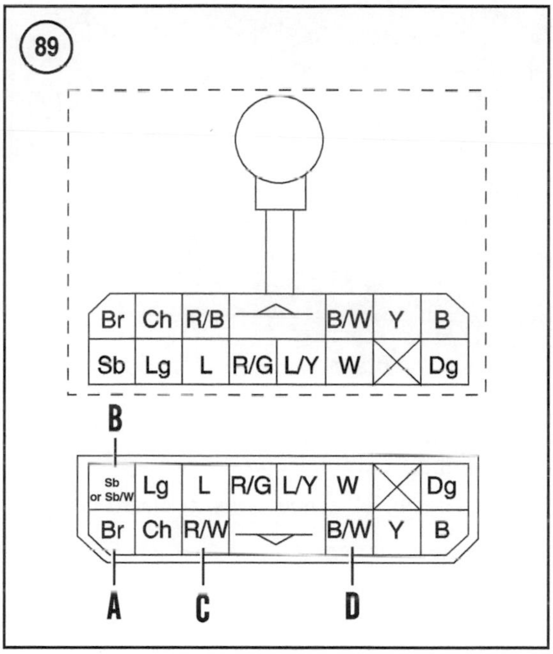

NOTE
On USA, California and Canada models, the flasher relay connector is a five-pin connector. A three-pin connector is used on Europe and Australia models.

1. Remove the fuel tank, air filter housing and surge tank as described in Chapter Eight.
2. Remove the frame neck covers (Chapter Fourteen).
3. Remove the starting circuit cutoff relay (A, **Figure 85**) from its mounting tang, then remove the flasher relay (B).
4. Disconnect the connector (C, **Figure 85**) from the flasher relay.
5. Installation is the reverse of removal.

NEUTRAL INDICATOR LIGHT

Circuit Test

1. Remove the fuel tank (Chapter Eight) and disconnect the meter assembly connector (**Figure 87**) from its harness mate.
2. Check the voltage in the circuit as follows:
   a. Set a voltmeter to the 20 DC volt range.
   b. Connect the voltmeter positive test lead to the brown terminal in the harness side of the meter assembly connector (A, **Figure 89**). Connect the voltmeter negative test lead to the connector's sky blue (or sky blue/white) terminal (B, **Figure 89**).
   c. Shift the transmission into neutral, turn the main switch to on, and check the voltage on the meter. It should read battery voltage.
   d. If the reading is less than battery voltage, the wiring between the main switch and the meter connector is faulty.

OIL LEVEL INDICATOR LIGHT

Indicator Test

1. Remove the oil level relay (A, **Figure 90**) as described in this chapter and disconnect the four-pin connector (B) from the relay.
2. Check the continuity of the oil level relay as follows:
   a. Use jumper wires to connect the battery positive terminal to the red/white terminal in the oil level relay (A, **Figure 91**). Connect the battery negative terminal to the relay's black/red terminal (B).
   b. Connect the ohmmeter's positive test probe to the black/white terminal in the oil level relay

(C, **Figure 91**). Connect the negative test probe to the black terminal (D).

c. If the relay does not have continuity, replace the oil lamp level relay.

3. Check the oil level indicator wiring as follows:

a. Remove the fuel tank (Chapter Eight) and disconnect the meter assembly connector (**Figure 87**) from its harness mate.

b. Set a voltmeter to the 20 DC volt scale.

c. Connect the voltmeter positive test lead to the red/white terminal on the harness side of the 14-pin meter connector (C, **Figure 89**); connect the negative test lead to the black/white terminal (D).

d. Turn the main switch on. The meter should indicate battery voltage. If the voltage is less than battery voltage, the wiring between the main switch and the meter connector is faulty. Make the necessary repairs.

Oil Level Relay Removal/Installation

1. Remove the battery cover as described in Chapter Fourteen.
2. Lift the oil level relay (A, **Figure 90**) from the mounting tang on the battery box.
3. Disconnect the four-pin connector (B, **Figure 90**) from the relay and remove the relay.
4. Installation is the reverse of removal.
   a. Pack the connector with dielectric grease.
   b. Make sure the connectors are clean and tight.

SPEED SENSOR

Removal/Installation

1. Remove the toolbox panel as described in Chapter Fourteen. Remove the side cover from the left side as described in Chapter Fourteen.
2. Roll the boot back from the electrical connector, and disconnect the three-pin speed sensor connector (A, **Figure 92**) from its harness mate.
3. Release any cable ties or holders that secure the speed sensor wire to the frame. Note the location of these ties and note how the speed sensor wire is routed along the motorcycle.
4. Remove the speed sensor screw and lift the sensor (**Figure 93**) from the middle drive housing.
5. Installation is the reverse of removal. Note the following:

91

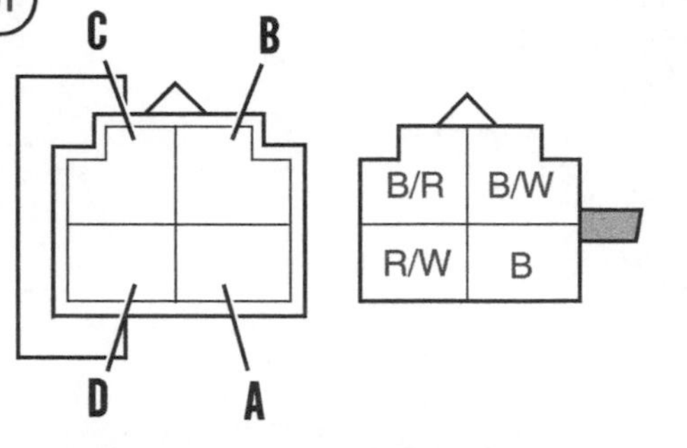

92

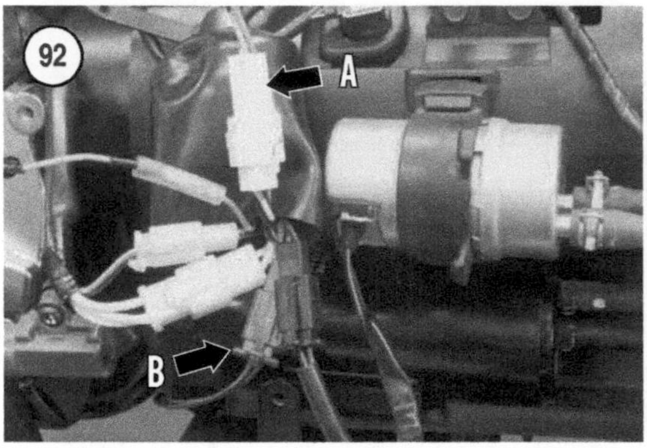

a. Torque the speed sensor screw to 7 N•m (62 in.-lb.).

b. Make sure all connections are clean and tight.

Test

Perform the following test whenever the self-diagnostic system flashes a speed sensor trouble code.

1. Securely support the motorcycle on a level surface with the rear wheel off the ground.
2. Check the wiring in the speed sensor circuit. All connections and grounds must be clean and tight.
3. Remove the toolbox cover and left side cover as described in Chapter Fourteen.
4. Roll the boot back from the electrical connector and locate the three-pin speed sensor connector (A, **Figure 92**).
5. Set a voltmeter to the 20 DC volt scale.
6. Back probe the connector and connect the voltmeter positive test lead to the white terminal (**Figure 94**) in the harness side of the connector. Connect the negative test lead to a good ground.

93

94

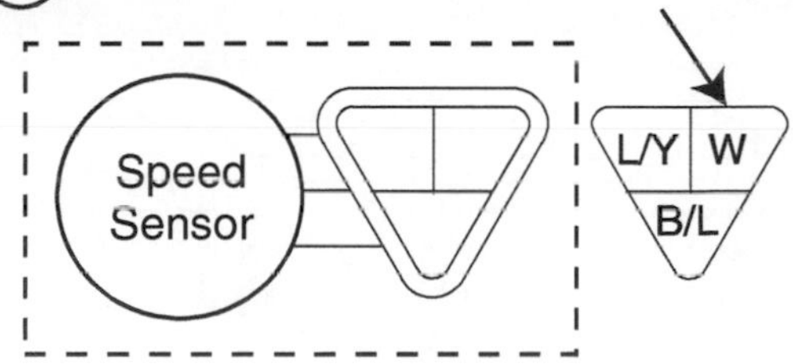

95

HORN SWITCH

Button position	Wire color	
	P	B
Push	●———	———●
Off		

7. Turn the main switch on and slowly rotate the rear wheel.
8. The speed sensor is operating properly if the voltage is 0-5 volts.
   a. If voltage is outside the specified range, replace the speed sensor.
   b. If the voltage is between zero and 5 volts, replace the ignitor unit.

SWITCHES

Switches can be tested for continuity with an ohmmeter or a test light (see Chapter One). The continuity diagrams for various switches are in the wiring diagrams at the back of this manual.

For example, **Figure 95** shows a continuity diagram for a horn button. It shows which terminals should have continuity when the horn switch is in a given position. The line on the continuity diagram indicates that there should be continuity between the black and pink terminals when the horn button is pressed. When the horn button is pressed, an ohmmeter connected to the black and pink terminals should indicate little or no resistance (a test lamp should light).

The horn switch diagram also indicates there should be no continuity between these terminals when the switch is free. When the horn button is released, the ohmmeter should indicate infinite resistance (a test lamp should not light) between the black and pink terminals.

When testing switches, note the following:
1. First check the fuses in the relevant circuit as described in this chapter.
2. Check the battery as described in Chapter Three. Charge the battery to the correct state of charge, if required.
3. Disconnect the negative cable from the battery if the switch connectors are not disconnected from the circuit.

CAUTION
Do not attempt to start the engine with the battery negative cable disconnected. This will damage the wiring harness.

4. When separating two connectors, pull the connector housings and not the wires.
5. After locating a defective circuit, check the connectors to make sure they are clean and properly connected. Check all wires going into a connector housing to make sure each wire is properly positioned and the wire end is not loose.
6. To properly connect connectors, push them together until they click into place.
7. When replacing a handlebar switch assembly, route the cables correctly so they will not be crimped when the handlebar is turned from side to side.

Left Handlebar Switch Replacement

1. Remove the fuel tank as described in Chapter Eight.

2. Remove the frame neck cover as described in Chapter Fourteen.
3. Disconnect the left handlebar switch white six-pin connector (A, **Figure 96**) and the blue six-pin connector (B) from their harness mates.
4. Remove the mounting screw (**Figure 97**) and remove the bottom cover from the handlebar switch.
5. Disconnect the choke cable end (**Figure 98**) from the choke lever.
6. Note how the switch cable is routed along the handlebar and through the fork legs. Release end clamps or cable ties that secure the switch cable to the motorcycle.
7. Remove the screws (**Figure 99**) and separate the housing halves from the handlebar.
8. Remove the switch housing and wiring cable.
9. Installation is the reverse of these steps. Note the following:
   a. Position the switch assembly so the edge of the switch half mating surface (A, **Figure 100**) aligns with the index mark (B) on the handlebar.
   b. Apply a dielectric grease to the electrical connector prior to reconnecting it.
   c. Make sure the electrical connectors are free of corrosion and the two sides are securely connected to each other.
   d. Adjust the choke cable free play as described in Chapter Three.

Right Handlebar Switch Replacement

1. Remove the fuel tank as described in Chapter Eight.
2. Remove the frame neck cover as described in Chapter Fourteen.
3. Disconnect the right handlebar switch black six-pin connector (C, **Figure 96**) from its harness mate.
4. Disconnect the two spade connectors (A, **Figure 101**) from the front brake switch.
5. Remove the cable ties that secure the switch cable to the motorcycle. Also note how the cable is routed. It must be rerouted along the same path.
6. Remove the screw that secures the pull cable to the switch housing.
7. Remove the two switch assembly screws (**Figure 102**) and separate the halves of the switch assembly from the handlebar.

96

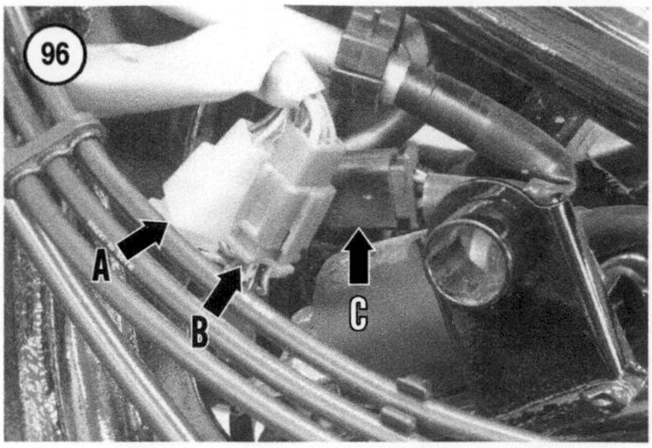

97

98

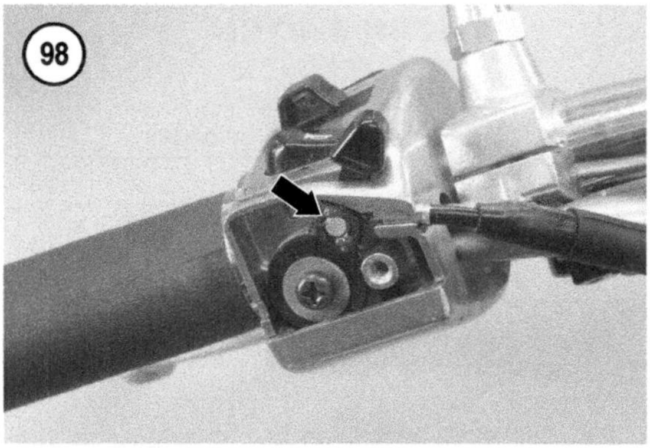

99

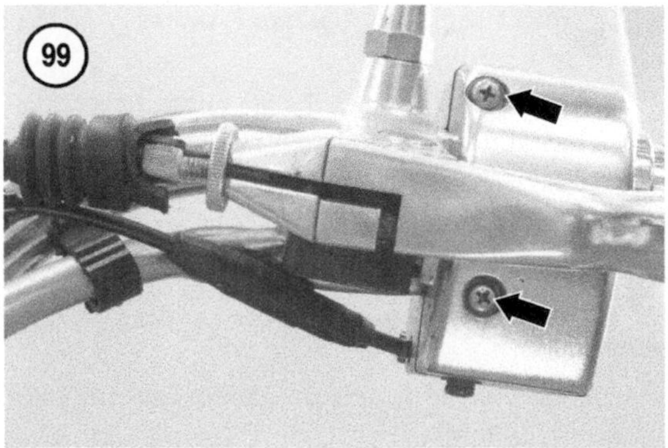

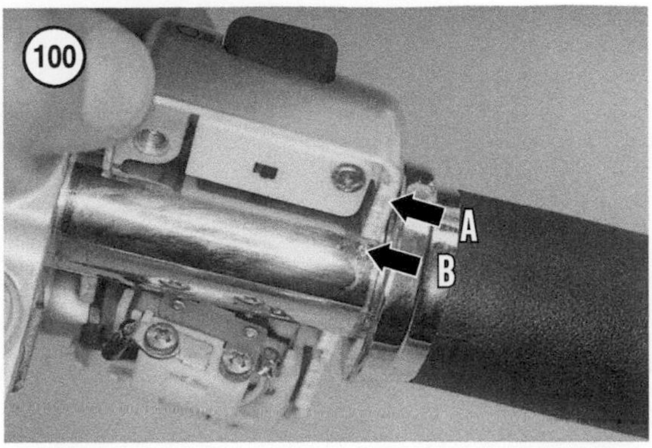

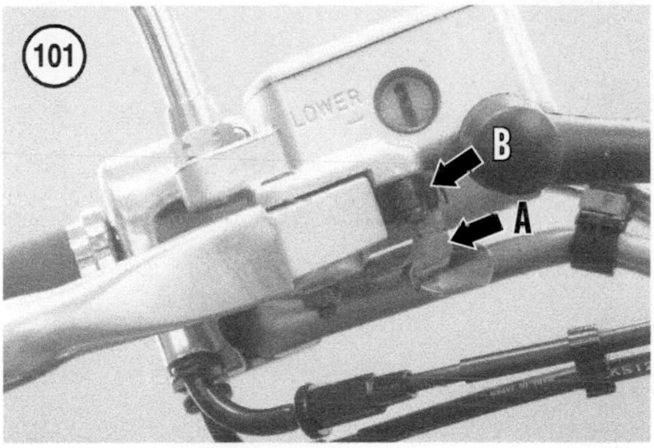

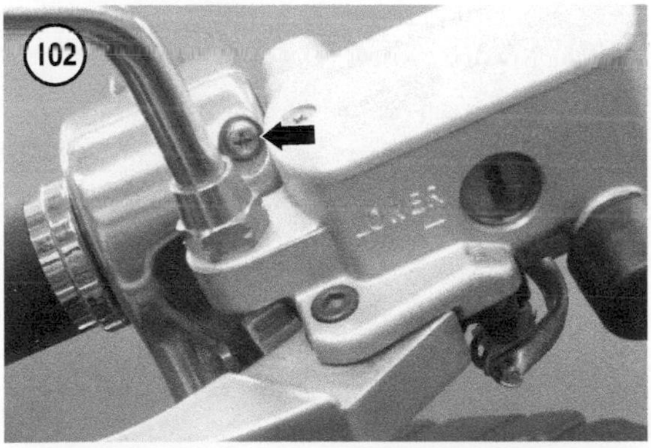

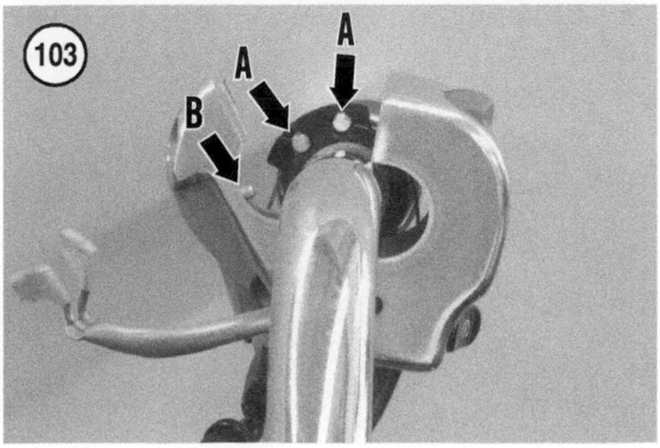

8. Disconnect the ends of the throttle cables from the throttle drum (A, **Figure 103**), and remove the pull and return cables from the switch housing.

9. Installation is the reverse of removal. Note the following:

   a. Apply lithium soap grease to the ends of the pull and return cables.
   b. The pin on the switch housing (B, **Figure 103**) must engage the hole in the handlebar.
   c. Apply a dielectric grease to the electrical connectors prior to reconnecting them.
   d. Make sure the electrical connectors are free of corrosion and the two sides are securely connected to each other.
   e. Use cable ties to secure the switch cable to the places noted during removal.
   f. Securely tighten the pull cable mounting screw.
   g. Adjust the throttle cable free play as described in Chapter Three.

Main Switch Replacement

1. Remove the fuel tank and surge tank as described in Chapter Eight.

2. Remove the frame neck covers as described in Chapter Fourteen.

3. Disconnect the switch's red, four-pin connector (A, **Figure 104**) from its mate on the wiring harness. On Europe models, disconnect the two-pin and four-pin connectors from their harness mates.

4. Remove the plug (B, **Figure 104**) from each mounting bolt and remove the two main switch mounting bolts. Remove the switch from the frame.

5. Installation is the reverse of these steps. Note the following:

a. Apply a dielectric compound to the electrical connector prior to reconnecting it.
b. Make sure the electrical connector is free of corrosion and both sides are completely coupled to each other.
c. Install all items removed.
d. When installing shear bolts, tighten the bolt until its head twists off.

Neutral Switch Replacement

The neutral switch sits in the bottom of the left crankcase half.

1. Drain the engine oil as described in Chapter Three.
2. Disconnect the electrical lead from the neutral switch (**Figure 105**).
3. Turn out and remove the neutral switch and its washer from the crankcase (**Figure 106**).
4. Installation is the reverse of removal. Note the following:
   a. Apply a medium strength threadlocking compound to the threads of the switch and torque the neutral switch to 20 N•m (15 ft.-lb.).
   b. Make sure the electrical connector is free of corrosion and is tight.

Sidestand Switch Replacement

1. Remove the toolbox cover and the left side cover as described in Chapter Fourteen.
2. Roll back the boot and disconnect the two-pin sidestand switch connector (B, **Figure 92**) from its harness mate.
3. Follow the sidestand switch electrical wire and disconnect any holders or cable ties that secure it in place. Note how the wire is routed along the frame. The new switch wire must be rerouted along the same path.
4. Remove the mounting screws (A, **Figure 107**) and remove the sidestand switch (B) from the sidestand bracket.
5. Installation is the reverse of these steps. Note the following:
   a. Apply dielectric grease to the electrical connector prior to reconnecting it.
   b. Make sure the electrical connector is free of corrosion and both sides are completely coupled to each other.

105

106

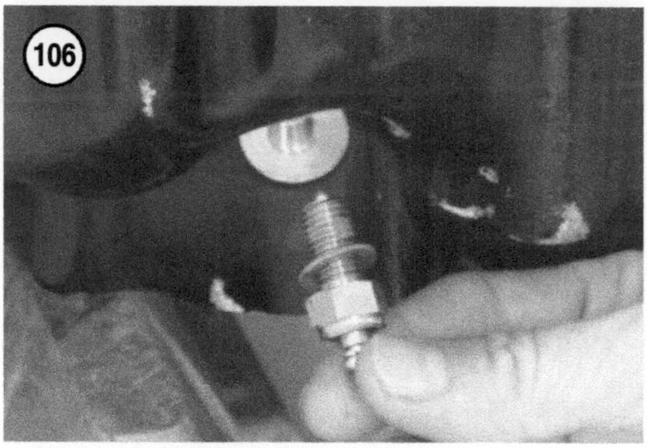

107

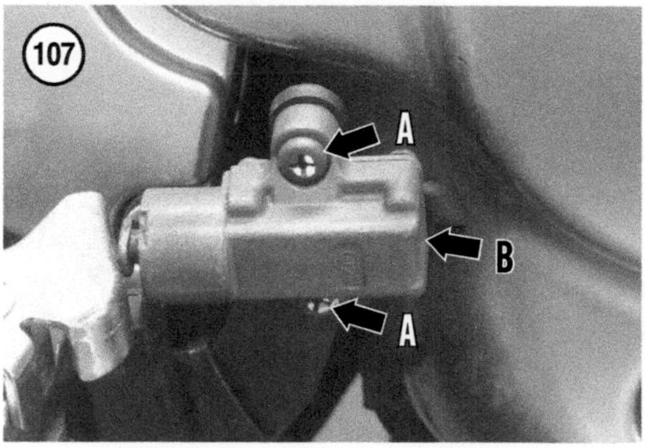

   c. Make sure the sidestand switch electrical wire is routed along the same path noted during removal.
   d. Secure the switch wire to the same points noted during removal.

Clutch Switch Replacement

1. Disconnect the electrical connector (A, **Figure 108**) from the clutch switch.

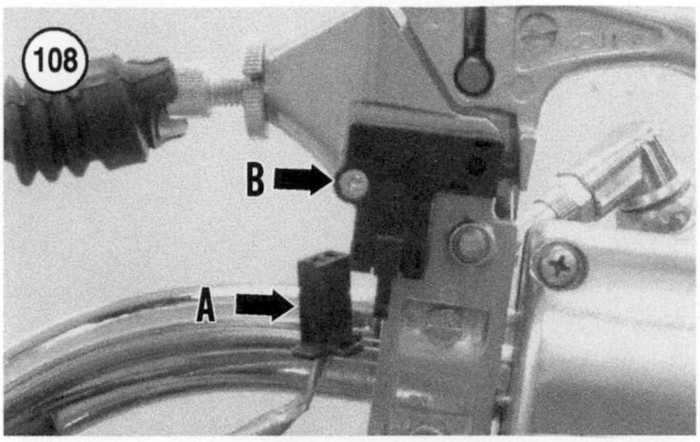

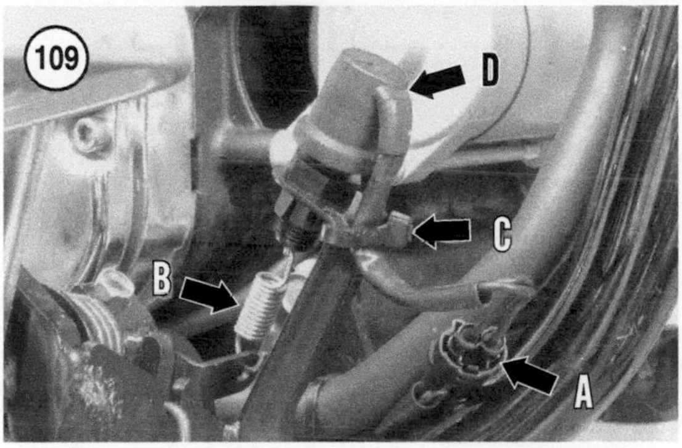

2. Remove the switch screw (B, **Figure 108**) and remove the switch from beneath the clutch lever.
3. Installation is the reverse of removal.

Front Brake Switch Replacement

The front brake light switch is mounted beneath the front brake master cylinder.

1. Disconnect the spade connectors (A, **Figure 101**) from the front brake switch.
2. Remove the switch mounting screw (B, **Figure 101**) and remove the switch.
3. Installation is the reverse of removal. Check switch operation. The rear brake light should come on when the front brake lever is applied.

Rear Brake Switch Replacement

The rear brake switch is mounted on the brake pedal/footrest bracket.

1. Disconnect the two-pin rear brake switch connector (A, **Figure 109**).
2. Disconnect the spring (B, **Figure 109**) from the boss on the brake pedal.
3. Release the cable tie (C, **Figure 109**) and remove the switch (D) from its mount.
4. Installation is the reverse of removal.
   a. Pack the connector with dielectric grease.
   b. Secure the electrical cable with a new cable tie (C, **Figure 109**).
5. Adjust the rear brake switch as described in Chapter Three.

SELF-DIAGNOSTIC SYSTEM

All models are equipped with a self-diagnostic system that checks the motorcycle electrical system before the engine is started. Immediately after turning on the main switch, the system is activated and turns on the engine indicator light. After 1.4 seconds, the light will turn off if no malfunctions are found in the system. If a malfunction is found, the light will go off after 1.4 seconds, then will flash a trouble code. If the engine is started when a malfunction is detected, the engine indicator light will remain on.

On early models, the self-diagnostic system checks only the throttle position sensor. On later models, the system checks the throttle position sensor, speed sensor, ignition coils and fuel cut solenoid valves.

On early models, if a malfunction is detected with the throttle position sensor, the engine indicator light will flash three times, pause, then repeat the trouble code.

On later models, if a malfunction is detected, there will be two digits in the trouble code. Count the number of long flashes to determine the first digit. After the light pauses, count the number of short flashes to determine the second digit.

Refer to **Table 5** for the trouble codes, detected conditions and fail-safe actions. The fail-safe actions allow the motorcycle to be started so it can be ridden to a service location, if necessary. The tests for the speed sensor and ignition coils are in this chapter. The tests for the throttle position sensor and fuel cut solenoid valves are in Chapter Eight.

FUSES

Whenever a fuse blows, determine the reason for the failure before replacing the fuse. Usually, the trouble is a short in the wiring. This may be caused by worn through insulation or a disconnected wire shorting to ground.

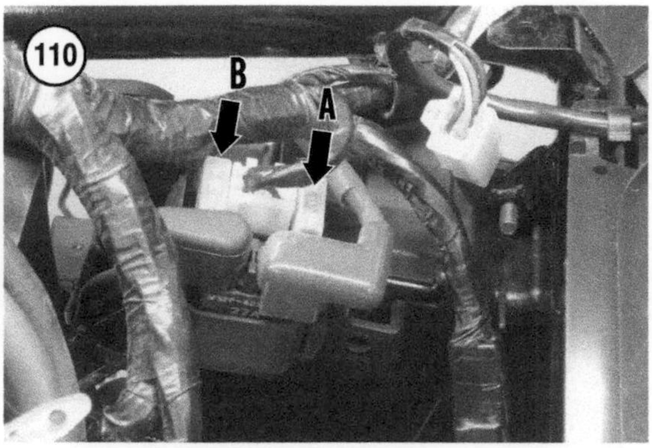

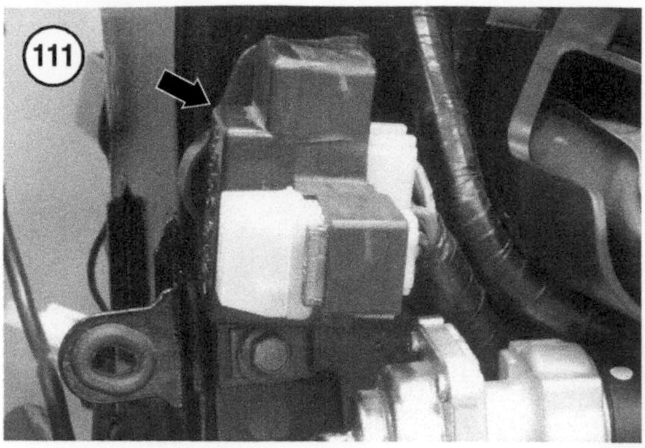

CAUTION
Never substitute metal foil or wire for a fuse. Never use a higher amperage fuse than specified. An overload could result in fire and complete loss of the bike.

Main Fuse Removal

The 30-amp main fuse (A, **Figure 110**) and a spare (B) are located on the starter relay. Access the main fuse as follows:

1. Remove the seats and ignitor panel as described in Chapter Fourteen.
2. Remove the fuse by pulling it out of the holder with needlenose pliers.

Fuse Removal

Fuses other than the main fuse are located in the fuse box (**Figure 111**), which sits on the toolbox panel. The fuses and their amperage ratings are listed in **Table 3**.

1. Remove the toolbox cover as described in Chapter Fourteen.
2. Remove the fuse box cover and open the box.
3. Remove the suspected fuse by pulling it out of the holder with needlenose pliers.

Fuse Testing

1. Remove the suspected fuse as described in this chapter.

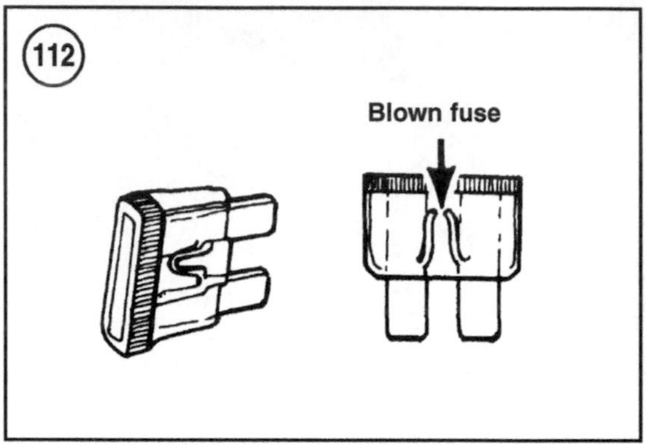

2. Visually inspect the fuse (**Figure 112**). Replace it if it is blown or cracked.
3. If necessary, check the continuity across the two spade connectors. Replace a fuse that does not have continuity.
4. A replacement fuse must have the same amperage rating as the original.

WIRING CONNECTORS

Many electrical troubles can be traced to damaged wiring or connectors that are contaminated with dirt and oil. Connectors can be serviced by disconnecting them and cleaning them with electrical contact cleaner. Multiple pin connectors should be packed with a dielectric compound (available at most automotive and motorcycle supply stores).

Table 1 ELECTRICAL SYSTEM SPECIFICATIONS

Cylinder numbering	Rear cylinder is No. 1
Battery	
Model	GT14B-4
Capacity	12 V 12 AH
Open circuit voltage	12.8 V or greater @ 20° C (68° F)
Charging voltage (output voltage)	14 volts, 340 watts @ 5000 rpm
Stator coil resistance	0.36-44 ohms @ 20° C (68° F)
Voltage regulator no-load output	14.1-14.9 volts
Rectifier	
Capacity	
1999-2003	18 A
2004-on	22 A
Withstand voltage	200 V
Pickup coil resistance	189-231 ohms @ 20° C (68° F)
Ignition minimum spark gap (air gap)	6 mm (0.24 in.)
Ignition coil	
Primary coil resistance	3.57-4.83 ohms @ 20° C (68° F)
Secondary coil resistance	10.7-14.5 k ohms @ 20° C (68° F)
Ignition timing	10° BTDC @ 1000 rpm
Ignition system	TCI
Recommended spark plug	NGK BPR7ES, Denso W22EPR-U
Spark plug gap	0.7-0.8 mm (0.028-0.031 in.)
Spark plug cap resistance	10 k ohms @ 20° C (68° F)
Starter	
Output	0.6 kW
Armature	
Armature coil resistance	0.026-0.034 ohms @ 20° C (68° F)
Armature insulation resistance	Greater than 1 meg ohm @ 20° C (68° F)
Brush length	10-12 mm (0.41-0.49 in.)
Brush length service limit	5 mm (0.20 in.)
Brush spring pressure	7.65-10.01 N (27.51-36.01 oz.)
Commutator diameter	28 mm (1.102 in.)
Commutator diameter service limit	27 mm (1.063 in.)
Mica undercut	0.7 mm (0.028 in.)
Starter relay amperage rating	180 A
Horn maximum amperage	3 A
Flasher relay	
Frequency	75-95 cycles/min
Wattage	21 W

Table 2 BULB SPECIFICATIONS

Item	Voltage/wattage
Headlight (high/low beam)	12 V 60/55 W
Taillight/brake light	12 V 8/27 W
Front turn signal/position light	
1999-2004	12 V 27/8 W
2005-on	12 V 23/8 W
Rear turn signal	
1999-2004	12 V 27 W
2005-on	12 V 21 W
License light	12 V 8 W

(continued)

9

Table 2 BULB SPECIFICATIONS (continued)

Item	Voltage/wattage
Meter light	14 V 1.4 W
Neutral indicator light	12 V 1.7 W
High beam indicator light	12 V 1.7 W
Turn signal indicator light	12 V 1.7 W
Oil level warning light	12 V 1.7 W
Engine warning light	12 V 1.7 W

Table 3 FUSES

Main	30 A
Headlight	15 A
Signal	10 A
Ignitor (2004-on)	5 A
Ignition	10 A
Odometer	5 A
Carburetor heater	15 A

Table 4 ELECTRICAL SYSTEM TORQUE SPECIFICATIONS

Item	N•m	in.-lb.	ft.-lb.
Alternator cover bolts	13	115	–
Flywheel nut	175	–	129
Headlight housing bolts	8	71	–
Ignition coil bolts	2.5	22	–
Neutral switch*	20	–	15
Pickup coil screws*	7	62	–
Stator mounting screws*	10	89	–
Starter mounting bolts	10	89	–
Spark plugs	20	–	15
Speed sensor screw	7	62	–
Starter relay terminal nuts	7	62	–

*Apply a medium-strength threadlocking compound.

Table 5 SELF-DIAGNOSTIC SYSTEM TROUBLE CODES

Code	Item	Condition	Fail-safe action
3 or 15	Throttle position sensor	Disconnected/shorted	Fixes sensor to fully open
3 or 16	Throttle position sensor	Locked/jammed	Fixes sensor to fully open
4 or 42	Speed sensor	Defective sensor pulse	–
33	Ignition coil No. 1	Primary lead shorted	Fuel cut solenoid valve No. 1 on
34	Ignition coil No. 2	Primary lead shorted	Fuel cut solenoid valve No. 2 on
57	Fuel cut solenoid valve No. 1	Disconnected/shorted	–
58	Fuel cut solenoid valve No. 2	Disconnected/shorted	–

CHAPTER TEN

WHEEL AND TIRES

This chapter describes repair and maintenance procedures for the wheels and tires. When inspecting any of the components described in this chapter, compare all measurements to the tire and wheel service specifications in the tables at the end of this chapter. Replace any component that is worn, damaged or out of specification. During assembly, tighten fasteners to the specified torque.

MOTORCYCLE STAND

Many procedures in this chapter require that the motorcycle be supported with a wheel off the ground. A quality motorcycle front end stand (**Figure 1**) or a swing arm stand does this safely and effectively. Before purchasing or using a stand, check the manufacturer's instructions to make sure the stand will work on the motorcycle. If the stand or motorcycle require any modifications or adjustment, perform the required service before lifting the motorcycle.

An adjustable centerstand can also be used to support the motorcycle with a wheel off the ground. Again, check the manufacturer's instructions and perform any necessary modifications before supporting the motorcycle with the adjustable centerstand.

When using a motorcycle stand, have an assistant standing by. Some means to tie down one end of the motorcycle is also needed. Regardless of the method used, make sure the motorcycle is properly supported before walking away from it.

BRAKE ROTOR PROTECTION

Be careful when removing, handling and installing a wheel with a disc brake rotor. Brake rotors are relatively thin in order to dissipate heat and to minimize unsprung weight. A rotor is designed to withstand tremendous rotational loads, but it can be damaged when subjected to side impact loads.

Protect the rotor when servicing a wheel. Never set a wheel down on the brake rotor. It may be bent or scratched. When a wheel must be placed on its side, support the wheel on wooden blocks (**Figure 2**). Position the blocks along the outer circumfer-

1

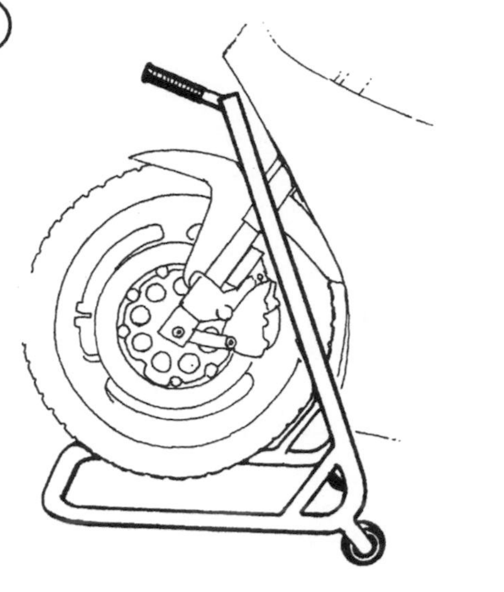

ence of the wheel so the rotor lies between the blocks and does not rest on them.

Also protect the rotor when transporting a wheel to a dealership or tire specialist. Do not place a wheel in a car trunk or truck bed without protecting the rotor from side impact.

If the rotor is knocked out of true by a side impact, a pulsation will be felt in the brake lever or pedal when braking. Since brake rotors are too thin to be trued, damaged rotors must be replaced.

WHEEL INSPECTION

During inspection, compare all measurement to the specification in **Table 1**. Replace any part that is damaged, out of specification or worn to the service limit.

1. Remove the wheel as described in this chapter.
2. Inspect the seals (A, **Figure 3**) for excessive wear, hardness, cracks or other damage. If necessary, replace the seals as described in the front or rear hub section later in this chapter.
3. Inspect the bearings as follows:
   a. Turn each bearing inner race (B, **Figure 3**) by hand. Each bearing must turn smoothly with no trace of roughness, binding or excessive noise. Some axial play (side-to-side) is normal, but radial play (up and down) must be negligible. See **Figure 4**. If either wheel bearing is damaged, replace them both as described later in this chapter.

2

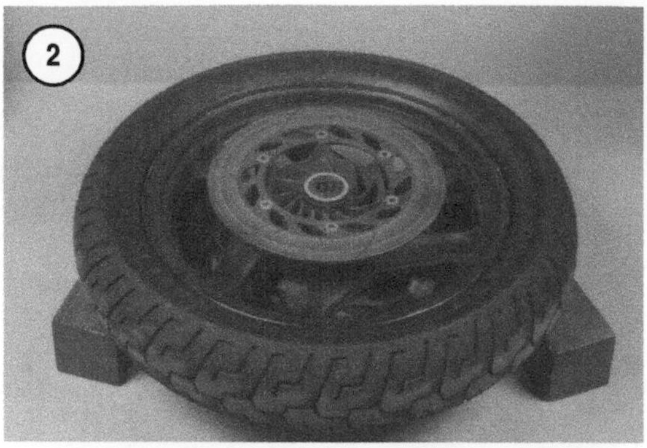

3

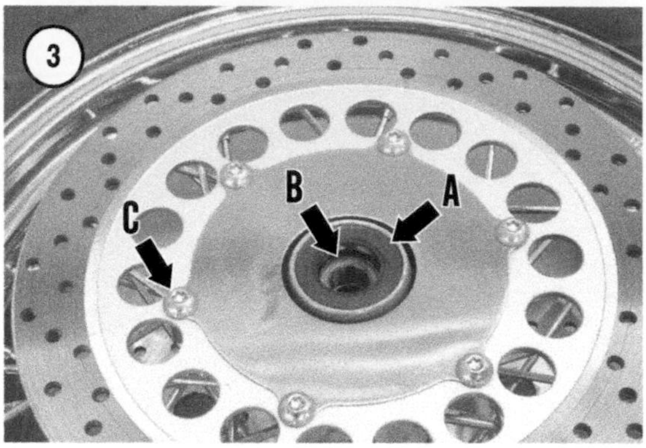

4

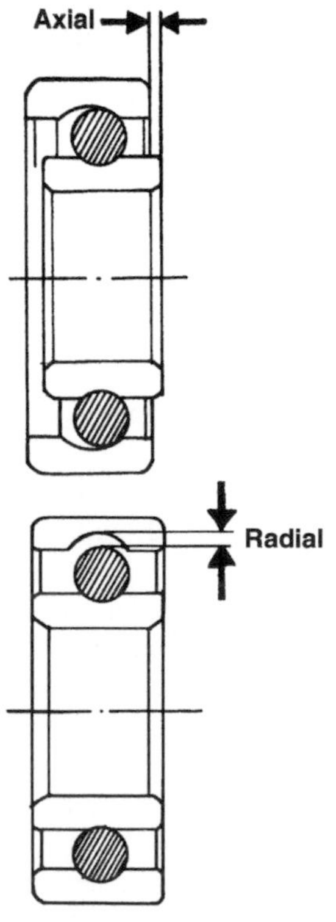

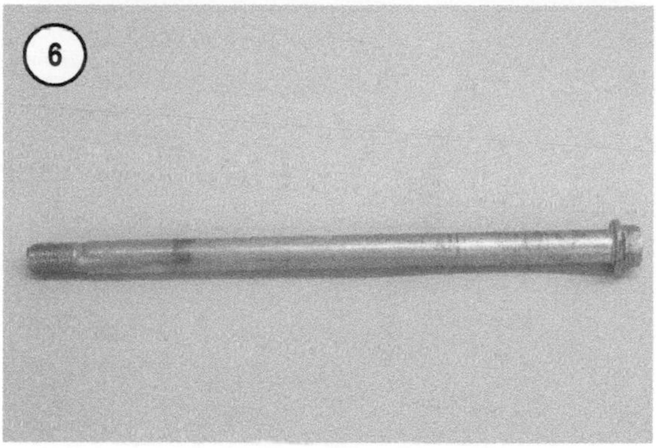

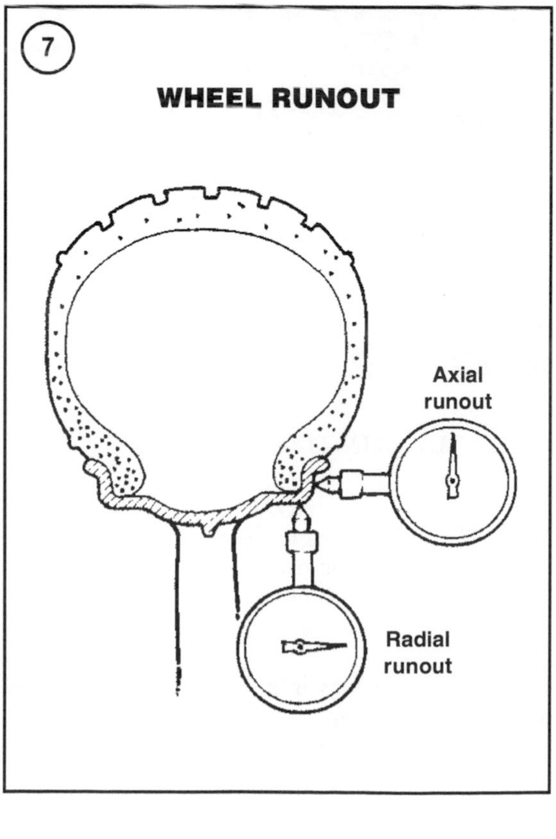

b. Check the bearing's outer seal (A, **Figure 5**) for buckling or other damage that would allow dirt to enter the bearing.

c. Check the bearing fit in the hub by trying to move the bearing laterally with your hand. The bearing should be tight in the bore. Loose bearings allow the wheel to wobble. If a bearing is loose, the bearing bore in the hub is probably worn or damaged.

4. Remove any corrosion from the axle (**Figure 6**) or collars with a piece of fine emery cloth.

WARNING
Do not attempt to straighten a bent axle.

5. Check axle runout by rolling the axle along a surface plate or a piece of glass. If the axle is not straight, replace it.

6. Install the wheel on a truing stand. Check wheel runout as follows:

a. Measure the radial (up and down) runout of the wheel rim with a dial indicator as shown in **Figure 7**.

b. Measure the axial (side to side) runout of the wheel rim with a dial indicator as shown in **Figure 7**.

7. If the wheel runout is out of specification (**Table 1**), inspect the wheel bearings as described earlier in this section.

a. If the wheel bearings are good, the wheel must be replaced.

b. If either wheel bearing is worn, disassemble the hub and replace both bearings as a set.

8. Check the tightness of the brake disc bolts (C, **Figure 3**). If a bolt is loose, remove and reinstall the bolts with medium-strength threadlocking compound. Clean any old threadlocking compound from the threads and torque the bolts to 23 N•m (17 ft.-lb.).

9. Visually inspect the brake discs and measure the brake disc deflection as described in Chapter Thirteen. If deflection is excessive, measure wheel runout. If wheel runout is within specification, replace the brake disc. Refer to the procedure in Chapter Thirteen.

10. Inspect the wheel rim for dents, bending or cracks. Check the rim and rim sealing surface (cast wheels) for scratches that are deeper than 0.5 mm (0.01 in.). If any of these conditions are present, replace the wheel.

11. On laced wheels, check the spoke tension as described in this chapter.

BEARING REMOVAL

NOTE
*The following procedure describes the use of the Kowa Seiki Wheel Bearing Remover shown in **Figure 8**. This set is available from K & L Supply Co., in Santa Clara, CA.*

1. Select the correct size remover head and insert it into the inner race of one bearing (**Figure 9**).
2. From the opposite side of the hub, insert the remover shaft through the hub bore and into the slot in the backside of the remover head. Position the hub with the remover head resting against a solid surface. Strike the remover shaft to force it into the slot in the remover head. This tightens the remover head against the bearing's inner race.
3. Reposition the wheel. Strike the end of the remover shaft with a hammer and drive the bearing out of the hub (**Figure 10**). Slide the bearing and tool assembly out of the hub.
4. Tap the remover head to release it from the bearing.
5. Remove the distance collar. Note how the collar is positioned in the hub. It will have to be reinstalled with the same orientation during assembly.
6. Repeat this procedure and remove the bearing from the other side. The left side of the rear hub has two bearings. Drive out each bearing as described in this procedure.

NOTE
If the Kowa Seiki tool, or its equivalent, is not available, remove the bearings as follows:

a. Use a long drift and hammer to tilt the distance collar away from one side of the right bearing (**Figure 11**), then drive the right bearing out of the hub.
b. Remove the distance collar and remove the left bearing.

7. Clean the hub and distance collar with solvent. Dry them with compressed air.
8. Install new bearings as described in this chapter.

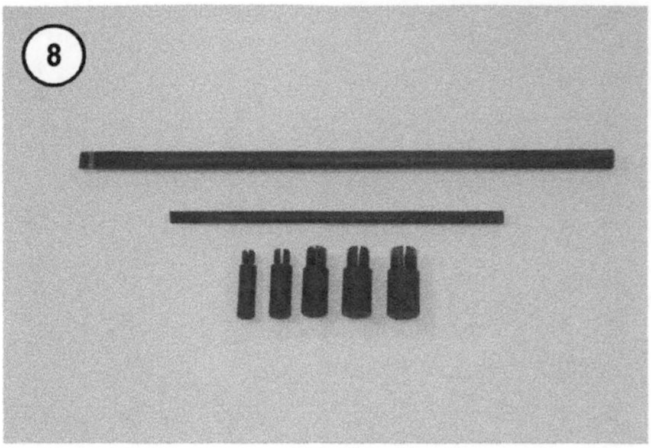

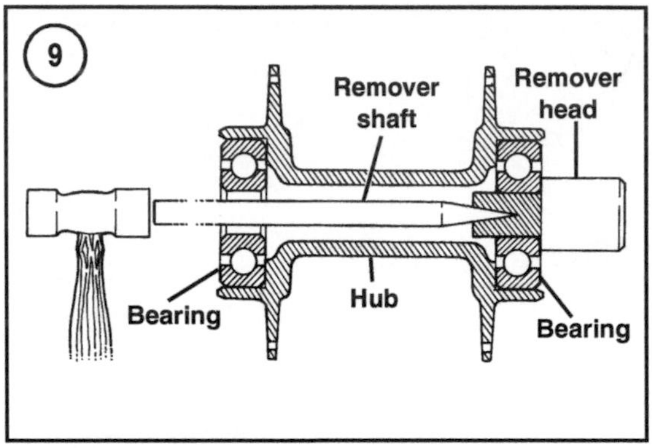

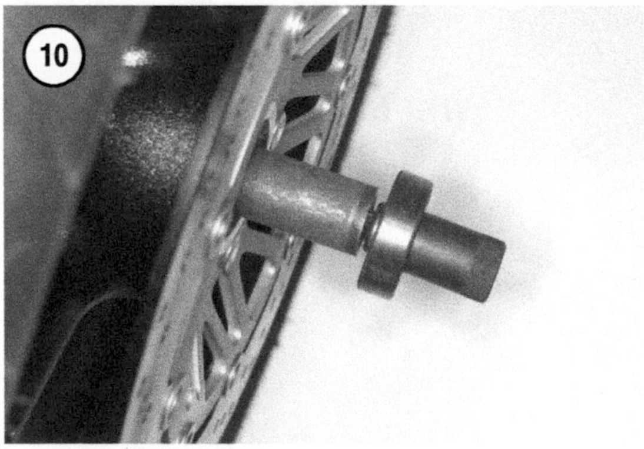

BEARING INSTALLATION

NOTE
The left and right bearings are not identical in the rear hub. Install each bearing into its proper location in the hub.

1. Place the bearings in a freezer overnight. This will ease installation.
2. Blow any dirt or foreign matter out of the hub before installing the bearing.

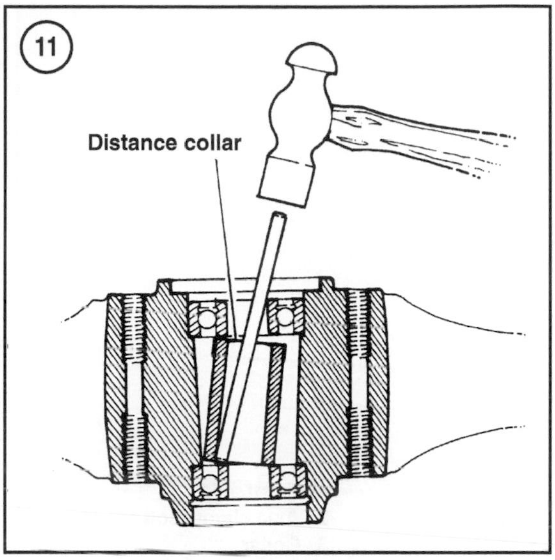

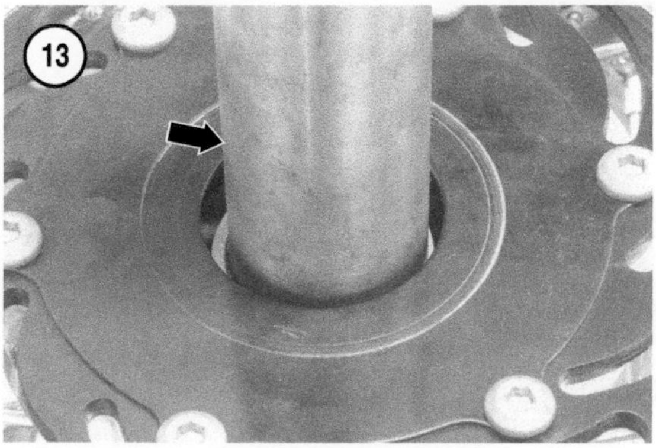

3. Pack the open side of each bearing with grease.

4A. Install rear hub bearings as follows:

   a. Position the left inner bearing with the manufacturer's marks facing out and place the bearing squarely on the bore opening on the left side of the hub. Select a bearing driver

(**Figure 12**) or socket with an outside diameter that matches (or is slightly smaller than) the outside diameter of the bearing. Drive the bearing into the bore until it bottoms.

   b. Repeat substep a and install the left outer bearing.

   c. Turn the hub over. Install the distance collar and center it against the left bearing's inner race.

   d. Place the right bearing (manufacturer's marks facing out) squarely against the bore opening. Select a bearing driver (**Figure 13**) or socket with an outside diameter that matches the outside diameter of the bearing and drive the bearing partway into the bore. Stop and check the distance collar. It must still be centered within the bearing. If it is not, install the axle partway through the hub and center the spacer. Remove the axle and continue installing the bearing until it bottoms.

4B. Install bearings in the front hub as follows:

   a. Position the bearing with the manufacturer's marks facing out and place the bearing squarely on the bore opening on the left side of the hub. Select a bearing driver or socket (**Figure 14**) with an outside diameter that matches the outside diameter of the bearing. Drive the bearing into the bore until it bottoms.

   b. Turn the hub over. Install the distance collar and center it against the left bearing's inner race.

   c. Place the right bearing (manufacturer's marks facing out) squarely against the bore opening. Using the same socket (**Figure 14**) or bearing driver, drive the bearing partway into the bore. Stop and check the distance collar. If it

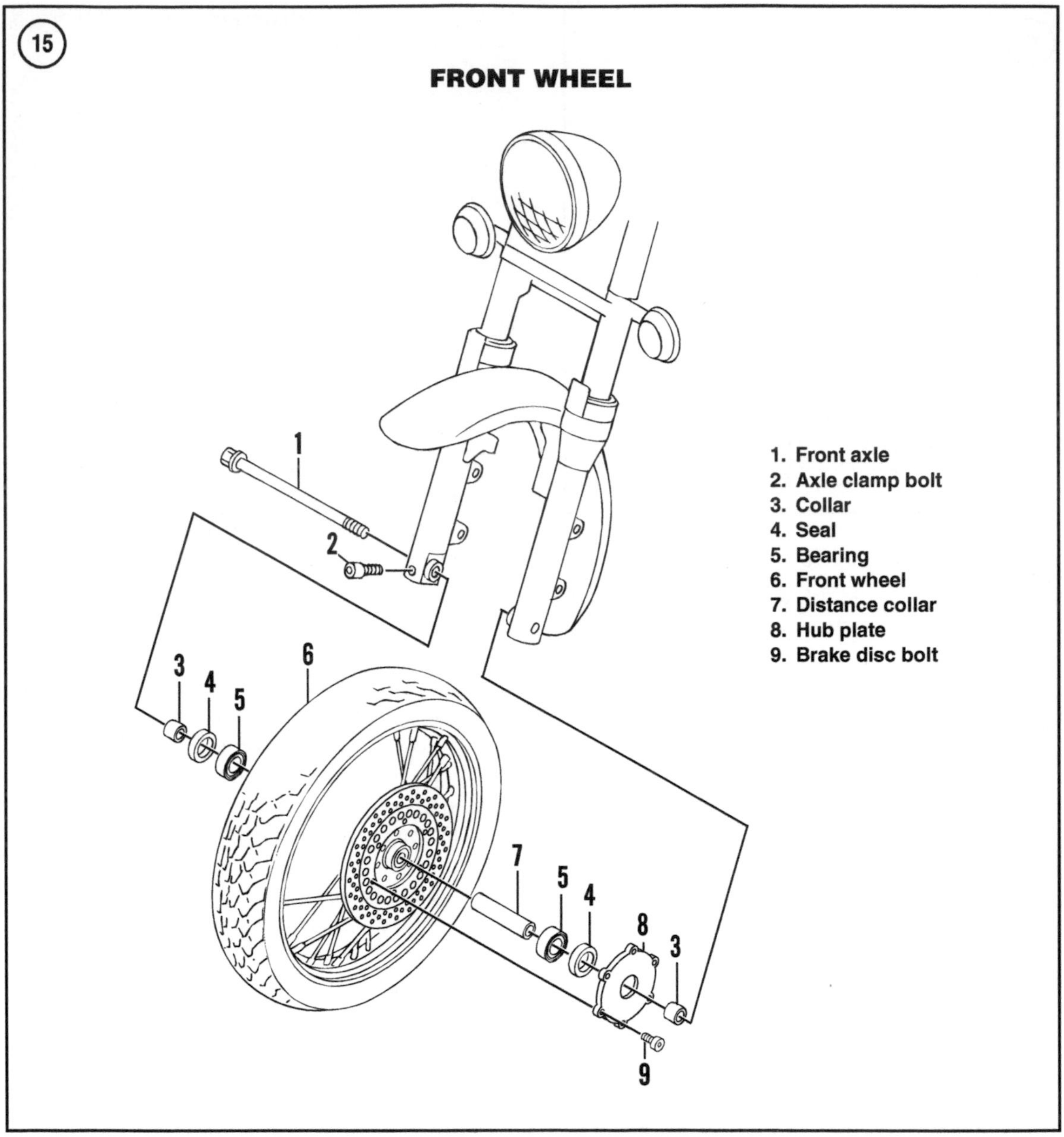

is still centered within the bearing, install the axle partway through the hub and center the spacer. Remove the axle and continue installing the bearing until it bottoms.

5. Reassemble the hub as described in this chapter.

FRONT WHEEL

Refer to **Figure 15**.

Removal

1. Securely support the motorcycle with the front wheel off the ground.
2. Test the wheel bearings as follows:
   a. Hold the wheel along its side (hands 180° apart), and try to rock it back and forth. If there is any play at the axle, the wheel bearings are worn or damaged.
   b. Have an assistant apply the brake while you rock the wheel again. On wheels with se-

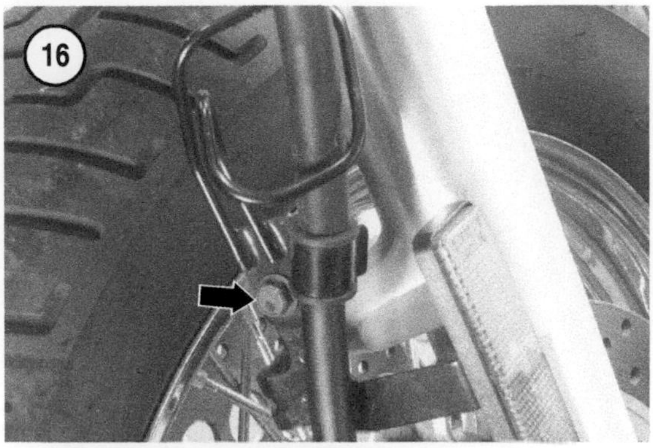

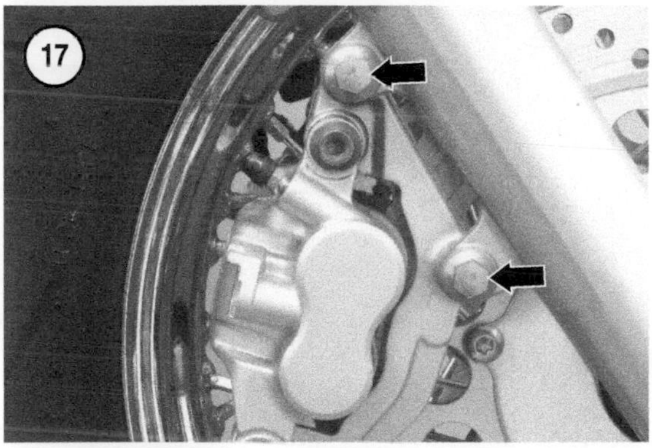

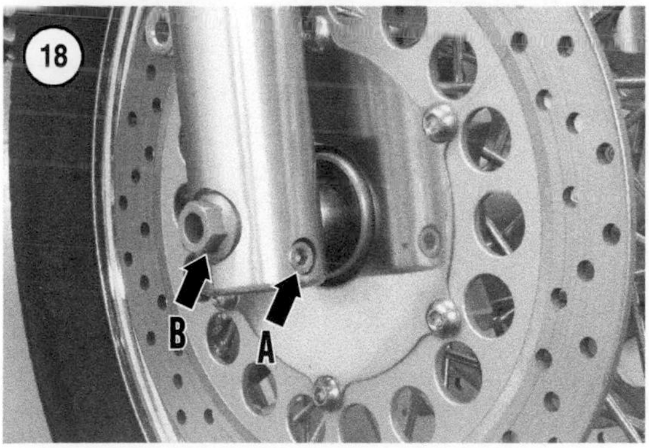

verely worn bearings, there will be play even though the wheel is locked in position.

c. Spin the wheel and listen for excessive wheel bearing noise. Grinding or catching noises indicate worn bearings.
d. If either bearing is worn or damage, replace both wheel bearings as a set. Refer to the hub disassembly procedures later in this chapter.

NOTE
If the brake lever is applied while the calipers are off the front wheel, both calipers must be disassembled to reseat the caliper pistons, and the brakes must be bled. A wooden block between the brake lever and the handlebar prevents the accidental application of the brakes.

3. Insert a wooden block between the brake lever and the handlebar grip. Use a rubber band to hold the block in place.
4. Remove a brake caliper as follows:
   a. Remove the brake hose holder bolt (**Figure 16**) and release the brake hose holder from the reflector bracket.
   b. Remove the caliper mounting bolts (**Figure 17**) from the caliper.
   c. Rotate the caliper off the brake disc. Use a stiff wire or bunjee cord to suspend the caliper from the motorcycle.
   d. Wrap a shop rag around the caliper so the front fender will not be scratched. Secure the rag with a zip tie or rubber band.
5. Repeat Step 4 for the other front brake caliper.

CAUTION
Since the full-size front fender on XVS1100A models wraps low over the front wheel, the motorcycle must be raised very high to remove the wheel with the fender installed. Unless you have a jack that can safely provide the necessary clearance, remove the front fender before removing the front wheel.

6. On XVS1100A models, remove the front fender as described in Chapter Fourteen.
7. Loosen the clamp bolt (A, **Figure 18**) in the right fork slider.
8. Loosen the front axle (B, **Figure 18**) and pull the axle from the right side.
9. Roll the wheel from between the fork legs. Watch for the collar (A, **Figure 19**) on each side of the hub.

CAUTION
Do not lay a wheel on the brake disc. The disc could be scratched or bent.

10. If the wheel must be set down on its side, place wooden blocks (**Figure 2**) beneath the tire.
11. Inspect the wheel as described earlier in this chapter.

Installation

1. Make sure the axle and the axle bearing surfaces of the fork sliders are free of burrs and nicks.
2. Lubricate the axle with lithium soap grease.
3. Apply a light coat of lithium soap grease to the lips of the oil seal in one side of the hub (B, **Figure 19**) and install the collar (A).
4. Repeat Step 3 on the other side.
5. Roll the wheel into place between the fork legs. Make sure the arrow on the tire points in the direction of forward rotation.
6. Lubricate the axle with lithium soap grease. Insert the front axle (B, **Figure 18**) through right fork slider and the hub, and turn the axle into the left fork slider. Do not tighten it to the final torque at this time.
7. Install a brake caliper as follows:
   a. Remove the shop rag from the caliper and route the brake hose along the path noted during removal.
   b. Lower the caliper onto the brake disc. Be careful not to damage the leading edge of each brake pad during installation.
   c. Install the caliper mounting bolts (**Figure 17**). Apply medium-strength threadlocking compound to the bolt threads and torque them to 40 N•m (30 ft.-lb.).
   d. Install the brake hose holder onto the reflector bracket and tighten the brake hose holder bolt (**Figure 16**) to 7 N•m (62 in.-lb.).
8. Repeat Step 7 to install the other brake caliper.
9. After the wheel is completely installed, rotate it several times to make sure it turns freely. Apply the front brake as many times as necessary to ensure the brake pads properly engage the brake disc.
10. Pump the fork several times to ensure the fork legs slide smoothly.
11. Torque the front axle (B, **Figure 18**) to 59 N•m (43 ft.-lb.), then torque the clamp bolt (A, **Figure 18**) to 20 N•m (15 ft-lb.).

FRONT HUB

Disassembly/Inspection/Assembly

1. Remove the front wheel as described in this chapter.
2. Remove the collar (A, **Figure 19**) from each side of the hub.

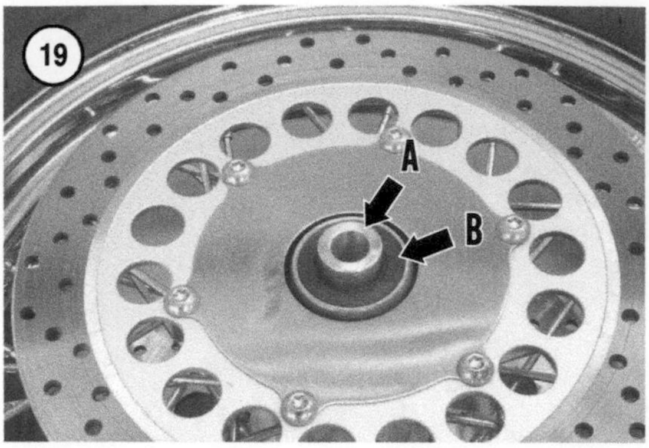

CAUTION
Do not lay a wheel on the brake disc. The disc could be scratched or bent.

3. Use wooden blocks to support the wheel. Place the tire on the blocks so the brake disc will not be damaged (**Figure 2**).

CAUTION
The hub can be serviced with the brake disc and hub plate in place. Nonetheless, consider removing the brake disc so it will not be damaged.

4. Remove the brake disc bolts (C, **Figure 3**). Lift the hub plate and brake disc from the hub. Repeat this on the opposite side.
5. Pry the seal (**Figure 20**) from each side of the hub. Place a shop rag beneath the pry tool so the hub will not be damaged.
6. Inspect the bearings as described in *Wheel Inspection* in this chapter. If necessary, remove and replace the bearings as described in this chapter.
7. Pack the lips of a new seal with lithium soap grease. Drive the seal in squarely with a seal driver or large diameter socket (**Figure 14**) seated on the outer portion of the seal. Drive the seal until it seats against the bearing or when the outer surface is flush with the hub.
8. Repeat for the seal on the opposite side of the hub.
9. Install the brake disc and hub cover, if it was removed, as follows:
   a. Set the brake disc in place so its arrow points in the direction of forward rotation.
   b. Set the hub plate on the hub.
   c. The brake disc bolts (C, **Figure 3**) are made from a harder material than similar bolts used on the motorcycle. When replacing the bolts, always use standard Yamaha brake disc bolts. Never compromise and use generic replacement bolts. They may not properly secure the disc to the hub.
   d. Use a small amount of medium-strength threadlocking compound on the brake disc bolts prior to installation.
   e. Evenly tighten the brake disc bolts in a crisscross pattern. Torque the bolts to 23 N•m (17 ft.-lb.).

REAR WHEEL

Removal

1. Securely support the motorcycle on a level surface. Drain the final gear oil as described in Chapter Three.
2. Remove the mufflers (Chapter Eight).

CAUTION
Since the rear fender on XVS1100A models wraps low around the rear wheel, the motorcycle must be raised very high to remove the rear wheel with the fender installed. Unless you have a jack that can safely provide the necessary clearance, remove the fender before removing the rear wheel.

3. On XVS1100A models, remove the rear fender as described in Chapter Fourteen.
4. Release the rear brake line from the clamp (**Figure 21**) on the swing arm.

NOTE
Insert vinyl tubing or a piece of wood between the brake pads in the caliper. That way, if the brake pedal is inadvertently applied, the pistons will not be forced out of the cylinders. If this does happen, the caliper might have to be disassembled to reseat the pistons and the system will have to be bled. By using the wood, bleeding the brake is not necessary when installing the wheel.

5. Remove the caliper mounting bolts (**Figure 22**). Lift the caliper from the brake disc and suspend it with wire or bunjee cord.
6. Remove the brake caliper bracket bolt (A, **Figure 23**).

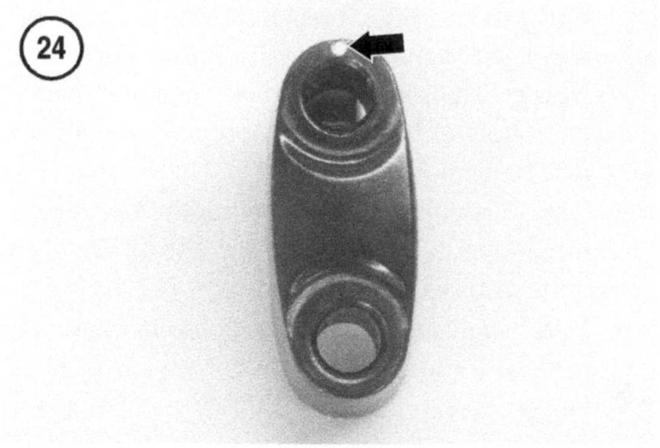

7. Remove the rear axle clamp bolts (B, **Figure 23**) and pull the rear axle clamp from the swing arm. Notice the dot (**Figure 24**) on the clamp. The clamp must be reinstalled with the dot facing up.

8. Remove the final gearcase bolts (A, **Figure 25**) and their washers.

9. Remove the axle nut (B, **Figure 25**) and washer (C) from the left side.

10. Pull the axle (A, **Figure 26**) from the right side, and remove the spacer (B) that sits between the swing arm and the caliper bracket.

11. Remove the rear caliper bracket (C, **Figure 26**) from the motorcycle.

12. Roll the wheel rearward until the driveshaft clears the universal joint (A, **Figure 27**) and remove the wheel assembly. Watch for the stepped collar in the right side of the hub.

13. If necessary, lift the final drive unit (**Figure 28**) from the hub. Watch for the distance collar (**Figure 29**). Reinstall it in the final drive unit (A, **Figure 30**) so it will not be misplaced.

14. Inspect the rear wheel as described in *Wheel Inspection* in this chapter.

15. If necessary, disassemble the hub as described later in this chapter.

Installation

1. If removed, install the final drive gearcase (**Figure 28**) onto the rear hub. Make sure the distance collar (A, **Figure 30**) is in place in the ring gear. Apply molybdenum disulfide grease to the splines in the ring gear (B, **Figure 30**) and the splines in the rear hub.

2. Apply a coat of lithium soap grease to the rear axle, bearings, seals and stepped collar.

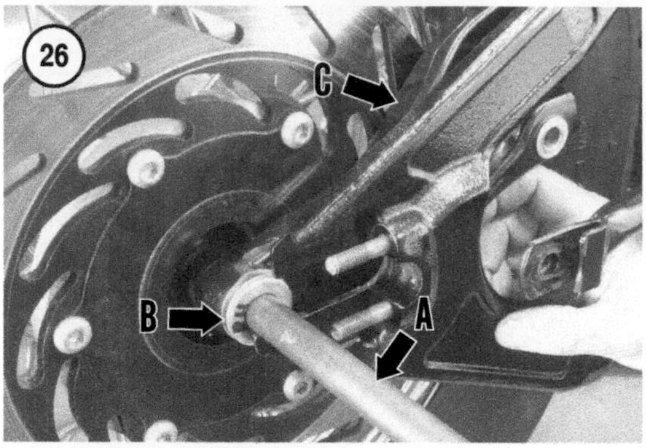

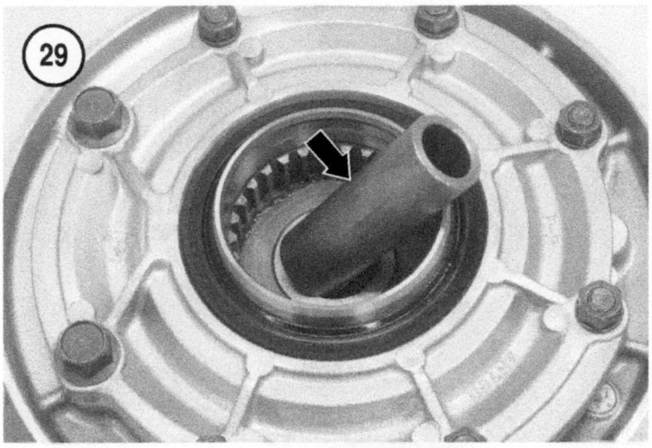
29

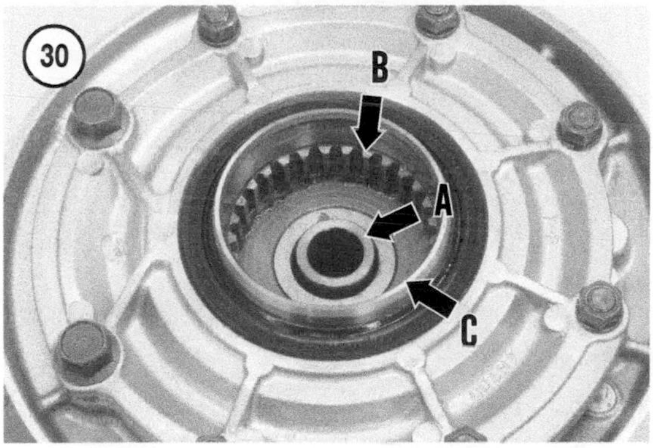

30

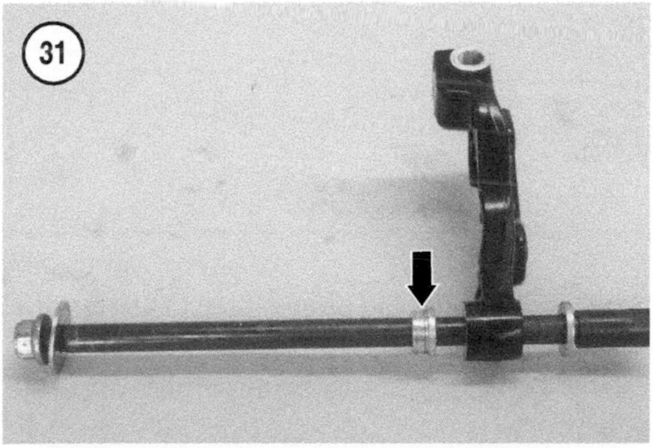
31

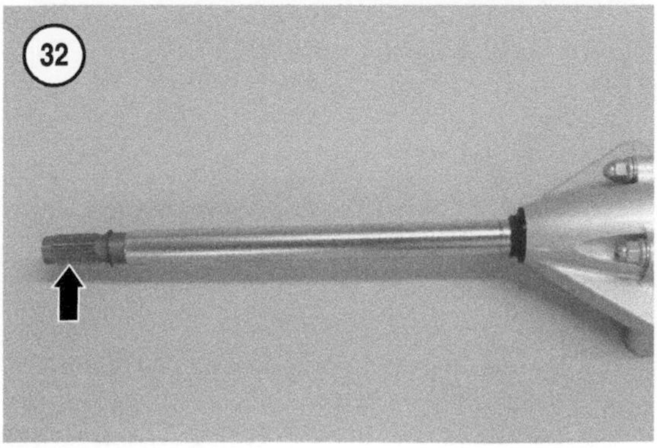
32

33

3. If removed, install the stepped collar (**Figure 31**) into the right side of the hub. Make sure the shouldered side of the collar faces out.
4. Apply molybdenum disulfide grease to the splines of the driveshaft (**Figure 32**) and the universal joint (**Figure 33**).
5. Roll the rear wheel into position in the swing arm so the splines of the driveshaft lightly engage those of the universal joint (A, **Figure 27**). Make sure the boot is in place on the driveshaft (B, **Figure 27**).
6. Set the caliper bracket (C, **Figure 26**) in position between the wheel and the swing arm.

WARNING

*The spacer (B, **Figure 26**) must sit between the caliper bracket (C) and the swing arm. If the washer is placed outside the swing arm, the inner brake pad will ride against the brake disc, causing excessive wear and overheating.*

7. Install the spacer (B, **Figure 26**) onto the axle (A), and slide the axle through the hub.
8. Roll the wheel forward so the driveshaft splines completely engage those of the universal joint. Hold the caliper bracket up so its cutout slides over the mount on the swing arm. Make sure the spacer (B, **Figure 26**) sits between the caliper bracket and the swing arm.
9. Install the clamp onto the studs and finger-tighten the clamp nuts (B, **Figure 23**). The indexing dot (**Figure 24**) on the clamp must face up.
10. Install the axle nut (B, **Figure 25**) and washer (C). Finger-tighten the nut.
11. Install and finger-tighten the final gearcase bolts (A, **Figure 25**).
12. Install and finger-tighten the rear caliper bracket bolt (A, **Figure 23**).

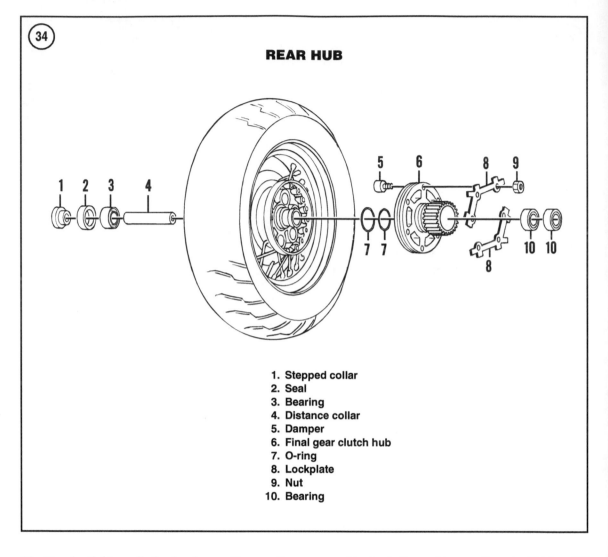

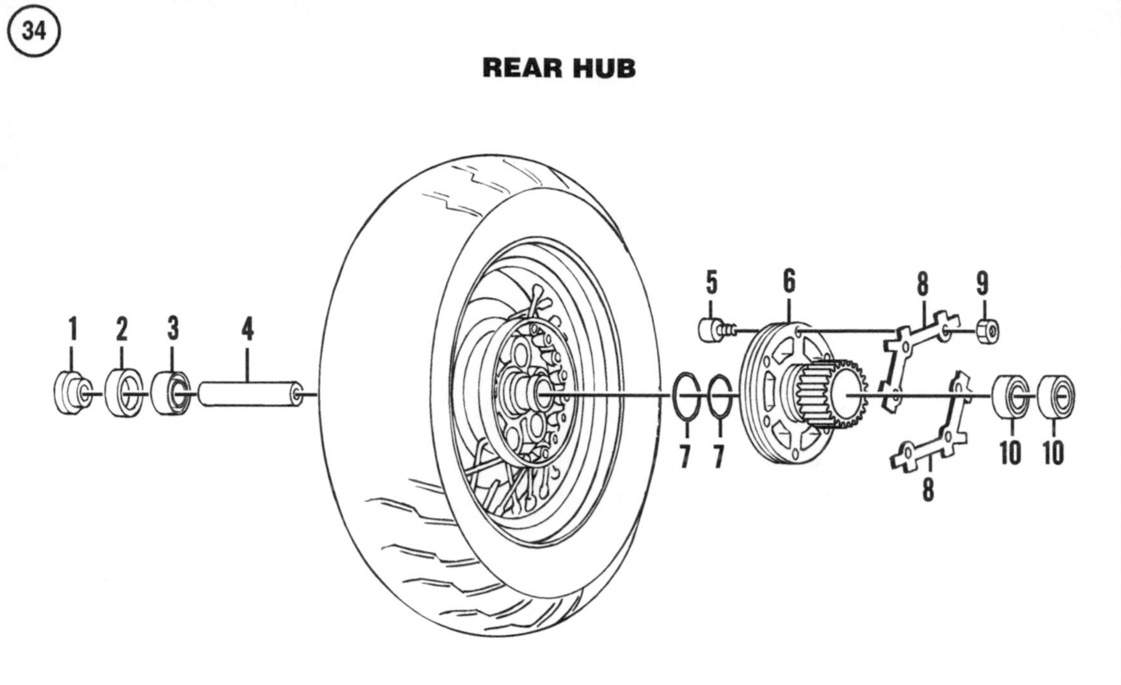

1. Stepped collar
2. Seal
3. Bearing
4. Distance collar
5. Damper
6. Final gear clutch hub
7. O-ring
8. Lockplate
9. Nut
10. Bearing

13. Evenly tighten all the hardware. Torque the hardware to the following specifications in their given order:
   a. Final gearcase bolts (A, **Figure 25**): 90 N•m (66 ft.-lb.).
   b. Axle nut (B, **Figure 25**): 107 N•m (79 ft.-lb.).
   c. Axle clamp nuts (B, **Figure 23**): 23 N•m (17 ft.-lb.).
   d. Rear caliper bracket bolt (A, **Figure 23**): 40 N•m (30 ft.-lb.).
14. Tap the boot (B, **Figure 27**) into the universal joint housing.
15. Install the caliper onto the brake disc. Be careful not to damage the leading edge of the brake pads during installation.
16. Install the two caliper mounting bolts (**Figure 22**). Apply medium-strength threadlocking compound to the bolts, and torque them to 40 N•m (30 ft.-lb.).
17. Secure the brake hose clamp (**Figure 21**) to the swing arm.
18. On XVS1100A models, install the rear fender as described in Chapter Fourteen.
19. Add final drive oil to the final gearcase (Chapter Three).

REAR HUB

Preliminary Inspection

CAUTION

Do not remove the wheel bearings for inspection purposes. The bearings are damaged during removal and

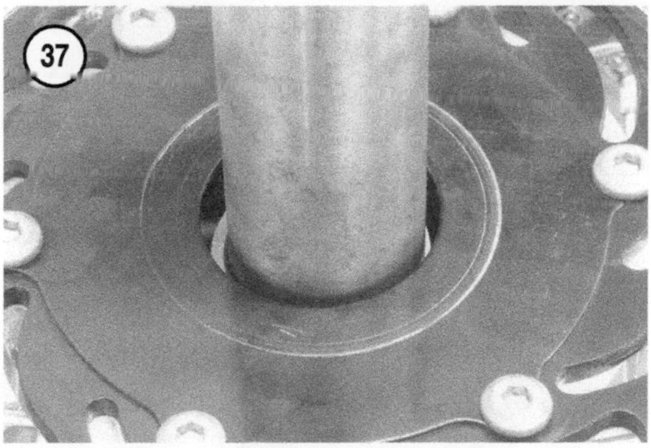

cannot be reused. Remove wheel bearings only if they must be replaced.

1. Remove the rear wheel as described in this chapter.

2. The condition of the rear wheel bearings is critical to the tracking and acceleration performance of the motorcycle. Check the wheel bearings whenever the wheel is removed or as one of the first steps when diagnosing handling or noise problems.

3. Insert the axle through the hub and turn the axle by hand. Each bearing should turn smoothly without noise or excessive play.

4. If the rear wheel was not inspected during removal, inspect it as described in *Wheel Inspection* in this chapter.

Disassembly/Assembly

Refer to **Figure 34** when servicing the rear hub.

1. Remove the rear wheel as described in this chapter. Set the wheel on wooden blocks.
2. Pull the final gearcase assembly (**Figure 28**) straight up and remove it from the hub. Watch for the distance collar (**Figure 29**). It should come out with the final gearcase. Reinstall the distance collar (A, **Figure 30**) in the final gearcase so the collar will not be misplaced.
3. Pry the seal (**Figure 35**) from the right side of the hub. Place a shop rag beneath the pry tool so the hub will not be damaged.
4. Inspect the bearings as described in *Wheel Inspection* in this chapter. If necessary, remove and replace the bearings as described in this chapter.
5. If the final gear clutch hub must be removed, remove it as follows:
   a. Bend the lock tab away from the flat of each clutch hub nut (A, **Figure 36**).
   b. Loosen and remove the clutch hub nuts.
   c. Remove and discard the two lockplates (B, **Figure 36**). New lockplates must be installed during assembly.
   d. Remove the clutch hub from the rear hub.
   e. Remove and discard the O-rings that sit behind the clutch hub. New O-rings must be installed during assembly.
   f. If necessary, remove the wheel dampers.
6. Inspect the rear hub as described in this chapter.
7. Pack the lips of a new oil seal with lithium soap grease. Drive the seal in squarely with a seal driver (**Figure 37**) or a large diameter socket seated on the outer portion of the seal. Drive the seal until it seats against the bearing or when the outer surface is flush with the hub.
8. Install the final gear clutch hub assembly, if it was removed, as follows:
   a. Install the dampers into the hub. If necessary, lubricate the dampers with a soap solution.

10

b. Install new O-rings. Lubricate the new O-rings with lithium soap grease and install them into the hub.
c. Install the clutch hub onto the damper studs and install two new lockplates (B, **Figure 36**).
d. Install and evenly tighten the nuts.
e. Torque the nuts to 62 N•m (45 ft.-lb.) and bend a locktab against a flat of each nut (A, **Figure 36**).

Inspection

1. Inspect the bearings as described in *Wheel Inspection* in this chapter. If necessary, remove and replace the bearings as described in this chapter.
2. Check the seal (**Figure 38**) for damage that might allow dirt to enter. If any seal is damaged, replace the bearings as a set.
3. Inspect the wheel dampers for cracks, wear or other signs of deterioration. Replace the dampers as a set.
4. Inspect the splines (C, **Figure 36**) on the final gear clutch hub. If there is damage, inspect the mating splines on the final gear assembly (**Figure 39**).

LACED WHEEL SERVICE

The laced or wire wheel assembly consists of a rim, spokes, nipples and hub (containing the wheel bearings, distance collars and seals).

Loose or improperly tightened spokes can cause hub damage. Periodically inspect the wheel assembly for loose, broken or missing spokes, rim damage and runout. Wheel bearing service is described in this chapter.

Component Condition

Wheels are subjected to a significant amount of punishment. Inspect the wheels regularly for lateral (side-to-side) and radial (up and down) runout, even spoke tension, and visible damage. When a wheel has a noticeable wobble, it is out of true. Loose spokes usually cause this, but it can be caused by impact damage.

Truing a wheel corrects the radial and lateral runout to bring the wheel back into specification. The condition of the individual wheel components will affect the ability to successfully true the wheel. Note the following:

38

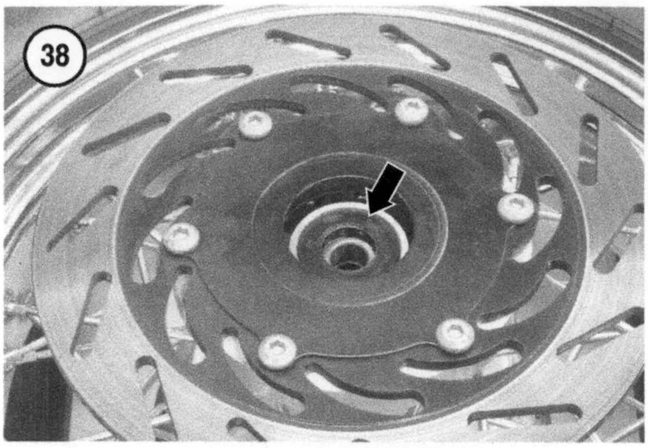

39

40

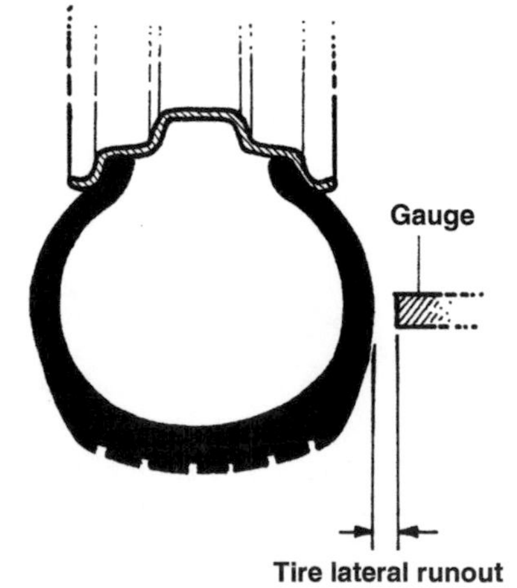

1. Spoke condition—Do not attempt to true a wheel with bent or damaged spokes. Doing so places an excessive amount of tension on the spoke and rim. The spoke may break and/or pull through the hole in the rim. Inspect spokes carefully and replace any that are damaged.

41

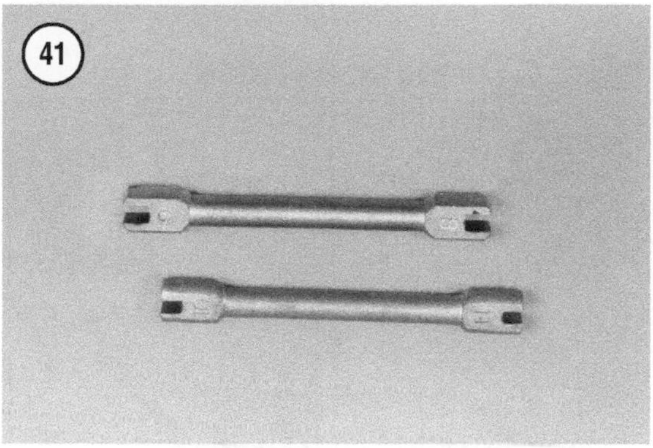

42

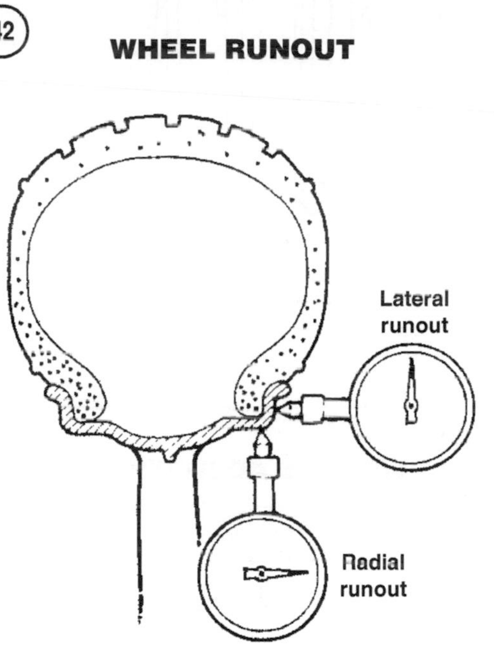

2. Nipple condition—When truing a wheel, the nipple should turn freely on the spoke. It is common for the spoke threads to become corroded and make turning the nipple difficult. Spray a penetrating liquid onto the nipple and allow sufficient time for it to penetrate. Use a spoke wrench and work the nipple in both directions and apply additional penetrating liquid. If the spoke wrench rounds off the nipple, remove the tire from the rim and cut the spoke(s) out of the wheel.

3. Rim condition—Minor rim damage can be corrected by truing; however, trying to correct excessive runout caused by impact damage will damage the hub and rim due to over-tightened spokes. Inspect the rims for cracks, flat spots or dents. Check the spoke holes for cracks or enlargement.

Wheel Truing Preliminaries

Before checking runout and truing the wheel, note the following:

1. Make sure the wheel bearings are in good condition.
2. A small amount of runout is acceptable, do not attempt to true the wheel to a perfect zero reading. Refer to **Table 1** for specifications.
3. Perform a quick runout check with the wheel on the motorcycle by placing a pointer against the fork (**Figure 40**) or swing arm and slowly rotating the wheel.
4. Perform major wheel truing with the tire removed and the wheel mounted in a wheel truing stand.
5. Use a spoke nipple wrench of the correct size. Using the wrong type of tool or one that is the incorrect size will round off the spoke nipples, making adjustment difficult. Quality wrenches (**Figure 41**) grip the nipple on four corners to prevent damage. Tighten spokes to 3 N•m (27 in.-lb.).

10

Wheel Truing Procedure

1. Set the wheel in a truing stand.

2A. When using a dial indicator, check rim runout as follows:

   a. Measure the radial runout with a dial indicator positioned as shown in **Figure 42**. If radial runout exceeds the service limit specified in **Table 1**, replace the rim.
   b. Measure the lateral runout with a dial indicator positioned as shown in **Figure 42**. If lateral runout exceeds the service limit specified in **Table 1**, replace the rim.

2B. If a dial indicator is not available, check rim runout as follows:

   a. Position a pointer facing toward the rim as shown in **Figure 43**. Spin the wheel slowly and check the lateral runout.
   b. Adjust the position of the pointer and check the radial runout.

3. If lateral runout is out of specification, the rim needs to be moved relative to the centerline of the wheel. See **Figure 44**. To move the rim to the left,

for example, tighten the spoke(s) on the left of the rim and loosen the opposite spoke(s) on the right.

NOTE
The number of spokes to loosen and tighten will depend on the amount of runout. As a minimum, always adjust two or three spokes in the vicinity of the rim runout. If runout affects a greater area along the rim, adjust a greater number of spokes.

4. If radial runout is excessive, the hub is not centered within the rim. The rim needs to move relative the centerline of the hub. See **Figure 45**. Draw the high point of the rim toward the centerline of the hub by tightening the spokes in the area of the high point and by loosening spokes on the low side. Tighten and loosen the spokes in equal amounts to prevent distortion.
5. Rotate the wheel and check runout. Continue adjusting the spokes until runout is within the specification listed in **Table 1**. Be patient and thorough, adjusting the position of the rim a little at a time.
6. After truing the wheel, seat each spoke in the hub by tapping it with a flat nose punch and hammer. Recheck the spoke tension and wheel runout. Readjust if necessary.
7. Check the ends of the spokes on the tube side of the rim. Grind off any spoke that protrudes from the nipple so it will not puncture the tube.

WHEEL BALANCE

An unbalanced wheel is unsafe. Depending on the degree of unbalance and the speed of the motorcycle, a rider may experience anything from a mild vibration to a violent shimmy that may result in loss of control.

Before balancing a wheel, thoroughly clean the wheel assembly. Make sure the wheel bearings are in good condition and properly lubricated. The wheel must rotate freely. Also make sure the balance mark on the tire (**Figure 46**) aligns with the valve stem. If not, break the tire loose from the rim and align it before balancing the wheel. Refer to *Tire Changing* in this chapter.

NOTE
Balance the wheels with the brake disc attached. The disc rotates with the wheel and affects the balance.

43

WHEEL RUNOUT

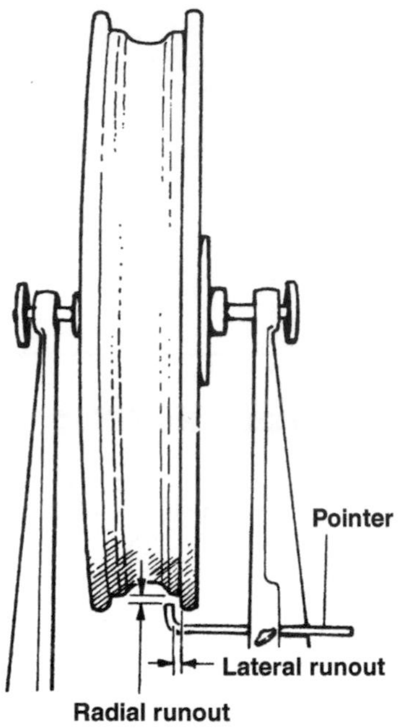

44

LATERAL ADJUSTMENT

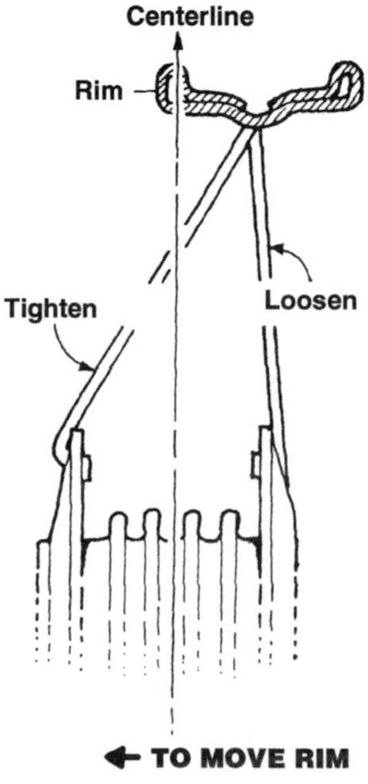

45

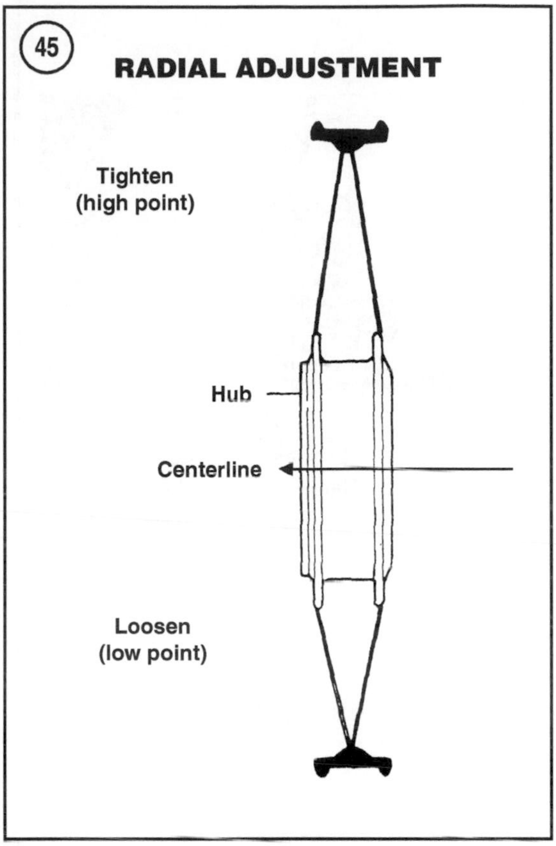

46

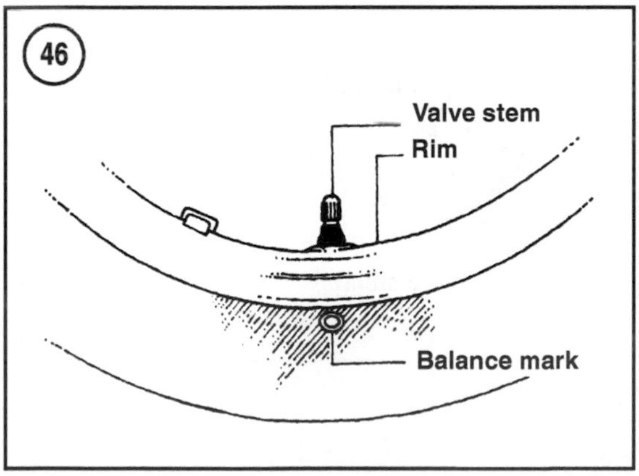

1. Remove the wheel as described in this chapter.

2. Make sure the valve stem and the valve cap are tight.

3. Mount the wheel on a stand such as the one shown in **Figure 47** so it can rotate freely.

4. Check the wheel runout as described in this chapter. Do not try to balance a wheel with excessive runout.

47

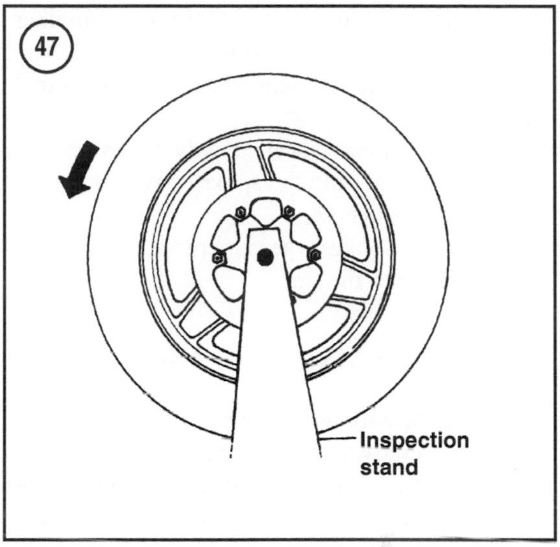

5. Remove any balance weights mounted on the wheel.

6. Give the wheel a spin and let it coast to a stop. Mark the tire at the highest point (12 o'clock). This is the wheel's lightest point.

7. Spin the wheel several more times. If the wheel keeps coming to rest at the same point, it is out of balance. If the wheel stops at different points each time, the wheel is balanced.

NOTE

Adhesive test weights are available from motorcycle dealerships. These are adhesive-backed weights that can be cut to the desired length and attached directly to the rim.

8. Loosely attach a balance weight (or tape a test weight) at the upper or light side (12 o'clock) of the wheel.

9. Rotate the wheel 1/4 turn (3 o'clock). Release the wheel and observe the following:

   a. If the wheel does not rotate (if it stays at the 3 o'clock position), the correct balance weight was installed. The wheel is balanced.

   b. If the wheel rotates and the weighted portion goes up, replace the weight with the next heavier size.

   c. If the wheel rotates and the weighted portion goes down, replace the weight with the next lighter size.

   d. Repeat this step until the wheel remains at rest after being rotated 1/4 turn. Rotate the wheel another 1/4 turn, another 1/4 turn, and another to see if the wheel is correctly balanced.

10. Remove the test weight and install the correct weight.

   a. On wire spoke wheels, firmly crimp the balance weight onto the spoke(s) with a pair of pliers (**Figure 48**).
   b. On cast spoke wheels, crimp the balance weight onto the rim.

TIRE CHANGING (LACED WHEELS)

The laced or wire wheels can easily be damaged during tire removal. Special care must be taken with tire irons to avoid scratches and gouges to the outer rim surface. Insert rim protectors (**Figure 49**) or scraps of leather between the tire iron and the rim.

Removal

CAUTION
While removing a front tire, support the wheel on two blocks of wood, so the brake disc does not contact the floor.

1. Remove the wheel as described in this chapter.
2. If the tire will be reinstalled, place a balance mark on the tire opposite the valve stem location (**Figure 46**) so the tire can be reinstalled in the same position for easier balancing.
3. Remove the valve core to deflate the tire.
4. Press the entire bead on both sides of the tire away from the rim and into the center of the rim.
5. Lubricate both beads with soapy water.

NOTE
*Use rim protectors (**Figure 49**) between the tire irons and the rim to protect the rim from damage. Also, use only quality tire irons without sharp edges (**Figure 50**). If necessary, file the ends of the tire irons to remove rough edges.*

6. Insert the tire iron under the upper bead next to the valve stem (**Figure 51**). Press the lower bead into the center of the rim and pry the upper bead over the rim with the tire iron.
7. Insert a second tire iron next to the first to hold the bead over the rim (**Figure 52**). Work around the

48

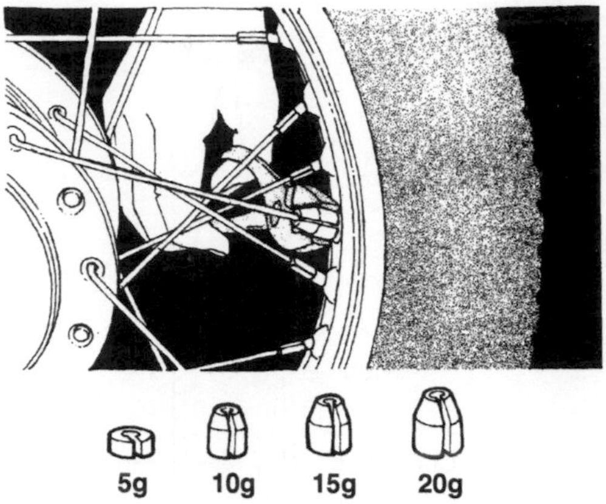

49

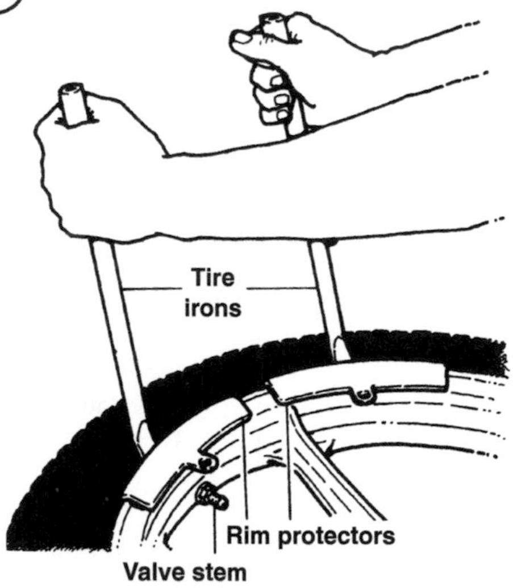

50

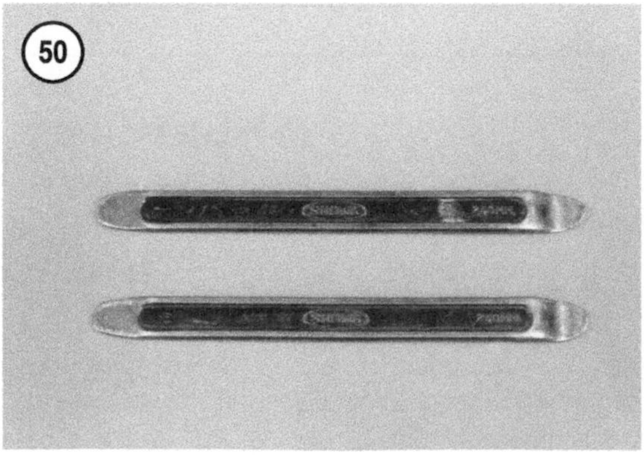

51

52

53

54

tire, prying the bead over the rim with the first tool. Be careful not to pinch the inner tube with the tire irons.

8. When the upper bead is off the rim, remove the nut from the valve stem. Remove the valve from the hole in the rim and remove the tube from the tire (**Figure 53**).

NOTE
Step 9 is required only if it is necessary to completely remove the tire from the rim.

9. Stand the wheel upright. Force the second bead into the center of the rim. Insert the tire iron between the second bead and the side of the rim that the first bead was pried over. Pry the second bead off the rim (**Figure 54**), working around the wheel with two tire irons as done earlier.

10. Inspect the rim as described in this chapter.

Installation

NOTE
Before installing the tire, place it in the sun or in a hot, closed car. The heat will soften the rubber and ease installation.

1. Reinstall the rubber rim band. Align the hole in the band with the valve hole in the rim.
2. Liberally sprinkle the inside of the tire with talcum powder to reduce chafing between the tire and tube.
3. Most tires have directional arrows on the sidewall. Install the tire so the arrow points in the direction of forward rotation.
4. If the tire was removed, lubricate the lower bead of the tire with soapy water and place the tire against the rim. Align the valve stem balance mark (**Figure 46**) with the valve stem hole in the rim.
5. Using your hand, push as much of the lower bead past the upper rim surface as possible (**Figure 55**). Work around the tire in both directions (**Figure 56**).
6. Install the valve core into the valve stem in the inner tube.
7. Put the tube into the tire and insert the valve stem through the hole in the rim. Inflate the tube just enough to round it out. Too much air will make tire installation difficult; too little air increases the chance of pinching the tube with the tire irons.

55

56

8. Lubricate the upper tire bead and rim with soapy water.

9. Press the upper bead into the rim opposite the valve stem. Pry the bead into the rim on both sides of this initial point with your hands and work around the rim to the valve stem. If the tire pulls up on one side, either use a tire iron or a knee to hold the tire in place. The last few inches are usually the toughest and also the place where most tubes are pinched. If possible, continue to push the tire into the rim with your hands. Relubricate the bead if necessary. If the tire bead pulls out from under the rim, use both of your knees to hold the tire in place. If necessary, use a tire iron and rim protector for the last few inches (**Figure 57**).

CAUTION

*Make sure the valve stem is not cocked in the rim (**Figure 58**).*

10. Wiggle the valve stem to make sure the tube is not trapped under the bead. Set the valve squarely in its hole.

WARNING

*In the next step, seat the tire on the rim by inflating the tire to approximately 10% above the recommended inflation pressure listed in **Table 3**. Do not exceed 10%. Never stand directly over a tire while inflating it. The tire could burst with enough force to cause severe injury.*

11. Check the bead on both sides of the tire for an even fit around the rim, then relubricate both sides of the tire. Inflate the tube to seat the tire on the rim. Check to see that both beads are fully seated and the tire rim lines (**Figure 59**) are the same distance from the rim all the way around the tire. If the beads will not seat, release air from the tire. Lubricate the rim and beads with soapy water, then reinflate the tube.

57

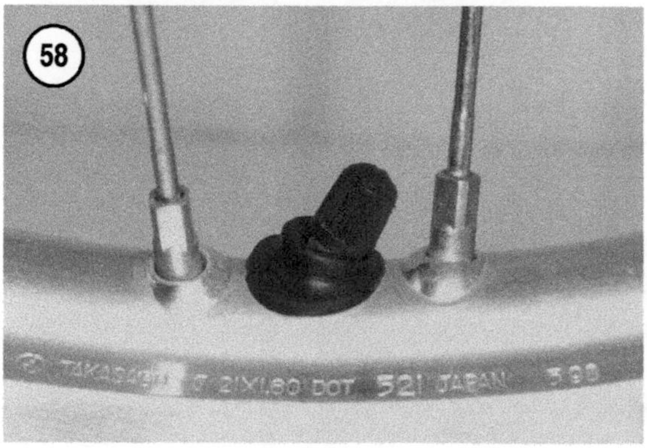
58

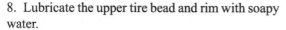

12. Bleed the tire pressure down to the recommended pressure listed in **Table 3**. Install the valve stem nut, tighten it against the rim, then install the valve stem cap.

13. Balance the wheel as described in this chapter.

59

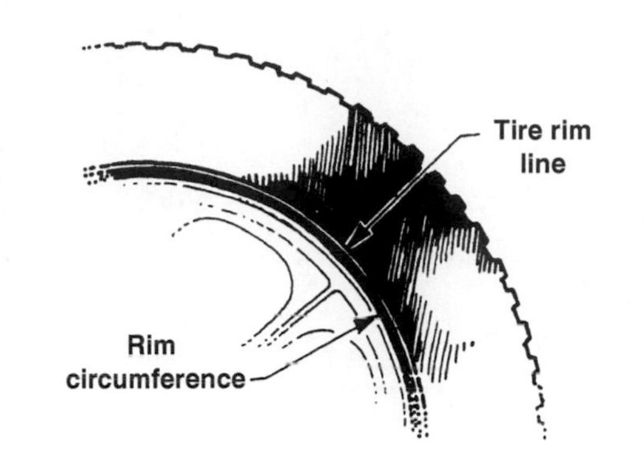

60

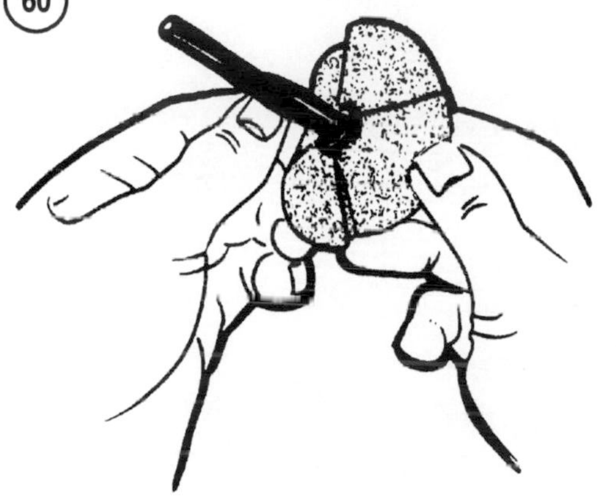

Inspection

1. Remove and inspect the rubber rim band. Replace the band if it is deteriorated or broken.
2. Clean the inner and outer rim surfaces of all dirt, rust, corrosion and rubber residue.
3. Inspect the valve stem hole in the rim. Remove any dirt or corrosion from the hole.
4. Inspect the rim profiles for any cracks or other damage.
5. The tube will be reused, reinstall the valve core, inflate the tube and check it for any leaks.
6. While the tube is inflated, clean it with water.
7. When reusing the tire, carefully check it inside and outside for damage. Replace the tire if there is any damage.
8. Make sure the spoke ends do not protrude from the nipples into the center of the rim.

TIRE CHANGING (CAST WHEELS)

WARNING
Do not install an inner tube inside a tubeless tire. The tube will cause an abnormal heat buildup in the tire.

Tubeless tires have the word TUBELESS molded in the tire sidewall and the rims have TUBELESS cast on them.

If the tire is punctured, remove it from the rim to inspect the inside of the tire and apply a combination plug/patch from inside the tire (**Figure 60**). A Plug applied from the outside of the tire should only be used as a temporary roadside repair.

Follow the repair kit manufacturer's instructions as to applicable repairs and any speed limitations. In most cases it is a good idea to consider replacing a patched or plugged tire as soon as possible.

Removal

The wheels can easily be damaged during tire removal. Special care must be taken with tire irons to avoid scratching and gouging the outer rim surface. Protect the rim by using rim protectors or scraps of leather between the tire iron and the rim (**Figure 49**). The stock cast wheels are designed for use with tubeless tires.

When removing a tubeless tire, be careful not to damage the tire beads, inner liner of the tire or the wheel rim flange. Use tire levers or flat-handled tire irons (**Figure 50**) with rounded ends.

NOTE
While removing a tire, support the wheel on two blocks of wood, so the brake disc does not contact the floor.

1. Place a balance mark opposite the valve stem (**Figure 46**) on the tire sidewall so the tire can be reinstalled in the same position for easier balancing.
2. Remove the valve core to deflate the tire.

CAUTION
Removal of tubeless tires from their rims can be very difficult because of the exceptionally tight bead/rim seal-

10

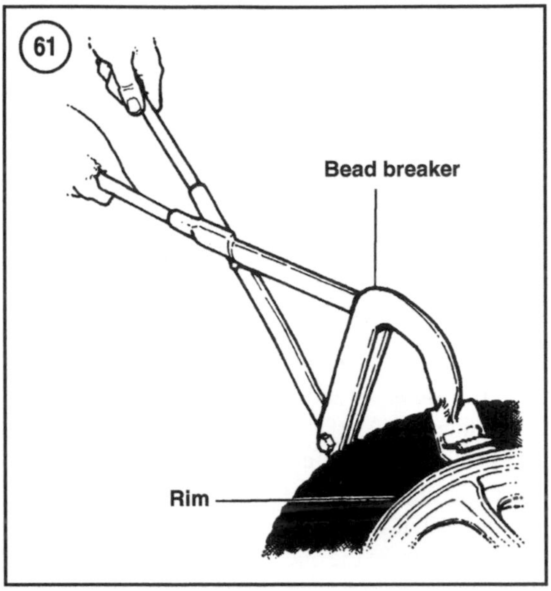

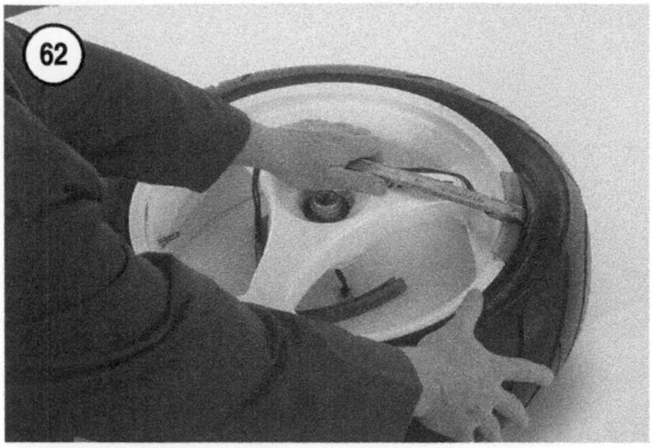

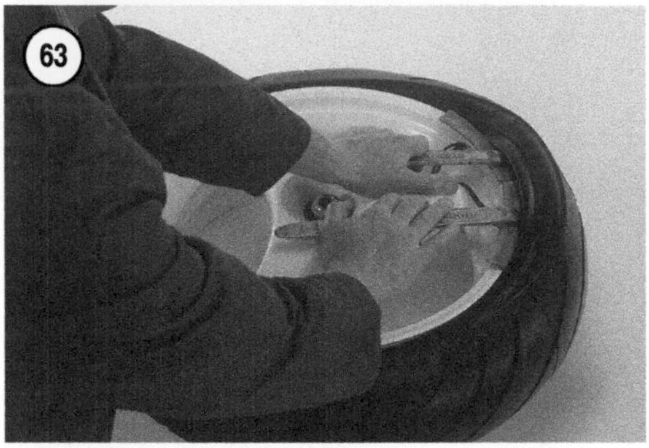

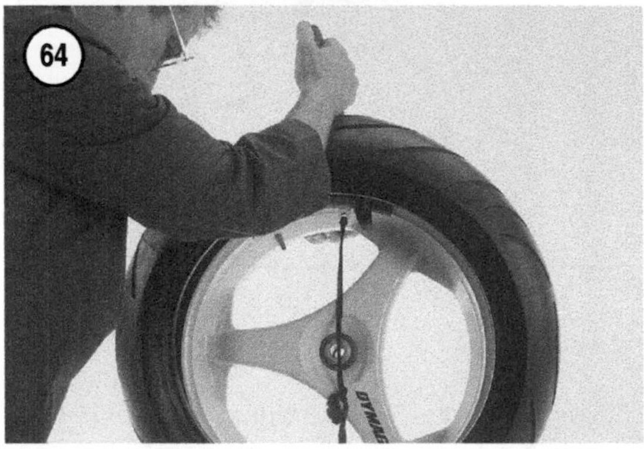

*ing surface. Breaking the bead seal may require the use of a bead breaker (**Figure 61**). Do not scratch the inside of the rim or damage the tire bead.*

3. Press the entire bead on both sides of the tire into the center of the rim.
4. Lubricate the beads with soapy water.

NOTE
*Use rim protectors (**Figure 49**) or insert scraps of leather between the tire irons and the rim to protect the rim from damage.*

5. Insert the tire iron under the bead next to the valve stem (**Figure 62**). Force the bead on the opposite side of the tire into the center of the rim and pry the bead over the rim with the tire iron.
6. Insert a second tire iron next to the first to hold the bead over the rim (**Figure 63**). Then work around the tire with the first tool prying the bead over the rim.

NOTE
Step 7 is required only if it is necessary to completely remove the tire from the rim.

7. Set the wheel on its edge. Insert a tire tool between the second bead and the same side of the rim that the first bead was pried over (**Figure 64**). Force the bead on the opposite side from the tool into the center of the rim. Pry the second bead off the rim, working around the wheel with two tire irons as with the first.
8. Inspect the valve stem seal. Because rubber deteriorates with age, it is advisable to replace the valve stem when replacing a tire.

Installation

1. Carefully inspect the tire for any damage, especially inside.

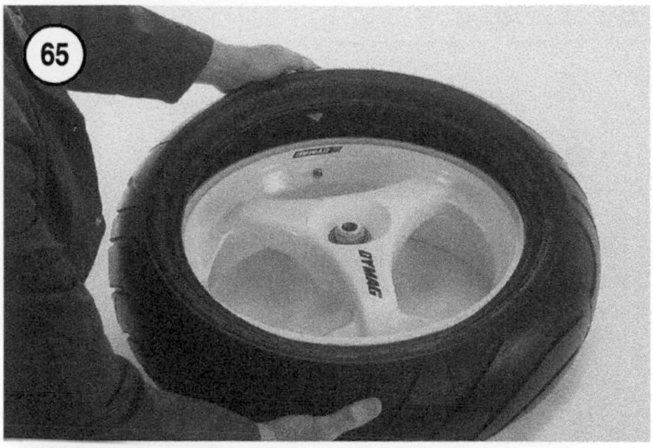
65

66

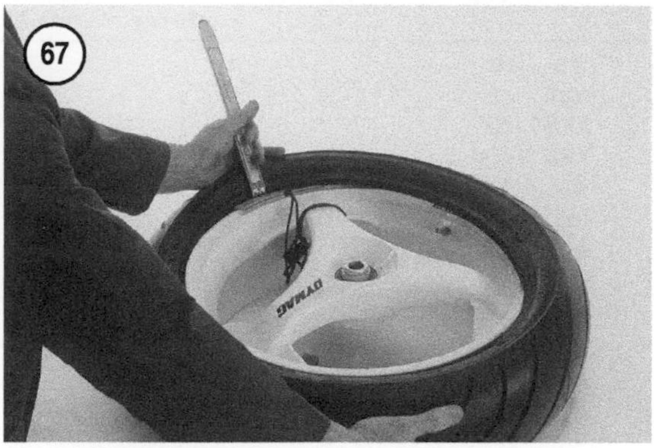
67

2. A new tire may have balancing rubbers inside. These are not patches and should not be disturbed.

3. Manufacturers place a colored spot near the bead, indicateing a lighter point on the tire. Install the tires so this balance mark (either the manufacturer's or the one made during removal) sits opposite the valve stem (**Figure 46**).

4. Most tires have directional arrows on the sidewall that indicate the direction of rotation. Install the tire so the arrow points in the direction of forward rotation.

5. Lubricate both beads of the tire with soapy water.

6. Place the backside of the tire into the center of the rim. The lower bead should go into the center of the rim and the upper bead outside. Work around the tire in both directions (**Figure 65**).

7. Starting at the side opposite the valve stem, press the upper bead into the rim (**Figure 66**). Pry the bead into the rim on both sides of the initial point with a tire tool, working around the rim to the valve (**Figure 67**).

8. Check the bead on both sides of the tire for an even fit around the rim.

9. Place an inflatable band around the circumference of the tire. Slowly inflate the band until the tire beads are pressed against the rim. Inflate the tire enough to seat it against the rim. Deflate and remove the band.

WARNING

Never exceed 56 psi (4.0 k/cm²) inflation pressure as the tire could burst causing severe injury. Never stand directly over the tire while inflating it.

10. After inflating the tire, check to see that the beads are fully seated and the tire rim lines are the same distance from the rim all the way around the tire (**Figure 59**). If the beads will not seat, deflate the tire, re-lubricate the rim and beads with soapy water, and re-inflate the tire.

11. Inflate the tire to the required pressure. Refer to tire inflation pressure specifications listed in **Table 3**. Screw on the valve stem cap.

12. Balance the wheel assembly as described in this chapter.

Tables 1-4 are on the following pages.

Table 1 WHEEL SPECIFICATIONS

Wheel size	
Front	
XVS1100	18 × 2.15
XVS1100A	16 × 3.00
Rear	15M/C × MT4.50
Wheel runout service limit	
Axial	0.5 mm (0.02 in.)
Radial	1.0 mm (0.04 in.)
Brake disc deflection	0.15 mm (0.006 in.)

Table 2 TIRE SPECIFICATIONS

XVS1100 models	
Front tire size	110/90-18 61S
Manufacturer	Bridgestone Exedra L309, Dunlop K555F
Rear tire size	170/80-15M/C 77S
Manufacturer	
USA, California and Canada models	Bridgestone Exedra G546G, Dunlop K555
Europe and Australia models	Bridgestone Exedra G546, Dunlop K555
XVS1100A models	
Front tire size	130/90-16 67S
Manufacturer	Dunlop D404F
Rear tire size	170/80-15M/C 77S
Manufacturer	Dunlop D404G
Minimum tread depth	1.6 mm (0.06 in.)

Table 3 TIRE INFLATION PRESSURE[1]

XVS1100 models	
0-90 kg (0-198 lb.) load[2]	
Front	200 kPa (28.5 psi)
Rear	225 kPa (32 psi)
90-200 kg (198-441 lb.) load[2]	
Front	225 kPa (32 psi)
Rear	250 kPa (36 psi)
XVS1100A models	
0-90 kg (0-198 lb.) load[2]	
Front	225 kPa (32 psi)
Rear	225 kPa (32 psi)
90-200 kg (198-441 lb.) load[2]	
Front	225 kPa (32 psi)
Rear	250 kPa (36 psi)
Maximum load[2]	200 kg (441 lb.)

1. Tire inflation pressures apply to original-equipment tires only. Aftermarket tires may require different pressures. Refer to the tire manufacturer's specifications.
2. Load equals the total weight of rider, passenger, accessories and all cargo.

Table 4 WHEEL TORQUE SPECIFICATIONS

Item	N•m	in.-lb.	ft.-lb.
Brake disc bolts*	23	–	17
Brake hose holder bolt	7	62	–
Final gear clutch hub nuts	62	–	45
Final gearcase bolts	90	–	66
Front brake caliper mounting bolt	40	–	30
Front caliper retaining bolt			
1999-on XVS1100 models	23	–	17
All models except 1999-on			
XVS1100 models	27	–	20
Front axle	59	–	43
Front axle clamp bolt	20	–	15
Hub plate bolt*	23	–	17
Muffler bracket bolt	30	–	22
Rear axle nut	107	–	79
Rear axle clamp nuts	23	–	17
Rear caliper bracket bolt	40	–	30
Rear caliper mounting bolts	40	–	30
Spokes	3	27	–

*Apply a medium-strength threadlocking compound.

CHAPTER ELEVEN

FRONT SUSPENSION AND STEERING

This chapter described the service procedures for the handlebar, front fork and steering components. When inspecting components, compare any measurements to the specifications in the tables at the end of this chapter. Replace any component that is worn, damaged or out of specification. During assembly, tighten hardware to the given torque specification.

FRONT FORK

Before concluding the front fork legs have major problems, drain the fork oil and refill the fork legs with the proper type and quantity of fork oil as described under *Assembly* in this section. If they are still having problems, such as poor damping, leaking around the seals or a tendency to bottom or top out, follow the service procedures in this section.

To simplify fork service and to prevent the mixing of parts, service each fork leg individually.

Removal (Fork Leg Will Not Be Serviced)

Refer to **Figure 1** or **Figure 2**.

NOTE

The photographs show the front fork removal and installation for an XVS1100A model. Except for the fork covers, these photographs also apply to XVS1100 models.

1. Securely support the motorcycle with a front end stand.
2. Remove the front wheel, along with the brake calipers and brake hose holders, as described in Chapter Ten.

NOTE

Insert vinyl tubing or a piece of wood between the pads in the caliper. That way, if the brake lever is inadvertently squeezed, the pistons will not be forced out of the caliper. If this does happen, the caliper might have to be disassembled to reseat the pistons. By

(1)

FORK ASSEMBLY (XVS1100 MODELS)

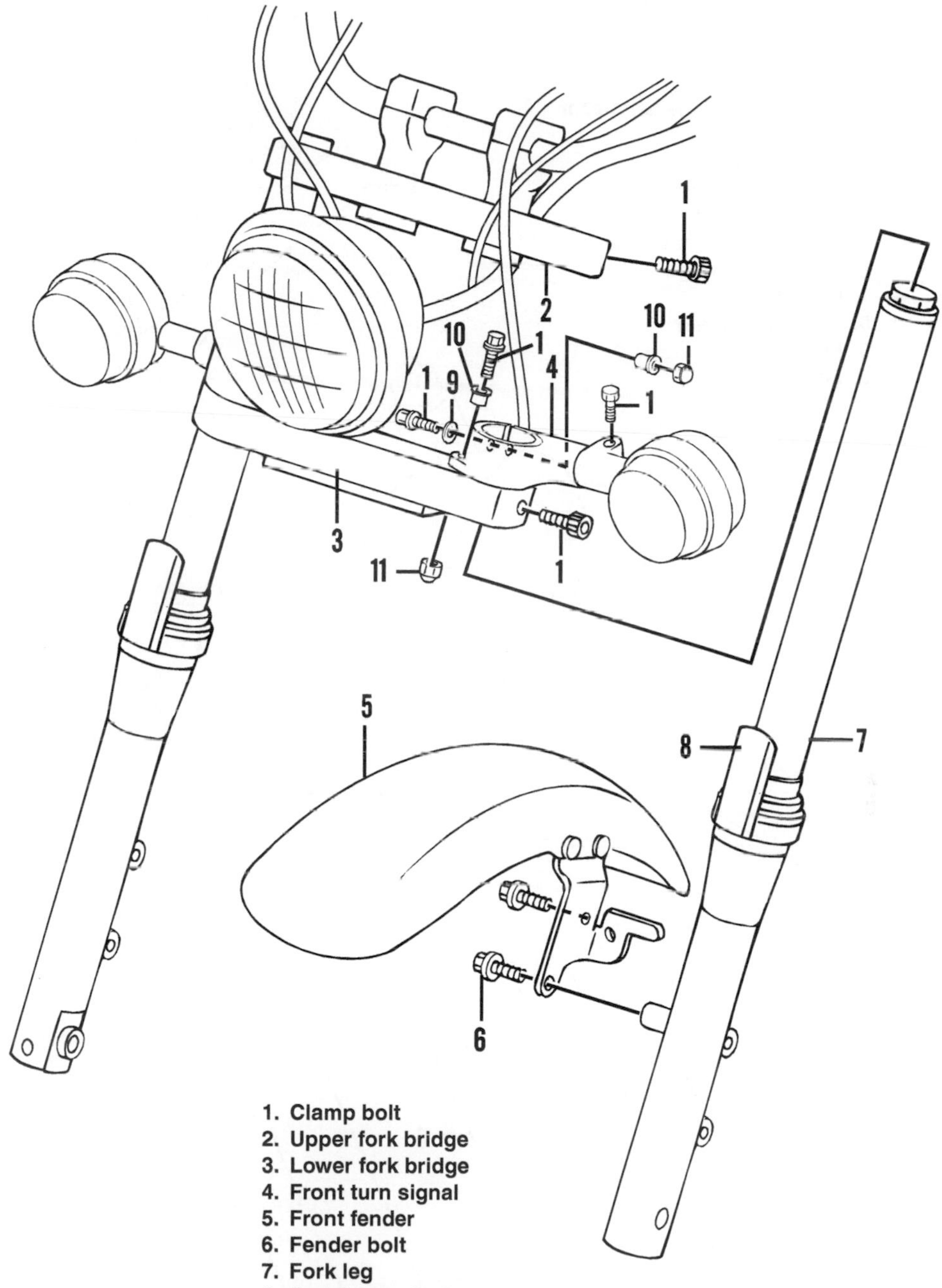

1. Clamp bolt
2. Upper fork bridge
3. Lower fork bridge
4. Front turn signal
5. Front fender
6. Fender bolt
7. Fork leg
8. Fork leg protector
9. Washer
10. Collar
11. Nut

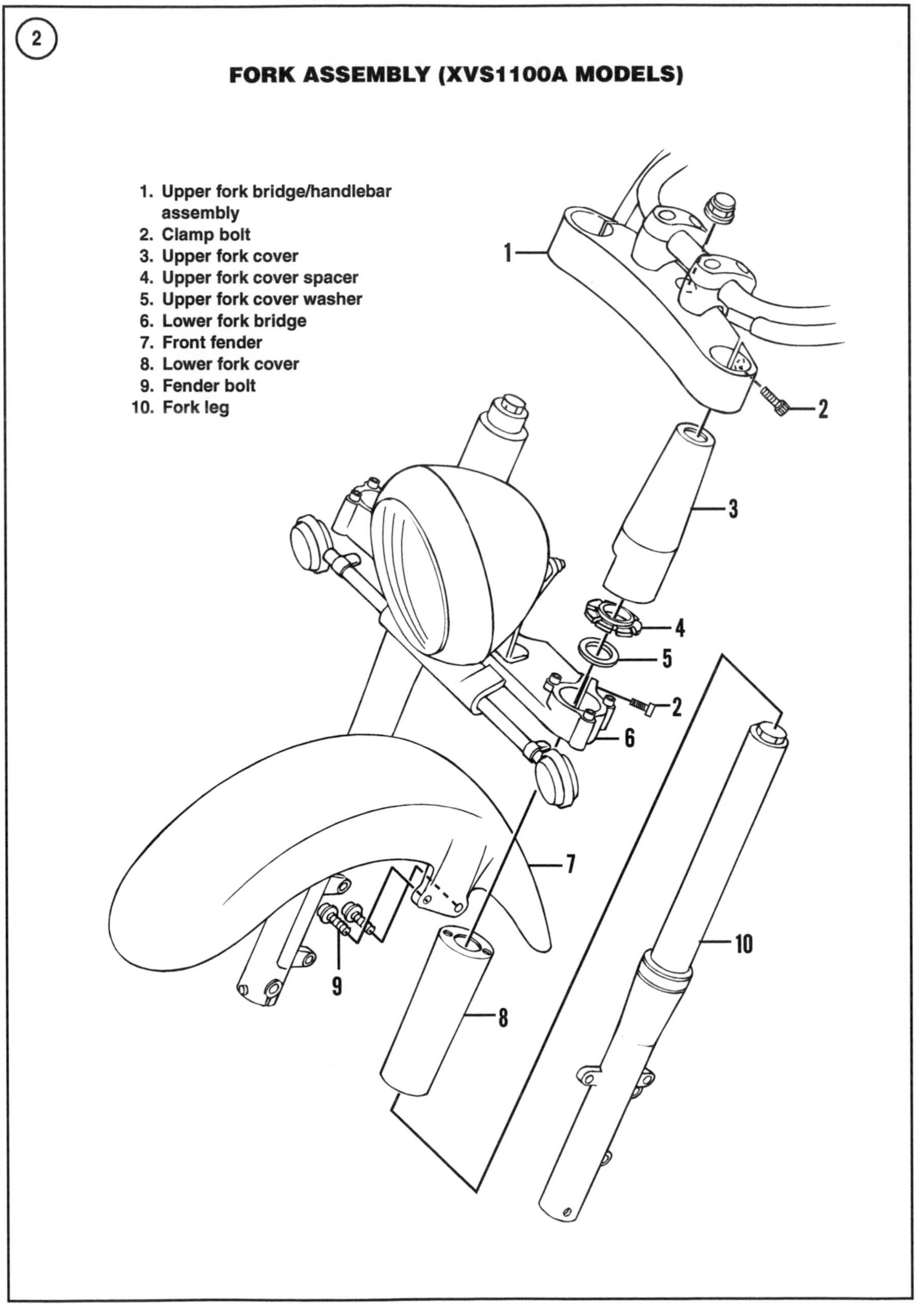
2
FORK ASSEMBLY (XVS1100A MODELS)
1. Upper fork bridge/handlebar assembly
2. Clamp bolt
3. Upper fork cover
4. Upper fork cover spacer
5. Upper fork cover washer
6. Lower fork bridge
7. Front fender
8. Lower fork cover
9. Fender bolt
10. Fork leg
1
2
3
4
5
2
6
7
10
9
8

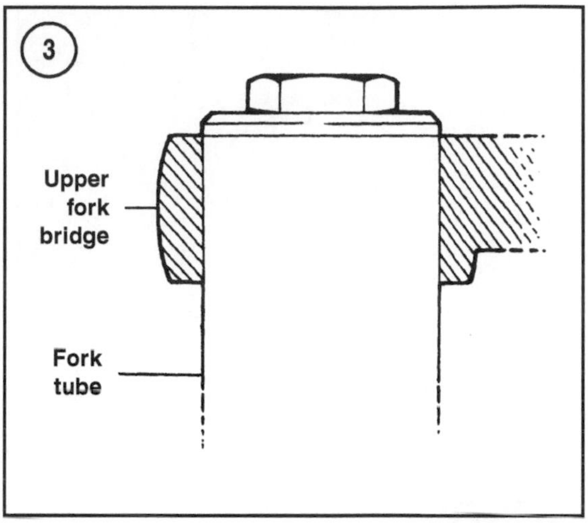

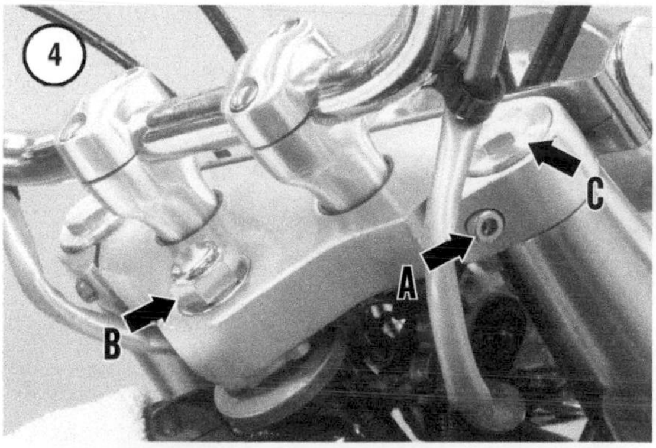

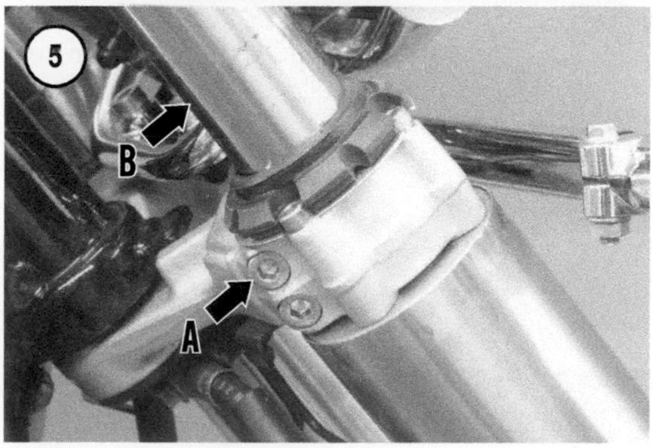

using the wood, bleeding the brake will not be necessary when installing the wheel.

3. Remove the front fender as described in Chapter Fourteen.

4. Note that the edge of the fork tube aligns with the top of the upper fork bridge (**Figure 3**). The fork

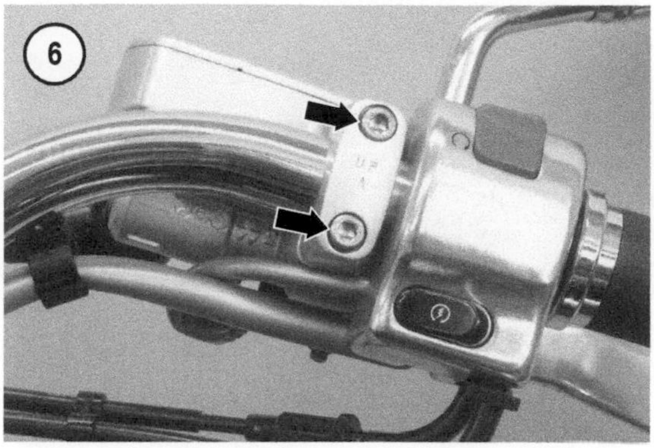

leg must be reinstalled to the same height during assembly.

5A. On XVS1100 models, perform the following:

a. Loosen the front flasher clamp bolt.

NOTE
A single clamp bolt is used on each side of the lower fork bridge on XVS1100 models.

b. Loosen the upper fork bridge clamp bolt (A, **Figure 4**) and the lower fork bridge clamp bolt (A, **Figure 5**).

c. Loosen the clamp bolt on the turn signal bracket.

d. Rotate the fork leg, and slide the fork tube (B, **Figure 5**) from the upper fork bridge, the turn signal bracket and the lower fork bridge.

5B. On XVS1100A models, perform the following:

CAUTION
The edges of the fork covers are sharp. Wear gloves and exercise caution to avoid injury.

a. Remove the fuel tank (Chapter Eight) so it will not be scratched or marred.

CAUTION
Cover the front fender and frame with a heavy cloth or plastic tarp to protect them from accidental brake fluid spills. Immediately wash spilled brake fluid off any painted or plated surface. Brake fluid will damage the finish. Use soapy water and rinse the area thoroughly.

b. Remove the front brake master cylinder clamp bolts (**Figure 6**) and remove the front

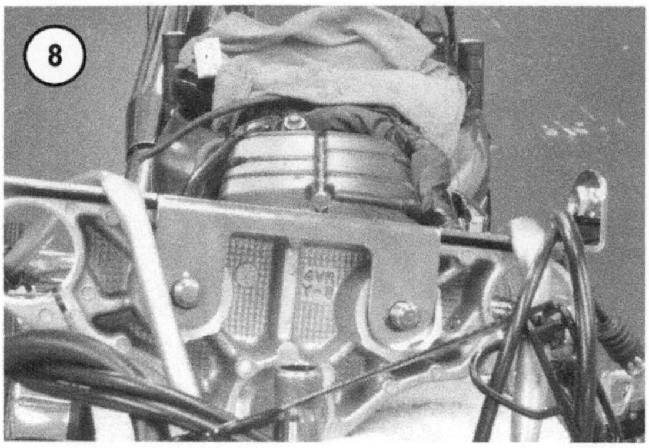

brake master cylinder from the handlebar. Suspend the master cylinder from the motorcycle with a bunjee cord. Keep the master cylinder reservoir upright. This prevents brake fluid spills and helps keep air out of the brake system. Do not disconnect the hydraulic brake line.

c. On models with a windshield, remove the two handlebar holder acorn nuts (**Figure 7**) under the upper fork bridge. These also secure the windshield bracket to the upper fork bridge.

d. Loosen the clamp bolt (A, **Figure 4**) on each side of the upper fork bridge.

e. Remove the steering head nut (B, **Figure 4**) and washer.

f. Place a tarp or plastic cover over the frame. Lift the upper fork bridge from the fork tubes, with the handlebar attached, and lay the fork bridge/handlebar assembly across the tarp (**Figure 8**).

g. Slide the upper fork cover (**Figure 9**) from the fork tube. Watch for the index pin (A, **Figure 10**) in the lower fork bridge.

h. Remove the indexing pin (A, **Figure 10**), upper fork cover spacer (B) and washer (C).

i. Loosen the two clamp bolts (A, **Figure 5**) on the lower fork bridge. Rotate the fork leg and lower it from the lower fork bridge and the lower fork cover.

j. If necessary, remove the mounting bolts (A, **Figure 11**) and remove the lower fork cover (B) from the lower fork bridge.

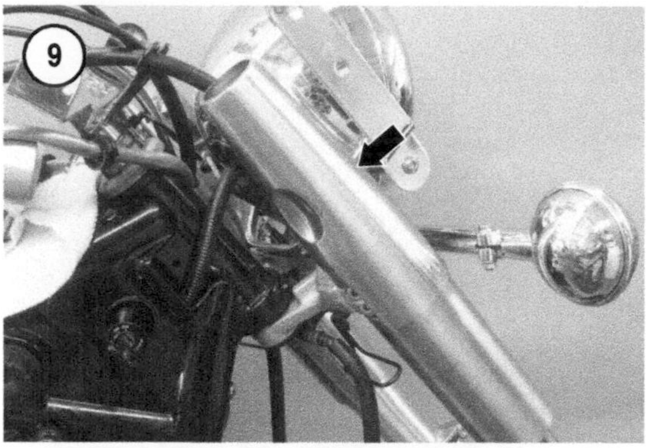

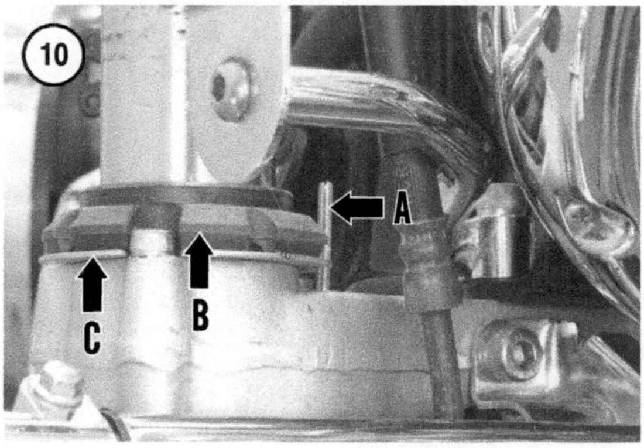

(12)

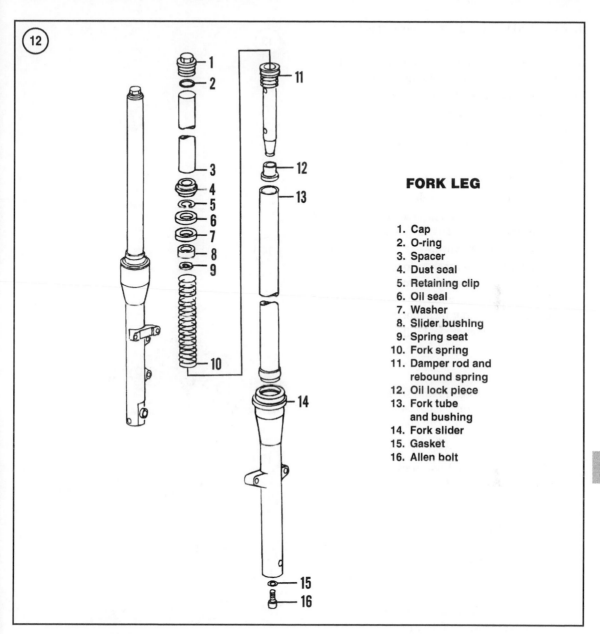

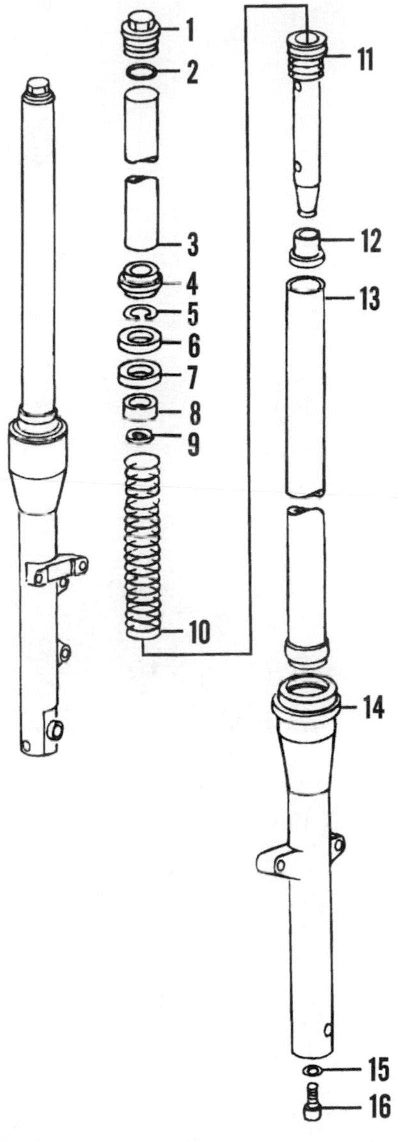

FORK LEG

1. Cap
2. O-ring
3. Spacer
4. Dust seal
5. Retaining clip
6. Oil seal
7. Washer
8. Slider bushing
9. Spring seat
10. Fork spring
11. Damper rod and rebound spring
12. Oil lock piece
13. Fork tube and bushing
14. Fork slider
15. Gasket
16. Allen bolt

Removal/Disassembly (Fork Leg Requires Service)

1. The following Yamaha special tools, or their equivalents, are needed to service a fork leg:

   a. Damper rod holder: part No. YM-1300-1 or 90890-01426.
   b. T-handle: part No. YM-01326 or 90890-01326.
   c. Fork seal driver: Yamaha part No. YM-33963 or 90809-01367.
   d. Driver adapter: part No. YM-33968 or 90890-01381.

NOTE

The fork leg Allen bolt can be removed without using the damper rod holder and T-handle. However, the Allen bolt cannot be torqued without these tools.

2. Securely support the motorcycle on a level surface.

3. Remove the front wheel (Chapter Ten) and the front fender (Chapter Fourteen).

4. On XVS1100A models, remove the fuel tank (Chapter Eight).
5. Note that the top of the fork tube aligns with the top of the upper fork bridge (**Figure 3**). The fork leg must be reinstalled in this same position.
6. Remove any cable ties or holders that secure a brake, clutch or electrical line to a fork leg. Note the location of these cable ties. They must be reinstalled during installation.
7. Remove the Allen bolt as follows:
   a. Place a drain pan beneath the fork leg.
   b. Insert the axle through both fork legs. The axle holds the sliders in position while the Allen bolt is loosened.
   c. Use an impact wrench to loosen and remove the Allen bolt from the bottom of the fork leg. If the Allen bolt cannot be loosened, use the damper rod holder mentioned in Step 10.
8. Pull out the fork slider and fully extend the fork leg.

CAUTION
*The cap bolt (C, **Figure 4**) is under spring pressure. Exercise caution when removing the cap bolt.*

NOTE
Use a 6-point, 22 mm socket on the cap bolt. A 12-point socket will mar the bolt.

9. Use a 6-point, 22 mm socket to break loose the cap bolt and slowly loosen the cap bolt (C, **Figure 4**). Remove the cap bolt, spacer, spring seat and the fork spring. Note that the closer wound coils on the spring face up.
10. If the Allen bolt could not be loosened during Step 7, remove it now as follows:
   a. Place a drain pan beneath the fork leg.
   b. Install the damper rod holder onto the T-handle.
   c. Insert the holder into the fork tube until the holder engages the damper rod (**Figure 13**). Hold the damper rod and remove the Allen bolt (**Figure 14**) along with its copper washer from the bottom of the slider.
   d. Let the oil drain from the fork. If necessary, pump the slider to expel any residual oil.
11. Remove the dust seal (A, **Figure 15**) and the retaining clip (B) from the fork slider.

13
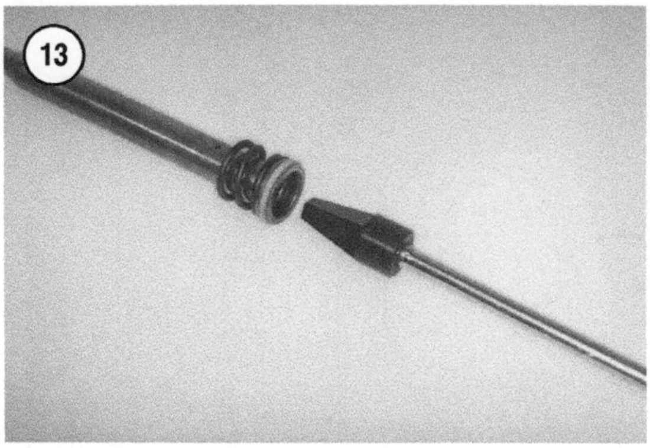

14
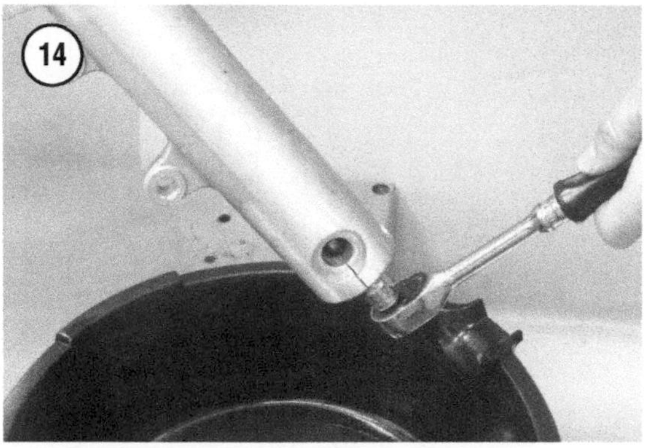

15
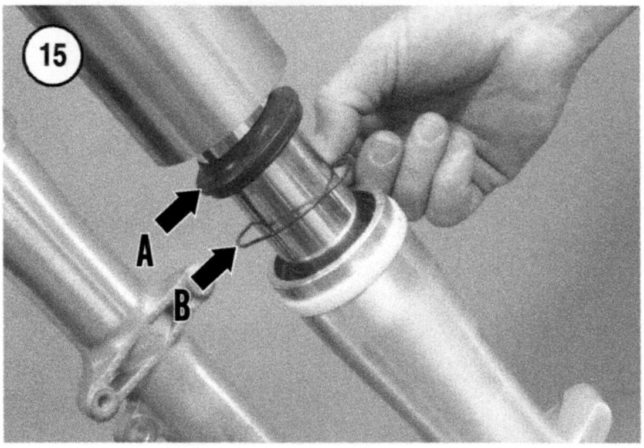

NOTE
It may be necessary to slightly heat the area on the slider around the oil seal prior to removal. Heat the area with a rag soaked in hot water. Do not apply a flame directly to the fork slider.

12. There is an interference fit between the bushing in the fork slider and the bushing on the fork tube. In order to remove the fork slider from the fork tube, firmly grasp the slider (**Figure 16**). Pull hard

16
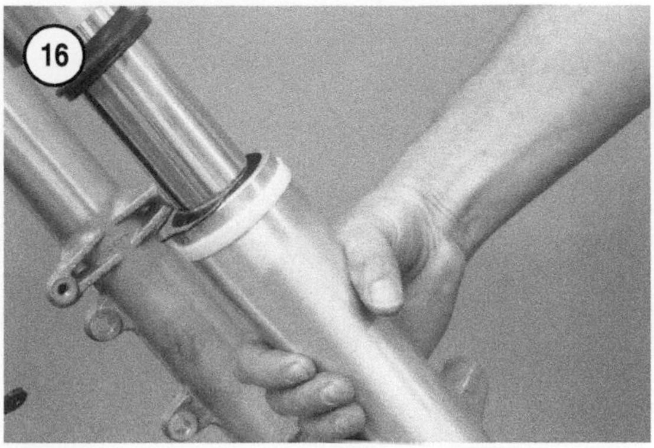

17
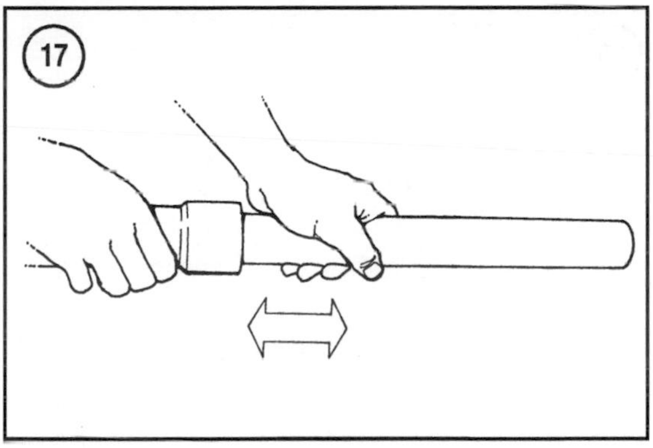

18
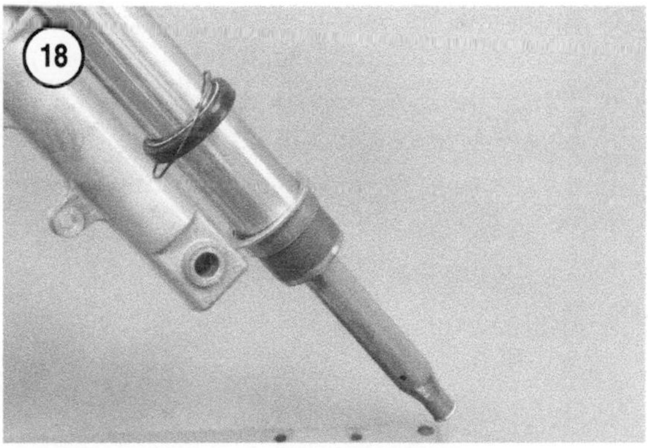

on the slider using quick in-and-out strokes (**Figure 17**), and pull the slider off the fork tube (**Figure 18**).

13. Pour any excess oil from the slider. Invert the slider and remove the oil lock piece.

14A. On XVS1100 models, perform the following:

a. Loosen the front flasher clamp bolt.

NOTE
On XVS1100 models, a single clamp bolt is used on each side of the lower fork bridge.

b. Loosen the upper fork bridge clamp bolt (A, **Figure 4**) and the lower fork bridge clamp bolt (A, **Figure 5**).

c. Rotate the fork tube (B, **Figure 5**), and slide it from the upper fork bridge, front turn signal assembly and the lower fork bridge. Take the fork tube to the bench for further disassembly.

d. Do not remove the fork leg protector from the slider unless it is damaged. The protector must be replaced once it has been removed.

14B. On XVS1100A models, perform the following:

CAUTION
The edges of the fork covers are sharp. Wear gloves and exercise caution to avoid injury.

a. On models with a windshield, remove the two handlebar holder acorn nuts (**Figure 7**) under the upper fork bridge. These also secure the windshield bracket to the upper fork bridge.

b. Remove the front brake master cylinder clamp bolts (**Figure 6**) and remove the front brake master cylinder from the handlebar. Suspend the master cylinder from the motorcycle with a bunjee cord. Keep the master cylinder reservoir upright. This prevents brake fluid spills and helps keep air out of the brake system. Do not disconnect the hydraulic brake line.

c. Loosen the clamp bolt (A, **Figure 4**) on each side of the upper fork bridge.

d. Remove the steering head nut (B, **Figure 4**) and washer.

CAUTION
Cover the front fender and frame with a heavy cloth or plastic tarp to protect them from accidental brake fluid spills. Immediately wash spilled brake fluid off any painted or plated surface. Brake fluid will damage the finish. Use soapy water and rinse the area thoroughly.

e. Place a tarp or plastic cover over the frame. Lift the upper fork bridge from the fork tubes, with the handlebar attached, and lay the fork bridge/handlebar assembly across the tarp (**Figure 8**).

f. Slide the upper fork cover (**Figure 9**) from the fork tube. Watch for the index pin (A, **Figure 10**) in the lower fork bridge.

g. Remove the index pin (A, **Figure 10**), upper fork cover spacer (B) and washer (C).
h. Loosen the two clamp bolts (A, **Figure 5**) on the lower fork bridge. Rotate the fork tube (B, **Figure 5**), and lower it from the lower fork bridge and the lower fork cover. Take the fork tube to the bench for further disassembly.
i. If necessary, remove the mounting bolts (A, **Figure 11**) and remove the lower fork cover (B) from the lower fork bridge.

15. Remove the damper rod (A, **Figure 19**) from the fork tube.

NOTE
*Do not remove the fork tube bushing (F, **Figure 20**) unless it is going to be replaced. Inspect it as described in this chapter.*

16. Slide the dust seal (A, **Figure 20**), retaining clip (B), oil seal (C), washer (D) and the slider bushing (E) from the fork tube.
17. Inspect all parts as described in this section.

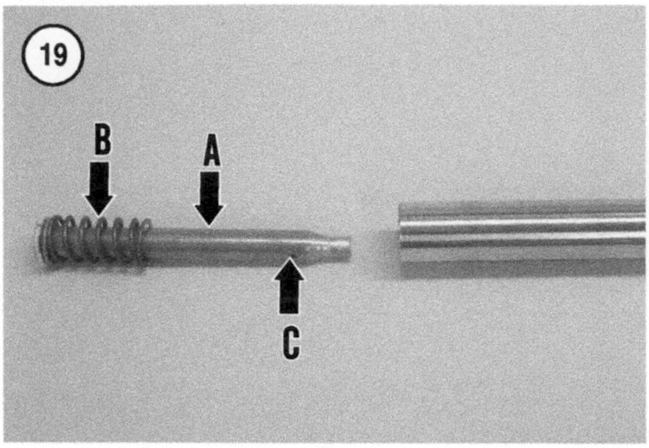

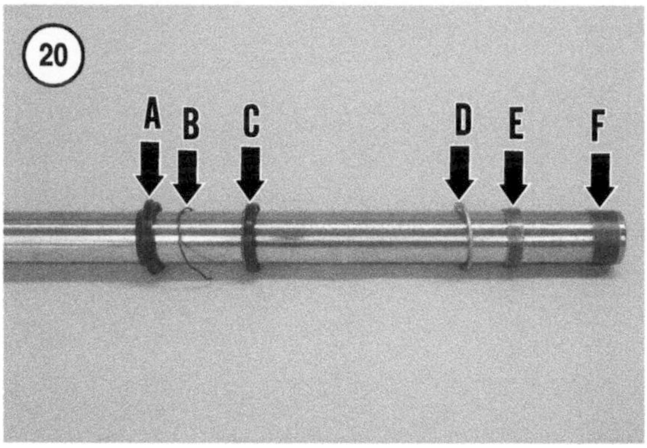

Installation

XVS1100 models

1. If the fork leg was disassembled, assemble it as described in this section.
2. Slide the fork tube up through the lower fork bridge, through the turn signal bracket, and into the upper fork bridge.
3. Align the top of the fork tube with the top edge of the upper fork bridge as shown in **Figure 3**.
4. Torque the fasteners in the following order:

NOTE
A single clamp bolt is used on each side of the lower fork bridge on XVS1100 models.

a. Torque the lower fork bridge clamp bolt (A, **Figure 5**) to 30 N•m (22 ft-lb.).
b. Torque the front turn signal clamp bolt to 7 N•m (62 in.-lb.).

NOTE
Use a 6-point, 22 mm socket on the cap bolt. A 12-point socket will mar the cap bolt.

c. If the fork leg was disassembled, torque the cap bolt (C, **Figure 4**) to 23 N•m (17 ft.-lb.).
d. Torque the upper fork bridge clamp bolt (A, **Figure 4**) to 20 N•m (15 ft.-lb.).

5. Install the front fender as described in Chapter Fourteen.
6. Install the front wheel as described in Chapter Ten.

XVS1100A models

1. If the fork leg was disassembled, assemble it as described in this chapter.
2. Install the lower fork cover (B, **Figure 11**) if it was removed. Secure the cover to the lower fork bridge with the mounting bolts (A, **Figure 11**).

CAUTION
The edges of the fork covers are sharp. Wear gloves and exercise caution. Hold the slider below the brake caliper mounting boss during installation so your hand will be protected if the fork leg slides abruptly.

21

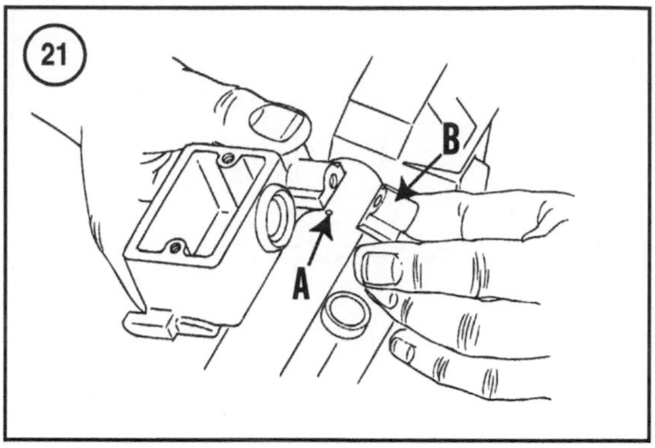

22

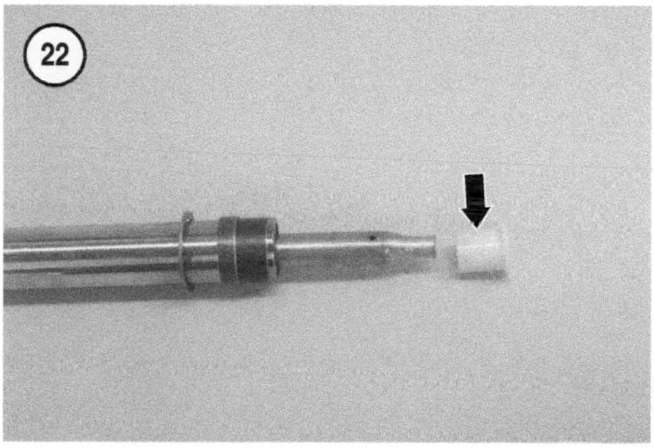

3. Install the washer (C, **Figure 10**), upper fork cover spacer (B) and the index pin (A).

4. Rotate the fork leg and install it (B, **Figure 5**) through the lower fork cover and the lower fork bridge. Tighten the lower fork bridge clamp bolts (A, **Figure 5**) enough to hold the fork leg in place.

NOTE
The upper fork bridge must be temporarily installed so the fork leg can set to the proper height.

5. Lower the upper fork bridge/handlebar assembly onto the front fork legs and the steering stem. Install the steering head nut and washer (B, **Figure 4**). Torque the steering head nut to 110 N•m (81 ft.-lb.).

6. Loosen the lower fork bridge clamp bolts. Reposition the fork leg and align the top edge of the fork tube with the top edge of the upper fork bridge as shown in **Figure 3**.

7. Torque the clamp bolts on the lower fork bridge (A, **Figure 5**) to 30 N•m (22 ft.-lb.).

NOTE
Use a 6-point, 22 mm socket on the cap bolt. A 12-point socket will mar the bolt.

8. If the fork leg was disassembled, torque the cap bolt (C, **Figure 4**) to 23 N•m (17 ft.-lb.).
9. Remove the steering head nut (B, **Figure 4**) and washer, and remove the upper fork bridge/handlebar assembly. Lay the assembly on the frame.
10. Slide the upper fork cover (**Figure 9**) down the fork tube. Make sure the cover engages the indexing pin (A, **Figure 10**) in the lower fork bridge.
11. Install the upper fork bridge/handlebar assembly onto the fork legs and steering stem. Install the steering head nut (B, **Figure 4**) and washer. Torque the steering head nut to 110 N•m (81 ft.-lb.).
12. Make sure the upper edge of each fork tube aligns with the upper edge of the upper fork bridge as shown in **Figure 3**. Adjust the fork leg as necessary.
13. Torque the upper fork bridge clamp bolt (A, **Figure 4**) on each side to 20 N•m (15 ft.-lb.).
14. Install the front brake master cylinder as follows:
   a. Position the master cylinder so the face of the clamp mating surface aligns with the mark on the handlebar. See A, **Figure 21**.
   b. Install the master cylinder clamp (B, **Figure 21**) so the UP stamped on the clamp faces up.
   c. Torque the clamp bolts (**Figure 6**) to 10 N•m (89 in.-lb.) Tighten the upper clamp bolt first, then the lower bolt. There should be a gap at the lower part of the clamp after tightening.
15. Install the front wheel (Chapter Ten).
16. Install the front fender (Chapter Fourteen).
17. Install the fuel tank (Chapter Eight).
18. If necessary, bleed the front brakes (Chapter Three).

Assembly

1. Coat all parts with fresh SAE 10W fork oil before installation.
2. If removed, install a new slider bushing (F, **Figure 20**) as described in *Inspection* in this section.
3. Slide the rebound spring (B, **Figure 19**) onto the damper rod (A).
4. Insert the damper rod into the fork tube until the rod emerges from the fork tube end. Install the oil lock piece (**Figure 22**) onto the damper rod.

23

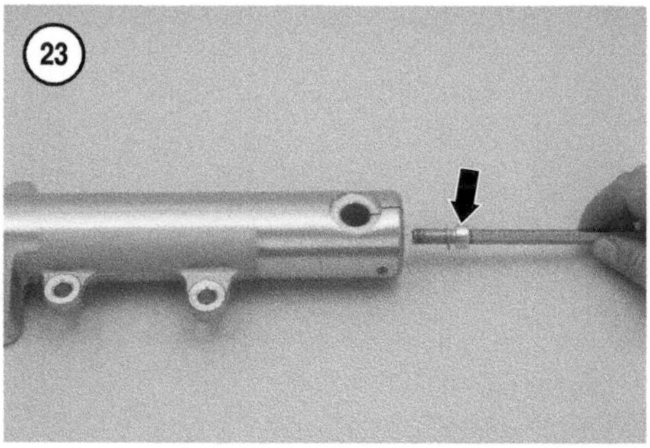

24

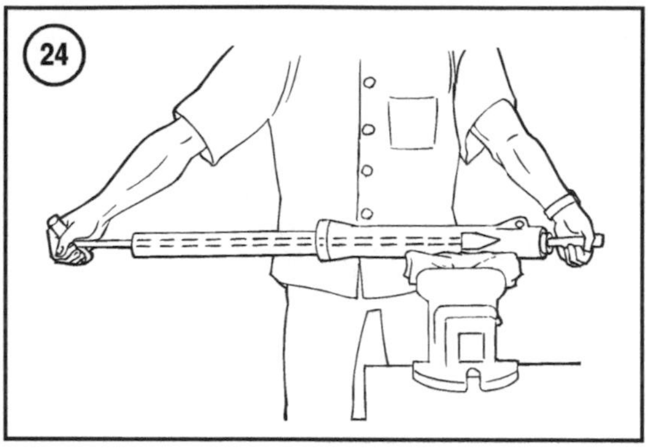

5. Install the fork tube into the slider until the tube bottoms.

6. Install the Yamaha damper rod holder onto the T-handle. Insert the tool into the fork tube so the rod holder engages the damper rod (**Figure 13**).

7. Slide a new copper washer onto the Allen bolt (**Figure 23**).

8. Apply a medium-strength threadlocking compound to the threads of the Allen bolt. Insert the Allen bolt through the bottom of the slider and thread it into the oil lock piece.

9. Hold the damper rod with the special tool (**Figure 24**) and tighten the Allen bolt to 30 N•m (22 ft.-lb.).

10. Slide a new slider bushing (E, **Figure 20**) and washer (D) down the fork tube.

NOTE

Use the Yamaha fork seal driver and adapter to install the slider bushing and washer in the next step. If these tools are not available, use a universal oil seal driver or a piece of galvanized pipe and a hammer. If both ends of the pipe are threaded, wrap one end with duct tape to prevent the threads from damaging the interior of the slider.

11. Drive the slider bushing and washer into place (**Figure 25**) until the bushing bottoms in the slider.

NOTE

The plastic wrap installed in the next step protects the oil seal and dust seal so they will not be torn during installation.

25

26

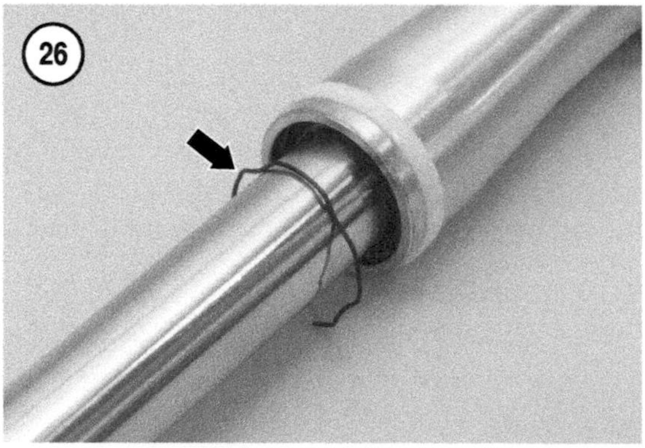

27

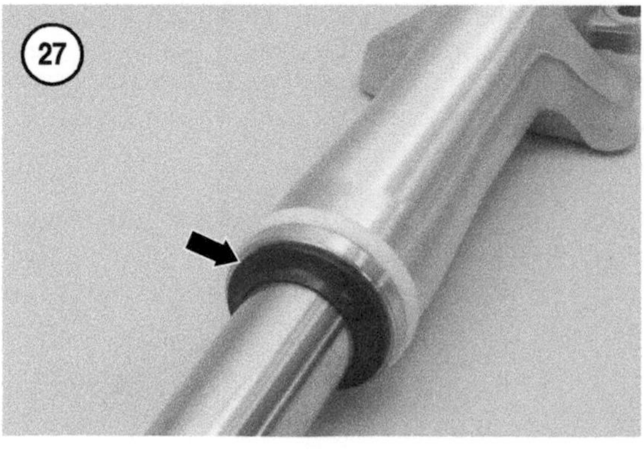

28

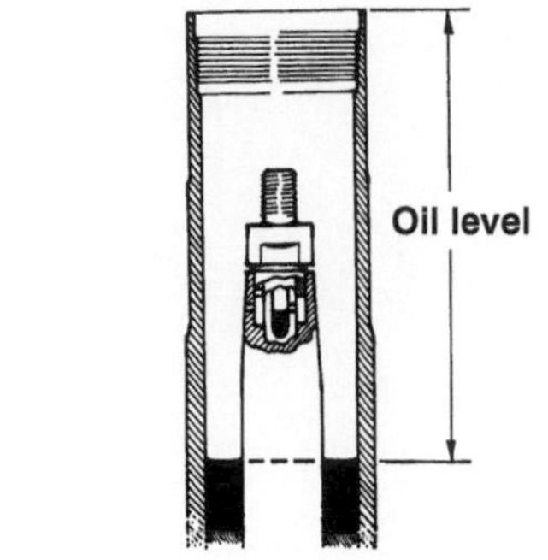

32

29

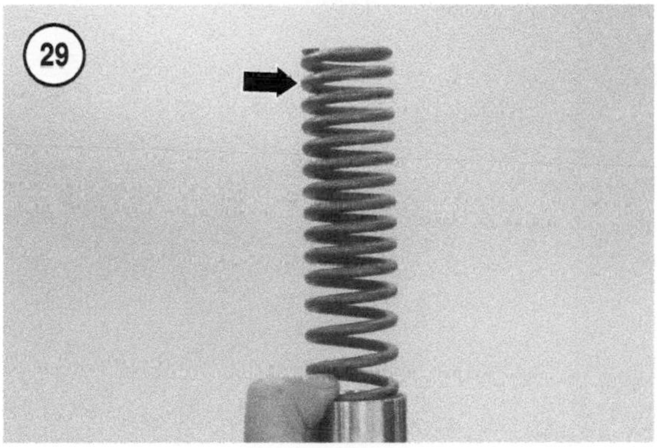

30

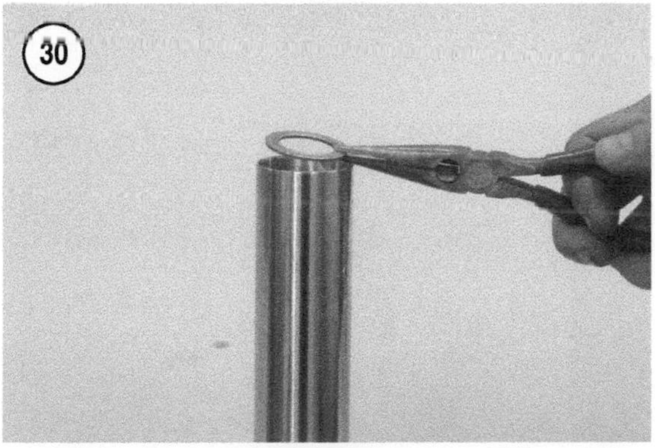

31

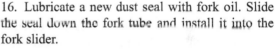

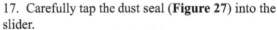

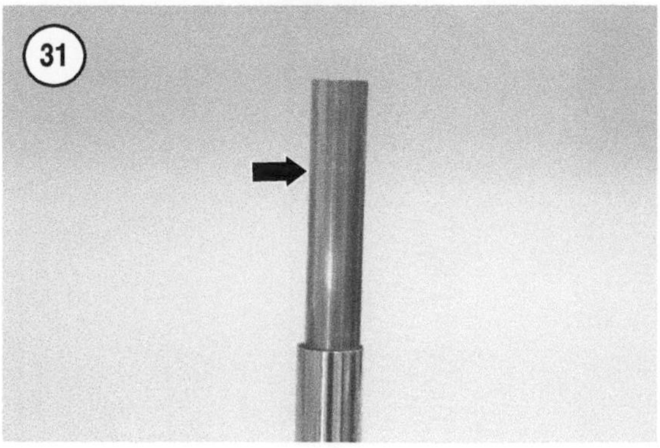

12. Wrap the end of the fork tube with plastic wrap. Liberally coat the plastic wrap with fork oil.
13. Lubricate the lips of the oil seal with fork oil and install the oil seal onto the fork tube (C, **Figure 20**). Make sure the side with the manufacturer's marks faces up, away from the fork slider.
14. Use the same tool used in Step 11 to drive the oil seal into the slider until the retaining clip groove in the slider can be seen above the top of the oil seal.
15. Install the retaining clip (**Figure 26**) down the fork tube and seat the clip into the groove in the slider. Make sure the clip is completely seated in the slider groove.
16. Lubricate a new dust seal with fork oil. Slide the seal down the fork tube and install it into the fork slider.
17. Carefully tap the dust seal (**Figure 27**) into the slider.
18. Secure the fork leg upright in a vise and fill the fork leg with the correct quantity of SAE 10W fork oil. Refer to **Table 1** for the specified quantity.
19. Slowly pump the fork up and down several times to distribute the fork oil.
20. Compress the fork completely and measure the fluid level after the fork oil settles. Use an oil level gauge to measure the fluid level from the top of the fork tube (**Figure 28**). If necessary, add or remove oil to set the fluid level to the value in **Table 1**.
21. Pull the fork tube out of the slider until it is fully extended.
22. Install the fork spring into the fork tube so the closer wound coils (**Figure 29**) face up.
23. Install the spring seat (**Figure 30**) onto the top of the fork spring. Make sure it is seated correctly.
24. Install the spacer (**Figure 31**).
25. Lubricate a new O-ring (**Figure 32**) with lithium soap grease and install it onto the fork cap.

11

26. Install the fork cap while pushing down on the spring. Start the bolt slowly; do not cross-thread it. Tighten the fork cap as tightly as possible. The cap bolt will be torqued to specification once the fork is installed on the motorcycle.
27. Install the fork leg as described earlier in this chapter.

Inspection

1. Thoroughly clean all parts in solvent and dry them with compressed air.
2. Blow out the oil holes in the damper rod with compressed air. Clean them if necessary.
3. Check the damper rod assembly for:
   a. Bent, cracked or otherwise damaged damper rod (A, **Figure 33**).
   b. Excessively worn rebound spring (B, **Figure 33**).
   c. Damaged oil lock piece (C, **Figure 33**).
   d. Excessively worn or damaged piston or piston ring (D, **Figure 33**).
4. Replace any worn part.
5. Check the fork tube for straightness and for signs of wear or scratches. If it is bent or severely scratched, replace it.
6. Check the fork tube for chrome flaking or creasing. This condition will damage the oil seal. Replace the fork tube if necessary.
7. Check the seal area of the slider (**Figure 34**) for dents, scratches or other damage that would allow oil leaks. Replace the slider if necessary.
8. Check the slider for dents or exterior damage that may cause the upper fork tube to hang up during riding. Replace it if necessary. Check for cracks or damage to the brake caliper (A, **Figure 35**) and fender mounting bosses (B).
9. Inspect the front axle threads in the slider for damage. If damage is slight, chase the threads with a metric tap. If damage is excessive, replace the slider.
10. Measure the uncompressed length of the fork spring (**Figure 36**). If the spring has sagged to the service limit (**Table 1**), replace it.
11. Inspect the fork tube bushing (F, **Figure 20**). If it is scratched or scored, replace it. If the Teflon coating is worn off so the copper base material is showing on approximately 3/4 of the total surface (**Figure 37**), replace the bushing.

33
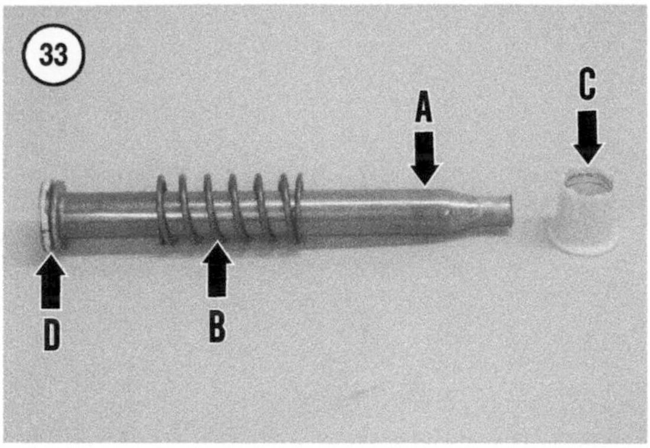

34

35
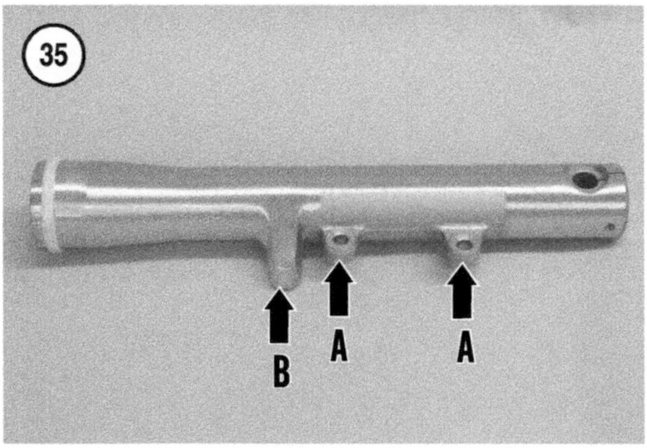

36
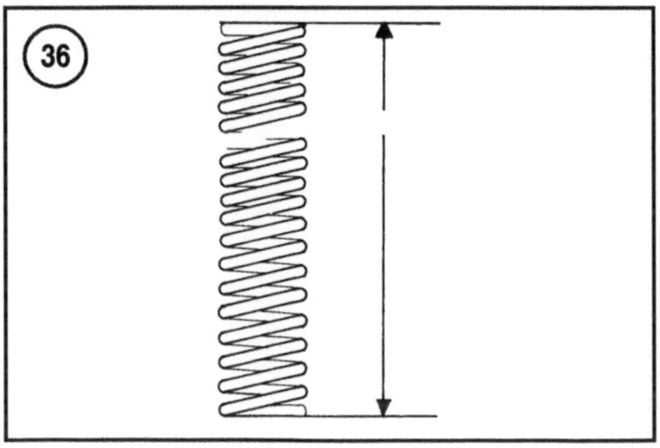

37

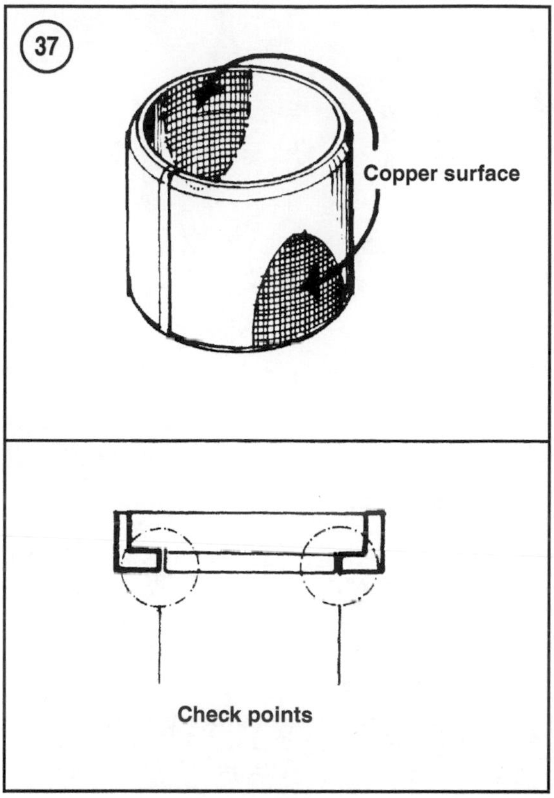

38

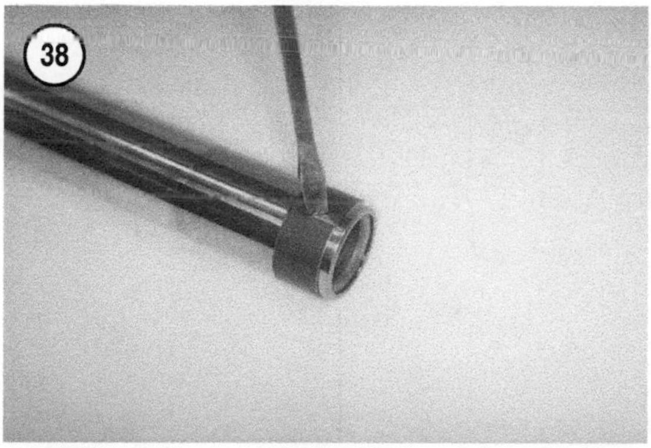

39

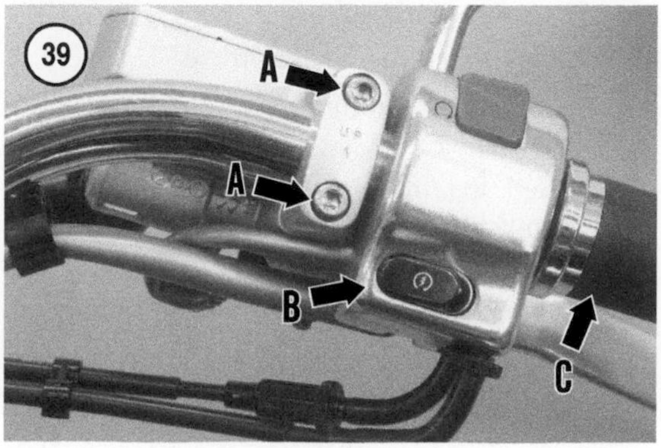

a. Open the bushing slot with a screwdriver (**Figure 38**) and slide the bushing off the fork tube.

b. Lubricate the new bushing with fork oil, open its slot and slide it onto the fork tube groove.

12. Replace the fork cap O-ring (**Figure 32**). Lubricate the O-ring with fork oil before installation.

13. Replace any parts that are worn or damaged. Simply cleaning and reinstalling unserviceable fork components will not improve performance of the front suspension.

HANDLEBAR

Removal/Installation

CAUTION

Cover the front fender, frame and fuel tank with a heavy cloth or plastic tarp to protect it from accidental brake fluid spills. Immediately wash spilled brake fluid off any painted or plated surface. Brake fluid will destroy the finish. Use soapy water and rinse the area thoroughly.

NOTE

If handlebar replacement is not required, proceed to Step 10.

1. Securely support the motorcycle on level ground.

2. Remove all cable ties that secure cables to the handlebar.

3. Remove the two master cylinder clamp bolts (A, **Figure 39**) and remove the master cylinder from the handlebar. Secure the master cylinder to the frame with a bunjee cord. Make sure the master cylinder reservoir remains upright. This prevents brake fluid spills and helps keep air out of the brake system. Do not disconnect the hydraulic brake line.

4. Remove the right handlebar switch assembly (B, **Figure 39**) as described in Chapter Nine.

5. Remove the handlebar weight and slide the throttle grip assembly (C, **Figure 39**) from the handlebar.

6. Remove the left handlebar switch as described in Chapter Nine.

7. Slide the clutch lever boot away from the adjuster and loosen the clutch cable locknut (A, **Figure 40**). Rotate the adjuster (B, **Figure 40**) to

provide maximum slack in the cable and disconnect the cable end from the clutch hand lever.

8. Disconnect the electrical lead (C, **Figure 40**) from the clutch switch.

9. Remove the clutch lever nuts and bolts (D, **Figure 40**), and remove the clutch lever assembly.

10. Remove the caps (A, **Figure 41**), then loosen the clamp bolts (B) on the upper handlebar holders.

11. Remove the upper handlebar holders from the handlebar and lift the handlebar from the lower handlebar holders.

12. If necessary, replace the left hand grip as follows:

a. Unscrew the handlebar end and remove it from the handlebar.

b. Insert a thin-bladed screwdriver under the hand grip.

c. Squirt electrical contact cleaner under the hand grip and twist it quickly to break its seal and remove it.

d. Apply a thin layer of rubber cement to the end of the handlebar when installing a new grip.

e. Check the hand grip after 10 minutes to make sure it is tight.

WARNING

Do not ride the motorcycle with a loose hand grip. Loss of control will occur.

13. Installation is the reverse of removal. Note the following:

a. Replace the handlebar if it is bent.

b. Make sure the punch mark on the handlebar (A, **Figure 42**) aligns with the top edge of the lower handlebar holder.

c. Install the upper handlebar holders so the punch mark on the top of each holder (B, **Figure 42**) faces forward.

d. Tighten the handlebar holder clamp bolts to 28 N•m (21 ft.-lb.). Tighten the front bolt first, then tighten the rear bolt for each handlebar holder.

e. Position the clutch lever bracket (A, **Figure 43**) so the gap in the bracket clamp aligns with the index mark (B) on the handlebar.

f. Install the left handlebar switch and the right handlebar switch as described in Chapter Nine.

40

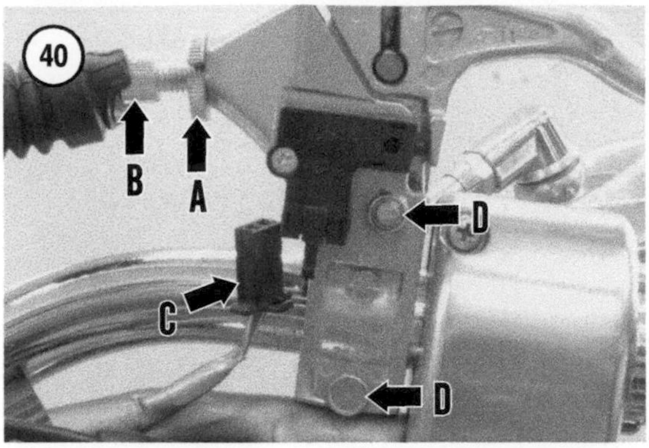

41

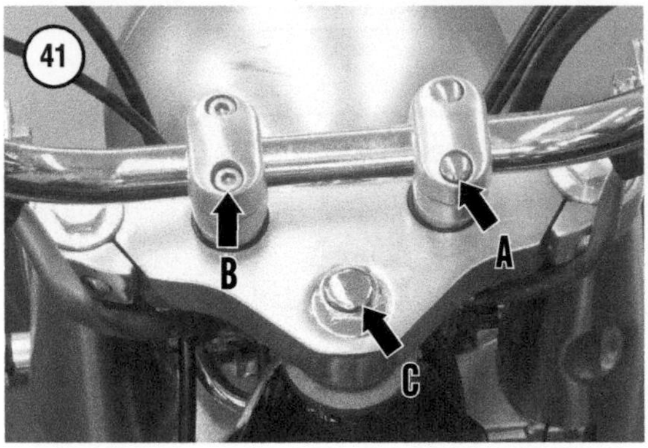

42

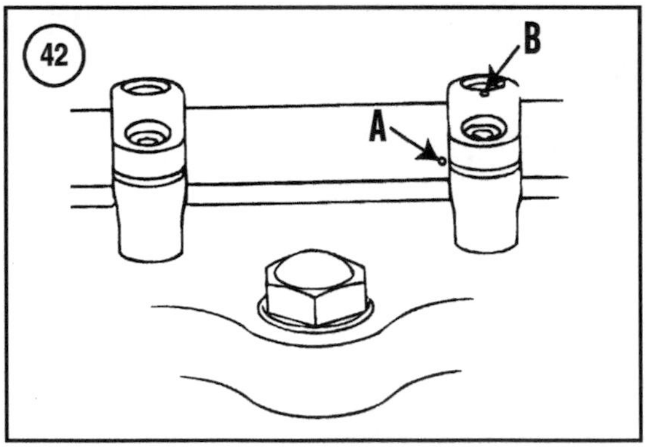

g. Lubricate the ends of the clutch and throttle cables with lithium soap grease.

h. Lubricate the throttle grip assembly with lithium soap grease.

i. Install the front brake master cylinder as follows:

j. Position the master cylinder so the face of the clamp mating surface aligns with the mark on the handlebar. See A, **Figure 44**.

k. Install the master cylinder clamp (B, **Figure 44**) so the UP stamped on the clamp faces up.

43

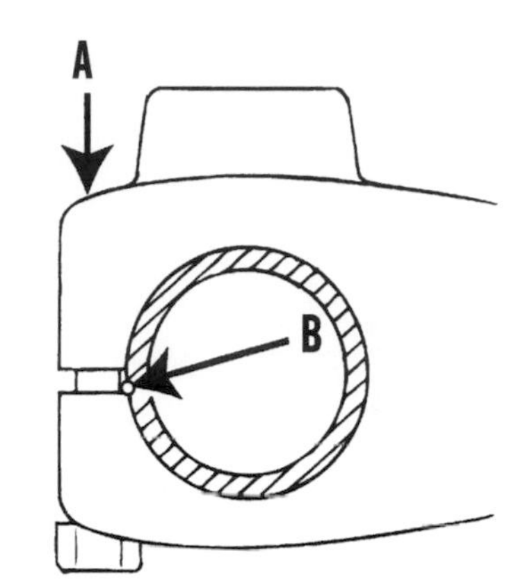

44

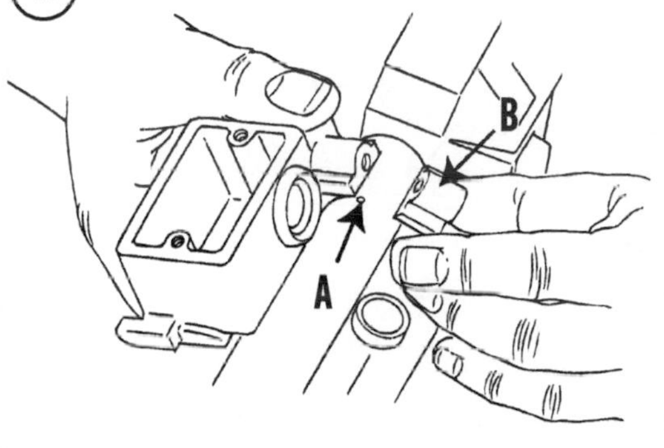

l. Torque the clamp bolts (A, **Figure 39**) to 10 N•m (89 in.-lb.). Tighten the upper clamp bolt first, then the lower bolt. There should be a gap at the lower part of the clamp after tightening.

Inspection

Check the handlebar along the entire mounting area for cracks or damage. Replace a bent or damaged handlebar immediately. If the bike is involved in a crash, examine the handlebar, steering stem and front fork legs carefully.

STEERING HEAD

Removal

The Yamaha ring nut wrench (part No. YU-33975 or 90890-01403) or an equivalent tool is needed to disassemble and reassemble the steering head.

Refer to **Figure 45** for this procedure.

1. Securely support the motorcycle on a level surface.
2. Remove the front wheel as described in Chapter Ten.
3. Remove the headlight housing as described in Chapter Nine.
4. Remove the each fork leg as described in this chapter.

5A. On XVS1100 models, perform the following:

a. Cover the frame and fuel tank with a heavy cloth or plastic tarp to protect them from accidental brake fluid spills.
b. Remove the steering head nut (C, **Figure 41**) and washer.
c. Lift the upper fork bridge from the steering stem, with the handlebar attached, and lay the fork bridge/handlebar assembly across the tank.
d. Remove the mounting nut and bolt, and remove each flasher assembly from the lower fork bridge.

5B. On XVZ1100A models, remove the flasher assembly as follows:

a. Pull the cover (A, **Figure 46**) from the flasher bracket.
b. Disconnect the flasher bullet connectors (B, **Figure 46**) from their harness mates.
c. Remove the mounting bolts (C, **Figure 46**). Pull the flasher wires through the hole in the headlight housing, and remove the flasher bracket from the lower fork bridge.

6. Remove the mounting bolts (A, **Figure 47**) and remove the brake hose union (B) from the lower fork bridge. Use a bunjee cord or wire to suspend the union from the motorcycle.
7. If necessary, remove the headlight housing bracket(s) from the lower fork bridge.
8. Remove the lockwasher (A, **Figure 48**) from the top of the ring nuts.
9. Remove the locking ring nut (B, **Figure 48**), then the rubber washer (**Figure 49**).
10. Hold onto the lower end of the steering stem assembly.
11. Loosen and remove the adjusting ring nut (A, **Figure 50**) and the bearing cap (B).
12. Remove the upper bearing inner race (A, **Figure 51**) and the bearing (B) from the steering head.

(45)

STEERING HEAD

1. Cap
2. Bolt
3. Upper handlebar holder
4. Lower handlebar holder
5. Washer
6. Damper
7. Cable holder
8. Nut
9. Steering head nut
10. Upper fork bridge
11. Lockwasher
12. Locking ring nut
13. Adjusting ring nut
14. Bearing cap
15. Bearing
16. Dust seal
17. Headlight housing bracket
18. Steering stem
19. Headlight housing bracket
20. Steering stem

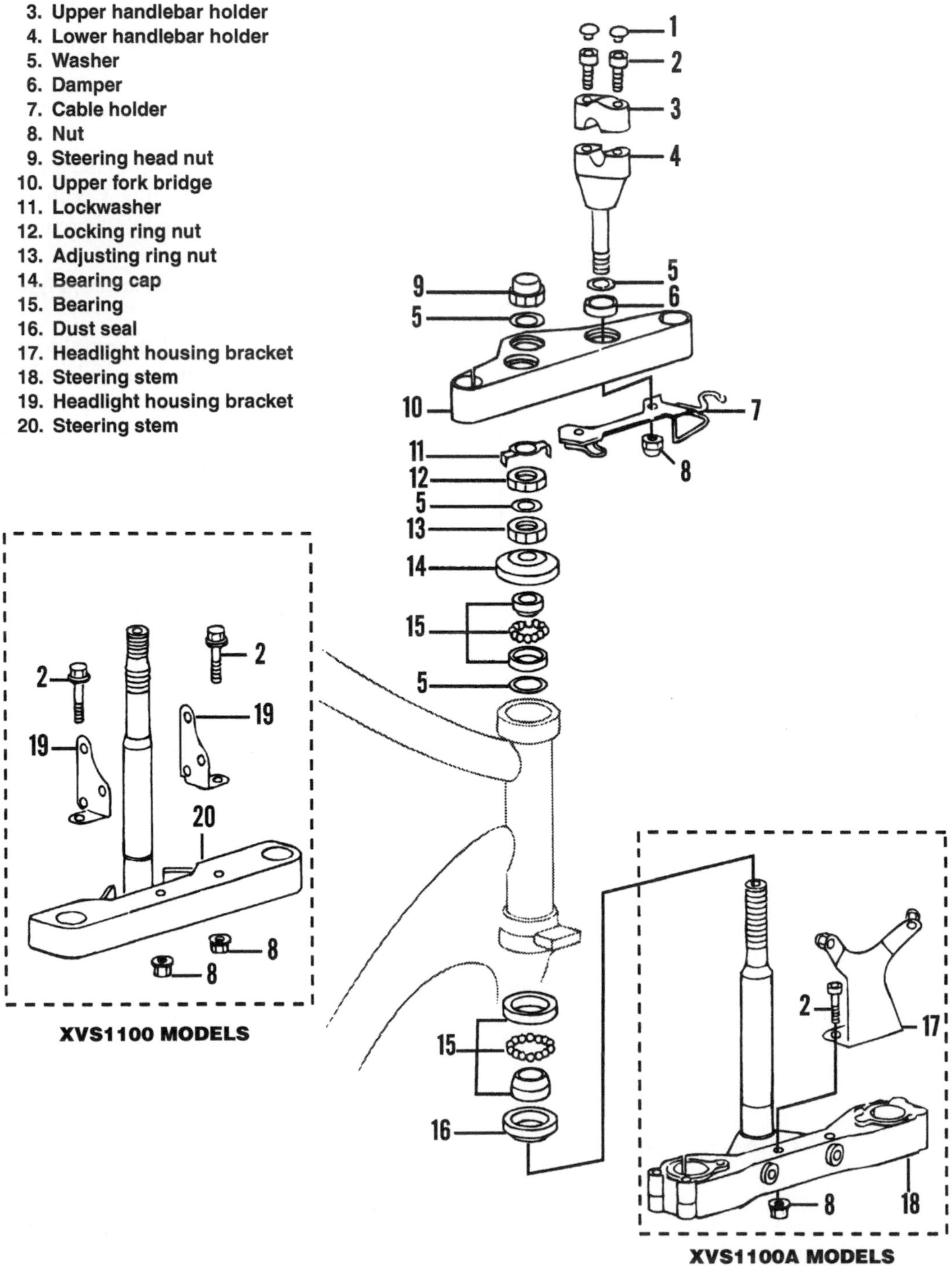

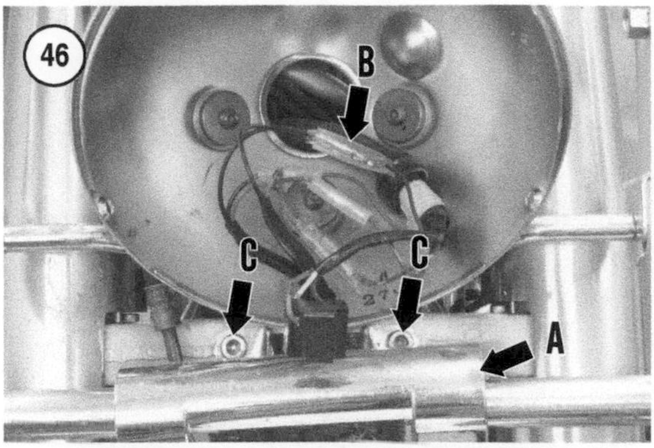

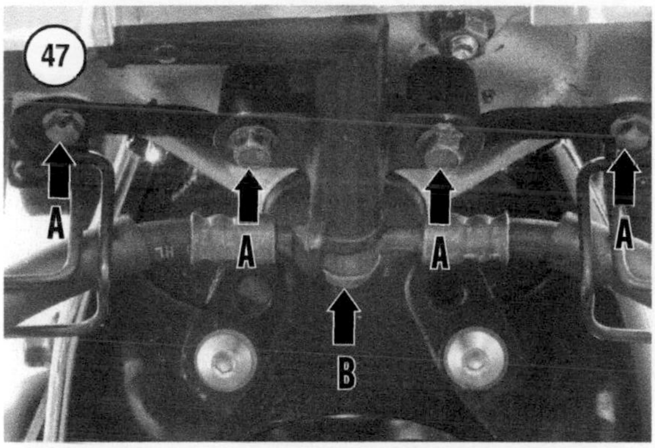

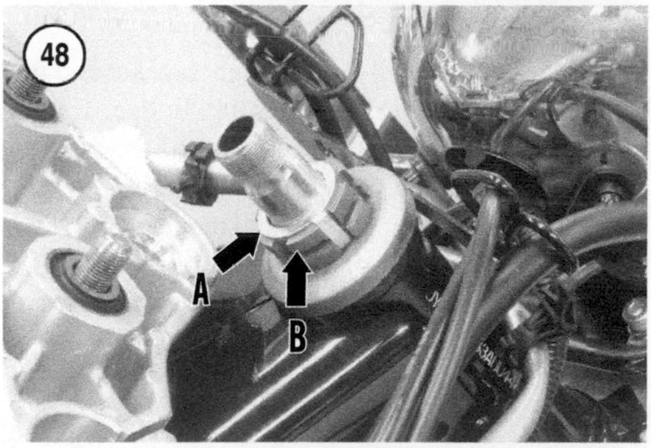

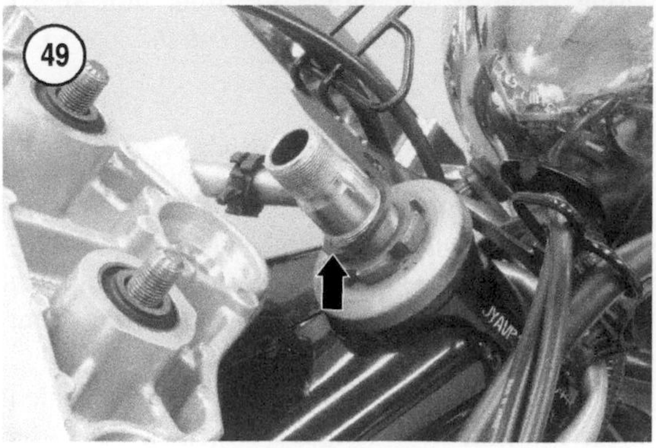

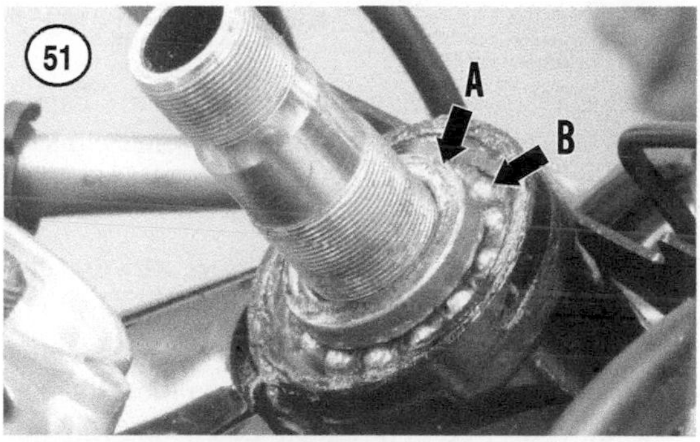

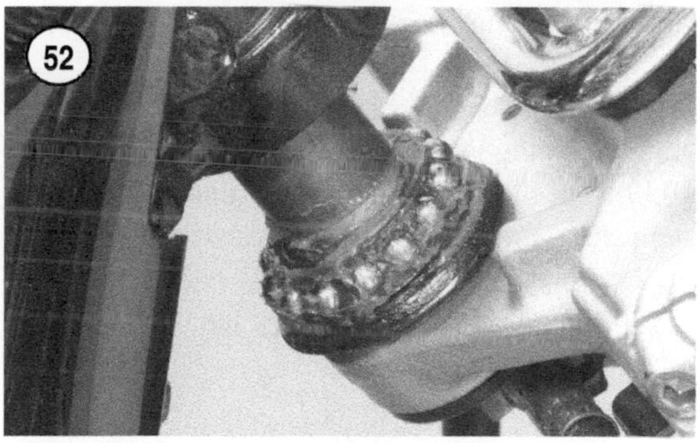

13. Lower the steering stem assembly down and out of the steering head.

14. Remove the lower bearing from the steering stem (**Figure 52**). The lower bearing race and the dust seal will remain on the steering stem.

Installation

1. Liberally apply lithium soap grease to the bearings and races.

2. Install a new dust seal and the lower bearing onto the steering stem.

3. Carefully slide the steering stem up through the frame steering head (**Figure 52**).

4. Install the upper bearing (B, **Figure 51**) and the bearing inner race (A).
5. Install the bearing cap (B, **Figure 50**) adjuster ring nut (A).
6. Adjust the steering head bearings as follows:
   a. Set a torque wrench at a right angle to the ring nut wrench (**Figure 53**).
   b. Seat the bearings within the steering head and tighten the adjuster ring nut to 52 N•m (38 ft.-lb.).
   c. Loosen the adjuster ring nut one turn.
   d. Tighten the adjuster ring nut to 18 N•m (13 ft.-lb.).
7. Inspect the steering head as described in Chapter Three. If there is any binding or looseness, disassemble and inspect the steering head as described in this chapter.
8. Install the rubber washer (**Figure 49**) and the locking ring nut (B, **Figure 48**). Tighten the locking ring nut finger-tight. Check the slots on the locking ring. They should align with those on the adjusting ring nut. If they do not, tighten the locking ring nut until alignment is achieved. If necessary, hold the adjusting ring nut so it does not move when the slots are aligned.
9. Install the lockwasher (A, **Figure 48**) so its fingers are seated in the ring nut slots.
10. Install the upper fork bridge onto the steering stem shaft. Loosely install the washer and the steering head nut (C, **Figure 41**).
11. Temporarily insert both fork tubes through the lower and upper fork bridges.

NOTE
On XVS1100 models, a single pinch bolt is used on each side of the lower fork bridge.

12. Tighten the lower bridge pinch bolts (**Figure 54**) to hold the fork tubes in position. Then tighten the steering head nut (C, **Figure 41**) to 110 N•m (81 ft.-lb.).
13. Turn the steering stem by hand to make sure it turns freely and does not bind. If the steering stem is too tight, the bearings can be damaged; if the steering stem is too loose, the steering will become unstable. Readjust if necessary.
14. Remove the fork tubes and install them as described in this chapter.
15. Install the front wheel (Chapter Ten) and fender (Chapter Fourteen).

53

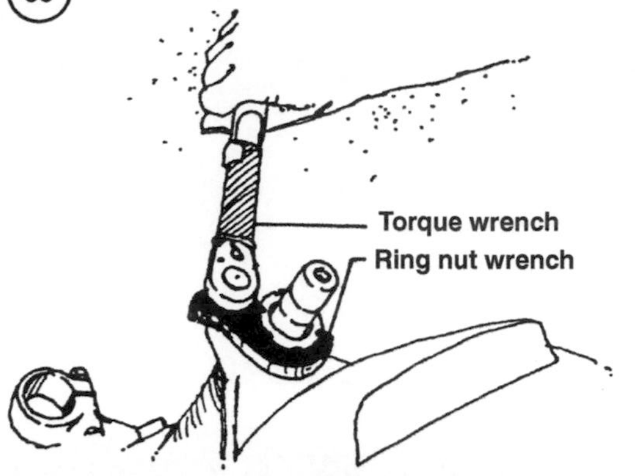

54

16. If removed, install the headlight bracket and the brake hose union onto the lower fork bridge. Torque the headlight bracket nuts to 7 N•m (62 in.-lb.) and the brake hose union bolts to 7 N•m (62 in.-lb.).
17. On XVZ1100 models, install the front turn signal flasher assemblies onto the lower fork bridge (if removed). Torque the front turn signal mounting nut to 7 N•m (62 in.-lb.).
18. On XVZ1100A models, install the front turn signal bracket and cover (A, **Figure 46**). Feed the flasher wires through the headlight housing as noted during removal and torque the front turn signal bracket bolts (B, **Figure 46**) to 7 N•m (62 in.-lb.).

Inspection

1. Clean the upper and lower bearings in a bearing degreaser. Thoroughly dry both bearings with compressed air. Make sure all solvent is removed from the lower bearing installed on the steering stem (**Figure 52**).
2. Wipe the old grease from the outer races located in the steering head (**Figure 55**), then clean the

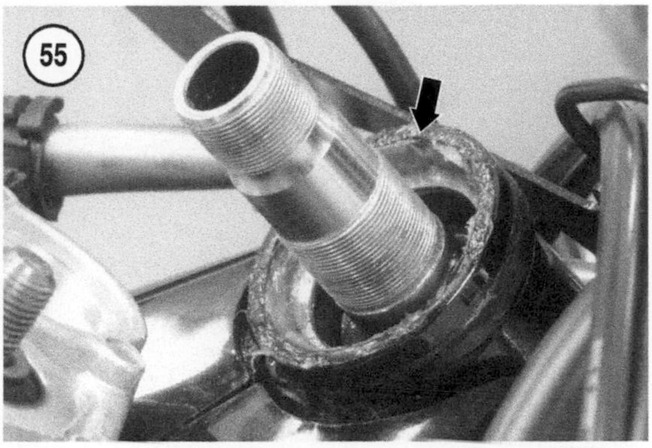

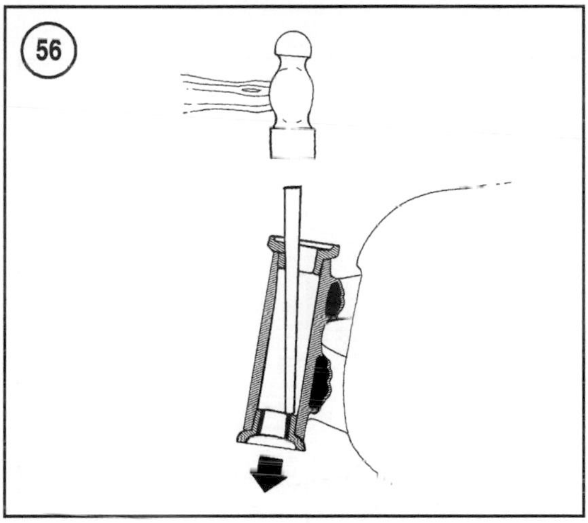

outer races with a rag soaked in solvent. Thoroughly dry the races with a lint-free cloth. Check the races for pitting, galling and corrosion. If any of these conditions exist, replace the races as described in this chapter.

3. If any race is worn or damaged, replace the race and bearing as an assembly as described in this chapter.

4. Check the welds around the steering head for cracks and fractures. If there is any damage, have the frame repaired at a competent frame or welding shop.

5. Check the bearings for pitting, scratches or discoloration indicating wear or corrosion. Replace the bearing if any ball is less than perfect.

6. If the bearings are in good condition, pack them thoroughly with grease. To pack the bearings, spread some grease in the palm of your hand and scrape the open side of the bearing cage across your palm until the bearing is completely full of grease.

7. Thoroughly clean all mounting parts in solvent. Dry them completely.

8. Inspect the ring nuts and washer for wear or damage. Inspect the nut threads. If necessary, clean them with an appropriate size metric tap or replace the nut(s). If the threads are damaged, inspect the appropriate steering stem thread(s) for damage. If necessary, clean the threads with an appropriate size metric die.

9. Check the underside of the steering head nut for damage. Replace the nut as necessary.

10. Inspect the steering stem and the lower fork bridge for cracks or other damage. Make sure the fork bridge clamping areas are free of burrs and the bolt holes are in good condition.

11. Inspect the upper fork bridge for cracks or other damage. Check both the upper and lower surface of the fork bridge. Make sure the fork bridge clamping areas are free of burrs and the bolt holes are in good condition.

STEERING HEAD BEARING RACES

The upper and lower bearing outer races must not be removed unless they are going to be replaced. These races are pressed into place and are damaged during removal. If removed, replace the outer races, the inner races, and the bearings as a set. Never reuse an outer race that has been removed. It is no longer true and will damage the bearings if reused.

1. Chill new bearing races overnight in a freezer to shrink the outside diameter.

2. Remove the steering stem as described in this chapter.

3. Insert a brass or aluminum drift into the steering head and carefully tap the lower race out from the steering head (**Figure 56**). Repeat this procedure for the upper race.

4. Clean the race seats in the steering head. Check for cracks or other damage.

5. Apply grease to a new upper race and insert the race into the steering head with the open side facing out. Square the race with the race bore. Tap it slowly and squarely with a block of wood (**Figure 57**).

LOWER BEARING REPLACEMENT

Do not remove the lower bearing and lower dust seal unless they are going to be replaced. The bearing can be difficult to remove. If it cannot be easily removed as described in this procedure, have a dealership service department replace the bearing and seal.

57

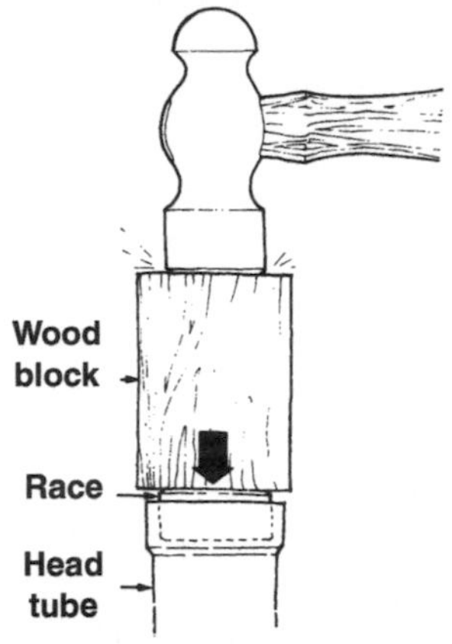

58

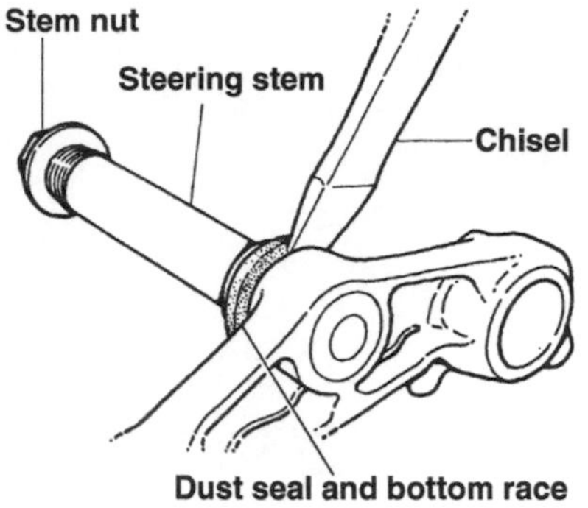

Never reinstall a bearing that has been removed. It is no longer true and will damage the rest of the bearing assembly if reused.

1. Install the adjusting and locking ring nuts as well as the steering head nut onto the top of the steering stem to protect the threads.
2. Use a chisel to drive the bearing/seal assembly from the steering stem (**Figure 58**). Work around in a circle and slowly drive the assembly from the shoulder on the steering stem. Remove the bearing and seal. Discard them both.
3. Clean the steering stem with solvent and dry it thoroughly.
4. Slide a new dust seal and the lower bearing onto the steering stem.
5. Align the bearing inner race with the machined shoulder on the steering stem.
6. Drive the bearing onto the steering stem shoulder with a piece of pipe that matches the diameter of the inner race (**Figure 59**).

59

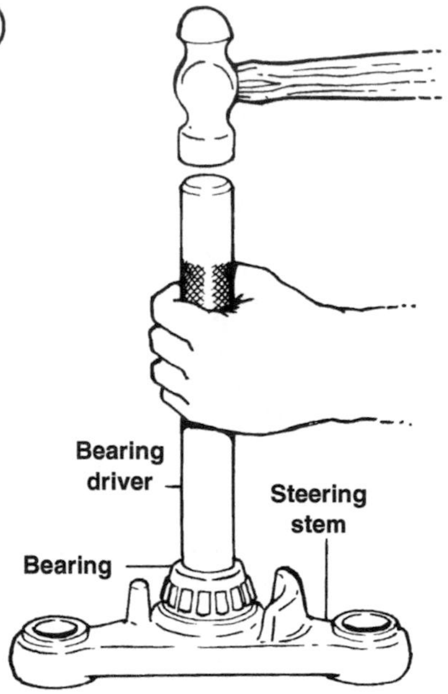

Table 1 FRONT SUSPENSION SPECIFICATIONS

Item	Specification
Fork oil	
Viscosity	SAE 10W fork oil
Capacity per leg	
1999-2003 XVS1100 models	464 cc (15.7 U.S. oz.)
2004-on XVS1100 models	467-481 cc (15.8-16.3 U.S. oz.)
2004-on XVS1100A models	488 cc (16.5 U.S. oz.)
Oil level (measured from top of the fully compressed fork tube with the fork spring removed)	
1999-2003 all models	108 mm (4.25 in.)
2004-on XVS1100 models	99-105 mm (3.89-4.13 in.)
2004-on XVS1100A models	99 mm (3.89 in.)

(continued)

Table 1 FRONT SUSPENSION SPECIFICATIONS (continued)

Item	Specification
Fork travel	
1999-2003	108 mm (4.25 in.)
2004-on	140 mm (5.51 in.)
Fork stroke	
K1	0-77.5 mm (0-3.05 in.)
K2	77.5-140 mm (3.05-5.51 in.)
Fork spring	
1999-2003 XVS1100 models	
Free length	356.9 mm (14.05 in.)
Service limit	350 mm (13.78 in.)
1999-2003 XVS1100A models	
Free length	361.9 mm (14.25 in.)
Service limit	350 mm (13.78 in.)
2004-on all models	
Free length	371.9 mm (14.64 in.)
Service limit	–
Installed length	
1999-2003 XVS1100 models	319.4 mm (12.57 in.)
1999-2003 XVS1100A models	324.4 mm (12.77 in.)
2004-on XVS1100 models	334.9 mm (13.19 in.)
2004-on XVS1100A models	334.4 mm (13.17 in.)
Spring rate	
1999-2003	
K1	8.8 N/mm (50.40 lb. in.)
K2	12.7 N/mm (72.80 lb./in.)
2004-on	
K1	4.4 N/mm (25.12 lb. in.)
K2	6.3 N/mm (35.97 lb./in.)
Spacer length	
2004-on XVS1100 models	177.5 mm (6.99 in.)
All other models	183 mm (7.20 in.)

11

Table 2 FRONT SUSPENSION AND STEERING TORQUE SPECIFICATIONS

Item	N•m	in.-lb.	ft.-lb.
Brake hose union bolt	7	62	–
Fork damper rod Allen bolt*	30		22
Fork cap bolt	23	–	17
Front axle	59	–	43
Front axle clamp bolt	20	–	15
Front brake caliper mounting bolt	40	–	29
Front brake master cylinder clamp bolts	10	89	–
Front turn signal clamp bolt			
(XVS1100A models)	7	62	–
Front turn signal mounting nut			
(XVS1100 models)	7	62	–
Handlebar holder clamp bolts	28	–	21
Handlebar holder mounting nut	32	–	24
Headlight bracket nuts	7	62	–
Lower fork bridge clamp bolts	30	–	22
Steering head nut	110	–	81
Steering stem adjuster ring nut		–	
First stage	52	–	38
Second stage	18	–	13
Upper fork bridge clamp bolt	20	–	15

*Apply Loctite 242 (blue) or an equivalent medium-strength threadlocking compound.

CHAPTER TWELVE

REAR SUSPENSION AND FINAL DRIVE

This chapter includes repair procedures for servicing the rear shock absorber and swing arm as well as the drive shaft and the final drive.

When inspecting rear suspension and final drive components, compare any measurements to the specifications in the tables at the end of this chapter. Replace parts that are damaged, worn or out of specification. During assembly, tighten fasteners to the specified torque.

SHOCK ABSORBER

Removal/Installation

Refer to **Figure 1**.

1. Securely support the motorcycle with the rear wheel off the ground so the rear shock absorber is not compressed.
2. Remove the muffler assembly (Chapter Eight).
3. Remove the battery box and toolbox panel as described in Chapter Fourteen.
4. Remove the rear wheel (Chapter Ten).

NOTE

The shock absorber preload adjuster sits at the top of the shock absorber. Locate the preload adjuster indicator and note which direction it faces. The shock absorber must be reinstalled in this same orientation.

5. Remove the nut and washer from the shock absorber lower bolt (A, **Figure 2**).
6. Remove the bolt (A, **Figure 2**) and lower the swing arm so the connecting arm separates from the shock absorber mount.
7. Remove the nut from the shock absorber upper bolt (A, **Figure 3**). Note that the bolt engages a receiver (B, **Figure 3**) in the frame.
8. Remove the shock absorber upper bolt and remove the shock absorber from the motorcycle.
9. Clean and dry the shock absorber mounts and hardware.
10. Inspect the shock absorber as described below.
11. Installation is the reverse of removal. Note the following:

1

SHOCK ABSORBER AND SWING ARM

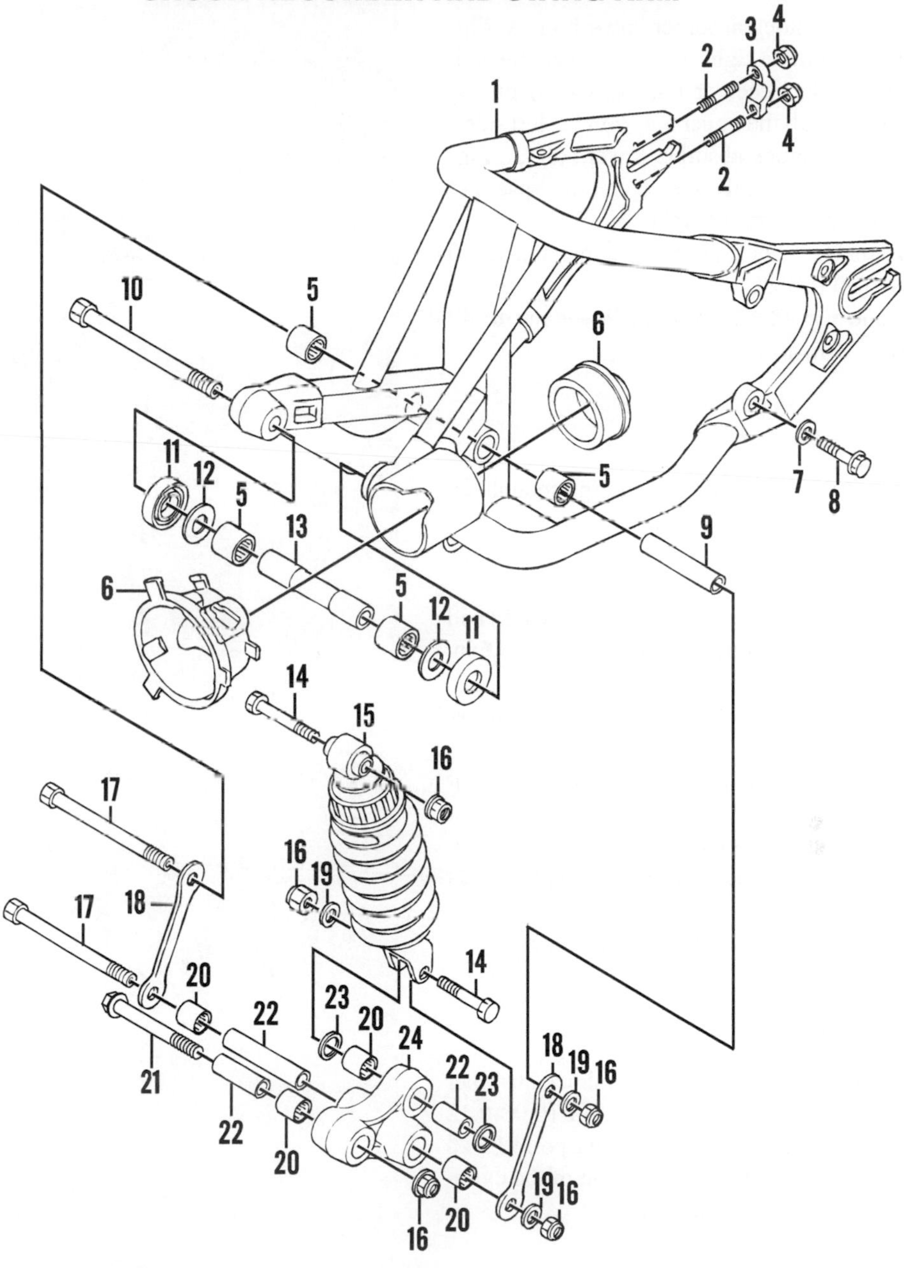

1. Swing arm	9. Collar	17. Connecting arm bolt
2. Stud	10. Pivot bolt	18. Connecting arm
3. Clamp	11. Thrust cover	19. Washer
4. Nut	12. Washer	20. Bearing
5. Bearing	13. Collar	21. Relay arm bolt
6. Boot	14. Shock bolt	22. Collar
7. Washer	15. Shock absorber	23. Seal
8. Bolt	16. Nut	24. Relay arm

a. Install the shock absorber with the preload adjuster indicator facing the direction noted during removal.
b. Install the shock absorber upper bolt (A, **Figure 3**) from the right side. Make sure the bolt engages the receiver (B, **Figure 3**) on the frame. Install the upper nut from the left side.
c. Install the shock absorber lower bolt (A, **Figure 2**) from the left side; install the lower nut and washer from the right.
d. Torque the shock absorber upper nut to 40 N•m (30 ft.-lb.).
e. Torque the shock absorber lower nut to 48 N•m (35 ft.-lb.).

Inspection

1. Clean and dry the hardware and the mounts on the shock absorber.
2. Inspect the shock absorber (**Figure 4**) for dents, damage or oil leaks.
3. Inspect the upper mount bushing (**Figure 5**) and inspect the lower mount. Look for elongation, cracks or other damage.
4. Inspect the spring for cracks or other signs of fatigue.
5. Inspect the shock absorber for signs of a gas or oil leak.
6. If any part of the shock absorber is worn or damaged, replace the shock absorber. Replacement parts are unavailable.
7. Replace the mounting hardware as necessary.

Shock Absorber Disposal

The gas must be released before a shock absorber is discarded in the trash. Drill a 2-3 mm (0.08-0.12 in.) hole through the shock absorber at a point 15-20 mm (0.59-0.79 in.) from the upper end of the cylinder (the end with the preload adjuster).

SUSPENSION LINKAGE

Removal/Installation

The rear suspension linkage consists of a relay arm and two connecting arms. Refer to **Figure 1** when servicing the suspension linkage.
1. Remove the shock absorber as described in this chapter.

2

3

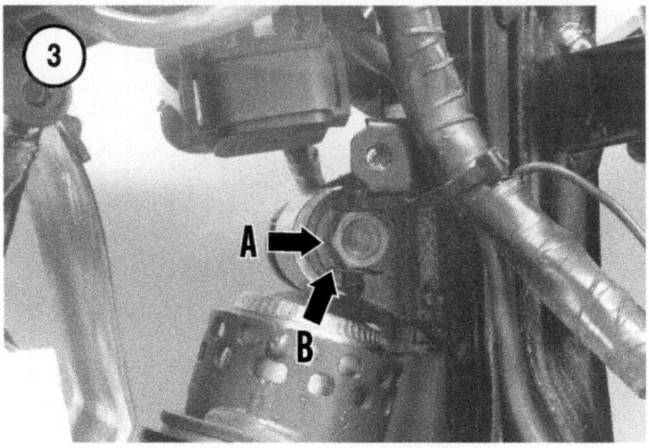

4

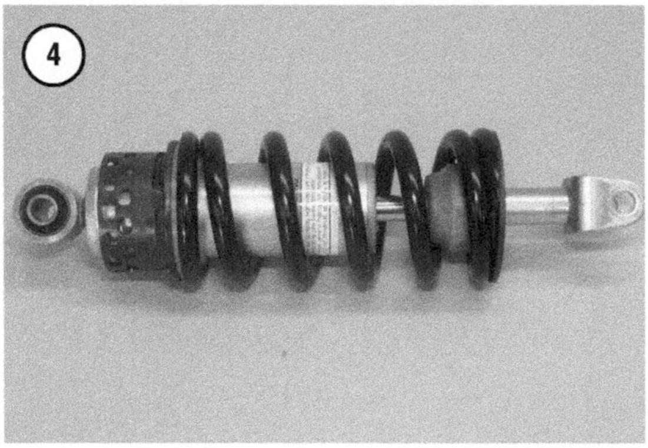

5

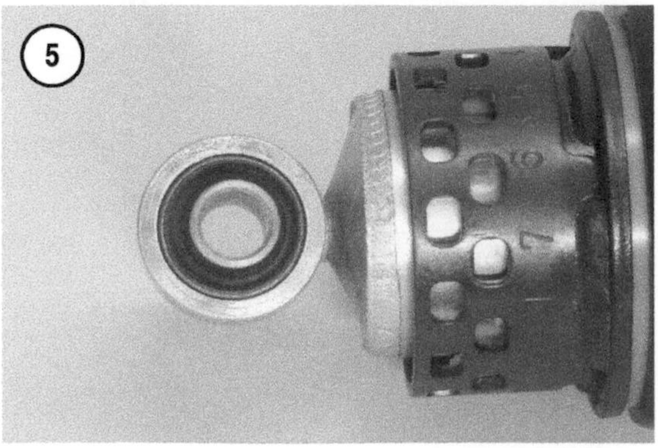

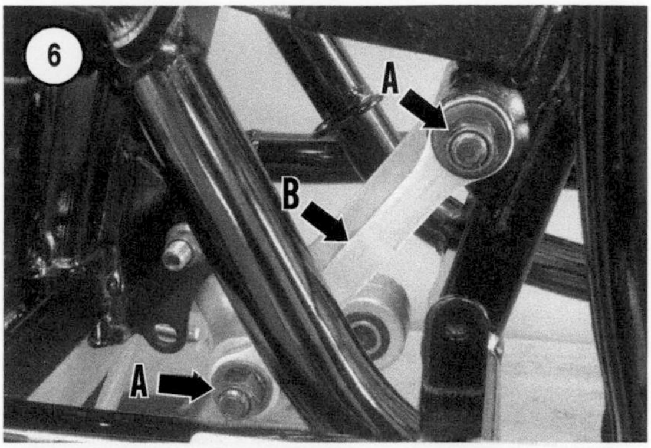

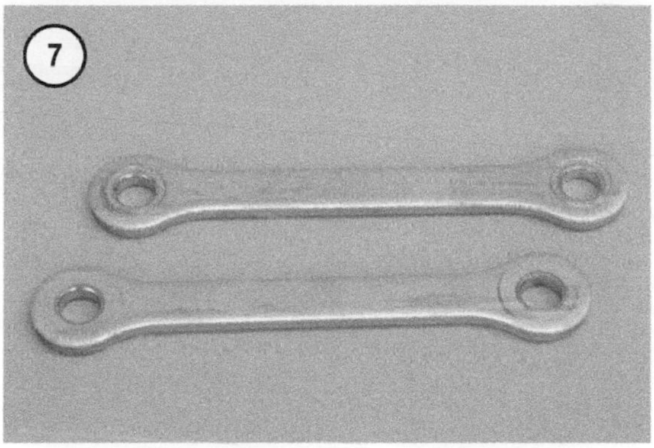

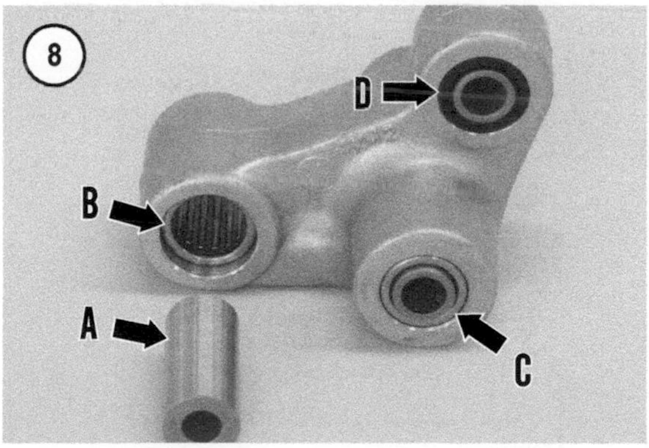

2. On the left side, remove the nut (A, **Figure 6**) and washer from the upper and lower connecting arm bolts.

3. Remove each connecting arm bolt and lower the connecting arms (B, **Figure 6**) from the motorcycle.

4. Remove the nut from the relay arm bolt (C, **Figure 2**).

5. Pull the relay arm bolt from the right side and lower the relay arm (B, **Figure 2**) from the frame mount.

6. Installation is the reverse of removal. Note the following:

a. Lubricate the bearings, collars and seals with molybdenum disulfide grease.

b. Install the bolts from the right side and the nuts from the left.

c. Torque the relay arm nut and the connecting arm nuts to 48 N•m (35 ft.-lb.).

Inspection

1. Clean and dry the hardware, collars and mounts.
2. Inspect the connecting arms (**Figure 7**) for dents or other damage. Also inspect the mounts for cracks or elongation.
3. Inspect the needle bearings in the relay arm as follows:

NOTE
A different length collar is used in each mount in the relay arm. Install each collar in the proper mount.

a. Remove the collar (A, **Figure 8**) from the relay arm mount.

b. Wipe excess grease from the bearing (B, **Figure 8**), and visually inspect the needles for pitting, wear or other damage.

c. Insert the collar into its bearing and turn the collar by hand. The bearing should turn smoothly without binding or excessive noise.

d. If any bearing is worn or damaged, replace it as described in *Needle Bearing Replacement* in this section.

e. Two bearings are used in the middle relay arm mount (C, **Figure 8**). If either bearing in this mount is worn, replace both bearings as a set. Install each bearing so it is flush with the outside edge of the bearing bore.

f. When installing a needle bearing in a mount that uses just one bearing, center the bearing in the mount.

4. Check the collar(s) for scoring or excessive wear. Replace collars as necessary.
5. Inspect the seal (D, **Figure 8**) in the rear mount. Replace the seal if it shows signs of leaking, is damaged or is becoming brittle.
6. Replace the mounting hardware as necessary.

Needle Bearing Replacement

1. Use a blind bearing remover to remove each bearing from the middle mount (C, **Figure 8**) on the relay arm. Follow the instructions from the tool's manufacturer.
2. Use a hydraulic press to remove a needle bearing from the front and rear relay mounts (B and D, **Figure 8**). Support the relay arm in the press and use a driver that matches the diameter of the needle bearing (**Figure 9**).
3. Thoroughly clean and dry the bearing bore.
4. Pack the new bearing with molybdenum disulfide grease.
5. Use a swing arm bearing installer, like the Motion Pro Swing Arm Bearing Tool (**Figure 10**, part No. 08-0213) to install the new bearing. Follow the manufacturer's instructions.
6. If a tool is not available, one can be fabricated from a socket, three large washers, a threaded rod and two nuts. Assemble the washers, threaded rod, bearing and nuts as shown in **Figure 11**. Hold the lower nut with a wrench and turn the upper nut to press the bearing into the bearing bore. Turn the nut slowly and watch the bearing carefully. Make sure it does not turn sideways.
7. Position a bearing in its bore as described in *Swing Arm* or *Suspension Linkage*.

9

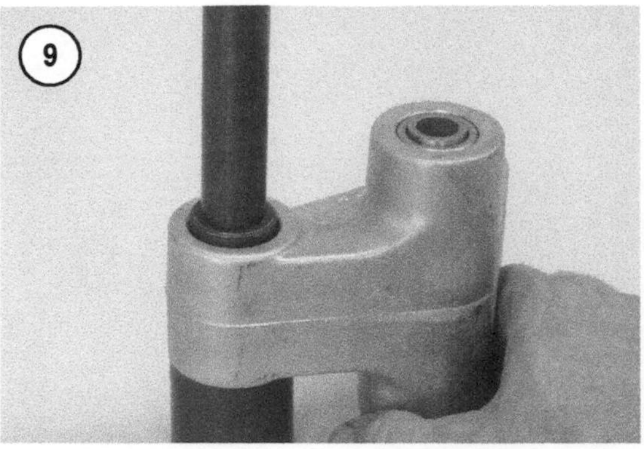

10

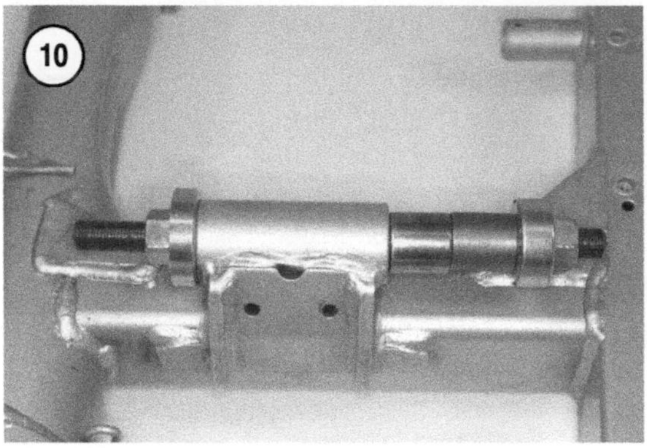

11

INSTALLING SWING ARM BEARINGS

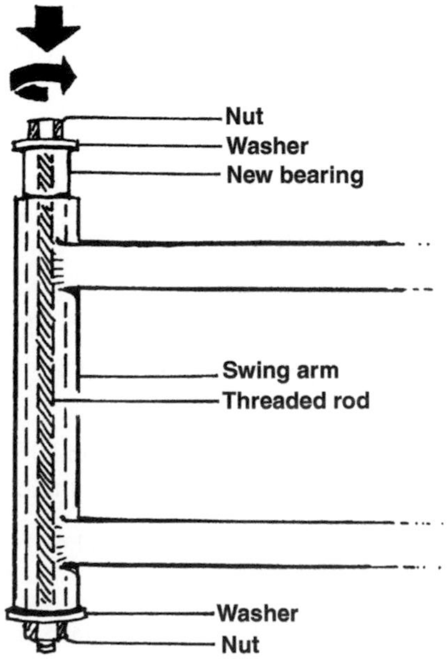

SWING ARM

Refer to **Figure 1**.

Removal

1. Securely support the motorcycle on level ground.
2. Remove the shock absorber and suspension linkage as described in this chapter.
3. Check the swing arm bearing as follows:
   a. Grasp both ends of the swing arm, and move it up and down. The swing arm should move smoothly with no binding or abnormal noise from the bearings. If there is binding or noise, the bearings are worn and must be replaced.
   b. Try to move the swing arm from side to side in a horizontal arc. If there is more than a slight amount of movement, the bearings are worn and must be replaced.

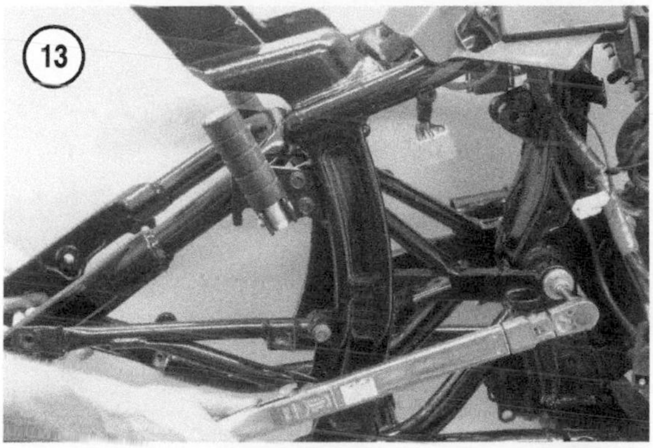

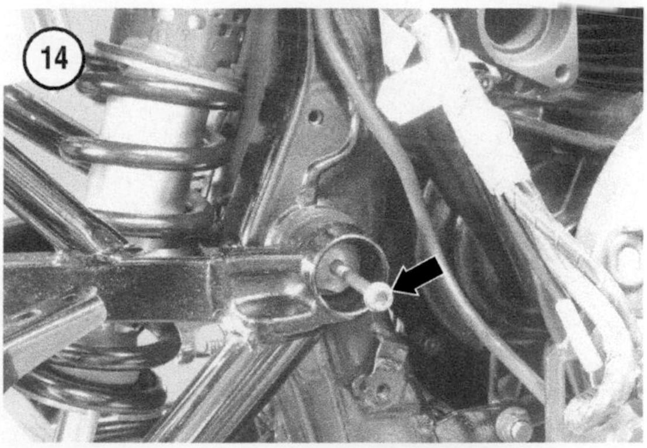

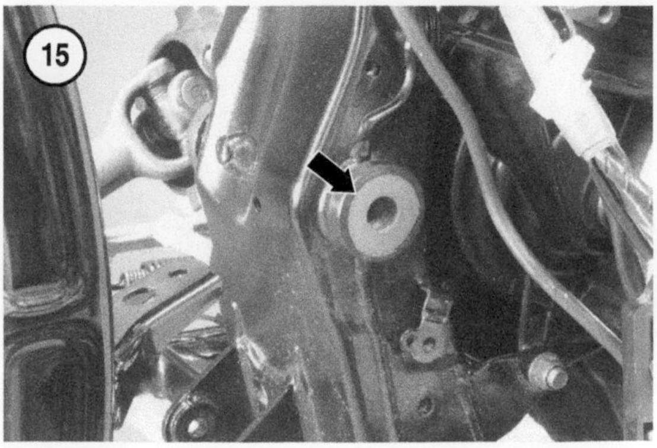

4. Note the location of any clamps (A, **Figure 12**) or cable ties that secure wires or the brake line to the swing arm. Release the lines from the clamps and suspend the rear brake caliper safely out of the way.

NOTE
*The swing arm pivot bolt threads into the swing arm. It does not go all the way through to the left side so it cannot be driven out of the pivot. The pivot bolt (B, **Figure 12**) has a 6-mm internal thread so it can be removed with a slide hammer.*

5. Loosen the swing arm pivot bolt (**Figure 13**).
6. Use a slide hammer to remove the pivot bolt. If a slide hammer is unavailable, turn the 6 mm bolt (**Figure 14**) into the threads in the pivot bolt. Grab the bolt with a pair of pliers and strike the pliers with a hammer to remove the pivot bolt.
7. Remove the swing arm from the frame.
8. Remove the thrust cover (**Figure 15**) and its washer from the swing arm pivot on the each side of the frame.
9. Remove the collar (A, **Figure 16**) from the frame.
10. Inspect the swing arm as described in this chapter.

Installation

1. Apply molybdenum disulfide grease to the bearing (B, **Figure 16**) on each side of the frame pivot.
2. Lubricate the thrust cover washer (A, **Figure 17**) with molybdenum disulfide grease and install the washer into the thrust cover (B).
3. Install the thrust cover (**Figure 15**) onto the frame pivot on the right side.

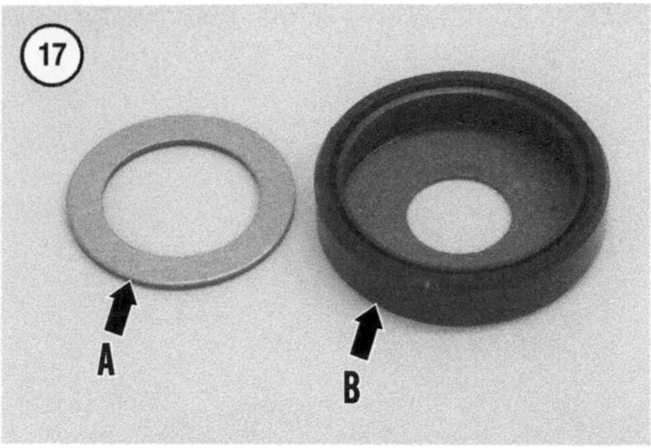

17

4. Lubricate the collar (A, **Figure 16**) with molybdenum disulfide grease and install the collar into the pivot bearings.
5. Repeat Step 2, and install the thrust cover (**Figure 15**) onto the frame pivot on the left side.
6. Raise the swing arm into position so its pivots are opposite the frame pivots. Insert the pivot bolt (**Figure 18**) from the right side and turn it into the threads in the left side of the swing arm.
7. Torque the swing arm pivot bolt to 90 N•m (66 ft.-lb.).
8. Install the suspension linkage and the shock absorber as described in this chapter.
9. Reinstall any removed components.

Inspection

1. Check the pivot bolt for straightness. A bent bolt will restrict the movement of the swing arm.
2. Inspect the pivot threads (A, **Figure 19**) in the swing arm. Clean and dress the threads as necessary.
3. Check the welds (B, **Figure 19**) on the swing arm for cracks or fractures.
4. Remove the collar (**Figure 20**) from the connecting arm pivot on the swing arm.
5. Check the swing arm pivot bearings (**Figure 21**) as follows:
   a. Turn each bearing (**Figure 21**) with a finger. A bearing should turn smoothly without excessive play or noise.
   b. Use a lint-free cloth to remove surface grease from the bearings.
   c. Check the bearing rollers for evidence of wear, pitting or rust.
   d. If either bearing is damaged, replace both bearings. Refer to *Needle Bearing Replace-*

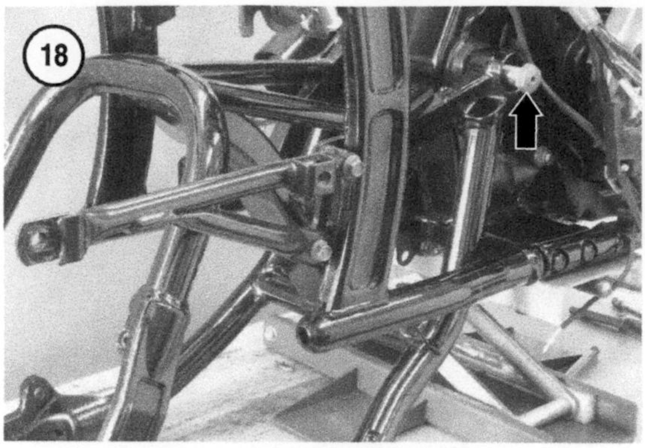
18

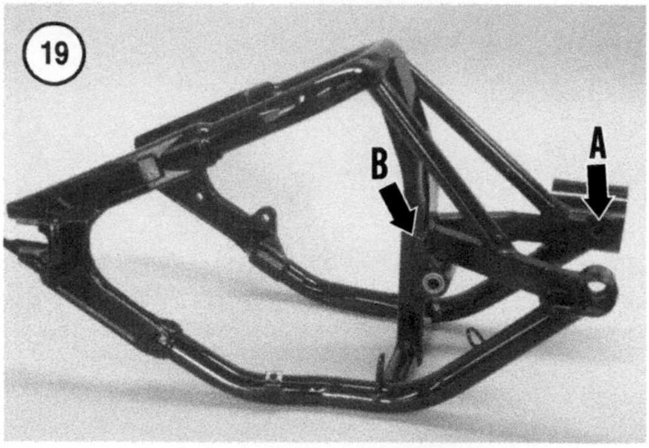

19

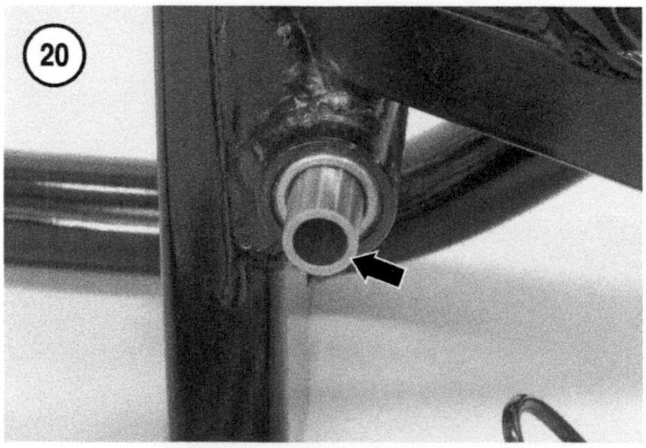
20

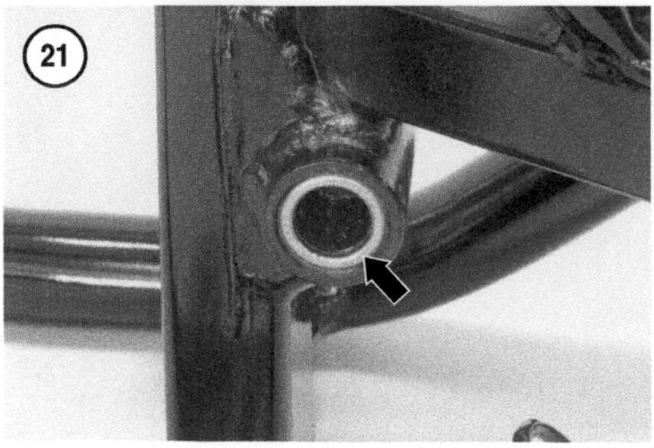
21

22

23

24
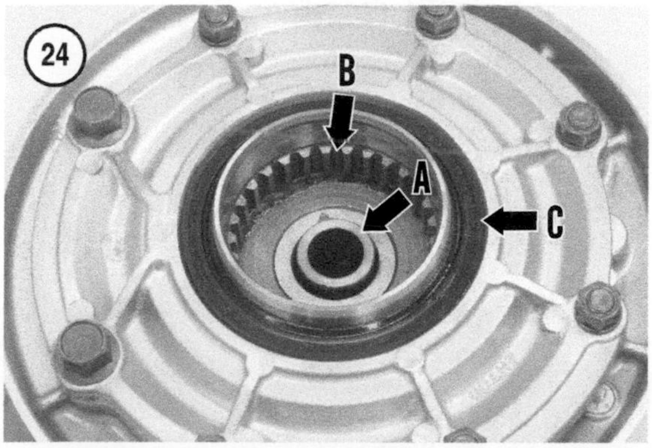

25

ment in this chapter. Install each swing arm bearing so it sits flush with the outside edge of the bearing bore.

6. Repeat Step 5 and inspect the swing arm bearing (B, **Figure 16**) in each side of the frame pivot.
7. Inspect the thrust cover (B, **Figure 17**) and washer (A) from each side of the frame pivot.

FINAL DRIVE ASSEMBLY

Removal/Installation

1. Securely support the motorcycle on a level surface.
2. If the final gearcase will be serviced, drain the gearcase oil as described in Chapter Three.
3. Remove the rear wheel (Chapter Ten) and set it on wooden blocks.
4. Grasp the drive shaft/final gearcase assembly and lift the final gearcase (A, **Figure 22**) from the rear wheel hub. Watch for the distance collar (**Figure 23**).
5. Installation is the reverse of removal.
   a. Make sure the distance collar (A, **Figure 24**) is seated within the ring gear.
   b. Apply molybdenum disulfide grease to the splines of the ring gear (B, **Figure 24**) and to the splines on the rear wheel hub (**Figure 25**).
   c. Make sure the ring gear splines mesh with those on the rear wheel.

12

Inspection

1. Inspect the gearcase for any external damage.
2. Make sure the oil filler bolt (B, **Figure 22**) and drain bolt (C) are in good condition.
3. Inspect the internal splines in the ring gear (B, **Figure 24**). If there is excessive wear or damage, replace the gear coupling.
4. Check the oil seal (C, **Figure 24**) for leaks or signs of damage. If necessary, replace the oil seal as follows:
   a. Pry the oil seal (**Figure 26**) from the ring gear bearing housing.
   b. Lubricate the new oil seal with lithium soap grease, and set it in place.
   c. Use a driver of a socket that matches the diameter of the oil seal (**Figure 27**) to drive the seal until it is seated in the housing.

26

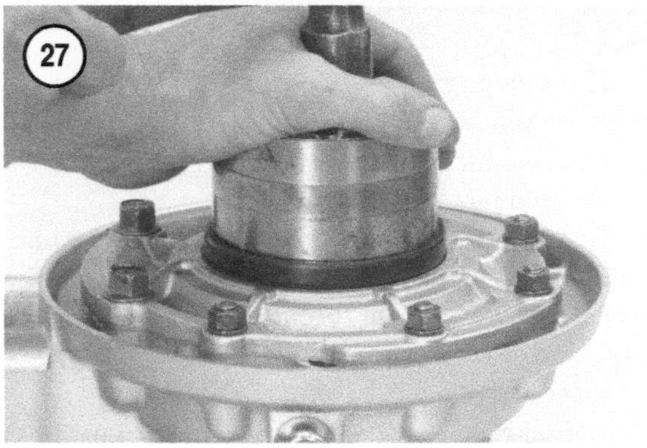
27

5. Check the operation of the final gear assembly by rotating the drive shaft. The shaft should turn smoothly and should transfer motion through the pinion gear assembly to the ring gear.
6. Drain the final gearcase oil as described in Chapter Three. Check the oil for metal particles. Small amounts of particles in the oil are normal. However, large amounts of particles indicate bearing problems.
7. Remove the drive shaft assembly. Measure the gear lash as described in this chapter.

Drive Shaft Removal/Installation

Refer to **Figure 28** when servicing the final drive assembly.
1. Remove the final drive assembly as described in this chapter.
2. Remove the final gearcase cover nuts (A, **Figure 29**).
3. Pull the cover/drive shaft assembly (B, **Figure 29**) from the final gearcase.
4. Remove the spring seated against the pinion gear nut.

28

FINAL DRIVE ASSEMBLY

1. Bearing housing nut
2. Washer
3. Distance collar
4. Bearing housing bolt
5. Bearing housing
6. O-ring
7. Oil seal
8. Ring gear shim
9. Bearing
10. Ring gear
11. Thrust washer
12. Bearing
13. Oil seal
14. Collar
15. Final gearcase
16. Spring
17. Pinion gear nut
18. Washer
19. Gear coupling
20. Oil seal
21. O-ring
22. Pinion bearing retainer
23. Bearing
24. Pinion gear shim
25. Pinion gear
26. Bearing
27. Stud
28. Snap ring
29. Dust seal
30. Nut
31. Washer
32. Gearcase cover
33. Washer
34. Oil seal
35. Drive shaft

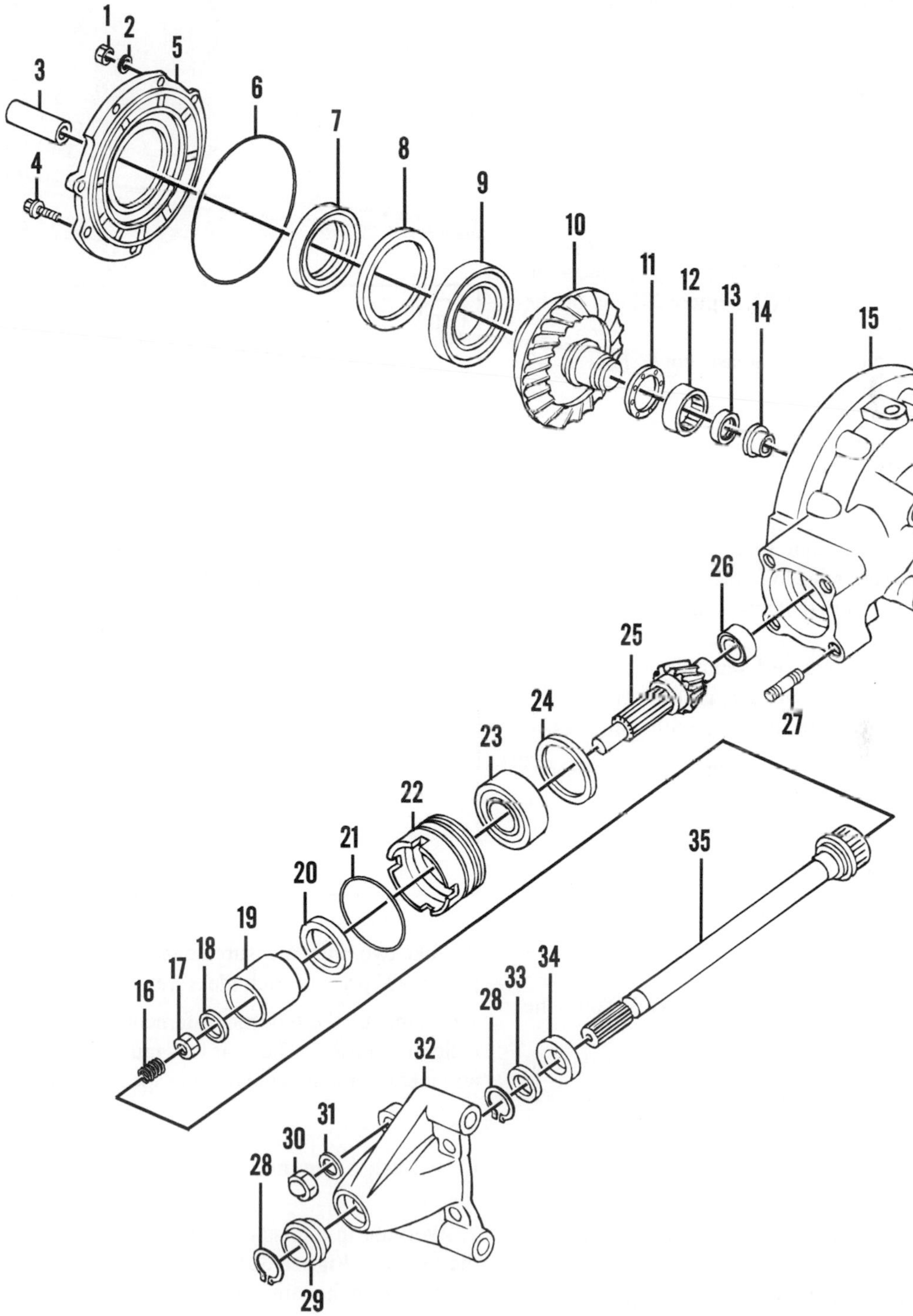
1
2
3
4
5
6
7
8
9
10
11
12
13
14
15
16
17
18
19
20
21
22
23
24
25
26
27
28
29
30
31
32
33
34
35

5. Inspect the splines of the drive shaft for wear or damage. Replace the drive shaft as necessary.
6. If necessary, remove the snap ring (C, **Figure 29**), and slide the cover (D) and its dust seal from the drive shaft.
7. Installation is the reverse of removal.
   a. Use a new snap ring when installing the gearcase cover on the drive shaft.
   b. Apply molybdenum disulfide grease to the drive shaft splines.
   c. Seat the spring against the pinion gear nut.
   d. Apply Yamaha Bond No. 1215 or an equivalent liquid gasket to the mating surfaces on the gearcase cover (D, **Figure 29**) and on the final gearcase.
   e. Torque the final gearcase cover nuts to 42 N•m (31 ft.-lb.).

Gear Lash Measurement

A dial indicator, a 14 × 100 mm bolt and the Yamaha gear lash tool (part No. YM-01230 or 90890-01230) are needed to measure backlash.

1. Drain the oil from the final gearcase as described in Chapter Three.
2. Remove the final drive assembly and remove the drive shaft from the final gearcase (this chapter).
3. Secure the gearcase in a vise with soft jaws so the drain hole faces up.

CAUTION
Do not overtighten the 14 × 100 mm bolt. Finger-tight is sufficient.

3. Turn a 14 × 100 mm bolt (**Figure 30**) into the drain hole and finger-tighten the bolt enough to hold the ring gear.
4. Install the gear lash tool (A, **Figure 31**) onto the gear coupling.
5. Secure a dial indicator (B, **Figure 31**) in place so the indicator's plunger rests against the gear lash tool. The plunger must rest at a point 54.5 mm (2.1 in.) from the center of the pinion gear assembly (C, **Figure 31**).
6. Gently rotate the gear coupling from tooth engagement to tooth engagement. Record the reading on the dial indicator.
7. Remove the gear lash tool and rotate the drive pinion 90°.
8. Measure the gear lash again and record this measurement.

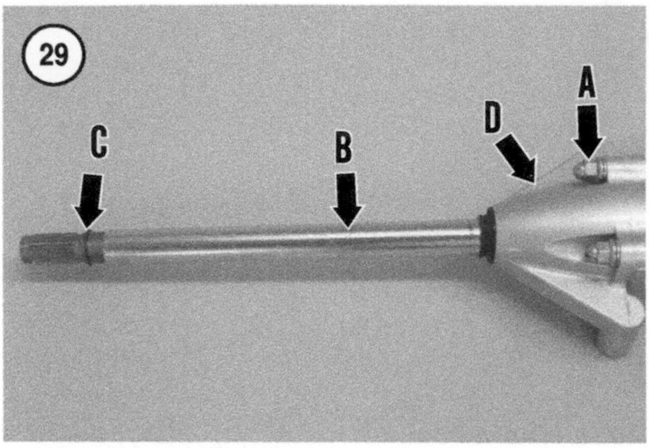

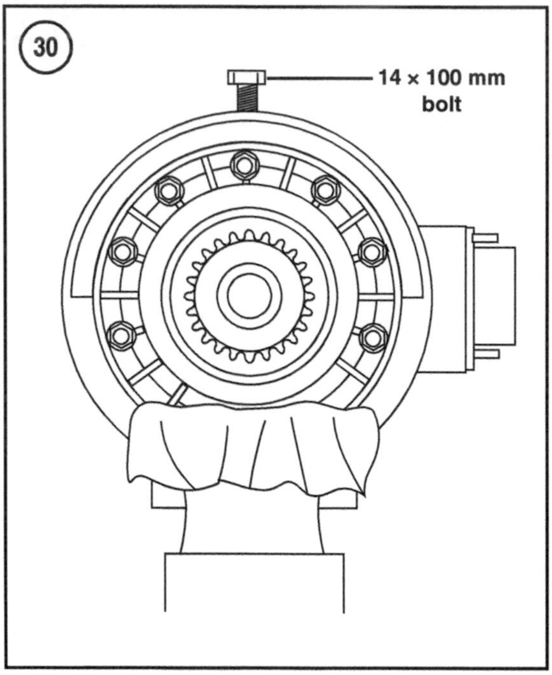

9. Take two more measurements, rotating the drive pinion 90° prior to each measurement.
10. Compare the four measurements to the specification in **Table 1**. The gear lash must be adjusted if any measurement is outside the specified range.

Gear Lash Adjustment

1. Working in a criss-cross pattern, evenly loosen the nuts (A, **Figure 32**) and bolts (B) on the ring gear bearing housing. Once all the hardware is loose, remove the nuts and bolts. Watch for the washer behind each nut.
2. Remove the bearing housing and the O-ring.

31

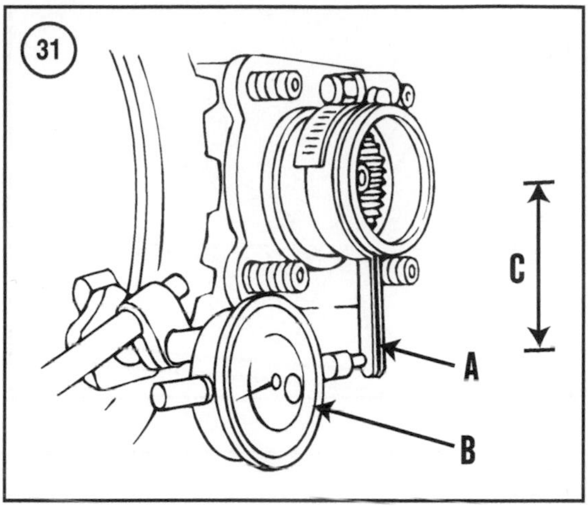

32

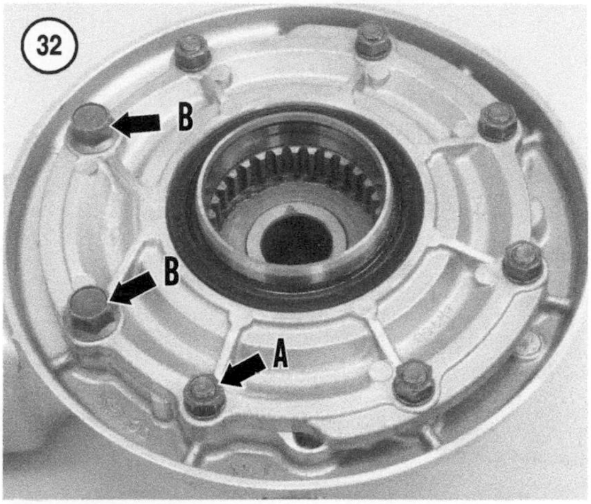

33

SHIMS AND THRUST WASHERS

Ring gear shim thickness (mm)	Thrust washer thickness (mm)	Pinion gear shim thickness (mm)
0.30, 0.40, 0.50	1.2, 1.4, 1.6, 1.8, 2.0	0.30, 0.40, 0.50

3. Remove the ring gear shim(s). Note the number of shims.

4. Remove the ring gear assembly and remove the thrust washer.

5. Ring gear shims and thrust washers are available in several sizes. See **Figure 33**.

6. Adjust gear lash by adding or removing ring gear shims as necessary to increase or decrease the gear lash. If the gear lash is less than specified, reduce ring gear shim thickness. If the gear lash exceeds specification, increase ring gear shim thickness.

   a. If the ring gear shim thickness is increased by more than 0.2 mm, reduce thrust washer thickness by 0.2 mm for each 0.2 mm increase in ring gear shim thickness.

   b. If the ring gear shim thickness decreases by more than 0.2 mm, increase thrust washer thickness 0.2 mm for each 0.2 mm decrease in ring gear shim thickness.

7. Reassemble the ring gear and bearing housing by reversing the disassembly procedures. Note the following:

   a. Pack the lips of the oil seal with lithium soap grease.

   b. Evenly tighten the bearing housing nuts (A, **Figure 32**) and bolts (B) in a criss-cross pattern. Apply a medium-strength threadlocking compound to the threads of the bearing housing bolts (B, **Figure 32**). Torque the nuts to 23 N•m (17 ft.-lb.). Torque the bolts to 40 N•m (30 ft.-lb.).

   c. Recheck the gear lash.

Final Gearcase Disassembly

Disassembly of the final gearcase assembly (**Figure 28**) requires considerable expertise as well as a number of Yamaha special tools. If the final gearcase requires service, general practice is to have the service performed by a Yamaha dealership or other qualified motorcycle shop.

1. The following tools, or their equivalents, are needed to disassemble and reassemble the final gear assembly:

   a. Gear coupling/shaft holder: part No. YM-01229 or 90890-01229.

b. Bearing retainer wrench: part No. YM-33214 or 90890-04077.

c. Crankshaft installer bolt adapter: part No. YM-90069 or 90890-91277.

d. Armature shock puller (slide hammer): part No. YU- 1047-3 or 90890-01290.

e. Weight: part No. YU-1047-4 or 90890-01291.

2. Remove the drive shaft from the final gearcase as described in this chapter.

3. Working in a criss-cross pattern, evenly loosen the nuts (A, **Figure 32**) and bolts (B) on the bearing housing. Once all the hardware is loose, remove the nuts and bolts.

4. Remove the bearing housing and the O-ring.

5. Remove the ring gear shim(s). Note the number of shims.

6. Remove the ring gear assembly and remove the thrust washer.

CAUTION
The pinion gear assembly should only be removed during ring gear replacement.

7. Hold the gear coupling with the Yamaha gear coupling/shaft holder (**Figure 34**) and remove the pinion gear nut along with its washer. Discard the self-locking pinion gear nut. A new nut must be installed during assembly.

8. Remove the gear coupling.

NOTE
The bearing retainer has left-hand threads. Turn the nut clockwise to remove it.

9. Use the Yamaha bearing retainer wrench (**Figure 35**) to remove the bearing retainer.

10. Use the crankshaft installer bolt adapter (A, **Figure 36**), armature shock puller (B) and the weight (C) to remove the pinion gear. Assemble the tools according to their instructions and remove the pinion gear. Discard the bearing inner race. A new one must be installed during assembly.

11. Replace the pinion gear bearing as described in *Gearcase Bearing Replacement* in this chapter.

34

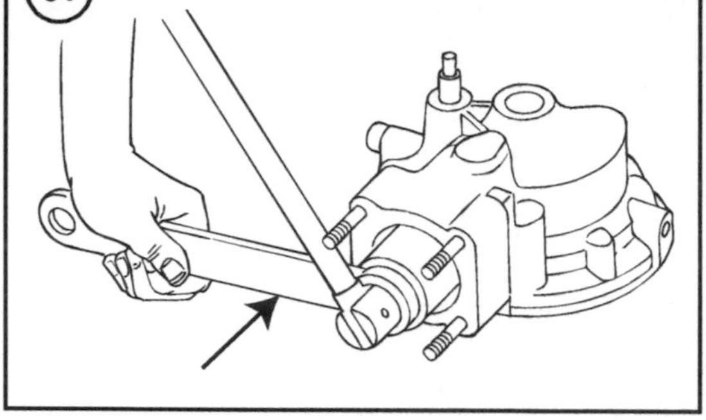

35

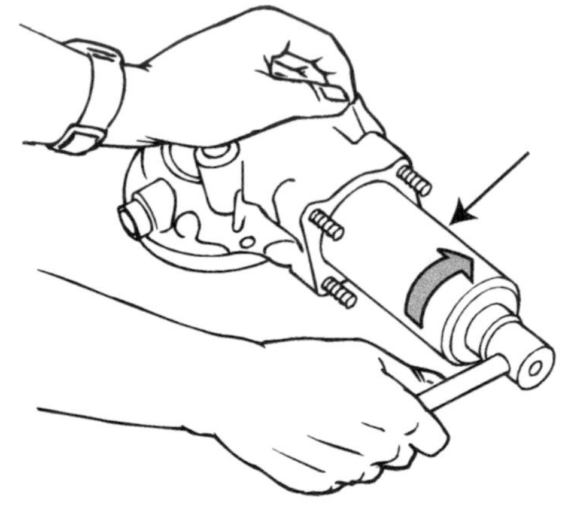

Final Gearcase Assembly

NOTE
The pinion and ring gears must be aligned whenever the final gearcase or any bearing is replaced. Refer to ***Pinion Gear/Ring Gear Alignment*** *and select new shims if the gearcase or a bearing has been replaced.*

1. Install the pinion gear shims into the gearcase.

2. Install the pinion gear assembly. Make sure the bearing inner race is in place on the inboard end of the pinion gear.

NOTE
The bearing retainer has left-hand threads. Turn it counterclockwise to tighten it.

36

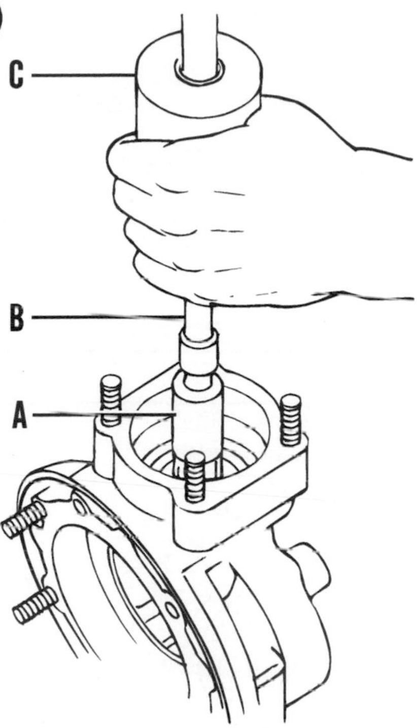

37

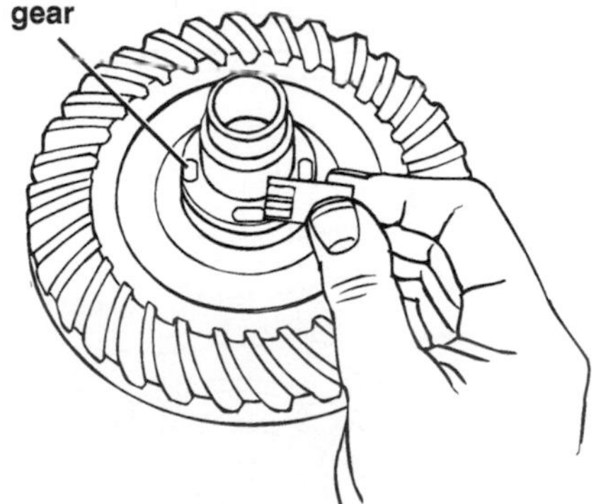

3. Install the bearing retainer with a new O-ring and new oil seal. Use lithium soap grease to pack the lips of the oil seal and to lubricate the O-ring.

4. Use the Yamaha bearing retainer wrench to torque the bearing retainer to 115 N•m (85 ft.-lb.). Turn the bearing retainer counterclockwise to tighten it.

5. Install the gear coupling and the washer.

6. Apply a medium-strength threadlocking compound to the new pinion gear nut and install it.

7. Hold the gear coupling with the Yamaha gear coupling/shaft holder (**Figure 34**) and torque the pinion gear nut 130 N•m (96 ft.-lb.).

NOTE
*At this point in the procedure, the ring gear assembly must be installed **without** the thrust washer.*

8. If the thrust washer is still in place, remove it from the ring gear. Install the ring gear assembly into final gearcase.

9. Install the bearing housing. Evenly tighten the bearing housing nuts (A, **Figure 32**) and bolts (B) in a criss-cross pattern. Torque the nuts to 23 N•m (17 ft.-lb.). Torque the bolts to 40 N•m (30 ft.-lb.).

10. Measure and adjust the gear lash as described in this chapter.

11. Measure the ring gear thrust clearance as follows:

   a. Remove the bearing housing and the ring gear as an assembly.

   b. Place four pieces of Plastigage around the inner circumference of the ring gear (**Figure 37**) and install the original thrust washer. Make sure the Plastigage is between the ring gear and the thrust washer.

CAUTION
Do not turn the pinion gear or the ring gear with Plastigage in the gearcase. This will produce an invalid measurement.

   c. Lower the final gearcase onto the ring gear/bearing housing assembly, and install the bearing housing nuts (A, **Figure 32**) and bolts (B). Evenly tighten the bearing housing nuts and bolts in a criss-cross pattern. Torque the nuts to 23 N•m (17 ft.-lb.). Torque the bolts to 40 N•m (30 ft.-lb.).

   d. Remove the bearing housing and the ring gear as an assembly, and measure the width of the Plastigage (**Figure 37**).

   e. If the thrust clearance is within the range specified in **Table 1**, remove the Plastigage. Reassemble the bearing housing and ring gear assembly with the original thrust washer. Tighten the bearing housing hardware as described in Step 9.

38

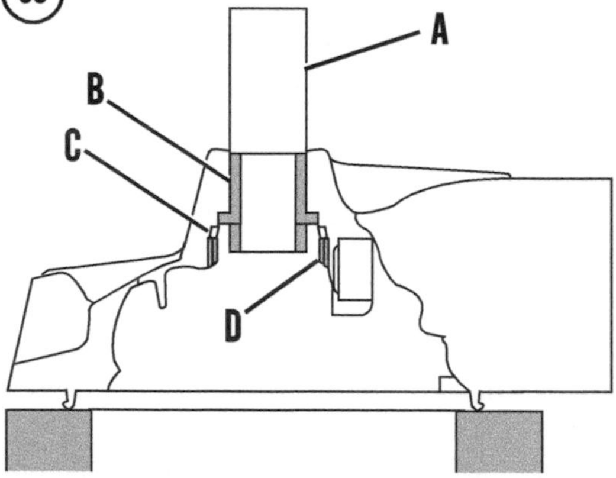

f. If the thrust clearance is outside the range specified in **Table 1**, refer to **Figure 33** and select a suitable thrust washer. Measure the thrust washer clearance with the new thrust washer. Repeat this process until the thrust washer clearance is within specification.

Gearcase Bearing Replacement

Removal

1. Disassemble the final gearcase as described in this chapter.
2. Securely support the final gearcase in a hydraulic press.
3. Use a driver (A, **Figure 38**) that matches the diameter of the collar (B), and press the collar, oil seal (C) and ring gear bearing (D) from the gearcase.

NOTE
Drive pinion bearing replacement is a difficult procedure. However, it is rarely necessary.

4. Place the final gearcase in an oven and heat the gearcase to 150° C (302° F).

WARNING
The final gearcase is extremely hot. Use welding gloves or other suitable means to protect your hands.

5. Remove the gearcase from the oven and set it on wooden blocks.

39

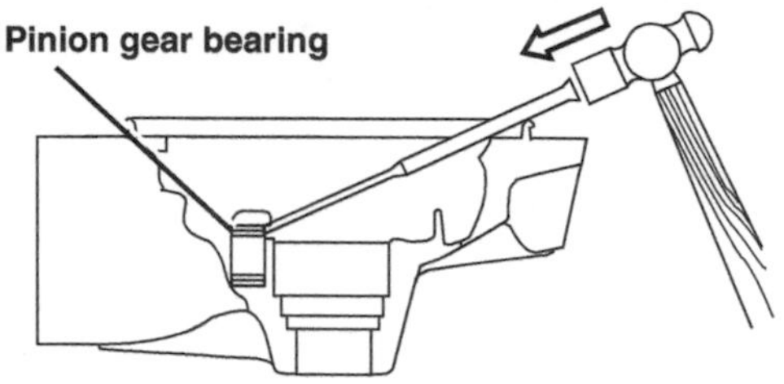

40

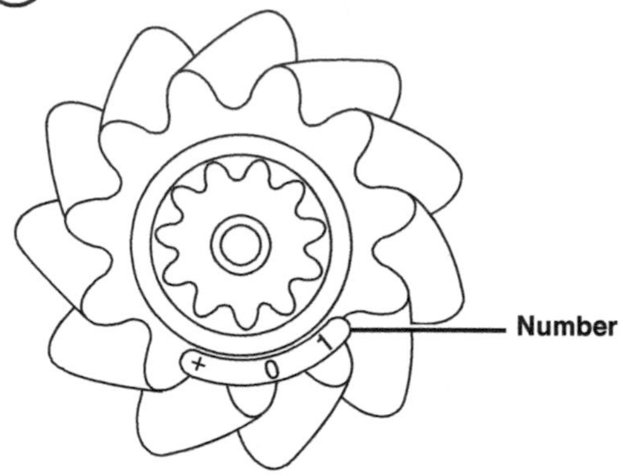

41

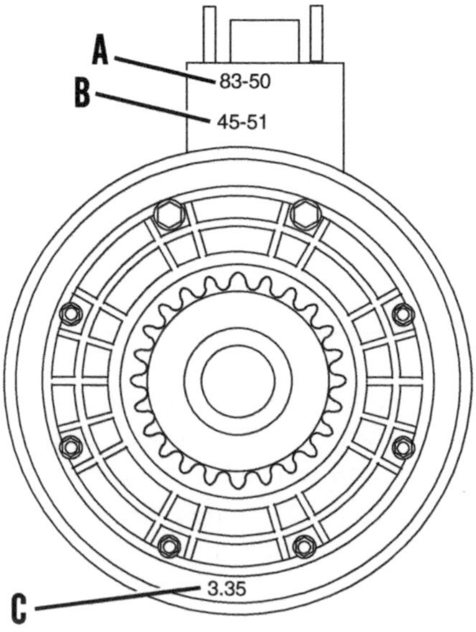

(42)

CALCULATED SHIM VALUE ROUNDED OFF

Hundredths digit	Rounded-off value
0, 1, 2, 3, 4	Round down to zero
5, 6, 7, 8, 9	Round up to the next tenth (10) digit.

6. Use a drift to tap around the circumference of the pinion gear bearing (**Figure 39**) and drive the race from the its boss in the gearcase.

Installation

1. Place the new ring gear bearing and the new pinion gear bearing outer race in a freezer overnight.
2. Heat the final gearcase to 150° C (302° F).

WARNING
The final gearcase is extremely hot. Use welding gloves or other suitable means to protect your hands.

3. Remove the gearcase from the oven and set it on wooden blocks.
4. Remove the pinion gear bearing outer race from the freezer and apply engine oil to the race.
5. Use a socket that matches the outside diameter of the race to drive the pinion gear bearing outer race (**Figure 39**) into its boss in the gearcase. Make sure the race bottoms in the boss.
6. Install the new pinion gear bearing inner race on the inboard end of the pinion gear.
7. Oil the ring gear collar and the boss in the final gearcase.
8. Support the gearcase housing in a hydraulic press.
9. Use a tool that matches the outside diameter of the collar to press the collar into place in the gearcase. Make sure the collar bottoms in the boss.
10. Pack the lips of a new oil seal with lithium soap grease, and press the oil seal over the collar and into the housing. The closed side of the oil seal should face down toward the lip of the collar.
11. Lubricate a new ring gear bearing, and press it over the collar and into the housing. Refer to the typical bearing procedures in Chapter One.

Pinion Gear/Ring Gear Alignment

The pinion gear and ring gear must be aligned whenever the final gearcase or any bearing is replaced. To align the pinion and rings gears, select and install new pinion gear shim(s) and new ring gear shim(s). See **Figure 28**.

1. Calculate the thickness of each shim as described in *Pinion Gear Shim Selection* and *Ring Gear Shim Selection*.
2. Refer to the chart (**Figure 33**) and select the shim(s) with the calculated thickness.
3. Assemble the final gearcase with these new shims.

Pinion Gear Shim Selection

Calculate the pinion gear shim thickness using the following formula:

Pinion gear shim thickness = (84 + A/100) – B.

A = the number on the pinion gear (**Figure 40**).

B = the number on the final gearcase (A, **Figure 41**).

For example: If the pinion gear number is +01 and the gearcase number is 83.50, then the pinion gear shim thickness:

= (84 + 1/100) – (83.50)

= (84 + 0.01) – (83.50)

= 84.01 – 83.50

= 0.51

Since shims are only available in increments of tenths of a millimeter, round off the hundredths digit according to the chart in **Figure 42**.

In the above example, round the calculated value of 0.51 mm down to 0.50 mm. Refer to the chart in **Figure 33** for available pinion gear shims and select a shim that is 0.50 mm thick.

12

Ring Gear Shim Thickness Selection

Calculate the ring gear shim thickness using the following formula:

Ring gear shim thickness = (C + D) – [(35.40 + or - E/100) + F].

C = the number on the final gearcase (B, **Figure 41**).

D = the number on the bearing housing (C, **Figure 41**).

E = the number on the ring gear (**Figure 43**).

F = the bearing thickness constant 13.00.

For example: If the final gearcase number (B, **Figure 41**) is 45.51, the bearing housing number (C) is 3.35, and the ring gear number (**Figure 43**) is -05, then the pinion gear shim thickness:

= (45.51 + 3.35) + [(35.40 - 05/100) + 13]

= (45.51) + (3.35) – [(35.40-0.05) + 13]

= (48.86) – (35.35 + 13)

= 48.86 – 48.35

= 0.51

43

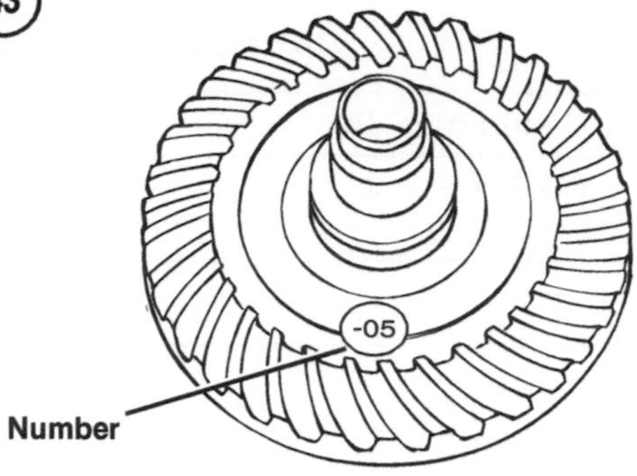

Since ring gear shims are only available in increments of tenths of a millimeter, round off the hundredths digit according to the chart in **Figure 42**.

In the above example, round the 0.51 mm thickness down to 0.50 mm. Refer to the chart in **Figure 33** for available ring gear shims and select the ring gear shim that is 0.50 mm thick.

Table 1 REAR SUSPENSION AND FINAL DRIVE SPECIFICATIONS

Item	Specification	Service limit
Final drive oil		
Viscosity	SAE 80 hypoid gear oil	–
Grade	API GL-4	–
Capacity	200 cc (6.8 U.S. oz.)	–
Shock absorber travel	113 mm (4.45 in.)	–
Shock spring free length	179.5 mm (7.07 in.)	
Shock spring installed length	163 mm (6.42 in.)	–
Spring rate (K1)	117.7 N•m (672 lb./in.)	–
Stroke (K1)	0-50 mm (0.00-1.97 in.)	–
Swing arm free play		
End	–	0.0 mm
Gear lash	0.1-0.2 mm (0.004-0.008 in.)	–
Ring gear thrust clearance	0.2 mm (0.008 in.)	–

Table 2 REAR SUSPENSION AND FINAL DRIVE TORQUE SPECIFICATIONS

Item	N•m	in.-lb.	ft.-lb.
Bearing retainer	115	–	85
Connecting arm nut	48	–	35
Final gearcase bolt	90	–	66
Final gearcase cover nuts	42	–	31
Final gearcase drain bolt	23	–	17
Final gearcase filler bolt	23	–	17
Final gearcase mounting stud[1]			
10 mm	18	–	13
8 mm	9	80	–
Pinion gear bearing retainer[2]	115	–	85
Pinion gear nut[1]	130	–	96
Rear axle clamp nut	23	–	17
Rear axle nut	107	–	79
Relay arm nut	48	–	35
Ring gear bearing housing bolts[1]	40	–	30
Ring gear bearing housing nuts	23	–	17
Shock absorber lower nut	48	–	35
Shock absorber upper nut	40	–	30
Swing arm pivot bolt	90	–	66

1. Apply a medium-strength threadlocking compound.
2. Uses left-hand threads.

12

CHAPTER THIRTEEN

BRAKES

The brake system on all models consists of dual front disc brakes and a single disc brake in the rear. This chapter describes the service procedures for brake system components.

When inspecting brake components, compare any measurements to the brake system specifications at the end of this chapter. Replace any part that is damaged, worn or out of specification. During assembly, tighten brake fasteners to the specified torque.

BRAKE SERVICE

When working on hydraulic brakes, all tools and the work area must be absolutely clean. Caliper or master cylinder components can be damaged by tiny particles of grit that enter the brake system. Do not use sharp tools inside the master cylinders, calipers or on the pistons. Sharp tools could damage these components and interfere with brake operation.

If there is any doubt about your ability to service a brake component safely and correctly, take the job to a Yamaha dealership or brake specialist.

Consider the following when servicing the front and rear brake systems.

1. Disc brake components rarely require disassembly. Do not disassemble them unless necessary.
2. When adding brake fluid, only use brake fluid clearly marked DOT 4 from a sealed container. Other grades of brake fluid may vaporize and cause brake failure.
3. Always use the same brand of brake fluid. One manufacturer's brake fluid may not be compatible with another's. Do not mix different brands of brake fluids.
4. Brake fluid absorbs moisture, which greatly reduces its ability to perform correctly. Purchase brake fluid in small containers and properly discard any small leftover quantities. Do not store a container of brake fluid with less than 1/4 of the fluid remaining. This small amount absorbs moisture very rapidly.

CAUTION

Do not use silicone based (DOT 5) brake fluid on the motorcycles covered in this manual. Silicone-based

1

fluid can damage these brake components leading to brake system failure.

CAUTION

Never reuse brake fluid (like the fluid expelled during brake bleeding). Contaminated brake fluid can cause brake failure.

5. Always keep the master cylinder reservoir cover installed to keep dust or moisture out of the system.

6. Use only DOT 4 brake fluid or isopropyl alcohol to wash parts. Never use petroleum-based solvents on the brake system's internal components. The seals will swell and distort, and have to be replaced.

7. Whenever any brake banjo bolt or brake line nut is loosened, the system is opened and must be bled to remove air. If the brakes feel spongy, this usually means air has entered the system. For safe operation, refer to *Brake Bleeding* in this chapter.

WARNING

*When working on the brake system, do **not** inhale brake dust. It may contain asbestos, which can cause lung injury and cancer. Wear a face mask that meets OSHA requirements for trapping asbestos particles, and wash your hands and forearms thoroughly after completing the work.*

WARNING

***Never** use compressed air to clean any part of the brake system. This releases harmful brake pad dust. Use an aerosol brake parts cleaner (**Figure 1**) to clean parts when servicing any component still installed on the motorcycle.*

PREVENTING BRAKE FLUID DAMAGE

Brake fluid will damage most surfaces on a motorcycle. To prevent brake fluid damage, note the following:

1. Protect the motorcycle before beginning any service requiring draining, bleeding or handling of brake fluid. Anticipate which parts are likely to leak brake fluid, and use a large tarp or piece of plastic to cover the areas beneath those parts. Even a few drops of brake fluid can extensively damage painted, plated or plastic surfaces.
2. Keep a bucket of soap and water close to the motorcycle while working on the brake system. If brake fluid spills on any surface, immediately wash the area with soap and water, then rinse it thoroughly.
3. To help control the flow of brake fluid when refilling the reservoirs, punch a small hole into the seal of a new container. Place this hole next to the edge of the pour spout.

BRAKE BLEEDING

Bleeding the brakes removes air from the brake system. Air in the brakes increases brake lever or brake pedal travel, and it makes the brakes feel soft or spongy. Under extreme circumstances, it can cause complete loss of brake pressure.

The brakes can be bled manually or with the use of a vacuum pump. Both methods are described here. Only use fresh DOT 4 brake fluid when bleeding the brakes. Do not reuse old brake fluid and do not use DOT 5 (silicone based) brake fluid.

1. Clean the bleed valve and area around the valve before beginning. Make sure the opening in the valve is clear.
2. Use a box-end wrench to open and close the bleed valve. This prevents damage to the valve, especially if the valve is rusted in place.

3. Replace a bleed valve with damaged threads or with a rounded hex head. A damaged valve is difficult to remove and it cannot be properly tightened.

NOTE
*The catch hose (**Figure 2**) is the hose installed between the bleed valve and the catch bottle.*

4. Use a clear catch hose so the fluid can be visually inspected as it leaves the bleed valve. Air bubbles in the catch hose indicate that air may be trapped in the brake system.

5. Open the bleed valve just enough to allow fluid to pass through the valve and into the catch bottle. If a bleed valve is too loose, air can be drawn into the system through the valve threads.

6. If air may be entering through the valve threads, apply silicone brake grease around the valve where it emerges from the caliper. The grease should seal the valve and prevent the entry of air. Wipe away the grease once the brakes have been bled. For a more permanent solution, remove the bleed valve and apply Teflon tape to the valve threads. Reinstall the valve and torque it to 6 N•m (53 in.-lb.). Make sure the Teflon tape does not cover the passage in the bleed valve.

7. If the system is difficult to bleed, tap the banjo bolt on the master cylinder a few times. This should dislodge air bubbles that may have become trapped at the hose connection. Also tap the banjo bolts at the brake hose union (beneath the lower fork bridge), at the calipers and any other hose connections in the brake line.

Manual Bleeding

1. Check all banjo bolts in the system. They must be tight.

2. Remove the dust cap from the bleed valve on the caliper assembly.

3. Connect a length of clear tubing to the bleed valve (**Figure 2**, typical). Place the other end of the tube into a clean container. Fill the container with enough fresh brake fluid to keep the end submerged. The tube should be long enough that its loop is higher than the bleed valve to prevent air from being drawn into the caliper during bleeding.

2

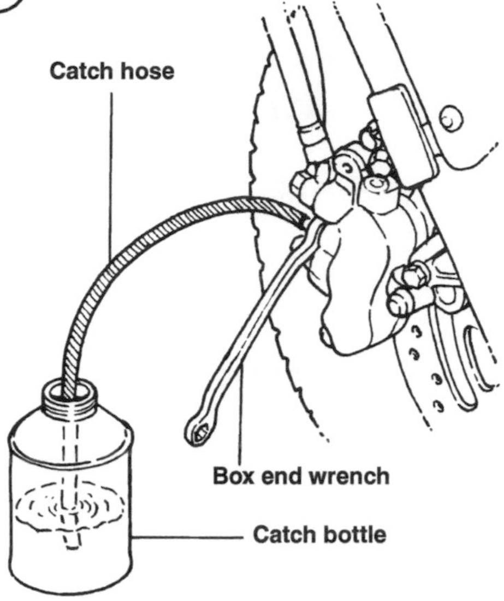

NOTE
When bleeding the front brakes, turn the handlebars to level the front master cylinder.

4. Clean all debris from the top of the master cylinder reservoir. Remove the top cover, diaphragm plate and the diaphragm from the reservoir.

5. Add brake fluid to the reservoir until the fluid level reaches the reservoir upper limit. Loosely install the diaphragm and the cover. Leave them in place during this procedure to keep dirt out of the system and so brake fluid cannot spurt out of the reservoir.

6. Pump the brake lever or brake pedal a few times, then release it.

7. Apply the brake lever or pedal until it stops and hold it in this position.

8. Open the bleed valve with a wrench (**Figure 2**, typical). Let the brake lever or pedal move to the limit of its travel, then close the bleed valve. Do not release the brake lever or pedal while the bleed valve is open.

NOTE
As break fluid enters the system, the level in the reservoir drops. Add brake fluid as necessary to keep the fluid level 10 mm (3/8 in.) below the reser-

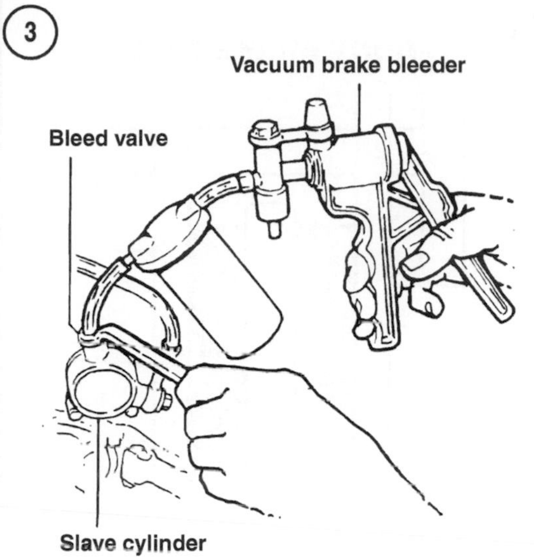

voir top so air will not be drawn into the system.

9. Repeat Steps 6-8 until the brake fluid flowing from the hose is clear and free of air. If the system is difficult to bleed, tap the master cylinder or caliper with a soft mallet to release trapped air bubbles.

10. Test the feel of the brake lever or pedal. It should feel firm and offer the same resistance each time it is operated. If the lever or pedal feels soft, air is still trapped in the system. Continue bleeding the system.

NOTE
The setting on the front brake lever adjuster affects bleeding. Initially bleed the front brakes with the adjuster turned to the softest setting. Once the brakes feel solid, check the feel with the adjuster in several different settings. If the lever feels soft at any setting or if the lever hits the handlebar, air is still trapped in the system. Continue bleeding the system.

11. When brake system bleeding is complete, disconnect the hose from the bleed valve. Torque the bleed valve to 6 N•m (53 in.-lb.).

12. When bleeding the front brakes, repeat Step 1-11 on the opposite front caliper.

13. Add brake fluid to the master cylinder to correct the fluid level.

14. Install the diaphragm, diaphragm plate and top cap. Make sure the cap is secured in place.

NOTE
Do not ride the motorcycle until the front and rear brakes, and the brake light are working properly.

15. Test ride the motorcycle slowly at first to make sure the brakes are operating properly.

Vacuum Bleeding

1. Check all banjo bolts in the system. They must be tight.

2. Remove the dust cap from the bleed valve on the caliper assembly.

NOTE
When bleeding the front brakes, turn the handlebars to level the front master cylinder.

3. Clean all dirt or foreign matter from the top of the master cylinder reservoir. Remove the top cover, diaphragm plate and the diaphragm from the reservoir.

4. Add brake fluid to the reservoir until the fluid level reaches the reservoir upper limit. Loosely install the diaphragm and the cover. Leave them in place during this procedure to keep dirt out of the system and so brake fluid cannot spurt out of the reservoir.

5. Assemble the vacuum tool following the manufacturer's instructions.

6. Connect the pump's catch hose to the bleed valve on the brake caliper (**Figure 3**).

NOTE
When using a vacuum pump, keep an eye on the brake fluid level in the reservoir. It will drop quite rapidly. This is particularly true for the rear reservoir, which does not hold as much brake fluid as the front. Stop often and check the brake fluid level. Maintain the level at 10 mm (3/8 in.) from the top of the reservoir so air will not be drawn into the system.

7. Operate the vacuum pump to create vacuum in the hose.

8. Use a wrench to open the bleed valve. The vacuum pump should pull fluid from the system. Close the bleed valve before the brake fluid stops flowing from the system or before the master cylinder reservoir runs empty. Add fluid to the reservoir as necessary.
9. Operate the brake lever or brake pedal a few times, and release it.
10. Repeat Steps 7 and 9 until the fluid leaving the bleed valve is clear and free of air bubbles. If the system is difficult to bleed, tap the master cylinder and caliper housing with a soft mallet to release trapped air bubbles.
11. Test the feel of the brake lever or brake pedal. It should feel firm and offer the same resistance each time it is operated. If the lever or pedal feels soft, air is still trapped in the system. Continue bleeding the system.
12. When bleeding is complete, disconnect the hose from the bleed valve and torque the valve to 6 N•m (53 in.-lb.).
13. When bleeding the front brakes, repeat Step 1-12 on the opposite front caliper.
14. Add brake fluid to the master cylinder to correct the fluid level.
15. Install the diaphragm, diaphragm plate and top cap. Make sure the cap is secured in place.

NOTE
Do not ride the motorcycle until both brakes and the brake light are working properly.

16. Test ride the motorcycle slowly at first to make sure the brakes are operating properly.

BRAKE FLUID DRAINING

Before disconnecting a front or rear brake hose, drain the brake fluid as described below. Draining the fluid reduces the amount of fluid that can spill out when system components are removed.

This section describes two methods for draining the brake system: manual and vacuum.

Manual Draining

An empty bottle, a length of clear hose and a wrench is required for this procedure.
1. Remove the dust cap from the bleed valve. Remove all dirt from the valve and its outlet port.

4

FRONT BRAKE PADS

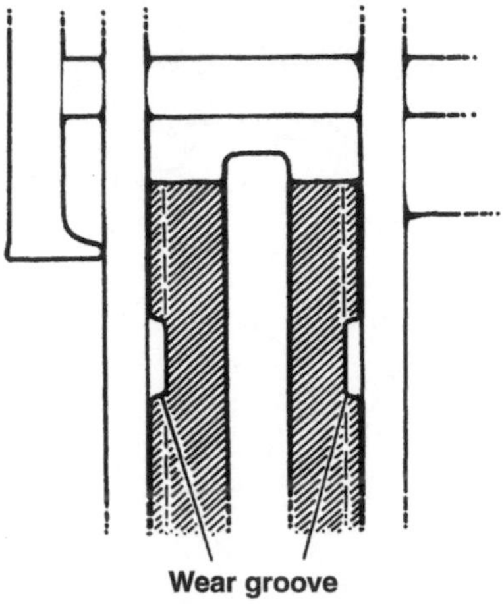

5

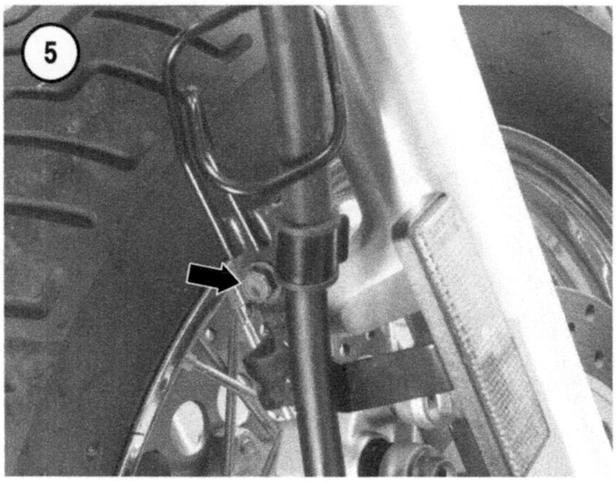

2. Connect a length of clear hose to the bleed valve on the caliper. Insert the other end into a container (**Figure 2**, typical).
3. Apply the front brake lever or the rear brake pedal until it stops. Hold the lever or pedal in this position.
4. Open the bleed valve with a wrench and let the lever or pedal move to the limit of its travel. Close the bleed valve.
5. Release the lever or pedal, and repeat Steps 3 and 4 until brake fluid stops flowing from the bleed valve.
6. When draining the front brakes, repeat this on the other brake caliper.
7. Discard the brake fluid.

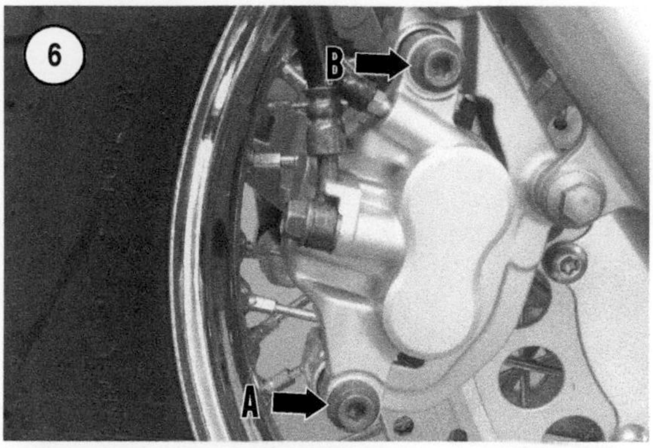

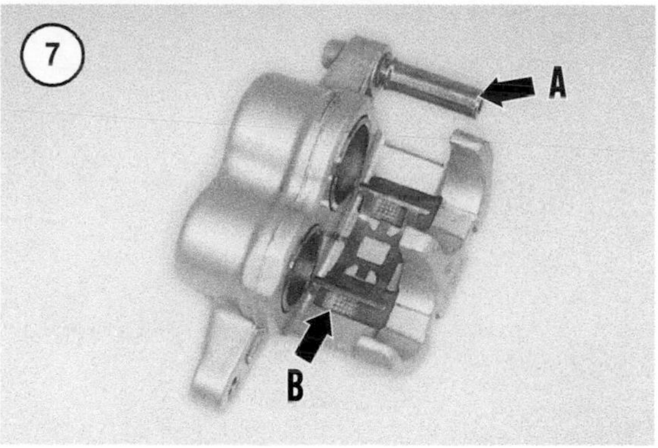

Vacuum Draining

A hand-operated vacuum pump is required when performing this procedure.

1. Connect the pump's catch hose to the bleed valve on the brake caliper (**Figure 3**).
2. Operate the vacuum pump to create vacuum in the hose.
3. Use a wrench to open the bleed valve. The vacuum pump should pull fluid from the system.
4. When fluid has stopped flowing through the hose, close the bleed valve.
5. Repeat Steps 2-4 until brake fluid no longer flows from the bleed valve.
6. When draining the front brake system, repeat this procedure on the opposite caliper.
7. Discard the brake fluid.

BRAKE PADS

Since brake pad wear depends greatly upon riding habits and conditions, manufacturers typically do not provide a recommended mileage interval for changing the brake pads. Check the brake pads for wear before each ride or when the wear indicator approaches the edge of the brake disc.

To maintain even brake pressure on the disc, always replace both pads in a caliper at the same time. When replacing the front brake pads, replace both pads in *both front calipers* at the same time. Refer to **Table 1** or **Table 2** for brake pad service limits.

Note that the brake hose does not need to be disconnected from the caliper during brake pad replacement. If the hose is removed, the brakes will have to be bled. Disconnect the hose only when servicing the brake caliper.

WARNING
Use brake fluid clearly marked DOT 4 from a sealed container. Other types may vaporize and cause brake failure. Always use the same brand of brake fluid. Do not mix brake fluids from different manufacturers. They may not be compatible.

WARNING
Do not ride the motorcycle until you are sure the brakes are operating correctly. If necessary, bleed the brake as described in this chapter.

CAUTION
*Check the pads more frequently when the wear limit lines (**Figure 4**) approach the disc. On some pads, the wear lines are very close to the metal backing plate. If pad wear happens to be uneven, the backing plate may contact the disc and cause damage.*

13

Front Brake Pad Replacement

1. Securely support the motorcycle on level ground.
2. Remove the mounting bolt (**Figure 5**) and release the brake hose holder from the fork leg.

3A. On 1999-2000 XVS1100 models, perform the following:

   a. Remove the single retaining bolt (A, **Figure 6**).
   b. Pivot the rear of the caliper up until the caliper clears the brake disc.
   c. Pull the caliper out until the caliper pin (A, **Figure 7**) clears the pivot (A, **Figure 8**) in the pad holder.

d. Use a bunjee cord to suspend the caliper from the motorcycle.

3B. On 2001-on XVS1100 models and all XVS1100A models, remove the two retaining bolts (A and B, **Figure 6**). Lift the caliper from the pad holder. Suspend it from the motorcycle with a bunjee cord.

4. Remove the outboard brake pad (A, **Figure 8**) and the inboard pad (**Figure 9**) from the pad holder.

5. Inspect the brake pads as described later in this section.

6. Remove the pad spring (A, **Figure 10**) from the caliper and check it for wear or fatigue. Replace the pad spring as necessary.

WARNING

The brake pads must be replaced as set. If any pad requires replacement, replace both pads in both front calipers.

7. When new pads are installed in the caliper, the master cylinder brake fluid level will rise as the caliper pistons are repositioned.

  a. Clean all dirt and foreign matter from the top of the master cylinder.
  b. Remove the screws securing the master cylinder cover. Remove the cover, diaphragm plate and the diaphragm from the master cylinder.
  c. Slowly push the caliper pistons (B, **Figure 10**) into the caliper. Constantly check the reservoir to make sure brake fluid does not overflow. Remove fluid if necessary.
  d. The pistons should move freely. If they do not move smoothly without sticking, the caliper should be removed and serviced as described in this chapter.

8. Push the caliper pistons until they bottom in the bore to allow room for the new pads.

NOTE

When purchasing new pads, make sure the friction compound of the new pads are compatible with the disc material. Remove any roughness from the backs of the new pads with a fine-cut file.

CAUTION

Position each brake pad so the friction material faces in toward the brake disc.

8

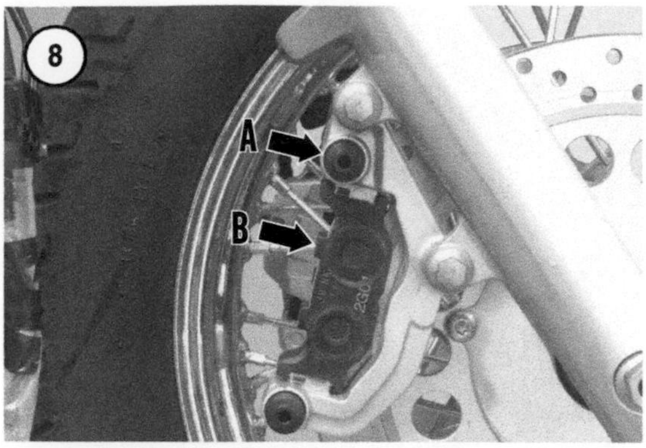

9

10

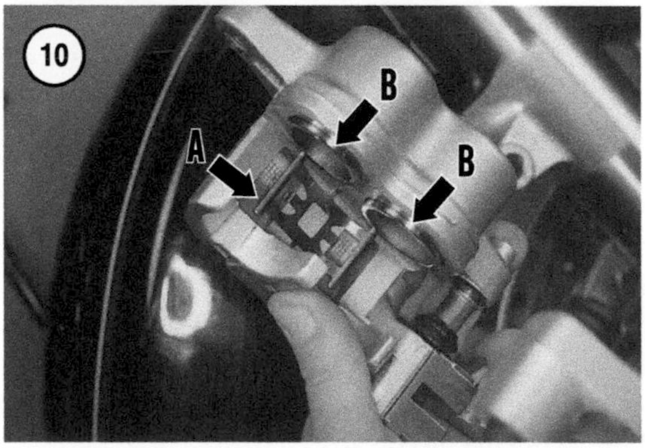

9. If removed, install the pad springs (**Figure 11**) onto the pad holder and into the caliper (B, **Figure 7**). Always use new pad springs when installing new brake pads.

10. Install the inboard (**Figure 9**), then outboard pad (B, **Figure 8**), into the pad holder. The ears on each pad must straddle the pad springs in the pad holder.

WARNING

Use just enough grease to lubricate the caliper pin and retaining bolt in

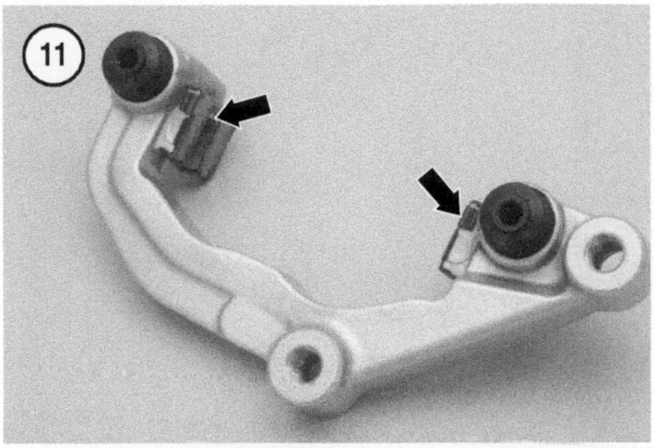

the next step. Excess grease could contaminate the brake pads. Do not get any grease on the brake pads.

11A. On 1999-2000 XVS1100 models, perform the following:

a. Lubricate the caliper pin (A, **Figure 7**) with lithium soap grease.
b. Insert the caliper pin into the pivot (B, **Figure 8**) in the pad holder.
c. Rotate the caliper and lower it onto the pad holder.
d. Lubricate the retaining bolt (A, **Figure 6**) with lithium soap grease and install it in the caliper. Torque the retaining bolt to 23 N•m (17 ft.-lb.).

WARNING
Use just enough grease to lubricate the retaining bolts in the next step. Excess grease could contaminate the brake pads. Do not get any grease on the brake pads.

11B. On 2001-on XVS1100 models and all XVS1100A models, perform the following:

a. Lower the caliper onto the pad holder.
b. Lubricate the retaining bolts with lithium soap grease.
c. Install the retaining bolts (A and B, **Figure 6**) and secure the caliper to the pad holder.
d. Torque the retaining bolts to 27 N•m (20 ft.-lb.).

12. Reinstall the brake hose holder. Torque the brake hose holder bolt (**Figure 5**) to 10 N•m (89 in.-lb.).
13. Repeat Steps 2-12 to replace the pads in the opposite front caliper.
14. Support the motorcycle with the front wheel off the ground. Spin the wheel and pump the brake lever until the pads are seated against the disc.

WARNING
Use brake fluid clearly marked DOT 4 from a sealed container. Other types may vaporize and cause brake failure. Always use the same brand of brake fluid. Do not intermix brake fluids. Many brands are not compatible with one another.

15. Refill the master cylinder reservoir, if necessary, to maintain the correct fluid level. Install the diaphragm and top cover. Tighten the cover screws securely.

WARNING
Do not ride the motorcycle until you are sure the brakes are operating correctly with full hydraulic advantage. If necessary, bleed the brake as described in this chapter.

16. Test ride the motorcycle slowly at first to make sure the brakes are operating properly.

Rear Brake Pad Replacement

1. Securely support the motorcycle on a level surface.
2. Remove the pad cover (**Figure 12**) from the caliper.
3. Remove the clip (**Figure 13**) from each retaining pin. Note that each clip sits between the caliper and the outboard brake pad. The clips must be reinstalled here during assembly.
4. Remove each retaining pin (A and B, **Figure 14**) and remove the pad spring (A, **Figure 15**). Note that

the side with the longer tang (**Figure 16**) points in the direction of forward wheel rotation. The spring must be installed with this same orientation during assembly.

5. Remove the inboard brake pad (**Figure 17**) and the outboard pad (**Figure 18**) from the caliper.

6. Inspect the brake pads as described in this section.

7. When new pads are installed in the caliper, the master cylinder brake fluid level rises as the caliper pistons are repositioned.

   a. Clean all dirt and foreign matter from the top of the master cylinder.
   b. Remove the master cylinder cover and the diaphragm from the master cylinder.
   c. Install the old outboard pad into the caliper (**Figure 18**). Use the pad to slowly push the piston into the caliper until the piston bottoms. Constantly check the reservoir to make sure brake fluid does not overflow. Remove brake fluid if necessary.
   d. Remove the outboard pad and repeat substep c with the inboard pad.
   e. The pistons should move freely. If they do not move smoothly without sticking, the caliper should be removed and serviced as described in this chapter.

NOTE

When purchasing new pads, check with the dealership to make sure the friction compound of the new pads is compatible with the disc material. Remove any roughness from the backs of the new pads with a fine-cut file.

CAUTION

Position each brake pad so the friction material faces in toward the brake disc.

8. Install a new outboard pad (**Figure 18**) into the caliper, then install the inboard pad (**Figure 17**).

9. Install the upper pad pin (**Figure 19**) and secure it with the clip (B, **Figure 15**). The clip must sit between the caliper body and the outboard pad.

10. Position the new pad spring so the side with the long tang (**Figure 16**) points in the direction of forward wheel rotation. See A, **Figure 15**.

13

14

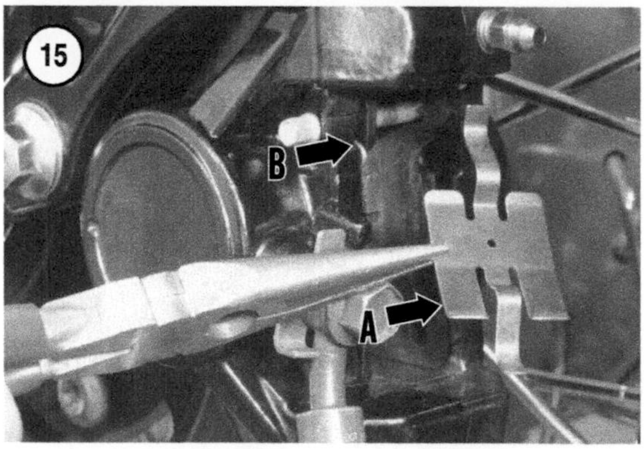

15

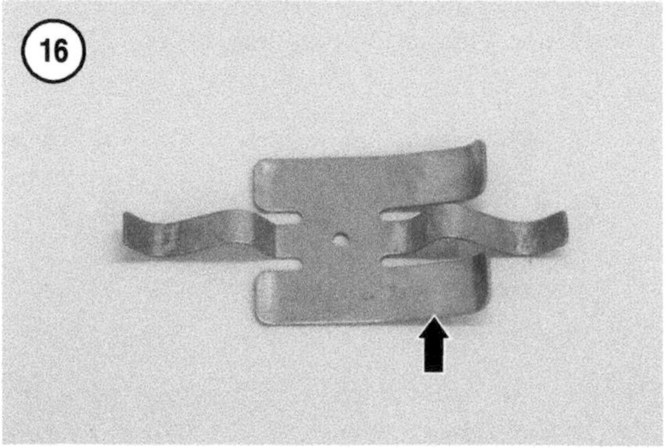
16

17

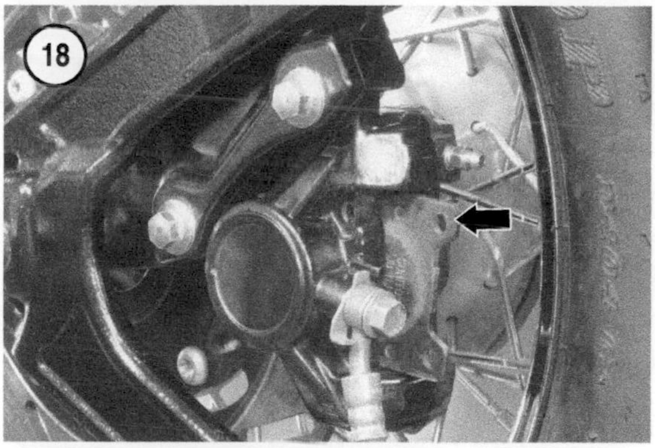

18

19

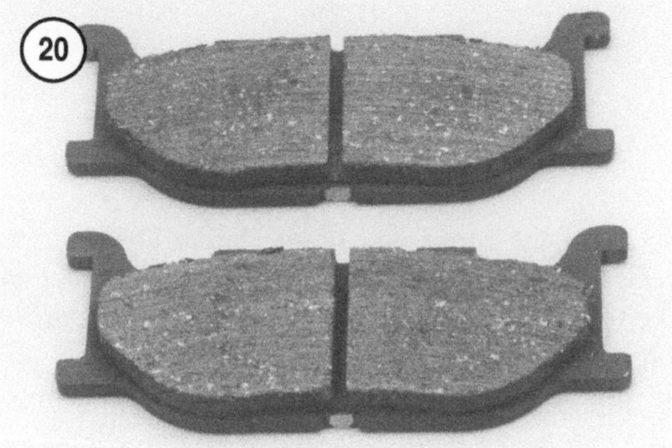

20

11. Slide the finger of the pad spring under the retaining pin (B, **Figure 14**) and seat the pad spring in the caliper.

12. Press the lower spring finger into the caliper and install the lower retaining pin (A, **Figure 14**). The pin must pass through the outboard pad, over the spring finger and through the inboard pad.

13. Install a clip through the hole in the retaining pin. The clip must sit between the caliper and the outboard brake pad as shown in **Figure 13**.

14. Install the pad cover (**Figure 12**) onto the caliper.

15. Support the motorcycle with the rear wheel off the ground. Spin the wheel and pump the brake pedal until the pads are seated against the disc.

WARNING

Use brake fluid clearly marked DOT 4 from a sealed container. Other types may vaporize and cause brake failure. Always use the same brand of brake fluid. Do not intermix brake fluids. Many brands are not compatible with one another.

16. Check the fluid level in the master cylinder reservoir. Add brake fluid as necessary to correct the fluid level. Install the diaphragm and top cover.

WARNING

Do not ride the motorcycle until you are sure the brakes are operating correctly with full hydraulic advantage. If necessary, bleed the brakes as described in this chapter.

17. Test ride the motorcycle slowly at first to make sure the brakes are operating properly.

Brake Pad Inspection

1. Inspect the brake pads (**Figure 20** for the front and **Figure 21** for the rear) as follows:
   a. Inspect the friction material for light surface dirt, grease and oil contamination. Remove light contamination with sandpaper. If contamination has penetrated the surface, replace the brake pads.
   b. Inspect the friction material for uneven wear, damage or contamination. Both pads should show approximately the same amount of

wear. If the pads are wearing unevenly, the caliper may not be operating correctly.

c. Measure the thickness of the friction material with a vernier caliper. Replace both brake pads if the thickness on either pad equals or is less than the service limit listed in **Table 1** or **Table 2**. When servicing the front brakes, replace both pads in both front calipers if any pad is worn to the service limit (**Figure 22**).
d. Inspect the metal plate on the back of each pad for corrosion and damage.

NOTE
Cleaning the brake disc is especially important if new pads are being installed. Many brake pad compounds are not compatible.

2. Use brake parts cleaner and a fine grade emery cloth to remove all road debris and brake pad residue from the brake disc surface.
3. Inspect the brake disc as described in this chapter.
4. Check the friction surface of the new pads for any foreign matter or manufacturing residue. If necessary, clean the pads with an aerosol brake cleaner.
5. Check the pad springs (**Figure 23** and **Figure 11** for the front, and **Figure 16** for the rear) for wear or fatigue. Replace a pad spring if it shows any sign of damage or excessive wear.
6. Install new pad spring(s) when installing new brake pads.

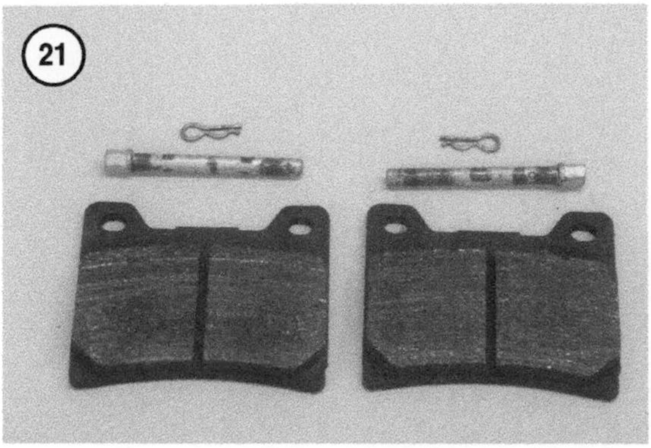
21

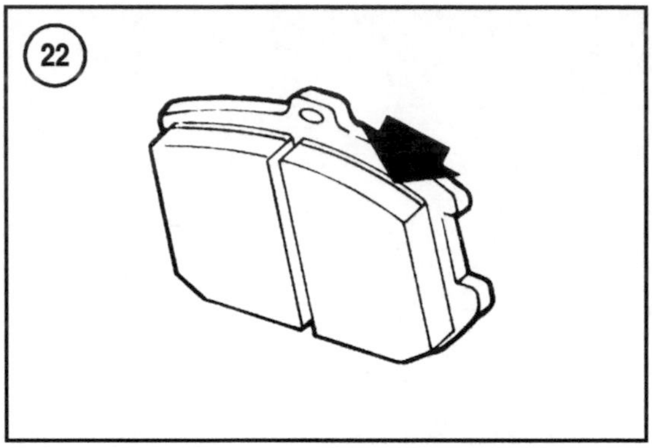
22

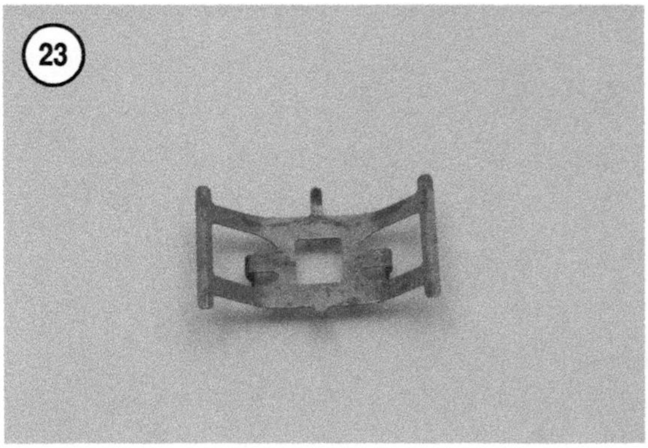
23

FRONT CALIPER

Refer to **Figure 24**.

Removal

1. Securely support the motorcycle on level ground.
2. Note that the brake hose neck sits against the inboard side of the indexing post (A, **Figure 25**) on the caliper. The hose must be installed on this side of the post during assembly.
3. Remove the banjo bolt (B, **Figure 25**) from the caliper. Remove the brake hose and the two copper washers from the caliper. Discard the washers. New ones must be installed during assembly.

CAUTION
Brake fluid damages paint and finish. Immediately wash any spilled brake fluid from the motorcycle. Use soapy water and rinse the area completely.

4. Place the loose end of the brake hose into a recloseable plastic bag to keep foreign matter out of the system and to protect the motorcycle from dribbling brake fluid.
5. Remove the brake pads as described in Steps 1-6 of *Front Brake Pad Replacement* in this chapter.

(24)

FRONT BRAKE CALIPER

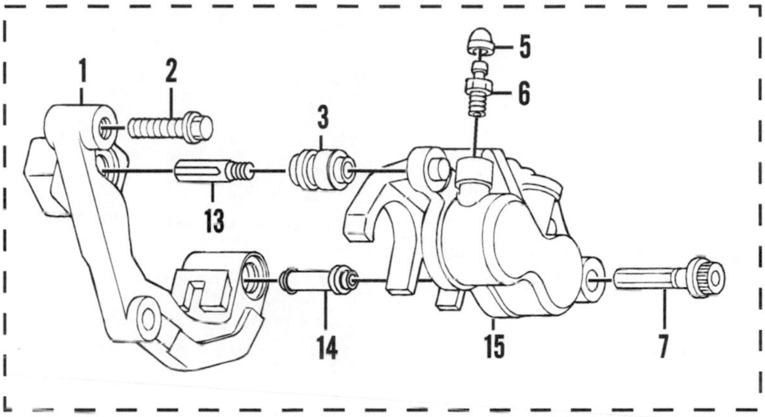

1999-2000 XVS1100

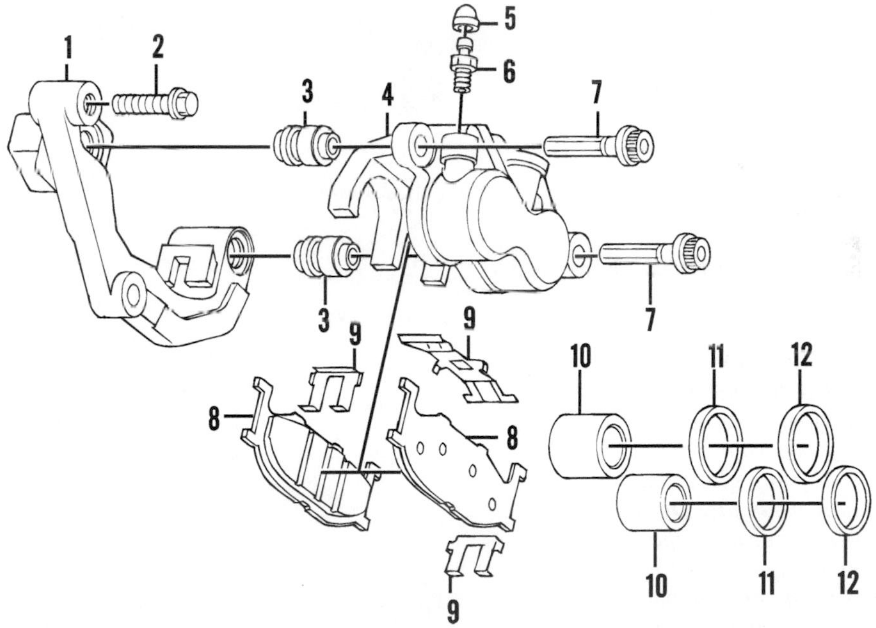

1. Pad holder
2. Caliper mounting bolt
3. Boot
4. Caliper (2001-on XVS1100 models, all XVS1100A models)
5. Cap
6. Bleed valve
7. Retaining bolt
8. Brake pad
9. Pad spring
10. Piston
11. Dust seat
12. Piston seal
13. Caliper pin (1999-2000 XVS1100 models)
14. Boot (1999-2000 XVS1100 models)
15. Caliper (1999-2000 XVS1100 models)

6. Remove the caliper mounting bolts (A, **Figure 26**) and lower the pad holder (B) from the fork slider.
7. Disassemble and inspect the caliper as described in this chapter.

Installation

1. Set the pad holder (B, **Figure 26**) into place on the fork slider and secure the holder with the caliper mounting bolts (A). Apply a medium-strength threadlocking compound to the threads of the caliper mounting bolts and torque the bolts to 40 N•m (30 ft.-lb.).
2. Install the brake pads as described in Steps 9-12 of *Front Brake Pad Replacement* in this chapter.
3. Place the brake hose against the caliper port so the brake hose neck sits on the inboard side of the indexing post (A, **Figure 25**). Install a new copper washer on each side of the brake hose fitting.
4. Install the banjo bolt (B, **Figure 25**). Torque the bolt to 30 N•m (22 ft.-lb.).
5. Add brake fluid to the reservoir and bleed the brakes as described earlier in this chapter.

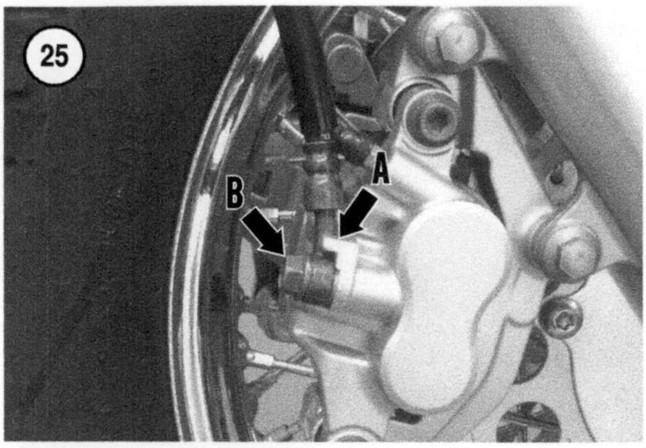

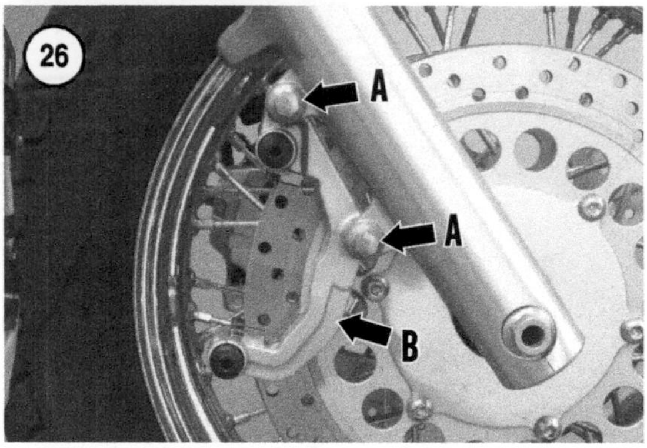

Disassembly/Assembly

1. Remove the brake pads and brake caliper as described in this chapter.

WARNING

In the next step, the pistons may shoot out of the caliper body with considerable force. Keep your fingers out of the way. Wear shop gloves and apply air pressure gradually. ***Do not*** *use high pressure air or place the air hose nozzle directly against the hydraulic line fitting in the caliper body. Hold the air nozzle away from the inlet, allowing some of the air to escape.*

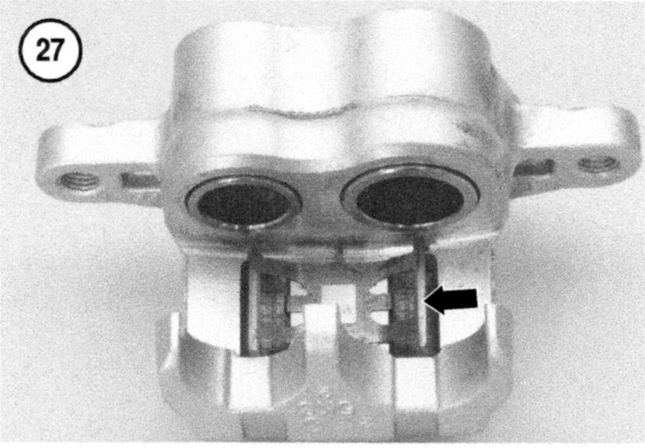

2. If still installed, remove the pad spring (**Figure 27**) from the caliper body.
3. Pad the pistons with shop rags or wood blocks as shown in **Figure 28**. Block the exposed housing fluid port holes on the caliper housing. Apply compressed air through the caliper hose port and blow the pistons out of the caliper. Remove the pistons from the caliper cylinders. See **Figure 29**.

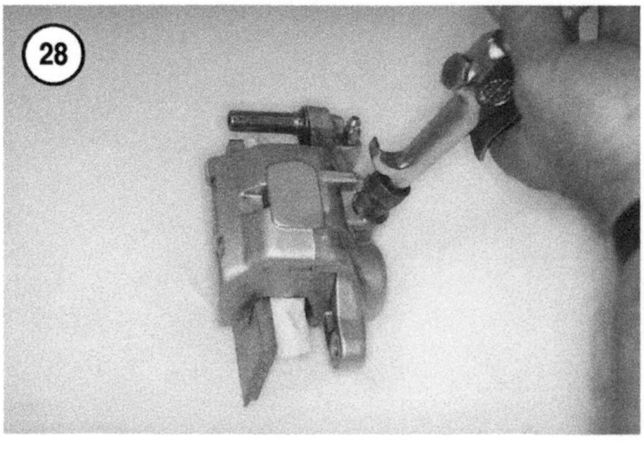

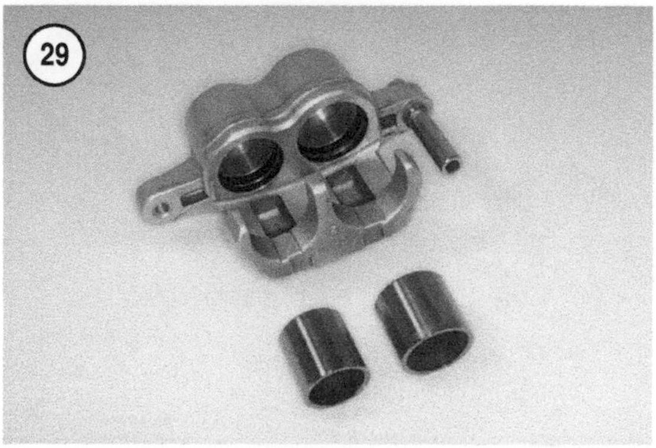

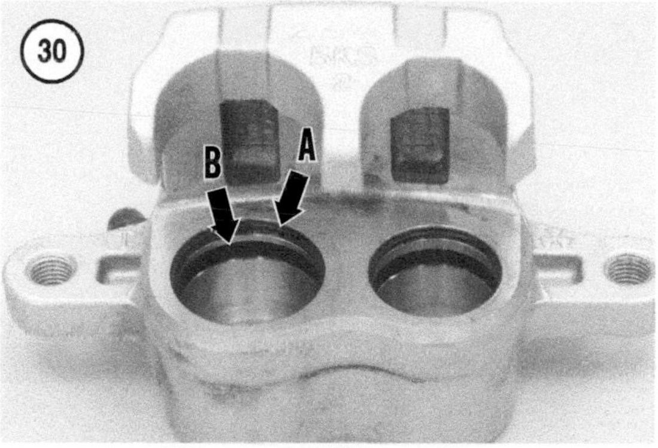

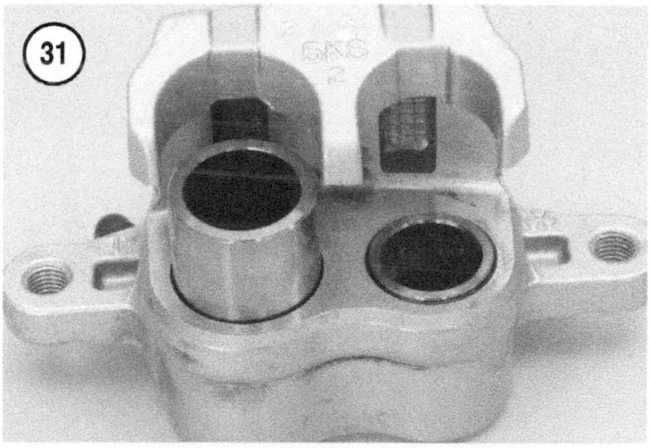

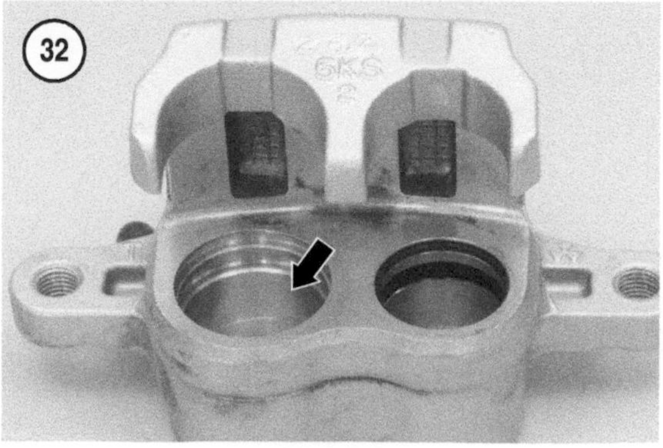

CAUTION

In the following step, do not use a sharp tool to remove the dust and piston seals from the caliper cylinder. Sharp tools could damage the cylinder surface. The caliper will have to be replaced if the cylinder surface is damaged.

4. Use a piece of plastic or wood to carefully remove the dust seal (A, **Figure 30**) and the piston seal (B) from their grooves in each caliper cylinder. Discard both seals.
5. Clean all caliper parts and inspect them as described in this section.

NOTE

Never reuse the old dust seals or piston seals. Very minor damage or age deterioration can make the seals useless.

6. Coat the new dust seal and piston seal with fresh DOT 4 brake fluid.
7. Carefully install the new piston seal (B, **Figure 30**) and dust seal (A) into the grooves in the caliper cylinders. Make sure the seals are properly seated in their respective grooves.
8. Coat the pistons and caliper cylinders with fresh DOT 4 brake fluid.
9. Position each piston with the open end facing toward the brake pads (**Figure 31**) and slide the pistons into the caliper cylinders. Push the pistons in until they bottom in the cylinders.
10. Install the pad spring. Make sure the spring is completely seated in the bottom of the caliper body as shown in **Figure 27**.
11. Install the caliper and brake pads as described in this chapter.

13

Inspection

1. Clean all parts, except brake pads, with clean DOT 4 brake fluid. Place the cleaned parts on a lint-free cloth while performing the following inspection procedures.
2. Inspect both seal grooves (**Figure 32**) in each cylinder for damage. If they are damaged or corroded, replace the caliper assembly.
3. Inspect the walls (**Figure 32**) in each cylinder for scratches, scoring or other damage. If there is rust or corrosion, replace the caliper assembly.

4. Measure the inside diameter of each cylinder bore with a bore gauge (**Figure 33**). Replace the brake caliper if the inside diameter of either bore exceeds the specification in **Table 1** or **Table 2**.
5. Inspect the pistons (**Figure 34**) for scratches, scoring or other damage. If they are rusty or corroded, replace the pistons.
6. Inspect the caliper body for scratches or other signs of damage. Replace the caliper assembly if necessary.
7. Inspect the pad spring(s) (**Figure 23** and A, **Figure 35** for the front caliper, and **Figure 36** for the rear caliper). Replace a spring if it is cracked, worn or shows signs of fatigue.
8. On front brake calipers, perform the following:
   a. Remove the pad springs (A, **Figure 35**) from the pad holder if they have not been removed, and inspect the springs. Replace both pad springs if either one is cracked, worn or shows signs of fatigue.
   b. Inspect the pad holder (B, **Figure 35**) for cracks or other signs of damage. Replace the caliper assembly if necessary.
   c. Inspect the boots (C, **Figure 35**) in the pad holder. Replace a boot that is torn or hard.
9. Inspect the mounting bolt holes (D, **Figure 35**) on the pad holder. If they are worn or damaged, replace the caliper assembly and inspect the bolt holes on the fork leg.
10. Remove the bleed valve from the caliper body. Apply compressed air to the opening and make sure it is clear. If necessary, clean out the bleed valve with fresh brake fluid. Install the bleed valve finger-tight. It will be torqued to specification during brake bleeding.
11. Inspect the fluid opening in the base of each cylinder bore. Apply compressed air to the opening and make sure it is clear. Clean out the opening with fresh brake fluid if necessary.
12. Inspect the threads of the banjo bolt and the retaining bolt for wear or damage. Clean up any minor thread damage. Replace the bolts and the caliper assembly if necessary.

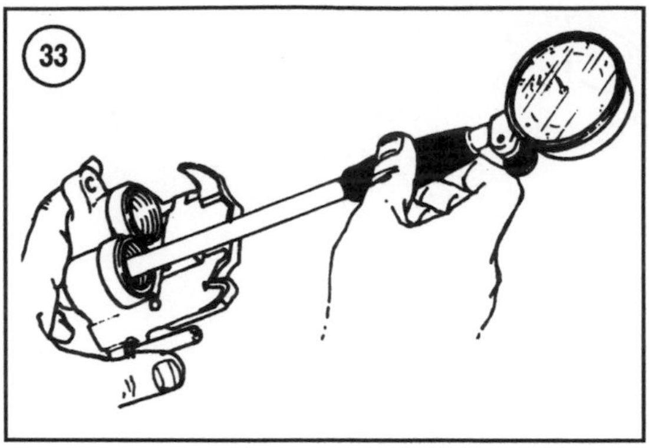

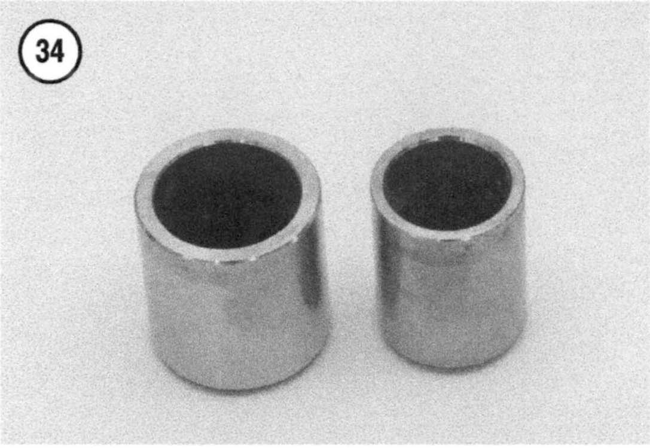

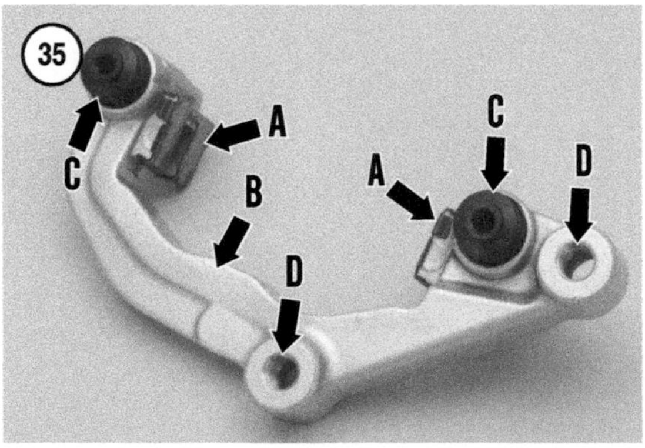

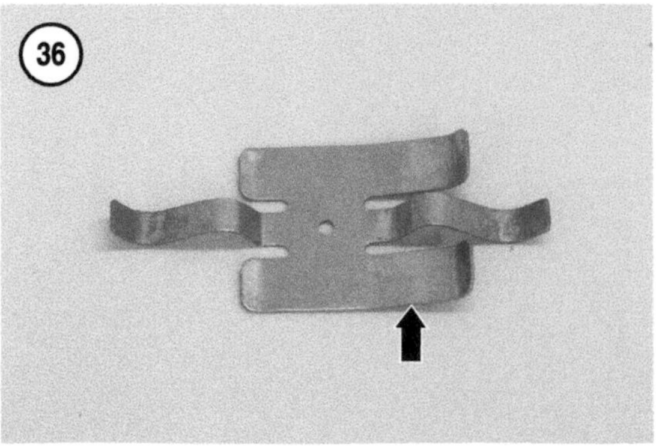

FRONT MASTER CYLINDER

Removal/Installation

CAUTION

Cover the fuel tank and front fender with a heavy cloth or plastic tarp to protect them from brake fluid spills.

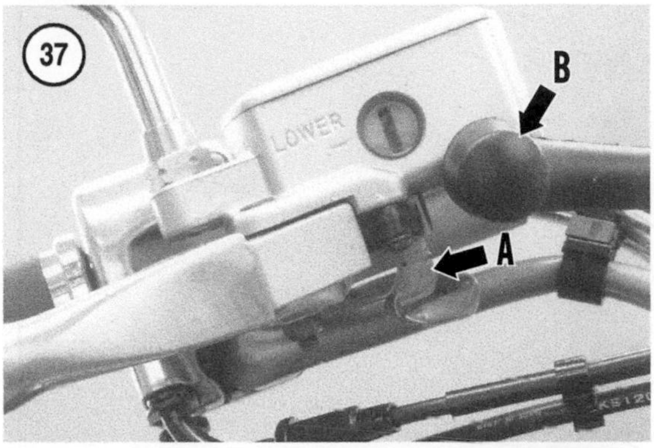

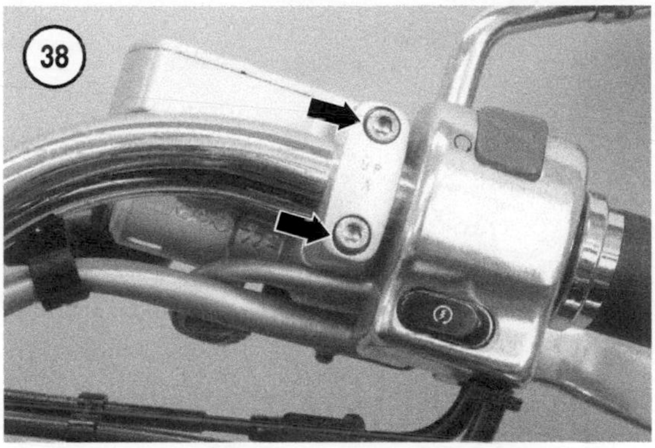

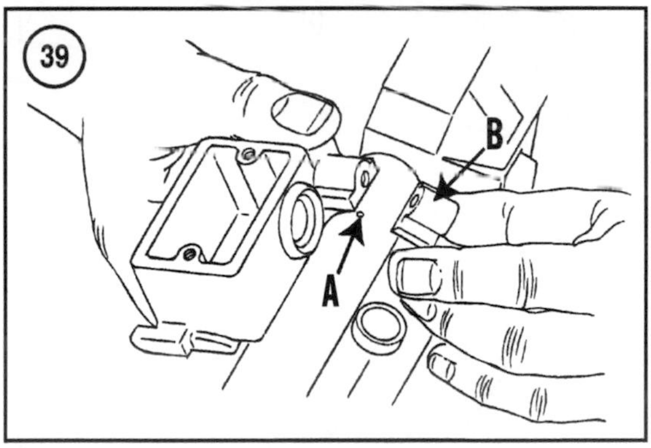

Wash spilled brake fluid off any painted or plated surfaces immediately. Brake fluid will damage the finish. Use soapy water and rinse the area completely.

1. Drain the master cylinder as described earlier in this chapter.
2. Remove the electrical connector from the brake switch (A, **Figure 37**) on the master cylinder.
3. Pull back the boot (B, **Figure 37**) and remove the banjo bolt securing the brake hose to the master cylinder. Remove the brake hose and both copper washers. Discard the washers. New ones must be installed during assembly.

CAUTION
Brake fluid damages paint and finish. Immediately wash any spilled brake fluid from the motorcycle. Use soapy water and rinse the area completely.

4. Place the loose end of the brake hose into a recloseable plastic bag to keep foreign matter from the system and to protect the motorcycle from leaking brake fluid.
5. Remove the two clamp bolts (**Figure 38**) and remove the master cylinder from the handlebar.
6. Remove the reservoir cap, diaphragm plate and diaphragm. Pour out and discard any remaining brake fluid. *Never* reuse brake fluid.
7. Installation is the reverse of removal. Note the following:
   a. Position the master cylinder so the face of the clamp mating surface aligns with the mark on the handlebar. See A, **Figure 39**.
   b. Install the master cylinder clamp so the UP stamp faces up.
   c. Torque the clamp bolts (**Figure 38**) to 10 N•m (89 in.-lb.). Tighten the upper clamp bolt first, then the lower bolt. There should be a gap at the lower part of the clamp after tightening.
   d. Install the brake hose on the master cylinder. Place a new copper washer on each side of the hose fitting and torque the banjo bolt to 30 N•m (22 ft.-lb.).
   e. Reconnect the brake switch electrical connector (A, **Figure 37**).
   f. Add fresh DOT 4 brake fluid to the master cylinder and bleed the brake system as described in this chapter.
   g. Install the diaphragm and diaphragm plate into the reservoir. Secure the reservoir cap in place with the two mounting screws.

WARNING
Do not ride the motorcycle until you are sure the brakes are operating correctly with full hydraulic advantage. If necessary, bleed the brakes as described in this chapter.

   h. Test ride the motorcycle slowly at first to make sure the brakes are operating properly.

13

40

MASTER CYLINDER

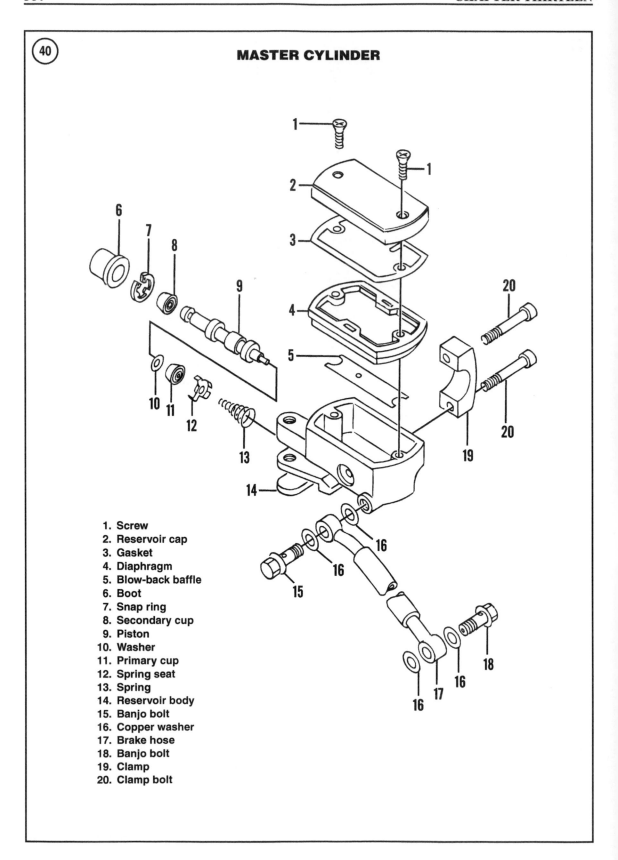

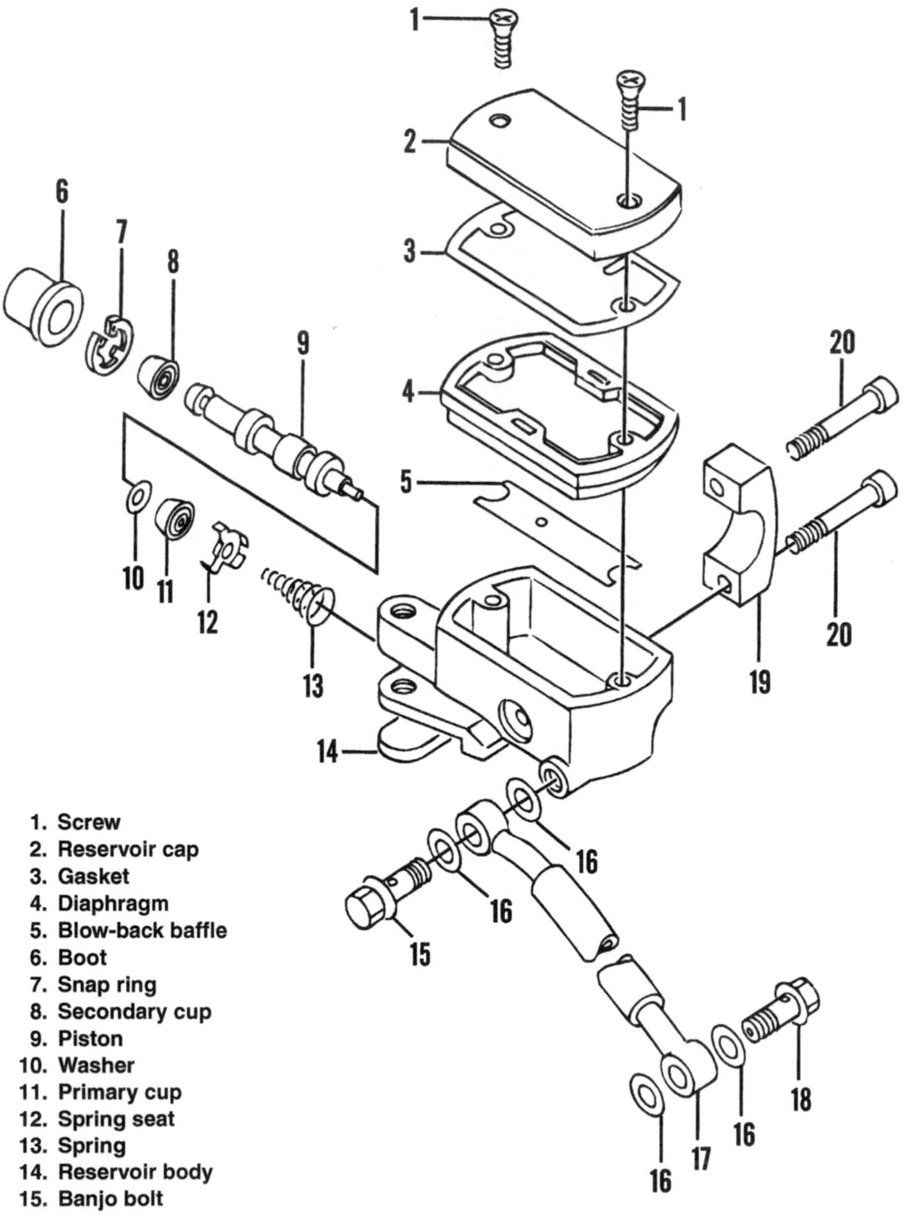

1. Screw
2. Reservoir cap
3. Gasket
4. Diaphragm
5. Blow-back baffle
6. Boot
7. Snap ring
8. Secondary cup
9. Piston
10. Washer
11. Primary cup
12. Spring seat
13. Spring
14. Reservoir body
15. Banjo bolt
16. Copper washer
17. Brake hose
18. Banjo bolt
19. Clamp
20. Clamp bolt

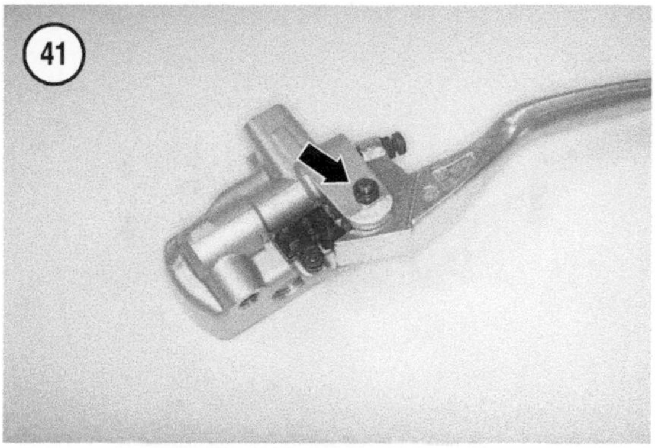

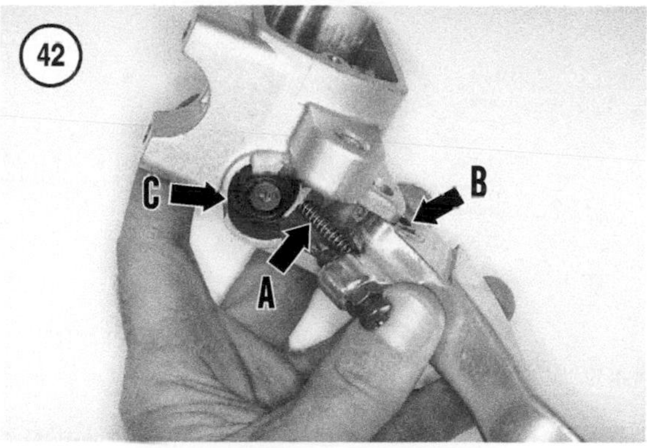

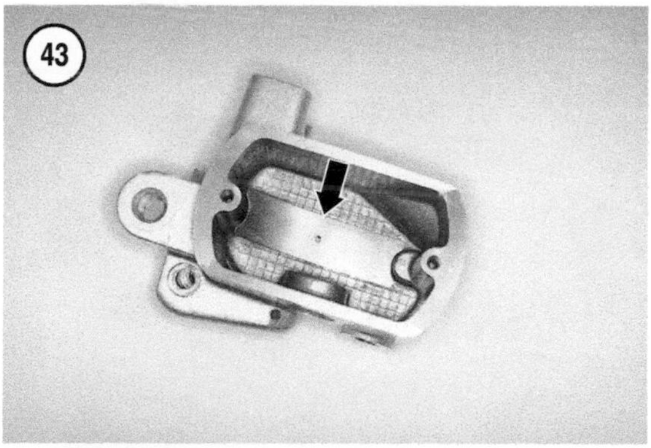

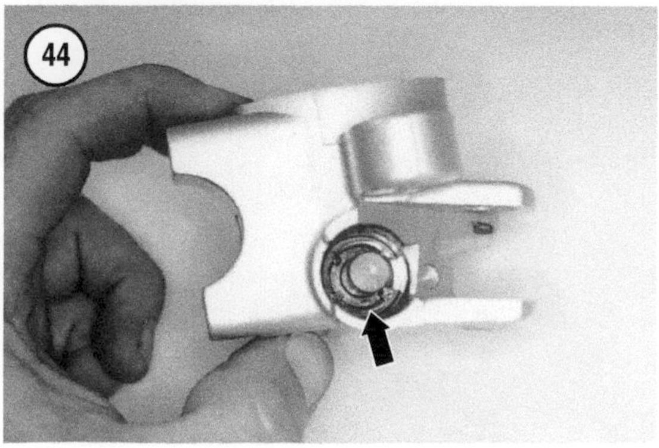

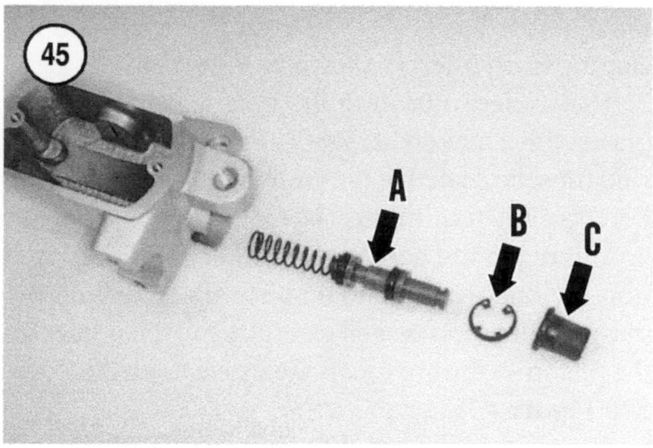

Disassembly

Refer to **Figure 40**.

1. Remove the master cylinder as described in this chapter.
2. Remove the reservoir cap, diaphragm plate and diaphragm. Pour out and discard any the remaining brake fluid. *Never* reuse brake fluid.
3. Remove the nut (**Figure 41**) from the lever bolt remove the bolt.
4. Carefully pull the brake lever from the master cylinder. Do not lose the compression spring (A, **Figure 42**) or the bushing (B) from the lever.
5. Remove the blow-back baffle (**Figure 43**) from the master cylinder.
6. Remove the boot (C, **Figure 42**) from the master cylinder bore.
7. Remove the snap ring (**Figure 44**) from its groove in the cylinder bore, then remove the piston assembly (A, **Figure 45**).

Assembly

Refer to **Figure 40**.

1. If removed, install the spring seat (A, **Figure 46**) onto the piston (B), and install the small end of the spring (C) onto the spring seat.

CAUTION
When installing the piston assembly, do not allow the cups to turn inside out. This will damage the cups and allow brake fluid to leak in the cylinder bore.

2. Lubricate the piston assembly (A, **Figure 45**) with fresh brake fluid, and install the assembly into

the cylinder bore. Make sure the master piston assembly is oriented as shown in **Figure 45**.

3. Press the piston into the bore and secure it in place with a new snap ring (B, **Figure 45**). The snap ring must be seated in the groove inside the cylinder bore as shown in **Figure 44**.

4. Lubricate the boot with fresh brake fluid. Position the boot so the end with the lip faces into the master cylinder bore and carefully roll the boot over the piston so the boot seals the master cylinder bore. See **Figure 47**.

5. If removed, install the compression spring (A, **Figure 42**) and the bushing (B) into place on the brake lever.

6. Slide the brake lever into place on the master cylinder body. Make sure the compression spring (A, **Figure 42**) engages the boss on the master cylinder body.

7. Lubricate the lever pivot bolt with lithium soap grease, and secure the brake lever to the body with the pivot bolt and nut (**Figure 41**).

8. Install the blow-back baffle (**Figure 43**) into the master cylinder reservoir.

9. Install the diaphragm, diaphragm plate and cover after the master cylinder has been installed onto the handlebar.

Inspection

1. Clean all parts in fresh DOT 4 brake fluid. Place the master cylinder components on a clean lint-free cloth when performing the following inspection procedures.

2. Inspect the cylinder bore (**Figure 48**) and piston contact surfaces for scratches, wear or other signs of damage. Replace the master cylinder body if necessary.

3. Inspect the inside of the reservoir for scratches, wear or other signs of damage. Replace the master cylinder body if necessary.

4. Make sure the passage in the bottom of the brake fluid reservoir (**Figure 49**) is clear.

NOTE

*The spring (C, **Figure 46**), spring seat (A), primary cup (D), secondary cup (E), piston (B), and snap ring (B, **Figure 45**) and boot (C) are sold as the master cylinder kit. If any part is worn or damaged, replace the master cylinder kit.*

46

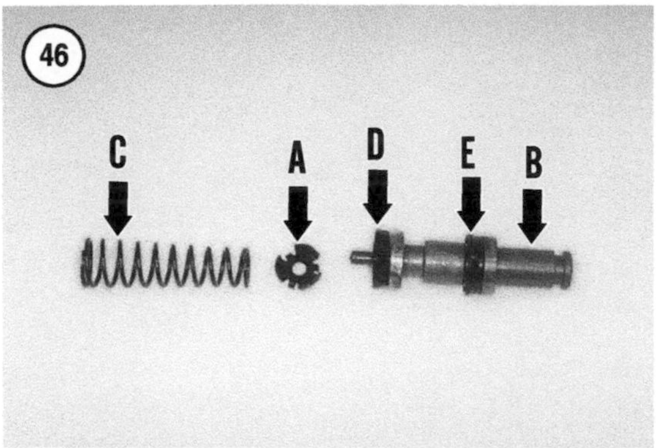

47

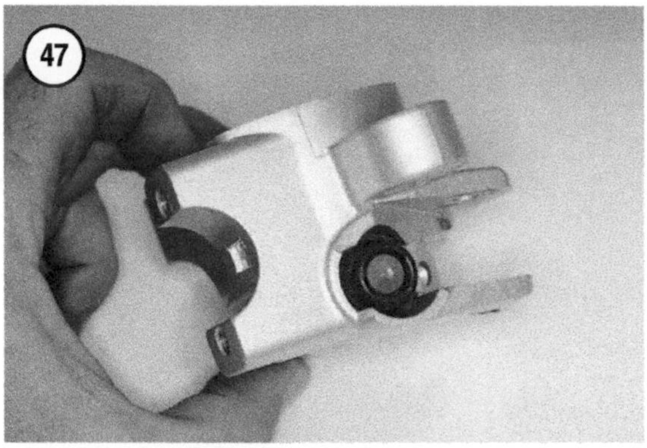

48

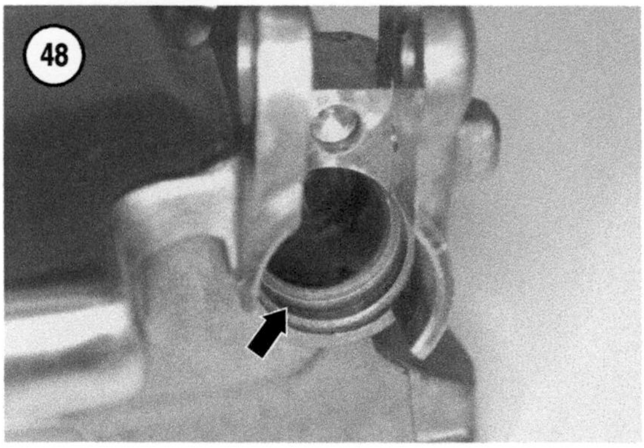

49

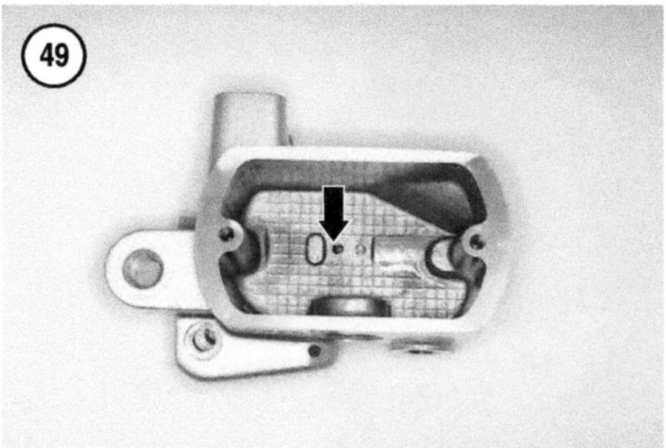

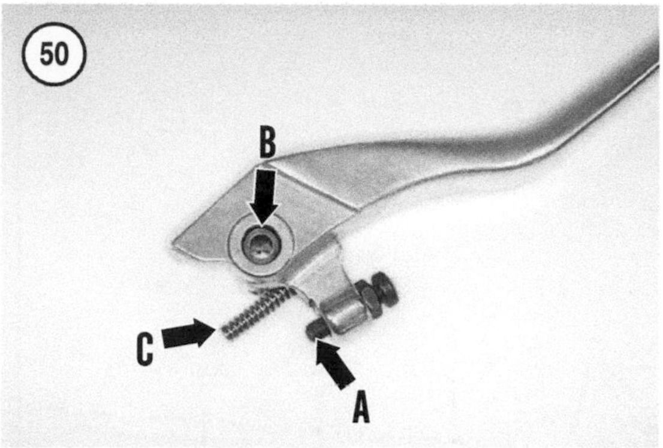

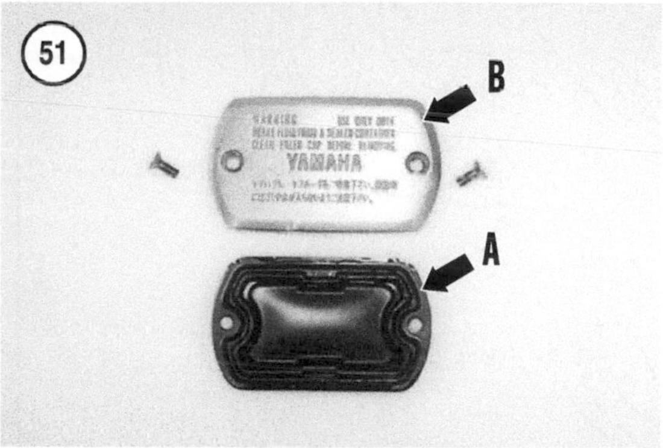

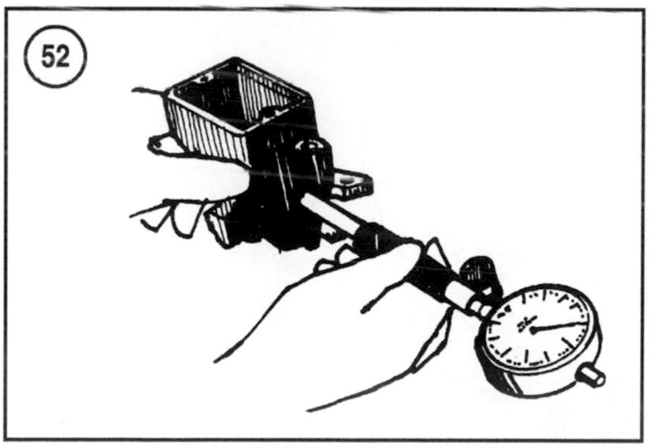

5. Check the end of the piston (B, **Figure 46**) for wear.

6. Check the primary cup (D, **Figure 46**) and secondary cup (E) on the master piston for damage, softness or swelling.

7. Check the end of the adjuster screw (A, **Figure 50**) for signs of wear. Replace the screw if necessary.

8. Remove and inspect the brake lever bushing (B, **Figure 50**). Replace the bushing if it is worn or elongated.

9. Remove the compression spring (C, **Figure 50**) from the hand lever. Replace the spring if it is worn or shows signs of fatigue.

10. Check the reservoir diaphragm (A, **Figure 51**) and cover (B) for damage and deterioration. Replace if necessary.

11. Inspect the threads in the master cylinder brake port. If the threads are damaged or partially stripped, replace the master cylinder body.

12. Measure the inside diameter of the master cylinder bore with a bore gauge (**Figure 52**). Replace the master cylinder if the inside diameter equals or exceeds the specification listed in **Table 1**.

REAR BRAKE CALIPER

Refer to **Figure 53**.

Removal

1. Securely support the motorcycle on level ground.

2. Drain the brake fluid from the rear brake as described in this chapter.

3. Remove the bolt and release the brake hose from the cable holder (**Figure 54**) on the swing arm.

NOTE

*Note the stop washer (A, **Figure 55**) that sits between the hose fitting and the caliper. This washer's upper stop sits against the outboard side of the caliper indexing post and the brake hose sits against the inboard side of the washer's lower stop. During assembly, the hose and stop washer must be installed as shown.*

4. Remove the banjo bolt (B, **Figure 55**) from the caliper and separate the brake hose from the caliper. Remove and discard the two washers. New washers must be used during installation.

5. Insert the end of the brake hose into a reclosable plastic bag so brake fluid will not leak onto the motorcycle.

6. Remove the brake pads as described in Steps 2-5 of *Rear Brake Pad Replacement* in this chapter.

7. Remove the caliper mounting bolts (C, **Figure 55**), and lift the caliper from the disc and caliper bracket.

(53)

REAR BRAKE CALIPER

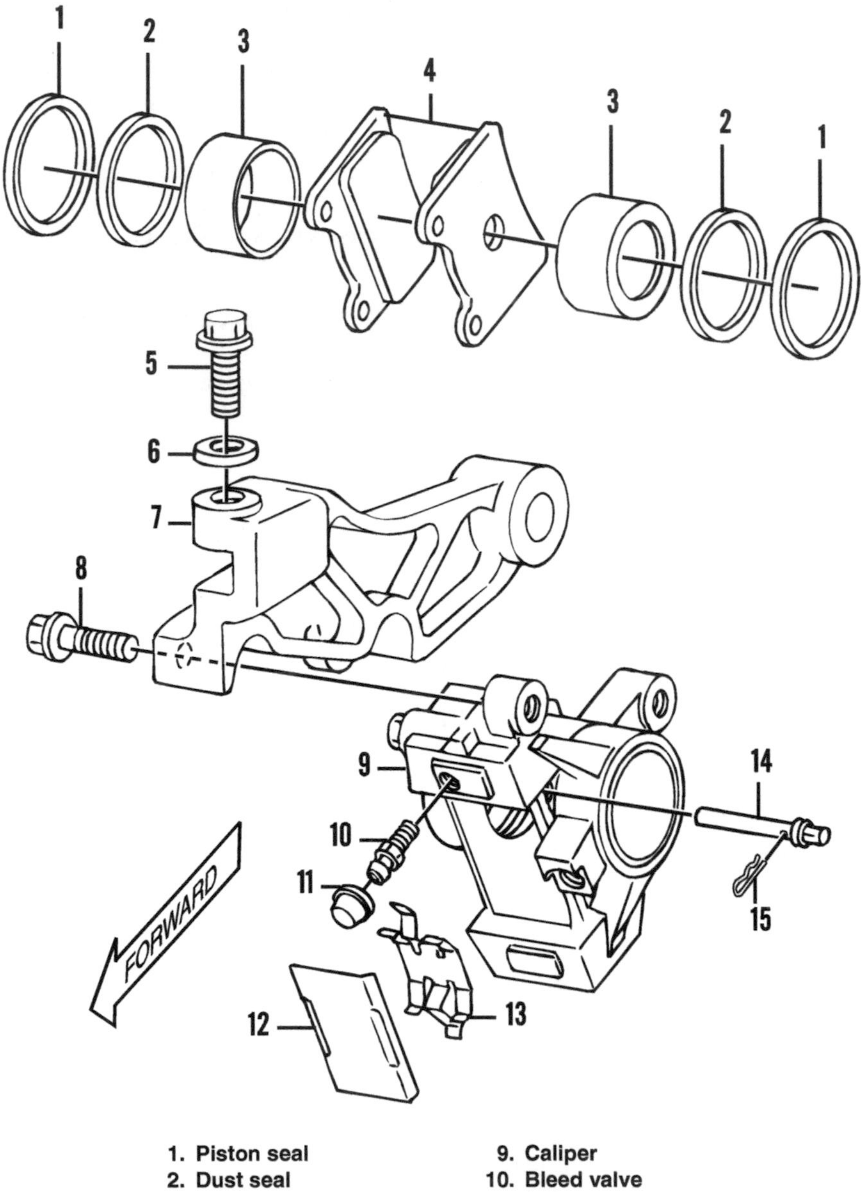

1. Piston seal
2. Dust seal
3. Piston
4. Brake pad
5. Bracket bolt
6. Washer
7. Caliper bracket
8. Caliper mounting bolt
9. Caliper
10. Bleed valve
11. Cap
12. Pad cover
13. Pad spring
14. Pad pin
15. Clip

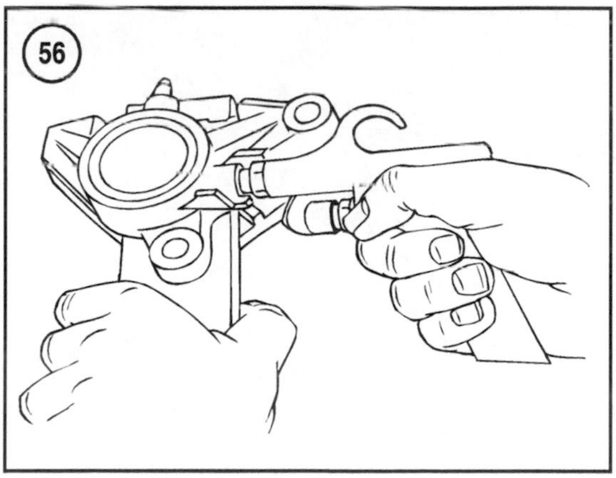

Installation

1. Lower the caliper over the brake disc and seat it on the caliper bracket.
2. Install the caliper mounting bolts (C, **Figure 55**). Apply a medium-strength threadlocking compound to the bolt threads and torque the caliper mounting bolts to 40 N•m (30 ft.-lb.).
3. Install the brake pads as described in Steps 7-14 of *Rear Brake Pad Replacement* in this chapter.
4. Secure the brake hose to the caliper with the banjo bolt (B, **Figure 55**). Place new washers on either side of the brake hose fitting, and torque the banjo bolt to 30 N•m (22 ft.-lb.). Note to the following:
   a. The stop washer (A, **Figure 55**) must sit between the caliper and the hose fitting.
   b. The upper stop on the washer must sit against the outboard side of the caliper indexing post.
   c. The brake hose neck must rest against the inboard side of the washer's lower stop (A, **Figure 55**).
5. Secure the brake hose to the swing arm with the cable holder (**Figure 54**).
6. Add brake fluid to the master cylinder and bleed the brakes as described earlier in this chapter.

Disassembly

Refer to **Figure 53**.

1. Remove the caliper and the brake pads as described in this chapter.
2. Set the caliper on a bench with the outboard side facing down.
3. Insert a piece of wood through the calipers and press the wood down against the outboard piston (**Figure 56**).

WARNING
In the next step, the piston may shoot out of the caliper body with great force. Keep your fingers out of the way. Wear shop gloves and apply air pressure gradually.

4. Apply compressed air through the brake hose fitting and blow the inboard piston out of its cylinder. Remove the piston from the caliper. Clearly label the piston so it can be reinstalled in its original cylinder in the caliper.
5. Turn the caliper so the outboard side faces up. Insert a rubber plug into the cylinder on the inboard side of the caliper. If a plug is not available, use a piece of rubber or wood that is large enough to cover the inboard cylinder.
6. Insert a piece of wood through the caliper and press the wood against the inboard cylinder. Apply compressed air to the brake hose fitting and blow the outboard piston (A, **Figure 57**) out of its cylinder.

13

7. Remove the outboard piston from the caliper. Label the piston so it can be reinstalled in its original location.

CAUTION
In the following step, do not use a sharp tool to remove the dust and piston seals from the caliper cylinder. Sharp tools could damage the cylinder surface. The caliper will have to be replaced if the cylinder surface is damaged.

NOTE
Yamaha recommends servicing the rear caliper without removing the caliper bolts (B, ***Figure 57****). The caliper halves have been separated for photographic clarity. This caliper can be serviced without separating the caliper halves.*

8. Use a piece of plastic or wood to carefully remove the dust seal (A, **Figure 58**) and the piston seal (B) from their grooves in each caliper cylinder. Discard all seals.
9. Clean all caliper parts and inspect them as described in this chapter.
10. Inspect the caliper as described in *Inspection* under *Front Caliper* in this chapter.

Assembly

NOTE
Never reuse the old dust seals or piston seals. Very minor damage or age deterioration can make the seals useless.

1. Coat the new dust seal and piston seal with fresh DOT 4 brake fluid.
2. Carefully install the new piston seal (B, **Figure 58**) and dust seal (A) into the grooves in each caliper cylinder. Make sure the seals are properly seated in their respective grooves.
3. Coat the pistons with DOT 4 brake fluid.

CAUTION
Install each piston into its original cylinder. The outboard piston must be installed in the outboard side of the caliper and vice versa.

4. Install the outboard piston as follows:

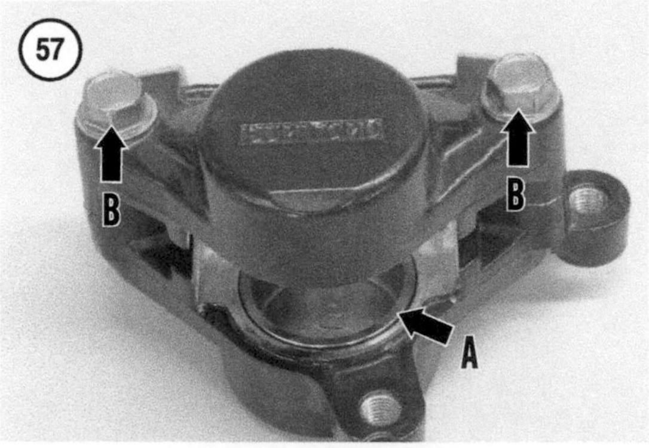

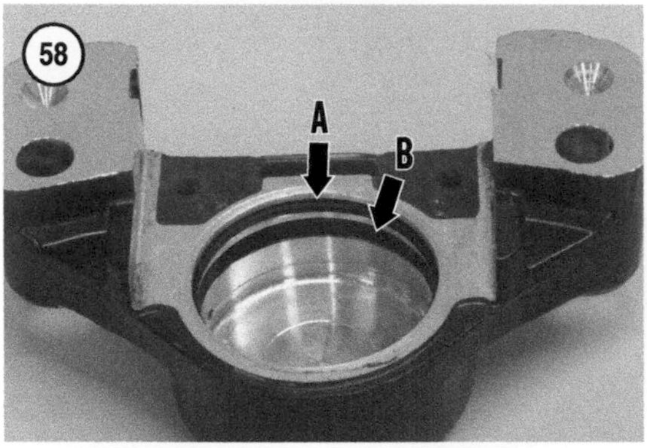

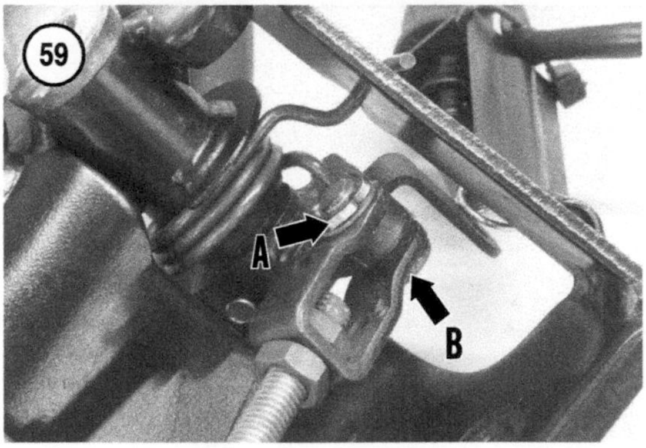

a. Lay the caliper on the bench with the outboard side facing down.
b. Position the piston with its open end up.
c. Insert a piston through the caliper body, set it in its cylinder and press the piston into the cylinder until it bottoms. See A, **Figure 57**.

5. Turn the caliper body over and install the inboard piston by performing the procedures in substeps b and c.
6. Install the bleed valve and torque it to 6 N•m (53 in.-lb.).

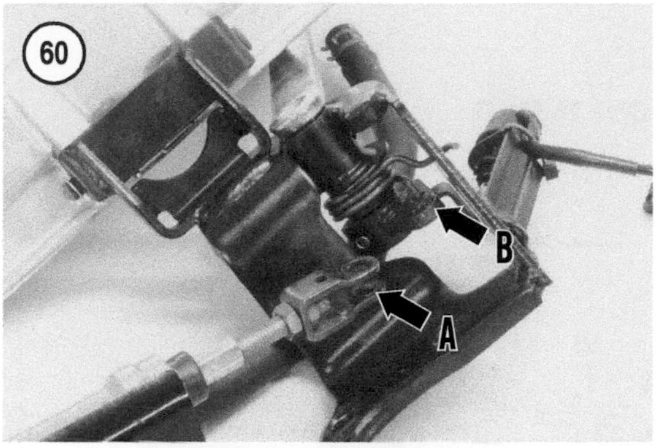

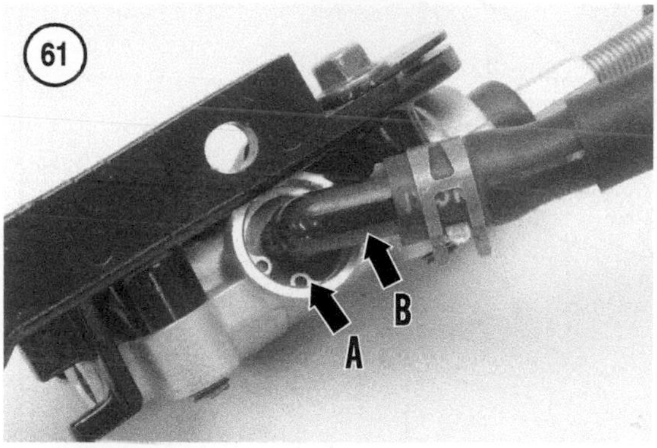

REAR BRAKE MASTER CYLINDER

Removal

1. Drain the brake fluid from the rear brakes as described in this chapter.

NOTE
The master cylinder can be removed while the brake pedal/footrest assembly is installed on the motorcycle. However, this is a very time consuming process. The master cylinder parts are more easily accessible if the brake pedal/footrest is removed first.

2. Remove the brake pedal/footrest assembly as described in this chapter.
3. Remove the cotter pin (A, **Figure 59**) and washer. Pull the clevis pin (B, **Figure 59**) from the master cylinder clevis.
4. Separate the master cylinder clevis (A, **Figure 60**) from the brake lever (B). Remove the master cylinder assembly and its bracket from the brake pedal/footrest assembly.
5. Remove the snap ring (A, **Figure 61**) and pull the reservoir hose fitting (B) from the port on the master cylinder. Discard the fitting O-ring. A new one must be installed during assembly.

Installation

1. Install a new O-ring onto the reservoir hose fitting and insert the fitting (B, **Figure 61**) into the master cylinder port.
2. Secure the fitting in place with the snap ring (A, **Figure 61**). The snap ring must be completely seated within the groove in the port.
3. Install the brake master cylinder clevis (A, **Figure 60**) onto the brake lever (B).
4. Secure it in place with the clevis pin (B, **Figure 59**).
5. Install the washer over the clevis pin and install a new cotter pin (A, **Figure 59**).
6. Install the brake pedal/footrest assembly as described in this chapter.
7. Add brake fluid and bleed the brakes as described in this chapter.

Disassembly

Refer to **Figure 62** when servicing the rear brake master cylinder.

1. Remove the rear brake master cylinder as described in this chapter.
2. Remove the bracket bolts (A, **Figure 63**). Lift the guard (B, **Figure 63**) and the master cylinder bracket (C) from the master cylinder body.
3. Roll back the dust boot (**Figure 64**) and remove the snap ring (**Figure 65**) from its groove in the cylinder bore.
4. Remove the pushrod assembly (A, **Figure 66**) and the piston assembly (B) from the cylinder bore. The spring should come out with the piston. If it does not, remove the spring from the cylinder bore.
5. Inspect the master cylinder and reservoir as described in this chapter.

Assembly

1A. When reinstalling the piston, soak the piston in fresh brake fluid for at least 15 minutes to make the primary (A, **Figure 67**) and secondary cups (B) pli-

13

(62)

REAR MASTER CYLINDER

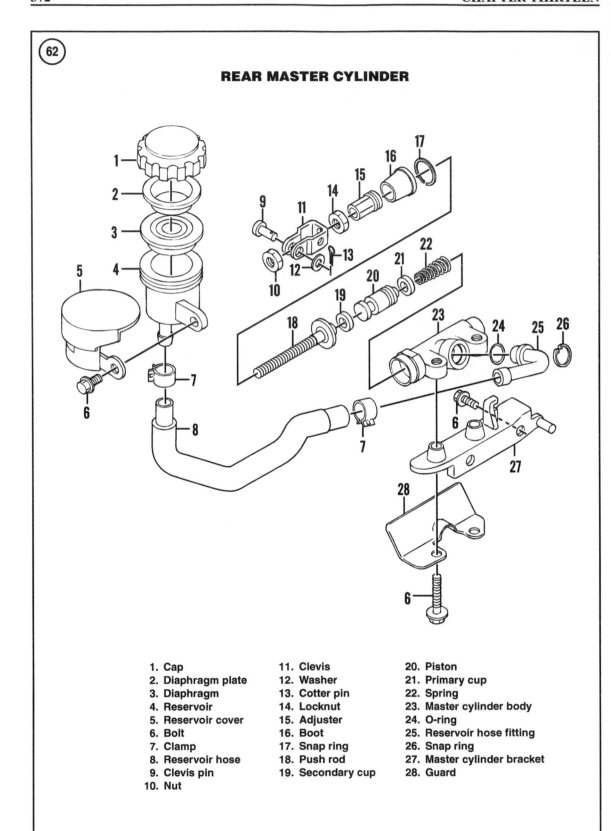

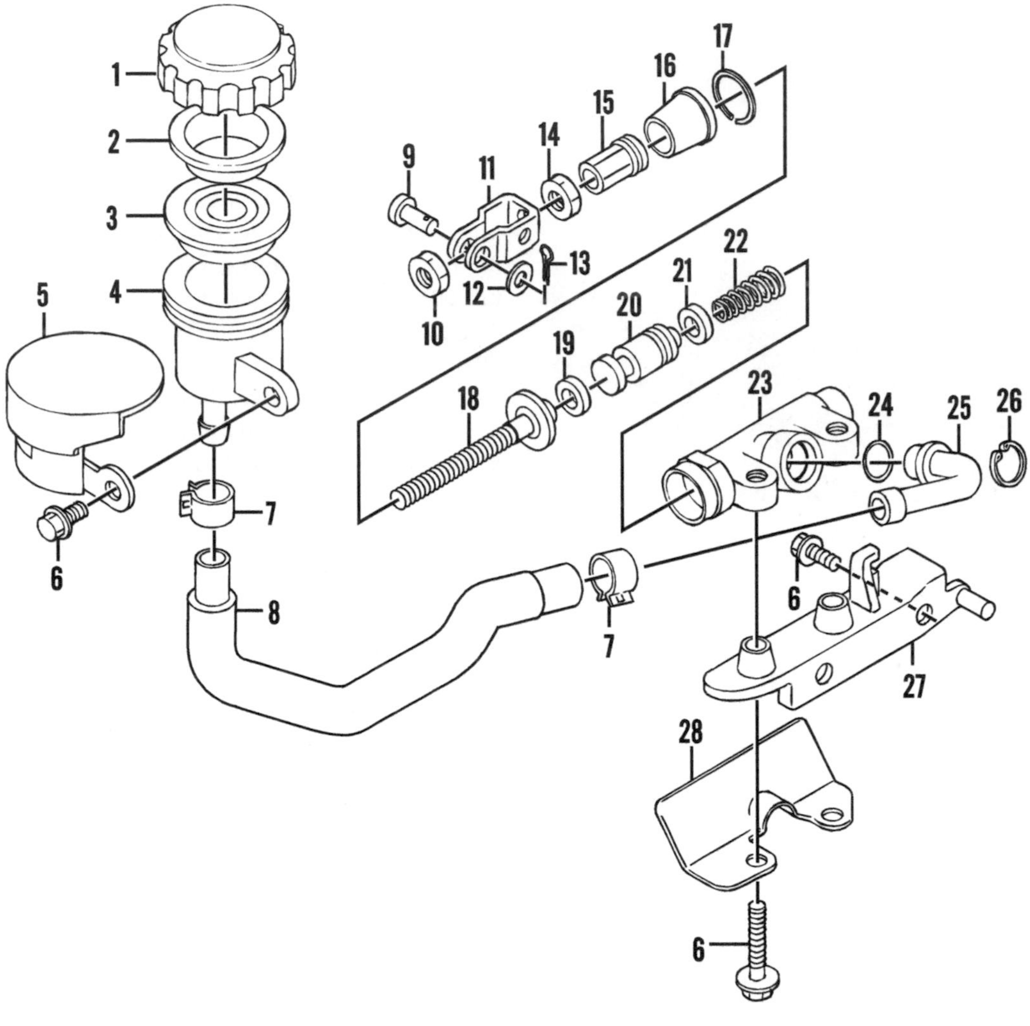

1. Cap
2. Diaphragm plate
3. Diaphragm
4. Reservoir
5. Reservoir cover
6. Bolt
7. Clamp
8. Reservoir hose
9. Clevis pin
10. Nut
11. Clevis
12. Washer
13. Cotter pin
14. Locknut
15. Adjuster
16. Boot
17. Snap ring
18. Push rod
19. Secondary cup
20. Piston
21. Primary cup
22. Spring
23. Master cylinder body
24. O-ring
25. Reservoir hose fitting
26. Snap ring
27. Master cylinder bracket
28. Guard

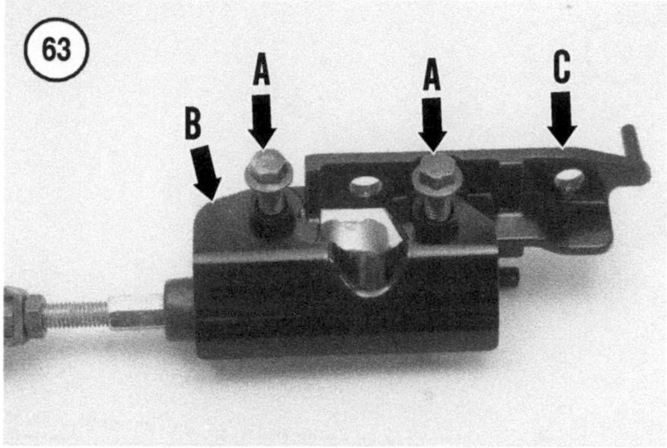

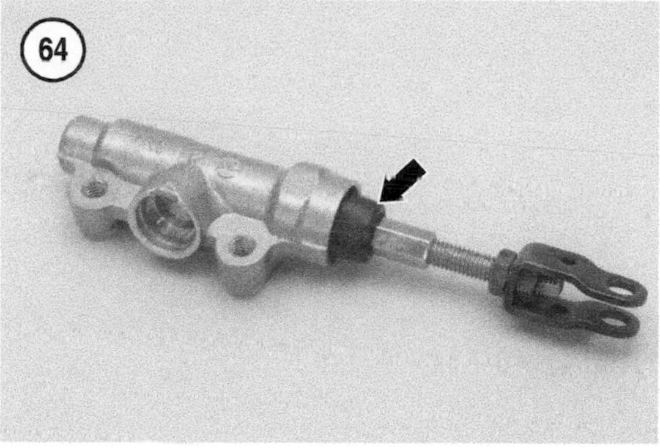

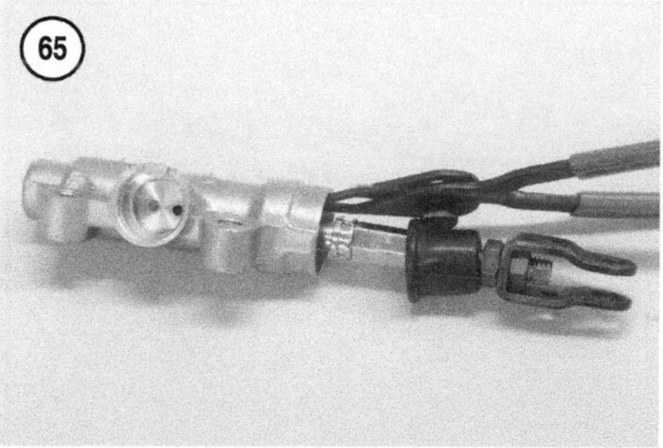

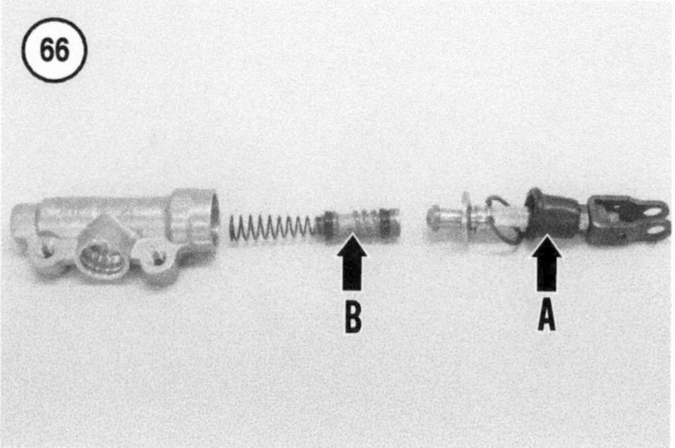

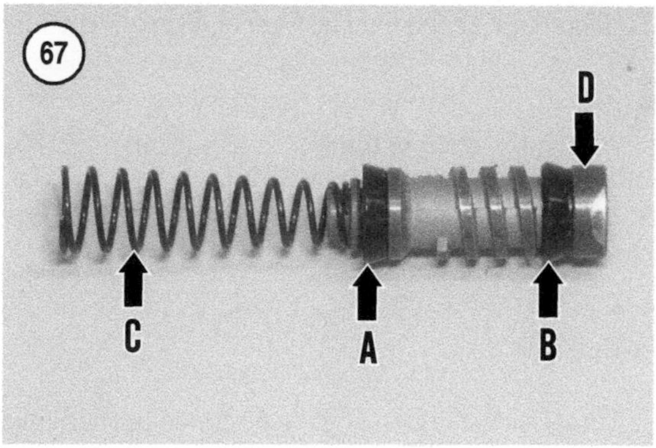

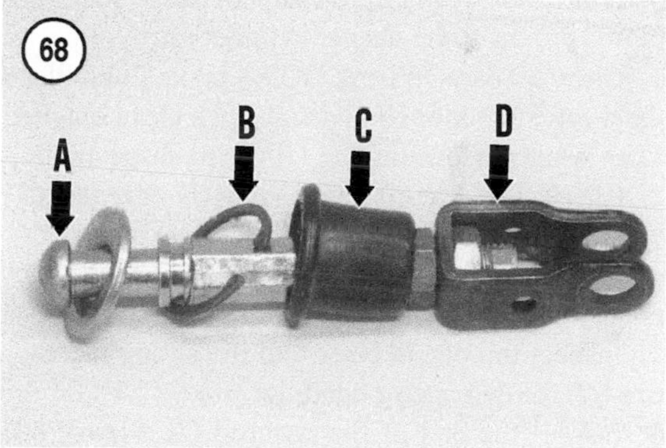

able. Coat the inside of the cylinder with fresh brake fluid prior to assembling the parts.

1B. When installing a new master cylinder kit, soak the new primary and secondary cups in brake fluid for at least 15 minutes. Roll the primary cup (A, **Figure 67**) onto the inboard end of the piston; roll the secondary cup (B) onto the outboard end.

2. If removed, install the narrow end of the spring (C, **Figure 67**) onto the master piston.

CAUTION

When installing the piston assembly, do not allow the cups to turn inside out. This will damage the cups and allow brake fluid to leak within the cylinder bore.

3. Install the piston/spring assembly (B, **Figure 66**) into the master cylinder.

4. Place a dab of grease on the end of the pushrod (A, **Figure 68**), and slowly push the spring and piston into the master cylinder with the pushrod assembly (A, **Figure 66**). Make sure the end of the push rod engages the seat in the master piston.

5. Install the snap ring (B, **Figure 68**) so it is completely seated in its groove in the master cylinder.
6. Check the operation of the master cylinder and slide the dust boot (**Figure 64**) into position. Make sure it is firmly seated against the master cylinder.
7. Install the master cylinder as described in this chapter.

Inspection

The piston, cups, spring, pushrod, boot, snap ring and adjuster bolt are all replaced as a kit. Individual parts are not available. If any of these parts are faulty, purchase the master cylinder kit.

1. Clean all parts in fresh DOT 4 brake fluid. Place the master cylinder components on a clean lint-free cloth when performing the following inspections.
2. Check the seat of the piston (D, **Figure 67**) where it contacts the pushrod for wear.
3. Check the primary cup (A, **Figure 67**) and secondary cup (B) for damage, softness or swelling.
4. Inspect the piston body and the spring (C, **Figure 67**) for damage or bending.
5. Inspect the end of the pushrod (A, **Figure 68**) where it contacts the piston for damage.
6. Inspect the pushrod dust boot (C, **Figure 68**) for tears or other signs of damage.
7. If any of the parts inspected in Steps 2-6 are worn, damaged or bent, replace all of them with the master cylinder kit.
8. Inspect the clevis (D, **Figure 68**) and brake lever (B, **Figure 60**) for cracks, bending or other signs of damage.
9. Inspect the cylinder bore (A, **Figure 69**) and piston contact surfaces for signs of wear or damage. If either part is less than perfect, replace the master cylinder.
10. Inspect the threads of the master cylinder mounting bosses (B, **Figure 69**) and the fluid outlet port (C). If the threads are damaged or partially stripped, replace the master cylinder assembly.
11. Make sure the passages in the inlet port (**Figure 70**) are clear.
12. Measure the master cylinder inside diameter with a bore gauge (**Figure 71**). If the inside diameter is out of specification (**Table 2**), replace the master cylinder body.
13. Inspect the reservoir diaphragm (A, **Figure 72**) and cap (B) for tears, cracks or other signs of damage. Replace as necessary.

69

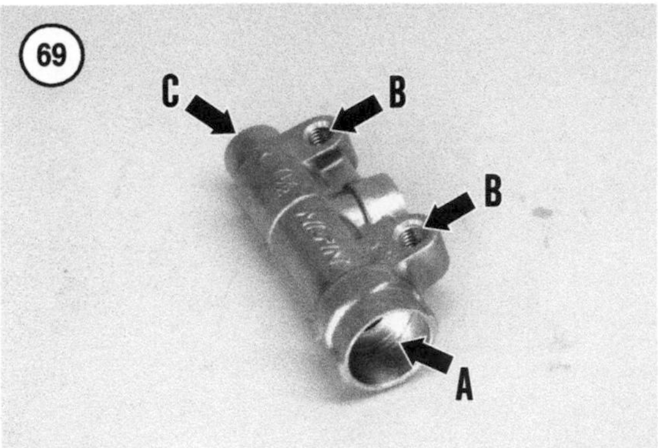

70

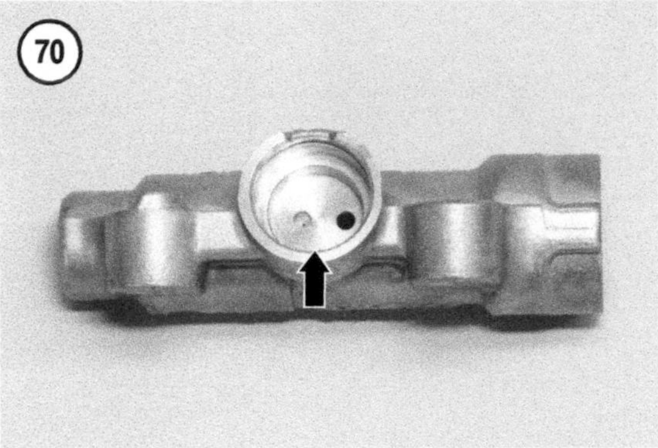

71

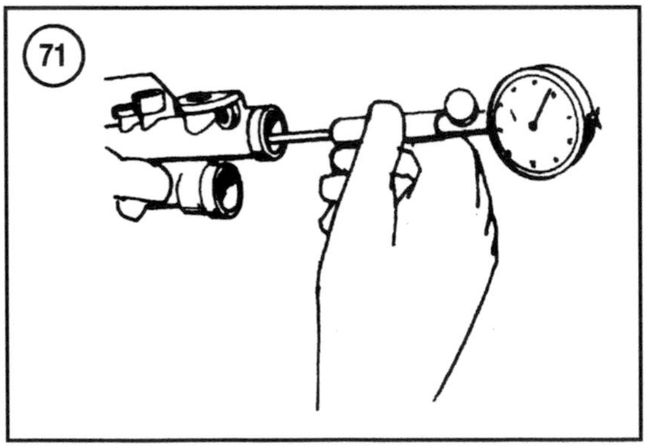

72

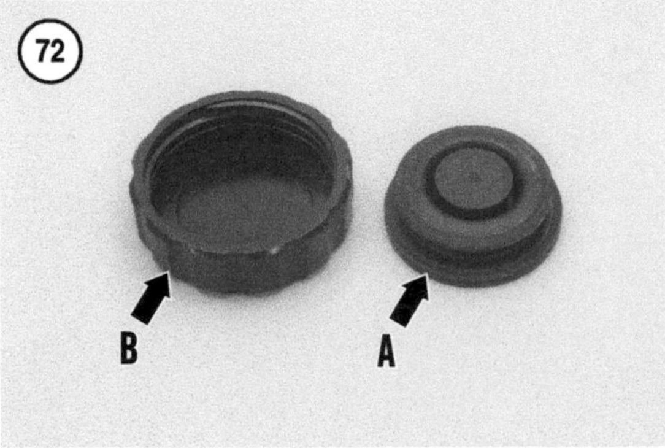

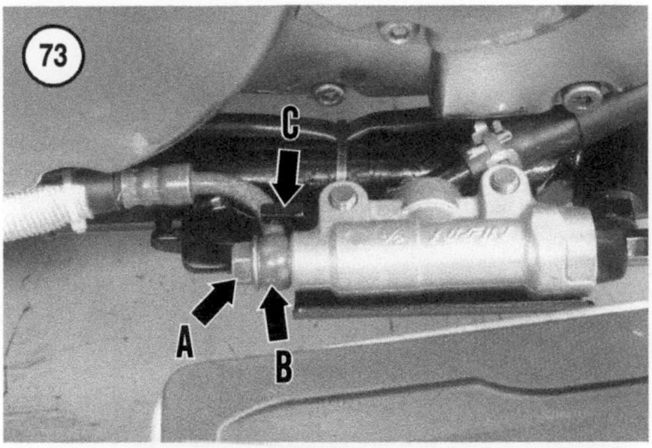

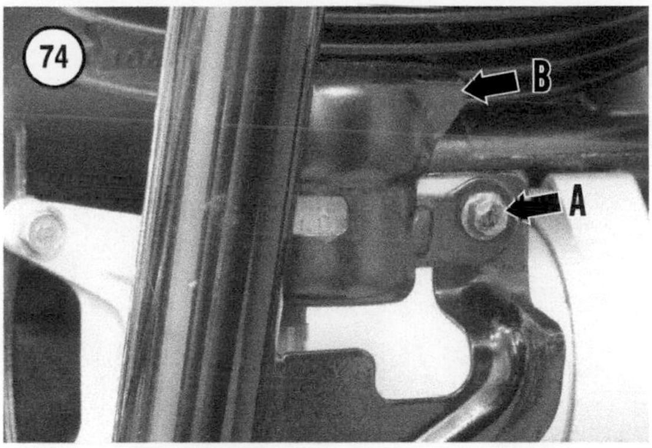

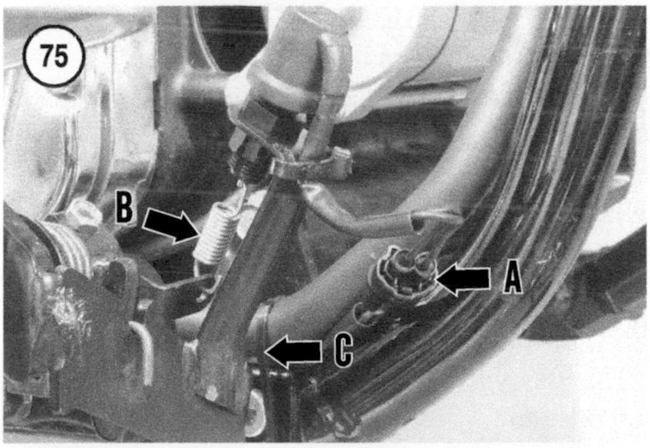

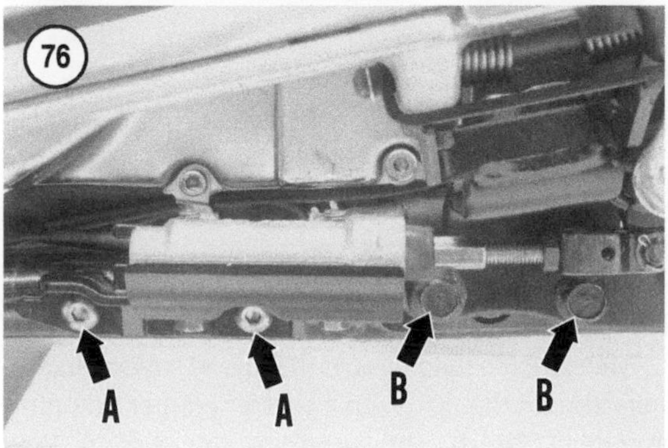

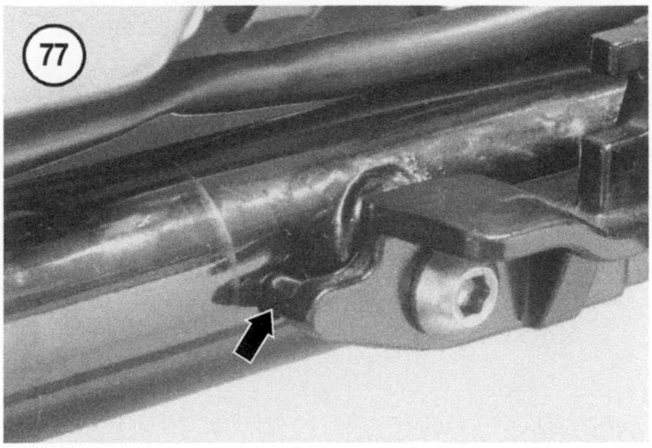

14. Inspect the reservoir and hose for cracks, wear or other signs of damage. Replace as necessary.

BRAKE PEDAL/FOOTREST ASSEMBLY

Removal

The rear brake master cylinder and reservoir come out with the brake pedal assembly.

1. Drain the brake fluid from the rear brakes as described in this chapter.
2. Remove the banjo bolt (A, **Figure 73**) and separate the brake hose (B) from the rear brake master cylinder. Insert the hose into a recloseable plastic bag so brake fluid will not leak onto the motorcycle.
3. Remove the mounting bolt (A, **Figure 74**) and lift the cover (B) from the rear brake reservoir.
4. Disconnect the rear brake switch connector (A, **Figure 75**) and unhook the switch spring (B) from the brake lever.
5. Remove the cable tie (C, **Figure 75**) that secures the brake reservoir hose to the frame downtube.
6. Loosen the master cylinder bracket bolts (A, **Figure 76**) and the footrest bracket bolts (B). Note that the indexing pin on the master cylinder bracket engages the hole in the frame (**Figure 77**).
7. Remove the bolts and lower the brake pedal/footrest assembly from the frame. Watch for the reservoir. It will come out with the assembly.
8. If necessary, remove the brake pedal as follows:
   a. Remove the cotter pin (A, **Figure 59**) and pull the clevis pin from the master cylinder clevis.
   b. Separate the master cylinder clevis (A, **Figure 60**) from the brake lever (B).
   c. Loosen the clamp bolt (A, **Figure 78**) on the brake lever. Note that the index mark on the

13

brake lever aligns with the mark on the pedal shaft (B, **Figure 78**). If these marks are not visible, make new ones.

d. Slide the brake lever from the brake pedal shaft.
e. Pull the return spring from the brake lever shaft.
f. Slide the brake pedal from the footrest bracket.

Installation

1. If the brake pedal was removed, perform the following:
   a. Lubricate the brake pedal shaft with lithium soap grease and slide the pedal onto the bracket.
   b. Install the return spring and the brake lever. Make sure the alignment dot on the brake lever aligns with the dot on the end of the brake pedal shaft (B, **Figure 78**).
   c. Tighten the lever clamp (A, **Figure 78**) bolt securely.
   d. Install the brake master cylinder clevis (A, **Figure 60**) on the brake lever (B).
   e. Secure it in place with the clevis pin (B, **Figure 59**).
   f. Install the washer over the clevis pin and install a new cotter pin (A, **Figure 59**).
2. Set the brake pedal/footrest assembly in place on the right side of the frame. Make sure the indexing pin on the master cylinder bracket engages the hole on the frame as shown in **Figure 77**.
3. Loosely install the master cylinder bracket bolts (A, **Figure 76**) and the footrest bracket bolts (B). Torque the footrest bracket bolts to 64 N•m (47 ft.-lb.) and torque the master cylinder bracket bolts to 23 N•m (17 ft.-lb.).
4. Hook the rear brake switch spring (B, **Figure 75**) to the brake lever and connect the switch connector (A) to its harness mate.
5. Place the cover over the master cylinder reservoir (B, **Figure 74**) and secure the reservoir in place with the mounting bolt (A).
6. Use a new cable tie (C, **Figure 75**) to secure the reservoir hose to the frame.
7. Install the brake hose (B, **Figure 73**) onto the port on the rear master cylinder. Install a new copper washer on each side of the hose fitting and torque the banjo bolt (A, **Figure 73**) to 30 N•m (22 ft.-lb.).

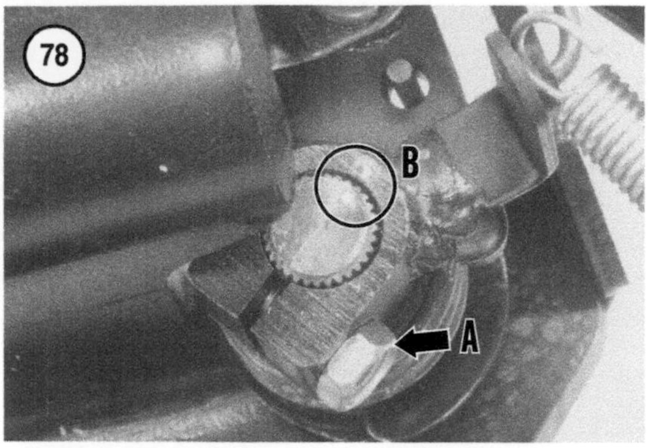

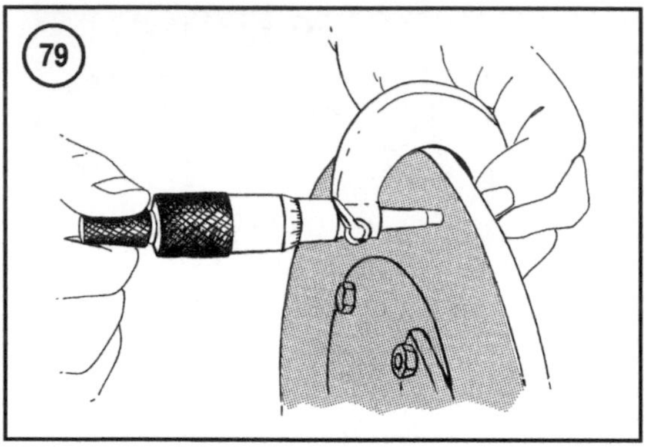

Make sure the brake hose neck sets below the index post (C, **Figure 73**) on the master cylinder.
8. Fill the master cylinder and bleed the system as described in this chapter.
9. Adjust the brake pedal height and rear brake switch as described in Chapter Three.

BRAKE DISC

The brake discs can be removed from the wheel hub once the wheel is removed from the motorcycle.

Inspection

A brake disc can be inspected while it is installed on the wheel.

1. Clean any rust or corrosion from the disc, and wipe it clean with brake parts cleaner. Never use oil-based solvents. They may leave an oil residue on the disc.
2. Measure the thickness of the disc at several locations around the disc with a vernier caliper or a micrometer (**Figure 79**). Replace the disc if the

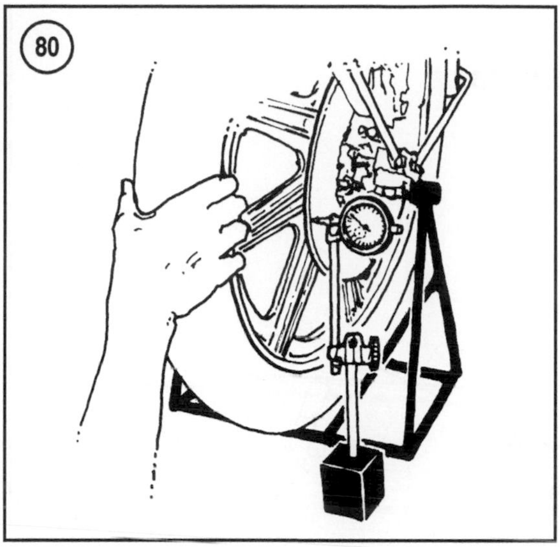

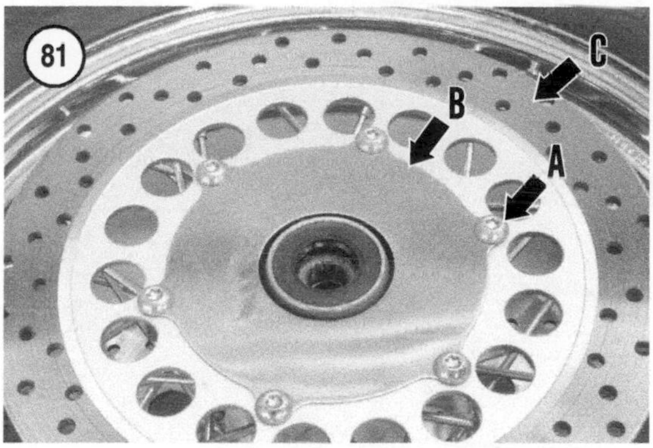

thickness in any area is equal to or less than the service limit specified in **Table 1** or **Table 2**.

3. Check the disc runout as follows:
   a. Make sure the disc mounting bolts are tight prior to running this check.
   b. Mount a dial indicator so its plunger sits 2-3 mm (0.09-0.12 in.) from the outside diameter of the disc (**Figure 80**).
   c. Slowly rotate the wheel and watch the dial indicator. Replace the disc if runout is out of specification.

4. A used disc will usually have some radial grooves. If these grooves are large enough to snag a fingernail, they will reduce brake effectiveness and increase brake pad wear. If large grooves are evident, consider replacing the disc. The discs cannot be machined, as the removal of material will reduce the disc thickness below the specification.

5. If there is evidence of disc overheating due to unequal pad pressure, inspect the following:
   a. The caliper may be binding on the caliper pin.
   b. The caliper piston is binding in the caliper.
   c. The master cylinder relief port is plugged.
   d. The master cylinder primary cup is worn or damaged.

Removal/Installation

1. Remove the front or rear wheel as described in Chapter Eleven.

CAUTION
Set the tire on two wooden blocks. Do not set the wheel down on the brake disc surface. It could be scratched or damaged.

NOTE
Insert a piece of wood or vinyl tube between the pads in the caliper(s). This way, if the brake lever or pedal is inadvertently applied, the pistons will not be forced out of the cylinders. If this does happen, the caliper(s) will have to be disassembled to reseat the pistons and the system will have to be bled. By using the wood or vinyl tube in place of the disc, the system will not have to be bled when installing the wheel.

2. Remove the brake disc bolts (A, **Figure 81**). On a front wheel, remove the hub cover (B).
3. Lift the brake disc (C, **Figure 81**) from the hub.
4. Clean the threaded holes in the hub.
5. Clean the brake disc mounting surface on the hub.
6. Installation is the reverse of removal. Note the following:
   a. The brake disc bolts are made from a harder material than similar bolts used on the motorcycle. When replacing the bolts, always use standard Yamaha brake disc bolts. Never compromise and use a generic replacement. They may not properly secure the disc to the hub.
   b. Install a disc so its arrow points in the direction of forward wheel rotation.
   c. Apply a small amount of a medium-strength threadlocking compound to the bolt threads.

(82)

FRONT BRAKE HOSE

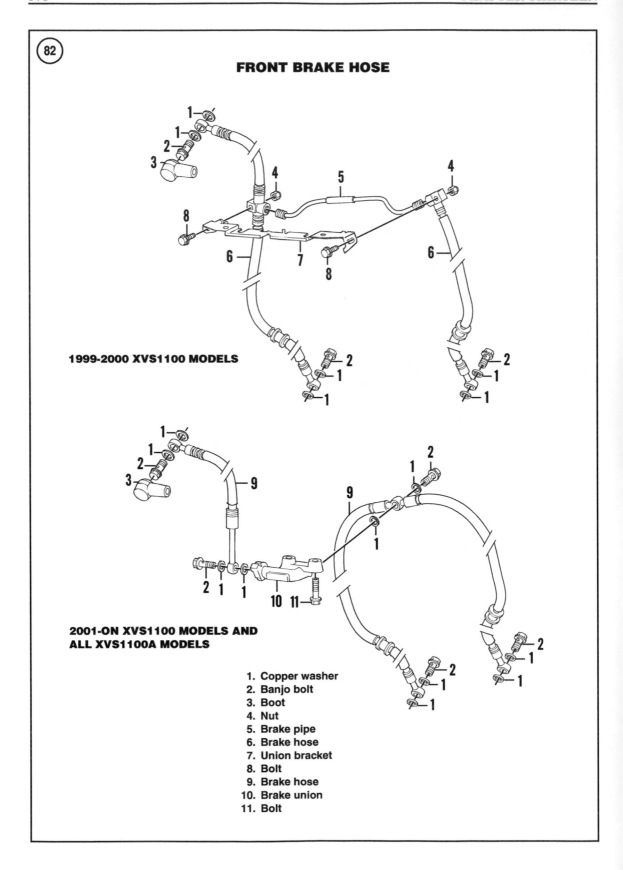

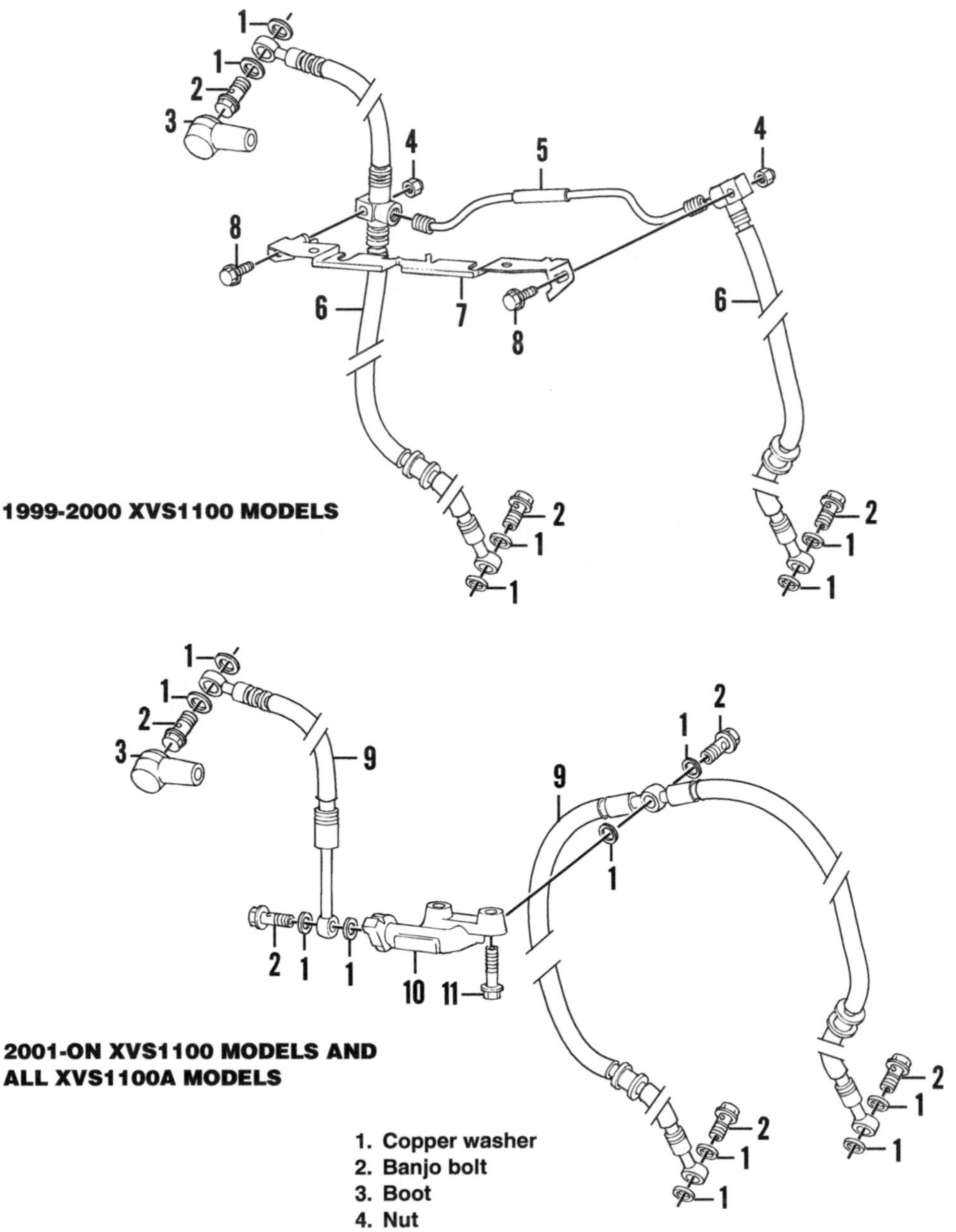

1. Copper washer
2. Banjo bolt
3. Boot
4. Nut
5. Brake pipe
6. Brake hose
7. Union bracket
8. Bolt
9. Brake hose
10. Brake union
11. Bolt

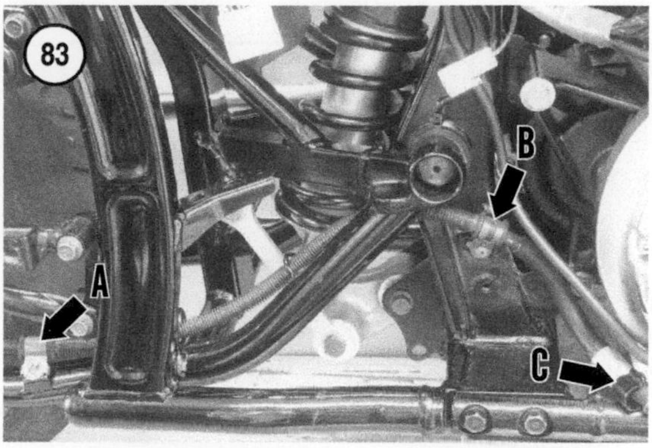

d. Evenly tighten the brake disc bolts in a criss-cross pattern. Torque the bolts to 23 N•m (17 ft.-lb.).

BRAKE HOSE REPLACEMENT

Check the brake hoses at the brake system inspection intervals listed in Chapter Three. Replace any brake hose that is cracked, bulging or shows signs of chafing, wear or other damage.

When replacing a brake hose, refer to **Figure 82** and perform the following:

1. Use a tarp or plastic drop cloth to cover areas of the motorcycle where brake fluid could spill.
2. Drain the brake fluid from the front or rear brake system as described in this chapter.
3. Note how the brake line is routed through the motorcycle. Make a drawing so the new line can be routed along the same path as the original hose or pipe.
4. Remove any holder or ties securing the line to the motorcycle. Also note the location of these clamps or ties. See A, B, and C **Figure 83** when replacing a rear brake hose.
5. Note how the hose fitting is installed on a master cylinder or caliper. Note whether the hose fitting sits on the inboard or outboard side of the indexing post. The new hose fitting must sit on the same side of the post.
6. Remove the banjo bolt from the hose fitting on each end of the hose. Discard the copper washer on each side of the hose fitting.
7. When replacing a rear brake reservoir hose, slide the hose clamps along the hose, and pull the hose from the fitting on the brake reservoir and on the master cylinder.
8. Installation is the reverse of removal. Note the following:
   a. Compare the new and old hoses. Make sure they are the same.
   b. Clean the banjo bolts and hose ends to remove any contamination.
   c. Refer to the notes made during removal and route the new hose along the same path as the original hose. Secure the hose to the motorcycle at the same locations noted during removal.
   d. Replace any banjo bolt with a damaged head or threads.
   e. Install a new copper washer on each side of a brake hose fitting.
   f. Torque the banjo bolt to 30 N•m (22 ft.-lb.).
   g. After replacing a front brake line, turn handlebars from side to side to make sure the hose does not rub against any part or pull away from its brake component.
   h. Refill the master cylinder and bleed the brakes as described in this chapter.

WARNING
Before riding the motorcycle, confirm that the brake lights work and that the front and rear brakes operate properly with full hydraulic advantage.

   i. Slowly test ride the motorcycle to confirm that the brakes are operating properly.

Tables 1-3 are on the following page.

13

Table 1 FRONT BRAKE SPECIFICATIONS

Item	Specification mm (in.)	Service limit mm (in.)
Brake pad thickness	6.2 (0.24)	0.8 (0.03)
Brake disc thickness	5.0 (0.2)	4.5 (0.177)
Brake disc outside diameter	298 (11.73)	–
Brake disc runout	–	0.15 (0.0059)
Caliper cylinder bore inside diameters		
	25.4 (1.0)	–
	30.1 (1.19)	–
Master cylinder bore inside diameter	14.0 (0.55)	–
Brake lever free play (at lever end)	5-8 (0.20-0.31)	–

Table 2 REAR BRAKE SPECIFICATIONS

Item	Specification mm (in.)	Service limit mm (in.)
Brake pad thickness	5.55 (0.219)	0.5 (0.020)
Brake disc thickness	6.0 (0.24)	5.5 (0.22)
Brake disc outside diameter	282 (11.10)	–
Brake disc runout	–	0.15 (0.0059)
Caliper cylinder bore inside diameter	42.9 (1.689)	–
Master cylinder bore inside diameter	12.7 (0.50)	–
Brake pedal height		
XVS1100 models (above footpeg)	81.8 (3.22)	–
XVS1100A models (above floorboard)	98.5 (3.88)	–
Brake pedal free play (at pedal end)	0	–

Table 3 BRAKE SYSTEM TORQUE SPECIFICATIONS

Item	N•m	in.-lb.	ft.-lb.
Brake hose banjo bolt	30	–	22
Brake hose holder bolt	10	89	–
Brake caliper bleed valve	6	53	–
Brake disc mounting bolts*	23	–	17
Brake pedal/footrest assembly bolts	64	–	47
Footrest bracket bolts	64	–	47
Front caliper retaining bolt			
1999-2000 XVS1100 models	23	–	17
2001-on XVS1100 models and all XVS1100A models	27	–	20
Front master cylinder clamp bolts	10	89	–
Rear caliper mounting bolts*	40	–	30
Rear caliper bracket bolt	40	–	30
Rear master cylinder bracket bolts	23	–	17

*Apply a medium-strength threadlocking compound.

CHAPTER FOURTEEN

FRAME

SEATS

Removal/Installation

Refer to **Figure 1**.

1. Remove the passenger seat nut (A, **Figure 2**) from the fender stud.
2. Pull the passenger seat rearward (B, **Figure 2**) until the seat tang disengages from the seat bracket (A, **Figure 3**), and remove the passenger seat.
3. On XVS1100 models, perform the following:
   a. Remove the seat bracket bolts and remove the seat bracket (A, **Figure 3**).
   b. Pull the rider's seat rearward until the seat tangs disengage from the fuel tank bracket (**Figure 4**), and remove the seat.
4. On XVS1100A models, perform the following:
   a. Remove the seat bolt (B, **Figure 3**) from the rider's seat.
   b. Pull the riders seat rearward until the seat tangs disengage from the fuel tank bracket (**Figure 4**), and remove the seat.
   c. If necessary, remove the seat bracket bolts and remove the seat bracket (A, **Figure 3**).
5. Installation is the reverse of removal.
   a. Slide the rider's seat forward until its tangs (**Figure 5**) engage the tank bracket.
   b. Torque the seat bracket bolts to 7 N•m (62 in.-lb.).
   c. Torque the rider's seat bolt to 7 N•m (62 in.-lb.) on models so equipped.
   d. Slide the passenger seat forward until its tang (**Figure 6**) engages the seat bracket (B, **Figure 3**).
   e. Torque the passenger seat nut to 7 N•m (62 in.-lb.).

IGNITOR PANEL

Removal/Installation

1. Remove the rider and passenger seats as described in this chapter.
2. Remove the fuel tank as described in Chapter Eight.
3. Release each push pin (A, **Figure 7**) by pressing the center pin.
4. Remove the push pins and lift the ignitor panel (B, **Figure 7**) from the frame.
5. Set the ignitor plate onto the frame.
6. Install each push pin as follows:
   a. From the locking side of the fastener, push the center pin until it extends out from the fastener head.
   b. Insert the push pin through the panel and frame mounts.

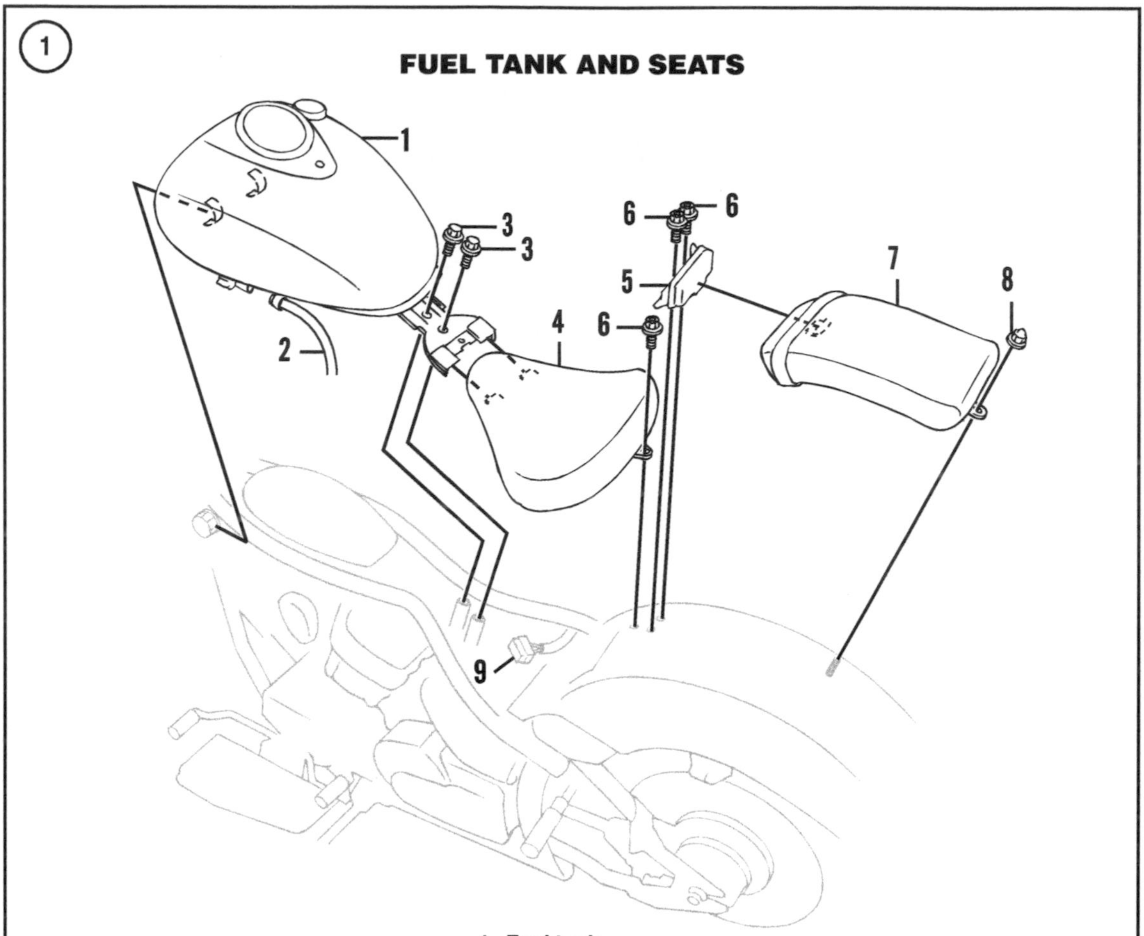

1. Fuel tank
2. Fuel hose
3. Fuel tank bolt
4. Rider's seat
5. Seat bracket
6. Seat bracket bolt
7. Passenger seat
8. Nut
9. Meter connector

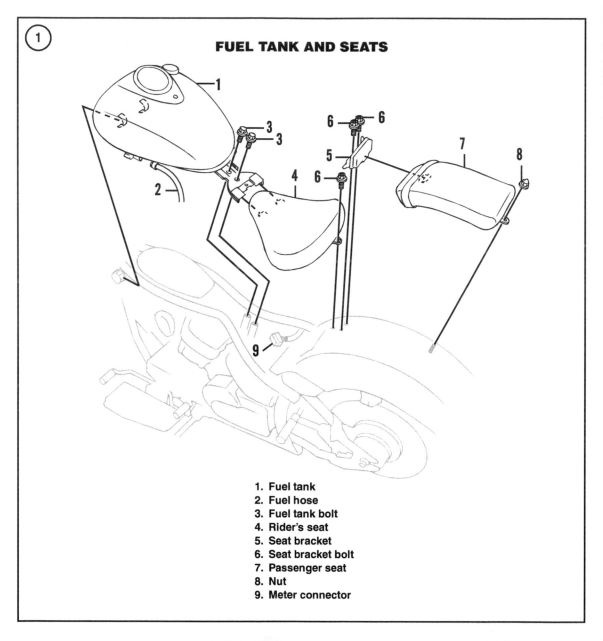

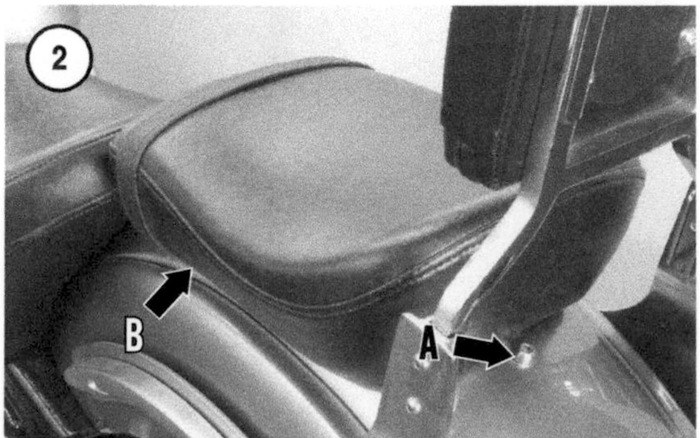

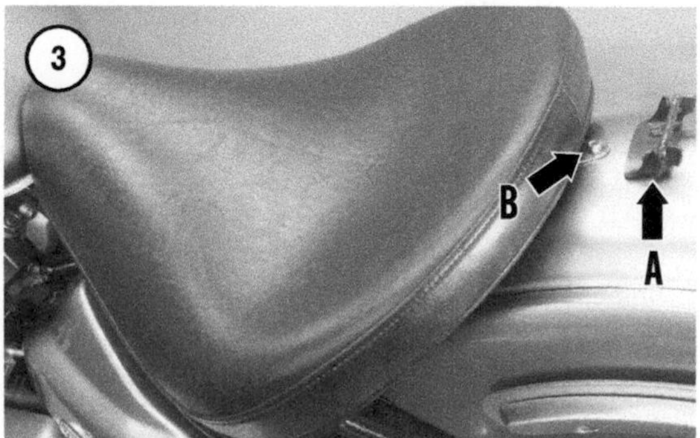

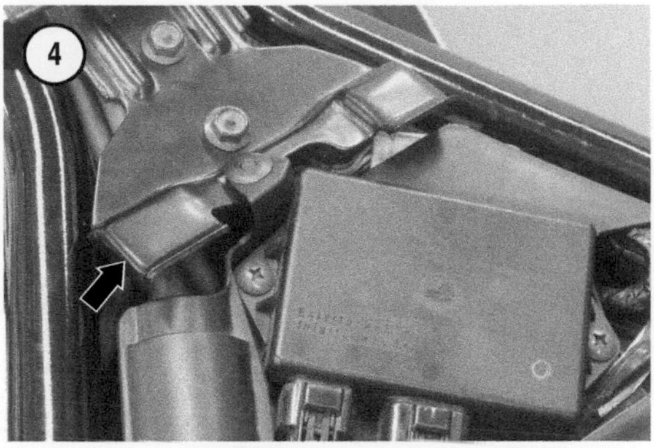

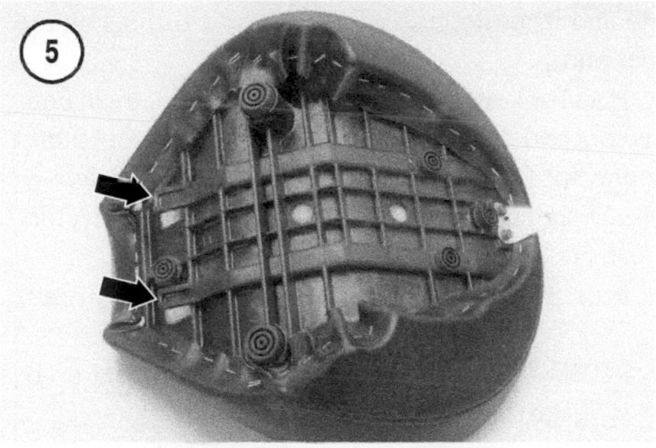

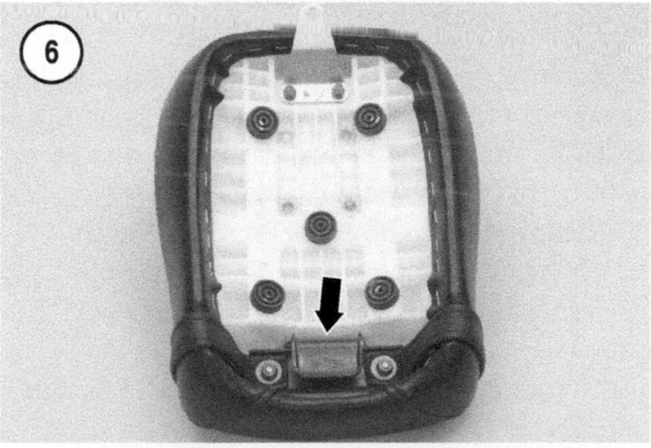

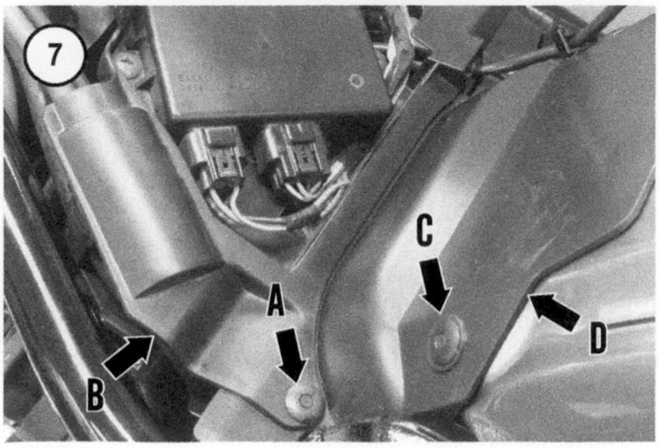

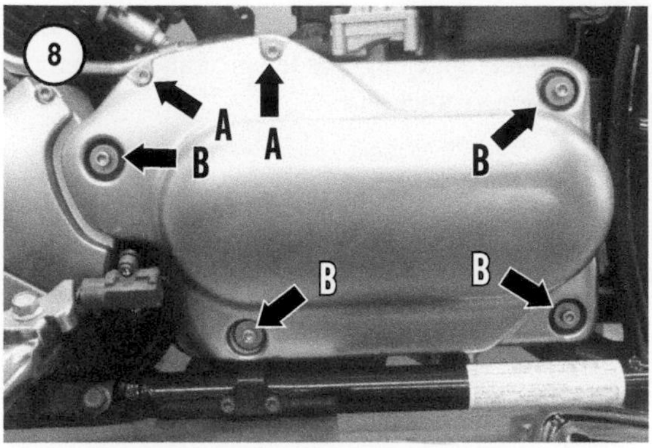

c. Push the center pin into the head. The fastener is locked when the pin is flush with the head. See A, **Figure 7**.

LEFT SIDE COVER

Removal/Installation

The left side cover is secured in place by six Allen bolts: two 6 × 54 mm bolts (A, **Figure 8**) and four 6 × 25 mm bolts (B). A grommet, collar and washer is used with four 6 × 25 mm bolts. Make sure these are in place at their respective locations during assembly.

1. Remove the shift pedal/footrest assembly from the motorcycle as described in Chapter Seven.
2. Remove each of the side cover bolts (A and B, **Figure 8**) along with their washers.
3. Lift the side cover from the motorcycle.
4. Installation is the reverse of removal. Note the following:
   a. Make sure a grommet and collar are in place on the four lower mounts (B, **Figure 8**).
   b. Install the long bolt (6 × 54 mm) at the upper forward mounts on the cover (A, **Figure 8**). Install a short bolt (6 × 25 mm) and washer at the four lower mounts (B, **Figure 8**).
   c. Install the shift pedal/footrest assembly as described in Chapter Seven.

TOOLBOX COVER

Removal/Installation

1. Insert the key into the lock (**Figure 9**) on the toolbox cover and turn the key clockwise.
2. Pull the cover out until the slot on the top left edge of the cover releases from the grommet. Remove the cover.

3. During installation, press the cover until the slot on the top left edge engages the grommet, then turn the key counterclockwise.

TOOLBOX PANEL

Removal/Installation

Refer to **Figure 10**.

CAUTION
Several hoses are disconnected from various fittings during this procedure. Label each hose and its fitting so they can be easily identified during assembly. Every hose must be reconnected to the correct fitting during assembly. The same applies to electrical connectors. Label each connector and its harness mate as they are disconnected. Also note how the hoses and wires are routed along the toolbox panel. They must be rerouted along the same path.

1. Remove the toolbox cover and the left side cover as described in this chapter.
2. Remove the battery as described in Chapter Three.
3. Remove the fuse holder as follows:
   a. Disconnect the electrical connectors (**Figure 11**) from the fuse holder.
   b. Pull the fuse holder damper from the tang on the toolbox panel.
4. On models with an AIS system, remove the AIS assembly as described in Chapter Eight.
5. Remove the tool kit from the toolbox.
6. Remove the mounting screws (A, **Figure 12**) and lift the hose bracket (B) from the toolbox panel. Note how the hoses behind this bracket are routed along the toolbox panel (**Figure 13**). They must be rerouted along the same path during assembly.
7. On California models, perform the following:
   a. Disconnect the purge hose (A, **Figure 13**) from the lower hose fitting on the EVAP solenoid valve.
   b. Disconnect the two-pin electrical connector from the EVAP solenoid valve (B, **Figure 13**).
   c. Disconnect the fuel tank hose (C, **Figure 13**) from the EVAP rollover valve. This valve can be found just forward of the EVAP solenoid valve, behind the fuel pump hoses. See D, **Figure 13**.

9

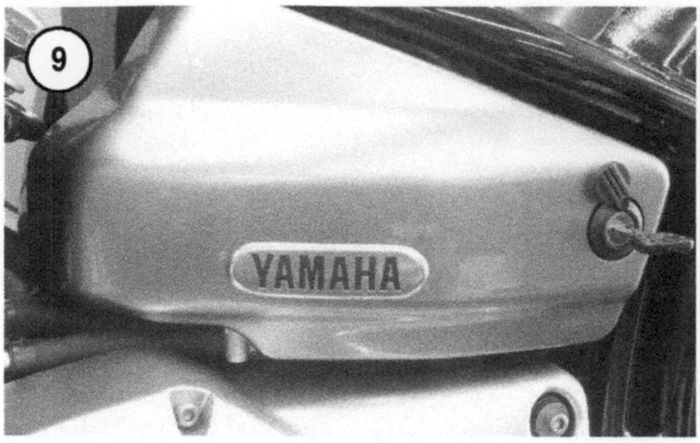

8. Disconnect the fuel pump inlet hose (A, **Figure 14**) and its outlet hose (B) from their fittings on the fuel pump.
9. Roll back the boot that protects the electrical connectors and disconnect the black, two-pin fuel pump connector (C, **Figure 14**) from its harness mate.
10. Disconnect the remaining electrical connectors from their harness mates:
    a. Three-pin speed sensor connector (A, **Figure 15**).
    b. Blue, two-pin sidestand switch connector (B, **Figure 15**).
    c. Three-pin stator connector (C, **Figure 15**).
    d. Two-pin pickup coil connector (D, **Figure 15**).
    e. Neutral switch bullet connector (E, **Figure 15**).
11. Remove the bolt (**Figure 16**) that secures the battery box to the toolbox panel.
12. Make sure all hoses and electrical connectors are released and moved safely out of the way.
13. Remove the toolbox panel screws (**Figure 17**) and lift the toolbox panel from the motorcycle.
14. Installation is the reverse of removal. Note the following:
    a. Make sure each electrical connector is connected to its original harness mate.
    b. Connect each hose to its original fitting.
    c. Route the hose and electrical cables along the original path noted during removal.

BATTERY COVER

Removal/Installation

1. Remove the battery cover bolt (**Figure 18**).
2. Grasp the cover and pull it out from the frame until the cover's slots in the rear side disengage from their grommets.
3. Remove the cover.

(10)

TOOLBOX PANEL

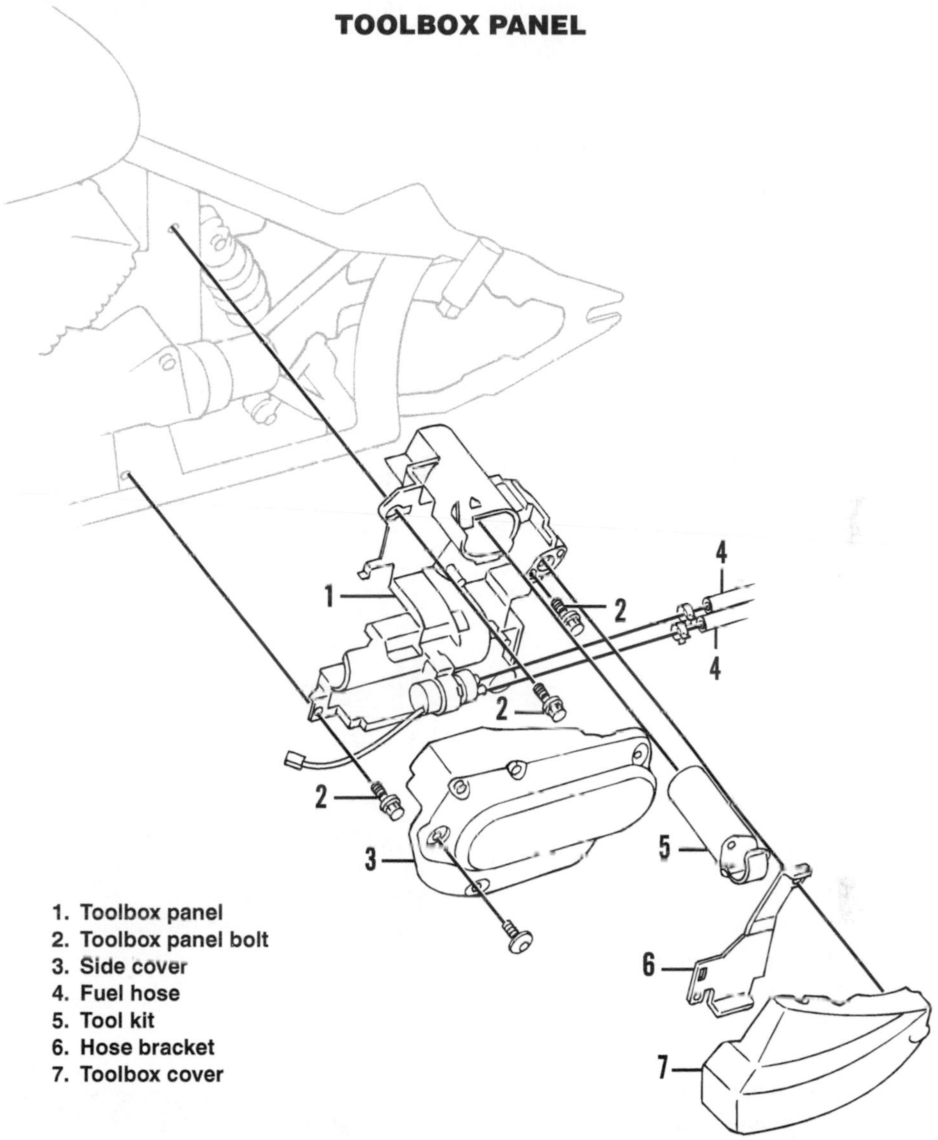

1. Toolbox panel
2. Toolbox panel bolt
3. Side cover
4. Fuel hose
5. Tool kit
6. Hose bracket
7. Toolbox cover

(11)

(12)

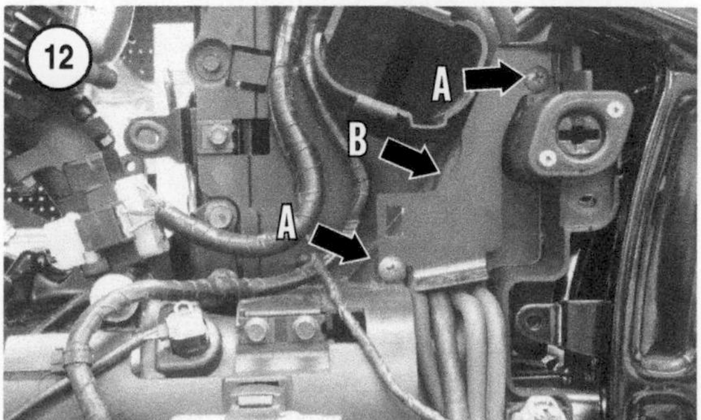

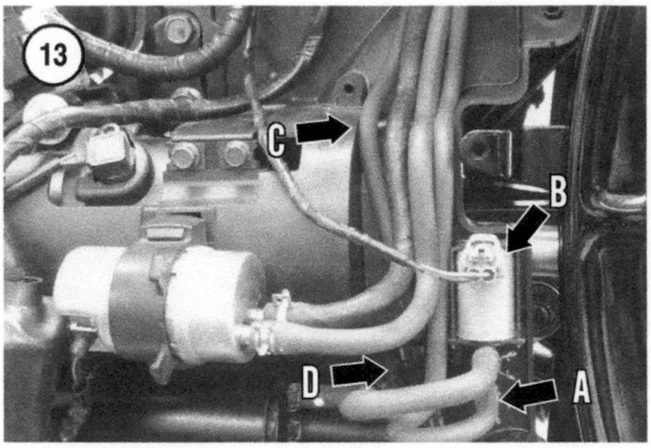

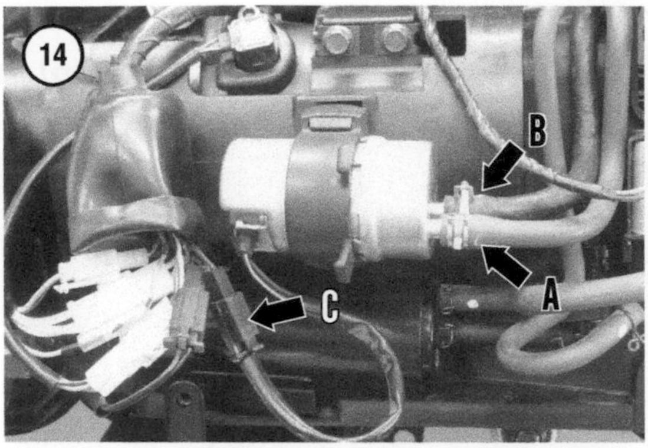

4. Install the cover as follows:
   a. Position the cover so the slots on the cover's rear side align with their respective grommets on the frame.
   b. Press the cover into the frame until the cover securely engages these grommets.
   c. Torque the battery cover bolt to 7 N•m (62 in.-lb.).

RIGHT SIDE COVER

Removal/Installation

1. Remove the mufflers and the rear exhaust pipe as described in Chapter Eight.
2. Remove the brake pedal/footrest assembly (Chapter Thirteen).
3. Remove the right side cover bolt (**Figure 19**).
4. Pull the rear of the cover out until the cover's post disengages from its grommet.
5. Installation is the reverse of removal. Note the following:
   a. Align the cover's post with its grommet and press the cover in until it snaps into place.
   b. Torque the side cover bolt (**Figure 19**) to 7 N•m (62 in.-lb.).
   c. Install the brake pedal/footrest assembly as described in Chapter Thirteen.
   d. Adjust the rear brake pedal height and the rear brake switch as described in Chapter Three.

BATTERY BOX

Removal/Installation

1. Remove the battery cover and the right side cover as described in this chapter.
2. Remove the battery as described in Chapter Three.

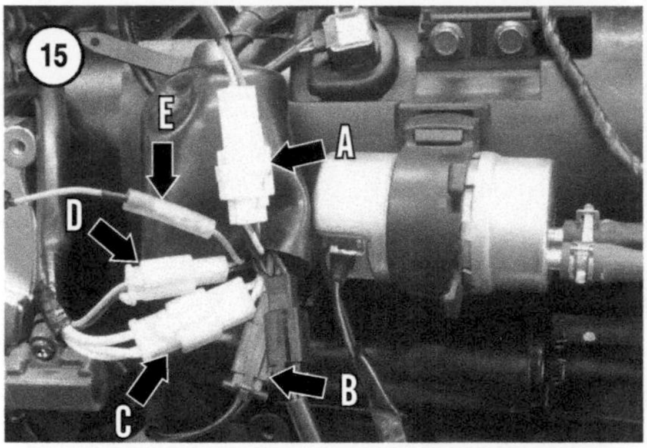

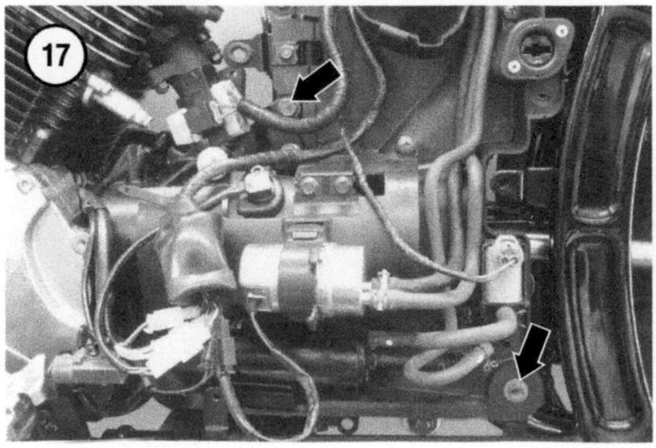

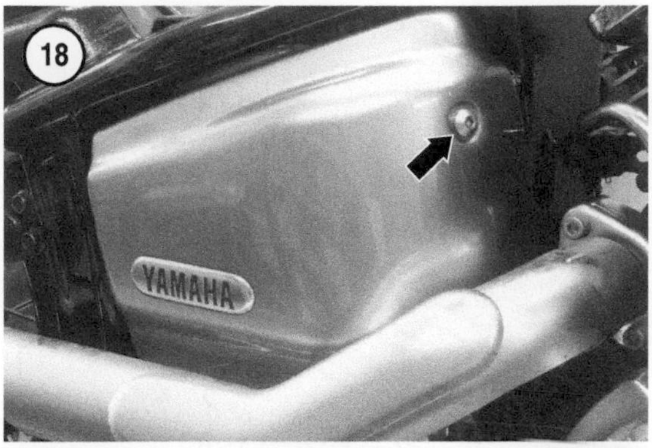

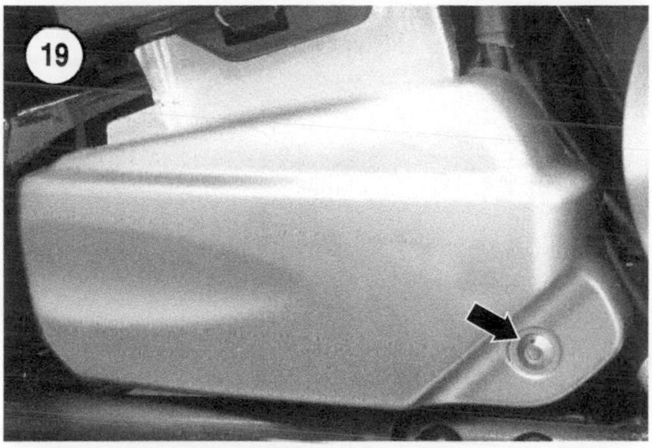

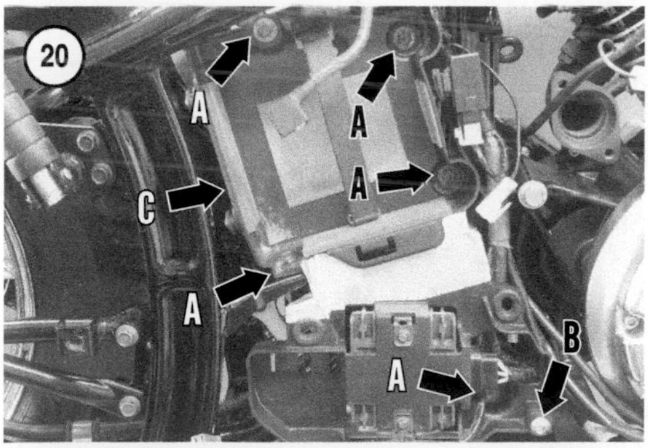

3. Remove the battery box bolts (A, **Figure 20**).

4. Remove the battery box screws (B, **Figure 20** and **Figure 16**).

5. Move the battery box (C, **Figure 20**) to the side, and suspend it from the frame with a bunjee cord.

6. If the battery box must be completely removed, disconnect the connector (**Figure 21**) from the voltage regulator/rectifier, and remove the battery box.

7. Installation is the reverse of removal.

FRAME NECK COVER

Removal/Installation

1. Remove the rider and passenger seats as described in this chapter.

2. Remove the fuel tank (Chapter Eight).

3. Remove the fuel tank grommet (A, **Figure 22**) from the post on each side of the frame.

4. Remove each neck cover screws **(Figure 23)** from the front of the frame.

5. Separate the top edges of the left and right covers, then remove each neck cover (B, **Figure 22**) from the frame.

24

RIDER FLOORBOARDS

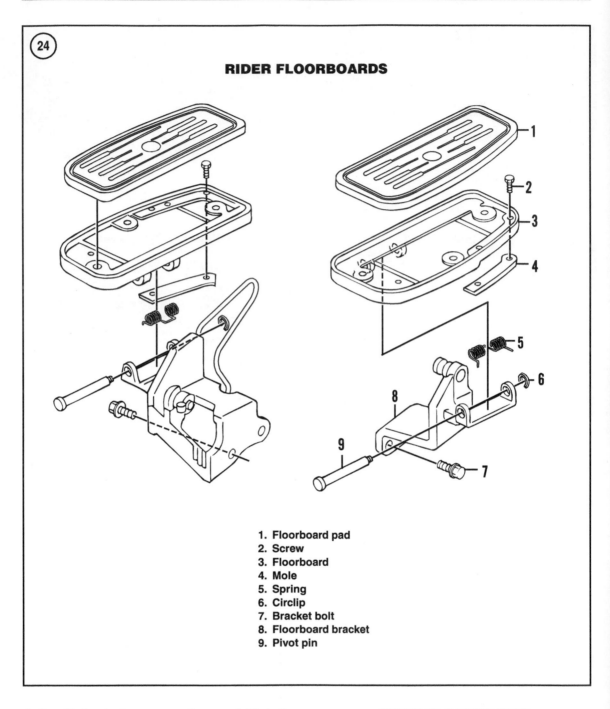

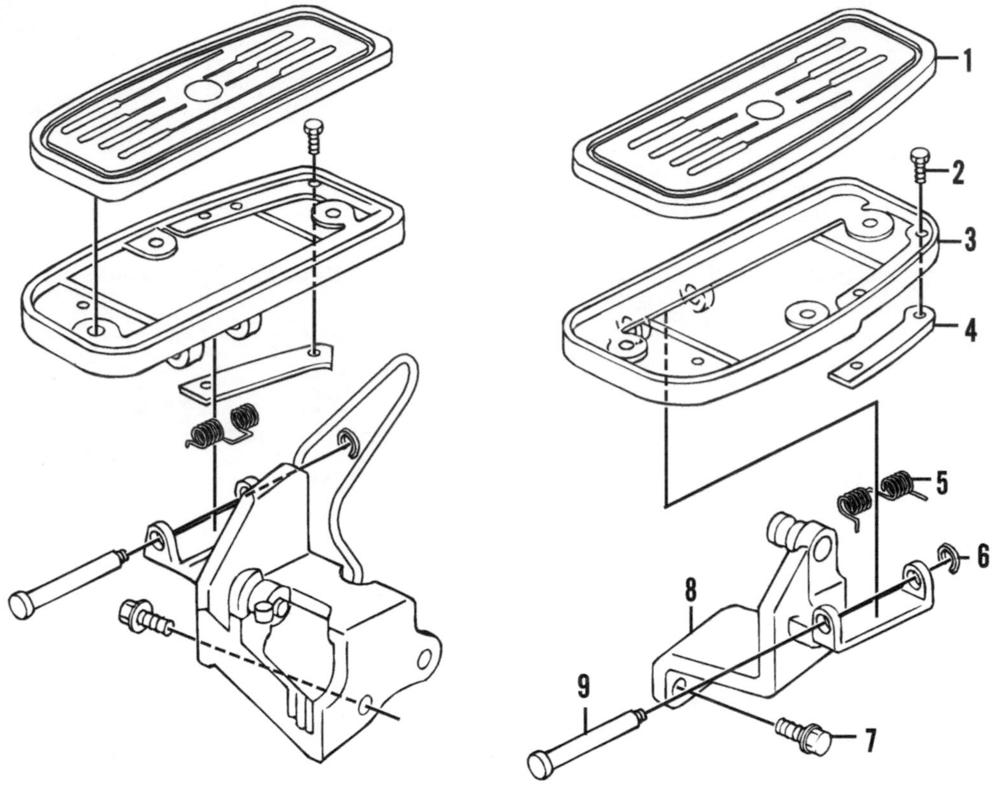

1. Floorboard pad
2. Screw
3. Floorboard
4. Mole
5. Spring
6. Circlip
7. Bracket bolt
8. Floorboard bracket
9. Pivot pin

6. Installation is the reverse of removal. Note the following:

   a. Set each neck cover into place on the frame. Make sure the top edges of the covers are properly mated.

   b. Once both covers are properly in place, install the neck cover screws (**Figure 23**) and the fuel tank grommets (A, **Figure 22**).

RIDER FLOORBOARDS

Disassembly/Assembly

Refer to **Figure 24**.

1A. Remove the brake pedal/footrest assembly as described in Chapter Thirteen.

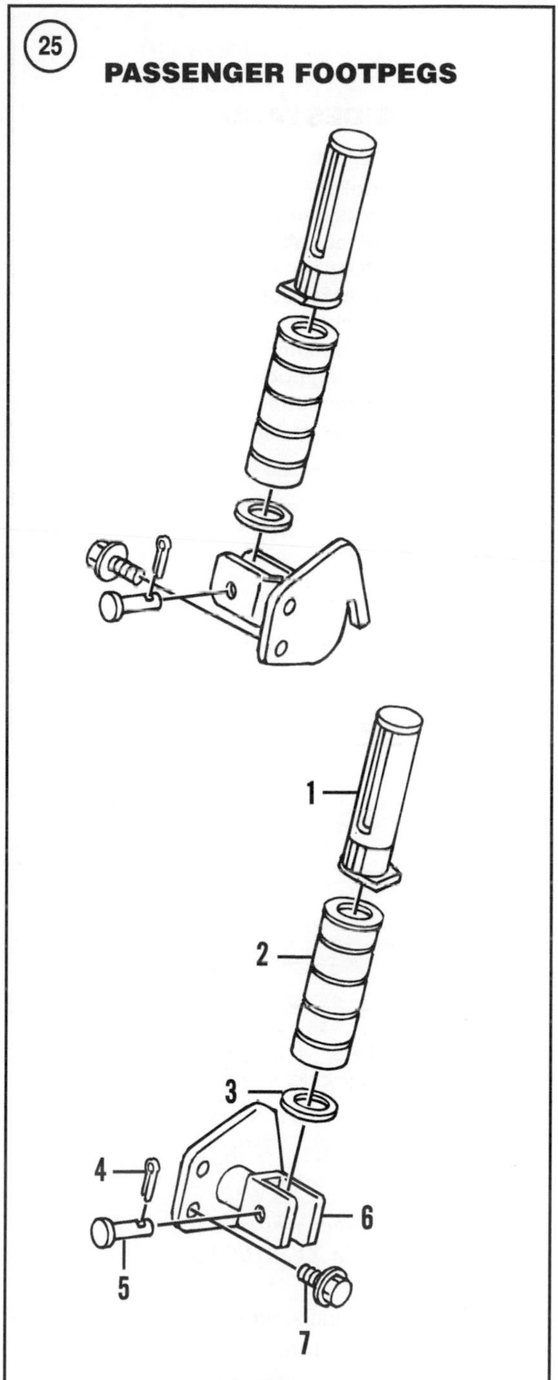

1. Footpeg
2. Footpeg cover
3. Washer
4. Cotter pin
5. Pivot pin
6. Footpeg bracket
7. Bolt

1B. Remove the shift pedal/footrest assembly as described in Chapter Seven.
2. Remove the circlip from the pivot pin.
3. Pull the pivot pin from the floorboard and separate the floorboard from the bracket. Watch for the spring.
4. Installation is the reverse of removal. Note the following.
   a. Make sure the pin passes through the spring.
   b. The circlip must be completely seated in its groove on the pivot pin.

PASSENGER FOOTPEGS

Removal/Installation

Refer to **Figure 25**.
1. Remove the passenger footpeg mounting bolts.
2. Lower the footpeg bracket from the frame.
3. If necessary, remove the footpeg as follows:
   a. Straighten the cotter pin and remove it from the pivot pin. Discard the cotter pin.
   b. Pull the pivot pin and separate the footpeg from the bracket.
4. Installation is the reverse of removal. Note the following:
   a. Install a new cotter pin.
   b. Torque the passenger footpeg bolts to 26 N•m (19 ft.-lb.)

SIDESTAND

Refer to **Figure 26**.

Removal/Installation

1. Securely support the motorcycle on a level surface.
2. Remove the bracket bolts (A, **Figure 27**) and lower the sidestand bracket (B) from the frame.
3. Installation is the reverse of removal. Torque the sidestand bracket bolts to 64 N•m (47 ft.-lb.)

Disassembly/Assembly

1. Raise the sidestand and use locking pliers to disconnect the spring hook (C, **Figure 27**) from the bracket boss.
2. Remove the self-locking nut from the pivot bolt (D, **Figure 27**).

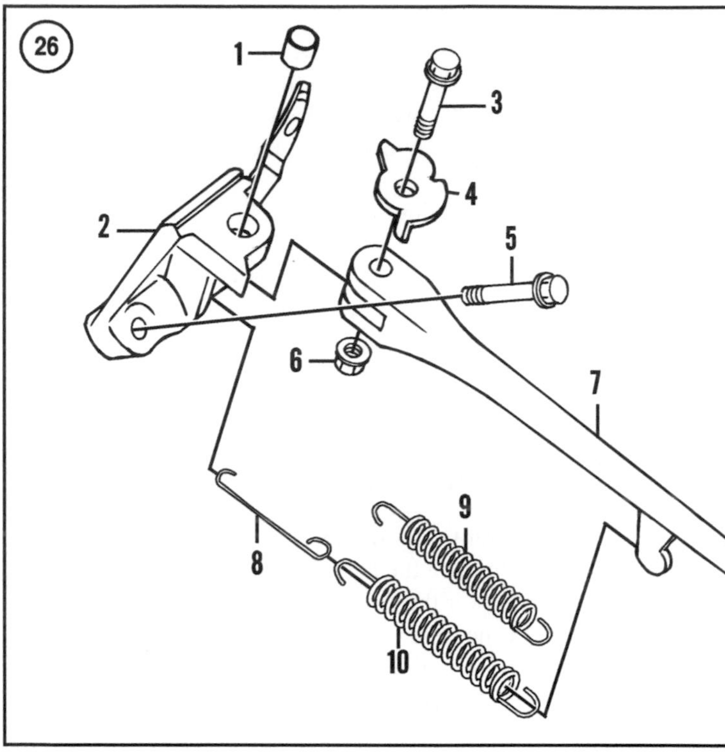

SIDESTAND

1. Collar
2. Sidestand bracket
3. Pivot bolt
4. Washer
5. Bracket bolt
6. Locknut
7. Sidestand
8. Spring hook
9. Inner spring
10. Outer spring

3. Remove the pivot bolt and washer.
4. Separate the sidestand from the bracket. Do not lose the collar from the bracket boss.
5. Installation is the reverse of removal. Note the following:
   a. Apply a light coat of lithium soap grease to the pivot surfaces of mounting bracket and sidestand.
   b. Make sure the collar is in place in the bracket boss.
   c. Install a new self-locking nut and torque the nut to the 56 N•m (41 ft.-lb.).

FRONT FENDER

Removal/Installation

Refer to **Figure 28**.

CAUTION

Since the full-size front fender on XVS1100A models wraps low over the front wheel, the motorcycle must be raised very high to remove the front wheel with the fender installed. Unless you have a jack that can safely provide the necessary clearance, remove the fender while the front wheel is installed on XVS1100A models.

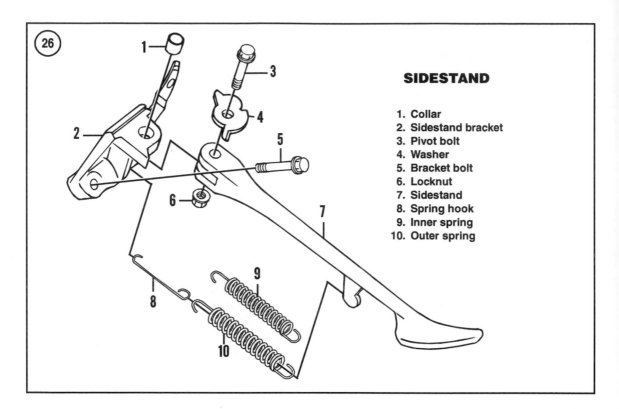

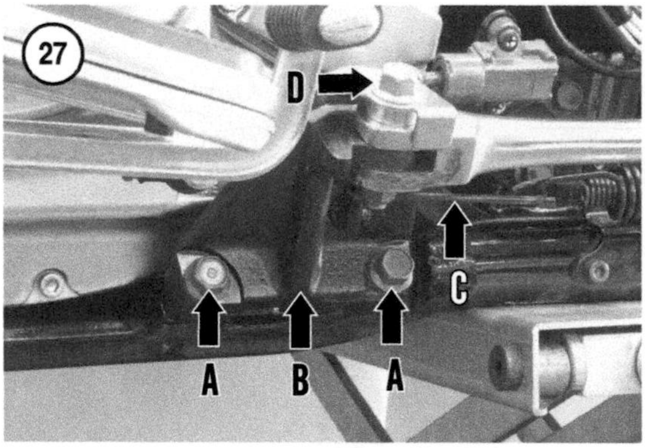

1. On XVS1100 models, remove the front wheel as described in Chapter Ten.
2. Remove the front fender bolts (**Figure 29**) from the inside of the fender and remove the fender from between the front fork legs. On XVS1100A models, watch for the reflector bracket that sits between the fender and fork leg boss.
3. Installation is the reverse of removal.
   a. On XVS1100A models, install the reflector bracket between the fender and the fork leg.

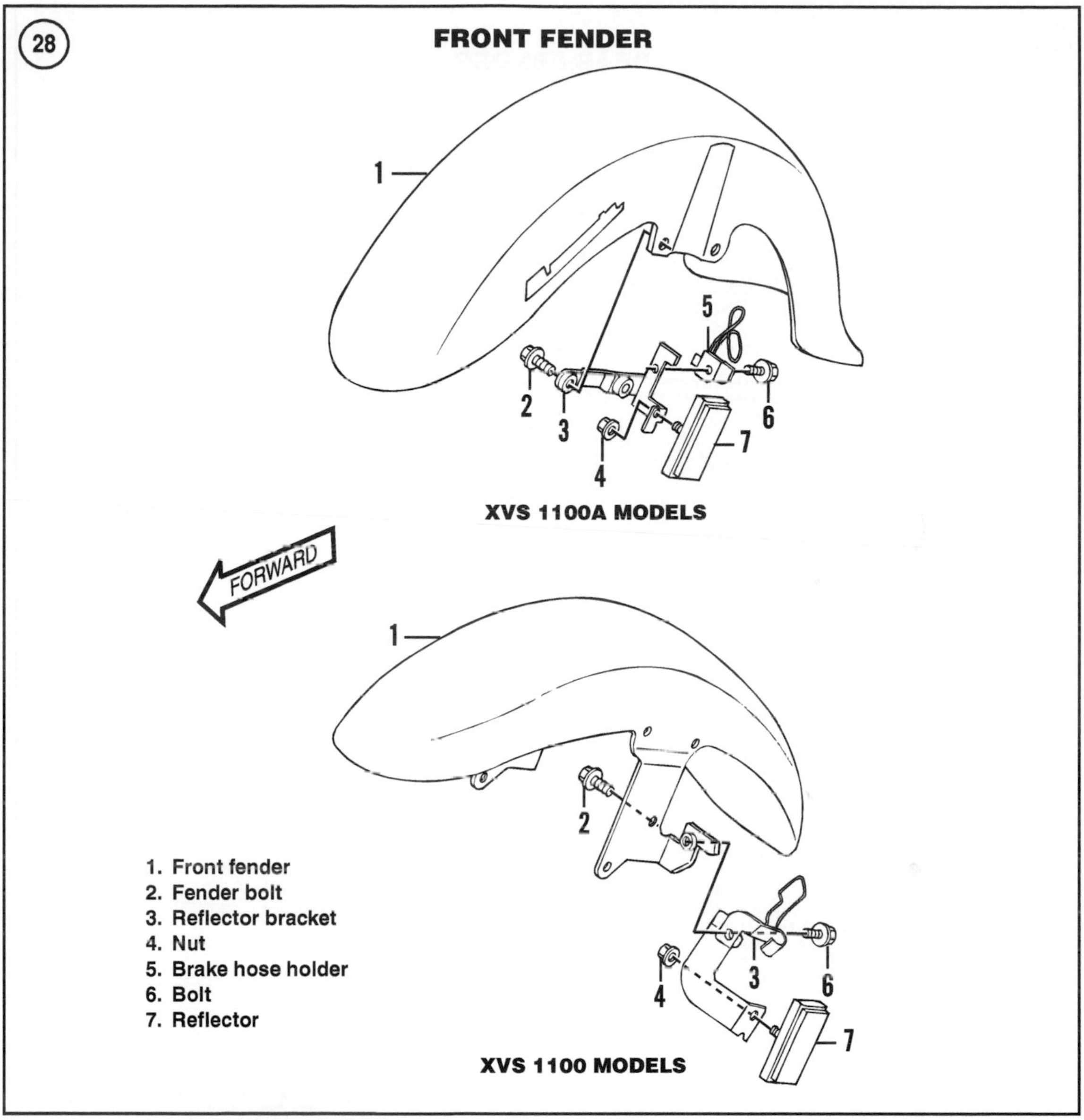

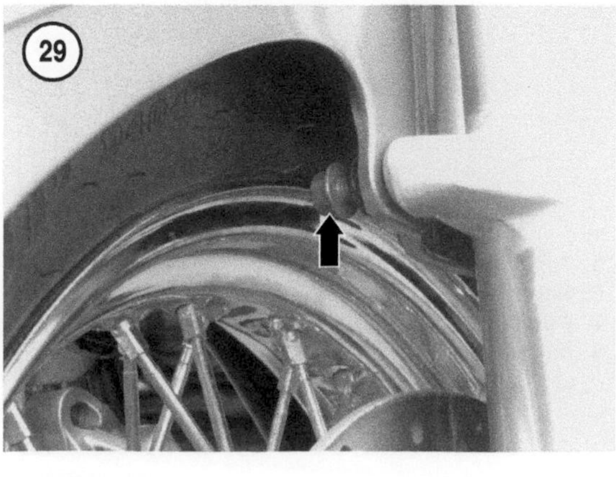

b. Torque the front fender bolts to 10 N•m (89 in.-lb.).

REAR FENDER

Refer to **Figure 30**.

Removal

1. Remove the rider and passenger seats, the battery cover, the toolbox cover and ignitor panel as described in this chapter.

30

REAR FENDER

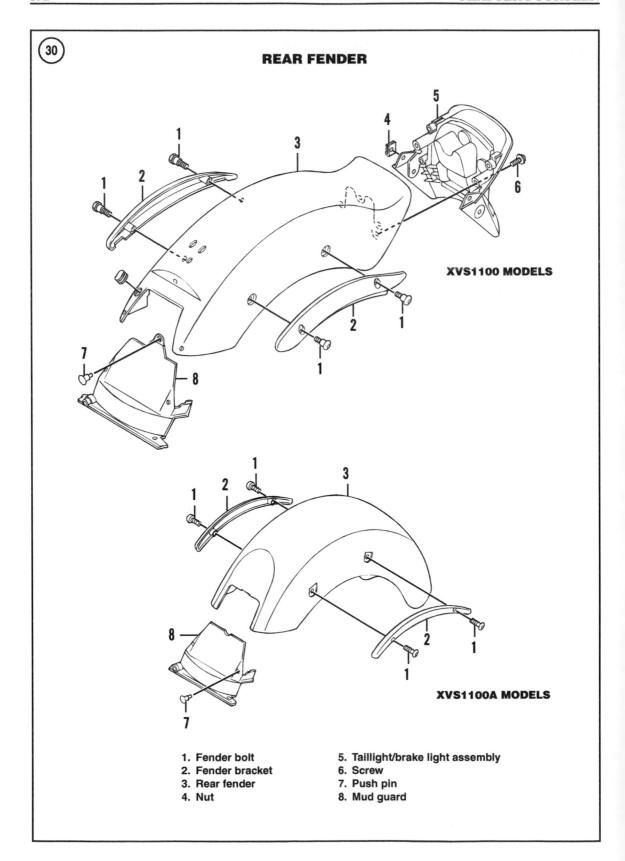

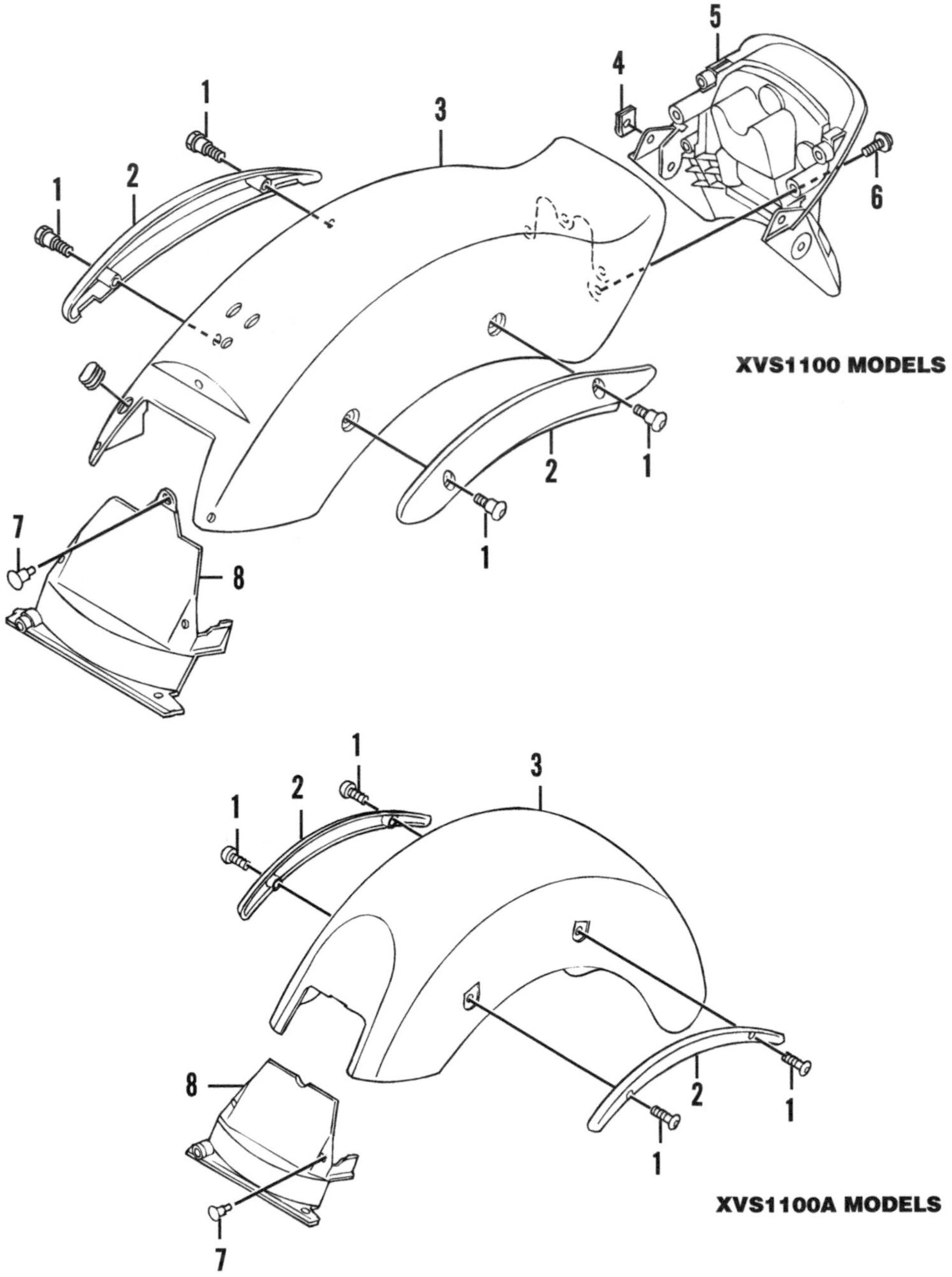

1. Fender bolt
2. Fender bracket
3. Rear fender
4. Nut
5. Taillight/brake light assembly
6. Screw
7. Push pin
8. Mud guard

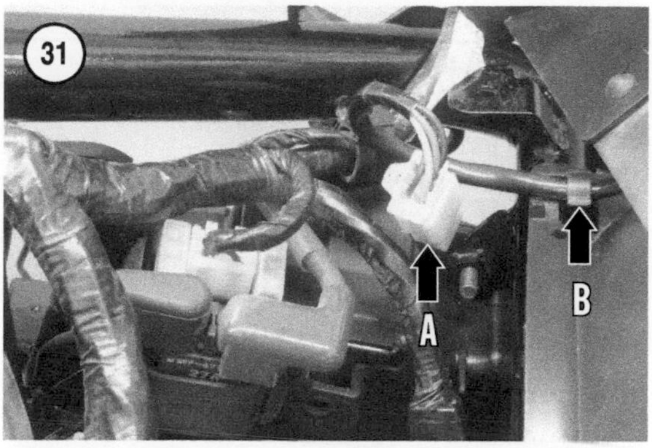

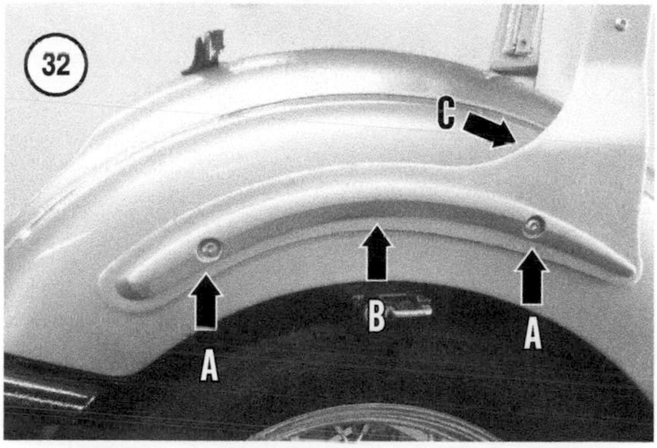

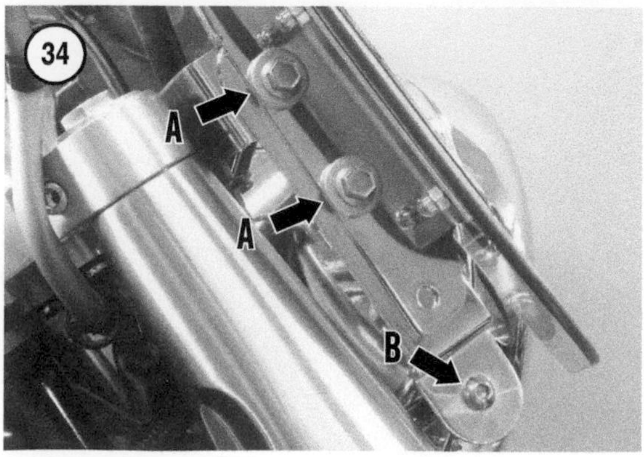

2. Remove the push pins (C, **Figure 7**) from each side of the mud guard (D) and lift the mud guard from the fender.

3. Disconnect the five-pin taillight/brake light connector (A, **Figure 31**) from its harness mate. Note how the wire is routed through the back of the frame. It will have to be rerouted along the same path during installation. Release the wire from any cable holders (B, **Figure 31**) that secure it to the frame.

NOTE

On Silverado models, have an assistant hold the backrest bracket so it will not scratch the fender as the fender bolts are removed.

4. Remove the fender bolts (A, **Figure 32**) and remove the fender bracket (B) from one side. On Silverado models, watch for the spacer between the fender and the backrest bracket (C, **Figure 32**).

5. Repeat Step 4 on the other side. On Silverado models, carefully lift the backrest from the fender. Since the spacers are removed, the backrest could scrape the fender and scratch the paint.

6. Remove the seat bracket bolts (**Figure 33**) and remove the seat bracket from the fender. This bracket is directional. Place an arrow on the bracket indicating forward.

7. Lift the fender from the motorcycle.

8. Installation is the reverse of removal. Note the following:

   a. Install the seat bracket so the arrow made during removal points forward.

   b. On Silverado models, install the spacers between the fender and the backrest.

   c. Torque the seat bracket bolts (**Figure 33**) to 7 N•m (62 in.-lb.)

   d. Torque the fender bolts (A, **Figure 32**) to 26 N•m (19 ft.-lb.).

   e. Route the taillight/brake light wire along the same path noted during removal.

WINDSHIELD (SILVERADO MODELS)

Removal/Installation

1. Remove the mounting bolts (A, **Figure 34**) on each side and remove the windshield from the windshield bracket.

2. Installation is the reverse of removal. If installing the two mounting bolts (A, **Figure 34**) is difficult, loosen the lower button head bolt (B) to make alignment easier.

Table 1 FRAME AND BODY TORQUE SPECIFICATIONS

Item	N•m	in.-lb.	ft.-lb.
Battery cover bolt	7	62	–
Front fender bolts	10	89	–
Passenger footpeg bolts	26	–	19
Passenger seat nut	7	62	–
Rear fender bolts	26	–	19
Seat bracket bolts	7	62	–
Side cover bolt	7	62	–
Sidestand bracket bolts	64	–	47
Sidestand nut	56	–	41

INDEX

A

Air
- filter housing 207-209
- induction system 231-234

Alternator cover 139-142

B

Battery 73-76, 84
- box 386-387
- cover 384-386

Bearing races 327-328

Body torque specifications 394

Brakes 80-82
- bleeding 349-352
- caliper
  - front 358-362
  - rear 367-370
- disc 376-379
- fluid
  - draining 352-353
  - preventing damage from 349
- hose replacement 379
- master cylinder
  - front 362-367
  - rear 371-375
- pads 353-358
- pedal/footrest assembly 375-376
- rotor protection 283-284
- service 348-349
- specifications
  - front 380
  - rear 380
  - torque 380
- troubleshooting 55

Break-in 166

C

Caliper
- front 358-362
- rear 367-370

Cam chain and drive assembly 104-110

Camshaft 99-102

Carburetor 211-222
- heater 225-227
  - troubleshooting 46-47
- specifications 239-240
- troubleshooting 45-46

Charging system 241-242

Choke cable 228-229

Clutch 170-177
- cable replacement 179-180
- cover 168-170
- release mechanism 177-179
- specifications 184
  - torque 184
- troubleshooting 43-44

Connecting rods 163-166
- bearing insert selection 166

Control cables 71-72

Conversion formulas 31

Crankcase 154-161

Crankshaft 162-163

Cylinder 120-122
- head 88-99
  - covers 87-88
- numbering 57

E

Electrical system
- charging 241-242
- diode 247
- fundamentals 19-20

fuses . . . 279-280
headlight . . . 258-261
horn . . . 268-269
ignition
coil. . . . 245-246
system . . . 243-245
ignitor unit . . . 247-248
indicator light
neutral . . . 273
oil level . . . 273-274
lighting system. . . . 258
meter assembly . . . 261-263
pickup coil . . . 246-247
self-diagnostic system . . . 279
signal system. . . . 268
spark plug cap . . . 245
specifications . . . 281
fuses . . . 282
replacement bulbs . . . 281-282
torque. . . . 282
speed sensor. . . . 274-275
starter . . . 248-255
relay . . . 257-258
starting
circuit cutoff relay . . . 255-257
system . . . 248
stator . . . 242-243
switches. . . . 275-279
taillight/brake light . . . 263-268
troubleshooting . . . 51-54
testing . . . 47-48
equipment . . . 48-50
procedures . . . 50-51
turn signals . . . 269-273
voltage regulator/rectifier . . . 243
wiring
connectors . . . 280
diagrams . . . 400
Emission system . . . 234-237
electrical specifications . . . 240
Engine
lower end . . . 133-139
alternator cover . . . 139-142
break-in. . . . 166
connecting rods . . . 163-166
crankcase . . . 154-161
crankshaft . . . 162-163
flywheel and starter clutch. . . . 142-148
oil pump . . . 148-154
servicing in-frame . . . 133
specifications. . . . 166
torque . . . 167
stator and pickup coils . . . 142
oil . . . 66-68
rotation . . . 57
top end
cam chain and drive assembly . . . 104-110
camshaft . . . 99-102
cylinder . . . 120-122
head . . . 88-99
covers . . . 87-88
pistons and rings. . . . 123-129
rocker arms . . . 102-104
servicing in-frame . . . 87
specifications . . . 130-131
torque . . . 131-132
valves and components . . . 110-120
troubleshooting
leak-down test . . . 43
lubrication . . . 42
noises . . . 41
performance . . . 39-41
Exhaust system . . . 237-239
evaporative emission control . . . 234-237
specifications . . . 240
torque. . . . 240

F

Fasteners . . . 4-6, 82
Fender
front. . . . 390-391
rear . . . 391-393
Final drive. . . . 79
assembly . . . 337-346
specifications . . . 346
torque. . . . 347
troubleshooting . . . 45
Floorboards. . . . 388-389
Flywheel and starter clutch . . . 142-148
Footpegs. . . . 389
Fork, front . . . 308-321
Frame
battery
box. . . . 386-387
cover. . . . 384-386
fender
front . . . 390-391
rear . . . 391-393
ignitor panel. . . . 381-383
neck cover. . . . 387-388

passenger footpegs 389
rider floorboards 388-389
seats 381
side cover
left 383
right 386
sidestand 389-390
toolbox
cover. 383-384
panel 384
torque specifications 394
windshield 393
Fuel
and exhaust systems 68-71
filter. 229-230
level. 222-223
pump 230-231
troubleshooting 47
system specifications, electrical 240
tank 203-205
valve 205-207
Fuses 279-280

G

Gearshift
linkage, troubleshooting. 44
specifications 202

H

Handlebar. 321-323
Headlight 258-261
Horn 268-269

I

Idle speed adjustment 224
Ignition
coil 245-246
system. 243-245
Ignitor
panel 381-383
unit 247-248
Internal shift mechanism 188-190

L

Laced wheel service 296-298
Lighting system 258
Linkage. 332-334
Lower bearing replacement 328
Lubricants and fluids recommendations. . . 84-85

M

Maintenance
battery 73-76
specifications 84
brakes 80-82
control cables 71-72
engine oil 66-68
fasteners. 82
final drive 79
fuel
and exhaust systems 68-71
type 56
intervals 56-57
lubrication recommendations 73, 84-85
schedule 82-83
sidestand. 72-73
spark plugs. 62-66
steering head. 77-78
suspension
front 77
rear 78-79
specifications
torque 86
tune-up. 85
tires 76-77
transmission. 72
Master cylinder
front. 362-367
rear 371-375
Meter assembly 261-263
Model names and numbers 1, 29

N

Neutral indicator light 273

O

Oil
level indicator light 273-274
pump 148-154
Operating requirements, troubleshooting 34

P

Pedal/footrest assembly. 375-376
Pickup coil 246-247
Pilot screw. 222
Pistons and rings 123-129

Primary drive gear 181-184
torque specifications 184

R

Rocker arms 102-104

S

Seats 381
Self-diagnostic system. 279
Serial numbers 3-4, 27-28
Shift mechanism torque specifications. 202
Shift pedal/footrest assembly 185-187
Shock absorber 330-332
Side cover
left 383
right 386
Sidestand 72-73, 389-390
Signal system 268
Spark plugs 62-66
cap 245
Specifications
brakes
front 380
rear 380
bulb replacement 281-282
carburetor 239-240
connecting rod bearing insert selection. . . . 166
dimensions and weight of vehicle 29
electrical system 281
emission system 240
engine
lower end 166
top end. 130-131
fuel system 240
fuses 282
maintenance and tune-up 85
suspension
front 329
rear, and final drive 346
tires 306
torque
brake system 380
electrical system 282
engine lower end 167
engine top end 131-132
frame and body 394
general. 30
maintenance and tune up 86
middle gear 202
suspension
front, and steering. 329
rear, and final drive 347
shift mechanism 202
transmission 202
wheels 307
transmission and gearshift 202
wheels 306
wiring diagrams 400
Speed sensor 274-275
Starter. 248-255
clutch 142, 147-148
relay. 257-258
Starting
circuit cutoff relay. 255-257
difficulties, troubleshooting 34-37, 37-39
system 248
Stator 242-243
and pickup coils 142
Steering
head 77-78, 323-327
bearing races. 327-328
handlebar 321-323
torque specifications 329
Storage 26
Surge tank 209-210
Suspension
front 77
fork 308-321
lower bearing replacement 328
rear. 78-79
linkage. 332-334
shock absorber. 330-332
swing arm 334-337
Swing arm 334-337
Switches 275-279

T

Taillight/brake light. 263-268
Technical abbreviations 31-32
Throttle
cable. 227-228
adjustment 224
position sensor 224-225
Tires 76-77
changing
cast wheels. 303-305
laced wheels 300-303
inflation pressure 306

specifications. 84, 306
Toolbox
cover 383-384
panel 384
Tools 9-14
precision measuring 14-19
Transmission 72, 190-195
middle
drive shaft 195-198
driven shaft 198-202
gear backlash 202
pedal/footrest assembly 185-187
shift mechanism
external 187-188
internal 188-190
specifications 202
torque. 202
troubleshooting 44-45
Troubleshooting
brakes 55
carburetor 45-46
heater. 46-47
clutch 43-44
electrical system 51-54
testing 47-48
equipment 48-50
procedures 50-51
engine
leakdown test 43
lubrication 42
noises 41
performance 39-41
final drive 45
fuel pump 47
gearshift linkage. 44
operating requirements 34
starting the engine 34-37
difficulties 37-39
suspension and steering 54-55
transmission 44-45
Tune-up 57-62
cylinder numbering 57
engine rotation. 57
Turn signals. 269-273

V

Valves and components 110-120
Voltage regulator/rectifier 243

W

Wheels
balance 298-300
bearing
installation. 286-288
removal. 286
front. 288-290
hub
front 290-291
rear 294-296
inspection 284-286
laced. 296-298
rear 291-294
specifications 306
torque. 307
Windshield 393
Wiring
connectors 280
diagrams 400

1999-2003 USA, CALIFORNIA AND CANADA MODELS
2004 AND 2005 MODELS EXCEPT CALIFORNIA

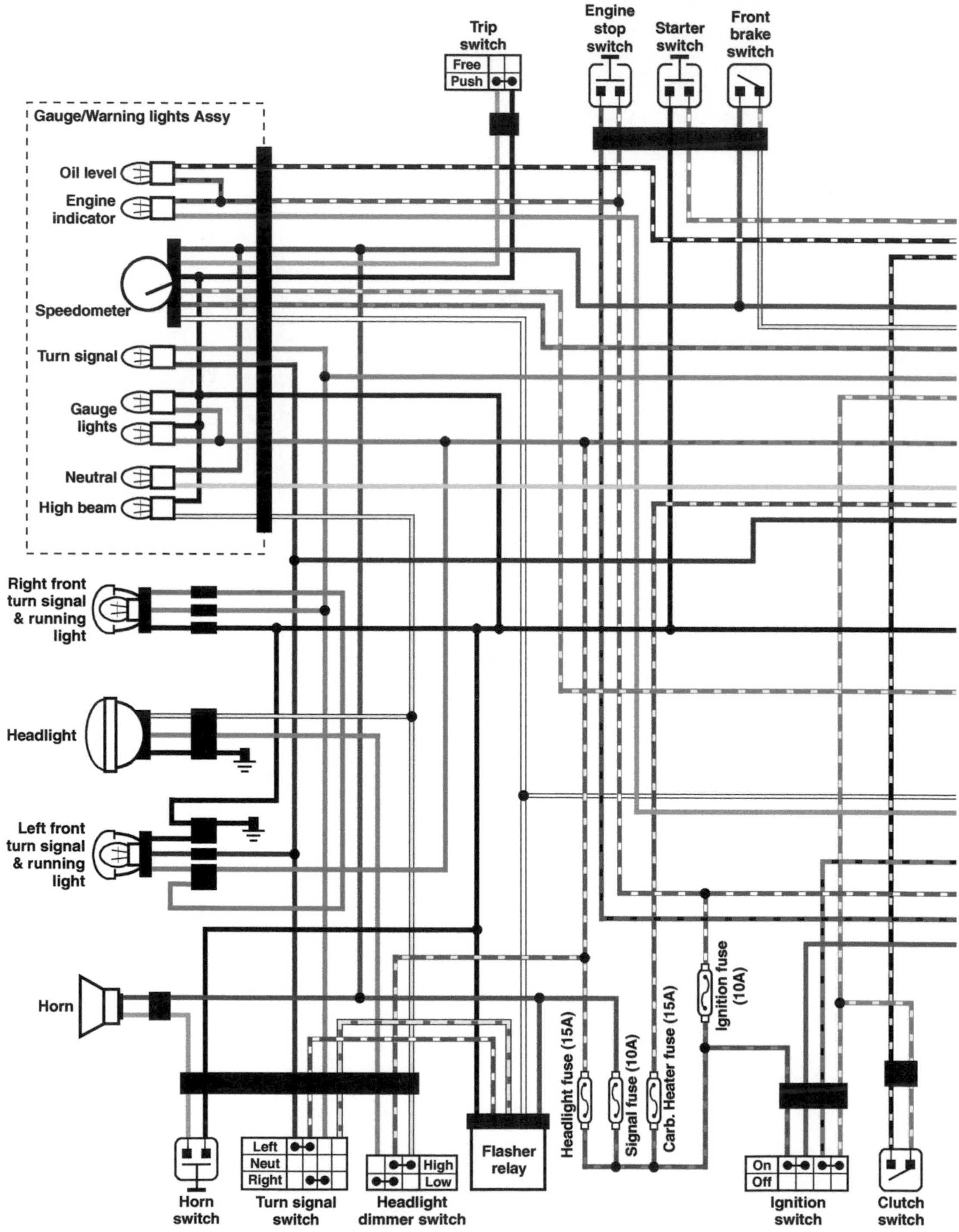

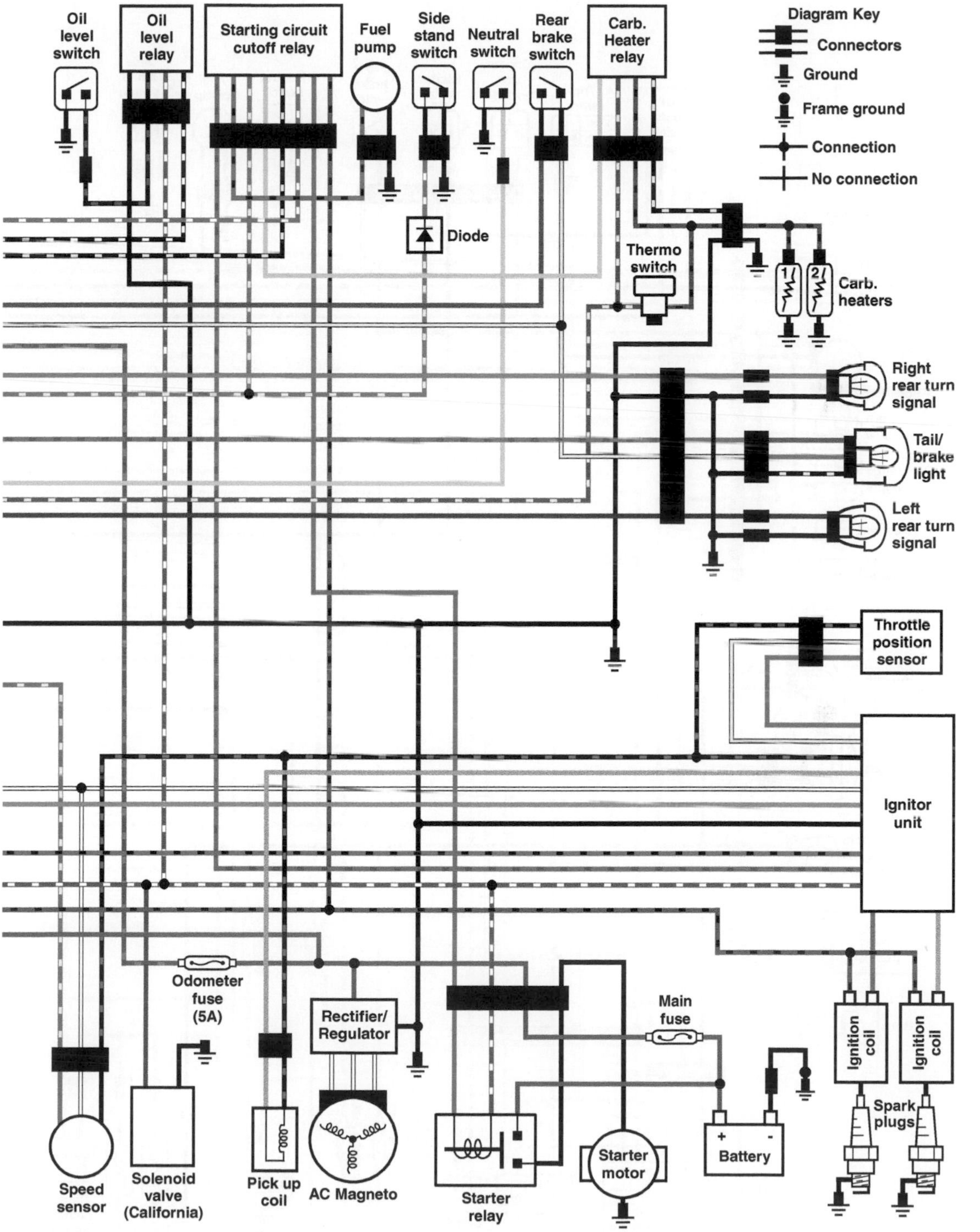
Oil level switch
Oil level relay
Starting circuit cutoff relay
Fuel pump
Side stand switch
Neutral switch
Rear brake switch
Carb. Heater relay
Diagram Key
Connectors
Ground
Frame ground
Connection
No connection
Diode
Thermo switch
Carb. heaters
Right rear turn signal
Tail/ brake light
Left rear turn signal
Throttle position sensor
Ignitor unit
Odometer fuse (5A)
Rectifier/ Regulator
Main fuse
Ignition coil
Ignition coil
Spark plugs
Speed sensor
Solenoid valve (California)
Pick up coil
AC Magneto
Starter relay
Starter motor
Battery

2004-2005 CALIFORNIA MODELS
2006-ON ALL MODELS

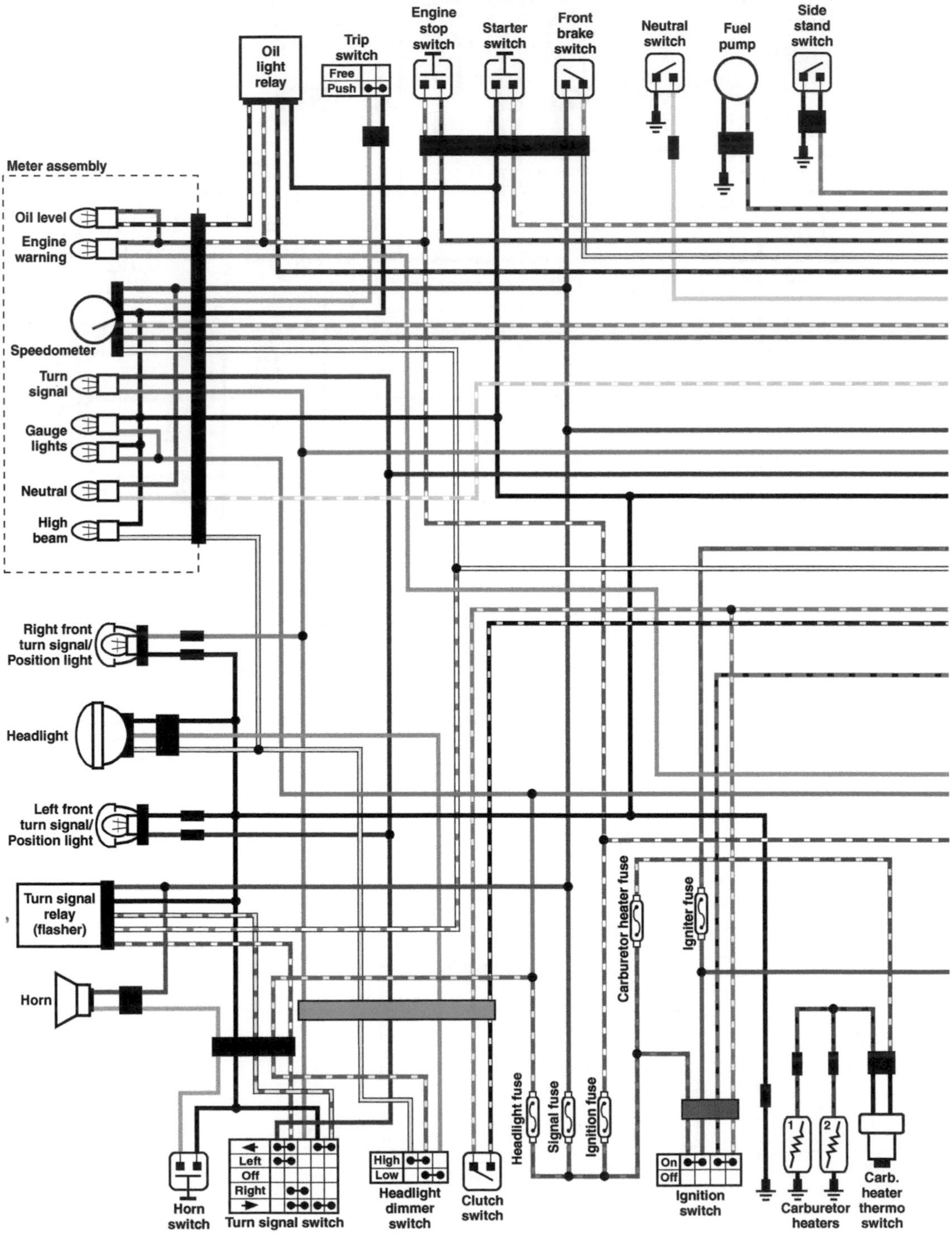

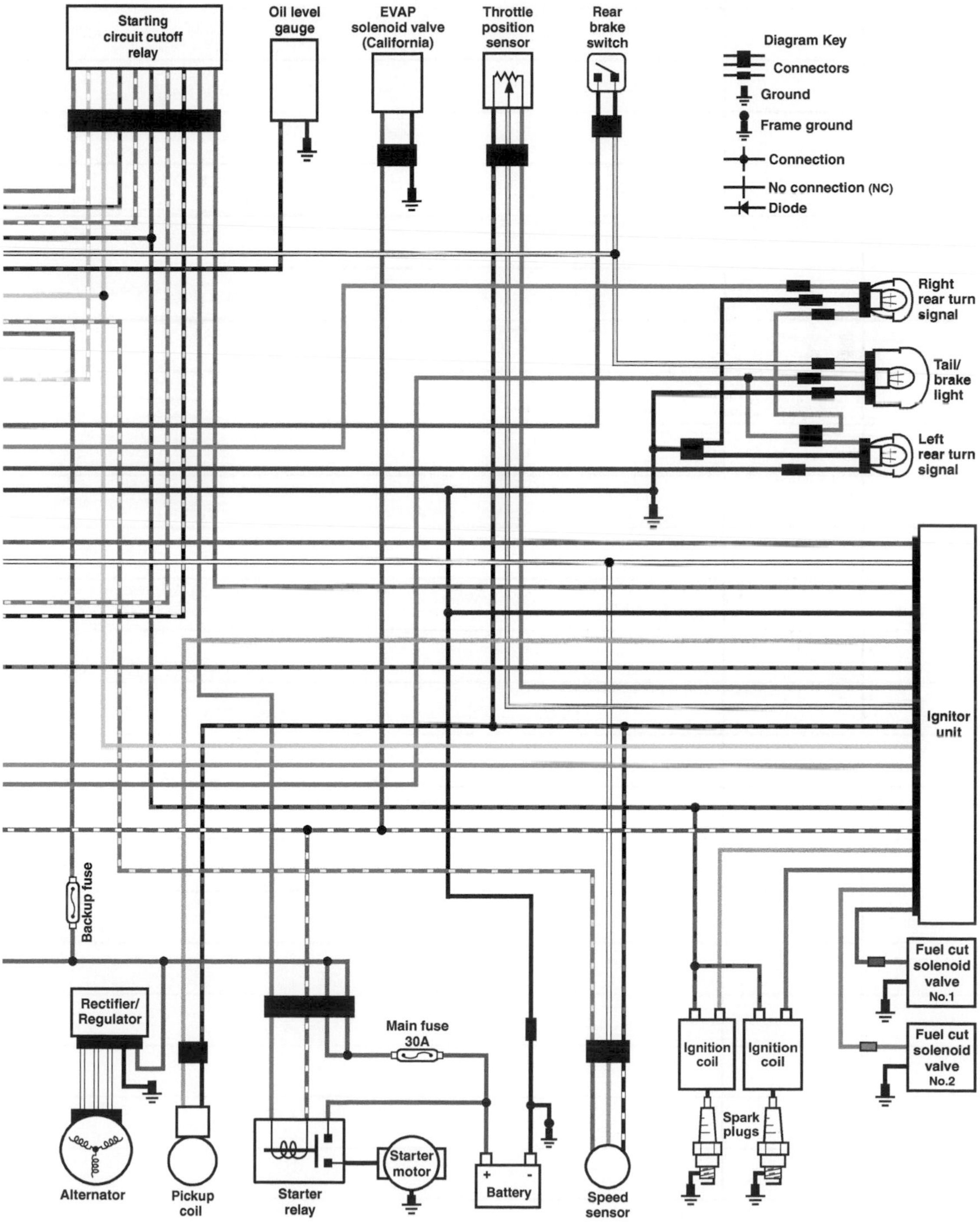
Starting circuit cutoff relay
Oil level gauge
EVAP solenoid valve (California)
Throttle position sensor
Rear brake switch
Diagram Key
Connectors
Ground
Frame ground
Connection
No connection (NC)
Diode
Right rear turn signal
Tail/ brake light
Left rear turn signal
Ignitor unit
Backup fuse
Fuel cut solenoid valve No.1
Fuel cut solenoid valve No.2
Rectifier/ Regulator
Main fuse 30A
Ignition coil
Ignition coil
Spark plugs
Alternator
Pickup coil
Starter relay
Starter motor
Battery
Speed sensor

MAINTENANCE LOG

Date	Miles	Type of Service